T0236147

Lecture Notes in Computer Science　　　10559

Commenced Publication in 1973
Founding and Former Series Editors:
Gerhard Goos, Juris Hartmanis, and Jan van Leeuwen

More information about this series at http://www.springer.com/series/7412

Yi Sun · Huchuan Lu
Lihe Zhang · Jian Yang
Hua Huang (Eds.)

Intelligence Science and Big Data Engineering

7th International Conference, IScIDE 2017
Dalian, China, September 22–23, 2017
Proceedings

Springer

Editors
Yi Sun
Dalian University of Technology
Dalian
China

Huchuan Lu
Dalian University of Technology
Dalian
China

Lihe Zhang
Dalian University of Technology
Dalian
China

Jian Yang
Nanjing University of Science
    and Technology
Nanjing
China

Hua Huang
Beijing Institute of Technology
Beijing
China

ISSN 0302-9743   ISSN 1611-3349   (electronic)
Lecture Notes in Computer Science
ISBN 978-3-319-67776-7   ISBN 978-3-319-67777-4   (eBook)
DOI 10.1007/978-3-319-67777-4

Library of Congress Control Number: 2017952884

LNCS Sublibrary: SL6 – Image Processing, Computer Vision, Pattern Recognition, and Graphics

Printed on acid-free paper

This Springer imprint is published by Springer Nature
The registered company is Springer International Publishing AG
The registered company address is: Gewerbestrasse 11, 6330 Cham, Switzerland

Preface

IScIDE 2017, the International Conference on Intelligence Science and Big Data Engineering, took place in Dalian, China, during September 22–23, 2017. As one of the annual events organized by the Chinese Golden Triangle ISIS (Information Science and Intelligence Science) Forum, this meeting was scheduled as the seventh in a series of annual meetings promoting the academic exchange of research on various areas of intelligence science and big data engineering in China and abroad. In response to the call for papers, a total of 121 papers were submitted and 61 papers were selected, yielding an acceptance rate of 51.2%. We would like to thank all the reviewers for spending their precious time on reviewing papers and for providing valuable comments that helped significantly in the paper selection process.

We would like to express our special thanks to the conference general co-chairs, Baocai Yin, Changyin Sun, and Yanning Zhang, for their leadership, advice, and help on crucial matters concerning the conference. We would like to thank all Steering Committee members, Program Committee members, Invited Speakers' Committee members, Organizing Committee members, and Publication Committee members for their hard work. We would like to thank Jingyi Yu, Fei Wu, Bin Hu, Tiejun Huang, and Kun Zhang for delivering the invited talks and sharing their insightful views on ISIS research issues. Finally, we would like to thank all the authors of the submitted papers, whether accepted or not, for their contribution to the high quality of this conference. We count on your continued support of the ISIS community in the future.

September 2017

Yi Sun
Huchuan Lu
Lihe Zhang
Jian Yang
Hua Huang

Organization

General Chairs

Baocai Yin   Dalian University of Technology, Dalian, China
Changyin Sun   Southeast University, Nanjing, China
Yanning Zhang   Northwestern Polytechnical University, Xi'an, China

Technical Program Committee Chairs

Jian Yang   Nanjing University of Science and Technology, Nanjing,
                            China
Hua Huang   Beijing Institute of Technology, Beijing, China
Yi Sun   Dalian University of Technology, Dalian, China
Huchuan Lu   Dalian University of Technology, Dalian, China

Local Arrangements Chairs

Xin Fan   Dalian University of Technology, Dalian, China
Yanqing Guo   Dalian University of Technology, Dalian, China
Wenda Zhao   Dalian University of Technology, Dalian, China

Publication Chairs

Lihe Zhang   Dalian University of Technology, Dalian, China
Wankou Yang   Southeast University, Nanjing, China

Publicity Chairs

XianPing Fu   Dalian Maritime University, Dalian, China
Haojie Li   Dalian University of Technology, Dalian, China

Tutorial Chairs

Yongri Piao   Dalian University of Technology, Dalian, China
Jinqing Qi   Dalian University of Technology, Dalian, China
Ming Li   Dalian University of Technology, Dalian, China

Registration Chairs

Risheng Liu   Dalian University of Technology, Dalian, China
Dong Wang   Dalian University of Technology, Dalian, China

Program Committee Members

Deng Cai	Zhejiang University, Hangzhou
Fang Fang	Peking University, Beijing
Jufu Feng	Peking University, Beijing
Xinbo Gao	Xidian University, Xi'an
Xin Geng	Southeast University, Nanjing
Ziyu Guan	Northwest University, Xi'an
Xiaofei He	Zhejiang University, Hangzhou
Kalviainen Heikki	Lappeenranta University of Technology, Finland
Akira Hirose	The University of Tokyo, Japan
Dewen Hu	National University of Defense Technology, Changsha
Hiroyuki Iida	Japan Advanced Institute of Science and Technology, Japan
Zhong Jin	Nanjing University of Science and Technology, Nanjing
Ikeda Kazushi	Nara Advanced Institute of Science and Technology, Japan
Andrey S. Krylov	Lomonosov Moscow State University, Russia
James Kwok	Hong Kong University of Science and Technology, Hong Kong
Jian-huang Lai	Sun Yat-sen University, Zhongshan
Shutao Li	Hunan University, Changsha
Xuelong Li	Xi'an Optics and Fine Mechanics, Chinese Academy of Scienses, Xi'an
Yi Li	Australian National University, Australia
Binbin Lin	University of Michigan, USA
Zhouchen Lin	Peking University, Beijing
Cheng Yuan Liou	National Taiwan University, Taiwan
Qingshan Liu	Nanjing University of Information Science and Technology, Nanjing
Yiguang Liu	Sichuang University, Chengdu
Bao-Liang Lu	Shanghai Jiao Tong University, Shanghai
Seiichi Ozawa	Kobe University, Japan
Yuhua Qian	Shanxi University, Taiyuan
Karl Ricanek	University of North Carolina Wilmington, USA
Shiguang Shan	Institute of Comp. Tec., Chinese Academy of Sciences, Beijing
Chunhua Shen	University of Adelaide, Australia
Changyin Sun	Southeast University, Nanjing
Dacheng Tao	University of Technology, Sydney, Australia
Vincent S. Tseng	National Cheng Kung University, Taiwan
Liang Wang	Institute of Automation, Chinese Academy of Sciences, Beijing
Liwei Wang	Peking University, Beijing
Yishi Wang	UNC Wilmington, USA
Jian Yang	Nanjing University of Science and Technology, Nanjing
Changshui Zhang	Tsinghua University, Beijing
Daoqiang Zhang	Nanjing University of Aeronautics and Astronautics, Nanjing

Lei Zhang   Hong Kong Polytechnic University, Hong Kong
Lijun Zhang   Nanjing University, Nanjing
Yanning Zhang   Northwestern Polytechnical University, Xi'an

Contents

Objects

Classification and Clustering

Imaging

Biomedical Signal Processing

Recommendation

Statistics and Learning

The Microphone Array Arrangement Method for High Order Ambisonics Recordings

Shan Gao, Xihong Wu, and Tianshu Qu^(✉)

Key Laboratory on Machine Perception (Ministry of Education),
Speech and Hearing Research Center, Peking University, Beijing, China
qutianshu@pku.edu.cn

Abstract. 3D sound field can be recorded using the method of Ambisonics. According to the vertices of Plato's polyhedrons, it is not possible to uniformly distribute arbitrary number of microphones on the sphere except five particular cases. Here we proposed a physical model to solve this problem. In this model, the microphones are assumed to be the charged particles that will move on the spherical surface according to the resultant force by other microphones. After several iteration times, the microphones will be at the equilibrium state, which means that they are distributed nearly uniformly. The emulation experiments were carried on for the 4th order of Ambisonics. The orthonormality errors of each number of the microphones were calculated and the results indicated that the proposed microphone arrangement method performs similar to the penatki-dodecahedron method when the microphone number is 32, but it can nearly uniformly distribute any number of microphones.

Keywords: Ambisonics · Microphone array · Physical model · Orthogonality error

1 Introduction

Spherical microphone arrays have the ability to record an arbitrary three dimensional sound field [1]. One common method to store and process the sound field is the Ambisonics system [2]. Ambisonics is a 3D sound spatialisation technology developed by Gerzon [3]. Ambisonics uses orthonormal spherical functions for describing the sound field in the area around the point of origin, also known as the sweet spot. One advantage of the Ambisonics representation is that the reproduction of the sound field can be adapted individually to any given loudspeaker arrangement.

The SoundField microphone, which is a practical 1st order Ambisonics [4], permits to render 3D sound fields in a flexible way from their 1st order directive components. Nevertheless, its low spatial resolution limits the correct sound field reconstruction in a small listening area, especially for high frequencies. Higher Order Ambisonics (HOA) [5] extends the 1st order directive components to higher resolution by means of spherical harmonic decomposition of the sound

© Springer International Publishing AG 2017
Y. Sun et al. (Eds.): IScIDE 2017, LNCS 10559, pp. 3–10, 2017.
DOI: 10.1007/978-3-319-67777-4_1

field, which results in enlarging the reproduction area. A 32 capsules microphone array (Eigenmike by mh acoustics) was used for acquisition the high spatial resolution [6,7].

During the acquisition of the sound field, microphones are distributed on the surface of the sampling ball discretely. There are three sampling method [8]: Equiangle Sampling, Gaussian Sampling and Nearly Uniform Sampling. Uniform sampling offers a tradeoff between the required number of microphones and the orthonormality of the array. Unfortunately it is not possible to uniformly sample the surface of a sphere, except in five particular cases according to the vertices of Plato's polyhedrons [9], tetrahedron, hexahedron, octahedron, dodecahedron and icosahedron. Therefore, the proposed method is to realize the uniform placement for arbitrary number of microphones on the surface of the sampling ball. With this method, the number of samples that have the minimum the orthonormality error can be founded.

The rest of the paper is arranged as follows. In Sect. 2, the Ambisonics encoding methods is introduced; in Sect. 3, the proposed microphone array arranged method is presented and in Sect. 4, the simulation experiments are carried out to evaluate the performance of the proposed method; lastly, conclusions based on the emulation experiments are summarized.

2   Ambisonics Encoding Methods

HOA is grounded on spatial harmonic representation of 3D sound field, the so-called Fourier-Bessel series, which comes from the resolution of the Acoustic wave equation in a source-free region of space. In the spherical coordinate system, the sound pressure defined by the Fourier-Bessel series is presented as follows [2]:

$$p(kr, \theta, \delta) = \sum_{m=0}^{+\infty} i^m j_m(kr) \sum_{n=0}^{m} \sum_{\sigma=\pm 1} B_{mn}^{\sigma} Y_{mn}^{\sigma}(\theta, \delta) \qquad (1)$$

Radial functions $j_m(kr)$ are spherical Bessel functions. Angular functions Y_{mn}^{σ} are the so-called spherical harmonics where m and n are the order of the spherical harmonics with $m \geq 0$, $0 \leq n \leq m$ and $\sigma = \pm 1$.

In order to realize a spherical HOA microphone arary, we must use a finite number of sensors. The order of the Ambisonics is limited to M, and the number of the microphones is Q, $Q \geq (M+1)^2$. The microphones are arranged on a spherical ball's surface, and the angle locations according to the center of the spherical ball are (θ_q, δ_q), $1 \leq q \leq Q$. As the sampling sphere is a rigid ball, the signal picked up by the microphone q on the surface of the sphere is a combination of direct and diffracted sound pressure [2]:

$$S_q(kR, \theta, \delta) = \sum_{m=0}^{+\infty} \frac{i^{m-1}}{(kR)^2 h_m^{-'}(kR)} \sum_{n=0}^{m} \sum_{\sigma=\pm 1} B_{mn}^{\sigma} Y_{mn}^{\sigma}(\theta_q, \delta_q), \qquad (2)$$

where R is the radius of the sphere, S_q is the signal picked up by the qth microphone, $1 \leq q \leq Q$ and $h_m^-(kR)$ is the divergent Hankel function according to kR.

The equations above can be expressed in matrix style,

$$s = T \cdot b \tag{3}$$

where

$$T = \begin{pmatrix} Y_{00}^1(\theta_1, \phi_1) & \cdots & Y_{M0}^1(\theta_1, \phi_1) \\ \vdots & \ddots & \vdots \\ Y_{00}^{-1}(\theta_Q, \phi_Q) & \cdots & Y_{M0}^{-1}(\theta_Q, \phi_Q) \end{pmatrix} \cdot diag[W_m(kR)] \tag{4}$$

$$W_m(kR) = \frac{i^{m-1}}{(kR)^2 h_m^{-'}(kR)} \tag{5}$$

$$s = (S_1, S_2, \ldots, S_Q)^T \tag{6}$$

$$b = (B_{00}^1, B_{11}^1, B_{11}^{-1}, B_{10}^1 \ldots, B_{M0}^1)^T \tag{7}$$

Due to the spherical geometry of the microphone array, the matrix T is a product of a real matrix Y whose columns are sampled spherical harmonics by a diagonal matrix of radial-dependent filters. The vector s contains the signals recorded by the Q sensors and b is the vector whose elements are the unknown HOA signals up to a finite order M.

In general, such a linear system cannot be solved exactly and the problem amounts to find an approximate solution which minimizes the square norm of the residual:

$$min \parallel s - Tb \parallel_2^2 \tag{8}$$

The least square solution b_{LS} to this minimization problem satisfies a new system of linear equations called normal equations.

$$(T^*T)b_{LS} = T^*s \tag{9}$$

So the solution of the system can be expressed as follows:

$$b_{LS} = (T^*T)^{-1}T^*s \tag{10}$$

By introducing the factorized expression of the matrix T in the equation, we finally obtain the HOA signal:

$$b_{LS} = diag[\frac{1}{W_m(kR)}]Es \tag{11}$$

where

$$E = (Y^TY)^{-1}Y^T \tag{12}$$

The discrete formulation of HOA recording introduces some limitations in the estimation process. The number of distinguishable harmonics depends on both the microphone number and their distribution. In order to estimate the harmonics up to order M, as least $(M+1)^2$ microphones are needed. Furthermore, the distribution method of the microphone array must satisfied the discrete orthonormality properties for sampled spherical harmonics. Therefore, we apply the uniformly sampling method to achieve the discrete orthonormality properties of the microphones array.

3   Methods

This section proposed a physical model to find a nearly uniform sampling scheme for arbitrary number of microphones. In the model, the microphones are regarded as charged particles and they can freely move on the surface of the sphere. There are repulsion force between particles, the value of which is inversely proportional to the square of the distance. Under the function of the repulsion force, the particle will move until the equilibrium state is reached, then the distribution of particles can be regarded as approximately uniform distribution. The calculate flow can be viewed in Fig. 1.

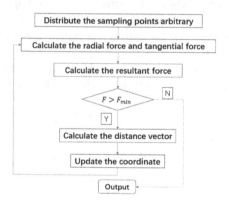

Fig. 1. Calculation flow of the method.

Firstly, the charged particles are randomly distributed on the surface of a unit sphere. The number of the particles is N and the value is set at $[0, 0, 0]$. The repulsion force constant is G.

Secondly, the distance vectors between each two points i and j are calcalted:

$$\overrightarrow{d_{ij}} = \overrightarrow{r_i} - \overrightarrow{r_j} \tag{13}$$

where the $\overrightarrow{r_i}$ and $\overrightarrow{r_j}$ are the coordiantes of the points i and j respectively. Then the size of the vector $\overrightarrow{d_{ij}}$ are calculated, denoted as L_{ij}. Assuming each particle carries the same charge, the repulse force between each two particles are inversely proportional to the square of their distance. The repulsion force of point i can be calculated as:

$$\vec{F_i} = \sum_{j \neq i} \vec{d_{ij}}/(L_{ij})^3 \tag{14}$$

Then for point i, the effect of the resultant force can be represented by the inner product of the resultant force and the coordinate vector. It consists of two parts: the radial component $\vec{F_{r_i}}$ which will not influence the particle motion and the tangential component $\vec{F_{v_i}}$ which lead to particle motion:

$$\vec{F_{r_i}} = \vec{r_i} \cdot (\vec{F_i} \cdot \vec{r_i}) \tag{15}$$

$$\vec{F_{v_i}} = \vec{F_i} - \vec{F_{r_i}} \tag{16}$$

Under the function of the tangential component of the resultant force, the particle has an displacement. The displacement $\vec{m_i}$ of the particle is the product of G and the acceleration.

$$\vec{m_i} = G \cdot \vec{F_{v_i}} \tag{17}$$

Then the coordinate of the particle $\vec{r_i}$ are changed to the sum of the original coordinate and the displacement.

$$\vec{r_i} = \vec{r_i} + \vec{m_i} \tag{18}$$

As the particles move along the tangential direction, they will fly out of the spherical surface. Normalize the new coordinate and let the particle move back to the surface of the spherical and the new coordinate can be expressed as

$$\vec{r_i} = \frac{\vec{r_i}}{|\vec{r_i}|} \tag{19}$$

Then the resultant force of each point is calculated. If each of them are less than the threshold value F_{min}, the iteration is finished, otherwise, continue the loop. When the cycle is over, the potential on the surface of the spherical is nearly uniform. So the particles are distributed uniformly.

Using the method mentioned above, given any number of microphones, we can obtain a uniform distribution of the samples on the surface of the sphere. Although this method yield a uniformly distribution of microphone array, the orthonormality condition for the sampled spherical harmonics must be analyzed in terms of the matrix D define by Eq. (20) which will be mentioned below.

4   Evaluation Experiments

Using the method mentioned above, the distribution scheme of 32 particles was obtained, which is shown in Fig. 2.

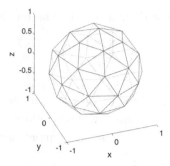

Fig. 2. Distribution diagram of the 32 microphones.

In order to verify the rationality of this distribution method, The orthonormality properties of induced sampled spherical harmonics should be examined. This can be done by considering the orthonormality error matrix defined by [2]:

$$D = I_K - \frac{1}{Q}Y^T Y \tag{20}$$

where I_K is the $K \times K$ identity matrix, K is the number of spherical harmonics of order M, Q is the number of sampling points, and the columns of the matrix Y are sampled spherical harmonics.

Figure 3 shows the matrix D of the distribution scheme generated by the method described above. Orthonormality error between two sampled spherical harmonics is represented by a small gray square. The coordinate axis of the figure represents the different orders of the spherical harmonic function. The lighter the color, the better the orthonormality of the spherical harmonics between different orders.

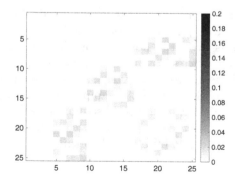

Fig. 3. The orthonormality error of the proposed method.

Figure 4 shows the orthonormality error associated to the pentaki-dodecahedron. As shown in the pictures, the method proposed in the paper performs similar to the pentaki dodecahedron distribution.

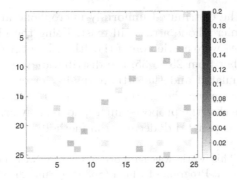

Fig. 4. The orthomormality error of the pentaki-dodecahedron method.

Since the Spherical harmonic function is space symmetry, the microphone array will satisfy the orthonormality only when its sampling mode is space symmetry. Though we can distribute the microphones uniformly, the symmetry error still exist which will result in the orthonormality error. Different number of microphones under the uniform distribution have different orthonormality error matrix. Assuming d_n is the sum of the elements in the orthonormality error matrix. When the number of microphones change from 25 to 35 under the 4th order of Ambisoncs, d_n are calculated and shown in the Fig. 5:

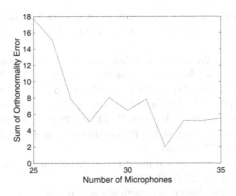

Fig. 5. The sum of orthonormality error for different number of microphones recording system.

Figure 5 indicated that the number of microphone that have the minimum orthonormality error is 32 when the order of Ambisonics is 4.

5   Conclusion

In this paper, a physical model is proposed with which we can distribute arbitrary number of microphone nearly uniformly on the surface of the sphere. However, as the spherical harmonic function is space symmetry which would not be strictly satisfied by the distributed uniformly microphone array, the orthonormality error of the microphone array will exist. Using the physical model, the evaluation experiments were carried on of the 4th order of Ambisonics for different microphone number from 25 to 35, the distribution scheme of each number of microphone is simulated and the orthonormality error is calculated. What should be noted is that when the microphone number is 32, the orhtonormality errors results indicate that the proposed microphone array arrangement method performs similar to the penatki-dodecahedron method.

Acknowledgments. This work was supported by the National High Technology Research and Development Program of China (863 Program: 2015AA016306), and the National Natural Science Foundation of China (Nos. 61175043, 61421062, 31170985).

References

1. Kordon, S., Batke, J.M., Krueger, A.: Method and Apparatus for Processing Signals of a Spherical Microphone Array on a Rigid Sphere Used for Generating an Ambisonics Representation of the Sound Field. US Patent App. 14/356,265 (2012)
2. Moreau, S., Daniel, J., Bertet, S.: 3D sound field recording with higher order ambisonics-objective measurements and validation of spherical microphone. In: 120th Audio Engineering Society Convention (2006)
3. Gerzon, M.A.: Periphony: with-height sound reproduction. J. Audio Eng. Soc. **21**(1), 2–10 (1973)
4. Craven, P.G., Gerzon, M.A.: Coincident Microphone Simulation Covering Three Dimensional Space and Yielding Various Directional Outputs. US Patent 4,042,779 (1977)
5. Daniel, J., Moreau, S., Nicol, R.: Further investigations of high-order ambisonics and wavefield synthesis for holophonic sound imaging. In: 114th Audio Engineering Society Convention, Amsterdam, The Netherlands, p. 5788 (2003)
6. Meyer, J., Elko, G.W.: A highly scalable spherical microphone array based on an orthonormal decomposition of the soundfield. In: 2002 IEEE International Conference on Acoustics, Speech, and Signal Processing (ICASSP), Orlando, Florida, USA, vol. 2, pp. 1781–1784. IEEE (2002)
7. Meyer, J., Elko, G.W.: A spherical microphone array for spatial sound recording. J. Acoust. Soc. Am. **111**(5), 2346 (2002)
8. Rafaely, B.: Analysis and design of spherical microphone arrays. IEEE Trans. Speech Audio Process. **13**(1), 135–143 (2005)
9. Verhaevert, J., Van Lil, E., Van de Capelle, A.: Uniform spherical distributions for adaptive array applications. In: VTC 2001 Spring. IEEE VTS 53rd Vehicular Technology Conference, Rhodes, Greece, vol. 1, pp. 98–102 (2001)

Learning an Alternating Bergman Network for Non-convex and Non-smooth Optimization Problems

Yiyang Wang[1], Risheng Liu[2,3,4(✉)], and Zhixun Su[1,5]

[1] School of Mathematical Science, Dalian University of Technology, Dalian, China
yywerica@gmail.com, zxsu@dlut.edu.cn
[2] DUT-RU International School of Information Science and Engineering,
Dalian University of Technology, Dalian, China
rsliu@dlut.edu.cn
[3] Shenzhen Key Laboratory of Media Security, Shenzhen University,
Shenzhen, China
[4] Key Laboratory for Ubiquitous Network and Service Software of Liaoning Province,
Dalian, China
[5] National Engineering Research Center of Digital Life, Guangzhou, China

Abstract. Recently, non-convex and non-smooth problems have received considerable interests in the fields of image processing and machine learning. The proposed conventional algorithms rely on carefully designed initializations, and the parameters can not be tuned adaptively during iterations with corresponding to various real-world data. To settle these problems, we propose an alternating Bregman network (ABN), which discriminatively learns all the parameters from training pairs and then is directly applied to test data without additional operations. Specifically, parameters of ABN are adaptively learnt from training data to force the objective value drop rapidly toward the optimal and then obtain a desired solution in practice. Furthermore, the basis algorithm of ABN is an alternating method with Bregman modification (AMBM), which solves each subproblem with a designated Bregman distance. This AMBM is more general and flexible than previous approaches; at the same time it is proved to receive the best convergence result for general non-convex and non-smooth optimization problems. Thus, our proposed ABN is an efficient and converged algorithm which rapidly converges to desired solutions in practice. We applied ABN to sparse coding problem with ℓ_0 penalty and the experimental results verify the efficiency of our proposed algorithm.

Keywords: Non-convex optimization · Alternating direction method · Sparse approximation · Learning-based algorithm · Optimization network

© Springer International Publishing AG 2017
Y. Sun et al. (Eds.): IScIDE 2017, LNCS 10559, pp. 11–27, 2017.
DOI: 10.1007/978-3-319-67777-4_2

1   Introduction

Recently, a variety of applications in the fields of data mining [1,2], signal and image processing [3–7], machine learning and statistical inference [8–12] have been modeled as non-convex and non-smooth optimization problems [13,14]. Among them, a class of non-convex and non-smooth problems has gained considerable attentions and many algorithms are designed for solving them, which can be formulated as a general problem with n variables as follows:

$$\min_{\mathbf{X}:=(\mathbf{x}_1,\ldots,\mathbf{x}_n)} \Psi(\mathbf{X}) = H(\mathbf{X}) + \sum_{i=1}^{n} f_i(\mathbf{x}_i), \tag{1}$$

where variables $\{\mathbf{x}_i\}_{i=1}^{n}$ can be either vectors or matrices throughout this paper. The objective function in problem (1) satisfies:

1. f_is are proper, lower semi-continuous functions.
2. H is a C^1 function; ∇H is Lipschitz continuous on a bounded set with constant M.
3. $\Psi(\mathbf{X})$ is a coercive, Kurdyka-Łojasiewicz (KL) function[1].

This problem (1) covers a variety of problems in various areas. For example, the non-negative matrix factorization problem framework [15,16] is always formed as a special two-block case of problem (1). The well-known sparse coding problems with various non-convex and non-smooth regularizers like ℓ_0 penalty [5,14] and MCP [17,18] belong to one-block case of problem (1).

Due to the non-convexity and non-smoothness of problem (1), there is not much work focusing on proposing converged and efficient algorithms for solving these complex problems. In 2009, the authors in [19] first propose a proximal algorithm (PA) for solving a one-block case of problem (1). This proposed algorithm is established with global convergence: *the algorithm generates a Cauchy sequence that converges to a critical point of the problem*, which as far as we know, is the best result in general non-convex and non-smooth optimization problems. Afterwards, they in [20,21] propose proximal alternating method (PAM) and proximal alternating linearized method (PALM) for solving problem (1). Following their pioneer work, plenty of algorithms for solving problem (1) have been proposed from different perspectives. The authors in [22] present a randomized/deterministic block prox-linear algorithm (BPA) for solving non-convex problems that may have block separable non-smooth terms. In [23], the authors introduce accelerated proximal gradient (APG) to general non-convex and non-smooth problems. They introduce a monitor in APG to ensure the objective function to have sufficient descent property. Moreover, the convergence rates of related algorithms are studied as well [19,24] by adding additional assumptions on KL functions.

All these proposed algorithms are demonstrated to converge to a critical point of the problem, however, in most cases, the converged critical points are not the desired solutions for specific problems. Since there always exists many

---

[1] The definition of KL function will be introduced afterwards.

critical points for non-convex and non-smooth problems, at the same time, the conventional algorithms converge to specific critical points by carefully selecting the algorithm parameters, thus it is extremely hard and time-consuming for conventional algorithms to converge to desired solutions in applications. One way to address this problem is to initialize the algorithm very close to the desired solution. For example, when solving the dictionary learning problem [21,25], the initialization of the dictionary is done via filling in local DCT transformation, which looks quite similar to the desired dictionary solution. With the carefully designed initializations, the process of tuning the algorithm parameters will be greatly simplified in practice.

On the other hand, in theory, the convergence rates of the previous algorithms are only affected by the objective function $\Psi(\mathbf{X})$ [23,24]. However, the truth is, carefully choosing appropriate parameters in the algorithms does affect the iteration time in practice. For example, the proximal parameters in all the algorithms mentioned above should not be set too large, otherwise the algorithm slowly converges with tiny step sizes. In the previous work, some strategies like line search with Barzilai-Borwein (BB) initialization [23,26] are proposed together with the algorithms to search appropriate parameters. However, the parameters in previously conventional algorithm framework can not be adaptively adjusted and tuned during iterations with corresponding to different data in various application problems. Adaptively selecting the parameters in every iteration does certainly shorten the whole number of iterations.

To address the above mentioned questions, we in this paper propose a data-driven general algorithm framework, which adaptively decides the algorithm parameters during iterations with the help of training data. Learning from training data give a guidance of the convergence direction so that it converges very close to the desired solution. Moreover, the algorithm parameters of the learning-based algorithm are adaptively decided during iterations to force the objective value drop rapidly to the optimal, thus it converges to the desired solution with much less iterations than conventional algorithms. It can be seen that learning an algorithm from data brings a new idea on designing an efficient and robust method for solving non-convex and non-smooth optimization problems, which is quite different from designing algorithms in conventional ways [19–23].

Further on, our learning-based algorithm framework named alternating Bregman network (ABN) can be regarded as a special network [27,28], which basis is an alternating method with Bregman modification (AMBM) designed in this paper for solving problem (1). Different from the previous work, our proposed AMBM solves each subproblem by adding a designated Bregman distance. Specifically, for different subproblems, AMBM adds different Bregman distances for the subproblems to simplify the solving processes. The previous algorithms, PA, PAM, PALM and BPA can all be seen as special cases of our AMBM. Thus our AMBM is more general, flexible and applicable than the previous proposed algorithms. At the same time, AMBM is established to retain the best convergence property and its convergence rate can also be theoretically analyzed for special objective function $\Psi(\mathbf{X})$, which shares the same conclusions with the previous algorithms. At last, our main contributions are summarized as follows:

1. We propose an efficient algorithm framework named AMBM for solving the non-convex and non-smooth optimization problems. In consideration of the complex subproblems in applications, AMBM appropriately solves each subproblem by adding a designated Bregman distance and is more general and flexible than existing algorithms. At the same time, AMBM has the same convergence property with the best result in general non-convex and non-smooth optimization.
2. On the basis of AMBM, we introduce an algorithm network named ABN, which is discriminatively learnt from training data. Different from conventional methods, ABN adaptively tune the algorithm parameters under the guidance of training data for rapidly converging to the desired solutions in practice.
3. We conduct experiments on solving sparse coding problem with ℓ_0 penalty. The experimental results in Sect. 5 demonstrate the efficiency of ABN and assist analyzing the convergence of our proposed algorithm.

The following part of this paper is organized as: we first give some necessary preliminaries in Sect. 2; then, in Sect. 3 we present the basis algorithm of ABN, i.e., AMBM on both detailed implementations and convergence analyses; follow that, we give the learning-based framework ABN in Sect. 4 and show experimental results in Sect. 5.

2   Preliminaries

To simplify the deduction, we analyze the convergence properties of AMBM and ABN on a special problem of (1):

$$\min_{\mathbf{z}:=(\mathbf{x},\mathbf{y})} \Psi(\mathbf{z}) = f(\mathbf{x}) + H(\mathbf{x},\mathbf{y}) + g(\mathbf{y}). \qquad (2)$$

For simplicity, we replace the notations in problem (1) with the new ones: correspondingly, f and g are proper, lower semi-continuous functions; H is a C^1 function and Ψ is a KL function. Though the convergence properties in this paper are analyzed on problem (2), it is straightforward to extend it to n-block problem (1) [21].

Definition 1 *(KL inequality and KL function). A proper, lower semi-continuous function σ has KL property at $\tilde{\mathbf{u}}$ if there exists $\eta, \varepsilon > 0$ and a function $\psi \in \Lambda_\eta$ such that for all $\mathbf{u} \in \mathcal{U}(\tilde{\mathbf{u}},\varepsilon) \cap [\sigma(\tilde{\mathbf{u}}) < \sigma(\mathbf{u}) < \sigma(\tilde{\mathbf{u}}) + \eta]$,*

$$\psi'(\sigma(\mathbf{u}) - \sigma(\tilde{\mathbf{u}}))\,dist(0, \partial\sigma(\mathbf{u})) \geq 1,$$

where $\mathcal{U}(\tilde{\mathbf{u}},\varepsilon)$ denotes a neighborhood of $\tilde{\mathbf{u}}$; $dist(\mathbf{u},S) := \inf\{\|\mathbf{v} - \mathbf{u}\|, \mathbf{v} \in S\}$ with $\|\cdot\|$ representing the ℓ_2 norm for vector and Frobenius norm for matrix *throughout the paper. Λ_η stands for a class of functions $\psi : [0, \eta) \to \mathbb{R}^+$ such that (1) $\psi(0) = 0$, $\psi'(s) > 0$ for all $s \in (0, \eta)$; (2) ψ is C^1 on $(0, \eta)$ and continuous at 0. Specially, $\psi(s)$ can be chosen as $cs^{1-\theta}$ with $c > 0$ and $\theta \in [0, 1)$ if Ψ is a semi-algebraic function [21]. Furthermore, if σ satisfies the KŁ property at each point in the domain of σ, then σ is called a KŁ function.*

KŁ functions include strongly convex functions, real analytic functions and semi-algebraic functions, e.g., capped-l_1 penalty [26], ℓ_p ($0 \leq p < 1$) norm [29], minimax concave penalty (MCP) [18] and smoothly clipped absolute deviation (SCAD) [30].

Definition 2 *(Bregman distance). φ is a convex differential function and its Bregman distance*

$$d_\varphi(\mathbf{u}, \mathbf{v}) = \varphi(\mathbf{u}) - \varphi(\mathbf{v}) - \nabla\varphi(\mathbf{v})^\top(\mathbf{u} - \mathbf{v}),$$

satisfies (1) $d_\varphi(\mathbf{u}, \mathbf{v}) \geq 0$, $\forall \mathbf{u}, \mathbf{v}$; (2) $d_\varphi(\mathbf{u}, \mathbf{v})$ is convex in $\mathbf{u}$, but not necessarily in $\mathbf{v}$.

We list the following φs according to the most commonly used Bregman distances [31,32], including

1. Euclidean distance: $\varphi(\mathbf{u}) = \|\mathbf{u}\|^2$;
2. Mahalanobis distance: $\varphi(\mathbf{u}) = \mathbf{u}^\top \mathbf{Q}\mathbf{u}$ with a symmetric positive definite $\mathbf{Q}$;
3. Kullback-Leibler divergence: $\varphi(\mathbf{u}) = \sum_j \mathbf{u}_j \log_2 \mathbf{u}_j$.

Assumption 3. *Suppose that $\{\mathbf{z}^k\}_{k\in\mathbb{N}}$ is the sequence generated by AMBM, then Bregman distances related to functions φ_1^k and φ_2^k are assumed to satisfy $d_{\varphi_1^k}(\mathbf{x}^{k+1}, \mathbf{x}^k) \geq \frac{\gamma_1^k}{2}\|\mathbf{x}^{k+1} - \mathbf{x}^k\|^2$, $d_{\varphi_2^k}(\mathbf{y}^{k+1}, \mathbf{y}^k) \geq \frac{\gamma_2^k}{2}\|\mathbf{y}^{k+1} - \mathbf{y}^k\|^2$ with bounded and positive parameters $\{\gamma_1^k\}_{k\in\mathbb{N}}$ and $\{\gamma_2^k\}_{k\in\mathbb{N}}$. Moreover, $\nabla\varphi_1^k$ and $\nabla\varphi_2^k$ are assumed to be Lipschitz continuous on bounded sets with bounded constants ζ_1^k and ζ_2^k.*

3   The Algorithmic Framework and Convergence Analyses

Before providing the learning-based algorithm network ABN, we in this section first introduce its basis algorithm AMBM, an alternating method with Bregman modification for solving problem (2). Firstly, we present the detailed implementation of AMBM and discuss its relationship with the previous algorithms. Then, we give convergence analyses, including the global convergence property and the convergence rate for our proposed AMBM. At last, we give examples to show the superiority of AMBM.

By adding Bregman distance, our proposed AMBM for solving problem (2) considers the following updates:

$$\mathbf{x}^{k+1} \in \arg\min_{\mathbf{x}} f(\mathbf{x}) + H(\mathbf{x}, \mathbf{y}^k) + d_{\varphi_1^k}(\mathbf{x}, \mathbf{x}^k),$$

$$\mathbf{y}^{k+1} \in \arg\min_{\mathbf{y}} g(\mathbf{y}) + H(\mathbf{x}^{k+1}, \mathbf{y}) + d_{\varphi_2^k}(\mathbf{y}, \mathbf{y}^k), \tag{3}$$

where $d_{\varphi_1^k}$ and $d_{\varphi_2^k}$ are the Bregman distances with respect to φ_1^k and φ_2^k that satisfy Assumption 3.

The proposed AMBM solves each subproblem of $\mathbf{x}^{k+1}$ and $\mathbf{y}^{k+1}$ in a flexible formulation with specific Bregman distances. Many existing methods [19, 21, 23] can be regarded as a special case of AMBM. For example, it is chosen as Euclidean distance in PA [19]. However, in PALM [21], the Bregman distances are chosen as Mahalanobis distances under special conditions to meet the requirement of symmetric, positive and definite Q for both subproblems of $\mathbf{x}$ and $\mathbf{y}$. All the subproblems in these previous work share the same Bregman distance. However, an appropriate Bremgan distance will certainly benefits to solving subproblems efficiently. For example, for solving the L_0 gradient minimization problem in [33]: $\min_{\mathbf{x},\mathbf{y}} \|\mathbf{y} - \mathbf{c}\|^2 + \lambda\|\mathbf{x}\|_0 + \|\mathbf{A}\mathbf{x} - \mathbf{y}\|^2$, the Bregman distance should be chosen as the Euclidean distance for $\mathbf{y}$-subproblem but the Mahalanobis distance for $\mathbf{x}$-subproblem so that each subproblem can be solved explicitly and efficiently. Thus our AMBM provides the chance for efficiently solving each subproblem with different Bregman distances. Further discussions on our proposed AMBM algorithm will be shown in Sect. 3.2. We would like to give the convergence analyses of our AMBM at first.

3.1 Convergence Analyses

Under the Assumption 3, we can obtain the key lemma as follows. The two assertions proposed in the following key lemma is the cornerstone for proving the main convergence theorem, i.e., Theorem 4.

Lemma 1. *Supposing* $\{(\mathbf{x}^k, \mathbf{y}^k)\}_{k\in\mathbb{N}}$ *be a bounded sequence generated by AMBM for solving the problem (2), then there exist positive integers* α *and* β *such that the following assertions hold:*

$$\Psi(\mathbf{z}^k) - \Psi(\mathbf{z}^{k+1}) \geq \alpha(\|\mathbf{x}^{k+1} - \mathbf{x}^k\|^2 + \|\mathbf{y}^{k+1} - \mathbf{y}^k\|^2), \tag{4}$$

$$dist(0, \partial\Psi(\mathbf{z}^k)) \leq \beta(\|\mathbf{x}^k - \mathbf{x}^{k-1}\| + \|\mathbf{y}^k - \mathbf{y}^{k-1}\|), \tag{5}$$

where constants are denoted: $\alpha = \min_{k\in\mathbb{N}}\{\frac{\gamma_1^k}{2}, \frac{\gamma_2^k}{2}\}$ *and* $\beta = \max_{k\in\mathbb{N}}\{\zeta_1^k, M + \zeta_2^k\}$.

Proof. Firstly, we prove the sufficiently descent property in Eq. (4). From the update of AMBM, i.e. Eq. (3), we have the following inequalities:

$$f(\mathbf{x}^{k+1}) + H(\mathbf{x}^{k+1}, \mathbf{y}^k) + d_{\varphi_1^k}(\mathbf{x}^{k+1}, \mathbf{x}^k) \leq f(\mathbf{x}^k) + H(\mathbf{x}^k, \mathbf{y}^k) + d_{\varphi_1^k}(\mathbf{x}^k, \mathbf{x}^k),$$

$$g(\mathbf{y}^{k+1}) + H(\mathbf{x}^{k+1}, \mathbf{y}^{k+1}) + d_{\varphi_2^k}(\mathbf{y}^{k+1}, \mathbf{y}^k) \leq g(\mathbf{y}^k) + H(\mathbf{x}^{k+1}, \mathbf{y}^k) + d_{\varphi_2^k}(\mathbf{y}^k, \mathbf{y}^k).$$

By adding the above two inequalities together, we have that

$$\Psi(\mathbf{x}^k, \mathbf{y}^k) - \Psi(\mathbf{x}^{k+1}, \mathbf{y}^{k+1})$$
$$\geq f(\mathbf{x}^k) + H(\mathbf{x}^k, \mathbf{y}^k) + g(\mathbf{y}^k) - f(\mathbf{x}^{k+1}) - H(\mathbf{x}^{k+1}, \mathbf{y}^{k+1}) - g(\mathbf{y}^{k+1})$$
$$\geq d_{\varphi_1^k}(\mathbf{x}^{k+1}, \mathbf{x}^k) - d_{\varphi_1^k}(\mathbf{x}^k, \mathbf{x}^k) + d_{\varphi_2^k}(\mathbf{y}^{k+1}, \mathbf{y}^k) - d_{\varphi_2^k}(\mathbf{y}^k, \mathbf{y}^k).$$

From the definition of the Bregman distance, we can see that $d_{\varphi_1^k}(\mathbf{x}^k, \mathbf{x}^k) = 0$ and $d_{\varphi_2^k}(\mathbf{y}^k, \mathbf{y}^k) = 0$. Together with the inequalities in Assumption 3, we conclude the proof of the sufficiently descent property Eq. (4):

$$\Psi(\mathbf{x}^k, \mathbf{y}^k) - \Psi(\mathbf{x}^{k+1}, \mathbf{y}^{k+1})$$

$$\geq \frac{\gamma_1^k}{2}\|\mathbf{x}^{k+1} - \mathbf{x}^k\|^2 + \frac{\gamma_2^k}{2}\|\mathbf{y}^{k+1} - \mathbf{y}^k\|^2 \geq \alpha(\|\mathbf{x}^{k+1} - \mathbf{x}^k\|^2 + \|\mathbf{y}^{k+1} - \mathbf{y}^k\|^2),$$

where $\alpha = \min_{k \in \mathbb{N}}\{\frac{\gamma_1^k}{2}, \frac{\gamma_2^k}{2}\}$.

Secondly, we prove the subgradient lower bound property, i.e. Eq. (5). We first write the first-order optimality condition of the update in Eq. (3):

$$0 = \mathbf{u}_1^k + \nabla_{\mathbf{x}} H(\mathbf{x}^k, \mathbf{y}^{k-1}) + \nabla d_{\varphi_1^k}(\mathbf{x}^k, \mathbf{x}^{k-1}),$$
$$0 = \mathbf{u}_2^k + \nabla_{\mathbf{y}} H(\mathbf{x}^k, \mathbf{y}^k) + \nabla d_{\varphi_2^k}(\mathbf{y}^k, \mathbf{y}^{k-1}),$$

where $\mathbf{u}_1^k \in \partial f(\mathbf{x}^k)$ and $\mathbf{u}_2^k \in \partial g(\mathbf{y}^k)$. Then we conclude that

$$(\nabla_{\mathbf{x}} H(\mathbf{x}^k, \mathbf{y}^k) - \nabla_{\mathbf{x}} H(\mathbf{x}^k, \mathbf{y}^{k-1}) - \nabla d_{\varphi_1^k}(\mathbf{x}^k, \mathbf{x}^{k-1}), -\nabla d_{\varphi_2^k}(\mathbf{y}^k, \mathbf{y}^{k-1}))$$
$$\in (\partial_{\mathbf{x}} \Psi(\mathbf{x}^k, \mathbf{y}^k), \partial_{\mathbf{y}} \Psi(\mathbf{x}^k, \mathbf{y}^k)).$$

Therefore, we have

$$\|\partial \Psi(\mathbf{x}^k, \mathbf{y}^k)\| \leq \|\partial_{\mathbf{x}} \Psi(\mathbf{x}^k, \mathbf{y}^k)\| + \|\partial_{\mathbf{y}} \Psi(\mathbf{x}^k, \mathbf{y}^k)\|$$
$$\leq \|\nabla_{\mathbf{x}} H(\mathbf{x}^k, \mathbf{y}^k) - \nabla_{\mathbf{x}} H(\mathbf{x}^k, \mathbf{y}^{k-1})\| + \|\nabla d_{\varphi_1^k}(\mathbf{x}^k, \mathbf{x}^{k-1})\| + \|\nabla d_{\varphi_2^k}(\mathbf{y}^k, \mathbf{y}^{k-1})\|$$
$$= \|\nabla_{\mathbf{x}} H(\mathbf{x}^k, \mathbf{y}^k) - \nabla_{\mathbf{x}} H(\mathbf{x}^k, \mathbf{y}^{k-1})\| + \|\nabla \varphi_1^k(\mathbf{x}^k) - \nabla \varphi_1^k(\mathbf{x}^{k-1})\| + \|\nabla \varphi_2^k(\mathbf{y}^k) - \nabla \varphi_2^k(\mathbf{y}^{k-1})\|$$
$$\leq M\|\mathbf{y}^k - \mathbf{y}^{k-1}\| + \zeta_1^k\|\mathbf{x}^k - \mathbf{x}^{k-1}\| + \zeta_2^k\|\mathbf{y}^k - \mathbf{y}^{k-1}\|.$$

The equality comes from the definition of Bregman distance, and the second inequality is derived from the assumptions on the Lipschitz continuity of ∇H, $\nabla \varphi_1^k$ and $\nabla \varphi_2^k$. Thus, we have finished the proofs of the two inequalities in Lemma 1. ∎

An additional hypothesis is added on the boundedness of $\{\mathbf{z}^k\}_{k \in \mathbb{N}}$. This boundedness holds in several scenarios such as the function Ψ has bounded level sets [21]. Obviously, the objective function Ψ is sufficiently descent (Eq. (4) in Lemma 1) during iterations. This non-increasing property is the key for proving the main theorem. Moreover, the main theorem can be proved in exactly the same way as [21]. Thus we only present the main theorem as follows and refer readers to [21] for detailed proof.

Theorem 4 (*Convergence result*). *Under the Assumption 3, $\{(\mathbf{x}^k, \mathbf{y}^k)\}_{k \in \mathbb{N}}$ is supposed to be a bounded sequence generated by AMBM for problem (2), we can conclude from Lemma 1 that $\{\mathbf{z}^k\}_{k \in \mathbb{N}}$ is a Cauchy sequence that converges to a critical point $\hat{\mathbf{z}}$ of Ψ.*

The convergence rate for non-convex and non-smooth optimization is quite complicated. The authors in [19] provide estimations when the function Ψ is semi-algebraic. Since the convergence rate of AMBM can be similarly estimated as [19], we only provide the conclusion in Theorem 5. Moreover, from the Theorem 5 we can tell that the convergence rate is not affected by the algorithm but the objective function Ψ.

Theorem 5 *(Convergence rate).* *If the function Ψ is semi-algebraic and the desingularizing function ψ is chosen as $\psi(s) = cs^{1-\theta}$ with $c > 0$ and $\theta \in [0, 1)$. Then by assuming the iterative sequence $\{(\mathbf{x}^k, \mathbf{y}^k)\}_{k\in\mathbb{N}}$ is bounded, we have the following estimations.*

(1) If $\theta = 0$, then $\exists K_1$ s.t. for $\forall k > K_1$, $\Psi(\mathbf{z}^k) = \Psi(\widehat{\mathbf{z}})$ and AMBM terminates in finite steps.

(2) If $\theta \in (0, 1/2]$, then there exists $\omega > 0$ and $\varrho \in [0, 1)$ such that $\|\mathbf{z}^k - \widehat{\mathbf{z}}\| \leq \omega\varrho^k$.

(3) If $\theta \in (1/2, 1)$, then there exists $\omega > 0$ such that $\|\mathbf{z}^k - \widehat{\mathbf{z}}\| \leq \omega k^{-\frac{1-\theta}{2\theta-1}}$.

The convergence property of AMBM can be summarized in a sentence: *if $\{\mathbf{z}^k\}_{k\in\mathbb{N}}$ is a bounded sequence and Assumption 3 holds, $\mathbf{z}^k$ always converges to a critical point $\widehat{\mathbf{z}}$ of Ψ no matter where the iteration starts.* The Bregman modification in AMBM provides a flexible framework but at the same time keeps the global convergence property (i.e., Theorem 4) and remains the convergence rate (i.e., Theorem 5) unchanged with the previous work [20,21,23,34], which is so far the best result identified by researchers as far as we know. It should be emphasized that a critical point of non-convex problem could be a local minimizer under some conditions, e.g. the second-order sufficient conditions [35].

3.2 Further Discussions on AMBM

It is apparent that not a few Bregman distances satisfy Assumption 3. However, different parameters γ lead to various Bregman distances. For clarity, we still take the L_0 gradient minimization problem [33]: $\min_{\mathbf{x},\mathbf{y}} \|\mathbf{y} - \mathbf{c}\|^2 + \lambda\|\mathbf{x}\|_0 + \|\mathbf{A}\mathbf{x} - \mathbf{y}\|^2$ as an example.

For efficiently updating the $\mathbf{x}^{k+1}$, we should add a special Mahalanobis distance with $\varphi(\mathbf{x}) = \mu\|\mathbf{x}\|^2 - \|\mathbf{A}\mathbf{x} - \mathbf{y}^k\|^2$ and $\mathbf{Q} = \mu\mathbf{I} - \mathbf{A}^\top\mathbf{A}$, where $\mathbf{I}$ denotes the identity matrix. Moreover, parameter μ should satisfy $\mu > \lambda_m$ (λ_m denotes the maximum eigenvalue of $\mathbf{A}^\top\mathbf{A}$) to make the Bregman distance $d_\varphi(\mathbf{x}, \mathbf{x}^k)$ satisfy $d_\varphi(\mathbf{x}^{k+1}, \mathbf{x}^k) \geq \frac{\gamma}{2}\|\mathbf{x}^{k+1} - \mathbf{x}^k\|^2$ with $\gamma = \mu - \lambda_m$. Hence, different μ in $\{\mu | \mu = \gamma + \lambda_m, \gamma > 0\}$ lead to different Bregman distances $d_\varphi(\mathbf{x}, \mathbf{x}^k)$ and then affect the step sizes of each iteration. In the following sections, we add the decisive parameter μ on $d_{\varphi(\mu)}(\mathbf{x}, \mathbf{x}^k)$ to specify particular Bregman distance.

It is well-known that an appropriate step size is beneficial to the convergence performance in first-order methods. Many strategies like line search [23] and BB rule [26] are proposed to search a good step size for each iteration. Inspired by data-driven methods, we in this paper learn the step size (i.e. parameter

μ) of each iteration adaptively and then produce an efficiently solving system, i.e., ABN for specific problems. The details of this learning-based algorithm are presented in the following section.

4   The Alternating Bregman Network and Theoretical Guarantee

To force the algorithm rapidly converging to a desired solution, we propose the learning-based algorithm network ABN on the basis of AMBM. The algorithm parameters of ABN is learnt from training data first and then is directly applied to all the test data with same distribution. This novel learning-based algorithm approaches to the desired solution rapidly and at the same time remains the same convergence property with its basis algorithm AMBM. In this section, at first, we present the motivation of learning an algorithm from data and give the general form of ABN. Then, detailed algorithm implementations are given for both learning and test part. At last, analyses on the theoretical guarantee of ABN are given with discussions.

4.1   Motivation and Detailed Implementation

As discussed in the previous section, the step size in each iteration plays an important role in first-order algorithms. Though many strategies like line search have been proposed to search a better step size, selecting appropriate parameters to force the algorithm converge to a desired solution is still extremely hard and time-consuming. Inspired by data-driven methods [27,28], we bring the assistance of the training data in our proposed AMBM and learn the step sizes adaptively.

Many existing data-driven algorithm network in machine learning [27,28] always contain two main stages. In the first stage, i.e., the training stage, the training data $\{z_1^*, \ldots, z_P^*\}$ are used to help train an efficient and universally applicable system for specific applications. Then in the second stage, i.e., the test stage, the pre-learnt algorithm net can be directly used for test data without manual operations. Inspired by the leaning-based process, we propose to learn an algorithm net to avoid manually setting step sizes in iterations. After learning the convergence algorithm with fixed parameters, prescribed expected error and certain iteration number, each test datum gets the result by applying the system directly without extra manually operations. But one must keep in mind that the learnt system is not available for all possible test samples, but only for *input variables drawn from the same distribution as the training samples* [28].

First, we would like to present the strategy for learning the algorithm parameters of ABN with the help of training data. By regarding the ground truth $z_p^* \in \{z_1^*, \ldots, z_P^*\}$ as a critical point of the optimization problem (2), the step size parameters μ_1^k and μ_2^k (corresponding to the Bregman distance $d_{\varphi_1^k(\mu_1^k)}(\mathbf{x}, \mathbf{x}^k)$ and $d_{\varphi_2^k(\mu_2^k)}(\mathbf{y}, \mathbf{y}^k)$) should be chosen to help ABN converges rapidly to the

---

Algorithm 1. ABN: the learning part.

1: Parameters setting: ε, $l_{\max}$, ϵ and $T = 0$.
2: Variables initialization: $\{\mathbf{z}_p^0\}_{p=1}^P$, μ_1^0 and μ_2^0
3: **while** $\sum_p \|\mathbf{z}_p^k - \mathbf{z}_p^*\| < \varepsilon$ **do**
4:     $T = T + 1$.
5:     Parameters setting: $l = 1$, $\{\nu_1^l\}_{l \in \mathbb{N}}$, $\{\nu_2^l\}_{l \in \mathbb{N}}$, $\mu_1^{k,0} = \mu_1^{k-1}$ and $\mu_2^{k,0} = \mu_2^{k-1}$.
6:     **while** $l < l_{\max}$ **do**
7:         (For example, use Projected Gradient Descent.)
8:         $\delta\mathbf{x}_p^{k,l} = \frac{\partial\mathbf{x}_p^{k+1}}{\partial\mu_1}(\mu_1^{k,l})$, $\delta\mathbf{y}_p^{k,l} = \frac{\partial\mathbf{y}_p^{k+1}}{\partial\mu_2}(\mu_2^{k,l})$.
9:         $\mu_1^{k,l+1} = \Pi_{\mathcal{X}_1^k}(\mu_1^{k,l} - \nu_1^l \Sigma_p \frac{\partial \mathcal{L}_p}{\partial \mathbf{x}_p^k}\delta\mathbf{x}_p^{k,l})$.
10:        $\mu_2^{k,l+1} = \Pi_{\mathcal{X}_2^k}(\mu_2^{k,l} - \nu_2^l \Sigma_p \frac{\partial \mathcal{L}_p}{\partial \mathbf{y}_p^k}\delta\mathbf{y}_p^{k,l})$.
11:        If $\|\mu_1^{k,l+1} - \mu_1^{k,l}\| + \|\mu_2^{k,l+1} - \mu_2^{k,l}\| \le \epsilon$ break;
12:        Set $l = l + 1$;
13:    **end while**
14:    Output $\mu_1^k = \mu_1^{k,l+1}$ and $\mu_2^k = \mu_2^{k,l+1}$.
15:    $\mathbf{x}_p^{k+1} \in \arg\min_{\mathbf{x}} f(\mathbf{x}) + H(\mathbf{x}, \mathbf{y}_p^k) + d_{\varphi_1^k(\mu_1^k)}(\mathbf{x}, \mathbf{x}_p^k)$.
16:    $\mathbf{y}_p^{k+1} \in \arg\min_{\mathbf{y}} g(\mathbf{y}) + H(\mathbf{x}_p^{k+1}, \mathbf{y}) + d_{\varphi_2^k(\mu_2^k)}(\mathbf{y}, \mathbf{y}_p^k)$
17: **end while**
18: Save $\{\mu_1^k\}_{k=1}^T$ and $\{\mu_2^k\}_{k=1}^T$.

Algorithm 2. ABN: the test part.

1: With fixed $\{\mu_1^k\}_{k=1}^T$ and $\{\mu_2^k\}_{k=1}^T$ learnt by Alg.1
2: Variable initialization: $\mathbf{z}^0$.
3: **for** $k = 1, 2, ..., T$ **do**
4:     $\mathbf{x}^{k+1} \in \arg\min_{\mathbf{x}} f(\mathbf{x}) + H(\mathbf{x}, \mathbf{y}^k) + d_{\varphi_1^k(\mu_1^k)}(\mathbf{x}, \mathbf{x}^k)$.
5:     $\mathbf{y}^{k+1} \in \arg\min_{\mathbf{y}} g(\mathbf{y}) + H(\mathbf{x}^{k+1}, \mathbf{y}) + d_{\varphi_2^k(\mu_2^k)}(\mathbf{y}, \mathbf{y}^k)$
6: **end for**
7: Output $\mathbf{z}^T$.

---

ground truth $\mathbf{z}_p^*$. Specifically, this intuitive command can be formulated as an optimization problem, that is,

$$\mu_1^k \in \min_{\mu_1} \mathcal{L}(\mathbf{x}_p^{k+1}(\mu_1), \mathbf{x}_p^*), \quad \mu_2^k \in \min_{\mu_2} \mathcal{L}(\mathbf{y}_p^{k+1}(\mu_2), \mathbf{y}_p^*), \qquad (6)$$

where we add subscript p to specify $\mathbf{x}_p^{k+1}$ and $\mathbf{y}_p^{k+1}$ as the iterative sequence related to $\mathbf{z}_p^* = (\mathbf{x}_p^*, \mathbf{y}_p^*)$. On the other hand, the step size μ_1^k and μ_2^k should ensure the Assumption 3 holds for $d_{\varphi_1^k(\mu_1^k)}(\mathbf{x}, \mathbf{x}^k)$ and $d_{\varphi_2^k(\mu_2^k)}(\mathbf{y}, \mathbf{y}^k)$, that is, $\mu_1^k \in \mathcal{X}_1^k$ and $\mu_2^k \in \mathcal{X}_2^k$ with

$$\mathcal{X}_1^k := \{\mu_1 | d_{\varphi_1^k(\mu_1)}(\mathbf{x}^{k+1}, \mathbf{x}^k) \ge \frac{\gamma_1^k}{2}\|\mathbf{x}^{k+1} - \mathbf{x}^k\|^2, \gamma_1^k > 0\},$$

$$\mathcal{X}_2^k := \{\mu_2 | d_{\varphi_2^k(\mu_2)}(\mathbf{y}^{k+1}, \mathbf{y}^k) \ge \frac{\gamma_2^k}{2}\|\mathbf{y}^{k+1} - \mathbf{y}^k\|^2, \gamma_2^k > 0\}.$$

$$(7)$$

Then by taking the average error of all the training data into consideration, we propose the learning process of ABN and present it as a bi-level optimization problem as follows to obtain μ_1^k and μ_2^k

$$
\begin{aligned}
\min_{\mu_1,\mu_2} \sum_{p=1}^{P} & \mathcal{L}(\mathbf{x}_p^{k+1}(\mu_1),\mathbf{x}_p^*) + \mathcal{L}(\mathbf{y}_p^{k+1}(\mu_2),\mathbf{y}_p^*), \\
\text{s.t. } \mathbf{x}_p^{k+1} & \in \arg\min_{\mathbf{x}} f(\mathbf{x}) + H(\mathbf{x},\mathbf{y}_p^k) + d_{\varphi_1^k(\mu_1^k)}(\mathbf{x},\mathbf{x}_p^k), \\
\mathbf{y}_p^{k+1} & \in \arg\min_{\mathbf{y}} g(\mathbf{y}) + H(\mathbf{x}_p^{k+1},\mathbf{y}) + d_{\varphi_2^k(\mu_2^k)}(\mathbf{y},\mathbf{y}_p^k), \\
\mu_1 & \in \mathcal{X}_1^k, \mu_2 \in \mathcal{X}_2^k.
\end{aligned}
\tag{8}
$$

Then μ_1^k and μ_2^k can be obtained by applying projected gradient descent, stochastic projected gradient descent [36] or other efficient algorithms to solve the bi-level problem (8). We take the projected gradient descent as an example, then μ_1^k and μ_2^k are obtained by iteratively updates the following equation with certain parameter ν_1^l and ν_2^l

$$
\begin{aligned}
\mu_1^{k,l+1} &= \Pi_{\mathcal{X}_1^k}(\mu_1^{k,l} - \nu_1^l \frac{\partial \mathcal{L}_P}{\partial \mu_1}(\mu_1^{k,l})), \ \mu_1^{k,0} = \mu_1^{k-1} \in \mathcal{X}_1^k, \\
\mu_2^{k,l+1} &= \Pi_{\mathcal{X}_2^k}(\mu_2^{k,l} - \nu_2^l \frac{\partial \mathcal{L}_P}{\partial \mu_2}(\mu_2^{k,l})), \ \mu_2^{k,0} = \mu_2^{k-1} \in \mathcal{X}_2^k,
\end{aligned}
\tag{9}
$$

with $\mathcal{L}_P = \sum_{p=1}^{P} \mathcal{L}(\mathbf{x}_p^{k+1}(\mu_1),\mathbf{x}_p^*) + \mathcal{L}(\mathbf{y}_p^{k+1}(\mu_2),\mathbf{y}_p^*)$. After learning the step size parameters adaptively from training data, these parameters together with the number of iterations T can be fixed due to a prescribed expected error.

Therefore, for the test data from the same distribution, we can automatically obtain the convergence solution after T-step iteration. Moreover, we summarize the procedures of "learning" the algorithm ABN in Algorithm 1 for solving problem (2). The "test" part of ABN with fixed step size parameters, number of iteration T is summarized in Algorithm 2.

Remark 1. The proposed ABN can be seen as a bi-level optimization problem [37]. The upper level optimization solving $\{\mu_1^k\}_{k=1}^T$ and $\{\mu_2^k\}_{k=1}^T$ is the leader which tries to anticipate the next move of the lower level problem for solving $\{\mathbf{z}^{k+1}\}_{k=1}^T$. The basic idea of learning an algorithm by using bi-level optimization has been proposed by others [38], but the lower level optimization in their work is a convex problem. Moreover, learning with bi-level optimization has also been used for applications in image processing [27,39]. However, no theoretical analysis has been made to guarantee the convergence of these bi-level problems. The convergence analysis is quite challenging even for convex objective functions [38].

4.2   Theoretical Guarantee

It is observed that the best step size parameters $\{\mu_1^k\}_{k=1}^T$ and $\{\mu_2^k\}_{k=1}^T$ are chosen from sets χ_1^k and χ_2^k respectively. Thus parameters $\{\mu_1^k\}_{k=1}^T$ and $\{\mu_2^k\}_{k=1}^T$ ensure that the corresponding Bregman distances satisfy the inequality in Assumption 3. In addition, ABN also has the sufficient descent property, i.e. Eq. (4) and subgradient lower bound property, i.e. Eq. (5). Hence ABN has the same convergence property of AMBM, that is, *suppose* $\{\mathbf{z}^k\}_{k\in\mathbb{N}}$ *is a bounded sequence and Assumption 3 holds,* $\mathbf{z}^k$ *always converges to a critical point* $\widehat{\mathbf{z}}$ *of* Ψ *no matter where the iteration starts.* However, ABN adaptively decides the step sizes during iterations, which ensures ABN perform better than AMBM with manually operating parameters. Though a faster convergence of ABN can not be guaranteed from theoretical analysis, it is verified through practical applications (see experiments in Sect. 5).

Remark 2. The iterative number T for the algorithm system is affected by the convergence rate of the basic algorithm. For example, we assume that the objective function Ψ satisfies the KL property with desingularizing function $\psi(s) = cs^{1-\theta}$ with $\theta \in (0, \frac{1}{2}]$. So from the Theorem 5, there exists $\omega > 0$ and $\varrho \in [0, 1)$ such that $\|\mathbf{z}^T - \mathbf{z}^*\| \le \omega\varrho^T < \varepsilon$. Then it can be deducted that $T \ln \varrho \le \ln \frac{\varepsilon}{\omega}$. With the fact that $\ln \varrho < 0$, we conclude that for achieving the error precision of ε, a theoretical T should satisfy $T \ge \frac{\ln \varepsilon - \ln \omega}{\ln \varrho}$.

5   Experiments

We conduct experiments to help analyze the convergence property of our proposed algorithm ABN. We first introduce the experimental setup and then give the analyses based on experimental results.

5.1   Experimental Setup

We evaluate our ABN algorithm by considering the Sparse Coding algorithm with ℓ_0 regularizer, that is, $f_i(\mathbf{x}_i) = \lambda|\mathbf{x}_i|_0$ and $g(\mathbf{x}) = \sum_i \|\mathbf{A}\mathbf{x}_i - \mathbf{y}_i\|^2$. We test the ABN on synthetic data $\mathbf{x}_i$s ($\mathbf{x}_i \in \mathbb{R}^m$) which are sparse and somehow high dimensional. In addition, the observe vector $\mathbf{y}_i = \mathbf{A}\mathbf{x}_i + noise$. The dictionary $\mathbf{A}$ in our experiments contains orthogonal bases. The regularizer parameter λ is set as $5e3$ in all the experiments and we carefully select λ to ensure the ground truth be the local minimizer of the problem. All algorithms are implemented in Matlab and executed on an Intel(R) Core(TM) $i5$-4200M CPU (2.50 GHz) with 8 GB memory.

5.2   Experimental Evaluations and Analyses

For solving sparse coding with ℓ_0 penalty, the Bregman distance $d_{\varphi(\mu^k)}(\mathbf{x}, \mathbf{x}^k)$ is chosen as a special Mahalanobis distance with $\mathbf{Q} = \mu\mathbf{I} - \mathbf{A}^\top\mathbf{A}$ and the $\mu^k > \lambda_m$ with λ_m denotes the maximum eigenvalue of $\mathbf{A}^\top\mathbf{A}$. Some fixed parameters are set as $\nu^l = 1$ for all l, $l_{\max} = 20$ and $\epsilon = 1e^{-3}$.

Firstly, the experimental results on applying ABN to recover the signal with different dimensions are shown in Fig. 1. The blue circles are the ground truth of

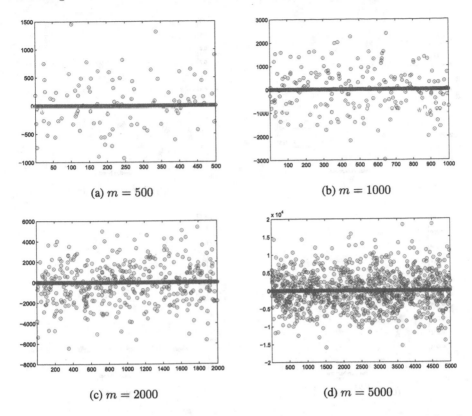

(a) $m = 500$ (b) $m = 1000$

(c) $m = 2000$ (d) $m = 5000$

Fig. 1. The recovered signals with different dimensions, which are recovered by the proposed ABN with learnt step sizes and iteration number. (Color figure online)

signals, and the red dots are the recovered signals by our proposed ABN method with learnt step sizes and iteration number. We give this result to demonstrate the accuracy on recovering signals solved by our algorithm.

Secondly, we give experiments to help verify the convergence property of ABN with learnt parameters calculated by Algorithm 1. It should be mentioned that some parameters in Fig. 2 are fixed with $m = 50$ and $\mu^{k,0} = 1.3\lambda_m$. By giving different initializations, the changes of the step sizes are given in Fig. 2(a). It can be seen from the figure that the step sizes changes a lot when the initialization (e.g., $\mathbf{x}_{ij}^0 = 1e5$) is far from the global minimizer. Then, we give the convergence performances on the test data in Fig. 2(b)–(d). Specifically, we give the relative error $(\|\mathbf{x}^{k+1} - \mathbf{x}^k\|^2)/\|\mathbf{x}^k\|^2$ in Fig. 2(b). The solid lines in Fig. 2(b) are corresponding to the relative error curves with learnt μ^ks while the dashed lines are with the curves of fixed $\mu^k = 1.3\lambda_m$. Obviously, the convergence property of ABN with learnt step sizes are quite similar on various test data. Moreover, it is faster and more robust of the learnt algorithm to converge to a small neighborhood of the local minimizer. The last two figures in Fig. 2 are the relative error with the ground truth $\mathbf{x}^*$, i.e. $(\|\mathbf{x}^{k+1} - \mathbf{x}^*\|^2)/\|\mathbf{x}^*\|^2$ and the changes of the

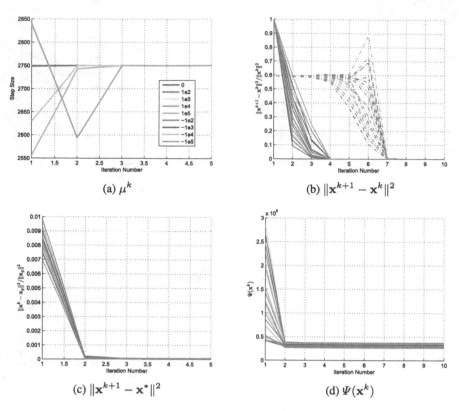

(a) μ^k

(b) $\|\mathbf{x}^{k+1} - \mathbf{x}^k\|^2$

(c) $\|\mathbf{x}^{k+1} - \mathbf{x}^*\|^2$

(d) $\Psi(\mathbf{x}^k)$

Fig. 2. Convergence property on (a) step size changes with various initial $\mathbf{x}^0$ and (b)–(d) the convergence properties of ABN on test data with adaptively step sizes learnt by Algorithm 1. The dotted curves in (b) can be regarded as one-block PALM [21] with fixed step sizes.

objective value $\Psi(\mathbf{x}^k)$. Both of them show the efficiency of our learnt algorithm ABN.

In the last, we show some evidences on $\mu^{k,0}$ in Fig. 3 We can obtain from the figures in Fig. 3 that the step size μ^k does affect the convergence rate. It seems like that a larger μ^k we have, a slower converge it is. For example, it takes 5 steps (dashed lines in Fig. 3(a)) for the case of $\mu^k = 1.1\lambda_m$ to converge to a relatively small error but it takes 9 steps for $\mu^k = 1.5\lambda_m$ in Fig. 3(d). However, the performances of the learnt algorithm ABN does not have such big differences. The solid lines are the convergence curves of the learnt algorithm ABN. The initial $\mu^{k,0}$s are set to the corresponding values of the fixed step sizes. It can be seen from the figures that it has 1 step difference between $\mu^{k,0} = 1.1\lambda_m$ and $\mu^{k,0} = 1.5\lambda_m$. This also demonstrates the robust and efficiency of ABN on deciding appropriate step sizes of iterations.

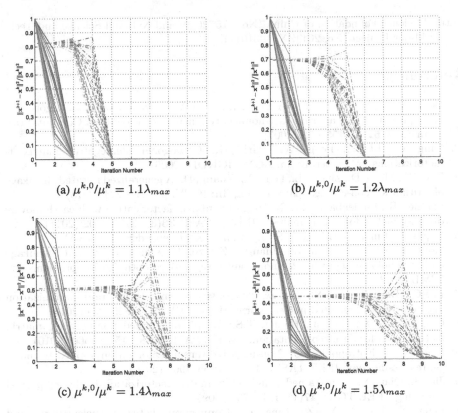

(a) $\mu^{k,0}/\mu^k = 1.1\lambda_{max}$

(b) $\mu^{k,0}/\mu^k = 1.2\lambda_{max}$

(c) $\mu^{k,0}/\mu^k = 1.4\lambda_{max}$

(d) $\mu^{k,0}/\mu^k = 1.5\lambda_{max}$

Fig. 3. The impact of $\mu^{k,0}$ and μ^k on ABN with learnt parameters (solid curves) and one-block PALM [21] with fixed step sizes (dotted curves).

6 Conclusion

We in this paper propose a learning-based algorithm network ABN for solving non-convex and non-smooth optimization problems. The basis algorithm of ABN is AMBM, which solves each subproblem with a specialized Bregman distance. Our proposed AMBM is more general and flexible than existing algorithm, and is proved to receive so far the best convergence result for general non-convex and non-smooth optimization problems. Different from the conventional algorithms, we propose the algorithm network ABN on the basis of AMBM to adaptively learn the algorithm parameters from training data to rapidly converge to desired solutions of the problems. Thus, our proposed ABN is an efficient and converged algorithm that adaptively tunes the algorithm parameters during iterations for fast converging performance in practice.

Acknowledgements. Risheng Liu is supported by the National Natural Science Foundation of China (Nos. 61672125, 61300086, 61572096, 61432003 and 61632019), the Fundamental Research Funds for the Central Universities (DUT2017TB02) and the Hong Kong Scholar Program (No. XJ2015008). Zhixun Su is supported by National

Natural Science Foundation of China (No. 61572099) and National Science and Technology Major Project (No. 2014ZX04001011).

References

1. Berry, M.W., Brown, M., Langvill, A.N., Pauca, V.P., Plemmons, R.J.: Algorithms and applications for approximate nonnegative matrix factorization. Comput. Stat. Data Anal. **52**(1), 155–173 (2007)
2. Ding, C., Li, T., Peng, W., Park, H.: Orthogonal nonnegative matrix t-factorizations for clustering. In: ACM SIGKDD (2006)
3. Zuo, W., Meng, D., Zhang, L., Feng, X., Zhang, D.: A generalized iterated shrinkage algorithm for non-convex sparse coding. In: ICCV, pp. 217–224 (2013)
4. Sandler, R., Lindenbaum, M.: Nonnegative matrix factorization with earth movers distance metric for image analysis. IEEE TPAMI **33**(8), 1590–1602 (2011)
5. Wang, Z., Ling, Q., Huang, T.S.: Learning deep ℓ_0 encoders. In: AAAI
6. Gong, P., Zhang, C., Lu, Z., Huang, J.Z., Ye, J.: A general iterative shrinkage and thresholding algorithm for non-convex regularized optimization problems. In: ICML (2013)
7. Lu, C., Tang, J., Yan, S., Lin, Z.: Nonconvex nonsmooth low rank minimization via iteratively reweighted nuclear norm. IEEE TIP **25**(2), 829–839 (2016)
8. Cai, D., He, X., Han, J., Huang, T.S.: Graph regularized non-negative matrix factorization for data representation. IEEE TPAMI **33**(8), 1548–1560 (2010)
9. Benetos, E., Kotropoulos, C.: Non-negative tensor factorization applied to music genre classification. IEEE TASLP **18**(8), 1955–1967 (2010)
10. Jia, S., Qian, Y.: Constrained nonnegative matrix factorization for hyperspectral unmixing. IEEE TGRS **47**(1), 161–173 (2009)
11. Peng, X., Lu, C., Yi, Z., Tang, H.: Connections between nuclear-norm and frobenius-norm-based representations. IEEE TNNLS
12. Deng, Y., Bao, F., Dai, Q.: A unified view of nonconvex heuristic approach for low-rank and sparse structure learning. In: Handbook of Robust Low-Rank and Sparse Matrix Decomposition: Applications in Image and Video Processing
13. Yuan, G., Ghanem, B.: A proximal alternating direction method for semi-definite rank minimization. In: AAAI (2016)
14. Wang, Y., Liu, R., Song, X., Su, Z.: Linearized alternating direction method with penalization for nonconvex and nonsmooth optimization. In: AAAI (2016)
15. Lee, D.D., Seung, H.S.: Algorithms for non-negative matrix factorization. In: NIPS, pp. 556–562 (2001)
16. Wang, Y.X., Zhang, Y.J.: Nonnegative matrix factorization: a comprehensive review. IEEE TKDE **25**(6), 1336–1353 (2013)
17. Shi, J., Ren, X., Dai, G., Wang, J.: A non-convex relaxation approach to sparse dictionary learning. In: CVPR (2011)
18. Zhang, C.H.: Nearly unbiased variable selection under minimax concave penalty. Ann. Stat. **38**(2), 894–942 (2010)
19. Attouch, H., Bolte, J.: On the convergence of the proximal algorithm for nonsmooth functions involving analytic features. MP **116**(1–2), 5–16 (2009)
20. Attouch, H., Bolte, J., Redont, P., Soubeyran, A.: Proximal alternating minimization and projection methods for nonconvex problems: an approach based on the kurdyka-lojasiewicz inequality. Math. Oper. Res. **35**(2), 438–457 (2010)
21. Bolte, J., Sabach, S., Teboulle, M.: Proximal alternating linearized minimization for nonconvex and nonsmooth problems. MP **146**(1–2), 459–494 (2014)

22. Xu, Y., Yin, W.: A globally convergent algorithm for nonconvex optimization based on block coordinate update, arXiv preprint arXiv:1410.1386
23. Li, H., Lin, Z.: Accelerated proximal gradient methods for nonconvex programming. In: NIPS (2015)
24. Frankel, P., Garrigos, G., Peypouquet, J.: Splitting methods with variable metric for kurdyka-łojasiewicz functions and general convergence rates. J. Optim. Theory Appl. **165**(3), 874–900 (2015)
25. Elad, M., Aharon, M.: Image denoising via sparse and redundant representations over learned dictionaries. IEEE TIP **15**(12), 3736–3745 (2007)
26. Pinghua, G., Zhang, C., Lu, Z., Huang, J., Jieping, Y.: A general iterative shrinkage and thresholding algorithm for non-convex regularized optimization porblems. In: ICML (2013)
27. Zuo, W., Ren, D., Gu, S., Lin, L.: Discriminative learning of iteration-wise priors for blind deconvolution. In: CVPR (2015)
28. Gregor, K., Lecun, Y.: Learning fast approximations of sparse coding. In: ICML (2010)
29. Foucart, S., Lai, M.-J.: Sparsest solutions of underdetermined linear systems via ℓ_q-minimization for $0 < q \leq 1$. ACHA **26**(3), 395–407 (2009)
30. Fan, J., Li, R.: Variable selection via nonconcave penalized likelihood and its oracle properties. J. Am. Stat. Assoc. **96**(456), 1348–1360 (2001)
31. Banerjee, A., Merugu, S., Dhillon, I.S., Ghosh, J.: Clustering with bregman divergences. JMLR **6**(4), 1705–1749 (2005)
32. Fischer, A.: Quantization and clustering with bregman divergences. J. Multivar. Anal. **101**(9), 2207–2221 (2010)
33. Xu, L., Lu, C., Xu, Y., Jia, J.: Image smoothing via l_0 gradient minimization. ACM TOG **30**(6), 174 (2011)
34. Kang, Y., Zhang, Z., Li, W.: On the global convergence of majorization minimization algorithms for nonconvex optimization problems, arXiv preprint arXiv:1504.07791
35. Nocedal, J., Wright, S.: Numerical Optimization. Springer Science & Business Media, New York (2006). doi:10.1007/978-0-387-40065-5
36. Sra, S., Nowozin, S., Wright, S.J.: Optimization for Machine Learning. MIT Press, Cambridge (2011)
37. Dempe, S.: Foundations of Bilevel Programming. Nonconvex Optimization & Its Applications, vol. 61
38. Ochs, P., Ranftl, R., Brox, T., Pock, T.: Bilevel optimization with nonsmooth lower level problems. In: Aujol, J.-F., Nikolova, M., Papadakis, N. (eds.) SSVM 2015. LNCS, vol. 9087, pp. 654–665. Springer, Cham (2015). doi:10.1007/978-3-319-18461-6_52
39. Schmidt, U., Roth, S.: Shrinkage fields for effective image restoration. In: CVPR (2014)

Sparse Multimodal Gaussian Processes

Qiuyang Liu and Shiliang Sun[✉]

Department of Computer Science and Technology, East China Normal University,
3663 North Zhongshan Road, Shanghai 200062, China
qiuyangliu2014@gmail.com, slsun@cs.ecnu.edu.cn

Abstract. Gaussian processes (GPs) are effective tools in machine learning. Unfortunately, due to their unfavorable scaling, a more widespread use has probably been impeded. By leveraging sparse approximation methods, sparse Gaussian processes extend the applicability of GPs to a richer data. Multimodal data are common in machine learning applications. However, there are few sparse multimodal approximation methods for GPs applicable to multimodal data. In this paper, we present two kinds of sparse multimodal approaches for multi-view GPs, the maximum informative vector machine (mIVM) and the alternative manifold-preserving (aMP), which are inspired by the information theory and the manifold preserving principle, respectively. The aMP uses an alternative selection strategy for preserving the high space connectivity. In the experiments, we apply the proposed sparse multimodal methods to a recent framework of multi-view GPs, and results have verified the effectiveness of the proposed methods.

Keywords: Classification · Kernel methods · Sparse Gaussian processes · Multimodal learning

1 Introduction

Gaussian processes (GPs) are widely used in machine learning and statistics as a powerful and flexible Bayesian nonparametric tool for probabilistic modeling [1]. However, computational requirements of the GPs grow as the cube of the size of the training set, impeding their widespread use to the scenario of scalable data. In order to address this limitation, researchers have recently proposed some sparse approximations [2–9]. They can be grouped into four classes. The first one uses only a subset of the data and focuses on the strategies of selecting the representative data points to form the subset [4]. The second kind of method concentrates on using a reduced-rank matrix to approximate the covariance matrix [2]. Another kind of method seeks to give a low rank approximation to the covariance matrix based on inducing points [6], while the fourth uses the method of variational inducing points [7,8]. These methods lead to a significant reduction of the computational complexity, which makes sparse Gaussian processes (SGPs) efficiently applied to a richer class of data [9–11].

© Springer International Publishing AG 2017
Y. Sun et al. (Eds.): IScIDE 2017, LNCS 10559, pp. 28–40, 2017.
DOI: 10.1007/978-3-319-67777-4_3

Typically, standard SGPs only pay attention to the scenario where data from a single modality are provided. In practice, multimodal data are common in applications of machine learning. They refer to the kind of data involving associated descriptive information from multiple domains, which are also called multi-view data. For instance, in speaker recognition, audio and visual data are correlated descriptions as phonemes and lip pose have correlations. In image classification, an image can be described by different features such as texture, shape, and color. As multiple modalities often provide complementary information, better performance is likely to be expected by utilizing multimodal instead of single-modal representations. Therefore, there has been a wealth of interest in multimodal learning recently [12–14]. However, SGPs, as popular and efficient methods in machine learning, are barely applied in multimodal learning. In this paper, our motivation is to study the sparse multimodal methods for GPs applicable to multimodal data.

We propose two kinds of sparse multimodal methods, the maximum informative vector machine (mIVM) and the alternative manifold-preserving (aMP), which are inspired by principles in information theory [4] and manifold learning [15], respectively. In the multimodal setting, the sparse multimodal methods need to consider all modalities together efficiently. On the one hand, the mIVM leverages a Gaussian process (GP) to model data from the same modality. Since data involve multiple related modalities, the mIVM uses multiple GPs, which are potentially correlated with each other as data from different modalities describe the same objective. For every example, it calculates the associated entropy reduction of each modality, and use the maximum entropy reduction among all the modalities as the overall entropy reduction of that data point. At each selection, the data point with the maximum overall entropy reduction is added to the sought sparse set. By using these strategies, the mIVM takes into consideration the entropy reduction of every modality for each data point. Overall, it tries to obtain the maximum of information among all the modalities with the minimum number of examples.

On the other side, for each modality, the aMP constructs a graph using the corresponding data. Vertices in different graphs are corresponding to each other if these vertices represent the same data point. Initially, the candidate set contains all the data points, while the sought sparse set is null. For each data point in the candidate set, the aMP calculates the degree of the corresponding vertex in each graph. To start the selection, it first chooses a modality randomly. Next, it selects a vertex with the maximum degree in the graph corresponding to the chosen modality. At the same time, all the vertices in other graphs corresponding to this chosen vertex are also selected. Then we include the data point associated with the chosen vertex into the sought sparse set and remove it from the candidate set. At the same time, we remove the chosen vertex and all the associated edges from each graph. Another round of selection will start with the alternative chosen modality. Overall, inspired by the manifold-preserving principle, the aMP makes use of data from all modalities by an alternative selection strategy for preserving the high space connectivity. Among the GP related multimodal learning methods [16–19], the two sparse multimodal approaches employ the recent framework of multi-view GPs [19] to evaluate the validity. It was a straightforward extension of GPs to multimodal learning with convenient implementation.

The contributions of our work are summarized as follows. First, we study the sparse multimodal methods for GPs from two different aspects and propose two kinds of sparse multimodal approaches. On the one hand, we present the mIVM to accommodate the multimodal data from the perspective of information theory. On the other hand, we present the aMP for multimodal sparsity from the manifold-preserving perspective. Secondly, we apply our two sparse multimodal methods to a recent multi-view GP framework. Finally, the proposed sparse multimodal methods can reduce the training time significantly with slight reduction of the accuracy, which extend the multimodal GPs to the scenario of scalable data.

The structure of the remainder of the paper is as follows. In Sect. 2, we briefly review GPs and propose the mIVM, our first sparse multimodal approach. Section 3 review the manifold-preserving principle, and present the other sparse multimodal method, aMP. A recent framework of multimodal GPs and our novel application are described in Sect. 4. Experimental results are reported in Sect. 5. Finally, we conclude this paper in Sect. 6.

2   Maximum Informative Vector Machine

This section first reviews the GP model, and then introduces the maximum Informative Vector Machine (mIVM), our first sparse multimodal approach. For the sake of clarity, in this section and Sect. 3, we take the case that data from two modalities are available as an example to illustrate our sparse multimodal methods. Similar algorithms can be mimicked if data concerning more than two modalities are adopted for the sparse multimodal GPs.

2.1   Gaussian Processes

GPs have proven their effectiveness as successful tools for classification and regression. They are frequently applied to describe a distribution over functions, and can be completely specified by its mean function and covariance function [1].

Suppose the training data are X, Y with N points, where $X = [x_1, x_2, ..., x_N]^T$, $x_i \in R^M$ is the ith input, $Y = [y_1, y_2, ..., y_N]^T$, and $y_i \in R$ is the ith output. The latent function of the data is denoted as f.

Following standard settings for GPs, the prior distribution for f is supposed to be Gaussian with a zero mean and a covariance matrix K, $f|X \sim \mathcal{N}(0, K)$, and the covariance function $k(x_i, x_j)$ determines the element K_{ij} of K. Numerous kernel functions can be applied in GPs. Since the squared exponential kernel is a frequently-used covariance function, we select it as the covariance function in this paper. The Gaussian likelihood for regression is $Y|f \sim \mathcal{N}(f, \sigma^2 I)$, and the marginal likelihood can be written as $Y|X \sim \mathcal{N}(0, K + \sigma^2 I)$. The posterior of the latent function is

$$f|Y \sim \mathcal{N}(\mu, \Sigma), \tag{1}$$

where $\mu = K(K + \sigma^2 I)^{-1}Y$ is the mean of the posterior distribution and $\Sigma = K - K(K + \sigma^2 I)^{-1}K$ is the covariance of the posterior distribution.

The prediction of a new point x^* is also Gaussian,

$$f^*|X, Y, x^* \sim \mathcal{N}(\bar{f}^*, \text{cov}(f^*)), \tag{2}$$

where $\bar{f}^* = k^{*\mathrm{T}}[K + \sigma^2 I]^{-1}Y$, $\text{cov}(f^*) = k(x^*, x^*) - k^{*\mathrm{T}}[K + \sigma^2 I]^{-1}k^*$, k is the covariance function, and k^* is the vector of covariance function values between x^* and the training data X.

Typically, if N is the size of training data, GPs need $\mathcal{O}(N^3)$ time and $\mathcal{O}(N^2)$ memory for training, and at least $\mathcal{O}(N)$ time for prediction on a test point.

2.2 Algorithm

Keeping the GP predictor only on a smaller subset of the data is a simple approximation to the full-sample GP predictor. This kind of approximation method makes sense if the information contained in points of the subset is sufficiently close to the information obtained by the full data set. Clearly, it is pivotal to select the points in the subset, which is called as the sparse set in this paper. Based on the information theory, [4] proposed to select the point with maximum differential entropy score to be included into the sparse set at every selection. In other words, they chose the point with the most information for inclusion. By the most information, it means that for a point the quantity

$$\Delta H_{in_i} = -\frac{1}{2} \log |\Sigma_{in_i}| + \frac{1}{2} \log |\Sigma_{i-1}| \tag{3}$$

is maximized, where Σ_{in_i} is the posterior covariance after choosing the n_ith point at the ith selection, and Σ_{i-1} is the posterior covariance after the $(i-1)$th choice. This quantity is the reduction in the posterior process entropy associated with selecting the n_ith point at the ith selection [4]. Inspired by these thoughts, we concentrate on the multimodal cases and propose the mIVM for sparse multimodal GPs.

The mIVM use a GP to model data from the same modality, which means that for each modality, the mIVM models data by a GP. In each selection, for each candidate point, it first calculates the entropy reduction associated with every modality. Next, the overall entropy reduction associated with the candidate is determined by the maximum among all the modalities. Then the candidate giving the largest overall reduction in the posterior process entropy is added to the sparse set.

Formally, assume that we have a data set $D = \{(x_i^1, x_i^2, y_i)\}_{i=1}^N$ with N examples, where $x_i^1 \in R^{M_1}$ is the ith observation from the first modality, $x_i^2 \in R^{M_2}$ is the ith observation from the second modality, and $y_i \in \{+1, -1\}$ is the corresponding label. Denote $X^1 = [x_1^1, ..., x_N^1]^{\mathrm{T}}$, $X^2 = [x_1^2, ..., x_N^2]^{\mathrm{T}}$, and $Y = [y_1, ..., y_N]^{\mathrm{T}}$. Let T denote the sparse set, I denote the candidate set, and t denote the number of points in the sparse set, namely the size of the sought sparse set.

The mIVM use two GPs to model two modalities of data. Specifically, it uses one GP to modal data from the first modality, i.e. $\{X^1, Y\}$, and uses the other GP to modal data from the second modality, i.e. $\{X^2, Y\}$. That is, the prior distributions for the latent functions f_1 on the first modality of data and f_2 on the second modality of data are assumed to be Gaussian, i.e. $p(f_1|X^1) = \mathcal{N}(0, K_1)$, and $p(f_2|X^2) = \mathcal{N}(0, K_2)$, where K_1 is the covariance matrix about the first modality of data and K_2 is the covariance matrix about the second modality of data. As for the likelihood, we use the Gaussian likelihood here. Although the Gaussian noise model is originally developed for regression, it has also been proved effective for classification, and its performance typically is comparable to the more complex probit and logit likelihood models used in classification problems [20]. Therefore, we also use Gaussian noise model for classification tasks in this paper.

Initially, T is a null set and I contains all the N examples. At the ith ($i = 1...t$) selection, the entropy reductions with the n_ith point for the first modality and the second modality are obtained by

$$\Delta H_{in_i}^1 = -\frac{1}{2} \log |\Sigma_{in_i}^1| + \frac{1}{2} \log |\Sigma_{i-1}^1|, \tag{4}$$

and

$$\Delta H_{in_i}^2 = -\frac{1}{2} \log |\Sigma_{in_i}^2| + \frac{1}{2} \log |\Sigma_{i-1}^2|, \tag{5}$$

respectively, where $\Sigma_{in_i}^1$ is the posterior covariance for the first modality of data after choosing the n_ith point at the ith selection, Σ_{i-1}^1 is the posterior covariance for the first modality of data after the $(i-1)$th choice, and $\Sigma_{in_i}^2$ and Σ_{i-1}^2 are defined analogously for the second modality. The overall entropy reduction associated with the n_ith point is given by

$$\Delta H_{in_i} = \max(\Delta H_{in_i}^1, \Delta H_{in_i}^2). \tag{6}$$

Then, at the ith selection, the n_i^*th data point is selected for inclusion at the sparse set T and removed from the candidate set I, where

$$n_i^* = \max_{n_i}(\{\Delta H_{in_i}\}_{n_i \in I}). \tag{7}$$

The selection procedure repeats until t points are added into the sparse set T.

The mIVM explores a sparse representation of multimodal data, which leverages the information from the input data and corresponding output labels. It attempts to obtain the maximum amount of information among all the modalities with the minimum number of data points. The computational complexity of the mIVM is $\mathcal{O}(t^2N)$, where t is the number of data points included in the sparse multimodal representation.

3   Alternative Manifold-Preserving

We first review the principle of manifold-preserving. Then we introduce our second sparse multimodal method, the alternative Manifold-Preserving (aMP).

3.1   Manifold-Preserving

Assume we are given a graph $G(V, E, W)$ corresponding to a manifold with vertex set $V = \{v_i\}_{i=1}^m$, edge set E, and weight matrix W, and the number of vertices reserved in the desired sparse graphs is s. Manifold-preserving seeks a sparse graph G', which is a subgraph of G with s vertices, having a high connectivity with G, that is to say, a candidate that maximizes the quantity

$$\frac{1}{m-s} \sum_{i=s+1}^{m} (\max_{j=1,..,s} W_{ij}), \tag{8}$$

where W_{ij} characterizes the similarity or closeness between the ith vertex and the jth vertex, and a small value denotes a low similarity [15].

The manifold-preserving sparse graph G' focuses on reducing the number of vertices, and the edge weights from the original graph G to sparse graph G' need not change. The high demand for space connectivity inclines to choose vital data points and thus remove outliers and noisy points, which can maintain the manifold structure. The maximum preservation of the manifold structure can be beneficial to machine learning tasks. Inspired by this thought, we propose the aMP in the following section.

3.2   Algorithm

To make this section self-contained, we repeat the data notations. We are given data $D = \{(x_i^1, x_i^2, y_i)\}_{i=1}^N$ with N examples, where $x_i^1 \in R^{M_1}$ is the ith observation from the first modality, $x_i^2 \in R^{M_2}$ is the ith observation from the second modality, and $y_i \in \{+1, -1\}$ is the corresponding output. Denote $X^1 = [x_1^1, ..., x_N^1]^{\mathrm{T}}$, $X^2 = [x_1^2, ..., x_N^2]^{\mathrm{T}}$, and $Y = [y_1, ..., y_N]^{\mathrm{T}}$.

We use $\{X^1, Y\}$ to construct the graph $G^1(V^1, E^1, W^1)$, where $V^1 = \{v_i^1\}_{i=1}^N$ is the vertex set of graph G^1. The graph $G^2(V^2, E^2, W^2)$ is constructed by using $\{X^2, Y\}$, where $V^2 = \{v_i^2\}_{i=1}^N$ is the vertex set of graph G^2. Clearly, graph G^1 is associated with the first modality of data, while graph G^2 is associated with the second modality. There are many methods to create the graphs. In this paper, we do not investigate the distinctions of properties of graphs constructed by different methods, but assume that a reasonable graph can be constructed.

Note that the vertex v_i^1 corresponds to the vertex v_i^2 since they are associated with the same example, namely the ith example. In fact, the ith example has observation (x_i^1, x_i^2) and label y_i, and is corresponding to the vertex v_i^1 in graph G^1, and vertex v_i^2 in graph G^2.

The degree $d^1(i)$ associated with vertex v_i^1 is defined to be $d^1(i) = \sum_{i \sim j} W_{ij}^1$, where $i \sim j$ denotes that there is an edge connecting the vertex v_i^1 and vertex

v_j^1 (if there is no edge between two vertices, their similarity is regarded as 0). For the degree $d^2(i)$ associated with vertex v_i^2, the definition is similar. Suppose that the number of retained examples is t. Our goal is to seek t examples to form a sparse set $T = \{x_i^1, x_i^2, y_i\}_{i \in T_I}$, where T_I is the index set of the sought sparse set, from the original N examples. Inspired by the manifold-preserving principle, we present the aMP whose details are described as follows.

The aMP first chooses a modality from the two modalities at random, for example, the second modality. Next, the vertex $v_{w_1}^2$ with the maximum degree in the graph associated with the chosen modality is selected. As we have mentioned above, the vertex $v_{w_1}^2$ is associated with the w_1th data point and vertex $v_{w_1}^1$ in the other graph. Thus, the vertex $v_{w_1}^1$ is also selected. All the edges and weights linked to the vertex $v_{w_1}^2$ from the original graph G_2 are then removed and all the edges and weights associated with the vertex $v_{w_1}^1$ in the other graph G_1 are also removed as they represent the same data point. At the same time, the chosen vertices and edges linking these vertices are added from the original graphs G^1 and G^2 to the corresponding sparse graphs G_s^1 and G_s^2 (which are null initially), respectively. Add the corresponding example w_1 to the index set T_I. Then a similar selection proceed on the resultant graphs with the first modality as the chosen modality. The alternative selection procedure repeats until t data points are added into the index set T_I. We summarize aMP in Algorithm 1.

Algorithm 1. Alternative Manifold-Preserving

Input: graphs $G^1(V^1, E^1, W^1)$, $G^2(V^2, E^2, W^2)$ with N vertices, t for the size of the sparse set, training data $\{(x_i^1, x_i^2, y_i)\}_{i=1}^N$.
Output: the index set T_I of sparse set, the sparse set T.

1: Initialize: a is randomly set in $\{1, 2\}$; $T_I = \emptyset$.
2: **for** $j = 1, ..., t$ **do**
3:     $b = a$, $a = (a \bmod 2) + 1$.
4:     compute degree $d^a(i)$ $(i = 1, ..., N - j + 1)$.
5:     pick one vertex v_w^a in graph G^a with the maximum degree.
6:     remove v_w^a and associated edges from graph G^a, remove v_w^b and associated edges from graph G^b.
7:     add w to the index set T_I.
8: **end for**
9: The sparse set is $T = \{x_i^1, x_i^2, y_i\}_{i \in T_I}$.

The aMP focuses on the sparse point selection for multimodal data. Motivated by the manifold preserving, it uses an alternative selection strategy to preserve the high space connectivity. Assume the maximum number of edges linked to a vertex in the original graphs G^1 and G^2 is d_E. The computational complexity of the aMP is

$$\mathcal{O}[d_E(N + (N - 1) + .. + (N - t + 1))] = \mathcal{O}(d_E N t), \tag{9}$$

Since the aMP is simple and efficient, it is quite straightforward to be applied to scalable data.

4 Application to Multi-view GPs

The framework of multi-view Gaussian processes (MvGPs) has recently been proposed as a straightforward extension of the GPs for multimodal data [19]. The core idea is to impose consistency between the posterior distributions of the functions across modalities.

Taking the data having two modalities as an example, the MvGP first models each modality of data by a GP. Then, it proposes the consistency criterion to regularize the objective function, and optimizes the hyperparameters collaboratively by the two modalities. The objective function of MvGP is

$$\min\{-[a\log p(\boldsymbol{Y}|\boldsymbol{X}^1) + (1-a)\log p(\boldsymbol{Y}|\boldsymbol{X}^2)]$$
$$+\frac{b}{2}[KL(p(\boldsymbol{f}_1|\boldsymbol{X}^1,\boldsymbol{Y})\|p(\boldsymbol{f}_2|\boldsymbol{X}^2,\boldsymbol{Y}))$$
$$+KL(p(\boldsymbol{f}_2|\boldsymbol{X}^2,\boldsymbol{Y})\|p(\boldsymbol{f}_1|\boldsymbol{X}^1,\boldsymbol{Y}))]\}, \tag{10}$$

where $\boldsymbol{X}^1 \in R^{N \times M_1}$ is the data matrix on the first modality, $\boldsymbol{X}^2 \in R^{N \times M_2}$ is the data matrix on the second modality, $\boldsymbol{Y}$ is corresponding label matrix, $\boldsymbol{f}^1$ and $\boldsymbol{f}^2$ are the associated latent functions for the two modalities of data, respectively, and a and b are parameters.

To demonstrate the performances of our proposed sparse multimodal methods, we apply them to the framework of MvGPs and use the combined models to solve the classification problem. For convenience, we denote the mIVM based MvGP as mMvGP, and the aMP based MvGP as aMvGP. The computational complexity of the mMvGP is

$$\mathcal{O}(t^2N + t^3) = \mathcal{O}(t^2N), \tag{11}$$

while the computational complexity of the aMvGP is

$$\mathcal{O}(d_ENt + t^3), \tag{12}$$

where N is the number of original training data points, t is the number of data points included in the sparse multimodal representation (usually, $t << N$), and d_E is the maximum number of edges linked to a vertex in the original graphs from all modalities. The original MvGP needs $\mathcal{O}(N^3)$ time, the same as the GP. From the analysis of computational complexity, it is clear that both the aMvGP and the mMvGP significantly reduce the training time. Thus, applying the mIVM and aMP to multimodal GPs would be quite efficient.

5 Experiment

In this section, experiments are conducted to assess the effectiveness of the two proposed sparse multimodal methods.

5.1 Data

Four Web-Page Data Sets. The web-page data sets, as widely used data sets in multimodal learning, consist of two-modalities web pages collected from computer science department websites of four universities: Cornell university, university of Washington, university of Wisconsin, and university of Texas. The two modalities are words occurring in a web page and words appearing in the links pointing to that page. We list the statistical information about the four data sets in Table 1. The web pages are classified into five classes: student, project, course, staff and faculty. In each data set, we set the category with the greatest size to be the positive class (denoted as "P class"), and all the other categories as the negative class (denoted as "N class").

Table 1. Statistical information of the data sets.

Data set	Size	Content dimension	Citation dimension	# P class	# N class
Cornell	195	1703	195	83	112
Washington	230	1703	230	107	123
Wisconsin	265	1703	265	122	143
Texas	187	1703	187	103	84
cora	2708	1433	2708	818	1890

Cora Data Set. The cora data set consists of 2708 scientific publications belonging to seven categories, of which the one with the most publications is set to be the positive class, and the rest the negative class. Each publication is represented by words in the content modality, and the numbers of citation links between other publications and itself in the citation modality. The dimensions are 1433 and 2708, respectively.

5.2 Setting

In the experiments, we select two-thirds of data in each data set as the training set, and the rest as the test set. For the four web-page data sets, the sizes of the sparse set are 40%, 60%, and 80% of the corresponding training sets, and we also conduct experiments without sparse approximation. For the cora data set, the sizes of the sparse set are 8%, 10%, and 12% of the size of the training data set. For comparison, we give a random sparse approximation for MvGP, which just randomly selects points to form the sparse set, and denote it as rMvGP. The kernel functions used in mIVM and aMP are the squared exponential kernel functions. After finding the sparse set, aMvGP, mMvGP, and rMvGP employ the similar hyperparameters learning and parameters setting as [19]. We repeat the experiments for all the data sets five times and record the average accuracies and the corresponding standard deviations. The average training times for each model on all the data sets are also reported.

5.3   Results on Four Web-Page Data Sets

We first evaluate mMvGP, aMvGP, and rMvGP on four web-page data sets in consideration of making comprehensive comparisons of the accuracies of such sparse methods. The results are shown in Fig. 1. Compared with the models without using sparse methods, the models leveraging sparse methods only reduce the classification accuracies slightly. For a range of t values, the classification results on the four data sets show that the aMvGP produce superior classification performance to other sparse models, which verifies the effectiveness of our proposed sparse multimodal method, the aMP. The performances of the mMvGP are not so well as aMvGP.

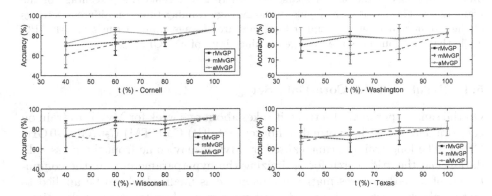

Fig. 1. Classification results for four web-page data sets. The x-axis corresponds to different settings of the size of sparse set, where t represents the percentage of the training set. The figures from top to bottom are results on the Cornell, Washington, Wisconsin, and Texas data set. Error bars represent standard deviations of the accuracies.

When the modalities are not necessarily compatible, a variant of the MvGP was given in [19]. We also combine the mIVM, the aMP, and random sparse approximation with this variant and denote the combinations as mMvGP2, aMvGP2, and rMvGP2, respectively. We evaluate mMvGP2, aMvGP2, and rMvGP2 on the four data sets. The corresponding classification results reflect that there is generally no improvement of the performances on accuracy.

The average training times of aMvGP and aMvGP2 on four data sets are presented in Fig. 2, which verify the significant reduction of computational complexity. It is shown that the training times increase rapidly with the size of the sparse set, which indicates that the sparse methods effectively reduce the training times. The average training times of other sparse models are comparable to aMvGP and aMvGP2.

Taken the computational complexity and the classification accuracy together, the sparse multimodal models significantly reduce the training times but without obvious loss of accuracies.

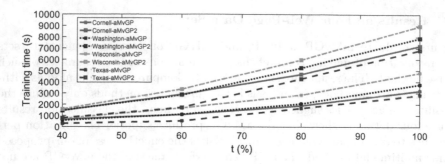

Fig. 2. Average training times on four web-page data sets. The x-axis corresponds to different settings of the size of sparse set, where t represents the percentage of the training set. The y-axis corresponds to the average training times on different settings of the size of sparse set. The upper four lines are average training times of aMvGP2, while the lower four lines are average training times of aMvGP.

5.4   Results on the Cora Data Set

On the same experimental setting, it takes about one week for MvGP to train on the cora data set with a normal computer (Intel(R) Core(TM) i7-6700 3.40 GHz CPU). The long training time caused by MvGP may come from the cross validation and the grid search for optimizing the hyperparameters. As the computation of MvGP involves matrix inversions, it is unaffordable to be applied to large-scale data sets, such as a dataset with more than ten thousands points. Considering those factors, we choose the cora data set. Since the performances of mMvGP, aMvGP, and rMvGP are generally better than mMvGP2, aMvGP2, and rMvGP2, we only evaluate mMvGP, aMvGP, and rMvGP on the cora data.

The classification results are demonstrated in Table 2. The average training times of mMvGP, aMvGP, and rMvGP are shown in Fig. 3. Taking into account of the results in the figure and table, we find that the aMP and mIVM greatly reduce the training times of the multimodal GPs with an acceptable performance on the accuracy.

Table 2. The accuracies on the cora data set (%).

Model	6%	8%	10%
rMvGP	72.46 ± 4.22	73.32 ± 10.04	78.51 ± 1.61
mMvGP	74.78 ± 3.09	76.42 ± 3.85	75.56 ± 2.89
aMvGP	**76.00** ± 3.00	**77.11** ± 2.08	**79.90** ± 2.22

The accuracy of the MvGP for the cora data set is around 92%. We can see the loss of the accuracy is slight though we only use a tiny proportion of the training set, such as 8%. Specifically, aMvGP can achieve the accuracy of

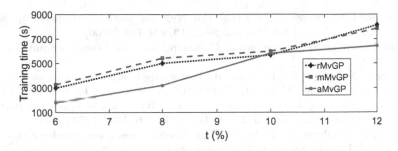

Fig. 3. Average training times on cora data. The x-axis corresponds to different settings of the size of sparse set, where t represents the percentage of the training set.

77.11% with only 8% of the whole training set, which is 83.82% of the accuracy of MvGP trained on the whole training set. It indicates that our sparse methods are scalable on large data sets.

Combining the results here and classification results on the four web-page data sets, we find that aMvGP achieves the best performance. We have also given the training times of aMvGP, which indicates that when it is unaffordable to train on the original full data set, we can use the aMvGP to approach a good approximation.

6   Conclusion

In this paper, we have proposed the mIVM and aMP as two kinds of sparse multimodal methods for the multimodal GPs. The mIVM is inspired by information theory, seeking the maximum amount of information from all the modalities with the same number of data points. The aMP is more intuitive, which adopts an alternative selection strategy to utilize data from all the modalities for preserving the high space connectivity. We apply the two proposed sparse multimodal methods to multi-view GPs to verify the effectiveness. The classification accuracies on four web-page data sets and the cora data set have shown that aMvGP outperforms other competitive methods. The scalability was also tested on preliminary experiments with tiny proportions of data for training. More experiments will be conducted in the future.

Acknowledgments. The corresponding author Shiliang Sun would like to thank supports from the National Natural Science Foundation of China under Projects 61673179 and 61370175, Shanghai Knowledge Service Platform Project (No. ZF1213), and the Fundamental Research Funds for the Central Universities.

References

1. Rasmussen, C.E., Williams, C.K.I.: Gaussian Processes for Machine Learning. MIT Press, Cambridge (2006)

2. Williams, C.K., Seeger, M.: Using the Nyström method to speed up kernel machines. Adv. Neural Inf. Process. Syst. **13**, 661–667 (2000)
3. Csató, L., Opper, M.: Sparse on-line Gaussian processes. Neural Comput. **14**, 641–668 (2002)
4. Lawrence, N., Seeger, M., Herbrich, R.: Fast sparse Gaussian process methods: the informative vector machine. Adv. Neural Inf. Process. Syst. **15**, 625–632 (2003)
5. Quiñonero-Candela, J., Rasmussen, C.E.: A unifying view of sparse approximate Gaussian process regression. J. Mach. Learn. Res. **6**, 1939–1959 (2005)
6. Snelson, E., Ghahramani, Z.: Sparse Gaussian processes using pseudo-inputs. Adv. Neural Inf. Process. Syst. **18**, 1257–1264 (2006)
7. Titsias, M.K.: Variational learning of inducing variables in sparse Gaussian processes. In: Proceedings of the 12th International Conference on Artificial Intelligence and Statistics, pp. 567–574 (2009)
8. Hensman, J., Fusi, N., Lawrence, N.D.: Gaussian processes for big data. In: Proceedings of the 29th Conference on Uncertainty in Artificial Intelligence, pp. 282–290 (2013)
9. Cheng, C.A., Boots, B.: Incremental variational sparse Gaussian process regression. Adv. Neural Inf. Process. Syst. **29**, 4410–4418 (2016)
10. Gal, Y., van der Wilk, M., Rasmussen, C.E.: Distributed variational inference in sparse Gaussian process regression and latent variable models. Adv. Neural Inf. Process. Syst. **27**, 3257–3265 (2014)
11. Deisenroth, M.P., Ng, J.W.: Distributed Gaussian processes. In: Proceedings of the 32nd International Conference on Machine Learning, pp. 1481–1490 (2015)
12. Ngiam, J., Khosla, A., Kim, M., Nam, J., Lee, H., Ng, A.Y.: Multimodal deep learning. In: Proceedings of the 28th International Conference on Machine Learning, pp. 689–696 (2011)
13. Sun, S.: A survey of multi-view machine learning. Neural Comput. Appl. **23**, 2031–2038 (2013)
14. Rao, D., De Deuge, M., Nourani-Vatani, N., Williams, S.B., Pizarro, O.: Multimodal learning and inference from visual and remotely sensed data. Int. J. Robot. Res. **36**, 24–43 (2016)
15. Sun, S., Hussain, Z., Shawe-Taylor, J.: Manifold-preserving graph reduction for sparse semi-supervised learning. Neurocomputing **124**, 13–21 (2014)
16. Shon, A.P., Grochow, K., Hertzmann, A., Rao, R.P.N.: Learning shared latent structure for image synthesis and robotic imitation. Adv. Neural Inf. Process. Syst. **19**, 1233–1240 (2005)
17. Yu, S., Krishnapuram, B., Rosales, R., Rao, R.B.: Bayesian co-training. J. Mach. Learn. Res. **12**, 2649–2680 (2011)
18. Xu, C., Tao, D., Li, Y., Xu, C.: Large-margin multi-view Gaussian process for image classification. In: Proceedings of the 5th International Conference on Internet Multimedia Computing and Service, pp. 7–12 (2013)
19. Liu, Q., Sun, S.: Multi-view regularized Gaussian processes. In: Kim, J., Shim, K., Cao, L., Lee, J.-G., Lin, X., Moon, Y.-S. (eds.) PAKDD 2017. LNCS, vol. 10235, pp. 655–667. Springer, Cham (2017). doi:10.1007/978-3-319-57529-2_51
20. Kapoor, A., Grauman, K., Urtasun, R., Darrell, T.: Gaussian processes for object categorization. Int. J. Comput. Vis. **88**, 169–188 (2010)

Document Analysis Based on Multi-view Intact Space Learning with Manifold Regularization

Zengrong Zhan[1,2](✉) and Zhengming Ma[1]

[1] School of Electronic and Information Engineering, Sun Yat-sen University,
Guangzhou, China
zhanpost@gmail.com, mazhengming@sysu.edu.cn
[2] School of Information Engineering, Guangzhou Panyu Polytechnic,
Guangzhou, China

Abstract. Document analysis plays an important role in our life, and traditional models like Latent Semantic Analysis (LSI) or Latent Dirichlet Allocation (LDA) cannot handle data from many sources. Multi-view learning technology like Multi-view Intact Space Learning (MISL), which integrates the complementary information on multiple views to discover a latent intact representation of the data, is effective for image or video application. But the model has not been applied to multi-lingual documents and has not considered the intrinsic geometrical and discriminating structure of the document data. To overcome this issue, we assume that if documents are close in the origin representation, they should also be close in the intact space representation. And we introduce a manifold regularization term to MISL so that the data is more smoothly in latent space. We conduct classification experiments on 10505 Wiki documents we crawled, and the result shows that it is outperforming TFIDF, LSI, LDA, and MISL.

Keywords: Multi-view learning · Manifold regularization · Document classification

1 Introduction

Document representation is the key problem for document analysis. The most common and natural way is using the Vector Space Model (VSM) [1] to represent every document as a bag of words. Distance measure methods like Cosine or inner product can be used as the similarity measure for documents. However, the Polysemy or Synonymy issues of documents make the VSM model unreliable or failed in real application. Hence, many dimension reduction techniques have been proposed to overcome this drawback. One of the most notable models is Latent Semantic Indexing (LSI) [2], which claims that using the representation in reduced latent space to measure the similarity between documents is more suitable and reliable than using the original representation. LSI has been followed by many researchers, and many variants versions have been proposed. The main

Y. Sun et al. (Eds.): IScIDE 2017, LNCS 10559, pp. 41–51, 2017.
DOI: 10.1007/978-3-319-67777-4_4

drawback of LSI is its unsatisfactory statistical formulation [3]. To address this issue, Hofman proposed a new model called Probabilistic Latent Semantic Indexing (PLSI) [4], which estimates the probability distribution of each document on the hidden topics. Using topics to represent the document is the main advantage of the PLSI. Unfortunately, the parameters of PLSI grow linearly with the size of the corpus, and it causes overfitting problems. To overcome this problem, Blei et al. [5] proposed the Latent Dirichlet Allocation (LDA) model, which uses K-parameter random variables for the probability distribution of each document over topics.

Despite the success of the existing algorithms like LSI or LDA in many areas, these algorithms are designed for only one view of the documents. While in real world application, many data are normally collected from different sources, single view data cannot comprehensively describe the features of documents. For example, so many multi-lingual documents have been generated in Wikipedia. In Wikipedia, people usually describe a term in many languages and some information may be missing in one language, but can be found in other languages. Since the multi-view model for data can better describe the features of data from different measuring methods, it has made great success in machine learning area [6].

Recently, a new method named Multi-view Intact Space Learning (MISL) [7] has been proposed and gotten so many attentions. The method assumes that each individual view only captures partial information, and all views together carry redundant information about the object. To module the problem, it assumes each view of the data is generated from an intact space by using a linear function, $z^v = W_v x$. Hence, the objective of the model is to learn a series of view generation matrix $\{W_v\}_{v=1}^m$. Although the model gets good performance on image data like Face Recognition, Human Motion Recognition, etc., it does not consider the special feature of documents for analysis. Especially, it does not consider the intrinsic geometrical and discriminating structure of the documents, which is very important to document classification.

In this paper, we proposed a method called Multi-view Intact Space Learning with Manifold Regularization (DMMR). It considers the multi-view intact space learning with document manifold. By discovering the local neighborhood structure, our algorithm can do better classification for documents. Our method assumes that two close documents in any view of the original representation will be close in the intact space in a high probability. We incorporate the nearest neighbor graph structure as the manifold regularization term into loss function of the multi-view intact space learning model.

The rest of this paper will be organized as follows: in Sect. 2, we introduce our proposed method by reviewing the multi-view intact space learning algorithm first and shows how to add the document manifold regularization term. Experimental results of document classification are presented in Sect. 3. Finally, we provide our conclusion in Sect. 4.

2   The Proposed Method

2.1   Problem Modeling

We assume that we are dealing with a set of multi-view documents, $Z = \{z_i^v | 1 \leqslant i \leqslant n, 1 \leqslant v \leqslant m\}$, where n is the document size and m is the view number. The information of each data point is composed of feature vectors of these m views, and since each view carries only some feature information of the documents, the information obtained from an individual view cannot completely describe all document information. Hence, combining all views information to better describe the documents become the popular method as a matter of course. The problem of multi-view intact space learning for documents is how to project those features of different views on an intact space to better describe the documents. For example, if we have a set of documents that describe terms such as "Service Economy" using different languages, then we can take each language as a view and the content of a certain item in a specified language will be the feature information on this view. And we assume there is an intact vector $x_i \in \Re^D$ for the i-th data point, therefore its v-th view data z_i^v can be reconstructed by a linear transformation,

$$z_i^v \leftarrow W_v x_i \tag{1}$$

Therefore, we can use the view generation matrix to transform the input documents to an intact space X that contain all information of the objects. And the reconstruction error over the latent intact space X can be measured using the Cauchy loss.

$$\min_{x,w} \frac{1}{mn} \sum_{v=1}^m \sum_{i=1}^n \log(1 + \frac{||z_i^v - W_v x_i||^2}{c^2}) \tag{2}$$

2.2   Multi-view Intact Space Learning with Manifold Regularization

By assuming that there is an intact vector for each data point in multi-view data, MISL method avoids "view insufficiency" problem with single view data information. However, the algorithm fails to discover the intrinsic geometrical and discriminating structure of the document space, which is essential to the real application. Hence, we introduce a manifold regularization to the objective function.

Recall that all documents can be represented in the intact space, one might hope that the knowledge of the relation for each document in single view may be exploited for better estimation of the representation of data in the intact space. Therefore, we make a specific assumption of the connection between z_i^v and x_i. We assume that if two documents $z_1^v, z_2^v \in Z$ in the v-th view are close in the intrinsic geometry, then they should be close to each other in the intact space, as we can see from Fig. 1. That means intact space representation x varies smoothly along the geodesics in the intrinsic geometry of Z. And this assumption is also referred to as manifold assumption [8], which is important to many algorithms in dimension reduction [8] and semi-supervised learning [9].

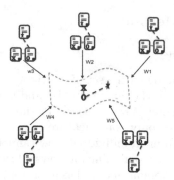

Fig. 1. Manifold for document space. The document X and O are close in the origin representation space and both of them are far from document ⋆. Hence, the geometrical features should be reserved in the intact space.

Here, we use $||x||_M^2$ to measure the smoothness of x along the geodesics in the intrinsic geometry of Z. In reality, the document manifold normally can not be computed, hence we use the studies on spectral graph theory and manifold learning theory in [10,11] to approximate $||x||_M^2$ through the nearest neighbor graph on a scatter of data points.

Consider a graph with n vertices, and each of which corresponds to a document in the document set Z. Define the edge weight matrix Ψ as follows:

$$\Psi_{ij} = \begin{cases} \min_v \ cos(z_i^v, z_j^v), \ if \ z_i^v \in N_k(z_j^v) \ or \ z_j^v \in N_k(z_i^v) \\ 0, \qquad\qquad\qquad\qquad\qquad Otherwise \end{cases} \tag{3}$$

where $N_k(z_i^v)$ denotes the set of k nearest neighbors of z_i in the v-th view. Thus the discrete approximation of $||x||_M^2$ can be computed as follows:

$$R_k = \frac{1}{2} \sum_{i,j=1}^{n} (x_i - x_j)^2 \Psi_{ij} \tag{4}$$

An intuitive explanation of minimizing R_k is that if two documents z_i and z_j are close in any view of data (i.e. Ψ_{ij} is close to 1), then their intact space representation x_i and x_j are similar to each other. Now we can redefine our intact space model as follows

$$\min_{x,w} \frac{1}{mn} \sum_{v=1}^{m} \sum_{i=1}^{n} \log(1 + \frac{||z_i^v - W_v x_i||^2}{c^2}) + \frac{1}{2}\alpha R_k \tag{5}$$

where α is the regularization parameter.

Moreover, we adopt the same regularization terms as those in MISL to penalize the latent data point x and the view generation matrix W, and we got the resulting objective function as follows:

$$\min_{x,w} \frac{1}{mn} \sum_{v=1}^{m} \sum_{i=1}^{n} log(1 + \frac{||z_i^v - W_v x_i||^2}{c^2}) + \frac{1}{2}\alpha R_k + C_1 \sum_{v=1}^{m} ||W_v||^2 + C_2 \sum_{i=1}^{n} ||x_i||^2$$
$$(6)$$

2.3 Opitmization

To solve the optimization problem in (6), we use the alternating optimization method which only considers updating one variable at a time and fixes the others. And the updated variable will be fixed in the next iteration when another variable is updating. In this section, we will discuss how to update each variable.

Updating x. When updating the $i-th$ point of data x in latent intact space X, we fix the view generation matrix W. And we obtain the following optimization problem.

$$\min_{x} J = \frac{1}{m} \sum_{v=1}^{m} log(1 + \frac{||z^v - W_v x||^2}{c^2}) + \frac{1}{2}\alpha(\sum_{j=1}^{n} ||x - x_j||^2 \psi_j + \sum_{i=1}^{n} ||x_i - x||^2 \psi_i) + C_2 ||x||^2$$
$$(7)$$

Setting the gradient of J over variable x to 0, and we obtain

$$\frac{1}{m} \sum_{v=1}^{m} \frac{-2W_v^T(z^v - W_v x)}{c^2 + ||z^v - W_v x||^2} + \alpha \sum_{j=1}^{n} (x - x_j)\psi_j + \alpha \sum_{i=1}^{n} (x - x_i)\psi_i + 2C_2 x = 0 \quad (8)$$

which is equal to

$$\frac{1}{m} \sum_{v=1}^{m} \frac{-2W_v^T(z^v - W_v x)}{c^2 + ||z^v - W_v x||^2} + 2\alpha \sum_{j=1}^{n} (x - x_j)\psi_j + 2C_2 x = 0 \qquad (9)$$

Let $\psi = [\psi_1, \psi_2, \ldots, \psi_n]$, the Eq. (7) can be rewritten as follows

$$\frac{1}{m} \sum_{v=1}^{m} \frac{-2W_v^T(z^v - W_v x)}{c^2 + ||z^v - W_v x||^2} + 2\alpha ||\psi|| x - 2\alpha X \psi^T + 2C_2 x = 0 \qquad (10)$$

Hence, we have

$$(\sum_{v=1}^{m} \frac{W_v^T W_v}{c^2 + ||z^v - W_v x||^2} + \alpha ||\psi|| (I + mC_2 I)) x = \alpha X \psi^T + \sum_{v=1}^{m} \frac{W_v^T z^v}{c^2 + ||z^v - W_v x||^2}$$
$$(11)$$

Let $r^v = z^v - W_v x$, and we define a weight function as follows

$$Q = [\frac{1}{C^2 + ||r^1||^2}, \ldots, \frac{1}{C^2 + ||r^m||^2}] \qquad (12)$$

By replacing Q to (9), and we get the final updating function for x.

$$x = (\sum_{v=1}^{m} W_v^T Q_v W_v + \alpha ||\psi|| (I + mC_2 I))^{-1} (\alpha X \psi + \sum_{v=1}^{m} W_v^T Q_v z^v) \qquad (13)$$

Updating W. Using all data points we got from (11) to fix the variables of x, the object function now is reduced to update the view generation function W, and it can be written as follows

$$\min_W J = \frac{1}{n} \sum_{i=1}^{n} log(1 + \frac{||z_i - Wx_i||^2}{c^2}) + C_1||W||_F^2 \tag{14}$$

Setting the gradient of J with respect to W,

$$W(\sum_{i=1}^{n} \frac{x_i x_i^T}{c^2 + ||z_i - Wx_i||^2} + nC_1 I) = \sum_{v=1}^{m} \frac{z_i x_i^T}{c^2 + ||z_i - Wx_i||^2} \tag{15}$$

Let $r_i = z_i - Wx_i$ and we define

$$Q = [\frac{1}{C^2 + ||r_1||^2}, \cdots, \frac{1}{C^2 + ||r_n||^2}] \tag{16}$$

we have the final updating function of W

$$W = \sum_{v=1}^{m} z_i Q_i x_i^T (\sum_{i=1}^{n} x_i Q_i x_i^T + nC_1 I)^{-1} \tag{17}$$

3   Experiments

3.1   Data Corpora

The data set we used for experiment compose of 10505 documents that are all crawled from Wikipedia. We write a python script based on the Scrapy [12] framework to crawl data from the following index URLs.

- https://en.wikipedia.org/wiki/Index_of_economics_articles
- https://en.wikipedia.org/wiki/Index_of_education_articles
- https://en.wikipedia.org/wiki/Index_of_law_articles
- https://en.wikipedia.org/wiki/Index_of_religion-related_articles
- https://en.wikipedia.org/wiki/Index_of_logic_articles
- https://en.wikipedia.org/wiki/Index_of_history_articles
- https://en.wikipedia.org/wiki/Index_of_health_articles

All items listed on content, which tags ID attribute with "mw-content-text", of the indexing page are crawled, and we drop those items whose title starts with "Index-of" or "Outline-of". Besides, when an item is crawled, its description in other languages (include German, Chinese, French, Spanish) are also crawled. All crawled data are stored in Mysql database, and we drop those items that don't have the description for all these 5 languages. And at last, there are 10505 items with 2101 items for each language, and they are separated into 7 categories according to their index URL, and we named the data set "W-2101".

After we got all items, we use the lxml html library [13] to extract the text of the page content of each item. For items in languages other than Chinese,

PunktSentenceTokenizer and WordPunctTokenizer from NLTK [14] are used to split the content to sentence and get the tokens of each term. Also, Stopwords and SnowballStemmer from NLTK library are used to remove the stop words and stem the tokens. In addition, tokens contain non-latin letters are dropped in these preprocess. For Chinese items, we use Stanford Segmenter [15] to get tokens for each document, and the punctuation, stop words or non-Chinese tokens are also removed.

At the end, we got 70–220 thousand features of items in each language, which are too large for us to conduct the experiments. Therefore, we use KMeans algorithm to cluster those futures and take the words of the center points as the features for each document. The number of the clusters is 500 and the features used in KMeans are all TDIDF values that are preprocessed by using Gensim library [16]. Besides, in order to conduct the experiment in different size of samples, we choose two subsets of the documents name W-200 and W-1000 which contains 200 samples and 1000 samples for each language respectively. The Statistics of these 3 data sets are listed in Table 1.

Table 1. Statistics of test data

	Religion	Economics	Education	Law	Logic	History	Health
W-2101	451	249	181	632	133	288	167
W-1000	197	109	88	278	52	199	77
W-200	62	19	24	41	9	25	20

3.2 Evaluation Measure

To measure the classification performance over the test, we use the classification accuracy which is defined as follows,

$$Accuracy = \frac{Number\ of\ Correctly\ Classified\ Test\ Data\ Points}{Number\ of\ Total\ Test\ Data\ Point} \tag{18}$$

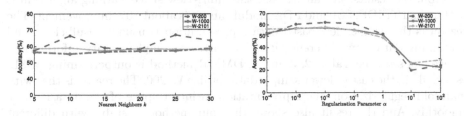

Fig. 2. Parameter analysis on k and α

3.3 Parameter Analysis

Our model has two essential parameters: the number of the nearest neighbors k in (3) and the regularization parameter α.

Parameter Analysis on Nearest Neighbors k. In order to show the robustness of our model, we conduct the experiments with different k in those data sets. Figure 2 shows the results of different choices, and it achieves consistent good performance with k varying from 5 to 30.

Parameter Analysis on Regularization Parameter α. The α controls the tradeoff of the Regularization Parameter, hence we perform the algorithm by using the values from 10^{-4} to 10^2 for parameter α. As we can see from Fig. 2, the α parameter will break down the accuracy when it raises from 1 to 10. From the experience, the best value for α is around $\frac{1}{n}$ where n is the total number of the instances.

3.4 Comparison Experiments

Classification is one of the most crucial techniques to organize the document in a supervised manner. Hence, we use the simplest method KNN to conduct the experiments. To demonstrate how the document classification performance can be improved by multi-view intact space learning approach. We also implement state-of-the-art algorithms includes LSI, LDA to extract the features for classification.

To conduct the experiments, we first run KNN algorithm on the features from TFIDF, LSI and LDA modules for documents in every language and the results are shown in Tables 2, 3 and 4. And the experimental protocol is that we randomly split the data set into 5 folds, and run 10 times for each round. The average accuracy is picked as the final result. To show the stability of different model, we run the KNN algorithm with 5–20 neighbors.

Besides, we use the same way to conduct the experiments on features from multi-view space learning algorithm and multi-view space learning algorithm with manifold regularization.

From the results, we can see that the multi-view space learning algorithm is better than TFIDF, LSI, and LDA module and it is about 10% improvement. The reason is that single view based data representation is usually insufficient, and combining data from different views can present complementary information. As we can see from Tables 2, 3 and 4, DMMR method is outperforming MISL, especially in the case of fewer sample data set like W-200. The reason is that with manifold regularization, the representation in intact space of the data is more smoothly. And the result also shows that our method is stable with different nearest neighbors K of KNN algorithm.

Table 2. Classification performance on W-2101

K	TFIDF EN	GE	ZH	FR	ES	Avg.	LSI EN	GE	ZH	FR	ES	Avg.	LDA EN	GE	ZG	FR	ES	Avg.	MISL	DMMR
5	0.44	0.38	0.42	0.41	0.46	**0.4211**	0.39	0.36	0.34	0.30	0.36	**0.3511**	0.45	0.43	0.42	0.40	0.44	**0.4305**	0.5419	0.5607
6	0.45	0.37	0.46	0.40	0.44	**0.4239**	0.39	0.35	0.37	0.30	0.37	**0.3577**	0.49	0.43	0.42	0.42	0.45	**0.4420**	0.5362	0.5693
7	0.47	0.36	0.43	0.41	0.45	**0.4231**	0.41	0.36	0.39	0.32	0.37	**0.3696**	0.48	0.44	0.45	0.41	0.47	**0.4512**	0.5383	0.5860
8	0.46	0.40	0.45	0.40	0.45	**0.4304**	0.40	0.36	0.38	0.34	0.39	**0.3725**	0.48	0.43	0.44	0.43	0.46	**0.4484**	0.5505	0.5824
9	0.46	0.41	0.45	0.40	0.43	**0.4312**	0.41	0.37	0.37	0.34	0.39	**0.3739**	0.47	0.46	0.43	0.41	0.47	**0.4501**	0.5519	0.5724
10	0.47	0.41	0.42	0.41	0.43	**0.4288**	0.42	0.36	0.36	0.34	0.39	**0.3722**	0.48	0.45	0.43	0.42	0.47	**0.4510**	0.5743	0.5826
11	0.44	0.42	0.45	0.41	0.43	**0.4296**	0.42	0.38	0.40	0.35	0.39	**0.3857**	0.47	0.47	0.46	0.42	0.48	**0.4607**	0.5633	0.5819
12	0.47	0.40	0.45	0.41	0.43	**0.4327**	0.40	0.37	0.38	0.33	0.40	**0.3770**	0.47	0.47	0.48	0.42	0.47	**0.4615**	0.5719	0.5829
13	0.46	0.41	0.46	0.40	0.44	**0.4336**	0.41	0.36	0.39	0.36	0.40	**0.3818**	0.48	0.45	0.45	0.43	0.46	**0.4541**	0.5693	0.5798
14	0.45	0.41	0.45	0.41	0.43	**0.4295**	0.39	0.36	0.40	0.35	0.40	**0.3810**	0.47	0.44	0.45	0.43	0.47	**0.4536**	0.5543	0.5776
15	0.46	0.39	0.46	0.42	0.42	**0.4289**	0.40	0.36	0.39	0.36	0.40	**0.3835**	0.48	0.46	0.49	0.43	0.47	**0.4650**	0.5633	0.5712
16	0.43	0.38	0.46	0.42	0.43	**0.4245**	0.41	0.38	0.41	0.35	0.38	**0.3871**	0.48	0.46	0.46	0.43	0.45	**0.4565**	0.5676	0.5840
17	0.44	0.39	0.46	0.42	0.41	**0.4247**	0.41	0.37	0.42	0.35	0.40	**0.3890**	0.49	0.45	0.48	0.44	0.47	**0.4646**	0.5555	0.5757
18	0.44	0.42	0.48	0.40	0.43	**0.4316**	0.40	0.37	0.41	0.35	0.39	**0.3841**	0.47	0.43	0.48	0.44	0.47	**0.4586**	0.5571	0.5898
19	0.43	0.42	0.47	0.42	0.42	**0.4322**	0.40	0.37	0.40	0.35	0.40	**0.3821**	0.48	0.45	0.48	0.42	0.46	**0.4596**	0.5679	0.5774
20	0.43	0.41	0.47	0.41	0.42	**0.4264**	0.41	0.37	0.41	0.35	0.40	**0.3875**	0.48	0.47	0.49	0.44	0.46	**0.4681**	0.5590	0.5800
Avg	0.45	0.40	0.45	0.41	0.43	**0.4283**	0.40	0.37	0.39	0.34	0.39	**0.3772**	0.48	0.45	0.46	0.42	0.47	**0.4547**	0.5576	0.5783

Table 3. Classification performance on W-1000

K	TFIDF EN	GE	ZH	FR	ES	Avg.	LSI EN	GE	ZH	FR	ES	Avg.	LDA EN	GE	ZG	FR	ES	Avg.	MISL	DMMR
5	0.41	0.35	0.42	0.35	0.39	**0.3848**	0.27	0.27	0.26	0.25	0.28	**0.2669**	0.46	0.41	0.48	0.39	0.45	**0.4373**	0.5225	0.5610
6	0.41	0.32	0.42	0.34	0.41	**0.3804**	0.26	0.28	0.28	0.27	0.29	**0.2785**	0.47	0.41	0.49	0.39	0.47	**0.4472**	0.5375	0.5685
7	0.40	0.36	0.43	0.35	0.41	**0.3920**	0.29	0.28	0.29	0.28	0.29	**0.2859**	0.46	0.41	0.49	0.38	0.47	**0.4427**	0.5315	0.5690
8	0.39	0.38	0.43	0.34	0.39	**0.3868**	0.29	0.30	0.28	0.28	0.30	**0.2876**	0.47	0.43	0.49	0.40	0.48	**0.4533**	0.5420	0.5905
9	0.40	0.38	0.45	0.34	0.38	**0.3907**	0.28	0.29	0.30	0.25	0.29	**0.2822**	0.49	0.43	0.50	0.40	0.49	**0.4624**	0.5420	0.5690
10	0.39	0.37	0.43	0.32	0.39	**0.3795**	0.30	0.27	0.30	0.27	0.29	**0.2836**	0.48	0.42	0.49	0.40	0.47	**0.4533**	0.5435	0.5955
11	0.38	0.38	0.43	0.32	0.38	**0.3788**	0.32	0.29	0.29	0.30	0.28	**0.2946**	0.48	0.43	0.51	0.41	0.49	**0.4643**	0.5550	0.5710
12	0.38	0.35	0.43	0.33	0.37	**0.3724**	0.29	0.30	0.30	0.27	0.29	**0.2910**	0.49	0.43	0.49	0.39	0.49	**0.4561**	0.5580	0.5650
13	0.39	0.37	0.47	0.35	0.35	**0.3834**	0.30	0.31	0.32	0.27	0.29	**0.2972**	0.47	0.44	0.49	0.38	0.49	**0.4541**	0.5600	0.5885
14	0.37	0.37	0.47	0.33	0.36	**0.3777**	0.29	0.29	0.32	0.26	0.29	**0.2892**	0.48	0.41	0.51	0.41	0.47	**0.4565**	0.5660	0.5895
15	0.37	0.38	0.49	0.34	0.34	**0.3842**	0.30	0.29	0.32	0.29	0.30	**0.2974**	0.45	0.43	0.49	0.39	0.46	**0.4421**	0.5490	0.5805
16	0.37	0.37	0.46	0.36	0.34	**0.3784**	0.30	0.31	0.32	0.27	0.28	**0.2949**	0.46	0.41	0.48	0.40	0.46	**0.4411**	0.5575	0.5895
17	0.36	0.35	0.47	0.37	0.35	**0.3774**	0.29	0.31	0.31	0.28	0.30	**0.2977**	0.46	0.41	0.49	0.41	0.47	**0.4465**	0.5590	0.5875
18	0.37	0.36	0.47	0.36	0.34	**0.3795**	0.29	0.31	0.32	0.28	0.29	**0.2970**	0.44	0.41	0.47	0.39	0.45	**0.4315**	0.5345	0.5650
19	0.36	0.36	0.49	0.38	0.35	**0.3887**	0.30	0.30	0.33	0.28	0.30	**0.3015**	0.45	0.41	0.50	0.41	0.46	**0.4472**	0.5555	0.5910
20	0.36	0.34	0.50	0.37	0.35	**0.3830**	0.29	0.29	0.32	0.29	0.29	**0.2958**	0.43	0.42	0.48	0.39	0.45	**0.4335**	0.5370	0.5705
Avg	0.38	0.36	0.45	0.35	0.37	**0.3824**	0.29	0.29	0.30	0.27	0.29	**0.2901**	0.46	0.42	0.49	0.40	0.47	**0.4481**	0.5469	0.5782

Table 4. Classification performance on W-200

	TFIDF						LSI						LDA						MISL	DMMR
K	EN	GE	ZH	FR	ES	Avg.	EN	GE	ZH	FR	ES	Avg.	EN	GE	ZG	FR	ES	Avg.	-	-
5	0.46	0.39	0.59	0.50	0.46	**0.4770**	0.28	0.19	0.25	0.18	0.29	**0.2375**	0.39	0.26	0.30	0.42	0.30	**0.3340**	**0.5150**	**0.6550**
6	0.46	0.46	0.60	0.57	0.45	**0.5065**	0.36	0.22	0.26	0.21	0.26	**0.2610**	0.40	0.31	0.31	0.42	0.34	**0.3560**	**0.5100**	**0.6325**
7	0.47	0.46	0.51	0.57	0.48	**0.4975**	0.32	0.24	0.28	0.20	0.34	**0.2735**	0.40	0.28	0.34	0.45	0.30	**0.3535**	**0.4875**	**0.6275**
8	0.54	0.45	0.55	0.57	0.51	**0.5215**	0.28	0.22	0.24	0.20	0.27	**0.2405**	0.37	0.28	0.37	0.42	0.33	**0.3540**	**0.4700**	**0.6550**
9	0.47	0.49	0.54	0.60	0.47	**0.5125**	0.31	0.24	0.30	0.18	0.29	**0.2620**	0.37	0.31	0.33	0.45	0.31	**0.3535**	**0.4825**	**0.6700**
10	0.53	0.47	0.55	0.58	0.50	**0.5235**	0.30	0.27	0.27	0.20	0.29	**0.2650**	0.43	0.29	0.38	0.45	0.32	**0.3730**	**0.4725**	**0.6175**
11	0.50	0.50	0.54	0.53	0.48	**0.5100**	0.28	0.32	0.27	0.20	0.31	**0.2745**	0.38	0.28	0.38	0.49	0.33	**0.3710**	**0.4975**	**0.5950**
12	0.50	0.50	0.53	0.58	0.49	**0.5200**	0.34	0.25	0.30	0.20	0.31	**0.2785**	0.40	0.33	0.35	0.44	0.28	**0.3595**	**0.4625**	**0.6250**
13	0.50	0.54	0.54	0.59	0.51	**0.5350**	0.30	0.25	0.31	0.24	0.32	**0.2825**	0.38	0.31	0.37	0.49	0.31	**0.3715**	**0.4750**	**0.6450**
14	0.52	0.48	0.53	0.56	0.50	**0.5165**	0.31	0.26	0.29	0.18	0.33	**0.2720**	0.43	0.29	0.38	0.47	0.26	**0.3645**	**0.4650**	**0.6300**
15	0.47	0.50	0.50	0.59	0.50	**0.5095**	0.30	0.25	0.27	0.23	0.29	**0.2655**	0.43	0.35	0.36	0.44	0.24	**0.3615**	**0.4825**	**0.6075**
16	0.48	0.52	0.55	0.56	0.46	**0.5135**	0.29	0.30	0.30	0.21	0.29	**0.2780**	0.36	0.33	0.36	0.51	0.25	**0.3615**	**0.4800**	**0.6400**
17	0.50	0.52	0.50	0.57	0.49	**0.5170**	0.31	0.25	0.31	0.25	0.29	**0.2845**	0.41	0.30	0.38	0.46	0.25	**0.3585**	**0.5050**	**0.6025**
18	0.49	0.49	0.56	0.56	0.51	**0.5205**	0.30	0.29	0.35	0.28	0.33	**0.3105**	0.38	0.35	0.35	0.53	0.28	**0.3760**	**0.4925**	**0.6125**
19	0.45	0.45	0.50	0.56	0.46	**0.4830**	0.30	0.26	0.31	0.28	0.28	**0.2855**	0.41	0.36	0.35	0.48	0.27	**0.3705**	**0.4450**	**0.6325**
20	0.47	0.48	0.52	0.57	0.47	**0.5020**	0.32	0.29	0.37	0.26	0.30	**0.3070**	0.43	0.34	0.37	0.50	0.23	**0.3720**	**0.4775**	**0.5900**
Avg	0.49	0.48	0.54	0.56	0.48	**0.5103**	0.30	0.26	0.29	0.22	0.30	**0.2736**	0.40	0.31	0.35	0.46	0.29	**0.3619**	**0.4825**	**0.6273**

4   Conclusions

We have presented a novel method of document analysis based on the multi-view intact space learning with manifold regularization (DMMR). DMMR models the document space from different views as multi-view intact space but considering the intrinsic geometrical and discriminating structure of the documents. As a result, DMMR can have more discriminating ability than MISL in document classification. Experimental results on document classification show that DMMR provides a better representation which leads to getting higher accuracy in the experiments.

References

1. Salton, G., Wong, A., Yang, C.-S.: A vector space model for automatic indexing. Commun. ACM **18**(11), 613–620 (1975)
2. Deerwester, S., Dumais, S.T., Furnas, G.W., Landauer, T.K., Harshman, R.: Indexing by latent semantic analysis. J. Am. Soc. Inf. Sci. **41**(6), 391 (1990)
3. Hofmann, T.: Unsupervised learning by probabilistic latent semantic analysis. Mach. Learn. **42**(1), 177–196 (2001)
4. Hofmann, T.: Probabilistic latent semantic indexing. In: Proceedings of the 22nd Annual International ACM SIGIR Conference on Research and Development in Information Retrieval, SIGIR 1999, pp. 50–57. ACM, New York (1999)
5. Blei, D.M., Ng, A.Y., Jordan, M.I.: Latent dirichlet allocation. J. Mach. Learn. Res. **3**(Jan), 993–1022 (2003)

6. Zhao, J., Xie, X., Xin, X., Sun, S.: Multi-view learning overview: recent progress and new challenges. Inf. Fusion **38**, 43–54 (2017)
7. Xu, C., Tao, D., Xu, C.: Multi-view intact space learning. IEEE Trans. Pattern Anal. Mach. Intell. **37**(12), 2531–2544 (2015)
8. Belkin, M., Niyogi, P.: Laplacian eigenmaps and spectral techniques for embedding and clustering. NIPS **14**, 585–591 (2001)
9. Belkin, M., Niyogi, P., Sindhwani, V.: Manifold regularization: a geometric framework for learning from labeled and unlabeled examples. J. Mach. Learn. Res. **7**(Nov), 2399–2434 (2006)
10. Chung, F.R.K.: Spectral Graph Theory, vol. 92. American Mathematical Society, Providence (1997)
11. Belkin, M.: Problems of learning on manifolds. Ph.D. thesis, The University of Chicago (2003). AAI3097083
12. A fast and powerful scraping and web crawling framework (2017). https://scrapy.org/
13. Processing xml and html with python (2017). https://lxml.de/
14. Natural language toolkit (2017). http://www.nltk.org/
15. The Stanford natural language processing group (2017). https://nlp.stanford.edu/
16. Efficient topic modelling of text semantics in python (2017). https://radimrehurek.com/gensim/index.html

Blind Image Quality Assessment: Using Statistics of Color Descriptors in the DCT Domain

Bingjie Lin, Wen Lu[✉], Lihuo He, and Xinbo Gao

School of Electronic Engineering, Xidian University, Xi'an 710071, China
bjlin@stu.xidian.edu.cn,
{luwen,lhhe,xbgao}@mail.xidian.edu.cn

Abstract. Our eyes receive the information of the images containing both the luminance information and chrominance information. However, the available blind image quality assessment (BIQA) criteria usually involve luminance information only. In this paper, we propose a novel efficient IQA metric via statistics of color descriptors in the DCT domain. Firstly, we calculate the saturation (S), hue (H), luminance (L) of the testing image simultaneously. Then the local DCT transform is implemented on each color descriptor, and the nature scene statistics (NSS) are extracted from the DCT coefficients. This is mainly based on the fact that the degradation of the image induces considerable deviation in the frequency domain characteristics of chromatic data in natural image. However, the deviation can be quantified by the DCT coefficients of the image's color descriptors effectively. Finally, we construct the mapping relation between the features and the image quality. Experimental results on several benchmarking databases (TID2013, LIVEII and CSIQ) show the proposed method is superior to other state-of-the-arts methods and reveal the rationality and the validity of the new approach.

Keywords: Blind image quality assessment · Color descriptor · DCT domain nature scene statistic

1 Introduction

When our eyes see an image, the luminance and chrominance information of the distorted image are received by the visual system, and then the image quality is formed in the brain. However, the majority of the existing visual perception simulation algorithms assess the image quality using only the luminance information, such as SSIM [1] and PSNR [2]. They both ignore the color information, use the pixel difference in the gray domain, and get unsatisfactory results. For example, there are four color images in the Fig. 1, and the changes of color can be easily noticed, indicating that different levels of distortions are involved. With the increase of the image's distortion levels, the mean opinion score (MOS) has a gradually decreasing trend, but the quality variation trend predicted by SSIM and PSNR is almost unchanged, showing that the image quality can't be evaluated effectively with only the luminance information. In recent years, researches have considered using color information to construct a reasonable

© Springer International Publishing AG 2017
Y. Sun et al. (Eds.): IScIDE 2017, LNCS 10559, pp. 52–63, 2017.
DOI: 10.1007/978-3-319-67777-4_5

IQA model. Xue and Jung [3] puts forward a chrominance just-noticeable-distortion model to evaluate the image quality. Dohyoung and Plataniotis [4] obtains the image quality by capture the correlation of chromatic data between spatially adjacent pixels by means of color invariance descriptors. Yuzhen et al. [5], color contrast and color value difference are employed in color space to derive image quality. Lee and Plataniotis [6], the angle, mean and standard deviation of luminance and chrominance in the CIELAB perceptual space are employed to derive a quality assessment score. In brief, mathematical statistics of color information [7], color features [8], and human visual system (HVS) to color information [9, 10] are utilized to design IQA models, which combine the luminance and chrominance of the image to improve the consistency of objective prediction and human perception. However, most of the features used are actually mainly on gray-level rather than color, and the results on the change of chrominance information have poor consistency with HVS.

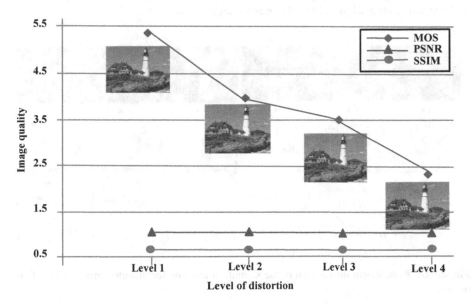

Fig. 1. Results of PSNR and SSIM on different distortion levels of distorted images (Color figure online)

In fact, the degradation of image causes considerable deviation in the frequency domain characteristics of chromatic data in natural image. We can use the DCT coefficients of the image's color descriptors to quantify the deviation effectively. Therefore, we propose a novel efficient IQA metric via statistics of color descriptors in the DCT domain. The proposed framework is divided into the following four parts: calculating the color descriptors, local DCT transform for the color descriptors, feature extraction using the nature scene statistics (NSS) model, constructing the mapping relation between the features and the image quality. Experimental results show the effectiveness and competitiveness of the proposed method with state-of-art technologies.

The rest of this paper is organized as follows. Section 2 gives an introduction of the DCT coefficient distribution of image descriptors. Section 3 details the proposed metric. Section 4 presents the experimental results and analysis on several databases. Finally, Sect. 5 concludes this paper and suggests directions for future research.

2 DCT Coefficient Distribution of Color Descriptors

The DCT coefficients of the image's chrominance information indicate the severity of the color change in the image. The low frequency part represents the smooth part of the color change, while the high frequency part is corresponding to the violent part of the color change. For example, the low frequency part in the sky is related to the overall red area or the overall white area whose color changes not seriously as shown in Fig. 2 (a_1), and the high frequency part is associated with the connection part of the red area and the white area whose color changes violently.

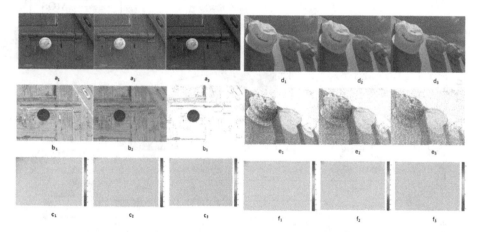

Fig. 2. DCT coefficient distribution of the saturation and hue for distorted image (Color figure online)

Figure 2 on the left side, shows the mean shift images with gradually increasing levels of distortion, their saturation maps and the DCT coefficient distribution of the saturation for distorted images. Figure 2 on the right side, shows the additive Gaussian noise images with gradually increasing levels of distortion, their hue maps and the DCT coefficient distribution of the hue for distorted images. It can be observed that the DCT coefficients of the saturation and hue of the images distribute regularly with the increasing of distortion level. The greater the degree of distortion is, the less low-frequency information the image has. It is suggested that the DCT coefficients of the saturation and hue of the images can be used to characterize the image distortion.

Where a_1, a_2, a_3 are the mean shift images with gradually increasing levels of distortion, b_1, b_2, b_3 are their saturation maps, c_1, c_2, c_3 are the DCT coefficient distribution of the saturation for distorted images. And d_1, d_2, d_3 are the additive

Gaussian noise images with gradually increasing levels of distortion, e_1, e_2, e_3 are their hue maps, f_1, f_2, f_3 are the DCT coefficient distribution of the hue for distorted images.

3 Blind Image Quality Assessment: Using Statistics of Color Descriptors in the DCT Domain

There are two important aspects in the process of image quality assessment: "look" image and "do" evaluation. The corresponding mathematical model process is as follows: extracting the image features highly correlated with the image quality and constructing the mapping relation between the features and the image quality. Among them, the use of NSS based on transform domain of IQA is a very typical way. In our IQA model, we focus on the DCT domain to extract features. Figure 3 illustrates the framework of our method. Image descriptors S, H, L are calculated from the input image, and then for each part, local DCT transform is implemented and the nature scene statistics (NSS) are extracted. Thereafter, a SVM is applied to predict visual quality given a set of extracted feature vector. By constructing the mapping relation between the features and the image quality, we propose a metric for blind image quality assessment. The detailed steps of the metric are provided in following sections.

3.1 Color Descriptor Computation

The image information perceived by the human eyes is represented by luminance and chrominance. Both of them can lead to the distortion of the image. Chrominance is generally characterized as the combination of hue and saturation, which describes the type and purity of a color, respectively. Color varies with the change of hue. As for the same kind of color, the more the white light it diluted with, the lower saturation it has and more faded it appears. In this paper, the formula of saturation, hue and luminance are as follows, S represents saturation, H represents hue, L represents luminance:

$$S = 1 - [(3 \times \min(R, G, B))/(R + G + B)] \tag{1}$$

$$H = \tan^{-1}\left(\frac{opp_1}{opp_2}\right) = \tan^{-1}\left(\frac{\sqrt{3}(R - G)}{R + G - 2B}\right) \tag{2}$$

$$L = 0.299R + 0.587G + 0.114B \tag{3}$$

where $opp_1 = (R - G)/\sqrt{2}$, $opp_2 = (R + G - 2B)/\sqrt{6}$ are two opponent channels of opponent color space, which corresponds to red – green and yellow – blue respectively.

3.2 Local DCT Transform and Features Extraction for Color Descriptors

The proposed approach relies on the fact that the statistical models for color descriptors in DCT domain change as introducing distortion in an image. The color descriptors S, H, L are divided into small non-overlapping blocks of size 5×5 and a local DCT is implemented on them over all blocks. Then the features are extracted from the DCT

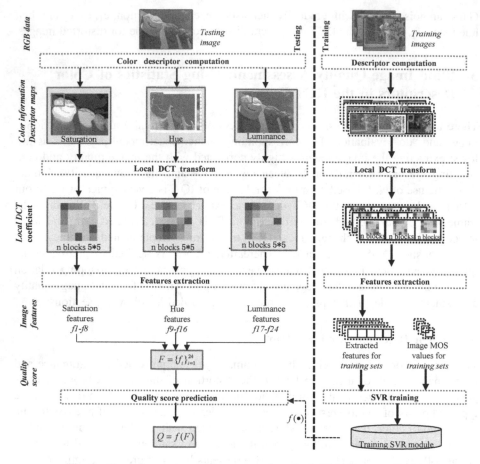

Fig. 3. General overview of the proposed method (Color figure online)

coefficients of color descriptors as described below. The NSS features are then used to predict perceptual image quality score.

3.2.1 Generalized Gaussian Model Shape Parameter for S, H, L

In image quality assessment methods based on NSS, the generalized Gaussian model is a very common statistical model because of its flexibility and robustness. Therefore, a generalized Gaussian fit is performed on the non-DC DCT coefficients of each color descriptor from 5×5 blocks. The shape parameter γ is one of the features that is computed over all blocks in the image using the following generalized Gaussian density in (3):

$$f_{x_S}(x_S; \mu_S, \sigma_S^2, \gamma_S) = a_S e^{-[b_S|x_S - \mu_S|]^{\gamma_S}} \, x_S \in R \tag{4}$$

where $\mu_S, \sigma_S, \gamma_S$ represent the mean, variance and shape parameter of the DCT coefficients of S. a_S and b_S are the normalizing and scale parameters given by:

$$a_S = \beta_S \gamma_S / 2\Gamma(1/\gamma_S) \tag{5}$$

$$b_S = (1/\sigma_S)\sqrt{\Gamma(3/\gamma_S)/(1/\gamma_S)} \tag{6}$$

The parameter γ_S is the effective feature extracted from the DCT coefficients of S. The shape parameter γ based H and L can be extracted from the same way, denoted by the variables γ_H and γ_L.

3.2.2 Frequency Variation for S, H, L

Variation coefficient is a parameter which used to measure the degree of data dispersion. In this section, the frequency variation of the DCT coefficients for S from 5×5 blocks in the image are given by:

$$\theta_S = \frac{\sigma_S|x|}{\mu_S|x|} \tag{7}$$

where x is the variable representing the DCT coefficients for S. $\sigma_S|x|, \mu_S|x|$ represent the standard deviation and mean of the DCT coefficients for S, which measures the center and the spread of the DCT coefficient magnitude distribution for S respectively. The frequency variation θ is one of the features that is computed over all blocks in the image. The frequency variation ϑ based H and L can be extracted from the same way, denoted by the variables θ_H and θ_L.

3.2.3 Energy Subband Ratio for S, H, L

In this section, we divide the local DCT coefficients of S, H, L into three bands from 5×5 blocks as shown in Fig. 4 which represent the lower, middle, higher frequency band separately. And different frequency bands have different energy distribution, so we compare the distribution of energy between different bands to measure energy subband ratio, denoted as R_n:

$$R_{ns} = \frac{|E_{ns} - \frac{1}{n-1}\sum_{j<n} E_{js}|}{E_{ns} + \frac{1}{n-1}\sum_{j<n} E_{js}} \quad n = 2, 3 \tag{8}$$

where $E_{ns} = \sigma_{ns}^2$, $ns = 1, 2, 3$ (lower, middle, higher), representing the DCT coefficient of S, R_S is the mean of R_{2s} and R_{3s} representing the energy subband ratio feature for S. The energy subband ratio feature R based H and L can be extracted from the same way, denoted by the variables R_H and R_L.

3.2.4 Orientation Feature for S, H, L

According to the human visual system is very sensitive to the changes with orientation energy in an image, we divide the local DCT coefficients of S, H, L along three orientation bands from 5×5 blocks as shown in Fig. 5.

A generalized Gaussian model is fitted to the coefficients within each orientation bands in Fig. 5, and θ is obtained from the model histogram fits for each orientation.

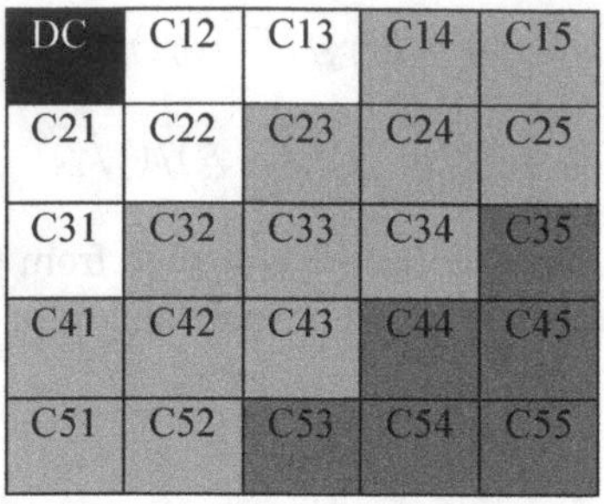

Fig. 4. DCT coefficients of frequency subbands

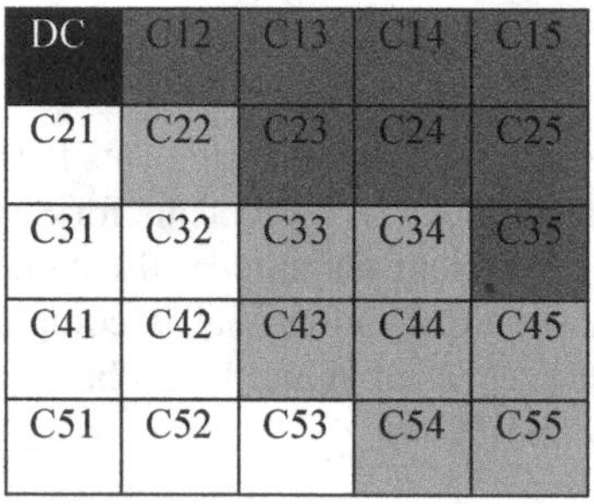

Fig. 5. DCT coefficients of orientation subbands

The variance of θ is computed along each of the three orientations. The variance of θ across the three orientations from all the blocks in the image is then pooled to obtain orientation feature O. O_S represents the orientation feature for S. The orientation feature O based H and L can be extracted from the same way, denoted by the variables O_H and O_L.

For each kind of feature, it's computed for all blocks resulting in a feature vector for the image. We would like to use the feature vector to obtain a scalar feature value for the image. So the features we extracted above are pooled in two ways. One is the lowest 10th percentile average of the local block feature across the image, and the other is the 100th percentile average of the local feature across the image.

3.3 Quality Score Prediction

So far, the image features we extract can be represented as $F = \{f_i\}_{i=1}^{24}$, as shown in Table 1, and superscript 1, 2 represent two different pooling ways. In the proposed framework the final quality is defined as a function of F:

Table 1. The extracted features, feature ID and image descriptors

Descriptor	Feature ID	Feature elements							
S	$f_1 - f_8$	γ_S^1	θ_S^1	R_S^1	O_S^1	γ_S^2	θ_S^2	R_S^2	O_S^2
H	$f_9 - f_{18}$	γ_H^1	θ_H^1	R_H^1	O_H^1	γ_H^2	θ_H^2	R_H^2	O_H^2
L	$f_{19} - f_{24}$	γ_L^1	θ_L^1	R_L^1	O_L^1	γ_L^2	θ_L^2	R_L^2	O_L^2

$$Q = f(F) \tag{9}$$

f is the mapping function from features to quality predicted by the support vector regression (SVR).

4   Experiment Results and Analysis

In this section, the evaluation criteria Pearson Linear Correlation Coefficient (PLCC) and Spearman Rank-order Correlation Coefficient (SRCC) would be used to compare the performance of the proposed metric with several existing IQA methods on the TID2013, CSIQ and LIVE II DB, and the value of PLCC or SRCC closer to 1 means a higher correlation with human perception.

4.1   Effectiveness Test of the Features

The features we used can be divided into three categories based S, H, L. In this section, we calculate PLCC with the single and joint contributions of these three kinds of features on TID2013, CSIQ and LIVE II DB. Figure 6 shows that the features luminance based, and saturation based contribute more than hue based, while the optimal performance of the algorithm would be obtained by combining the three kinds of features together. Therefore, the statistics of color descriptors in the DCT Domain proposed in our framework is impactful to blind image quality assessment.

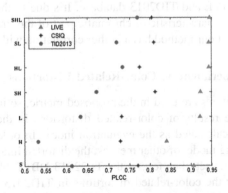

Fig. 6. PLCC performance of feature combinations

4.2   Performance Dependency on Training Set Size

In order to verify the effect of training set size on our metric, we have analyzed the changing of PLCC using varying ratio of train/test sets from 0.1 to 0.9 for individual databases in Fig. 7. It can be observed that the larger the ratio of train/test sets is, the higher the proposed method performance is. Meanwhile with the decrease of the ratio of train/test sets, the performance of the algorithm decreases gently, rather than sharply, which indicates that the proposed method can deal with the small sample problem well.

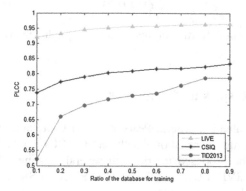

Fig. 7. Variation of metric performance for different training set ratio

4.3   Consistency Experiment on the Overall Database

The databases used in this experiment are TID2013, CSIQ, LIVE II. In the experiment, 80% of the images in the database are randomly selected to form the training set, and the remaining 20% is used as a test set. We repeat this test procedure 1000 times for each database, the median of the PLCC and SRCC values are calculated as the evaluation index. Since the algorithm doesn't use any information of the original image, the comparison methods of this experiment adopt the no-reference (NR) approaches, such as BIQI [11], BRISQUE [12], DIIVINE [13], CORNIA [14], NIQE [15], QAC [16], IL-NIQE [17], BLIINDS2 [18], and IDEAL [4]. Table 2 shows the experimental results. The CC values on each database show the algorithm performance clearly. Our metric outperforms all other methods on LIVE and TID2013 database. It's due to the fact that our metric considers the frequency characteristic of chromatic part of the image. The consistency experiment indicates that our method has a higher correlation with human perception.

4.4   Consistency Experiment on Color-Related Distortions

Due to lots of color features are used in the proposed metric, so in this experiment, we focus on analyzing the results of color-related distortions in the TID2013 database. The PLCC values are calculated as the evaluation index. In order to keep coincident with the original training model of other metrics, the distorted images in LIVE database II are used as the training set, then test on the TID2013 DB. Table 3 shows the PLCC of different metrics for the color-related distortions in TID2013. It can be seen from Table 3 that the proposed method can assess the color distortion effectively.

Table 2. PLCC and SRCC of different metrics on TID2013, CSIQ and LIVE II databases

Datasets	Methods	SRCC	PLCC
TID2013	BIQI	0.349	0.366
	BRISQUE	0.573	0.651
	DIIVINE	0.549	0.654
	CORNIA	0.549	0.613
	NIQE	0.317	0.426
	QAC	0.39	0.495
	IL-NIQE	0.521	0.648
	BLIINDS2	0.536	0.628
	IDEAL	0.719	0.767
	Proposed	**0.769**	**0.787**
CSIQ	BIQI	0.092	0.237
	BRISQUE	0.775	0.817
	DIIVINE	0.757	0.795
	CORNIA	0.714	0.781
	NIQE	0.627	0.725
	QAC	0.486	0.654
	IL-NIQE	0.822	0.865
	BLIINDS2	0.78	0.832
	IDEAL	**0.868**	**0.891**
	Proposed	**0.804**	**0.828**
LIVE	BIQI	0.825	0.84
	BRISQUE	0.933	0.931
	DIIVINE	0.884	0.893
	CORNIA	0.94	0.944
	NIQE	0.908	0.908
	QAC	0.874	0.868
	IL-NIQE	0.902	0.906
	BLIINDS2	0.924	0.927
	IDEAL	0.941	0.946
	Proposed	**0.945**	**0.963**

Table 3. PLCC of different metrics for the color-related distortions in TID2013

TID2013	BIQI	BRISQUE	BLIINDS2	DIIVINE	CORNIA	NIQE	QAC	Ours
#2	0.5405	0.7090	0.6497	0.7120	**0.7498**	0.6699	0.7184	0.7011
#7	0.3894	0.7984	0.7816	0.1650	0.0156	**0.8500**	0.7089	0.6544
#10	0.8567	0.8521	0.8643	0.6628	**0.8743**	0.8402	0.8369	0.8431
#16	0.0346	0.1241	0.1284	0.0104	0.0645	0.1626	0.3060	**0.3882**
#17	0.4125	0.0403	0.1505	**0.4601**	0.1823	0.0180	0.2067	0.2656
#18	0.1418	0.1093	0.0178	0.0684	0.0807	0.2460	0.3691	**0.4778**
#22	0.6983	0.7640	0.7358	0.4362	0.2717	0.7832	**0.8733**	0.7641
#23	0.5435	0.6160	0.5397	0.6608	**0.7922**	0.5612	0.6249	0.7888

5 Conclusions

A novel image quality assessment metric using statistics of color descriptors in the DCT domain is proposed. In the proposed framework, the image descriptors L, S, H are calculated from the input images, and then for each part, the local DCT transform is implemented and the nature scene statistics (NSS) are extracted. Thereafter, a SVM is applied to predict visual quality given a set of extracted feature vector. Finally, the objective image quality assessment is acquired by constructing the mapping relation between the features and the image quality. Although our metric achieves a desirable performance on the overall database, it is still limited in assessing color-related distortions, such as quantization noise, more effective color features are needed to be developed to improve the performance of IQA algorithms.

Acknowledgments. This research was supported partially by the National Natural Science Foundation of China (Nos. 61372130, 61432014, 61501349, 61571343).

References

1. Wang, Z., Bovik, A.C., Sheikh, H.R., Simoncelli, E.P.: Image quality assessment: from error visibility to structural similarity. IEEE Trans. Image Process. **13**(4), 600–612 (2004)
2. Eskicioglu, A.M., Fisher, P.S.: Image quality measures and their performance. IEEE Trans. Image Commun. **43**(12), 2959–2965 (1996)
3. Xue, F., Jung, C.: Chrominance just-noticeable-distortion model based on human color perception. Electron. Lett. **50**(22), 1587–1589 (2014)
4. Dohyoung, L., Plataniotis, K.N.: Toward a no-reference image quality assessment using statistics of perceptual color descriptors. IEEE Trans. Image Process. **25**(8), 3875–3889 (2016)
5. Yuzhen, N., Haifeng, Z., Wenzhong, G.: Image quality assessment for color correction based on color contrast similarity and color value difference. IEEE Trans. Circ. Syst. Video Technol. (2016)
6. Lee, D., Plataniotis, K.: Towards a full-reference quality assessment for color images using directional statistics. IEEE Trans. Image Process. **24**(11), 3950–3965 (2015)
7. Omari, M., Abdelouahad, A.A., Hassouni, M.E., Cherifi, H.: Color image quality assessment measure using multivariate generalized Gaussian distribution. In: International Conference on Signal-Image Technology and Internet-Based Systems, Japan, pp. 195–200 (2013)
8. Redi, J.A., Gastaldo, P., Heynderickx, I., Zunino, R.: Color distribution information for the reduced-reference assessment of perceived image quality. IEEE Trans. Circ. Syst. Video Technol. **20**(12), 1757–1769 (2010)
9. Lissner, I., Preiss, J., Urban, P., Lichtenauer, M.S.: Image-difference prediction: from grayscale to color. IEEE Trans. Image Process. **22**(2), 435–446 (2013)
10. Preiss, J., Fernandes, F., Urban, P.: Color-image quality assessment: from prediction to optimization. IEEE Trans. Image Process. **23**(3), 1366–1378 (2014)
11. Moorthy, A.K., Bovik, A.C.: A two-step framework for constructing blind image quality indices. IEEE Sig. Process. **17**(5), 513–516 (2010)
12. Mittal, A., Moorthy, K., Bovik, A.C.: No-reference image quality assessment in the spatial domain. IEEE Trans. Image Process. **21**(12), 4695–4708 (2012)

13. Moorthy, A.K., Bovik, A.C.: Blind image quality assessment: from natural scene statistics to perceptual quality. IEEE Trans. Image Process. **20**(12), 3350–3364 (2011)
14. Ye, P., Kumar, J., Kang, L.: Unsupervised feature learning framework for no-reference image quality assessment. In: Conference on Computer Vision and Pattern Recognition, RI, USA, pp. 1098–1105 (2012)
15. Mittal, A., Soundararajan, R., Bovik, A.C.: Making a 'completely blind' image quality analyzer. IEEE Sig. Process. Lett. **20**(3), 209–212 (2013)
16. Xue, W., Zhang, L., Mou, X.: Learning without human scores for blind image quality assessment. In: Conference on Computer Vision and Pattern Recognition, Portland, USA, pp. 995–1002 (2013)
17. Zhang, L., Zhang, L., Bovik, A.C.: A feature-enriched completely blind image quality evaluator. IEEE Trans. Image Process. **24**(8), 2579–2591 (2015)
18. Saad, M.A., Bovik, A.C., Charrier, C.: Blind image quality assessment: a natural scene statistics approach in the DCT domain. IEEE Trans. Image Process. **21**(8), 3339–3352 (2011)

Location Dependent Dirichlet Processes

Shiliang Sun[1(✉)], John Paisley[2], and Qiuyang Liu[1]

[1] Department of Computer Science and Technology, East China Normal University,
3663 North Zhongshan Road, Shanghai 200062, China
slsun@cs.ecnu.edu.cn, qiuyangliu2014@gmail.com
[2] Department of Electrical Engineering, Columbia University, New York City, USA
jpaisley@columbia.edu

Abstract. Dirichlet processes (DP) are widely applied in Bayesian non-parametric modeling. However, in their basic form they do not directly integrate dependency information among data arising from space and time. In this paper, we propose location dependent Dirichlet processes (LDDP) which incorporate nonparametric Gaussian processes in the DP modeling framework to model such dependencies. We develop the LDDP in the context of mixture modeling, and develop a mean field variational inference algorithm for this mixture model. The effectiveness of the proposed modeling framework is shown on an image segmentation task.

Keywords: Bayesian nonparametric model · Dirichlet process · Infinite mixture model · Variational inference

1 Introduction

For many practical problems, nonparametric models are often chosen over alternatives to parametric models that use a fixed and finite number of parameters [1]. In contrast, Bayesian nonparametric priors are defined on infinite-dimensional parameter spaces [2], but fitting such models to data allows for an adaptive model complexity to be learned in the posterior distribution. In theory, this mitigates the underfitting and overfitting problems faced by model selection for parametric models [3].

Dirichlet processes (DPs) [4] are a standard Bayesian nonparametric prior for modeling data, typically via mixtures of simpler distributions. In this scenario, each draw of a DP gives a discrete distribution on an infinite parameter space that can be used to cluster data into a varying number of groups. While DPs have this flexibility as prior models for generating and modeling data, common additional data-specific markers such as time and space are often not incorporated in the mixture modeling formulation, and are simply ignored [5,6]. However, for many problems this additional information can be an important part of data clustering. For example, in text models articles nearby in time may be more likely to be clustered together by topics, and in image segmentation, neighboring pixels are more likely to fall in the same category. DP-based mixture models can often be improved by incorporating such information.

© Springer International Publishing AG 2017
Y. Sun et al. (Eds.): IScIDE 2017, LNCS 10559, pp. 64–76, 2017.
DOI: 10.1007/978-3-319-67777-4_6

Such dependencies among the data are addressed in the literature through dependent Dirichlet processes and their generalizations [7–11]. For example, the distance dependent Chinese restaurant process (ddCRPs) of Blei and Frazier [12] is a clustering framework that uses a distance function between locations attached to data points to encourage clustering by their "proximity." In the generative definition, it first partitions data points by sequentially creating a linked network between observations, rather than by assigning data to clusters. The cluster assignments are then obtained as a by-product of the partition of the data according to the cliques in the ddCRP network.

While shown to be useful for spatial modeling [13], the non-exchangeability of the ddCRP means that a mixing measure for such a process cannot be found along the lines of the stick-breaking construction for the DP. Therefore, there exists no distribution that makes all observations conditionally independent. As a result, the order of the data crucially matters for the ddCRP, which is often arbitrary and leads to local optimal issues. Since variational inference is a significant challenge as a result, Gibbs sampling was used for posterior inference of the ddCRP, which is computationally demanding and difficult to scale. Related exchangeable dependent random processes based on beta process and probit stick-breaking processes have been recently proposed, but for the mixed-membership model setting [14, 15].

In this paper, we propose location dependent Dirichlet processes (LDDPs) as a general dependent Dirichlet process modeling framework. Since a mixing measure for the ddCRP does not exist, our motivation is to define such a mixing measure for a model that achieves the same end goal, but is not equivalent to the ddCRP. To this end, we adapt ideas from the discrete infinite logistic normal (DILN) model [16] by combining Gaussian processes (GP) with Dirichlet processes. However, whereas DILN is a mixed-membership model that uses a single GP across latent cluster locations, the LDDP is a mixture model in which cluster-specific GP's interact directly with the data to capture distance dependencies. The direct definition of the LDDP mixing measure immediately allows for a variational inference algorithm to be derived. While the LDDP framework is general, we apply it to the Gaussian mixture model for image segmentation.

2 LDDPs and an Inference Algorithm

We first review the connection between Dirichlet and gamma processes and define the generative process of the location dependent Dirichlet process (LDDP). We then derive mean field variational inference with the general LDDP and discuss a proposed model for Gaussian data. We note that the term "location" refers to any auxiliary information connected to the primary data, such as time or space.

2.1 DPs and the Gamma Process

The DP is a prior widely used for Bayesian nonparametric mixture modeling. A draw G from a DP with concentration parameter α_0 and base distribution G_0, written as $G \sim DP(\alpha_0 G_0)$, is a discrete probability distribution on the support of G_0. Suppose

$$v_i \overset{iid}{\sim} Beta(1, \alpha_0), \quad \theta_i^* \overset{iid}{\sim} G_0, \tag{1}$$

and define $\pi_i = v_i \prod_{j=1}^{i-1}(1 - v_j)$. A way of constructing the infinite distribution G is [17]

$$G = \sum_{i=1}^{\infty} \pi_i \delta_{\theta_i^*}. \tag{2}$$

With the DP mixture, data are generated independently as,

$$\theta_n | G \overset{iid}{\sim} G, \quad x_n | \theta_n \sim p(x | \theta_n). \tag{3}$$

A partition of the data is naturally formed according to the repeating of atoms $\{\theta_i^*\}$ among the parameters $\{\theta_n\}$ that are used by the data.

It is well-known that the DP can be equivalently represented as an infinite limit of a finite mixture model, and through normalized gamma measures [4,18]. In this case, suppose there are K_0 components in the finite mixture model, and

$$\theta_i^* \overset{iid}{\sim} G_0, \quad z_i \overset{iid}{\sim} Gamma(\tfrac{\alpha_0}{K_0}, 1) \quad G_{K_0} = \sum_{i=1}^{K_0} \frac{z_i}{\sum_{j=1}^{K_0} z_j} \delta_{\theta_i^*}.$$

Then as $K_0 \to \infty$, $G_\infty \sim DP(\alpha_0 G_0)$. For computational convenience, we can form an accurate approximation to the DP by using G_{K_0} with a large value of K_0 [18].

2.2 Location Dependent Dirichlet Processes

We extend the DP to the LDDP by associating with the atom θ_i^* of each cluster a Gaussian process

$$f_i(\ell) \sim \mathcal{GP}(0, k(\ell, \ell')), \quad i = 1, 2, \ldots$$

on a particular space of interest, $\ell \in \Omega$. For example, the ℓ indicates geographic location or is a time stamp. We note that the Gaussian process of each cluster is defined on, e.g., all time or space. Our goal is to allow the associated location ℓ_n for observation x_n to *increase* the probability of using cluster parameter θ_i^* when $f_i(\ell_n) > 0$, and *decrease* that probability when $f_i(\ell_n) < 0$. The kernel of the Gaussian process $k(\cdot, \cdot)$ ensures that each cluster marks off contiguous regions in space or time. For example, we use the common Gaussian kernel in our experiments,

$$k(\ell, \ell') = \sigma_f^2 \exp\left[-\|\ell - \ell'\|^2 / \sigma_\ell^2\right]. \tag{4}$$

We observe that GPs generated with such a kernel will be continuous and are flexible enough to be positive or negative in various regions of space [19], which provides more modeling capacity than the ddCRP.

The LDDP uses these GPs in combination with a gamma process representation of the DP to generate an *observation-specific* distribution on clusters.

Employing the finite-K_0 approximation to the DP above, we again let G_0 be the base distribution for $\{\theta_i^*\}$. Suppose c_n is a discrete latent variable which indicates the atom assigned to x_n, so that $\theta_n = \theta_{c_n}^*$. We first generate

$$\theta_i^* \overset{iid}{\sim} G_0, \quad z_i \overset{iid}{\sim} Gamma(\tfrac{\alpha_0}{K_0}, 1)$$

exactly as before. Then, for observation n our construction of the LDDP distribution on the clusters is

$$P(c_n = k | \mathbf{z}, \mathbf{f}, \ell_n) \propto z_k e^{f_k(\ell_n)}, \tag{5}$$

for each observation $n = 1, \ldots, N$. This is a trade-off between how prevalent cluster k is globally—z_k—and how appropriate cluster k is for the nth observation—$e^{f_k(\ell_n)}$. Using the previous notation, this can also be written

$$G_n = \sum_{i=1}^{K_0} \frac{z_i e^{f_i(\ell_n)}}{\sum_{j=1}^{K_0} z_j e^{f_j(\ell_n)}} \delta_{\theta_i^*}, \tag{6}$$

from which we generate data

$$\theta_n | G_n \sim G_n, \quad x_n | \theta_n \sim p(x | \theta_n). \tag{7}$$

Since the atoms θ_i^* are shared among each distribution G_n, a partition of the data is formed according to the values of the indicator variables $c_1, \ldots, c_N$. However, as is clear from Eq. (6), each observation x_n does not use these atoms i.i.d. as in the standard DP. Instead, the Gaussian processes encourages those x_n that have auxiliary information ℓ_n in positive regions of the Gaussian processes to cluster together. These will tend to cluster x_n with ℓ_n that are close (e.g., in time or space). We note that we do not define a generative model for these ℓ_n, but only $x_n | \ell_n$. In posterior inference, clustering will be a trade-off between how similar two ℓ_n are according to the Gaussian process, and how similar two x_n are according to the data distribution $p(x | \theta)$.

2.3   Mean-Field Variational Inference

We let the data-generating distribution $p(x | \theta)$ be generic for the moment and discuss a variational inference algorithm for the LDDP in general. Given N observations with corresponding location variables $\{(x_n, \ell_n)\}$, the joint distribution of the model variables and data factorizes as

$$p(\mathbf{x}, \mathbf{c}, \mathbf{z}, \mathbf{f}, \boldsymbol{\theta} | \boldsymbol{\ell}) = p(\mathbf{z}, \mathbf{f}, \boldsymbol{\theta}) \prod_n p(x_n | \theta_{c_n}) p(c_n | \mathbf{z}, \mathbf{f}, \ell_n). \tag{8}$$

We derive a variational inference algorithm for the sets of variables $\mathbf{z}$, $\mathbf{f}$ and $\mathbf{c}$, which occur in all potential LDDP models. We recall that with mean-field variational inference [20,21], we define a factorized approximation to the posterior distribution,

$$p(\mathbf{c}, \mathbf{z}, \mathbf{f}, \boldsymbol{\theta} | \boldsymbol{\ell}, \mathbf{x}) \approx \big[\textstyle\prod_n q(c_n) \big] \big[\textstyle\prod_k q(z_k) q(f_k) q(\theta_k) \big].$$

After choosing specific distributions for each q,[1] we then tune the parameters of these distributions to maximize the variational objective function

$$\mathcal{L} = \mathbb{E}_q[\ln p(\boldsymbol{x}, \boldsymbol{c}, \boldsymbol{z}, \boldsymbol{f}, \boldsymbol{\theta}|\boldsymbol{\ell})] - \mathbb{E}_q[\ln q].$$

Coordinate ascent is usually adopted to optimize the objective by cycling through optimizing each q within each iteration. For the LDDP model, we choose,

$$q(z_k) = Gam(a_k, b_k), \quad q(c_n) = Mult(\phi_n), \quad q(f_k) = \delta_{f_k}.$$

The last choice is out of convenience, since a distribution of f_k (an N dimensional vector) has computationally-intensive tractability issues. A delta q distribution amounts to a point estimate of the variable in the objective function $\mathcal{L}$.

The joint distribution $p(\boldsymbol{x}, \boldsymbol{c}, \boldsymbol{z}, \boldsymbol{f}, \boldsymbol{\theta}|\boldsymbol{\ell})$ presents further difficulties, which can be seen by expanding it as

$$\prod_{n=1}^{N} \prod_{i=1}^{K_0} \left(p(x_n|c_n) \frac{z_i e^{f_i(\ell_n)}}{\sum_{j=1}^{K_0} z_j e f_j(\ell_n)} \right)^{1(c_n=i)} \left[\prod_{i=1}^{K_0} z_i^{\frac{\alpha_0}{K_0}-1} e^{-z_i} \right] \left[\prod_{i=1}^{K_0} e^{-\frac{1}{2} f_i^\top K^{-1} f_i} \right].$$

$$(9)$$

The normalization of $z_i e^{f_i}$ makes directly calculating $\mathcal{L}$ intractable, since $\mathbb{E}_q[-\ln \sum z_j e^{f_j(\ell_n)}]$ is not in closed form when integrating over each z_j. We therefore use a lower bound of this term found useful in similar situations, e.g., [16]. Introducing an auxiliary parameter $\xi_n > 0$, by a simple first order Taylor expansion of the convex function $-\ln(\cdot)$ we have

$$-\ln \sum_j z_j e^{f_j(\ell_n)} \geq -\ln \xi_n - \frac{\sum_j z_j e^{f_j(\ell_n)} - \xi_n}{\xi_n}.$$

$$(10)$$

Therefore, in the joint likelihood we replace

$$\frac{1}{\sum_j z_j e^{f_j(\ell_n)}} \geq \frac{1}{\xi_n} e^{-\xi_n^{-1} \sum_j z_j e^{f_j(\ell_n)}}.$$

$$(11)$$

Differentiating the new objective with respect to ξ_n and setting to zero, we see that the lower bound is tightest at

$$\xi_n = \sum_j \mathbb{E}_q[z_j] e^{f_j(\ell_n)}.$$

$$(12)$$

Thus, ξ_n becomes a new parameter in the model that is set to this value at the end of each iteration. In this and all following equations, the expectations are calculated using the most recent parameters of the relevant q distribution.

For the remaining q distributions, following the steps in [20] for $q(c_n)$ and $q(z_k)$, the multinomial distribution $q(c_n)$ can be updated at each iteration by setting its discrete distribution parameter

$$\phi_n(k) \propto \exp\{\mathbb{E}_q[\ln p(x_n|\theta_k)] + \mathbb{E}_q[\ln z_k] + f_k(\ell_n)\}.$$

$$(13)$$

---

[1] $q(\theta_k)$ is problem-specific, so we ignore it here.

The first expectation is problem-specific and depends on the data x_n and the distributions chosen for modeling it and θ_k.

The parameters for the gamma distribution $q(z_k)$ can be updated by setting them to

$$a_k = \frac{a}{K_0} + \sum_{n=1}^{N} \mathbb{E}_q[1(c_n = k)], \quad b_k = 1 + \sum_{n=1}^{N} \frac{e^{f_k(\ell_n)}}{\xi_n}. \tag{14}$$

To update each Gaussian process f_k at the N locations, we use gradient ascent. The gradient $\nabla_{f_k}\widetilde{\mathcal{L}}$, where the tilde indicates the lower bound approximation to $\mathcal{L}$, is

$$\nabla_{f_k}\widetilde{\mathcal{L}} = \left[\mathbb{E}_q[1(c_n = k)] - \frac{1}{\xi_n}\mathbb{E}_q[z_k]e^{f_k(\ell_n)} \right]_n - \mathbf{K}^{-1}f_k. \tag{15}$$

We take a gradient step using $\mathbf{K}$ as a convenient preconditioner (discussed more below)

$$f_k \leftarrow f_k + \rho\mathbf{K}\nabla_{f_k}\widetilde{\mathcal{L}}. \tag{16}$$

2.4 Computational Considerations

When the number of observations N is large, the $N \times N$ kernel $\mathbf{K}$ can be massive. Not only is the inverse not feasible in this situation, but calculating $\mathbf{K}$ itself results in memory issues. We use a simple approach based on the Nyström method to address this issue [22, 23].

Specifically, let ℓ^* be a set of $N_2 \ll N$ locations in the same space as ℓ. These N_2 locations can be different from those in the data set, and should be spread out in the space. For example, in an image these might be N_2 evenly spaced grid points. Then let $\mathbf{K}^*$ be the kernel restricted to these N_2 locations, and $\mathbf{K}^{**}$ the kernel between ℓ^* and ℓ, so that

$$\mathbf{K}^*_{i,j} = k(\ell^*_i, \ell^*_j), \quad \mathbf{K}^{**}_{i,j} = k(\ell^*_i, \ell_j).$$

Then it is well-known that for appropriately chosen ℓ^*, an accurate approximation to the $N \times N$ kernel $\mathbf{K}$ is

$$\mathbf{K} \approx (\mathbf{K}^{**})^{\top}(\mathbf{K}^*)^{-1}\mathbf{K}^{**} \tag{17}$$

As a result, when updating each f_k as in Eq. (16), we only need to work with $N_2 \times N_2$ and $N_2 \times N$ matrices. These matrices are much smaller and can be calculated in advance and stored for re-use, and so an $N \times N$ matrix never needs to be constructed. We also note that this approximation is being performed (in principle) after multiplying $\mathbf{K}\nabla_{f_k}\widetilde{\mathcal{L}}$, and so we do not need to approximate $\mathbf{K}^{-1}$.

2.5   LDDP Mixtures of Gaussian Distributions

We apply the LDDP prior to mixture models for which the data are Gaussian. In this case, $\theta_i^* = \{\mu_i, R_i\}$ where μ_i and R_i are the mean and inverse covariance of a multivariate Gaussian distribution. We also specify the priors for μ_i and R_i as normal and Wishart distributions as $\mu_i \sim \mathcal{N}(\mu_0, R_0^{-1})$, $R_i \sim \mathcal{W}(W_0, \nu_0)$. The graphical model for the LDDP mixture of Gaussian distributions is given in Fig. 1, where the hyperparameters are not shown. Inference details for this model are standard, and thus omitted here.

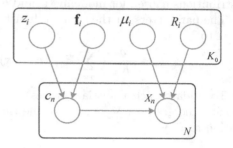

Fig. 1. The graphical model for the LDDP mixture of Gaussian distributions. This model extends the GMM by including a Gaussian process f_i in the cluster assignment prior, which encourages region-based clustering.

Some Applications: An example we consider in our experiments is image segmentation. In this setting, each x_n could be the 3-D RGB vector of pixel n. The location ℓ_n would then be the 2-D coordinates of this pixel in the image. Each cluster would consist of a 3-D Gaussian distribution on RGB, which would cluster similar colors, and a Gaussian process that would indicate which regions of the image this cluster would be more likely to be active. This GP would be intended to improve the segmentation over a direct Gaussian mixture model (GMM). Another example for future consideration would be audio segmentation. The setup would be almost identical to image segmentation, however in this case x_n could be a short-time frequency content vector, such as an MFCC, and ℓ_n would be the time stamp within the audio. A third example could capture geographic information in ℓ_n, and a feature vector x_n for the person or business with index n having the location ℓ_n.

3   Discussion

The LDDP is a type of dependent DP where Gaussian processes are involved to adapt the generating probabilities of the atoms in a nonparametric manner. In addition to ddCRPs, our model is also related to but still different from several other dependent DPs which we briefly discuss here.

The kernel stick-breaking process [24] is constructed by introducing a countable sequence of mutually independent random variables

$$\{\Gamma_h, V_h, G_h^*, h = 1, \ldots, \infty\}, \tag{18}$$

where $\Gamma_h \sim H$ is a location, $V_h \sim Beta(a_h, b_h)$, and $G_h^* \sim Q$ is a probability measure. Then, the process is defined as follows:

$$G_{x_n} = \sum_{h=1}^{\infty} U(x_n, V_h, \Gamma_h) \prod_{i<h} \{1 - U(x_n, V_i, \Gamma_i)\} G_h^*,$$

$$U(x_n, V_h, \Gamma_h) = V_h k(x_n, \Gamma_h), \tag{19}$$

where $k(\cdot, \cdot)$ is a kernel function. The kernel stick-breaking process accommodates dependency since for close x_n and $x_{n'}$, the G_{x_n} and $G_{x_{n'}}$ will assign similar probabilities to the elements of $\{G_h^*\}_{h=1}^{\infty}$. By inspection, we find that the table assignments induced from the kernel stick-breaking process are generally not exchangeable but marginally invariant. This process was later extended to hierarchical kernel stick-breaking process [25] for multi-task learning.

Foti and Williamson [26] introduced a large class of dependent nonparametric processes which are also similar to LDDPs, but uses parametric kernels to weight dependency. Foti et al. [27] presented a general construction for dependent random measures based on thinning Poisson processes, which can be used as priors for a large class of nonparametric Bayesian models. In contrast to our LDDP, the proportion variable of the thinned completely random measures comes from the global measure, and the rate measures involve parametric formulations.

4   Experiments

4.1   Setup

We define the kernel function to be the radial basis function (RBF) in Eq. (4) with settings for σ_f and σ_ℓ described below. For hyperparameters, we set $\alpha_0 = 1$ and μ_0 and R_0 are set to the empirical mean and inverse covariance of the training data x. The Wishart parameter ν_0 is set to the dimensionality d of the training data, with $d = 3$ for our problems. W_0 is set to R/d such that the mean of R_i under the Wishart distribution is R. For the q distributions, we initialize each f_i to be the zero vector, and set each $a_i = b_i = 1$. We define $q(\theta_i) = q(\mu_i)q(R_i)$ to be normal-Wishart and initialize them to be equal to the prior, with the important exception that the mean of $q(\mu_i)$ is initialized using K-means.

We apply the Gaussian LDDP to a segmentation problem of natural scene images. The size of the images we used is 128×128, and thus the size of the kernel matrix of the Gaussian process is 16384×16384. We use the Nyström method here with approximately 5% of these locations evenly spaced in the image.

For comparison, we use the K-means clustering algorithm as a baseline for performance comparison. We also compare our method with normalized cut spectral clustering [28], dependent Pitman-Yor processes (DPYP) [10] and hierarchical Pitman-Yor (HPY) [29], as well as special cases of the LDDP such as the GMM. We run all algorithms for 1000 iterations, which empirically were sufficient for convergence.

Fig. 2. Original images (first row) and segmentation results (other rows). The second row is obtained by K-means with RGB and pixel locations. The third and fourth rows are DPYP and HPY, respectively. The fifth and sixth rows are the proposed LDDP using $K_0 = 5$ and $K_0 = 100$.

4.2   Image Segmentation Results

We show segmentation results using images of different scenes from the LabelMe data set [30] in Fig. 2. In our experiments, we consider a parametric version of the LDDP in which $K_0 = 5$ and a nonparametric approximation, where $K_0 = 100$. These two cases were differentiated by the posterior cluster usage, where all were used in the first case and only a subset used in the second.

We compare with K-means segmentation in which we use a 5-D vector, three for RGB and two for the
pixel location in the image. These
results are shown in the second row
of Fig. 2, where five clusters are used
(i.e., $K_0 = 5$). The dependent Pitman-
Yor processes (DPYP) segmentation
results for the images are shown in the
third row. The DPYP uses thresholded
Gaussian processes to generate spatially coupled Pitman-Yor processes.
The hyperparameters involved in the

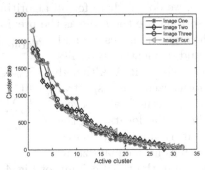

Fig. 3. #pixels assigned to active clusters

DPYP are set analogously to the ones in [10]. For the covariance functions involved in the DPYP, we use the distance-based squared exponential covariance, which has been shown to give good results [10]. The DPYP includes the hierarchical Pitman-Yor (HPY) model as a special case when the Gaussian processes involved have identity covariance functions. The HPY mixture segmentation results for the images are shown in the fourth row of Fig. 2.

For the LDDP mixture, when $\sigma_\ell = 0.1$ and $\sigma_f = 1$, the segmentation results are given in the fifth ($K_0 = 5$) and sixth ($K = 100$) rows of Fig. 2. We further modified σ_f to $\sqrt{10}$ in our experiments, and found that the segmentation results are quite similar to the setting $\sigma_f = 1$ and thus not provided here. As is evident, using more clusters creates a finer segmentation, though the results are still similar. Subjectively, we see that LDDP outperforms K-means, while the two Pitman-Yor models do not have as clear a segmentation.

With most DP-based models, including the LDDP, the number of used clusters is expected to grow logarithmically with the number of observations [3]. Therefore, it is not surprising that more clusters are used by the LDDP model when $K_0 = 100$. We show this in Fig. 3 for the four images considered. These plots give an ordered histogram of the number of pixels assigned to a cluster. We see that far fewer than 100 clusters contain data, highlighting the nonparametric aspect of the LDDP, but still more than the (possibly) desired number of segments. Therefore it is arguable that for image segmentation a nonparametric model is not ideal and K_0 should be set to a small number such as 5. We note that the LDDP can easily make this shift to parametric modeling as presented and derived above.

4.3   Results with Ground Truth Segmentation

We further compare our LDDP mixture model with the normalized cut spectral clustering method [28], Pitman-Yor based models, and the GMM using human-segmented images. We use the Rand index [31] to quantitatively evaluate the results. For these images [32], we know the number of true clusters from the human segmentations and set K_0 to this number. While this is not possible in practice, we do this here for all algorithms as a head-to-head comparison. The other settings are the same as the previous experiments.

The original images, ground truth and segmentation results are all shown in Fig. 4. Although the normalized cut spectral clustering method also leverages the spatial location of the pixels, it is implicitly biased towards regions of equal size, as we can see in the third column of Fig. 4. The LDDP appears to perform qualitatively better than other Bayesian methods. We particularly note the improvement over the GMM, a special case of LDDP, which is due to the addition of Gaussian processes to each cluster.

Table 1. Comparison of Rand index values (%).

	Rock	Mountain	Hut	Building
Normalized cut	78.63	79.62	78.35	**81.86**
HPY	58.61	56.82	56.02	58.68
DPYP	59.01	58.71	58.57	54.28
GMM	**95.79**	80.98	81.63	76.84
LDDP	95.44	**88.82**	**85.74**	74.00

The Rand index is a standard quantitative measurement of the similarity between a segmentation and the ground truth. The corresponding Rand index

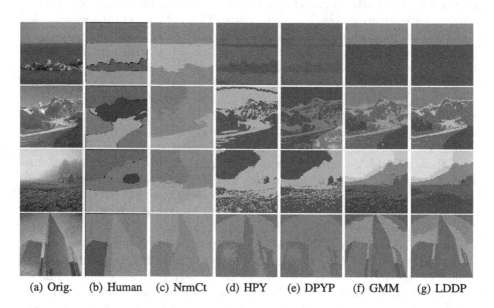

(a) Orig.   (b) Human   (c) NrmCt   (d) HPY   (e) DPYP   (f) GMM   (g) LDDP

Fig. 4. Example segmentations when ground truth is known. The number of clusters is set to ground truth for all experiments to facilitate a head-to-head comparison of the modeling structure. LDDP performs better than other Bayesian methods and comparable to normalized cuts. We see the improvement of LDDP over GMM as a result of the added Gaussian processes.

values for the images segmentation results are given in Table 1. These results indicate the overall competitiveness of LDDP for segmentation. We again highlight the general improvement over other general Bayesian methods, which like LDDP are also applicable to a broader set of modeling problems.

5 Conclusion

We have proposed location dependent Dirichlet processes (LDDP), a general mixture modeling framework for clustering data using additional location information. We derived a general variational inference algorithm for both parametric and approximately nonparametric settings. We presented a case study of a Gaussian LDDP for an image segmentation task where we saw competitive results. Future research will focus on exploring more applications of the proposed framework beyond the current image data.

Acknowledgments. The corresponding author Shiliang Sun would like to thank supports from the National Natural Science Foundation of China under Projects 61673179 and 61370175, Shanghai Knowledge Service Platform Project (No. ZF1213), and the Fundamental Research Funds for the Central Universities.

References

1. Hjort, N., Holmes, C., Müller, P., Walker, S. (eds.): Bayesian Nonparametrics: Principles and Practice. Cambridge University Press, Cambridge (2010)
2. Orbanz, P., Teh, Y.W.: Bayesian nonparametric models. In: Sammut, C., Webb, G.I. (eds.) Encyclopedia of Machine Learning, pp. 81–89. Springer, New York (2010). doi:10.1007/978-0-387-30164-8_66
3. Teh, Y.W.: Dirichlet processes. In: Sammut, C., Webb, G.I. (eds.) Encyclopedia of Machine Learning, pp. 280–287. Springer, New York (2010). doi:10.1007/978-0-387-30164-8_219
4. Ferguson, T.: A Bayesian analysis of some nonparametric problems. Ann. Stat. **1**, 209–230 (1973)
5. Sun, X., Yung, N., Lam, E.: Unsupervised tracking with the doubly stochastic Dirichlet process mixture model. IEEE Trans. Intell. Transp. Syst. **17**, 2594–2599 (2016)
6. Zhu, F., Chen, G., Hao, J., Heng, P.A.: Blind image denoising via dependent Dirichlet process tree. IEEE Trans. Pattern Anal. Mach. Intell. **38**, 1–14 (2016)
7. MacEachern, S.: Dependent nonparametric processes. In: ASA Proceedings of the Section on Bayesian Statistical Science, pp. 50–55 (1999)
8. Griffin, J., Steel, M.: Order-based dependent Dirichlet processes. J. Am. Stat. Assoc. **101**, 179–194 (2006)
9. Duan, J., Guindani, M., Gelfand, A.: Generalized spatial Dirichlet process models. Biometrika **94**, 809–825 (2007)
10. Sudderth, E., Jordan, M.: Shared segmentation of natural scences using dependent Pitman-Yor processes. Adv. Neural Inf. Process. Syst. **21**, 1585–1592 (2008)
11. Griffin, J.: The Ornstein-Uhlenbeck Dirichlet process and other time-varying processes for Bayesian nonparametric inference. J. Stat. Plann. Infer. **141**, 3648–3664 (2011)

12. Blei, D., Frazier, P.: Distance dependent Chinese restaurant processes. J. Mach. Learn. Res. **12**, 2461–2488 (2011)
13. Ghosh, S., Ungureunu, A., Sudderth, E., Blei, D.: Spatial distance dependent Chinese restaurant processes for image segmentation. In: Advances in Neural Information Processing Systems, pp. 1476–1484 (2011)
14. Ren, L., Wang, Y., Dunson, D., Carin, L.: The kernel beta process. Adv. Neural Inf. Process. Syst. **24**, 963–971 (2011)
15. Rodríguez, A., Dunson, D.: Nonparametric Bayesian models through probit stick-breaking processes. Bayesian Anal. **6**, 145–178 (2011)
16. Paisley, J., Wang, C., Blei, D.: The discrete infinite logistic normal distribution. Bayesian Anal. **7**, 235–272 (2012)
17. Sethuraman, J.: A constructive definition of Dirichlet priors. Stat. Sin. **4**, 639–650 (1994)
18. Ishwaran, H., Zarepour, M.: Dirichlet prior sieves in finite normal mixtures. Stat. Sin. **12**, 941–963 (2002)
19. Rasmussen, C.E.: Gaussian Processes for Machine Learning. MIT Press, Cambridge (2006)
20. Bishop, C.: Pattern Recognition and Machine Learning. Springer, New York (2006)
21. Blei, D.: Build, compute, critique, repeat: data analysis with latent variable models. Ann. Rev. Stat. Appl. **1**, 203–232 (2014)
22. Williams, C., Seeger, M.: Using the Nyström method to speed up kernel machines. Adv. Neural Inf. Process. Syst. **13**, 682–688 (2011)
23. Kumar, S., Mohri, M., Talwalkar, A.: Sampling methods for the Nyström method. J. Mach. Learn. Res. **13**, 981–1006 (2012)
24. Dunson, D., Park, J.H.: Kernel stick-breaking processes. Biometrika **95**, 307–323 (2008)
25. An, Q., Wang, C., Shterev, I., Wang, E., Carin, L., Dunson, D.: Hierarchical kernel stick-breaking process for multi-task image analysis. In: Proceedings of the International Conference on Machine Learning, pp. 17–24 (2008)
26. Foti, N., Williamson, S.: Slice sampling normalized kernel-weighted completely random measure mixture models. Adv. Neural Inf. Process. Syst. **25**, 2240–2248 (2012)
27. Foti, N., Futoma, J., Rockmore, D., Williamson, S.: A unifying representation for a class of dependent random measures. In: Proceedings of the International Conference on Artificial Intelligence and Statistics, pp. 20–28 (2013)
28. Shi, J., Malik, J.: Normalized cuts and image segmentation. IEEE Trans. Pattern Anal. Mach. Intell. **22**, 888–905 (2000)
29. Teh, Y.: A hierarchical Bayesian language model based on Pitman-Yor processes. In: Proceedings of the 21st International Conference on Computational Linguistics and the 44th Annual Meeting of the Association for Computational Linguistics, pp. 985–992 (2006)
30. Uetz, R., Behnke, S.: Large-scale object recognition with CUDA-accelerated hierarchical neural networks. In: Proceedings of the IEEE International Conference on Intelligent Computing and Intelligent Systems, pp. 1–6 (2009)
31. Unnikrishnan, R., Pantofaru, C., Hebert, M.: Toward objective evaluation of image segmentation algorithms. IEEE Trans. Pattern Anal. Mach. Intell. **29**, 929–944 (2007)
32. Oliva, A., Torralba, A.: Modeling the shape of the scene: a holistic representation of the spatial envelope. Int. J. Comput. Vis. **42**, 145–175 (2001)

Frequency Recognition Based on Optimized Power Spectral Density Analysis for SSSEP-Based BCIs

Xing Han, Yadong Liu$^{(\boxtimes)}$, Yang Yu, and Zongtan Zhou

College of Mechatronic and Automation,
National University of Defense Technology,
Changsha 410073, Hunan, People's Republic of China
liuyadong1977@163.com

Abstract. This paper presents a novel three-class Steady-state somatosensory evoked potentials (SSSEPs)-based BCI paradigm for target identification. The improved stimulation pattern accompanied by rhythmic pulses (i.e., 'Tic-Tic-Toc') was provided to the arm, waist and thigh of three healthy subjects by tactile tactors vibrating at different frequencies. The subjects were asked to selectively focus their attention on flutter sensation derived from stimulation of one site among three. To improve classification accuracy, we added a posterior processing after the power spectral density (PSD) analysis to reduce the inter-frequency variation and named the new method D-PSD. Experimental results for three subjects suggested that D-PSD method and the 'Tic-Tic-Toc' pattern increased accuracy (60.41%) compared with traditional PSD (42.56%) and absence of 'Tic-Tic-Toc' pattern (54.72%), respectively. These indicate that our BCI paradigm deserves being further explored as an alternative to SSVEP-based BCI applications.

Keywords: Brain-computer interface · Electroencephalogram · Steady-state somatosensory evoked potentials · Power spectral density

1 Introduction

An electroencephalogram (EEG)-based brain-computer interface (BCI) technology provides a direct and non-invasive channel between the human brain and artificial devices which can translate brain activities to computer commands without the participation of peripheral nerves and muscles [1]. Robot arm control is one of popular applications of BCI. Because of the high information rate (ITR), unsophisticated system configuration and less mental effort, steady-state evoked potentials have become one of the most promising paradigms for real BCI applications.

However, one apparent drawback of visual method is that great visual pressure imposed on subjects who are supposed to focus on the flicker characters or sudden flash of light during the whole stimulus presentation process. Specifically, patients with some kinds of optical illness cannot possibly control external devices based on SSVEP or visual P300. Besides, in case system manipulation requires additional visual interaction, the communication function of the system relied on visual pathways is decreased

© Springer International Publishing AG 2017
Y. Sun et al. (Eds.): IScIDE 2017, LNCS 10559, pp. 77–87, 2017.
DOI: 10.1007/978-3-319-67777-4_7

dramatically. To guarantee the practicality of BCI system in above situations, new paradigm based on steady-state evoked potentials other than SSVEP should be considered and adopted. It is expected, accordingly, somatosensory system remains to be available for individual who suffers eye-related ailments or whose vision channel is already occupied.

Steady-state somatosensory evoked potential (SSSEP) is a sinusoidal Electrophysiological brain response to tactile stimulation delivered to the glabrous skin in a wide range of particular frequencies [2]. Induced SSSEP amplitudes at the driving frequency of the vibratory stimulation can be modulated by selective attention to the stimulated location compared to when the location is ignored [4, 5]. Decoding the EEG recording unveils the right target that subject focused on, which is similar to the basic principle in SSVEP (Steady-state visual evoked potentials)-based BCIs.

However, there is a significant physiological superiority of SSVEP against SSSEP: SSVEPs are much easier decoded from EEG segments because participant using SSVEP-based BCI can direct their gaze to a single visual stimulus patch and maintain fixation to that one for a while [6].

Despite the above-mentioned challenge, enhancement of SSSEPs via purely 'shift' attention (i.e., attention without a gaze conversion) to a tactile stimulus has been proved to be feasible in [3]. To enhance classification accuracy, several signal processing methods have been proposed, such as power spectral density (PSD) analysis based on Fast Fourier Transform (FFT) [7] and time/frequency maps [8]. [9] introduced the common spatial pattern (CSP) method as employed to the SMR-based BCI into SSSEPs processing to help enhance signal separabiliy. Recent studies have focused on selecting suitable stimulus frequencies dependent on subjects [10], which was not studied intensively in this paper.

In this study, we proposed a novel multi-class SSSEP-based BCI paradigm. The tactile stimuli were mounted at the subjects' arm, waist and thigh on the same side of the body, which improve the practicability of the BCI devices at the cost of increasing signal separability possibly. First, we introduced the 'Tic-Tic-Toc' stimulation pattern into SSSEP BCI to help keep users' attention to the target stimulation unit. Second, we employed a novel approach called D-PSD (D means differentiation) to SSSEP signal discrimination based on the degree of deviation of PSD between reference period and stimulation period, and strategy to pick multiple channels in a local area was based on that.

The remainder of this paper is organized as follows: the stimulation pattern, experiment procedures and signal processing are described in the Materials and Methods section, the experimental results are presented in the Results section, Discussion and Conclusions section provides an overall discussion of the results and concludes the paper.

2   Materials and Methods

2.1   Experiment Procedure

Each multi-class identification task consisted of 2 blocks, one of which was training session and another was testing session. Each block included 8 runs. Pseudo-Randomly distributed attention-modulated task was executed in every 9 trials, which were run successively in single run, lasting for 120 s in whole. Subjects were supposed to wear earphones listening to white noise during the experiment process to avoid the interference by vibration noise. Before the start of the first trial, a 3 break was put in. A 'start' sound from earphones cued the beginning of the trial. After that was a 4.5-s long reference period with applying stimulation to stimulus sites. Following the stimulation duration, a short ticking sound like 'beep' signalized the end of one trial. Subsequently, a 1 s break was inserted, followed by a cue in form of vibration lasting 1 s. After 0.5 s of interval. 4.5 s of stimulus presentation occurred again. This time subject had to pay close attention to the only vibrator instructed. 3-s long break was inserted and after that the trial started anew. There was a break of 30 s between runs and a 2-min rest every four runs. The sequence of one trial is shown in Fig. 1(c).

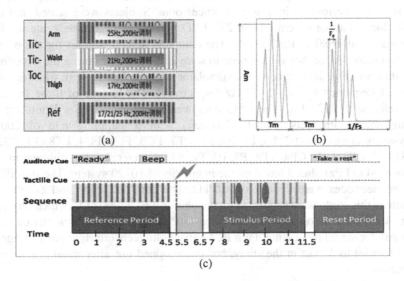

(a)

(b)

(c)

Fig. 1. (a) 'Tic-Tic-Toc' pattern for stimulation period and normal pattern for reference period, where thick bars for 'normal' pattern and light-colored bars for 'Tic' pattern. The circles are for 'Toc' pattern. (b) Schematic diagram of amplitude-modulated stimulus. In our study, the carrier frequency (F_c) of motors was 200 Hz, and modulation frequencies denoted as stimulus frequency F_s hereafter, varied with sites where vibrators shook. (c) Sequence diagram of the 'Tic-Tic-Toc' pattern for one trial.

2.2 Paradigm Description

To enhance attention to the task and to minimize the effects of weariness and boredom, participants finished 'Tic-Tic-Toc' detection tasks. The rhythmic pulse (i.e., 'Tic-Tic-Toc') was dissimilar to that proposed in [2]. As Fig. 1(b) showed, the tactor vibrated at medium amplitude, denoted as A_m, lasting T_m seconds long (i.e., $0.5/F_s$) and kept out of work for another T_m. This cycle happened during most of vibration time, and we took it 'normal' pattern. To generate the transient pulses of the 'Tic-Tic-Toc' pattern, tactor vibrated at 75% of A_m and kept closed for 2 times of T_m

for the first two beats ('Tic-Tic') while vibrated at 150% of A_m and turned down for 2 times of T_m for the third beat ('Toc'). The 'Tic-Tic-Toc' pattern was arranged to randomly generate one to three times in the second half of one trial. Participants were instructed to mentally count the exact number of occurrences of rhythmic pattern. After each trial, the participants were asked about the number they have remarked in mind to just test their concentration level.

Amplitude-modulated (AM) tactile waveforms was used to restrict the power to the carrier frequency and its sidebands, which ensured that any response at the modulation frequency originated from attention-based modulation [11].

2.3 Subjects and Experimental Setup

SSSEP-based BCI systems were designed and offline analyses were employed. The experiment was performed in a normal office room. Subjects were seated in a chair naturally located about 60 cm from a 27" LED monitor with a refresh rate of 60 Hz and resolution of 1680 × 1080 pixels. The computer screen was set for the operator rather than subjects, for they do not need to stare at the monitor during the experiment.

As shown in Fig. 2(c), three tactile stimulators mounted at the arm, waist and thigh vibrated at frequency of 25, 21, 17 Hz respectively. The choose of these frequency values referred to [12, 13]. The EEG data were recorded using a BrainAmp DC Amplifier (Brain Products GmbH, Germany) with a sampling rate of 200 Hz. Twenty-seven electrodes (F7, F3, Fz, F4, F8, FT7, FC3, FCz, FC4, FT8, T7, C3, Cz, C4, T8, TP7, CP3, CPz, CP4, TP8, P7, P3, Pz, P4, P8, referenced to right ear mastoid and grounded to FPz) placed based on the international 10–20 system. The impedances of all the electrodes were kept below 10 kΩ. The data collection and experimental procedure in this study were supported by the platform BCI2000 [14], which provides a Python Application Programming Interface (API) for stimulus presentation and a Matlab API for signal processing. Four BCI-naïve volunteers (all right-handed, age 23–26, one female) took part in the study, but one dropped out after training session for some reason.

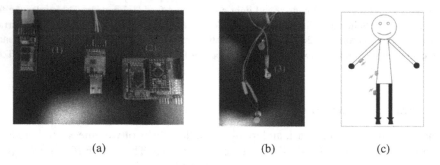

(a)   (b)   (c)

Fig. 2. (a) Communication Unit for host computer (1) and Micro Control Unit (2), (b) Stimulation Unit, and (c) sketch indicating the stimulator placement.

2.4 Stimulation Unit

The amplitude-modulated mechanical stimulation unit generating rectangular waves with 50% duty cycle consisted of three parts: (1) the wireless communication module based on Bluetooth HC-05. It received signals from another Bluetooth connected the host computer and sent the signal to slave computer. (2) The slave computer was a programmed Arduino Pro mini, which drove motors to generate amplitude-modulated stimulus. (3) Tactile actuators were composed of ULN2003A driver chips and vibrators (C0834B002F), which were electrically safe to subjects. These vibrators were taped to the subjects' epidermis with medical adhesive tape. Vibrators at different places were set the same magnitude for comparison. Once receiving driving signals, vibrators started to shake within 30 ms, which guaranteed the time synchronization of stimuli and EEG signal recording. All these three components were integrated in a self-made Printed Circuit Board (PCB), as Fig. 2(a) and (b) showed.

2.5 Signal Processing

The EEG signals were sampled at 200 Hz and filtered using a 50-Hz notch filter. To weaken the artifacts and noise, the data were band-pass filtered between 15–27 Hz using a FIR filter with order 96. These data were then parsed into trials. For each trial, the 4.5 reference period and the 4.5 stimulation period were separated. To reduce the effect of evoked potentials caused by the stimulation onset, the first 0.5 s was omitted as described in [10].

2.5.1 D-PSD Method

Traditional SSSEP-based EEG system choose the target frequency by calculating the PSD of SSSEP response directly. Because the SSSEP response is likely to be contaminated by background noise in the subjects' brain, the amplitude of the SSSEP response exhibits a complex inter-frequency variation among subjects. Therefore, some targets corresponding to less noisy SSSEP responses may be detected more easily than others. In view of the above-mentioned facts, we introduce reference period into the SSSEP-based multi-class identification task and achieve better classification performance than that without reference period. Though few mistakes happen occasionally, D-PSD did reduce the inter-frequency variation of the SSSEP response (see Results).

For the purpose of conditioning (to obtain a normal distribution), the logarithm was applied. The score of EEG data X(t) corresponding to each stimulus frequency is defined as follows:

$$\text{score}_f(t) = \frac{1}{K}\sum\nolimits_{k=1}^{K} 10\log_{10}\text{PSD}_f^{(k)}(t). \tag{1}$$

PSD is denoted as:

$$\text{PSD}_f^{(k)}(t) = \left\|\text{FFT}\left(X_{(t)}^{(k)}, N\right)\right\|. \tag{2}$$

where f is the stimulus frequency, k is the channel index, K is the number of selected channels (see Sect. 2.5.3), t is the time duration of EEG data, which is treated as constant value of 4.5 throughout the analysis and N is constant value 256. $\|\cdot\|$ implies the absolute value. Specifically, we obtained reference period scores $\left(\text{score}_f^{RP}(t)\right)$ of each frequency from training session. During testing session, the SSSEP response scores were calculated as following:

$$\text{Score}_f(t) = \frac{\text{score}_f(t) - \text{score}_f^{RP}(t)}{\text{score}_f(t) + \text{score}_f^{RP}(t)}. \tag{3}$$

2.5.2   Recognition Strategy

The main purpose of attention-modulated SSSEP-based BCI systems is to detect the frequency of the SSSEP component. Let us suppose that a signal with t seconds long acquired from K channels is to be recognized from F stimulus frequencies $f_1, f_2, f_3, \ldots, f_F$. Our tactics for identifying the stimulus frequency f_s is as follows:

$$f_s = \max_f \text{Score}_f, \quad f = f_1, f_2, f_3, \ldots, f_F. \tag{4}$$

The definition of Score_f is indicated in (3). Here symbol for time t is omitted.

2.5.3   Channel Selection

As SSSEP is a localized potential related to stimulation site, taking too large an area for PSD analysis is more likely to introduce less relevant EEG features (i.e., the frequency of the SSSEP). Therefore, signal processing is carried out on 2 channels in our study. In [15], Muller et al. simply chose sensorimotor cortex (i.e., C3, Cz, C4 referred to 10–20 system) acquiring signal. In view of that the spatial distance between stimulation locations in our study is large, the activated area involved SSSEP may vary obviously. Besides, attention modulation is a complex process, not only primary sensory cortex, but somatic sensory association area is responsive to tactile stimulation. Consequently, we take 25 electrodes into account (listed in Sect. 2.3).

We proposed a method for channel selection based on D-PSD, which describes the degree of variation of PSD before and during the attention-converted task. For each of three stimulation class related to stimulation sites and frequency, the average score_f of each electrode across trials, in form of scalp topography, is depicted for both reference period and its subsequent stimulation period. Specially, K is 1 in formula (1) while calculating on single electrode. The channels with the 2 largest Score_f are selected for stimulus recognition with frequency f. In our analysis, we empirically set K equals 2 and eliminate the repeated channel selected in more than one class if such condition exists, which characterizes the CSP feature extraction method.

2.5.4   Evaluation Methods

A ten-fold cross-validation was used to evaluate the D-PSD method. The data acquired from session 1 and session 2 was randomly divided into 2 groups for selecting channel and testing the performance respectively. Each multi-class recognition task in one session had 24 trials, each of that included 4.5-second-long reference period and 4.5-second-long stimulation period. The analysis was repeated 10 times. The mean accuracy p and single classification accuracy p_i were defined as follows:

$$p = \frac{N_{trial(f=F)}}{N_{trial}} * 100\%. \tag{5}$$

$$p_i = \frac{n_{trial(f_i=F_i)}}{n_{trial}} * 100\%, \, i = 1, 2, 3. \tag{6}$$

3 Results

3.1 Performance of the BCI Paradigm

Four subjects participated in the training experiment for channel selection, and three among whom finished testing procedure. Testing experiment includes two sessions, one of which is with 'Tic-Tic-Toc' paradigm and another is not. These two conditions are denoted as 'Y' and 'N' in Table 1. The mean offline separability with 'Tic-Tic-Toc' pattern was 60.41% across subjects and stimulation types, which is a little higher than that without 'Tic-Tic-Toc' pattern. The highest accuracy of classification exists in arm for all subjects, among whom subject B reaches the highest ratio by almost 70%.

Table 1. Ten-fold cross-validation classification accuracy for multi-class SSSEP tasks

Subject	Pattern	Accuracy (%)			
		Arm	Waist	Thigh	Avg.
A	Y	64.27	59.13	52.44	58.61
	N	59.10	50.28	48.16	52.51
B	Y	69.74	67.32	62.59	66.88
	N	55.13	63.20	60.71	59.68
C	Y	60.37	55.63	52.25	56.08
	N	57.09	56.87	41.93	51.96
Mean	Y	64.79	60.69	55.76	60.41
	N	57.11	56.81	50.26	54.72

To improve the validity of proposed D-PSD pattern recognition method, we take a simple test contrasting performance between D-PSD and PSD analysis introduced in [7], which tried to recognize the target by picking peak value on power spectrum from the several choices. The results statics can be seen from Fig. 3.

3.2 Channel Selection Based on D-PSD

As discussed in Sect. 2.5.3, the channel selected to analyze SSSEPs motivated by tactile stimulator with one frequency is different from another. Figure 4 depicts average

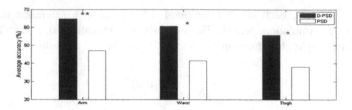

Fig. 3. Average accuracy across subjects and paired t-test result. Black and white bars represent average value of D-PSD and PSD method respectively. The significance of paired t-test is symbolized by pentacles, where ★★ stands for p < 0.05 and ★ the PSD method for each of stimulation targets.

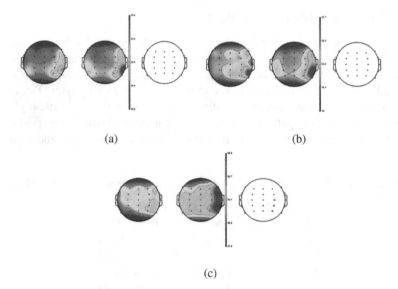

(a)   (b)

(c)

Fig. 4. The average distribution topographies of log spectral power density of subject B. The left diagram of each row is average topography about stimulation period. The middle diagram of each row is average topography about reference period. The solid red circle on the right diagram reflect the channel whose Score is larger than threshold (0.02) we set empirically: graph (a), (b) and (c) are average distribution topography at 17 Hz, 21 Hz and 25 Hz respectively. (Color figure online)

topographies of the 25 channels' log spectral power density $(10 * \log_{10}(\mu v^2/\text{Hz}))$ of frequency at 17, 21, 25 Hz respectively in subject B derived from SSSEP data analysis on training session.

For target stimulus on thigh with 17 Hz, the channels to analyze were T7 and P7. For target stimulus on waist with 21 Hz, the channels that we focused are TP7 and P8. For target stimulus on arm with 25 Hz, the channels chose on account that Score was larger than empirical value 0.02 were FT8, T7, C4, CP4, P7, P4 and P8. Getting rid of T7, P7 and P8 (see Sect. 2.5.3 for reasons), the remaining two channels with largest

Score value were CP4 and FT8, and these two were used for analysis of SSSEP derived from arm stimulation.

Figure 5 is one illustration of PSD comparison between stimulation period and reference period of subject A's arm over sum of signals obtained from CP4 and FT8.

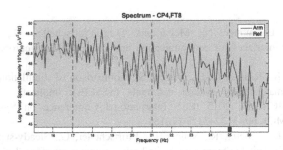

Fig. 5. The average spectrum of CP4 and FT8 between 15 and 27 Hz. The blue and green line indicated the stimulation period and the reference period of arm respectively. The red vertical lines instructed the stimulation frequency of 17,21 and 25 Hz respectively. The red square in the under panel implies the maximum of Score among the three frequencies above. (Color figure online)

4 Discussion and Conclusion

The results of this study showed that SSSEPs can be enhanced via purely focused attention. Comparing the amplitude variation between reference period and stimulation period based on proposed D-PSD method, we selected the item whose corresponding Score was the largest as the target. And this method has been proved to be valid by comparing with traditional PSD analysis. Three healthy subjects participated in our experiments, and the average accuracy of classification for the three-class SSSEPs tasks was 60.41%. Subject B achieved the best performance, with an average accuracy of 66.88%. Besides, employed 'Tic-Tic-Toc' pattern helps to improve the ability of command separability. However, whether this paradigm has a remarkable difference in statistics needs further discussed.

In order to make the best of the stimulus signals from different parts of the body, we proposed a specific channel selection method based on D-PSD. From the distribution topographies as shown in Fig. 4, we inferred that most area related to SSSEPs distributed in the back of the central sulcus, which was a little different from previous findings that the main regions located at primary sensory cortex contralateral to the stimulated sites. One possible reason for that is the duration time of stimulation. In our study, the reference and stimulation period were both 4 s (the first 0.5 s was removed), during which time the shift of distribution of cortical potentials may have occurred. The scalp topographies, as expected, reflected the spatially variation of SSSEPs induced by different parts of body, which form the basis of proposed channel selection method.

In this paper, the length of time window to analyze SSSEPs is fixed, but the relationship between window length and task recognition performance will be investigated in our future work. We will extend this BCI paradigm to online test with practical applications, for example, humanoid robot or neuroprosthesis control, providing practical assistance to the disabled especially patients with eye disease.

References

1. Wolpaw, J.R., Birbaumer, N., McFarland, D.J., Pfurtscheller, G., Vaughan, T.M.: Brain-computer interfaces for communication and control. Clin. Neurophysiol. **113**, 767–791 (2002)
2. Choi, I., Bond, K., Krusienski, D., Nam, C.S.: Comparison of stimulation patterns to elicit steady-state somatosensory evoked potentials (SSSEPs): implications for hybrid and SSSEP-based BCIs. In: 2015 IEEE International Conference on Systems, Man, and Cybernetics (SMC), pp. 3122–3127 (2015)
3. Muller-Putz, G.R., Scherer, R., Neuper, C., Pfurtscheller, G.: Steady-state somatosensory evoked potentials: suitable brain signals for brain-computer interfaces? IEEE Trans. Neural Syst. Rehabil. Eng. **14**, 30–37 (2006)
4. Adler, J., Giabbiconi, C.-M., Müller, M.M.: Shift of attention to the body location of distracters is mediated by perceptual load in sustained somatosensory attention. Biol. Psychol. **81**, 77–85 (2009)
5. Giabbiconi, C.M., Dancer, C., Zopf, R., Gruber, T., Müller, M.M.: Selective spatial attention to left or right hand flutter sensation modulates the steady-state somatosensory evoked potential. Cogn. Brain. Res. **20**, 58–66 (2004)
6. Smith, D.J., Varghese, L.A., Stepp, C.E., Guenther, F.H.: Comparison of steady-state visual and somatosensory evoked potentials for brain-computer interface control. In: 2014 36th Annual International Conference of the IEEE Engineering in Medicine and Biology Society, pp. 1234–1237 (2014)
7. Severens, M., Farquhar, J., Desain, P., Duysens, J., Gielen, C.: Transient and steady-state responses to mechanical stimulation of different fingers reveal interactions based on lateral inhibition. Clin. Neurophysiol. Off. J. Int. Fed. Clin. Neurophysiol. **121**, 2090–2096 (2010)
8. Spitzer, B., Wacker, E., Blankenburg, F.: Oscillatory correlates of vibrotactile frequency processing in human working memory. J. Neurosci. **30**, 4496–4502 (2010)
9. Nam, Y., Cichocki, A., Choi, S.: Common spatial patterns for steady-state somatosensory evoked potentials. In: 2013 35th Annual International Conference of the IEEE Engineering in Medicine and Biology Society (EMBC), pp. 2255–2258 (2013)
10. Breitwieser, C., Kaiser, V., Neuper, C., Muller-Putz, G.R.: Stability and distribution of steady-state somatosensory evoked potentials elicited by vibro-tactile stimulation. Med. Biol. Eng. Comput. **50**, 347–357 (2012)
11. Noss, R.S., Boles, C.D., Yingling, C.D.: Steady-state analysis of somatosensory evoked potentials. Electroencephalogr. Clin. Neurophysiol./Evoked Potentials Sect. **100**, 453–461 (1996)
12. Wolpaw, J., Wolpaw, E.W.: Brain-Computer Interfaces: Principles and Practice. OUP, New York (2012)
13. Muller, G.R., Neuper, C., Pfurtscheller, G.: „Resonance-like "frequencies of sensorimotor areas evoked by repetitive tactile stimulation-resonanzeffekte in sensomotorischen arealen, evoziert durch rhythmische taktile stimulation. Biomed. Technik/Biomed. Eng. **46**, 186–190 (2001)

14. Schalk, G., McFarland, D.J., Hinterberger, T., Birbaumer, N., Wolpaw, J.R.: BCI2000: a general-purpose brain-computer interface (BCI) system. IEEE Trans. Biomed. Eng. **51**, 1034–1043 (2004)
15. Breitwieser, C., Kaiser, V., Neuper, C., Muller-Putz, G.R.: Stability and distribution of steady-state somatosensory evoked potentials elicited by vibro-tactile stimulation. Med. Biol. Eng. Comput. **50**, 347–357 (2012)

A Hybrid Particle Swarm Optimization Algorithm Based on Migration Mechanism

Ning Lai[✉] and Fei Han

School of Computer Science and Communication Engineering,
Jiangsu University, Zhenjiang, Jiangsu, China
2211508048@stmail.ujs.edu.cn, hanfei@ujs.edu.cn

Abstract. Standard Particle Swarm Optimization 2011 is the latest version of standard PSO algorithm, with an adaptive random topology and rotational invariance. However, it needs to be improved on non-separable and asymmetrical problems. Migration mechanism is inspired by biogeography-based optimization, which is insensitive to dense swarms, and could effectively develop the local solution space. In this study, in order to avoid the algorithm has been in a local optimal state, when the optimal value of the algorithm does not improve within the threshold value, the introduction of migration mechanism used to jump out of that state. In addition, topological migration is used in the migration mechanism, in order to more effectively share the solution features and improve the search capabilities of the algorithm. Finally, six different types of benchmark functions are selected to compare the proposed algorithm with the above two algorithms, which verifies the effectiveness of the proposed algorithm.

Keywords: Particle swarm optimization · Standard PSO 2011 · Biogeography-based optimization · Migration mechanism · Topological migration

1 Introduction

Particle swarm optimization (PSO) is an effective optimization tools, firstly introduced by Kennedy and Eberhart [1], inspired from the foraging behavior of birds, and using biological population model proposed by Heppner and Grenander [2]. The algorithm has some similarities with other evolutionary algorithms, such as genetic algorithm (GA) [3]. Due to strong distributed capabilities, global search capabilities and the ease of parallel implementation, PSO has received a surge of attention in recent years. Since its inception in 1995, many ideas were introduced thereafter for trying to improve the original version of PSO, aimed at avoiding the algorithm fall into local minimum. The particles fly too fast or too slow in the original PSO, and exploration and development capabilities are uncontrolled, thus introducing inertia weights to balance this behavior [4]. After that, the PSO algorithm with a contraction factor model is proposed by Clerc, which is equivalent to the PSO algorithm with inertia weight as long as the parameters adjust appropriately [5]. In addition to the improvement of the PSO algorithm on improving convergence speed, many improvements focused on the domain structure of particles [6], the improvement of objective function [7], the mixing of PSO algorithm

© Springer International Publishing AG 2017
Y. Sun et al. (Eds.): IScIDE 2017, LNCS 10559, pp. 88–100, 2017.
DOI: 10.1007/978-3-319-67777-4_8

with other optimization algorithms [8] and the study of the discrete version of the PSO algorithm [9] and so on.

With the advent of variants of different PSO versions, which use different "standard" PSO algorithms in the corresponding literatures, some researchers have attempted to define the standard version of the PSO algorithm [10, 11]. So far, three versions of the standard PSO algorithm have been recognized by most researchers, known as SPSO2006, SPSO2007 and SPSO2011 [11]. SPSO2011 is the latest version of PSO algorithm, which has been applied to many standard benchmark functions, and its results provide a baseline for future improvement of SPSO2011 performance [12]. By analyzing the stability, local convergence and rotation sensitivity of SPSO2011 algorithm and its variants, it is shown that the algorithm has good stability and rotational invariance, and yet the algorithm does not converge locally [13, 14].

In [15], a meta-optimization environment is used to tune the parameters of the SPSO2011 algorithm and the SPSO2006 algorithm to improve the performance of algorithms. The results show that the PSO algorithm needs a larger population size than is usually set in the standard versions. In [16], Snehal and Varsha discussed the impact of acceleration coefficients and random neighborhood topology modification threshold on the SPSO2011 algorithm. Although the tuning of the SPSO2011 algorithm has an effect on the performance of the algorithm, the ability of the algorithm to jump out of the local optimum does not improve. Due to the performance of the SPSO2011 algorithm is affected by the distribution of the center of the search range and its global search ability slowly disappear because of the update rule of the center, a new update rule is propose to improve the global search ability of algorithm and it has the characteristic of maintaining the diversity [17]. Modifying an update rule can achieve the effect of maintaining the diversity of the swarm without increasing the complexity of the algorithm. However, the increase in the diversity of the swarm will only make the SPSO2011 algorithm converge more slowly to the local optimum, due to the lower diversity of the swarm in the later period of the algorithm.

The migration mechanism is a method for each candidate solution to exchange information according to migration rate, and its basic idea comes from the BBO algorithm [18]. On the one hand, due to insensitive to dense swarms, it is usually used for local search. Therefore, the migration mechanism is proposed to improve the local search ability of the SPSO2011 algorithm. In [19], a hybrid algorithms of mixed BBO and PSO are proposed, which combines the exploration of PSO with the development of the BBO algorithm. In addition to having a strong local search capability, the migration mechanism has no a so-called particle velocity term like the PSO algorithm. Therefore, when the migration mechanism performs a local search, there is no fear that the solution velocity is too large to find the optimal value. On the other hand, because the SPSO2011 algorithm has an adaptive topology, only the SPSO2011 algorithm is simply used for global search, and the migration mechanism is used for local search, which will destroy the original topology of the SPSO2011 algorithm and make the convergence speed of the hybrid algorithm slow. Therefore, in order to avoid the algorithm has been in a local optimal state, an improved SPSO2011 algorithm is proposed by introducing topological migration aim to jump out of that state. In the remainder of the paper, improved SPSO2011 algorithm is simply represented by ISPSO2011 algorithm. The experiment results show that it has a higher success rate and faster convergence speed, compared with the SPSO2011 algorithm and the migration mechanism.

2 Preliminaries

2.1 Standard PSO 2011

The PSO algorithm is similar to other evolutionary algorithms, also uses the conception of the "swarm" and the "evolution". Each individual in the swarm is regarded as particle in PSO terminology, and each particle dynamically adjusts its position according to its individual cognition and swarm cognition. Each particle of the PSO algorithm represents a candidate solution of the solution space, containing three vectors: position, velocity and previous best. Assuming that the solution space is a D-dimensional search space, the position and velocity of the ith particle are represented by $X_i = (x_{i1}, x_{i2}, \ldots, x_{iD})$ and $V_i = (v_{i1}, v_{i2}, \ldots, v_{iD})$ respectively. Previous best is known best position of the ith particle, as a memory for storing information of the best found solution, is represented by $P_i = (p_{i1}, p_{i2}, \ldots, p_{iD})$, whereas L_i^t is the local best position within that particle's field, $i = 1, 2, \ldots, N$, where N is the number of swarm size. In the iterative process, the velocity and position of the ith particle are updated according to the following equation:

$$V_i^{t+1} = \omega V_i^t + \emptyset_1 R_{1,i}^t * \left(P_i^t - X_i^t\right) + \emptyset_2 R_{2,i}^t * \left(P_g^t - X_i^t\right) \tag{1}$$

$$X_i^{t+1} = X_i^t + V_i^{t+1} \tag{2}$$

where ω is the inertia weight coefficient, $\emptyset_1$ and $\emptyset_2$ are the two acceleration coefficients. In iteration t, $R_{1,i}^t$ and $R_{2,i}^t$ are randomly generated diagonal matrix, in which each element is a number distributed uniformly in interval $[0, 1]$. Also, P_i^t and P_g^t are the previous best position vectors of the ith particle and the global best position vectors of all particles, respectively, at iteration t.

It is well known that there are biases in dimension by dimension method, such as the canonical PSO algorithm, whose performance is affected by the rotation of the coordinate system [20]. Due to SPSO2011 algorithm use a structure analogous to the "sphere model" with rotational invariance, attracting many researchers to study its property, such as stability, local convergence and rotation sensitivity [13, 14]. The velocity update rule of the SPSO2011 algorithm is as follows:

$$V_i^{t+1} = \omega V_i^t + H_i\left(G_i^t, \left\|G_i^t - X_i^t\right\|\right) - X_i^t \tag{3}$$

where $H_i\left(G_i^t, \left\|G_i^t - X_i^t\right\|\right)$ is a spherical distribution with the center G_i^t and radius $\left\|G_i^t - X_i^t\right\|$. G_i^t is the center of gravity of three points, denoted by $\frac{X_i^t + p_i^t + l_i^t}{3}$, where $p_i^t = X_i^t + c_1\left(P_i^t - X_i^t\right)$ and $l_i^t = X_i^t + c_2\left(L_i^t - X_i^t\right)$. Whereas the position update rules of the SPSO2011 algorithm follow the Eq. (2), which can also be simplified as:

$$X_i^{t+1} = \omega V_i^t + x_i' \tag{4}$$

when calculating the new position, a random point x_i' needs to be defined in the hyper-sphere. Figure 1 shows that the support of the distribution of all possible next positions is a D-sphere, and the distribution itself is not uniform.

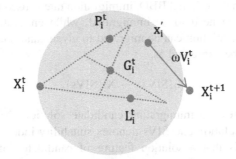

Fig. 1. Geometrical interpretation of the SPSO2011 algorithm

In addition, the SPSO2011 algorithm uses an adaptive random topology to connect all particles, controls the exchange of information between particles, which uses a matrix $L(N, N)$ to store this information. The set of particles "informing" the ith particle is called the field of that particle, where includes the particle itself (a particle informs itself). The number of the set of particles, called the field size of the ith particle, denoted by K, which follows a probability distribution. In this case, with a probability equal to $1 - \left(1 - \frac{1}{N}\right)^K$ to choose K in the swarm (except itself), because of the probability of the particle informs itself is equal to 1 [21]. If the global best position does not improve, then regenerate a new topology, that is, modify the connection between all particles.

2.2 Biogeography-Based Optimization

Biogeography-based optimization (BBO) is also an evolutionary algorithm, firstly proposed by Simon [18]. As name suggests, the BBO algorithm derives from mathematical modeling of biogeography, just as genetic algorithm derives from mathematical modeling of biological genetics. In the BBO algorithm, mainly contains two probabilistic operators. The one of probabilistic operator is migration operator, and the rate of immigration and emigration are usually expressed by τ and μ. The former represents the probability that the new species will reach a habitat and the latter represents the probability of leaving the native region. The value of τ and μ are given as:

$$\tau_i = I\left(1 - \frac{f(i)}{N}\right), \mu_i = E\left(\frac{f(i)}{N}\right) \tag{5}$$

where I denotes the maximum immigrated rate and E denotes the maximum emigration rate, both are usually equal to 1. $f(i)$ can be either the fitness-based ranking of the ith habitat or based on the ranking of the number of species of the ith habitat, and i is the ascending order of N (1 is worst and N is best). The behavior of the probabilistic migration of the species in habitats imply that good solutions tend to share their features with other solutions, while poor solutions are more likely to accept features from other solutions. In the original BBO, immigration rate is used for decide whether a given candidate solution need to be immigrated, while emigration rate is used for decide which candidate solution can be migrated to given candidate solution described above. Migration can be expressed as:

$$X_i(SIVs) \leftarrow X_j(SIVs) \tag{6}$$

where $X_i(SIVs)$ denotes the immigrating candidate solution, $X_j(SIVs)$ denotes the emigrating candidate solution and SIVs denotes suitability index variables. Note that the above migration is that a solution feature of candidate solution X_i is simply replaced by a solution feature from candidate solution X_j. In [22], a blended migration is presented as follows:

$$X_i(SIVs) \leftarrow \alpha X_i(SIVs) + (1 - \alpha)X_j(SIVs) \tag{7}$$

where α is a random number in the interval $[0, 1]$, which is usually 0.5. It may be a determined number or proportional to the relative fitness of the candidate solutions X_i and X_j.

In addition to the migration operator, the migration mechanism also includes the mutation operator, which is used for increasing the biological diversity by modifying the values of some randomly selected SIVs of each habitat.

3   The Improved Standard PSO 2011 Algorithm

Due to the limited local search ability of the SPSO2011 algorithm and easily fall into the local optimum for complicated and combined function, an ISPSO2011 algorithm based on the migration mechanism is proposed to improve the performance of the SPSO2011 algorithm. The migration mechanism is derived from the BBO algorithm, which has a strong ability of local search. As a result, the ISPSO2011 algorithm not only introduced the migration mechanism for local search, but also improved the migration mechanism, the topological migration. Topological migration uses the topology in the SPSO2011 algorithm, in order to avoid the slow convergence speed of the SPSO2011 algorithm to the rotational or symmetrical function and more effectively develop the optimal solution. In the basic migration mechanism, each particle migrates according to the information of all particles in the particle swarm, while the topological migration is based only on the information of the best particles in the field. The pseudo-codes of the topological migration are given in Algorithm1. When the optimal value of the algorithm is in the local optimal state, and the number of times reaches the threshold T, then the topology migration is introduced into the algorithm to jump out of the state.

Define a topological structure
For each particle X_i
    Finding the all particle of the field of X_i
    For each decision solution SIVs
        Use τ to decide whether to immigrate to X_i
        If immigrating then
            Use μ to decide whether the best particle X_j in the field to emigrate
            $X_i(SIVs) \leftarrow \alpha X_i(SIVs) + (1 - \alpha)X_j(SIVs)$
        End If
    End For
End For

Algorithm1: the topological migration

The flow diagram of the ISPSO2011 algorithm is shown in Fig. 2, and the steps of the algorithm are as follows:

Step 1: Initialize some of the parameters and initialize the position and velocity of each particle in the swarm. Calculate the fitness values of each particle in the swarm, and initialize the previous best and global best.

Step 2: Execute the SPSO2011 algorithm. If the number of times that the global best has not been improved reaches the designated threshold value, then go to Step 3.

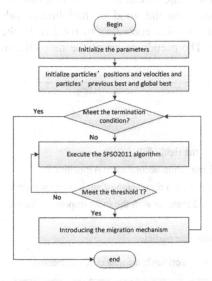

Fig. 2. The flow diagram of ISPSO2011 algorithm

Step 3: Generate a topology matrix to construct the structure of the topological migration between the eigenvalues of the particles in the swarm, and initialize the immigration rate and the migration rate of each particle in the swarm according to the ranking of the number of particles.

Step 4: According to the topological field where the particle is located, each particle in the swarm is subjected to the corresponding topological migration and each dimension of each particle in the swarm is subjected to mutation operation according to the mutation rate.

Step 5: Record the number of times each dimension of each particle in the swarm changes, and if the particle is improved, then the number of times plus one, otherwise minus one. The number of times is greater, and the probability of particle improvement is greater. Then, there are more probabilistic improved particles to replace those particles that are less likely to improve.

Step 6: Replacing those particles of lower fitness value with elite particles. Regardless of whether the particles have improved, go to Step 2.

Step 7: If the termination condition is not reached, go to Step 2.

In this paper, the migration mechanism use a blended migration in [22], and migration model is a linear model based on the ranking of the number of particles. Most of the mixed evolutionary algorithm and BBO algorithm is that the former is used for global search while the latter is used for local search or both of them are used separately in real-world optimization problems [19]. The demarcation point of the global search and local search process are not based on the diversity density used by most algorithms, but judge whether the global optimal value is improved in the set threshold value. The threshold value is expressed by T. A new elite mechanism is used in the ISPSO2011 algorithm, which is a double elite mechanism. The first elite mechanism performs the replacement operation based on the good or bad fitness value, while another one performs the replacement operation according to the size of the improvement of each dimension of the particle. The pseudo-codes of the new elite mechanism are given in Algorithm2.

Define a topological structure

Store elite particle

For each particle X_i

    Finding the all particle of the field of X_i

    Execute the topological migration according to Algorithm1

    Record the number of times each dimension of each improved particle in the swarm

    Replace the least improved features with the most improved features in particles

End For

Replace the worst particle with elite particle

Algorithm2: the new elite mechanism

---

4 Experiment Results and Discussion

In order to verity the performance of ISPSO2011 algorithm, six well-known bench-mark functions are chosen to compare with the SPSO2011 algorithms and the BBO algorithm. The details of benchmark functions are shown in Table 1, where the theoretical global minimum value of all benchmark functions is equal to 0. F1 is separable and unimodal function, and in addition to the global minimum, also has a local minimum of the same number as the dimension. F2 is complicated function, with many local minimum, but the global minimum is away from the second better local minimum. F3 and F4 is a highly multimodal function with several local minimum. F5 is characterized by a nearly flat outer region, and a large hole at the center, which makes the search algorithms are easy to be trapped into its local minimum. F6 has many widespread local minimum regularly distributed, but only one global minimum.

Table 1. The six benchmark functions

Test function	Formula	Search range
Sphere (F1)	$f_1(x) = \sum_{i=1}^{D} x_i^2$	$[-5.12, 5.12]^D$
Schwefel (F2)	$f_2(x) = 418.9828D - \sum_{i=1}^{D} x_i \sin\left(\sqrt{|x_i|}\right)$	$[-500, 500]^D$
Rastrigin (F3)	$f_3(x) = 10D + \sum_{i=1}^{D} \left(x_i^2 - 10\cos(2\pi x_i)\right)$	$[-5.12, 5.12]^D$
Alpine (F4)	$f_4(x) = \sum_{i=1}^{D} |x_i \sin(x_i) + 0.1 x_i|$	$[-10, 10]^D$
Ackley (F5)	$f_5(x) = -20e^{\left(-0.2\sqrt{\frac{1}{D}\sum_{i=1}^{D} x_i^2}\right)} - e^{\left(\frac{1}{D}\sum_{i=1}^{D}\cos(2\pi x_i)\right)} + 20 - e^1$	$[-32.768, 32.768]^D$
Griewank (F6)	$f_6(x) = \frac{1}{4000}\sum_{i=1}^{D} x_i^2 - \prod_{i=1}^{D} \cos\left(\frac{x_i}{\sqrt{i}}\right) + 1$	$[-600, 600]^D$

All simulation experiments are run on a PC with Intel(R) Core(TM) i5-2450 M 2.50 GHz CPU and MATLAB R2013a. All the algorithms have the same particle random initialization setting. The parameters are set as follows: The particle size is equal to 40, the maximum iteration is equal to 5000 and each algorithm experiment was repeated 50 times. The success rate (SR) indicates the ratio of the number of finding theoretical global minimum value to the total number of trials. According to [12], in SPSO2011 algorithm, the acceleration coefficients $\varnothing_1$ and $\varnothing_2$ are $0.5 + \log 2$; the inertia weight coefficient ω is $1/(2 * \log 2)$; particles velocity is limited to the interval $[-X_{min}, X_{max}]$; and the random topology with $K = 3$ informants. In BBO algorithm, the mutation rate is 0.05; the number of habitat elites is 3. The ISPSO2011 algorithm has the same parameters as the two algorithms described above.

The details of the three algorithms are described in Table 2. From Table 2, it can be seen that ISPSO2011 has better average performance than the other two algorithms on all of test functions. The performance of the BBO algorithm is far less than the SPSO2011 algorithm, but after mixing, it shows a stronger performance. The ISPSO2011 algorithm can find the theoretical global minimum when running the F1 and F4 functions on 10 dimensions, however, for the F3 and F6 functions, no matter which dimension can find the theoretical global minimum. In finding the theoretical global minimum, the ISPSO2011 algorithm has a stronger search capability than the

Table 2. Summary statistics for the 10, 30 and 50 dimensional case in fifty times experiments

Test function	Dimension	SPSO2011		BBO		ISPSO2011	
		Mean ± std	SR	Mean ± std	SR	Mean ± std	SR
F1	10	1.21E −301 ± 0.00E+00	0%	4.17E −03 ± 2.48E−03	0%	9.97E −319 ± 0.00E+00	64%
	30	3.16E −133 ± 1.35E−132	0%	5.72E −01 ± 1.29E−01	0%	3.20E −150 ± 1.18E−149	0%
	50	3.67E−79 ± 1.05E −78	0%	3.29E +00 ± 5.91E−01	0%	9.71E−91 ± 4.38E −90	0%
F2	10	8.77E+02 ± 2.48E +02	0%	1.51E +01 ± 6.72E+00	0%	1.42E+01 ± 7.43E +01	0%
	30	4.86E+03 ± 6.46E +02	0%	1.98E +03 ± 3.25E+02	0%	8.36E+02 ± 1.28E +03	0%
	50	9.15E+03 ± 1.18E +03	0%	6.49E +03 ± 5.01E+02	0%	1.78E+03 ± 2.35E +03	0%
F3	10	4.27E+00 ± 2.01E +00	0%	4.33E +00 ± 1.56E+00	0%	3.13E−15 ± 7.75E −15	84%
	30	2.23E+01 ± 6.34E +00	0%	7.11E +01 ± 1.24E+01	0%	6.85E−01 ± 2.47E +00	32%
	50	5.08E+01 ± 1.04E +01	0%	2.11E +02 ± 2.22E+01	0%	3.19E+00 ± 9.63E +00	22%
F4	10	3.71E−04 ± 3.28E −04	0%	5.44E −02 ± 4.64E−02	0%	6.28E−05 ± 1.08E −03	62%
	30	8.32E−02 ± 5.06E −02	0%	4.67E +00 ± 1.27E+00	0%	4.89E−02 ± 8.87E −02	0%
	50	6.67E−01 ± 9.62E −01	0%	1.92E +01 ± 2.95E+00	0%	4.80E−01 ± 7.13E −01	0%
F5	10	4.44E−15 ± 0.00E +00	0%	1.24E +00 ± 4.82E−01	0%	3.66E−15 ± 1.49E −15	0%
	30	9.68E−01 ± 7.88E −01	0%	5.39E +00 ± 5.23E−01	0%	7.92E−15 ± 1.34E −15	0%
	50	2.28E+00 ± 3.49E −01	0%	8.83E +00 ± 4.87E−01	0%	3.59E−01 ± 5.78E −01	0%
F6	10	1.97E−02 ± 1.64E −02	12%	8.08E −01 ± 1.52E−01	0%	1.20E−06 ± 8.45E −06	96%
	30	7.00E−03 ± 1.06E −02	28%	3.05E +00 ± 5.38E−01	0%	2.22E−18 ± 1.57E −17	98%
	50	5.40E−03 ± 7.50E −03	0%	1.18E +01 ± 2.10E+00	0%	8.16E−09 ± 5.77E −08	94%

SPSO2011 algorithm, and the SR value in Table 2 can explain it. In particular, for F6 function, the success rate of the ISPSO2011 algorithm is higher than ninety percent in all dimensions. For F5 function on 10 dimensions, The SPSO2011 algorithm is obviously trapped into local optimal, and the ISPSO2011 algorithm can jump out of this local optimal point.

Figure 3 shows the mean error convergence curve for selected functions of the three algorithms on ten dimensions.

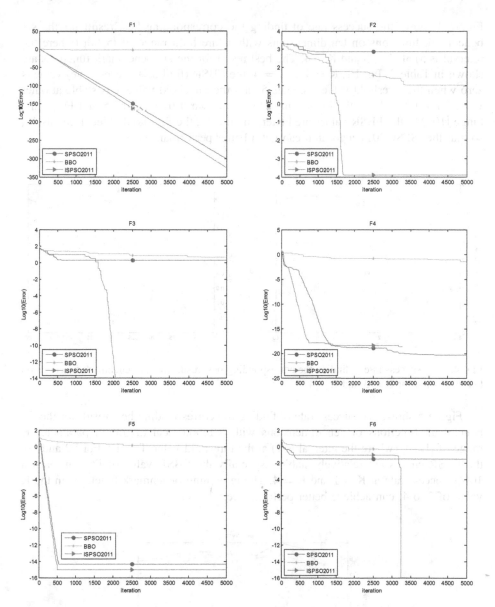

Fig. 3. The mean error convergence curve for selected functions of the three algorithms on ten dimensions

Below, we will discuss the two parameters of the ISPSO2011 algorithm for the threshold value T and topological migration parameter K, respectively. The threshold value T indicates the degree of implementation of the migration mechanism. The higher threshold value T corresponds to less implementation of the migration mechanism. Topological migration parameter K is the number of informants in the topology of the migration mechanism, similar to parameter K that in the SPSO2011 algorithm.

Figure 4 shows the success rate of finding the corresponding best result for the six benchmark functions on ten dimensions with a threshold range of [5, 50] (where the interval is 5) at K = 3 and K = 4. The best result for the six benchmark functions are shown in Table 2. For F1, K = 3 or K = 4, the FBSR (find best success rate) value is zero when the threshold value T equals 5, and the threshold value T is stable at other values. For F2 and F3, the FBSR curve to a downward trend. For F5 and F6, in the range [10, 24], the FBSR values are better. In all case, the threshold value T equals 10, so that the ISPSO2011 algorithm can get a better performance.

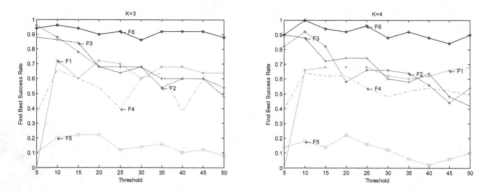

Fig. 4. The success rate of finding the corresponding best result for six benchmark functions on ten dimensions at K = 3 and K = 4

Figure 5 shows the success rate of finding the corresponding best result for the six benchmark functions on ten dimensions with a topological migration parameter K range of [1, 10] (where the interval is 1) at the threshold value T = 10. For F2 and F6, the FBSR curves are relatively stable, especially the FBSR value of F6 can reach a 100% success rate at K = 1 and K = 4. The remaining benchmark functions in the K value of 2 to 4, can achieve better performance.

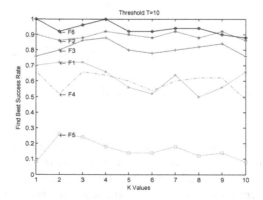

Fig. 5. The success rate of finding the corresponding best result for the six benchmark functions on ten dimensions at the threshold value T = 10

5   Conclusions

In this paper, an improved SPSO2011 algorithm is proposed, which combines the global search of SPSO2011 algorithm with the local search of the migration mechanism to balance the capability of SPSO2011 algorithm to explore and develop. In addition, the introduction of topological migration and a new elite mechanism, so that the algorithm to avoid trapping into the local optimum. In order to verify the performance of the ISPSO2011 algorithm, six benchmark functions are employed. Experimental results demonstrate the good performance of our approach by comparing with the SPSO2011 algorithm and the BBO algorithm. In this work, we just consider the unconstrained function optimization and the local optimum of some benchmark functions are difficult to jump out. In our further work, we will solve these problems and apply the improved algorithm to neural networks, pattern recognition and extreme learning.

Acknowledgments. This work was supported by the National Natural Science Foundation of China (Nos. 61572241 and 61271385), the Foundation of the Peak of Six Talents of Jiangsu Province (No. 2015-DZXX-024), and the Fifth "333 High Level Talented Person Cultivating Project" of Jiangsu Province.

References

1. Kennedy, J., Eberhart, R.: A new optimizer using particle swarm theory. In: Proceedings of the Sixth International Symposium on Micro Machine and Human Science, Australia, pp. 39–43. IEEE (1995)
2. Heppner, F., Grenander, U.: The Ubiquity of Chaos: A Stochastic Nonlinear Model for Coordinate Bird Flocks. AAAS Publications, American (1990)
3. Eberhart, R.C., Shi, Y.: Comparison between genetic algorithms and particle swarm optimization. In: Porto, V.W., Saravanan, N., Waagen, D., Eiben, A.E. (eds.) EP 1998. LNCS, vol. 1447, pp. 611–616. Springer, Heidelberg (1998). doi:10.1007/BFb0040812
4. Shi, Y., Eberhart, R.: A modified particle swarm optimizer. In: Evolutionary Computation Proceedings, Anchorage, pp. 69–73. IEEE (1998)
5. Clerc, M.: The swarm and the queen: towards a deterministic and adaptive particle swarm optimization. In: Evolutionary Computation, Washington, pp. 1951–1957. IEEE (1999)
6. Kennedy, J.: Small worlds and mega-minds: effects of neighborhood topology on particle swarm performance. In: Evolutionary Computation, Washington, pp. 1931—1938. IEEE (1999)
7. Parsopoulos, K.E., Plagianakos, V.P., Magoulas, G.D., Vrahatis, M.N.: Stretching technique for obtaining global minimizers through particle swarm optimization. In: Proceeding of the Particle Swarm Optimization Workshop, PSOW, Indianapolis, pp. 22–29. (2001)
8. Liu, Q., Han, F.: A hybrid attractive and repulsive particle swarm optimization based on gradient search. In: Huang, D.-S., Jo, K.-H., Zhou, Y.-Q., Han, K. (eds.) ICIC 2013. LNCS, vol. 7996, pp. 155–162. Springer, Heidelberg (2013). doi:10.1007/978-3-642-39482-9_18
9. Kennedy, J., Eberhart, R.C.: A discrete binary version of the particle swarm algorithm. In: Computational Cybemetics and Simulation, Orlando, pp. 4104–4108. IEEE (1997)
10. Bratton, D., Kennedy, J.: Defining a standard for particle swarm optimization. In: Swarm Intelligence Symposium, Honolulu, pp. 120—127. IEEE (2007)

11. Clerc, M.: Beyond Standard Particle Swarm Optimisation. IJSIR **1**(4), 46–61 (2010)
12. Zambrano-Bigiarini, M., Clerc, M., Rojas, R.: Standard particle swarm optimisation 2011 at CEC-2013: a baseline for future PSO improvements. In: 2013 IEEE Congress on Evolutionary Computation (CEC), Cancun, Mexico, pp. 2337–2344. IEEE (2013)
13. Bonyadi, M.R., Michalewicz, Z.: SPSO 2011: analysis of stability; local convergence; and rotation sensitivity. In: Proceedings of the 2014 Annual Conference on Genetic and Evolutionary Computation, Vancouver, pp. 9–16. ACM (2014)
14. Bonyadi, M.R., Michalewicz, Z.: Analysis of stability, local convergence, and transformation sensitivity of a variant of particle swarm optimization algorithm. IEEE Trans. Evol. Comput. **20**(3), 370–385 (2016)
15. Ugolotti, R., Cagnoni, S.: Automatic tuning of standard PSO versions. In: Proceeding of the Companion Publication of the 2015 Annual Conference on Genetic and Evolutionary Computation, pp. 1501–1502. ACM, New York (2015)
16. Kamalapur, S.M., Patil, V.H.: Impact of acceleration coefficients and random neighborhood topology modification threshold on SPSO. In: 2015 International Conference on Computing Communication Control and Automation (ICCUBEA), Pune, India, pp. 437–441. IEEE (2015)
17. Hariya, Y., Shindo, T., Jin'No, K.: An improved rotationally invariant PSO: a modified standard PSO-2011. In: 2016 IEEE Congress on Evolutionary Computation (CEC 2016), Vancouver, BC, Canada, pp. 1839–1844. IEEE (2016)
18. Simon, D.: Biogeography-based optimization. IEEE Trans. Evol. Comput. **12**(6), 702–713 (2008)
19. Cheng, G., Lv, C., Yan, S., Xu, L.: A novel hybrid optimization algorithm combined with BBO and PSO. In: 2016 Chinese Control and Decision Conference, China, pp. 1198–1202. IEEE (2016)
20. Spears, W.M., Green, D.T., Spears, D.F.: Biases in particle swarm optimization. Int. J. Swarm Intell. Res. **1**(2), 34–57 (2010)
21. Clerc, M.: Back to random topology. Relatório Técnico (2007)
22. Ma, H., Dan, S.: Blended biogeography-based optimization for constrained optimization. Eng. Appl. Artif. Intell. **24**(3), 517–525 (2011)

A Hybrid Multi-swarm PSO Algorithm Based on Shuffled Frog Leaping Algorithm

Hongfei Bao[✉] and Fei Han

School of Computer Science and Communication Engineering,
Jiangsu University, Zhenjiang 212013, Jiangsu, China
rainbow_eris@sina.com, hanfei@ujs.edu.cn

Abstract. As an effective swarm intelligence algorithm, multi-swarm particle swarm optimization (PSO) has better search ability than single-swarm PSO. In order to enhance the ability of group communication as well as improve the ability of local search, this paper proposes a hybrid multi-swarm PSO algorithm. Three strategies have been proposed, which are multi-swarm strategy, update strategy and cooperation strategy. A new way of grouping the particle swarms is put forward by calculating the fitness value of particles. In each group, the particles updates according to the formula which is morphed from the shuffled frog leaping algorithm. Moreover, a new information communication strategy is proposed. The cooperation of these three strategies maintains the diversity of algorithm and improves the ability of searching the optimal solution. Finally, the experimental results on the benchmark functions verify the effectiveness of the proposed PSO.

Keywords: Particle swarm optimization · Global optimization · Shuffled frog leaping algorithm

1 Introduction

Particle swarm optimization (PSO) algorithm, as a swarm intelligence algorithm, is a new optimization technique which has attracted more and more researchers' attention. Particle swarm optimization (PSO) was firstly proposed by Kennedy and Eberhart [1]. By observing flock of birds' behaviors, the researchers found that sharing information in groups was beneficial to gain advantage during evolution. This made up the basis of PSO [2]. However, particle swarm optimization algorithm has some shortcomings. So, in recent years, many researchers have been working to improve the PSO algorithm.

In [3], a multi-strategy adaptive particle swarm optimization (MAPSO) was proposed which changed dynamically the inertia weight and introduced an elitist leaning strategy to enhance the diversity of population. Two modified PSO algorithms were introduced to solve large scale continuous optimization problems [4, 5]. The three algorithms above emphasized the update of particles, but they were lack of communication between groups which needs great attention. A novel parallel multi-swarm algorithm based on comprehensive learning particle swarm optimization (PCLPSO) was introduced which solved the problem of communication [6]. This algorithm based on the master-slave paradigm had multiple swarms which work cooperatively.

© Springer International Publishing AG 2017
Y. Sun et al. (Eds.): IScIDE 2017, LNCS 10559, pp. 101–112, 2017.
DOI: 10.1007/978-3-319-67777-4_9

However, as iteration increased, the diversity of population lost rapidly. In [7], the dynamic multi-swarm particle swarm optimizer (DMS-PSO) and a new cooperative learning strategy (CLS) were hybridized. The algorithm selected the two worst particles of groups to be updated. But with this strategy, the complexity of algorithm becomes larger and the convergence speed becomes slower which was not expected in this paper. A new hybrid algorithm was proposed. In this algorithm, the particles update, and then the best particles form a new group [8]. This method improved the diversity of population. Therefore, in the new algorithm, the update of particles and the communication between subgroups are noticed. At the same time, the implementation of the algorithm is not difficult.

Shuffled frog leaping algorithm (SLFA) was a new swarm intelligence optimization algorithm proposed in [9]. In its update strategy, particles regenerate when it cannot update to a better solution. The strategy maintains the diversity of the population. So, in this paper, a hybrid multi-swarm PSO algorithm was proposed by combining with the shuffled frog leaping algorithm. Firstly, a new way of grouping the particle swarms is put forward by calculating the fitness value of particles. This way reorders the particles in a descending order, which is first come up in the shuffled frog leaping algorithm. Secondly, in each group, the particles updates according to the formula which is morphed from the shuffled frog leaping algorithm. Through the two strategies above, the particles of each group are searching in the global range, which improve the diversity of particles. Thirdly, a new information communication strategy is proposed in this paper.

2   Preliminaries

2.1   Particle Swarm Optimization

Particle swarm optimization (PSO) is a stochastic optimization algorithm which uses an iterative model. Each particle represents a potential solution which initializes the position and speed degrees within the solution space in a random way. In the process of evolution, each particle exchanges information with other particles. The process based on its own experience and the "experience" of neighboring particles which is called personal best and global best to determine their flight.

In order to obtain better performance, adaptive particle swarm optimization (APSO) algorithm was proposed [10]. Assume that the dimension of the search space is D, and swarm is M which consists of particles without weight and volume. Each particle represents a position in the D dimension. The velocity of the i-th particle is expressed as $v_i = (v_{i1}, v_{i2}, \cdots, v_{iD})$; the position of the i-th particle in the search space can be denoted as $x_i = (x_{i1}, x_{i2}, \cdots, x_{iD})$; the best position of the i-th particle being searched until now is called *pbest* which is expressed as $p_i = (p_{i1}, p_{i2}, \cdots, p_{iD})$. The best position of the all particles is called *gbest* which is denoted as $p_g = (p_{g1}, p_{g2}, \cdots, p_{gD})$. The APSO was described as [1]:

$$v_{ij}(t+1) = wv_{ij}(t) + c_1r_{1j}(t)\left(p_{ij}(t) - x_{ij}(t)\right) + c_2r_{2j}(t)\left(p_g(t) - x_{ij}(t)\right). \tag{1}$$

$$x_i(t+1) = x_i(t) + v_i(t+1). \tag{2}$$

where w called the inertia weight factor, which is used to adjust the global and local search ability; c_1 and c_2 are the acceleration constants with positive values which make the particles themselves have the ability to learn from the excellent individual in groups; r_1, r_2 are random number ranged from 0 to 1.

In APSO, w can be computed by the following equation:

$$w = w_{max} - t * (w_{max} - w_{min})/N_{pso}. \tag{3}$$

where w_{max}, w_{min} and N_{pso} are the initial inertial weight, the final inertial weight and the maximum iterations, respectively.

In the multi-swarm PSO [11], the way of updating is same as the standard particle swarm optimization. The difference is that, in the multi-swarm PSO, a certain number particle swarm is subdivided into several groups, and then they evolve with the rule of particle swarm optimizer. It not only keeps the independent of the particle swarm and the superiority of the optimizer, but also not increases the complexity of algorithm.

2.2 Shuffled Frog Leaping Algorithm

Shuffled Frog Leaping Algorithm was a new swarm intelligence optimization algorithm firstly proposed by Muzaffar et al. [12, 13]. It has the simple concept, the less number of parameters, the faster computing speed and the better global optimization ability.

The dimension of the search space is D. In the search space, the number of frog is P which generated randomly. The i-th frog represents that the solution of the problem is $X_i = (x_{i1}, x_{i2}, \cdots, x_{iD})$. According to the fitness value of frog individuals, the whole group can be divided into M subgroups. The first frog points into the first group; The second frog points into the second group; The i-th frog points into the M group and the $(i + 1)$-th frog points into the first group and so on, until all the frogs divided.

Each group searches deeply in the local space. The worst of a subgroup in the current iteration is called X_w, the best individual is called X_b and the best solution in the space is called X_g. The algorithm updates the current worst X_w as follows:

Formula for updating step:

$$\Omega_i = rand() \cdot (X_b - X_w).$$
$$(\|\Omega_{min}\| \le \|\Omega_i\| \le \|\Omega_{max}\|) \tag{4}$$

Formula for updating position:

$$new\, X_w = X_w + \Omega_i. \tag{5}$$

The algorithm updates strategy according to the formula. If the fitness of new X_w does not improve, the algorithm generates a new X_w randomly.

When all subgroups of local search finished, all of the frogs mixed again and divided into groups again. Then the algorithm repeats the steps above.

3   The Improved Multi-swarm PSO Based on SFLA

In order to enhance the ability of group communication as well as improve the ability of local search, this paper puts forward a hybrid multi-swarm PSO algorithm based on shuffled frog leaping algorithm. In the improved multi-swarm PSO, three strategies has been proposed, which are multi-swarm strategy, update strategy and cooperation strategy.

In the multi-swarm strategy, the population is divided into several subpopulations. Main swarm and sub-swarm are two types of swarms. Each swarm $s_i, i \in \{1, 2, \cdots, n\}$, independently runs the algorithm. Thus, each swarm explores the whole search space.

This strategy places particles in descending order according to the fitness value of the particle individuals. The whole group can be divided into M subgroups. The first particle is divided into the first group; The second particle is divided into the second group; The i-th particle is divided into the M group and the $(i + 1)$-th particle is divided into the first group and so on. The group of the global optimal particle is the main group and the other is the subgroup.

In the main group and subgroups, the algorithm uses different strategies to update. The main group is mainly used for convergence of the algorithm where the standard PSO is used for the update. The subgroups are mainly used for the searching in space. In each subgroup, particles update themselves according to the update rule of shuffled frog leaping algorithm. The main formulas are as follows:

$$v_i(t+1) = wv_i(t) + c_1 r_1(t)(p_i(t) - x_i(t)) + c_2 r_2(t)(p_s(t) - x_i(t)) \qquad (6)$$

where p_i is the best position of individual and p_s is the best position in the each subgroup. If the particles do not update to a better position, the particle updates by the following formula:

$$v_i(t+1) = wv_i(t) + c_1 r_1(t)(p_i(t) - x_i(t)) + c_2 r_2(t)(p_g(t) - x_i(t)) \qquad (7)$$

where p_g is the global best position of particles. If the particle is unable to get a better update after the operations, regenerate the velocity and position. Also, deep searching is added in the subgroups, which accelerates the convergence speed of the particles. Through the two strategies above, the particles of each group are searching in the global space, which keep the diversity of particles (Fig. 1).

In the cooperation strategy, the swarms in algorithm periodically cooperate to exchange solutions found up-to date and to guide the search towards solutions of better quality. There are several proposed cooperation strategies in the literature. Some of them are Independent Runs, Ring, 2D Mesh and Complete Graph [14].

In [6], some important questions encountered were the followings: Firstly, when should the migration occur? Secondly, how should the exchange topology be? Thirdly, which information should be exchanged between swarms? Finally, how should the integration policy be? In order to solve these questions, a strategy carried out.

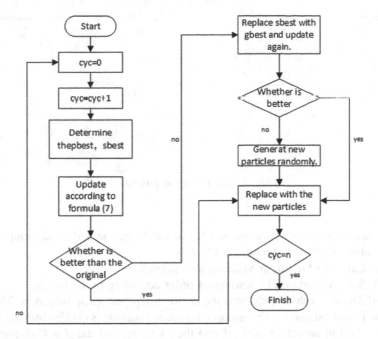

Fig. 1. The flowchart of the algorithm for updating particles in each subgroup.

The first factor in the cooperation strategy is the migration period. There is always a tradeoff at the migration process between the solution quality and computational efficiency. If information exchanges too frequently, then the solution quality may be better, but the computational efficiency deteriorates. If the information exchange slowly, then the computational efficiency is better, but the solution quality may be worse. There are two kinds of the strategies of determining the migration period as blind and adaptive [14]. In this paper, the blind strategy where migration process occurs periodically after a certain number of iterations is preferred.

The second factor is the topology where the neighbor swarms to share the information are determined. If the information is shared between too many swarms, the computational efficiency naturally deteriorates. In addition, with the convergence of particles, the algorithm may fall into the local minima and lost the ability of jumping out the local minima. If the information is shared between too few swarms, then the computational efficiency increases, but the algorithm may result in an unsatisfactory result. In this paper, there is not any directly communication between subgroups. Subgroups are responsible for the global search. When the subgroups get to a better solution, the subgroup communicates with the main group by replacing the poor particles at a designated spot in the main group, as shown in the Fig. 2. At the same time, subgroups remain updating with the original speed and position.

The basic steps of the improved PSO are as follows:

Step 1: Initialize particles randomly and set the parameters of the algorithm: the total of particles is N, the dimension is d, the individual number of each group is m,

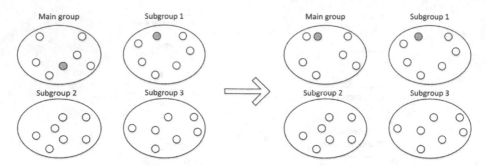

Fig. 2. The migration process.

the number of subgroups is n, the number of local iteration within subgroups is cyc, the number of global generation is *MaxIter*.

Step 2: Calculate the fitness value of each particle.

Step 3: Sort the particles in descending order according to the fitness.

Step 4: Divide each particle into the main group and each subgroup. The first particle is divided into the first group; The second particle is divided into the second group; The i-th particle is divided into the M group and the $(i + 1)$-th particle is divided into the first group and so on. The group of the global optimal particle is the main group and the other is the subgroup.

Step 5: Repeat the following steps in the subgroups, until the number of deep search is met.

(1) Determine the optimal solution for each subgroup which is called *sbest* according to the fitness value.

(2) Update each particle with the formula (2) and the formula (6). If the particles do not update to a better position, the particle updates by the formulas (2) and (7). If the particle is unable to get a better update after the operations, regenerate the velocity and position.

(3) Update the optimal solution in each subgroup. If the *sbest* is better the global best, replace the poor particles at a designated spot in the main group.

Step 6: If the iteration meets the migration period, all the particles mix and go to the step 2 to calculate. Otherwise, the algorithm goes to the step 5.

Step 7: If the algorithm meets the iteration, the process of evolution finish. Otherwise, the algorithm goes to the step 5.

4   Numerical Simulation and Analysis

In this section, the computational results of the proposed algorithm are presented. Two unimodal and four multimodal benchmark functions are selected to test the performance of the hybrid algorithm and to compare with other work.

Test functions are shown in Table 1. In this experiment, six high dimensional functions are chosen to test the search efficiency of the algorithm we proposed. Functions f_1 and f_2 are unimodal. Unimodal functions have only one optimum and no local minima. Function f_1 is the Sphere function which has been studied by many researchers. Function f_2 is the Rosenbrock function. The classical Rosenbrock function is a two-dimensional unimodal function. It has been extended to higher dimensions in recent years. Many researchers take the high-dimensional Rosenbrock function as a unimodal function. Function $f_3 - f_6$ are multimodal.

Table 2 shows the mean best solution for the six test functions on different dimensions by using six PSO variants. Three of the PSO variants are single group and

Table 1. The six test functions

Test function	Expression	Space search	Global minima
Sphere (f_1)	$\sum_{i=1}^{D} x_i^2$	$(-100,100)^n$	0
Rosenbrock (f_2)	$\sum_{i=1}^{D-1}\left[100\left(x_i^2 - x_{i+1}\right)^2 + (x_i - 1)^2\right]$	$(-5.12, 5.12)^n$	0
Ackley (f_3)	$-20exp\left(-0.2\sqrt{\frac{1}{D}\sum_{i=1}^{D} x_i^2}\right) - exp\left(\frac{1}{D}\sum_{i=1}^{D} \cos(2\pi x_i)\right) + 20 + e$	$(-32, 32)^n$	0
Griewank (f_4)	$\sum_{i=1}^{D} \frac{x_i^2}{4000} - \prod_{i=1}^{D} \cos\left(\frac{x_i}{\sqrt{i}}\right) + 1$	$(-100,100)^n$	0
Rastrigin (f_5)	$f_5(x) = 10D + \sum_{i=1}^{D}\left[x_i^2 - 10\cos(2\pi x_i)\right]$	$(-100,100)^n$	0
Quadric (f_6)	$f_6(x) = \sum_{i=1}^{D}\left(\sum_{j=1}^{i} x_j^2\right)$	$(-100,100)^n$	0

Table 2. Mean best solution for the six test functions by using six PSOs

Test functions	Dimension	SPSO	APSO	PSOPC	MSCPSO	SFLA	MSLPSO
Sphere (f_1)	10	3.4964e−03	6.2670e−03	4.8414e−27	8.0783e−14	1.0472e−12	9.6713e−49
	20	6.7322e−04	2.0192e−06	7.1524e−30	2.2870e−10	3.0907e−13	1.3367e−44
	30	0.0263	6.3582e−03	1.1550e−29	3.3160e−10	2.9812e−13	1.1818e−44
Rosenbrock (f_2)	10	9.0582e+03	3.8753e+03	48.6985	22.4092	4.1661e−12	9.1454e−15
	20	9.2505e+03	5.1605e+03	6.5012e+02	2.0727e+02	3.2681e−11	2.8679e−19
	30	4.6146e+03	4.9595e+03	1.7273e+03	7.1262e+02	2.2247e−09	3.6101e−20
Ackley (f_3)	10	3.8832	4.4151	2.2723	1.1747	5.0865e−07	4.4136e−12
	20	4.4909	4.2862	2.9699	2.7213	6.5464e−06	1.4599e−12
	30	4.4242	4.4121	3.4866	3.1960	5.3903e−05	6.4424e−10
Griewank (f_4)	10	0.5425	0.5941	0.1243	0.0457	9.3440e−12	2.2204e−17
	20	0.5526	0.5561	0.2868	0.1835	1.3527e−12	2.4425e−16
	30	0.6072	0.6710	0.3866	0.3355	1.2653e−12	6.6613e−16
Rastrigin (f_5)	10	2.7143e+02	2.6456e+02	32.2989	17.7443	9.8340e−12	0
	20	2.6785e+02	2.7433e+02	1.0809e+02	85.9334	2.6998e−10	0
	30	2.7224e+02	2.6777e−02	1.8053e+02	1.5895e+02	1.3489e−10	0
Quadric (f_6)	10	25.0789	67.6534	1.5853	1.1141	6.3365e−10	1.1870e−19
	20	1.1532e+02	1.0054e+02	7.7726	8.8357	2.7199e−09	4.8961e−19
	30	1.2097e+02	78.3611	19.7630	19.4801	2.5127e−08	5.9126e−20

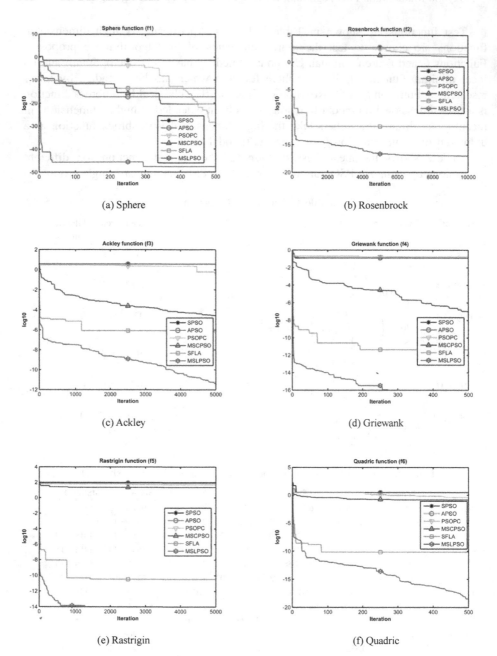

Fig. 3. Best solution versus iteration number for the six test functions by using six PSO variants

the population for them is 20. The others are multi-swarm optimization algorithm and the population for them is 100. The values of inertia weight w and the acceleration coefficient c in experiments are same with each other. Namely, the inertia weight w linearly decreases from 0.8 to 0.4 during the iterations. The acceleration coefficient c is equal to 1.4962. All the programs are carried out in MATLAB 2012a environment on an Intel Core(TM) i5-3210 M CPU.

The migration period in the algorithm is 50. Table 2 shows that the mean best solution for the six test functions by using six PSO variants.

In the Fig. 3, this paper presents the best solution of each function with ten dimensions for six algorithms. It is obvious that the proposed algorithm has an excellent performance on the most functions. The proposed algorithm had a higher efficiency than others on the functions and a faster convergence speed. Moreover, the algorithm performed very well in the f_4 and f_5, especially.

Figure 4 shows the result of the best solution of some functions with ten dimensions by using different number of subgroups. In the following experiment, the numbers of subgroups are ranged from 3 to 7. As shown in the result, when the number of subgroups is 5, the algorithm gets a better optimal value.

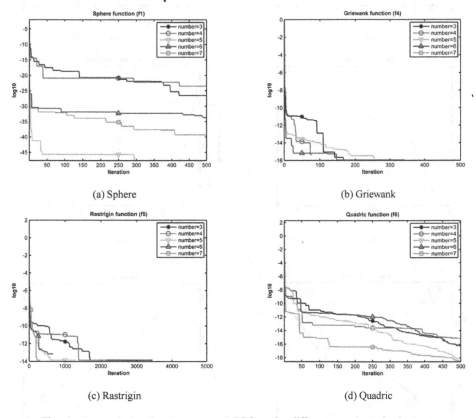

(a) Sphere

(b) Griewank

(c) Rastrigin

(d) Quadric

Fig. 4. Best solution for the proposed PSO under different number of subgroups

Figure 5 shows the result of the best solution and running time of each function with ten dimensions by using different number of deep search. The number of deep search is called *cyc*. With the increased of *cyc*, the result could be better, but the running time of algorithm would increase. As a result, in this paper, the number of deep search is set up to 50.

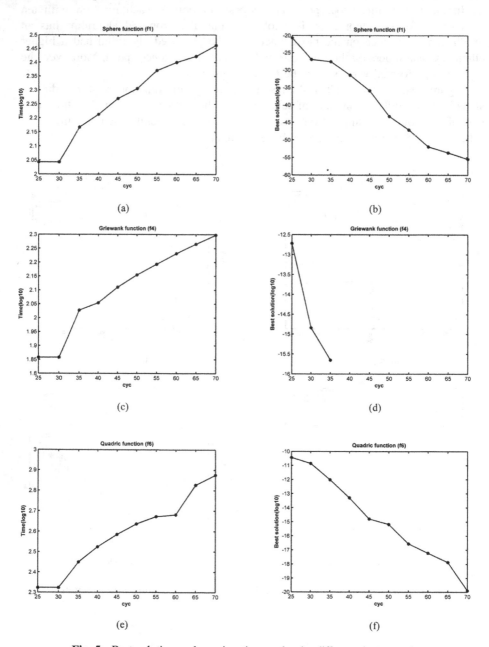

Fig. 5. Best solution and running time under the different deep search.

5 Conclusions

In order to improve the ability of group communication as well as the ability of local search, three strategies had been proposed, which were multi-swarm strategy, update strategy and cooperation strategy. A new way of grouping the particle swarms is put forward. The particles updates according to the formula which is morphed from the shuffled frog leaping algorithm. Subgroups communicated with the main group by replacing the poor particles at a designated spot in the main group. These three strategies guaranteed the ability of global search and kept the diversity of particles. The results were provided to verify that the proposed algorithm performs better than the other algorithms in most cases. In the future, we should continue to improve its ability of jumping out local minima.

Acknowledgements. This work was supported by the National Natural Science Foundation of China (Nos. 61271385 and 61572241), the Foundation of the Peak of Six Talents of Jiangsu Province (No. 2015-DZXX-024), and the Fifth "333 High Level Talented Person Cultivating Project" of Jiangsu Province.

References

1. Kennedy, J., Eberhart, R.C.: Particle swarm optimization. IEEE Int. Conf. Neural Netw. **4**, 1942–1948 (1995)
2. Xie, X.F., Zhang, W.J., Yang, Z.L.: Overview of particle swarm optimization. J. Kongzhi Yu Juece/Control Decis. **18**(2), 129–134 (2003). Beijing
3. Tang, K., Li, Z., Luo, L., et al.: Multi-strategy adaptive particle swarm optimization for numerical optimization. J. Eng. Appl. Artif. Intell. **37**, 9–19 (2015)
4. Oca, M.A.M.D., Aydın, D., Stützle, T.: An incremental particle swarm for large-scale continuous optimization problems: an example of tuning-in-the-loop (re) design of optimization algorithms. J. Soft Comput. **15**(11), 2233–2255 (2011)
5. García-Nieto, J., Alba, E.: Restart particle swarm optimization with velocity modulation: a scalability test. J. Soft Comput. **15**(11), 2221–2232 (2011)
6. Gülcü, S., Kodaz, H.: A novel parallel multi-swarm algorithm based on comprehensive learning particle swarm optimization. J. Eng. Appl. Artif. Intell. **45**, 33–45 (2015)
7. Xu, X., Tang, Y., Li, J., et al.: Dynamic multi-swarm particle swarm optimizer with cooperative learning strategy. J. Appl. Soft Comput. **29**, 169–183 (2015)
8. Li, J., Sun, H., Shi, X., et al.: The research on multi-swarm particle swarm optimization algorithm and hybrid frog leaping algorithm. J. J. Chin. Mini-Micro Comput. Syst. **34**(9), 2164–2168 (2013)
9. Eusuff, M.M., Lansey, K.E.: Optimization of water distribution network design using shuffled frog leaping algorithm. J. Water Resour. Plan. Manag. **129**(3), 210–225 (2003)
10. Zhang, X., Du, Y., Qin, Z., Qin, G., Lu, J.: A modified particle swarm optimizer for tracking dynamic systems. In: Wang, L., Chen, K., Ong, Y.S. (eds.) ICNC 2005. LNCS, vol. 3612, pp. 592–601. Springer, Heidelberg (2005). doi:10.1007/11539902_72
11. Luo, D.X., Zhou, Y.Q., Huang, H.J., et al.: Multi-colony particle swarm optimization algorithm. Comput. Eng. Appl. **46**(19), 51–54 (2010)
12. Muzaffar, E., Kevin, L., Fayzul, P.: Shuffled frog-leaping algorithm: a memetic meta-heuristic for discrete optimization. J. Eng. Optim. **38**(2), 129–154 (2006)

13. Rahimi-Vahed, A., Mirzaei, A.H.: A hybrid multi-objective shuffled frog-leaping algorithm for a mixed-model assembly line sequencing problem. J. Comput. Ind. Eng. **53**(4), 642–666 (2007)
14. Talbi, E.G.: Metaheuristics: from design to implementation. J. Proc. SPIE – Int. Soc. Opt. Eng. **42**(4), 497–541 (2009)

Similarity Degree for Multi-Attribute Decision Making with Incomplete Dual Hesitant Fuzzy Sets

Xin Liu[2], Yuanyuan Shi[2], Li Zou[1(✉)], and Siyuan Luo[2]

[1] School of Computer and Information Technology,
Liaoning Normal University, Dalian, Liaoning, China
zoulicn@163.com
[2] School of Mathematics, Liaoning Normal University, Dalian, Liaoning, China
liuxin67812@163.com, shiyuanyuanwww@163.com,
luosiyuanlz@163.com

Abstract. Due to the uncertainty and fuzziness of the real world, the dual hesitant fuzzy set (DHFS) has been proposed to express uncertain information during the process of multi-attribute decision making (MADM). In order to process the information with incomplete dual hesitant fuzzy elements (IDHFEs) in MADM, a new similarity degree for MADM with incomplete dual hesitant fuzzy sets (IDHFSs) is proposed. The concept of similarity degree of IDHFEs and similarity aggregation matrix are introduced. Then a complete dual hesitant fuzzy matrix (CDHFM) is obtained by using the maximum similarity to complement the data. An investment selection example is provided to illustrate the validity and applicability of the proposed method.

Keywords: Incomplete dual hesitant fuzzy element · Maximum similarity · Multi-attribute decision making

1 Introduction

MADM is a common process to human beings. In classical MADM, the assessments of alternatives are precisely known [1]. However, because of the inherent vagueness of human preferences and the uncertainty of objects, the attributes involved in decision making problems are better denoted by the fuzzy concept. Fuzzy sets have been introduced by Zadeh [2]. Subsequently, many scholars have done a lot of achievements on fuzzy decision making [3, 4, 5]. Hesitant fuzzy set (HFS), which has been introduced by Torra and Narukawa as an extension of fuzzy set [6, 7], describes the situations that permit the membership of an element to a given set having a few different values, which is a useful means to describe and deal with uncertain information in the process of MADM. However, the HFS only depicts people's membership and ignors the people's non-membership. Therefore, Zhu et al. have proposed the DHFS [8]. Comparing with the classical fuzzy set theory, DHFS increase the non-membership degree to describe the fuzzy nature of the objective world comprehensively and show a strong flexibility and practicality.

Y. Sun et al. (Eds.): IScIDE 2017, LNCS 10559, pp. 113–122, 2017.
DOI: 10.1007/978-3-319-67777-4_10

In recent years, hesitant fuzzy MADM has received more and more attention. Herrera-Videma et al. have emphasized fuzzy preference matrix and studied different forms of expression in fuzzy preference matrix approach [9]. Xu et al. have defined the distance and similarity measure of the HFSs [10, 11]. Zhang et al. have proposed hesitant fuzzy power aggregation operator to aggregate MADM evaluation values [12]. Xia and Xu also have developed some aggregation operators for hesitant fuzzy information and applied them in MADM problems under the hesitant fuzzy environment [13]. In the real life, people often judge the truth-values of a fuzzy proposition with language. Lin R has defined the hesitant fuzzy linguistic set by combining the advantages of linguistic evaluation values and HFS [14]. On the basis of multi-hesitant fuzzy sets, literature [15] has proposed the multi-hesitant fuzzy linguistic term sets.

The above studies are based on the assumption that the evaluation information is complete. However, due to the complexity of the decision making environment, the hesitation of the decision makers and the improper operation in the decision making process, the information is often incomplete. In order to solve the problem of MADM with information loss, the information is usually complemented by some methods. Therefore, this paper proposes a multi-attribute decision making method based on the maximum similarity of IDHFSs. We define the similarity of the IDHFEs in the IDHFSs and compare the similarity among individuals to complement the maximum data. An example is then provided to illustrate the proposed approach is more flexible and effective.

2 Preliminaries

DHFS, as a generalization of fuzzy set, permit the membership degree and the non-membership degree of an element to a set presented as several possible values between 0 and 1, which can describe the situations where people have hesitancy in providing their preferences over objects in the process of decision making.

Definition 1 [17]. Let X be a fixed set, then DHFS on X is defined as:

$$D = \{\langle x, h(x), g(x) | x \in X \rangle\}$$

where $h(x)$ and $g(x)$ are two sets of several values in $[0, 1]$, representing the possible membership degrees and the non-membership degrees for $x \in X$. Also, there is

$$0 \leq \gamma, \eta \leq 1, 0 \leq \gamma^+ + \eta^+ \leq 1$$

where $\gamma \in h(x), \eta \in g(x), \gamma^+ \in h^+(x) = \cup_{\gamma \in h(x)} \max\{\gamma\}$ and $\eta^+ \in g^+(x) = \cup_{\eta \in g(x)} \max\{\eta\}$. The DHFS is composed of dual hesitant fuzzy elements (DHFEs), which is denoted by $d(x) = (h(x), g(x))$ ($d = (h, g)$ for short).

To compare the DHFSs, Zhu et al. have gave the following comparison laws:

Definition 2 [8]. Let $A_i = \{h_{A_i}, g_{A_i}\}(i = 1, 2)$ be any two DHFSs, $s_{A_i} = \frac{1}{l_h}\sum_{\gamma \in h}\gamma - \frac{1}{l_g}\sum_{\eta \in g}\eta(i = 1, 2)$ the score function of $A_i(i = 1, 2)$ and $p_{A_i} = \frac{1}{l_h}\sum_{\gamma \in h}\gamma + \frac{1}{l_g}\sum_{\eta \in g}\eta(i = 1, 2)$.

The accuracy function of $A_i (i = 1, 2)$, where l_h and l_g be the number of elements in h and g, respectively, then

(i) if $s_{A_1} > s_{A_2}$, then A_1 is superior to A_2, denoted by $A_1 \succ A_2$;
(ii) if $s_{A_1} = s_{A_2}$, then
    (1) if $p_{A_1} = p_{A_2}$, then A_1 is equivalent to A_2, denoted by $A_1 \sim A_2$;
    (2) if $p_{A_1} > p_{A_2}$, then A_1 is superior to A_2, denoted by $A_1 \succ A_2$.

3 Similarity of Incomplete Dual Hesitant Fuzzy Elements

In the real life, due to the preference of experts and the influence of external factors, some values of the information can't be given. So there is a case where the DHFEs are incomplete when the values of the information are represented by DHFEs. Therefore, we divide the DHFEs into complete double hesitation fuzzy elements and incomplete double hesitation fuzzy elements. The corresponding definitions as follows:

Definition 3. Let $U = \{u_1, u_2, \ldots, u_n\}$ be a fixed set and $A_i = (h_i, g_i)$ be any DHFEs on U, $i = 1, 2, \ldots n$.

(1) if $n_i^h = (n_i)^* = n_i^g$, then the DHFEs are called complete dual hesitant fuzzy elements, denoted by CDHFEs; The matrix(s) composed of CDHFEs are called complete double hesitation fuzzy matrix(s), denoted by CDHFM(s); The set(s) composed of CDHFEs are called complete double hesitation fuzzy set(s), denoted by CDHFS(s).

(2) if $n_i^h \neq (n_i)^* \neq n_i^g$, then the DHFEs are called incomplete dual hesitant fuzzy elements, denoted by IDHFEs. The matrix(s) composed of IDHFEs are called incomplete double hesitation fuzzy matrix(s), denoted by IDHFM(s); The set(s) composed of IDHFEs are called incomplete double hesitation fuzzy set(s), denoted by IDHFS(s).

Where n_i^h and n_i^g are the number of values in h_i and g_i, respectively. $(n_i)^*$ is the largest value of n_i^h and n_i^g.

Singh [16] has proposed the similarity measures of CDHFSs. However, it is no longer applicable for IDHFS(s). We improve the definition of similarity measures in [16] and give the definition of similarity degree of IDHFEs.

Definition 4. Let $A(u_\mu) = (h(u_\mu), g(u_\mu))$, and $B(u_v) = (h(u_v), g(u_v))$ be two IDHFEs, the similarity degree $s_{IDHFSs}(A, B)$ between A and B is defined as:

$$s_{IDHFEs}(A, B) = 2 \frac{\frac{1}{p} \sum_{k=1}^{p} h_{ik} \cdot \frac{1}{q} \sum_{s=1}^{q} h_{js} + \frac{1}{\alpha} \sum_{l=1}^{\alpha} g_{il} \cdot \frac{1}{\beta} \sum_{r=1}^{\beta} g_{jr}}{(\frac{1}{p} \sum_{k=1}^{p} h_{ik})^2 + (\frac{1}{q} \sum_{s=1}^{q} h_{js})^2 + (\frac{1}{\alpha} \sum_{l=1}^{\alpha} g_{il})^2 + (\frac{1}{\beta} \sum_{r=1}^{\beta} g_{jr})^2} \quad (1)$$

If we take weight of each element $x \in X$ for membership function and non-membership function into account, then

$$s_{IDHFEs}(A, B) = 2 \frac{\sum\limits_{k=1}^{p} \omega_{\mu k} h_{\mu k} \cdot \sum\limits_{s=1}^{q} \omega_{vs} h_{vs} + \sum\limits_{l=1}^{\alpha} \omega_{\mu l} h_{\mu l} \cdot \sum\limits_{r=1}^{\beta} \omega_{vr} h_{vr}}{(\sum\limits_{k=1}^{p} \omega_{\mu k} h_{\mu k})^2 + (\sum\limits_{s=1}^{q} \omega_{vs} h_{vs})^2 + (\sum\limits_{l=1}^{\alpha} \omega_{\mu l} h_{\mu l})^2 + (\sum\limits_{r=1}^{\beta} \omega_{vr} h_{vr})^2} \qquad (2)$$

where $\sum\limits_{k=1}^{p} \omega_{\mu k} h_{\mu k}, \sum\limits_{s=1}^{q} \omega_{vs} h_{vs}, \sum\limits_{l=1}^{\alpha} w_{\mu l} g_{\mu l}, \sum\limits_{r=1}^{\beta} w_{vr} g_{vr}$ are called the weighted summation of each element $x \in X$ for membership function and non-membership function, respectively. Particularly, if each element has same importance, then (2) is reduced to (1).

According to the Definition 2, we can easily proof Definition 4 satisfies the following properties:

Proposition 1. Let A, B and C be any IDHFSs, then $s_{IDHFEs}(A, B)$ is the similarity degree, which satisfies the following properties:

(i) $0 \leq s_{IDHFEs}(A, B) \leq 1$;
(ii) $s_{IDHFEs}(A, B) = 1$ if and only if $A = B$;
(iii) $s_{IDHFEs}(A, B) = s_{IDHFEs}(B, A)$;
(iv) Let C be any IDHFS, if $A \subseteq B \subseteq C$, then $s_{IDHFEs}(A, B) \geq s_{IDHFEs}(A, C)$, $s_{IDHFEs}(B, C) \geq s_{IDHFEs}(A, C)$.

Definition 5. Let $A = (h_D(u_\mu), g_D(u_\mu))$ and $B = (h_D(u_v), g_D(u_v))$ be two IDHFEs, then the similarity matrix of A and B on attribute c_j is defined as:

$$R_{c_j} = (x_{\mu v})_{m \times m}, \text{where } x_{\mu v} = s_{c_j}(A, B), 1 \leq j \leq n$$

Definition 6. Let $A = (h_D(u_\mu), g_D(u_\mu))$ and $B = (h_D(u_v), g_D(u_v))$ be two IDHFEs, then the similarity aggregation matrix $R_\aleph$ is defined as:

$$R_\aleph = (r_{\mu v})_{m \times m} = \frac{\sum\limits_{j=1}^{n} R_{c_j}(x_{\mu v})}{n}, \text{where } r_{\mu v} = x_{c_1(\mu v)} \cdot x_{c_2(\mu v)} \cdots x_{c_j(\mu v)}$$

4   Application of Proposed Similarity Degree in MADM

4.1   Method of Proposed Similarity Degree in MADM

When we complement the data to IDHFEs based on the maximum similarity, we need to operate and organize the system of the information many times. So we make the following signs:

(1) Denote the similarity matrix as $R_{c_j}^t$ after the tth operations for the information system;

(2) Denote the similarity aggregation matrix as $R^t_{\aleph(c_j)}$ after the tth operations for the information system;

(3) Denote the information system as D^t after the tth operations;

(4) Denote the integral information system as D.

Note: Two IDHFEs, which similarity degree is the greatest, we complement the two IDHFEs by adding to the weighted summation each other until the IDHFEs get to CDHFEs, then ordering the membership degree and non-membership degree of the CDHFEs from the largest to the smallest.

Based on the above analysis, when the information of system is incomplete, we use the following method to find the best alternative in MADM problem.

Let $U = \{u_1, u_2, \ldots, u_n\}$ be a set of alternatives and $C = \{c_1, c_2, \ldots, c_m\}$ be a set of attributes, we construct the decision matrix $D^0 = (d_{\mu v})_{m \times n}$ where $d_{\mu v} = (h_{\mu v}, g_{\mu v})$ $(1 \le \mu \le m, 1 \le v \le n)$. The method involves the following steps:

Step 1. Let $t = 1$ and calculate the similarity degrees $s_{c_j}(u_\mu, u_v)$ of any two alternatives u_μ and u_v of attribute c_j in $U^{t-1} \in D^{t-1}$, we obtain the similarity matrix $R^t_{c_j} = (x_{\mu v})_{n \times n}$ and the similarity aggregation matrix $R^t_{\aleph(c_j)} = (r_{\mu v})_{n \times n}$;

Step 2. Scan similarity aggregation matrix $R^t_{c_j} = (x_{\mu v})_{n \times n}$ and find the maximum of the similarity of any two alternatives with IDHFEs. According to the *Note*, we arrange the matrix D^{t-1} to get matrix D^t:

(i) If $D^t \ne D, t = t + 1$, return **Step 1**,

(ii) If $D^t = D$, then to **Step 3**;

Step 3. Calculate the similarity degrees among the alternatives $u^i_{c_j}$ and the ideal alternatives $u^*_{c_j}$ by using formula (2) $(i = 1, 2, \ldots, m, j = 1, 2, \ldots, n)$;

Step 4. Rank the alternatives according to the results of **Step 3**;

Step 5. Select the best alternative according to the **Step 4**.

4.2 Practical Example

Here, we take the example [16], to illustrate the utility of the proposed method. Also, we show that the results obtained using the proposed method are same as the results of Ref. [16].

There is an investment company, which wants to invest a sum of money in the best option. There is a panel with four possible alternatives to invest the money: u_1 is a car company; u_2 is a food company; u_3 is a computer company; u_4 is an arms company. The investment company must make a decision according to the following three attributes: c_1 is the risk analysis; c_2 is the growth analysis; c_3 is the environmental impact analysis. The attribute weight vector for membership degree and nonmembership degree is given as $w = (0.35, 0.25, 0.40)^T$ and $z = (0.30, 0.40, 0.30)^T$, respectively. The four possible alternatives u_i $(i = 1, 2, 3, 4)$ are to be evaluated using the dual hesitant fuzzy information by three decision makers under three attributes c_j $(j = 1, 2, 3)$, as listed in the following dual hesitant fuzzy decision matrix D^0 ("$\widehat{\Delta}$" represent that the missing value):

$$D^0 = \begin{pmatrix} \{\{0.5, 0.4, 0.3\}, \{0.4, 0.3\}\} & \{\{0.6, 0.4, \widehat{\Delta}\}, \{0.4, 0.2\}\} & \{\{0.3, 0.2, 0.1\}, \{0.6, 0.5\}\} \\ \{\{0.7, 0.6, 0.4\}, \{0.3, 0.2\}\} & \{\{0.7, 0.6, \widehat{\Delta}\}, \{0.3, 0.2\}\} & \{\{0.7, 0.6, 0.4\}, \{0.2, 0.1\}\} \\ \{\{0.6, 0.4, 0.3\}, \{0.3, \widehat{\Delta}\}\} & \{\{0.6, 0.5, \widehat{\Delta}\}, \{0.3, \widehat{\Delta}\}\} & \{\{0.6, 0.5, \widehat{\Delta}\}, \{0.3, 0.1\}\} \\ \{\{0.8, 0.7, 0.6\}, \{0.2, 0.1\}\} & \{\{0.7, 0.6, \widehat{\Delta}\}, \{0.2, \widehat{\Delta}\}\} & \{\{0.4, 0.3, \widehat{\Delta}\}, \{0.2, 0.1\}\} \end{pmatrix}$$

Step 1. Let $t = 1$ and calculate the similarity degrees $s_{c_j}(u_\mu, u_\nu)$ of any two alternatives u_μ and u_ν of attribute c_j to obtain the similarity matrix $R_{c_j}^1 = (x_{\mu\nu})_{4 \times 4}$, where $1 \le \mu, \nu \le 4, 1 \le j \le 3$. Then we get the similarity aggregation matrix $R_{\aleph(c_j)}^1 = (r_{\mu\nu})_{4 \times 4}$ of c_j.

$$R_{c_1}^1 = \begin{pmatrix} 1 & 0.942 & 0.968 & 0.852 \\ 0.942 & 1 & 0.959 & 0.976 \\ 0.968 & 0.959 & 1 & 0.895 \\ 0.852 & 0.976 & 0.895 & 1 \end{pmatrix} \quad R_{c_2}^1 = \begin{pmatrix} 1 & 0.965 & 0.971 & 0.926 \\ 0.965 & 1 & 0.981 & 0.983 \\ 0.971 & 0.981 & 1 & 0.983 \\ 0.926 & 0.983 & 0.983 & 1 \end{pmatrix}$$

$$R_{c_3}^1 = \begin{pmatrix} 1 & 0.599 & 0.763 & 0.729 \\ 0.599 & 1 & 0.924 & 0.728 \\ 0.763 & 0.924 & 1 & 0.908 \\ 0.729 & 0.728 & 0.908 & 1 \end{pmatrix} \quad R_{\aleph(c_j)}^1 = \begin{pmatrix} 1 & 0.835 & 0.900 & 0.835 \\ 0.835 & 1 & 0.954 & 0.895 \\ 0.900 & 0.954 & 1 & 0.928 \\ 0.835 & 0.895 & 0.928 & 1 \end{pmatrix}$$

Step 2. We can see $r_{23} = r_{32}$ is the largest from $R_{\aleph(c_j)}^1$ easily, that is, the similarity of u_2 and u_3 is the largest. Arranging matrix D^0, we can get matrix D^1.

$$D^1 = \begin{pmatrix} \{\{0.5, 0.4, 0.3\}, \{0.4, 0.3\}\} & \{\{0.6, 0.4, \widehat{\Delta}\}, \{0.4, 0.2\}\} & \{\{0.3, 0.2, 0.1\}, \{0.6, 0.5\}\} \\ \left\{ \begin{array}{l} \{0.7, 0.6, 0.4\}, \\ \{0.3, 0.2, 0.105\} \end{array} \right\} & \left\{ \begin{array}{l} \{0.7, 0.6, 0.38\}, \\ \{0.3, 0.2, 0.105\} \end{array} \right\} & \left\{ \begin{array}{l} \{0.7, 0.6, 0.4\}, \\ \{0.2, 0.13, 0.1\} \end{array} \right\} \\ \left\{ \begin{array}{l} \{0.6, 0.4, 0.3\}, \\ \{0.3, 0.155, 0.155\} \end{array} \right\} & \left\{ \begin{array}{l} \{0.6, 0.5, 0.45\}, \\ \{0.3, 0.155, 0.155\} \end{array} \right\} & \left\{ \begin{array}{l} \{0.6, 0.57, 0.5\}, \\ \{0.3, 0.1, 0.095\} \end{array} \right\} \\ \{\{0.8, 0.7, 0.6\}, \{0.2, 0.1\}\} & \{\{0.7, 0.6, \widehat{\Delta}\}, \{0.2, \widehat{\Delta}\}\} & \{\{0.4, 0.3, \widehat{\Delta}\}, \{0.2, 0.1\}\} \end{pmatrix}$$

Thus, we can get $D^1 \ne D$, return **Step 1**.

Let $t = 2$ and calculate the similarity degrees $s_{c_j}(u_\mu, u_\nu)$ of any two alternatives u_μ and u_ν of attribute c_j to obtain the similarity matrix $R_{c_j}^2 = (x_{\mu\nu})_{4 \times 4}$, where $1 \le \mu, \nu \le 4, 1 \le j \le 3$. Then we get the similarity aggregation matrix $R_{\aleph(c_j)}^2 = (r_{\mu\nu})_{4 \times 4}$ of c_j.

$$R^2_{c_1} = \begin{pmatrix} 1 & 0.948 & 0.997 & 0.852 \\ 0.948 & 1 & 0.966 & 0.968 \\ 0.997 & 0.966 & 1 & 0.882 \\ 0.852 & 0.968 & 0.882 & 1 \end{pmatrix} \quad R^2_{c_2} = \begin{pmatrix} 1 & 0.901 & 0.932 & 0.926 \\ 0.901 & 1 & 0.996 & 0.948 \\ 0.932 & 0.996 & 1 & 0.956 \\ 0.926 & 0.948 & 0.956 & 1 \end{pmatrix}$$

$$R^2_{c_3} = \begin{pmatrix} 1 & 0.650 & 0.682 & 0.729 \\ 0.650 & 1 & 0.998 & 0.730 \\ 0.682 & 0.998 & 1 & 0.737 \\ 0.729 & 0.730 & 0.737 & 1 \end{pmatrix} \quad R^2_{\aleph(c_j)} = \begin{pmatrix} 1 & 0.833 & 0.871 & 0.835 \\ 0.833 & 1 & 0.987 & 0.882 \\ 0.871 & 0.987 & 1 & 0.858 \\ 0.835 & 0.882 & 0.858 & 1 \end{pmatrix}$$

Due to the similarity of u_2 and u_3 are complete, overlook $s(u_2, u_3)$. We can see $r_{24} = r_{42}$ is the largest from $R^2_{\aleph(c_j)}$ easily, that is, the similarity of u_2 and u_4 is the largest. Arranging matrix D^1, we can get matrix D^2.

$$D^2 = \begin{pmatrix} \{\{0.5,0.4,0.3\},\{0.4,0.3\}\} & \{\{0.6,0.4,\widehat{\Delta}\},\{0.4,0.2\}\} & \{\{0.3,0.2,0.1\},\{0.6,0.5\}\} \\ \left\{\begin{matrix}\{0.7,0.6,0.4\}, \\ \{0.3,0.2,0.105\}\end{matrix}\right\} & \left\{\begin{matrix}\{0.7,0.6,0.38\}, \\ \{0.3,0.2,0.105\}\end{matrix}\right\} & \left\{\begin{matrix}\{0.7,0.6,0.4\}, \\ \{0.2,0.13,0.1\}\end{matrix}\right\} \\ \left\{\begin{matrix}\{0.6,0.4,0.3\}, \\ \{0.3,0.155,0.155\}\end{matrix}\right\} & \left\{\begin{matrix}\{0.6,0.5,0.45\}, \\ \{0.3,0.155,0.155\}\end{matrix}\right\} & \left\{\begin{matrix}\{0.6,0.57,0.5\}, \\ \{0.3,0.1,0.095\}\end{matrix}\right\} \\ \left\{\begin{matrix}\{0.8,0.7,0.6\}, \\ \{0.2,0.197,0.1\}\end{matrix}\right\} & \left\{\begin{matrix}\{0.7,0.6,0.564\}, \\ \{0.2,0.197,0.197\}\end{matrix}\right\} & \left\{\begin{matrix}\{0.4,0.57,0.3\}, \\ \{0.2,0.1,0.1425\}\end{matrix}\right\} \end{pmatrix}$$

Thus, we can get $D^2 \neq D$, return **Step 1**.

Let $t = 3$ and calculate the similarity degrees $s_{c_j}(u_\mu, u_\nu)$ of any two alternatives u_μ and u_ν of attribute c_j to obtain the similarity matrix $R^3_{c_j} = (x_{\mu\nu})_{4\times4}$, where $1 \le \mu, \nu \le 4, 1 \le j \le 3$. Then we get the similarity aggregation matrix $R^3_{\aleph(c_j)} = (r_{\mu\nu})_{4\times4}$ of c_j.

$$R^3_{c_1} = \begin{pmatrix} 1 & 0.948 & 0.997 & 0.871 \\ 0.948 & 1 & 0.966 & 0.979 \\ 0.997 & 0.966 & 1 & 0.898 \\ 0.871 & 0.979 & 0.898 & 1 \end{pmatrix} \quad R^3_{c_2} = \begin{pmatrix} 1 & 0.901 & 0.932 & 0.864 \\ 0.901 & 1 & 0.996 & 0.996 \\ 0.932 & 0.996 & 1 & 0.985 \\ 0.864 & 0.996 & 0.985 & 1 \end{pmatrix}$$

$$R^3_{c_3} = \begin{pmatrix} 1 & 0.650 & 0.682 & 0.754 \\ 0.650 & 1 & 0.998 & 0.968 \\ 0.682 & 0.998 & 1 & 0.973 \\ 0.754 & 0.968 & 0.973 & 1 \end{pmatrix} \quad R^3_{\aleph(c_j)} = \begin{pmatrix} 1 & 0.833 & 0.871 & 0.829 \\ 0.833 & 1 & 0.987 & 0.981 \\ 0.871 & 0.987 & 1 & 0.973 \\ 0.829 & 0.981 & 0.973 & 1 \end{pmatrix}$$

Due to the similarity of u_2, u_3, u_4, are complete, overlook $s(u_2, u_3), s(u_2, u_4), s(u_3, u_4)$. We can see $r_{13} = r_{31}$ is the largest from $R^3_{\aleph(c_j)}$ easily, that is, the similarity of u_1 and u_3 is the largest. Arranging matrix D^2, we can get matrix D^3.

$$
D^3 = \begin{pmatrix}
\left\{\begin{array}{l} \{0.5, 0.4, 0.3\}, \\ \{0.4, 0.3, 0.2058\} \end{array}\right\} & \left\{\begin{array}{l} \{0.6, 0.515, 0.4\}, \\ \{0.4, 0.2058, 0.2\} \end{array}\right\} & \left\{\begin{array}{l} \{0.3, 0.2, 0.1\}, \\ \{0.6, 0.5, 0.168\} \end{array}\right\} \\
\left\{\begin{array}{l} \{0.7, 0.6, 0.4\}, \\ \{0.3, 0.2, 0.105\} \end{array}\right\} & \left\{\begin{array}{l} \{0.7, 0.6, 0.38\}, \\ \{0.3, 0.2, 0.105\} \end{array}\right\} & \left\{\begin{array}{l} \{0.7, 0.6, 0.4\}, \\ \{0.2, 0.13, 0.1\} \end{array}\right\} \\
\left\{\begin{array}{l} \{0.6, 0.4, 0.3\}, \\ \{0.3, 0.155, 0.155\} \end{array}\right\} & \left\{\begin{array}{l} \{0.6, 0.5, 0.45\}, \\ \{0.3, 0.155, 0.155\} \end{array}\right\} & \left\{\begin{array}{l} \{0.6, 0.57, 0.5\}, \\ \{0.3, 0.1, 0.095\} \end{array}\right\} \\
\left\{\begin{array}{l} \{0.8, 0.7, 0.6\}, \\ \{0.2, 0.197, 0.1\} \end{array}\right\} & \left\{\begin{array}{l} \{0.7, 0.6, 0.564\}, \\ \{0.2, 0.197, 0.197\} \end{array}\right\} & \left\{\begin{array}{l} \{0.4, 0.57, 0.3\}, \\ \{0.2, 0.1, 0.1425\} \end{array}\right\}
\end{pmatrix}
$$

Thus, we can get $D^3 = D$, turn to **Step 3**.

Step 3. Calculate the similarity degrees among the alternatives $u^i_{c_j}$ and the ideal alternatives $u^*_{c_j}$ by using formula (2) ($i = 1, 2, \ldots, m, j = 1, 2, \ldots, n$);

(1) The ideal alternatives $u^*_{c_j} : u^*_{c_1} = (\{0.8, 0.7, 0.6\}, \{0.2, 0.155, 0.1\}); u^*_{c_2} = (\{0.7, 0.6, 0.564\}, \{0.2, 0.155, 0.105\}); u^*_{c_3} = (\{0.7, 0.6, 0.5\}, \{0.2, 0.1, 0.095\})$

(2) The similarity degrees:

$$s_{DHFEs}(u^1_{c_1}, u^*_{c_1}) = 0.852, s_{DHFEs}(u^2_{c_1}, u^*_{c_1}) = 0.978, s_{DHFEs}(u^3_{c_1}, u^*_{c_1}) = 0.897,$$
$$s_{DHFEs}(u^4_{c_1}, u^*_{c_1}) = 0.999;$$
$$s_{DHFEs}(u^1_{c_2}, u^*_{c_2}) = 0.962, s_{DHFEs}(u^2_{c_2}, u^*_{c_2}) = 0.993, s_{DHFEs}(u^3_{c_2}, u^*_{c_2}) = 0.980,$$
$$s_{DHFEs}(u^4_{c_2}, u^*_{c_2}) = 0.997;$$
$$s_{DHFEs}(u^1_{c_3}, u^*_{c_3}) = 0.598, s_{DHFEs}(u^2_{c_3}, u^*_{c_3}) = 0.998, s_{DHFEs}(u^3_{c_3}, u^*_{c_3}) = 0.955,$$
$$s_{DHFEs}(u^4_{c_3}, u^*_{c_3}) = 0.955.$$

Step 4. Rank the alternatives according to the results of **Step 3**;

$$\hat{s}_{u_2} > \hat{s}_{u_4} > \hat{s}_{u_3} > \hat{s}_{u_1}$$

Step 5. Select the best alternative according to the **Step 4**.

$$u_2 \succ u_4 \succ u_3 \succ u_1$$

Therefore, u_2 is the best alternative to choose.

4.3 Advantages of the Proposed Distance Measures

(i) Singh has proposed the distance and similarity measures for MADM with DHFSs in [16]. The result of the proposed method is matched with existing method of Singh. But the method in [16] based on decision maker's preference parameter is known to make decision. The approach we proposed is to complement the missing data when the decision maker's risk preference is unknown, extracting rules from the existed information to solve the problem of missing information. Thus the proposed method is more practical and general.

(ii) As mentioned above, we give the definition of the similarity degree for IDHFEs and give some related properties. Using the method of maximum similarity degree to complement the IDHFEs not only keeps the maximum similarity among the hesitant fuzzy information, but also solves the problem of information loss in IDHFSs.

5 Conclusions

In the real world, the HFS or DHFS is adequate for dealing with the vagueness of DM's judgment. However, due to the preference of experts and the influence of external factors, some information values can't be given, which results in the fuzzy information is not complete. In order to solve this problem, we propose a completion method of information values based on the maximum similarity degree in IDHFSs. An important advantage of the proposed method is that it not only solves the problem of information loss in IDHFSs, but also keeps the maximum similarity among the hesitant fuzzy information.

The next step of the research work is to extend the proposed method to the wider areas such as machine learning, clustering, reasoning and so on.

Acknowledgments. This work is partially supported by the National Natural Science Foundation of P. R. China (Nos. 61772250, 61673320, 61672127), the Fundamental Research Funds for the Central Universities (No. 2682017ZT12), and the National Natural Science Foundation of Liaoning Province (No. 2015020059).

References

1. Dyer, J.S., Fishburn, P.C.: Multiple criteria decision making, multiattribute utility theory: the next ten years. Manag. Sci. **38**(5), 645–654 (1992)
2. Zadeh, L.A.: Fuzzy sets. Inf. Control **8**(3), 338–353 (1965)
3. Cao, Q., Wu, J.: The extended COWG operators and their application to multiple attributive group decision making problems with interval numbers. Appl. Math. Model. **35**(5), 2075–2086 (2011)
4. Wan, S., Li, D.: Fuzzy mathematical programming approach to heterogeneous multiattribute decision-making with interval-valued intuitionistic fuzzy truth degrees. Inf. Sci. **325**, 484–503 (2015)

5. Joshi, D., Kumar, S.: Interval-valued intuitionistic hesitant fuzzy Choquet integral based TOPSIS method for multi-criteria group decision making. Eur. J. Oper. Res. **248**(1), 183–191 (2016)
6. Torra, V., Narukawa, Y.: On hesitant fuzzy sets and decision. In: IEEE International Conference on Fuzzy Systems, pp. 1378–1382 (2009)
7. Torra, V.: Hesitant fuzzy sets. Int. J. Intell. Syst. **25**(6), 529–539 (2010)
8. Zhu, B., Xu, Z., Xia, M.: Dual hesitant fuzzy sets. J. Appl. Math. **2012**(11), 2607–2645 (2012)
9. Herreraviedma, E., Herrera, F., Chiclana, F., Luque, M.S.: Some issues on consistency of fuzzy preference relations. Eur. J. Oper. Res. **154**(1), 98–109 (2004)
10. Xu, Z., Xia, M.: On distance and correlation measures of hesitant fuzzy information. Int. J. Intell. Syst. **26**(5), 410–425 (2011)
11. Xu, Z., Xia, M.: Distance and similarity measures for hesitant fuzzy sets. Inf. Sci. **181**(11), 2128–2138 (2011)
12. Zhang, Z.: Hesitant fuzzy power aggregation operators and their application to multiple attribute group decision making. Inf. Sci. **234**(10), 150–181 (2013)
13. Xia, M., Xu, Z.: Hesitant fuzzy information aggregation in decision making. Int. J. Approx. Reason. **52**(3), 395–407 (2011)
14. Lin, R., Zhao, X., Wei, G.: Models for selecting an ERP system with hesitant fuzzy linguistic information. J. Intell. Fuzzy Syst. **26**(5), 2155–2165 (2014)
15. Wang, J., Wang, J., Zhang, H., Chen, X.: Multi-criteria group decision-making approach based on 2-tuple linguistic aggregation operators with multi-hesitant fuzzy linguistic information. Int. J. Fuzzy Syst. **18**(1), 81–97 (2015)
16. Singh, P.: Distance and similarity measures for multiple-attribute decision making with dual hesitant fuzzy sets. Comput. Appl. Math. **36**, 1–16 (2015)
17. Zhao, N., Xu, Z., Liu, F.: Group decision making with dual hesitant fuzzy preference relations. Cogn. Comput. **8**, 1–25 (2016)

A Unified Confidence Measure Framework Using Auxiliary Normalization Graph

Zhehuai Chen, Yanmin Qian, and Kai Yu$^{(\boxtimes)}$

Key Laboratory of Shanghai Education Commission for Intelligent Interaction
and Cognitive Engineering, SpeechLab, Department of Computer Science
and Engineering, Brain Science and Technology Research Center,
Shanghai Jiao Tong University, Shanghai, China
{chenzhehuai,yanminqian,kai.yu}@sjtu.edu.cn

Abstract. Due to the distinct search space and efficiency demands in
different ASR applications, the state-of-the-art confidence measures and
their decoding frameworks are heterogeneous among keyword spotting,
domain-specific recognition and LVCSR. Inspired by the success in applying a phone level language model to replace the word lattice in discriminative training, the *auxiliary normalization graph* is proposed in
this work, and it is constructed to model the observation probability in
hypothesis posterior based confidence measure. In this way, confidence
measure normalizing term modelling can be independent from the original search space and the confidence measure can be grouped into an
unified framework. Experiments on three typical ASR applications show
that the proposed method using a unified confidence measure framework
achieves comparable performance to the separately optimized system on
each task.

Keywords: Confidence measure · Auxiliary normalization graph · Con-
nectionist temporal classification · Phone synchronous decoding

1 Introduction

In automatic speech recognition (ASR), *confidence measure* (CM) is used to evaluate the reliability of recognition results. CM is taken as a further verification
stage to different ASR applications, e.g., *keyword spotting* (KWS) [1] to guarantee low false acceptance rates, grammar [2] or class [3] language model based
domain-specific recognition to verify in-domain recognition result [4,5], and
large vocabulary continuous speech recognition (LVCSR) [2] to support semantic processing [6]. Due to distinct search space and efficiency demands in above
applications, confidence measures are usually heterogeneous [7–10], which will
be reviewed in Sect. 2.

K. Yu—This work was supported by the Shanghai Sailing Program No. 16YF1405
300, the China NSFC projects (No. 61573241 and No. 61603252) and the Interdisciplinary Program (14JCZ03) of Shanghai Jiao Tong University in China. Experiments
have been carried out on the PI supercomputer at Shanghai Jiao Tong University.

© Springer International Publishing AG 2017
Y. Sun et al. (Eds.): IScIDE 2017, LNCS 10559, pp. 123–133, 2017.
DOI: 10.1007/978-3-319-67777-4_11

It is challenging to propose an unified framework across all above ASR applications with high performance in both CM and accurary. The key challenge of the unified framework includes two sides: (i) how to normalize the best ASR result with proper overall evaluation, i.e., the normalizing term in CM. (ii) how to keep the computational efficiency in resource limited scenarios, e.g., spoken term detection in some personal digital assistant. *Keyword-filler* based method [9] and *utterance verification* [11] can be viewed as such kind of trials. For instance, in KWS, a context independent (CI) linguistic unit, called `filler` is proposed to model all the non-keyword elements, which is imperfect. In domain-specific ASR, the framework even suffers from weakness in context dependency (CD) modelling, which results in worse `filler` recognition ability. In LVCSR, a theoretically better method, i.e., hypothesis posterior based CM [10], is proposed. ASR is formulated as *maximum a posterior* (MAP) in the framework. The posterior probability of ASR output given the whole utterance can be served as CM. The observation probability is modelled by the summation of probability of all the hypothesis from the ASR search space. Because ASR search space is always tremendous, the word lattice recorded in the decoding process is used to constrain the hypothesis. In LVCSR, hypothesis posterior based CM is significantly better than the unified framework, although other applications can't benefit from it, e.g. KWS.

Modeling the search space with an elaborately optimized phone level language model to replace the word lattice, recently shows competitive performance in discriminative training [12,13]. In the paper, similar idea is adopted to the normalizing term modelling, i.e., *auxiliary normalization graph*. Because of the phone level acoustic modelling, such method is theoretically sound in all above applications. Therefore, the CM normalizing term modelling can be independent with both the original search space and the acoustic modelling. To reduce the computational cost from the search space modelling, CTC-based *phone synchronous decoding* (PSD) [14] is further adopted, which shows great efficiency in phone level decoding.

In the paper, an unified and efficient confidence measure framework using auxiliary normalization graph and phone synchronous decoding is proposed to provide consistent performance among all types of ASR applications within a single framework, for the first time. The whole paper is arranged as follow. In Sect. 2, the state-of-the-art confidence measures and decoding frameworks in different ASR search space are compared and summarized. In Sect. 3, the auxiliary normalization graph is proposed to form an unified and efficient framework for varieties of search space. Section 4 describes experiments and analysis, followed by the conclusion in Sect. 5.

2   Confidence Measure and Search Space

Regarding to the difference in search space and decoding frameworks, ASR applications can mainly be divided into three types, and the state-of-the-art confidence measures are heterogeneous among these various ASR search space.

2.1   Keyword Spotting

KWS is to accurately and efficiently detect words or phrases of interest, i.e., keywords, in continuous speech. Therefore, the search space of KWS is all the keyword sequences[1]. False acceptance reveals to falsely recognize speech spans with interested keywords, which is undesirable. i.e., false alarm segments are not in the original search space. To treat this problem, a branch of methods [8] include a post-processing algorithm to provide a word level CM with specific threshold to potentially model the non-keyword elements. Another branch of methods [9] add non-keyword units into acoustic modelling. Due to the difficulty in non-keyword modelling, the previous branch shows better performance especially in restricted KWS [8].

2.2   Domain-Specific Recognition

Recent focus on assistant products has increased the need for users to make voice commands referencing their own personal data, such as favorite songs, names and contacts. In the scenarios, grammar [2] or class [3] based language model with *slots* to dynamically indexing personal words or phrases, is the most suitable search space of the domain-specific recognition [4]. The false recognition in the task includes: (i) falsely recognizing out-domain utterance to the in-domain result; (ii) correctly recognizing the domain but with false slot values. To verify the recognition result, a sentence level CM should be effective in both cases. For in-domain utterances, the language patterns and *slots* can be viewed as the complete search space. Therefore, the CM discussed in Sect. 2.3 is proper. However for out-domain utterances, the out-domain elements need to be modelled specifically as the methods in Sect. 2.1. To our knowledge, there hasn't been careful research conducted in this aspect.

2.3   Large Vocabulary Continuous Speech Recognition

In *large vocabulary continuous speech recognition* (LVCSR) [2], the search space is modelled by a n-gram language model. To support semantic post-processing [6], CM is proposed to model the reliability of spontaneous speech recognition results. Hypothesis posterior based CM [10], is commonly used in LVCSR. In this framework, ASR is formulated as the *maximum a posterior* (MAP) decision process. The posterior probability of ASR output given the whole feature sequence can be served as a word level or sentence level CM. The normalizing term of MAP, i.e., the observation probability, is modelled by the summation of probability of all the hypothesis from the ASR search space. Because the ASR search space is always tremendous, the word lattice recorded through the decoding process is used to constrain the hypothesis.

---

[1] The LVCSR based KWS is not included in the discussion because it's mostly a problem to enhance the acoustic model performance and keyword indexing algorithm. Besides, the computational burden is not suitable for resource-limited scenarios.

3 Auxiliary Normalization Graph Based Confidence Measure

In this paper, an unified framework among all above ASR applications with good performance in both CM and accurary is proposed using *auxiliary normalization graph* and CTC-based *phone synchronous decoding*.

3.1 Unified Confidence Measure Framework

The posterior probability of ASR output given the whole utterance is served as CM in MAP framework,

$$CM = P(\mathbf{w}|\mathbf{x}) = \frac{P(\mathbf{x}|\mathbf{w}) \cdot P(\mathbf{w})}{P(\mathbf{x})} \tag{1}$$

Here, $P(\mathbf{w})$ is the language model probability and $P(\mathbf{x}|\mathbf{w})$ is the acoustic part. $P(\mathbf{x})$ is the probability of observing $\mathbf{x}$ and can be modelled as below,

$$P(\mathbf{x}) = \sum_H P(\mathbf{x}, H) = \sum_H P(H) \cdot P(\mathbf{x}|H) \tag{2}$$

Here, H denotes all the alternative competing hypotheses of the recognition results. H is distinct between different ASR applications and always infinite.Accordingly, the modelling on H is the bottleneck of the performance.

The key challenge of the unified framework includes two sides: (i) how to model the task-specific set H in an unified method. (ii) How to keep the computational efficiency in modelling the theoretically infinite set H.

3.2 Auxiliary Normalization Graph

To solve this problem, a *auxiliary normalization graph* is proposed to be integrated with the original ASR search space. The architecture of the proposed method is compared with traditional methods in Fig. 1.

In lattice based method, $P(\mathbf{x})$ is obtained from the lattice recorded from a subset of decoding graph. In filler based method, $P(\mathbf{x})$ is modelled by phone-loop graph. In the proposed method, $P(\mathbf{x})$ can be obtained from the modified search space as,

$$P(\mathbf{x}) \approx \max_H P(H) \cdot P(\mathbf{x}|H) \tag{3}$$

Here, three types of auxiliary normalization graphs are proposed for comparison.

- Phone loop graph (AX1). All the phone-loop can be combined together as the auxiliary normalization graph. Similar to the traditional keyword-filler CM [9], an all-phone recognition can be conducted to normalize the word based decoding result.
- Lexicon free graph (AX2). Innovated by recent progress in lattice-free discriminative training [13], the auxiliary normalization graph can be constructed from a phone level language model to approximate the search space.

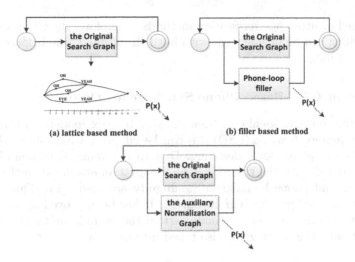

Fig. 1. Architecture comparison. The original search graph can be from KWS, domain-specific recognition and LVCSR.

- Lexicon based graph (**AX3**). In some language, e.g., Mandarin, the mapping between acoustics to characters is always many-to-one[2]. Therefore, given the limited number of characters, the expected pronunciations are also limited. All the possible pronunciations can be combined together as the auxiliary normalization graph.

Besides, the proposed method can be served as a word level CM. In this case, Eqs. (1) and (3) can be transformed to Eqs. (4) and (5),

$$CM = P(w|\mathbf{x}) = \frac{P(\mathbf{x}|w) \cdot P(w)}{P(\mathbf{x}^w)} \tag{4}$$

$$P(\mathbf{x}^w) \approx \max_{H^w} P(H^w) \cdot P(\mathbf{x}^w|H^w) \tag{5}$$

Here, $P(\mathbf{x}|w)$ is the acoustic model probability within the time span of the word w and $P(w)$ is the language model probability of w. $\mathbf{x}^w$ is the feature sequence within the time span of w and H^w is the alternative hypothesis of w. $P(\mathbf{x}^w)$ can be specifically obtained from the auxiliary normalization graph in decoding stage to form the word level CM.

Decoding results in the auxiliary normalization graph is a good approximation of the observing probability to form the CM in MAP framework. Because the acoustic model unit is always phone level, such method of search space modelling is theoretically sound in all above ASR applications. Compared to the traditional lattice based method, the proposed method is more stable among all types of ASR search space, which will be revealed in the experiment part. Besides, the method doesn't need to further include a series of non-keyword model units as

---

[2] While in language like English, the mapping is many-to-many.

`filler` based or utterance verification methods. Therefore, the CM normalizing term modelling can be independent with both the original search space and the acoustic modelling.

3.3 Efficient CTC-Based Phone Synchronous Decoding

To reduce the computational cost from the search space modelling, CTC-based *phone synchronous decoding* (PSD) [14] can be adopted, which shows the great efficiency in the phone level decoding. Due to the removal of non-phonemic frames, compared with the traditional *frame synchronous decoding* (FSD), less search errors and phone boundary disambiguity are made [15]. This phenomenon results in less hypothesis H in Eq. (2). It has been proved in [14,16] that the decoding process becomes a small part of the overall computation. In the experiment part, the efficiency is also taken into account.

3.4 Empirical Implementation

The search space construction of PSD based system is discussed in [16]. In KWS, the search graph G is a series of linear acceptors of keywords [1]. G can be grammar or class based language model in domain-specific recognition, and n-gram [3,17] language model in LVCSR.

The proposed auxiliary normalization graph is finally unified with the original search space. The output symbol of the auxiliary normalization graph is the symbol $\langle$fil$\rangle$ as in `filler` based methods. In the decoding stage, the probability of $\langle$fil$\rangle$ symbols can be obtained specifically to form Eq. (5).

4   Experiments

To evaluate the unified framework in all ASR search space, experiments are conducted on three typical ASR applications, i.e., KWS, domain-specific recognition[3] and LVCSR. A 5000 h Mandarin corpus is used to train the CTC model[4] and the training configuration is the same as [14].

In KWS, the word level CM is evaluated by the false alarm and false rejection of the keyword sequences in each utterance. In domain-specific recognition, the sentence level CM is used to filter out two types of recognition errors discussed in Sect. 2.2. *Equal error rate* (EER) is taken as the sentence level metric in KWS and domain-specific recognition, which reflects the average error rate of the false alarm and false rejection. The lower EER is the better. *Normalised cross entropy* (NCE) [15] is taken as the metric of the word level CM quality in LVCSR. The higher NCE is the better. To ensure that the ASR precision isn't

---

[3] Grammar language model based decoding is taken, as the in-domain and out-domain evaluation discussed in Sect. 2.2 are similar between grammar and class based model.

[4] The comparison between CMs in CTC and HMM frameworks has been conducted in previous research [15], all the comparisons below are within the CTC.

deteriorated by the unified framework, the sentence level recalling rate of the positive examples is provided in KWS and domain-specific recognition, denoted as *snt recall*. In LVCSR, *character error rate* (CER) is used. To evaluate the efficiency of proposed unified framework, the portion of the decoding time except the acoustic model computation, versus all the decoding time is also measured, denoted as *portion of time except acoustic model* (PEA). Because experiments are all conducted with the same acoustic model, the lower PEA shows the less time taken in the other process, e.g., graph searching, lattice generation and post-processing[5]. i.e., the higher PEA indicates the more computational cost from the CM framework.

The baseline CM methods include the predictor feature based CM and the hypothesis posterior based CM, denoted as AC and CN. They show the best reported result in CTC framework and outperform their HMM competitors [15]. The proposed methods are compared in the experiments, i.e., AX1, AX2 and AX3 in Sect. 3.2. The auxiliary normalization graph AX2 is generated from a tri-gram phone language model with *145K* grams. AX3 is generated from a lexicon graph L mapping phone to syllable, and an all syllable-loop graph G, by $L \circ G$. filler based method is not included as it's theoretically similar to AX1.

4.1 Keyword Spotting

In the task, 398 home appliance keywords are chosen and tested in 17332 utterances (11789 positive and 5543 negative examples). Table 1 shows the CM quality of different methods.

Table 1. KWS task

CM	Setup	EER (%)	snt recall (%)	PEA (%)
Phonemic	AC	11.65	88.4	10
Hypothesis	CN	12.55	88.4	11
	AX1	11.60	88.0	10
	AX2	10.16	88.2	16
	AX3	10.10	88.2	15

Result shows that AC outperforms CN. The reason is that the overall search space is very limited in KWS, resulting imperfect observing probability modeling from decoding lattice. The imprecision of normalizing term in MAP results in worse CM quality. The proposed auxiliary normalization graph can alleviate the problem and bring about better CM. Concretely, AX1 can't benefit much. We suspect it is because the end-to-end model CTC intrinsically shares similar

---

[5] As in [14], the acoustic model is a small size one applied in the embedded application. Therefore, computation time is comparable between all above portions in the three tasks.

normalization of all-phone recognition in the model. AX2 and AX3 are significantly better than traditional methods, while AX3 is slightly better. The reason is from better characterization of the linguistic search space in Mandarin, discussed in Sect. 3.2.

Besides, the recalling rate shows that proposed unified confidence measure and its decoding framework slightly affects the model precision. Regarding to the improvement in false alarm revealed by EER, the side effect is tolerable.

In aspect of efficiency, the computation of CN is around 10% more than AC because of lattice and confusion network generation. Although PEA of proposed method is notably more than both AC and CN, the total time except the acoustic model computation still takes a very small portion in the task. The reason is from application of PSD. Detail comparison of decoding time versus search space size in PSD can be referred to [14].

4.2 Domain-Specific Recognition

A task of larger ASR search space is examined in the section, i.e., grammar based language model ASR decoding. The test-set includes 13186 voice assistant utterances (7923 positive examples and 5263 negative examples), e.g., phone calls, voice commands and etc. The grammar contains several supported speaking styles and different contact information. The negative examples include both in-domain and out-domain situations discussed in Sect. 2.2.

Table 2. Grammar based ASR task

CM	Setup	EER (%)	snt recall (%)	PEA (%)
Phonemic	AC	19.86	87.4	38
Hypothesis	CN	15.78	87.4	43
	AX1	19.80	86.0	38
	AX2	16.23	87.2	41
	AX3	16.12	87.2	40

Table 2 shows the result. In the task of larger ASR search space, CN significantly outperforms AC, because AC doesn't harness the competitive relationship between hypotheses and results in false acceptance. AX2 and AX3 are similar to CN. The reason is that with larger ASR search space, both the decoding lattice and the auxiliary normalization graph can be a good approximation of observing probability. AX3 is still slightly better than AX2, so we only use AX3 in the latter experiment.

In the task, AX1 notably does harm to the recalling rate. It reveals that a series of CI model units is hard to model all the non-keyword and out-domain elements. Besides, the weakness in modelling context dependency also affects the recognition results. There is no such weakness in AX2 and AX3.

Regarding to the efficiency, the proposed framework only slightly affects the computation time compared with the previous task. The reason is compared with original larger search space in grammar based decoding, the search space increment from the proposed auxiliary normalization graph is more ignorable.

4.3 LVCSR

In the section, a mandarin spontaneous conversation test-set (about 25 h) is taken. A tri-gram language model with *118K* words and *1.9M* grams is used in the decoding stage (Table 3).

Table 3. LVCSR task

CM	Setup	NCE	CER (%)	PEA (%)
Phonemic	AC	0.182	10.2	45
Hypothesis	CN	0.302	10.2	50
	AX3	0.260	10.1	46

As in Sect. 4.2, CN outperforms AC in LVCSR. AX3 shows worse but comparable result compared with CN. The reason is that in large ASR search space, the auxiliary normalization graph can't provide extra competing information. Meanwhile, in AX3, only the best decoding path is taken to simulate the search space, which is worse than CN from the decoding lattice.

The CER of proposed framework is slightly better than the baseline. We believe it's because the auxiliary normalization graph fitters out a portion of false decoding paths, which reduces the insertion and substitution errors. And regarding to the efficiency, it's similar to the previous task.

5   Conclusion

In the paper, the unified confidence measure and efficient decoding framework using auxiliary normalization graph and CTC-based phone synchronous decoding achieves comparable performance to the separately optimized systems on three typical ASR applications. The proposed unified framework can be independent with both the original search space and the acoustic modelling. Future work includes extending the proposed framework to other state-of-the-art acoustic models and language models, e.g., LFMMI and NNLM.

6   Relation to Prior Work

To form an unified confidence measure framework for different ASR applications, prior trial on *keyword-filler* [9] and *utterance verification* [11] are not very successful. Besides, they both need to further include a series of non-keyword units into the acoustic model. The proposed auxiliary normalization graph is inspired by the success in applying an elaborately optimized phone level language model to replace the word lattice in discriminative training [12,13]. In

[18], the word level n-gram language model is combined with the keyword search graph to improve its recognition. The work shares similar profits from the auxiliary graph added into the original search space, but it's different in: (i) the language model in the auxiliary normalization graph is phone level; (ii) the proposed auxiliary normalization graph is combined with varieties of search space; (iii) most importantly, the motivation is different, i.e., the proposed graph is to model the normalization term in CM. Compared with separately optimized confidence measures [7–10] in different ASR applications, the proposed method achieves consistent performance within a single framework for the first time.

References

1. Weintraub, M.: Keyword-spotting using SRI's DECIPHER large-vocabulary speech-recognition system. In: 1993 IEEE International Conference on Acoustics, Speech, and Signal Processing, vol. 2, pp. 463–466. IEEE (1993)
2. Woodland, P.C., Odell, J.J., Valtchev, V., Young, S.J.: Large vocabulary continuous speech recognition using HTK. In: 1994 IEEE International Conference on Acoustics, Speech, and Signal Processing, vol. 2, pp. II–125. IEEE (1994)
3. Ward, W., Issar, S.: A class based language model for speech recognition. In: 1996 IEEE International Conference on Acoustics, Speech, and Signal Processing, ICASSP-1996, Conference Proceedings, vol. 1, pp. 416–418. IEEE (1996)
4. Vasserman, L., Haynor, B., Aleksic, P.: Contextual language model adaptation using dynamic classes. In: 2016 IEEE Spoken Language Technology Workshop (SLT), pp. 441–446. IEEE (2016)
5. Cleveland, J., Thakur, D., Dames, P., Phillips, C., Kientz, T., Daniilidis, K., Bergstrom, J., Kumar, V.: Automated system for semantic object labeling with soft-object recognition and dynamic programming segmentation. IEEE Trans. Autom. Sci. Eng. 14(2), 820–833 (2017)
6. Hakkani-Tür, D., Béchet, F., Riccardi, G., Tur, G.: Beyond ASR 1-best: using word confusion networks in spoken language understanding. Comput. Speech Lang. 20(4), 495–514 (2006)
7. Hu, W., Qian, Y., Soong, F.K.: A new DNN-based high quality pronunciation evaluation for computer-aided language learning (CALL). In: INTERSPEECH, pp. 1886–1890 (2013)
8. Chen, G., Parada, C., Heigold, G.: Small-footprint keyword spotting using deep neural networks. In: 2014 IEEE International Conference on Acoustics, Speech and Signal Processing (ICASSP), pp. 4087–4091. IEEE (2014)
9. Young, S.R.: Detecting misrecognitions and out-of-vocabulary words. In: 1994 IEEE International Conference on Acoustics, Speech, and Signal Processing, ICASSP-1994, vol. 2. pp. II–21. IEEE (1994)
10. Wessel, F., Schluter, R., Macherey, K., Ney, H.: Confidence measures for large vocabulary continuous speech recognition. IEEE Trans. Speech Audio Process. 9(3), 288–298 (2001)
11. Rose, R.C., Juang, B.-H., Lee, C.-H.: A training procedure for verifying string hypotheses in continuous speech recognition. In: 1995 International Conference on Acoustics, Speech, and Signal Processing, ICASSP-1995, vol. 1, pp. 281–284. IEEE (1995)

12. Chen, S.F., Kingsbury, B., Mangu, L., Povey, D., Saon, G., Soltau, H., Zweig, G.: Advances in speech transcription at IBM under the DARPA EARS program. IEEE Trans. Audio Speech Lang. Process. **14**(5), 1596–1608 (2006)
13. Povey, D., Peddinti, V., Galvez, D., Ghahrmani, P., Manohar, V., Na, X., Wang, Y., Khudanpur, S.: Purely sequence-trained neural networks for ASR based on lattice-free MMI. In: Submitted to Interspeech (2016)
14. Chen, Z., Deng, W., Xu, T., Yu, K.: Phone synchronous decoding with CTC lattice. In: Interspeech 2016, pp. 1923–1927 (2016). http://dx.doi.org/10.21437/Interspeech.2016-831
15. Chen, Z., Zhuang, Y., Yu, K.: Confidence measures for CTC-based phone synchronous decoding. In: 2017 IEEE International Conference on Acoustics, Speech and Signal Processing. IEEE (2017)
16. Chen, Z., Zhuang, Y., Qian, Y., Yu, K.: Phone synchronous speech recognition with CTC lattices. IEEE/ACM Trans. Audio Speech Lang. Process. **25**(1), 86–97 (2017)
17. Stolcke, A., et al.: SRILM-an extensible language modeling toolkit. In: Interspeech 2002, vol. 2002, 2002 p. (2002)
18. Chen, I.-F., Ni, C., Lim, B.P., Chen, N.F., Lee, C.-H.: A novel keyword+ LVCSR-filler based grammar network representation for spoken keyword search. In: 2014 9th International Symposium on Chinese Spoken Language Processing (ISCSLP), pp. 192–196. IEEE (2014)

Decentralized Pinning Synchronization of Colored Time-Delayed Networks via Decentralized Pinning Periodically Intermittent Control

Guoliang Cai[1,2(✉)], Wenjun Shi[1], Yuxiu Li[2], Zhiyin Zhang[1], and Gaihong Feng[1]

[1] Institute of Applied Mathematics,
Zhengzhou Shengda University, Zhengzhou 451191, China
glcai@ujs.edu.cn
[2] Nonlinear Scientific Research Center, Jiangsu University,
Zhenjiang 212013, China

Abstract. This paper investigates synchronization of colored time-delayed networks via decentralized pinning periodically intermittent control. Unlike most pinning periodically intermittent control applying the pinned nodes only from 1 to l or a centralized adaptive control, one proposes a decentralized pinning periodically intermittent control to weak the restrictions. Based on Lyapunov stability theorem, sufficient conditions are derived to guarantee the realization of colored delayed networks' synchronization. Finally, the corresponding numerical simulations have been presented to verified effectiveness and correctness of the theoretical results.

Keywords: Colored time-delayed networks · Decentralized pinning control · Intermittent control · Lyapunov stability theorem

1 Introduction

Various control and synchronization of complex networks have been widely studying, which have many potential applications in many areas such as biology system, physics, communication, traffic and so on [1–4]. In view of the ubiquitous synchronization phenomena, the study of synchronization and control has been attracted increasing attention [5–7]. Many works mainly study the outer relationship between the nodes. However, the inner relationship ignored in many literatures plays important roles for the study of the whole networks.

In many realistic systems, another relationship may exist in the social networks consisting of N individuals, e.g., schoolmates, relatives and collaborative relationship. For individuals i and j, they may be either schoolmate or relatives but have no collaborative relationship, while for individual i and k ($k \neq j$), they may only have collaborative relationship. To depict this phenomenon more clearly, the graph theory in mathematics is introduced to solve such problem [8–11]. In the colored networks, nodes with different color signify that they have different properties, and a pair of nodes

© Springer International Publishing AG 2017
Y. Sun et al. (Eds.): IScIDE 2017, LNCS 10559, pp. 134–143, 2017.
DOI: 10.1007/978-3-319-67777-4_12

connected by different color edges means that they have different mutual interactions [12]. In particular, networks of coupled nonidentical dynamical systems with identical inner coupling matrixes can be deem as node colored networks, while networks of coupled identical dynamical systems with nonidentical inner coupling matrixes can be regarded as edge colored networks.

To our best knowledge, many complex networks have a large number of nodes, which means it is difficult to apply control technique to every node. Based on this issue, pinning control, in which is an effective control only applied to a small fraction of networks nodes, is proposed. Much literature has been presented on synchronization of complex networks via pinning control [13–15]. There are also many works concerned the synchronization of complex works via pinning periodically intermittent control. However, a common feature of the works presented in [13–15] is that there are fixed pinned nodes from 1 to l or a centralized adaptive control. Although there is research related to this problem, few works concerns the decentralized pinning periodically intermittent control applied to the colored networks [16]. Therefore, novel techniques applied to the synchronization of the colored networks should be explored and put into use.

Therefore, this letter investigates the problems of synchronization of colored time-delayed networks via decentralized pinning periodically intermittent control. By using decentralized pinning periodically intermittent control techniques, one consider a colored time-delayed networks consisting of N linearly and diffusively coupled identical nodes, which little colored networks involves the time delay. Based on Lyapunov stability theory, sufficient conditions for ensuring the synchronization of colored time-delayed networks are derived through designing appropriate controllers.

This paper is organized as follow: In Sect. 2, a general colored time-delayed networks consisting of N linearly and diffusively coupled identical nodes is considered. At the same time, assumption and lemma are stated. In order to reach the decentralized pinning synchronization with general colored time-delayed networks, a sufficient criterion is presented in Sect. 3. In Sect. 4, several simulations are illustrated to verify the effectiveness of the theory proposed. Finally, conclusions are gained in Sect. 5.

2 Problem Statement

In this paper, one considers a general colored time-delayed networks consisting of N linearly and diffusively coupled identical nodes described as follows:

$$\dot{x}_i(t) = F_i(t, x_i(t), x_i(t - \tau(t))) + \varepsilon \sum_{j=1, j \neq i}^{N} a_{ij} H_{ij}(x_j(t) - x_i(t)), \quad i = 1, 2, \ldots, N. \quad (1)$$

where $x_i(t) = (x_{i1}(t), x_{i2}(t), \ldots, x_{in}(t))^T \in R^n$ is the state vector of the ith node. $F_i(t, x_i(t), x_i(t - \tau(t))) : R \times R^n \times R^n$ represents the local dynamic of node ith, which is continuous differentiable. The time delay: $\tau(t)$ only need bound. The matrix $A = (a_{ij})_{N \times N}$ is outer-coupling matrix, which denotes the networks topology. If there is a connection between node i and node j ($i \neq j$), then $a_{ij} > 0$, otherwise $a_{ij} = 0$, and the entire diagonal element $a_{ii} = 0$. $H_{ij} = diag(h_{ij}^1, h_{ij}^2, \ldots, h_{ij}^n)$ is the inner coupling matrix, which represents the mutual interactions between nodes i and j, which is defined as the following: if the ζ th component of node i is affected by that of node j, then $h_{ij}^\xi \neq 0$, otherwise $h_{ij}^\xi = 0$.

Figure 1 indicates that $F_1 = F_3 = F_4$, $F_2 = F_5 = F_6$, $H_{16} = H_{23}$, $H_{12} = H_{35}$, $H_{24} = H_{36}$. When n = 3, and $H_{16} = diag\{1,1,0\}$ and $H_{56} = diag\{1,0,1\}$, then the first and second components of node 1 are affected by those of node 6, and the first and third components of node 6 are affected by that of node 5, which is shown by Fig. 2.

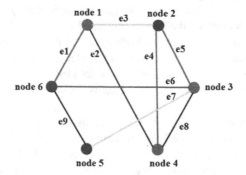

Fig. 1. A colored networks consisting of 6 colored nodes and 9 colored edges (Color figure online)

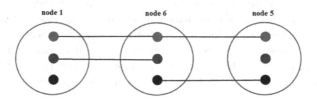

Fig. 2. The red, blue and black stand for the first, second, and third components of each individual node, respectively (Color figure online)

Let $c_{ij} = diag(c_{ij}^1, c_{ij}^2, \ldots, c_{ij}^n)$, where $c_{ij}^k = a_{ij}h_{ij}^k$ for $i \neq j$ and $c_{ii}^k = -\sum_{j=1,j\neq i}^{N} c_{ij}^k$,

Then, the colored time-delayed networks (1) can be represented as:

$$\dot{x}_i(t) = F_i(t, x_i(t), x_i(t - \tau(t))) + \varepsilon \sum_{j=1}^{N} c_{ij}H_i(x_j(t)), \quad i = 1, 2, \ldots, N. \tag{2}$$

Let $C_\xi = (c_{ij}^\xi) \in R^{N \times N}, \xi = 1, 2, \ldots, N$, then we regard the colored networks (2) as a combination of n component sub-networks with a topology determined by $C_\xi, \xi = 1, 2, \ldots, n$.

The main objective of this letter is to apply decentralized pinning periodically intermittent control to make the colored time-delayed networks (1) synchronize with $s(t)$, i.e.,

$$\lim_{t \to \infty} \|x_i(t) - s(t)\| = 0, \quad i = 1, 2, \ldots, N$$

where $s(t) = s(t; t_0, x_0) \in R^n$ be a solution of dynamics of the isolated node $\dot{x}(t) = f(t, x(t), x(t - \tau(t)))$. So $s(t)$ may be an equilibrium point, a limit cycle, or even a chaotic attractor. On the premise of the proposed scheme, some periodically intermittent controls are added to the colored time-delayed networks (1), described as:

$$\dot{x}_i(t) = F_i(t, x_i(t), x_i(t - \tau(t))) + \varepsilon \sum_{j=1}^{N} c_{ij} H_i(x_j(t)) + h_i d_i(t)(x_i(t) - s(t)), \quad (3)$$

If node i is selected to be pinned, then $h_i = 1$; otherwise $h_i = 0$, where $d_i(t)$ is the intermittent feedback control gain, which defined as:

$$d_i(t) = \begin{cases} -d_i, & t \in [mT, mT + \delta) \\ 0, & t \in [mT + \delta, (m+1)T) \end{cases} \quad m = 0, 1, 2, \ldots \qquad (4)$$

where $d_i > 0$ is a positive constant control gain, $T > 0$ is control period, $\delta > 0$ is called control width. Then derive the error dynamical system represented as:

$$\begin{cases} \dot{e}_i(t) = \hat{F}_i(t, x_i(t), x_i(t - \tau(t))) + \varepsilon \sum_{j=1}^{N} c_{ij} \hat{H}_i(x_i(t)) - h_i d_i(t) e_i(t) \\ mT \leq t < (m+\theta)T \\ \dot{e}_i(t) = \hat{F}_i(t, x_i(t), x_i(t - \tau(t))) + \varepsilon \sum_{j=1}^{N} c_{ij} \hat{H}_i(x_i(t)) \\ (m+\theta)T \leq t < (m+1)T \end{cases}$$

where $\hat{F}_i(t, x_i(t), x_i(t - \tau(t))) = F_i(t, x_i(t), x_i(t - \tau(t))) - F_i(t, s(t), s(t - \tau(t)))$, and $\hat{H}_i(x_i(t)) = H_i(x_i(t)) - H_i(s(t))$, which satisfies $\sum_{j=1}^{N} c_{ij} H_i(s(t)) = 0$, The following assumptions and lemmas are required for proving the proposed results.

Assumption 1. For any $x_i, y_i \in R^n$, the vector-valued function $F_i(t, y_i(t), y_i(t - \tau(t)))$, $H_i(y_i(t))$ is uniformly continuous and there exists constant $L_1, L_2, L_3 > 0$ satisfying

$$[y_i(t) - x_i(t)]^T [F_i(t, y_i(t), y_i(t - \tau(t))) - F_i(t, x_i(t), x_i(t - \tau(t)))] \leq$$
$$L_1[y_i(t) - x_i(t)]^T [y_i(t) - x_i(t)] + L_2[y_i(t - \tau(t)) - x_i(t - \tau(t))]^T [y_i(t - \tau(t)) - x_i(t - \tau(t))]$$
$$[y_i(t) - x_i(t)]^T [H_i(y_i(t)) - H_i(x_i(t))] \leq L_3[y_i(t) - x_i(t)]^T [y_i(t) - x_i(t)]$$

Lemma 1 [16]. Let $0 \leq \tau(t) \leq \tau$, $y(t)$ is a continuous and nonnegative function. If $t \in [-\tau, \infty]$, and satisfies the following conditions

$$\begin{cases} \dot{y}(t) \leq -\gamma_1 y(t) + \gamma_2 y(t - \tau(t)) & nT \leq t < (n+\theta)T \\ x(t) = \varphi(t), & -\tau \leq t \leq 0 \\ \dot{y}(t) \leq -\gamma_3 y(t) + \gamma_2 y(t - \tau(t)) & (n+\theta)T \leq t < (n+1)T \\ y_i(t) = \Phi(t), & -\tau \leq t \leq 0 \end{cases}$$

where $\gamma_1, \gamma_2, \gamma_3$, is constant, $n = 1, 2, \ldots, N$, if the condition $\gamma_1 > \gamma_2 > 0$, $\delta = \gamma_1 + \gamma_3 > 0$ and $\eta = \lambda - \delta(1-\theta) > 0$, so we can get $y(t) \leq \sup_{-\tau \leq s \leq 0} y(s) \exp(-\eta t)$, $t \geq 0$, in which $\lambda > 0$ is the only positive solution of function $\lambda - \gamma_1 + \gamma_2 \exp(\lambda \tau) = 0$.

3   Main Results

In this section, we will present our main results to make the colored time-delayed networks (2) achieve general projective synchronization with general colored time-delayed networks (3).

Theorem 1. Suppose the Assumption 1 holds. The drive system and response system can achieve synchronization if the following condition holds:

(1) $a_1 = -\lambda_{\max}(L_1 - h_{il}d_{il} + \varepsilon L_3(C_l + C_l^T)) > 0$
(2) $a_2 = -h_i d_i$
(3) $a_3 = L_2 > 0$
(4) $a_1 - a_2 > 0$ (5)

Proof. Construct a Lyapunov function as the following

$$V(t) = e^T(t)e^T(t) = \frac{1}{2}\sum_{i=1}^{N} e_i^T(t)e_i(t)$$

Then differentiating $V(t)$ along the error systems with respect to t, given as follows: When $mT \leq t < (m+\theta)T$, $m = 1, 2, \ldots$

$$\dot{V}(t) = \sum_{i=1}^{N} e_i^T(t)\dot{e}_i(t)$$

$$= \sum_{i=1}^{N} e_i^T(t)[\hat{F}_i(t, x_i(t), x_i(t - \tau(t))) + \varepsilon \sum_{j=1}^{N} c_{ij}\hat{H}_i(x_j(t)) - h_i d_i(t)e_i(t)]$$

$$\leq (L_1 - h_i d_i)\sum_{i=1}^{N} e_i^T(t)e_i(t) + L_2 \sum_{i=1}^{N} e_i^T(t - \tau(t))e_i(t - \tau(t)) + \varepsilon L_3 \sum_{i=1}^{N}\sum_{j=1}^{N} c_{ij}e_i^T(t)e_j(t)$$

According to Assumption 1 and Lemma 1, the derivation of $V(t)$ gives:

$$\dot{V}(t) \leq \sum_{i=1}^{N} \hat{e}_i^{\mathrm{T}}(t)[L_1 - h_{il}d_{il} + \varepsilon L_3(C_i + C_i^T)/2]\hat{e}_i(t)$$

$$+ L_2 \sum_{i=1}^{N} \hat{e}_i^{\mathrm{T}}(t - \tau(t))\hat{e}_i(t - \tau(t))$$

$$\leq -2a_1 V(t) + 2a_3 V(t - \tau(t))$$

When $(m + \theta)T \leq t < (m+1)T$, $m = 1,2,\ldots$

$$\dot{V}(t) = \sum_{i=1}^{N} e_i^{\mathrm{T}}(t)\dot{e}_i(t) = \sum_{i=1}^{N} e_i^{\mathrm{T}}(t)[\hat{F}_i(t, x_i(t), x_i(t - \tau(t))) + \varepsilon \sum_{j=1}^{N} c_{ij}\hat{H}_i(x_i(t - \tau(t)))]$$

$$\leq L_1 \sum_{i=1}^{N} e_i^{\mathrm{T}}(t)e_i(t) + L_2 \sum_{i=1}^{N} e_i^{\mathrm{T}}(t - \tau(t))e_i(t - \tau(t)) + \varepsilon L_3 \sum_{i=1}^{N} c_{ij}e_i^{\mathrm{T}}(t)e_j(t)$$

$$\leq 2(a_2 - a_1)V(t) + 2a_3 V(t - \tau(t))$$

From Lemma 1, Theorem 1 and on the basis of the Lyapunov stability theorem, one has $V(t) \leq \sup\limits_{-\tau \leq s \leq 0} V(s)\exp(-\bar{w}t)$, $t \geq 0$, which means the drive system (1) can achieve the synchronization with $s(t)$. Thus, one completes the proof.

Remark 1. It should emphasized that decentralized pinning periodically intermittent control techniques are adopted to guarantee general synchronization of colored time-delayed networks, while little of form paper has been done on this work, which can applied to many practical areas.

4   Numerical Simulation Examples

In this section, two illustrative examples are adopted to demonstrate the validity and reduce conservatism of the above theory.

Example 1. Consider an edge-colored networks with 10 coupled time-delayed Lorenz systems:

$$F(t, x(t), t(t - \tau(t))) = Ax(t) + g_1(x(t)) + g_2(x(t - \tau(t)))$$

where $x(t) = (x_1, x_2, x_3)^{\mathrm{T}}$, $g_1(x(t)) = (0, -x_1x_3, x_1x_2)^{\mathrm{T}}$, $g_2(x(t)) = (0, \sigma_0 x_2(t), 0)^{\mathrm{T}}$,

$$A = \begin{pmatrix} -a_0 & a_0 & 0 \\ r_0 & \sigma_0 - 1 & 0 \\ 0 & 0 & -b_0 \end{pmatrix},$$

and $a_0 = 10$, $b_0 = 8/3$, $r_0 = 28$, $\sigma_0 = 5$, $\tau = 0.1$.

In numerical simulation, the initial values of drive-response system are chosen as $x_i(0) = (0.3 + 0.1i, 0.3 + 0.1i, 0.3 + 0.1i)^{\mathrm{T}}$. $H_i(x_j(t) = x_j(t)$. The control period, control width and control strength are set as $T = 0.7$, $\delta = 0.97$, $d_i = 8$, and the pinned control nodes are the last four ($N = 5, 6,...,$ 10). The Fig. 3 shows the synchronization errors of the edge-colored networks.

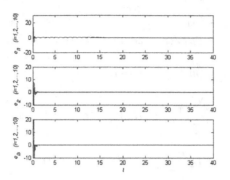

Fig. 3. Synchronization errors of the edge-colored networks coupled with 10 Lorenz systems.

Example 2. Two general colored time-delayed networks, whose topology coupled with 4 time-delayed Lorenz systems and 4 time-delayed Chua systems are considered. The Chua system is given as follows:

$$x(t) = F(t, x(t), t - \tau(t)) = Ax(t) + h_1(x(t)) + h_2(x(t - \tau(t)))$$

where $x(t) = (x_1, x_2, x_3)^{\mathrm{T}}$,

$$h_1(x(t)) = (-1/2\alpha(m_1 - m_2)(|x_1(t) + 1| - |x_1(t) - 1|), 0, 0)^{\mathrm{T}}$$
$$h_2(x(t)) = (0, 0, -\beta\rho_0 \sin(vx_1(t - \tau(t))))^{\mathrm{T}}$$
$$A = \begin{pmatrix} -\alpha(1 + m_2) & \alpha & 0 \\ 1 & -1 & 1 \\ 0 & -\beta & -\omega \end{pmatrix}$$

and $\alpha = 10$, $\beta = 19.53$, $\omega = 0.1636$, $m_1 = -14.325$, $m_2 = -0.7831$, $v = 0.5$, $\rho_0 = 0.2$ and $\tau(t) = 0.02$. The control period, control width and control strength are set as $T = 1.0$, $\delta = 0.90$, $d_i = 10$ and the pinned control nodes are the first five ($N = 1, 2, ..., 5$). The initial values of drive-response system are chosen as the Example 1. The synchronization error of the general colored networks is shown in Fig. 4.

Remark 2. Pinning synchronization of colored networks has been extensively studied, in which all the nodes synchronized each other in a common manner. However, in real complex networks, different communities usually synchronize with each other in a different manner. So in this paper, one considers decentralized pinning synchronization by using decentralized pinning periodically intermittent control techniques. If the

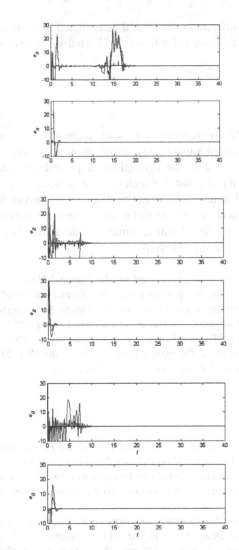

Fig. 4. Synchronization errors of the colored networks coupled with 4 time-delayed Lorenz system and 4 time-delayed Chua system, respectively.

intermittent feedback control gain $d_i(t) = 1$, the general pinning projective synchronization can be realized.

Remark 3. In the existing research of synchronization of the colored networks, certain networks are often considered [17]. However, information may be not available in many practical cases. The uncertain networks (1) can be seen as the special case of the colored networks.

Remark 4. The topology of colored network coupled with one Chen system, two Rössler systems and two Lorenz systems in Fig. 5 (in ref. [17]) are not accurate. In fact, different nodes of colored networks indicate different dynamical behavior. The dynamical behavior of different nodes should be assigned to different colors. However, different systems have the same color in ref. [17]. That is the reason why the figure in [17] is incorrect.

5 Conclusions

In this paper, by periodically intermittent control, some new methods are employed to investigate the decentralized pinning synchronization of colored time-delayed networks. This application weakens the restrictions of pinned nodes only from 1 to l or central nodes via intermittent control. Specially, one considers a colored time-delayed networks model, on which there is light works about colored network concerned. Based on Lyapunov stability theorem, simple and useful criteria for the colored time-delayed networks have been established. Finally, numerical simulation is provided to support the effectiveness of the theoretical results.

Acknowledgments. This work was supported by the National Nature Science foundation of China (No. 61603157), the Society Science Foundation from Ministry of Education of China (No. 15YJAZH002), the Society Science Foundation of Henan Province (No. 2018BJJ008), the Key Technologies R&D Program of Henan Province (No. 172102210553), the Key Scientific Research Projects of Higher Education Institutions of Henan Province (No. 18A120013), and the Advanced Talents' Foundation of Zhengzhou Shengda University (No. 2016SDKY001). Especially, thanks for the support of Zhengzhou Shengda University.

References

1. Salarieh, H., Alasty, A.: Adaptive synchronization of two chaotic systems with stochastic unknown parameters. Commun. Nonlinear Sci. Numer. Simul. **14**(2), 508–519 (2009)
2. Cai, G.L., Yao, Q., Shao, H.J.: Global synchronization of weighted cellular neural network with time-varying coupling delays. Commun. Nonlinear Sci. Numer. Simul. **17**(10), 3843–3847 (2012)
3. Li, Y., Cai, G.: Adaptive cluster synchronization for weighted cellular neural networks with time-varying delays. In: Deng, Z., Li, H. (eds.) Proceedings of the 2015 Chinese Intelligent Automation Conference. LNEE, vol. 338, pp. 21–28. Springer, Heidelberg (2015). doi:10.1007/978-3-662-46466-3_3
4. Ding, W.: Synchronization of delayed fuzzy cellular neural networks with impulsive effects. Commun. Nonlinear Sci. Numer. Simul. **14**, 3945–3952 (2009)
5. Jiang, S.Q., Cai, G.L., Cai, S.M., Tian, L.X., Lu, X.B.: Adaptive cluster general projective synchronization of complex dynamic networks in finite time. Commun. Nonlinear Sci. Numer. Simul. **28**(10), 194–200 (2015)
6. Zhu, Q.X., Cao, J.D.: Adaptive synchronization of chaotic cohen-crossberg neural networks with mixed time delays. Nonlinear Dyn. **61**(3), 517–534 (2010)
7. Cai, G.L., Yao, Q., Fan, X.H., Ding, J.: Adaptive projective synchronization in an array of asymmetric neural networks. J. Comput. **7**(8), 2024–2030 (2012)
8. Cai, G.L., Jiang, S.Q., Cai, S.M., Tian, L.X.: Cluster synchronization of overlapping uncertain complex networks with time-varying impulse disturbances. Nonlinear Dyn. **80**(1), 503–513 (2015)

9. Li, T., Wang, T., Yang, X.: Cluster synchronization in hybrid coupled discrete-time delayed complex networks. Commun. Theor. Phys. **56**(5), 686–696 (2012)
10. Wu, X.J., Lu, H.T.: Cluster anti-synchronization of complex networks with nonidentical dynamical nodes. Phys. Lett. A **375**(14), 1559–1565 (2011)
11. Cai, S.M., Hao, J.J., He, Q.B., Liu, Z.R.: New results on synchronization of chaotic systems with time-varying delays via intermittent control. Nonlinear Dyn. **67**, 393–402 (2012)
12. Cai, G.L., Jiang, S.Q., Cai, S.M., Tian, L.X.: Finite-time analysis of global projective synchronization on coloured networks. Pramana J. Phys. **86**(3), 545–554 (2016)
13. Zhou, J., Lu, J., Lü, J.: Pinning adaptive synchronization of a general complex dynamical network. Automatica **44**(4), 996–1003 (2008)
14. Yu, W., Chen, G., Lü, J.: On pinning synchronization of complex dynamical networks. Automatica **45**(2), 429–435 (2009)
15. Liu, X.W., Chen, T.P.: Cluster synchronization in directed networks via intermittent pinning control. IEEE Trans. Neur. Netw. **22**(7), 1009–1020 (2011)
16. Cai, S.M., Hao, J.B., He, Q.B., Liu, Z.R.: Exponential synchronization of complex delayed dynamical networks via pinning periodically intermittent control. Phys. Lett. A **375**, 1965–1971 (2011)
17. Sun, M., Li, D.D., Han, D., Jia, Q.: Synchronization of colored networks via discrete control. Chin. Phys. Lett. **30**(9), 90202–90206 (2013)

Deep Neural Networks

Image Fusion Using Pulse Coupled Neural Network and CNN

Weiwei Kong$^{(\boxtimes)}$, Wenzhun Huang, and Yang Lei

Xijing University, Xi'an 710123, China
{kongweiwei,huangwenzhun,leiyang}@xijing.edu.cn

Abstract. Image fusion has been a hotspot in the area of image processing. How to extract and fuse the main and detailed information as accurately as possible from the source images into the single one is the key to resolving the above problem. Convolutional neural network (CNN) has been proved to be an effective tool to cope with many issues of image processing, such as image classification. In this paper, a novel image fusion method based on pulse-coupled neural network (PCNN) and CNN is proposed. CNN is used to obtain a series of convolution and linear layers which represent the high-frequency and low-frequency information, respectively. The traditional PCNN is improved to be responsible for selecting the coefficients of the sub-images. Experimental results indicate that the proposed method has obvious superiorities over the current main-streamed ones in terms of fusion performance and computational complexity.

Keywords: Image fusion · Pulse coupled neural network · Convolutional neural network · Time matrix

1 Introduction

Image fusion is a hotspot in the field of information processing, which has attracted a lot of attentions in both domestic and abroad. Due to the significant improvements of the visual performance, the fused image is very beneficial for the following computer processing so that the technology of image fusion has been widely used in a lot of areas, such as medical imaging [1], remote sensing [2], and so on.

Recently, a variety of methods [3–30] have been proposed to deal with the issue of image fusion, which can be mainly classified into three types including transform-domain-based methods, spatial-domain-based ones, and neural-network-based ones.

As for the first type, the core idea is composed three steps as follows. (a) The source image is decomposed into a series of sub-images. (b) Certain fusion rules are adopted to complete the choice of coefficients in the sub-images. (c) The final image is reconstructed. As a result, the mechanism and effects of decompositions and recon-structions are very crucial in the whole course. Discrete wavelet transform (DWT) was regarded as an ideal pioneer before. However, further researches indicate that DWT still has its inherent limitations. First, it is merely good at capturing point-wise sin-gularities, but the edge expression performance is poor. Second, it captures limited directional information only along vertical, horizontal and diagonal directions [3]. In

© Springer International Publishing AG 2017
Y. Sun et al. (Eds.): IScIDE 2017, LNCS 10559, pp. 147–160, 2017.
DOI: 10.1007/978-3-319-67777-4_13

order to overcome the drawbacks of DWT, a series of extensive improved models have been proposed, such as quaternion wavelet transform (QWT) [4], ridgelet transform (RT) [5], curvelet directional transform (CDT) [6], quaternion curvelet transform (QCT) [7], contourlet transform (CT) [8] and shearlet transform (ST) [9]. However, the performance of the above models is severely limited because of the absence of the shift-invariance property introduced by the down-sampling procedure. The shift-invariance extension of CT, namely non-subsampled contourlet transform (NSCT) [1, 10] has been explored and used, but its computational complexity is rather higher compared with aforementioned transform-domain-based methods. Easley *et al.* proposed an improved version of ST called non-subsampled shearlet transform (NSST) [11] which not only has higher capability of capturing the feature information of the input images, but also costs much lower computational resources compared with NSCT. In spite of relatively good performance of preserving the details of the source images, transform-domain-based methods may produce brightness distortions since spatial consistency is not well considered in the fusion process.

Compared with the above ones, the spatial-domain-based methods are much easily implemented. A pioneer in SD is the weighted technique (WT) [12], and the final fused image is estimated as the weighted compromise among the pixels with the same spatial location in the corresponding inputs. WT is resistant to the existent noise in the inputs to a certain degree, but it always results in the decline of the contrast level of the fused image because it always treats all of the pixels without distinction. Furthermore, the theories of principal component analysis (PCA) [13] and independent component analysis (ICA) [14] have been used for image fusion as well. However, the methods based on PCA and ICA both put forward high requirements to the component selection.

Recently, the neural-network-based methods have become the research hotspot during the area of image fusion. As the third generation of the artificial neural networks (ANN), pulse coupled neural network (PCNN) [15] as well as its extensive versions, e.g., intersecting cortical model (ICM) [16] has been successfully proposed and widely used to deal with the issue of image fusion. In essence, ICM is the improved version of the traditional PCNN model. The above two models are able to simulate the process of biological pulse motivation to capture the inherent information of source images. Unfortunately, the traditional models of PCNN and ICM have so many parameters requiring setting. Furthermore, the mechanisms are not obvious enough so that it may bring troubles for further processing of the human or computers.

On the other hand, recently, deep learning (DP) [31] has been successfully used in the fields of image classification and natural language processing. Different from past neural network models, DP with multi-hidden layers has an remarkable feature-learning ability which is helpful to describe and represent the nature of data, and the difficulty during the training course can effectively decline via the mechanism of "gradually initialization". Although the great progress DP has made, its application in the area of image fusion has not been involved. As a typical branch of DP, convolution neural network (CNN) has been widely utilized in the area of information processing.

Based on the content mentioned above, a novel technique for image fusion base on PCNN and CNN is proposed in this paper. The main idea consists of several following stages. On the one hand, CNN is responsible for decomposing the source images into several layers with different functions. On the other hand, PCNN is used to conduct the coefficients choice in each hidden layer.

The rest of this paper is organized as follows. The CNN model is briefly reviewed in Sect. 2. An improved PCNN model and the fusion framework based on the proposed technique are proposed in Sect. 3. Experimental results with relevant analysis are reported in Sect. 4. Conclusions are summarized in Sect. 5.

2 Convolution Neural Network

Convolutional neural network is a kind of artificial neural network, which has become a hot topic in the field of speech analysis and image recognition. The weight sharing network structure makes it more similar to the biological neural network, which reduces the complexity of the network model and the number of weights. The advantage is more obvious when the input of the network is a multi-dimensional image, so that the image can be directly used as the input of the network, so as to avoid the complex feature extraction and data reconstruction process in the traditional recognition algorithm. The convolutional network is a multilayer perceptron which is designed to recognize the shape of the two dimensions. The network structure is highly invariant to translation, scaling, inclination, or deformation.

CNN is affected by the early delayed neural network (TDNN). The time-delay neural network reduces the complexity of learning by sharing the weights on the time dimension, which is suitable for the processing of speech and time series signals. CNN is the first truly successful learning algorithm for multilayer network architecture. It reduces the number of parameters that need to be studied by the use of spatial relations in order to improve the training performance of general forward BP algorithm. CNN as a deep learning architecture is proposed to minimize the data preprocessing requirements. In CNN, a small part of the image (local receptive field) as the minimum level structure of the input information, and then transmitted to different layers, each layer by a digital filter to obtain the most significant features of observed data. This method can obtain the remarkable characteristics of the translation, zoom and rotation invariant observation data, because the characteristics of the local area of the image feel neurons or processing unit allowed access to the most basic, such as directional edge or corner.

CNN is a multilayer neural network, and each layer is composed of a number of two-dimensional planes each of which consists of a number of independent neurons. The concept demonstration of CNN is shown in Fig. 1.

The input image can be convoluted with three training filters and biases. After convolution, three feature maps appear in C1 layer, then four pixels in each group of the feature map is summed plus bias get. Three feature maps in S2 layer can be obtained through a sigmoid function. These maps are then filtered into the C3 layer. Similar to S2 layer, S4 layer can be obtained via the above hierarchical structure. Ultimately, these pixel values are rasterization and connected into a vector input to be the traditional neural network.

In general, C layer is responsible for feature extractions. The inputs of each neuron are connected to the local receptive field of the previous layer to extract the local features. Once the local feature is extracted, its position relationships to other

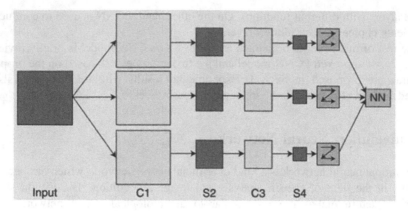

Fig. 1. Concept demonstration of convolutional neural network

characteristics are determined. S layer is the feature mapping layer, and each computing layer of the network is composed of multiple feature maps. Each feature is mapped to a plane where the weights of neurons are equal. The sigmoid function is used as the activation function of the convolution network, so that the feature mapping has the invariance of the displacement.

In addition, since the neurons within a map share the weights, the number of free parameters of the network and the complexity of the network parameter selection both decrease. Each feature extracting layer (C layer) in CNN is followed by a computing layer (S layer) for computing the local average and twice extraction. This unique two feature extraction structure allows the network to identify the input sample with a higher tolerance to distortion.

3   Improved Pulse Coupled Neural Network

3.1   The Basic PCNN Model

As the famous third generation of artificial neural network, PCNN is a model based on the cat's primary visual cortex which is formed by the connection of lots of neurons. A pulse coupled neuron commonly denoted by N_{ij} is composed of three units: the receptive field, the modulation field and the pulse generator. Figure 2 shows the structure of a basic pulse coupled neuron, whose corresponding discrete mathematical expressions can be described as follows:

$$F_{ij}[n] = \exp(-\alpha_F)F_{ij}[n-1] + V_F \sum_{kl} M_{ijkl}Y_{kl}[n-1] + I_{ij} \qquad (1)$$

$$L_{ij}[n] = \exp(-\alpha_L)L_{ij}[n-1] + V_L \sum_{kl} W_{ijkl}Y_{ij}[n-1] \qquad (2)$$

$$U_{ij}[n] = F_{ij}[n](1 + \beta L_{ij}[n]) \qquad (3)$$

$$Y_{ij}[n] = \begin{cases} 1, & if \ U_{ij}[n] \geq \theta_{ij}[n] \\ 0, & else \end{cases} \tag{4}$$

$$\theta_{ij}[n] = \exp(-\alpha_\theta)\,\theta_{ij}[n-1] + V_\theta Y_{ij}[n] \tag{5}$$

As shown in Fig. 2, N_{ij} receives input signals via other neurons and external sources by two channels in the receptive field. One channel is the feeding input F_{ij} and the other one is the linking input L_{ij}, both of which correspond to Eqs. (1) and (2), respectively. U_{ij} combines the information of the above two channels in a second order mode to form the total internal activity, as shown in Eq. (3). Equation (4) indicates that U_{ij} is then compared with the dynamic threshold θ_{ij} to decide the value of the output Y_{ij}. If U_{ij} is larger than θ_{ij}, then the neuron N_{ij} will be activated and generate a pulse, which is characterized by $Y_{ij} = 1$, else $Y_{ij} = 0$. According to Eq. (5), the dynamic threshold θ_{ij} will decline with the iterative number n increasing. However, if $Y_{ij} = 1$, θ_{ij} will immediately raise with the function of V_θ whose value is relatively large, so that the behavior of firing of N_{ij} will stop at once and $Y_{ij} = 0$. Later, if x_{ij} reduces to be equal to or less than U_{ij}, N_{ij} will fire again and make an impulse sequence. On the other hand, the relations between N_{ij} and its surrounding neurons exist, therefore, if N_{ij} is activated, those neurons having similar gray values around it may also be activated at the next iteration. The result is an auto wave expanding from an active neuron to the whole region.

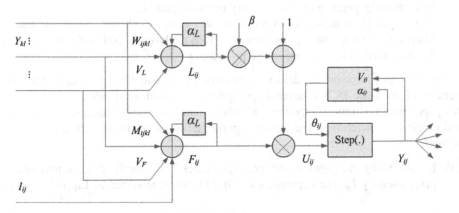

Fig. 2. The basic model of PCNN neuron

Apart from the parameters mentioned above, there are still nine ones required explaining. I_{ij} is the input signal; β is the linking strength; V_F and V_L are the magnitude scaling terms. α_F, α_L and α_θ are the time decayed constants associated with F_{ij}, L_{ij} and θ_{ij}, respectively. M_{ijkl} and W_{ijkl} are both the linking matrices, which respectively correspond to F channel and L channel.

The PCNN used for image fusion is a single layer two-dimensional array of laterally linked pulse coupled neurons. The number of neurons in PCNN is equal to that of pixels in the input image, and all neurons in the network are considered to be identical. There exists a one-to-one correspondence between the image pixels and network neurons. Commonly, I_{ij}, the gray value of each pixel, is directly referred to as the external stimulus of N_{ij}.

3.2   IPCNN and Its Time Matrix

The basic PCNN provides us with a novel viewpoint to deal with the issue of image fusion; however, its several inherent drawbacks have already been exposed to us which are listed in detail as follows.

(a) There are a large number of parameters required settings in the basic PCNN, so it is often difficult for us to determine the near optimal parameters of the network to achieve satisfactory performance for image fusion. Up to now, during the current main-streamed methods involving PCNN, the parameters are commonly adjusted manually according to personal experience or too much repeated, unnecessary and time-consuming experiments. Even worse, a group of parameters producing good effects on certain occasions may not be suitable for other applications.

(b) The mechanism that the dynamic threshold θ_{ij} is exponential decayed and varies periodically is not an effective mode of controlling synchronous pulse bursting. It doesn't consist with the reality of human optical response to intensity variety. Meanwhile, it costs a great deal of computational resources. Moreover, unlike the linear decay, the mode of exponential decay directly causes the treatments of both high intensity pixels and low intensity ones are partial.

(c) The output Y_{ij} in basic PCNN has only two values to choose, namely 0 and 1. Therefore, the impulse sequence resulted from the basic PCNN is binary, which is not beneficial to subsequent image processing.

In order to overcome the defects mentioned above and further enhance the efficiencies of the basic PCNN, several appropriate improved measures are required. In this paper, we tend to modify the basic PCNN structure in terms of parameters reducing and efficiencies increasing. The concrete improved measures of IPCNN are given as follows.

(a) For simplicity, the signal of the feeding channel F_{ij} is simplified to the normalized pixel intensity I_{ij}. The expression of Eq. (1) can be rewritten as Eq. (6).

$$F_{ij}[n] = I_{ij} \tag{6}$$

(b) In Ref. [32], a new PCNN model called unit-linking PCNN was proposed. In the linking channel of the unit-linking PCNN, let $N(ij)$ denote the neighborhood with the neuron N_{ij} as its center. Note that when calculating the linking input L_{ij}, we should exclude N_{ij} from $N(ij)$. If any neuron except N_{ij} in $N(ij)$ fires, the linking input L_{ij} will be set as 1, otherwise L_{ij} is 0. In other words, if N_{ij} fires, any other neuron which has not fired yet in the region $N(ij)$ but has a similar input to N_{ij} may be also encouraged to fire. Obviously, this model reduces the number of undetermined parameters a lot and makes the linking inputs of unit-linking neurons uniform. As a result, the mechanism of unit-linking PCNN can be utilized to complete the task of linking channel in IPCNN, whose expression is given as follows.

$$L_{ij}[n] = \begin{cases} 1, & \text{if } \sum\limits_{k \in N(ij)} Y_k(n) > 0 \\ 0, & \text{otherwise} \end{cases} \tag{7}$$

(c) How to determine the proper iterative number n is a knotty problem all the time. If n is too small, the neurons can't be activated adequately to make use of the characteristics of the synchronous impulse, so that the performance of image processing is not satisfactory commonly. On the other hand, if n is exceedingly large, it will not only sharply increase computational complexity, but result in an adverse influence on the visual effects of the final image. Consequently, it is very necessary to develop an efficient scheme of setting the iterative number.

In order to settle the above problem, the time matrix model [33], whose size is the same as that of the image, is adopted in IPCNN in this paper. With the help of the time matrix T, the iterative number can be determined adaptively according to the intensity distribution of pixels in images. The mechanism of time matrix T can be described as:

$$T_{ij}[n] = \begin{cases} n, & \text{if } Y_{ij} = 1 \text{ for the first time} \\ T_{ij}[n-1], & \text{otherwise} \end{cases} \tag{8}$$

With regard to Eq. (8), there are several aspects required to be explained and noted. (i) T_{ij} will keep invariable if N_{ij} does not fire all the time. (ii) If N_{ij} fires for the first time, T_{ij} will be set as the ordinal value of corresponding iteration. (iii) Once N_{ij} has already fired, T_{ij} will not alter again. Its value will be saved as the ordinal value of iteration during which N_{ij} fired for the first time even N_{ij} may fire later. As known to us, the pixels having similar intensity values often share the same or approximate firing times. Accordingly, their corresponding values of the elements in T are also near. Once all pixels have fired, the whole iteration process is over, and the value of the largest element in T is the total iteration times.

(d) In order to eliminate the irrationality of exponential decay in the basic PCNN, the mode of linear decay is utilized in IPCNN as a substitute for the former. Its mathematical expression is listed as follows.

$$\theta_{ij}[n] = \theta_{ij}[n-1] - \Delta + V_\theta Y_{ij}[n] \tag{9}$$

Where step Δ is a positive constant. It guarantees that the dynamic threshold θ_{ij} decreases linearly with the iterative number n increasing. V_θ is usually set as a relatively large value to ensure that the firing times of each neuron will not exceed one at most.

Moreover, we have to note that the expression of the total internal activity in the basic PCNN is still in use in IPCNN.

In conclusion, IPCNN has overcome the drawbacks of the basic PCNN to a great extent.

(a) There are only four parameters required settings in IPCNN, while the original basic PCNN has ten parameters in all. Obviously, the number of parameters has been reduced a lot in IPCNN.

(b) As shown in Eq. (9), the decayed mechanism of the dynamic threshold θ_{ij} is linear in IPCNN, which not only conforms to the human visual characteristics, but is more helpful to enhance the computational efficiency compared with the basic PCNN.

(c) In IPCNN, the original function of Y has been replaced by the time matrix T, and the element T_{ij} commonly involves too many values. In comparison to the basic PCNN, the time matrix in IPCNN can provide rich temporal and spatial information simultaneously, which is more beneficial to subsequent image processing.

3.3   Parameters Determination of IPCNN

In the basic PCNN, there are ten different parameters in all, namely I_{ij}, β, M_{ijkl}, W_{ijkl}, α_F, α_L, α_θ, V_F, V_L and V_θ. In comparison, IPCNN proposed in this paper has much fewer parameters to determine. It only has four variables including I_{ij}, β, V_θ and Δ. Where I_{ij} is set as the normalized intensity of neuron N_{ij}; the function of V_θ is to ensure each neuron can fire only once at most, so letting V_θ be a certain comparatively positive and large value can easily satisfy the requirement. As a result, the settings of I_{ij} and V_θ are not difficult at all. With regard to the step Δ, we can set it as 0.01 to guarantee that the decayed speed of the dynamic threshold θ_{ij} is moderate and acceptable. In conclusion, the linking strength β comes to be the only stress of the entire task of parameters determination.

Currently, during too many applications of image processing, the values of β are commonly set as a constant. However, according to the human visual characteristics, the responses to the region with remarkable features are supposed to be stronger than those with inconspicuous features to a certain extent. Thus, the mode of setting β as a constant is not reasonable and required to be modified. In this paper, the model of local directional contrast (LDC) is established and introduced into IPCNN to decide the value of β, which is defined as follows.

$$\beta_X^{K,r}(i,j) = \frac{|X^{K,r}(i,j)|}{\overline{X_K^0(i,j)}} \tag{10}$$

where X denotes the source images required fusing; $X^{K,r}(i, j)$ is the coefficients located at (i, j) in the rth directional sub-image at the Kth NSP decomposition level; "$\|$" is the symbol of absolute value; $\overline{X_K^0(i,j)}$ denotes the local average value of the low-frequency coefficients from image X at the Kth level. The expression of $\overline{X_K^0(i,j)}$ is given like this.

$$\overline{X_K^0(i,j)} = \frac{1}{M \times N} \sum_{r=-(M-1)/2}^{(M-1)/2} \sum_{c=-(N-1)/2}^{(N-1)/2} X_K^0(i+r,j+c) \tag{11}$$

where M is commonly assumed to be equal to N. $M \times N$ is the size of the neighborhood within the center at (i, j).

Note that Eq. (10) is mainly utilized to decide the values of β in high-frequency directional sub-images. When required to determine the values of β in low-frequency sub-images, we can rectify Eq. (10) as following:

$$\beta_X^{low}(i,j) = \frac{|X^{low}(i,j)|}{\overline{X^{low}(i,j)}} \qquad (12)$$

As known to us, there is only one low-frequency sub-image left when the course of multi-scale decompositions ends. In Eq. (12), $\overline{X^{low}(i,j)}$ denotes the local average value of the coefficients in the low-frequency sub-image from X, whose expression can be represented like this.

$$\overline{X^{low}(i,j)} = \frac{1}{M \times N} \sum_{r=-(M-1)/2}^{(M-1)/2} \sum_{c=-(N-1)/2}^{(N-1)/2} X^{low}(i+r,j+c) \qquad (13)$$

Obviously, the larger $\beta_X^{K,r}(i,j)$ is, the more remarkable the characteristics of the corresponding pixel (i, j) from image X are; furthermore, N_{ij} is prone to be activated much earlier than others.

The basic framework of the proposed technique is as follows. To begin with, decompose the source images into several layers via CNN. Then, PCNN is responsible for coefficients choice in each pairs of layers. The final fused image can be reconstructed by CNN.

4 Experimental Results and Analysis

In order to demonstrate the effectiveness of the proposed technique, two pairs of simulation experiments are conducted in this section. The experimental platform is a PC with Intel Core i7/2.6 GHz/4G and MATLAB 2013a. The whole section can be classified into two main sections. (a) Methods introduction and performance evaluation. (b) Subjective and objective evaluation on the experimental results.

4.1 Methods Introduction and Parameters Setting

The parameters setting of the proposed method is as follows: $W = [0.707\ 1\ 0.707;\ 1\ 0\ 1;\ 0.707\ 1\ 0.707]$, $\Delta = 15$, $h = 500$. It is noteworthy that it is not necessary for us to modify the parameters manually during the following simulation experiments. For simplicity, we term the proposed method M6.

In addition, Five current typical fusion algorithms are adopted to compare with the proposed one in this paper, which are PCNN-based algorithm (M1), NSCT-based algorithm (M2), algorithm in Ref. [34] (M3), NMF-based algorithm (basic NMF model, M4), WNMF-based algorithm [35] (M5). In M1, the parameters are initialized as follows: $\alpha_F = +\infty$, $\alpha_L = 1.0$, $\alpha_\theta = 0.2$, $V_F = 0.5$, $V_L = 0.2$, $V_\theta = 20$, $W = M =$

[0.707 1 0.707; 1 1 1; 0.707 1 0.707], β is a constant as 0.2, the number of iterations is initialized as 50, all of the fused coefficients are determined by the firing times of neurons in PCNN. In M2 and M3, the stage number of the multi-scale decomposition in NSCT is set as 3, and the levels of the multi-direction decomposition are 4, 3 and 2 respectively from fine resolution to coarse resolution, besides, the size of the neighborhood is 3×3. As for concrete fused schemes, the rule of the selection of maximum coefficients is adopted at the high-frequency level and the average fusion rule is used at the low-frequency level in M2, the initialization of parameters in M3 is the same as that in Ref. [34]. In M4 and M5, the number of iterations is set as 50, and the group of weight coefficients is given as (0.5, 0.5). In addition, what is necessarily needed to note is that, since the random initialization of parameters W and H commonly has a great influence on the performance of the ultimate fused effect, the fused images, whose information entropy value is the largest in three random simulations, will be chosen to be the final results in M4 and M5 respectively.

In order to testify the superior performance of M6, extensive fusion experiments with two pairs of source images have been performed. The related source images with the size of 512×512 have been already accurately registered. The source images used in the following experiments cover 256 gray levels and can be downloaded on the web at http://www.imagefusion.org/. Subjective evaluation system can be adopted to provide direct comparisons. However, it is easily prone to be affected by lots of personal factors, such as eyesight level, mental state, even the mood, and so on. As a result, it is very necessary for us to evaluate the fusion effects based on both subjective vision and objective quality assessment. In this paper, we choose the information entropy (IE), root mean square cross entropy (RCE), standard deviation (SD), and average grads (AG) as evaluation metrics. IE is one of the most important evaluation indexes, whose value directly reflects the amount of average information in the image. The larger the IE value is, the more abundant the information amount of the image is. RCE is used to express the extent of the difference between source images and ultimate fused image. The value of RCE is in inverse proportion to the fusion performance. SD indicates the deviation extent between the gray values of pixels and the average of the fused image. In a sense, the fusion effect is in direct proportion to the value of the SD. AG, which is the last metric, is able to illuminate the clarity extent of the fused image. Being similar to SD, the clarity extent will be better with the AG value increasing. The expressions of the above indexes can be referred to in Ref. [36].

4.2   Subjective and Objective Evaluation on the Experimental Results

Experimental results of multi-focus image fusion are shown in Fig. 3. Figure 3(a) and (b) are the corresponding source images. The fused images based on M1–M6 are given in Fig. 3(c)–(h).

From the visual angle, we observe that the intensity of the whole fused image based on M1–M3 is not sufficient enough especially the image based on M1 whose indexes of the two clocks are greatly blurred. On the contrary, the overall performances of the final fused image based on M4, M5 and the proposed algorithm are much better. Further

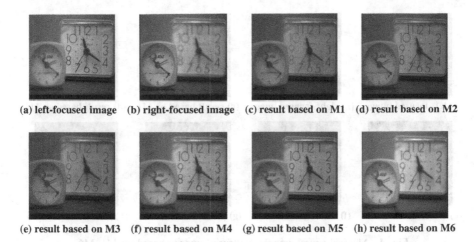

(a) left-focused image   (b) right-focused image   (c) result based on M1   (d) result based on M2

(e) result based on M3   (f) result based on M4   (g) result based on M5   (h) result based on M6

Fig. 3. Multi-focus source images and fused images based on M1–M6

compared with M4 and M5, although the definition of the right clock is slightly low, the whole image fusion result is superior to others. The above visual effect is verified in Table 1.

Table 1. Comparison of the fusion methods for multi-focus images

Method	M1	M2	M3	M4	M5	M6
IE	6.967	7.025	7.041	7.331	7.320	**7.598**
MCE	0.035	**0.027**	0.145	0.505	0.495	0.374
SD	26.70	27.62	26.72	34.83	34.84	**39.45**
AG	2.960	3.881	4.013	3.798	3.949	**4.187**

Figure 4(a) and (b) show a medical CT image and a MRI image, whose sizes are 256×256, respectively. The fused images based on M1–M6 are given in Fig. 4(c)–(h).

As revealed in Fig. 4, the effects of fused images based on M4 and M5 are not as good as others, whose drawbacks mainly lie in that the information of the source MRI image are not fully described; despite the fact that the performances of the fused images based on M1 and M3 are relatively satisfactory, it is undeniable that the external outlines of these images are not clear enough; A2 overcomes the deficiencies emerged in M1 and M3, but it still has its own problems such as the low intensity of the whole fused image, and the undesirable depiction of the middle part of the image. On the contrary, the proposed method not only has clear external outlines and a rational intensity level, but protects and enhances the details information well. Table 2 reports an objective evaluation of the above mentioned six methods.

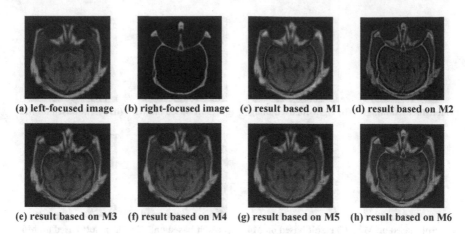

(a) left-focused image   (b) right-focused image   (c) result based on M1   (d) result based on M2

(e) result based on M3   (f) result based on M4   (g) result based on M5   (h) result based on M6

Fig. 4. Medical source images and fused images based on M1–M6

Table 2. Comparison of the fusion methods for medical images

Method	M1	M2	M3	M4	M5	M6
IE	5.801	5.443	5.965	5.770	5.755	**6.243**
MCE	6.547	**2.099**	4.778	6.735	6.701	3.523
SD	28.13	19.24	26.98	25.77	25.88	**30.42**
AG	4.795	4.443	4.310	3.806	3.815	**5.002**

5   Conclusions

In this paper, a new technique for image fusion based on PCNN and CNN is proposed. Experimental results demonstrate that the proposed method has obvious superiorities over current typical ones. The optimization of the proposed method will be the focus in our future work.

Acknowledgements. The authors thank all the reviewers and editors for their valuable comments and works. The work was supported in part by the National Natural Science Foundations of China under Grant 61309008 and 61309022, in part by Natural Science Foundation of Shannxi Province of China under Grant 2014JQ8349, in part by Foundation of Science and Technology on Information Assurance Laboratory under Grant KJ-15-102, in part by the Natural Science Foundation of Shannxi Provincial Department of Education under Grant 16JK2246, and the Foundation of Xijing University Under Grant XJ16T03.

References

1. Yang, Y., Que, Y., Huang, S., Lin, P.: Multimodal sensor medical image fusion based on type-2 fuzzy logic in NSCT domain. IEEE Sens. J. **16**, 3735–3745 (2016)
2. Ghahremani, M., Ghassemian, H.: Remote sensing image fusion using ripplet transform and compressed sensing. IEEE Geosci. Remote Sens. Lett. **12**, 502–506 (2015)

3. Ali, F.E., El-Dokany, I.M., Saad, A.A., El-Samie, F.E.A.: Curvelet fusion of MR and CT images. Prog. Electromagnet. Res. C **3**, 215–224 (2008)
4. Pertuz, S., Puig, D., Garcia, M.A., Fusiello, A.: Genaration of all-in-focus images by noise-robust selective fusion of limited depth-of-field images. IEEE Trans. Image Process. **22**, 1242–1251 (2013)
5. Do, M.N., Vetterli, M.: The finite ridgelet transform for image representation. IEEE Trans. Image Process. **12**, 16–28 (2003)
6. Candes, E.J., Donoho, D.L.: Curvelets: A Surprisingly Effective Non-adaptive Representation for Objects With Edges. Stanford University, Stanford (1999)
7. Cao, L., Jin, L., Tao, H., Li, G., Zhang, Z., Zhang, Y.: Multi-focus image fusion based on spatial frequency in discrete cosine transform domain. IEEE Signal Process. Lett. **22**, 220–224 (2015)
8. Do, M.N., Vetterli, M.: The contourlet transform: an efficient directional multi-resolution image representation. IEEE Trans. Image Process. **14**, 2091–2106 (2005)
9. Miao, Q.G., Shi, C., Xu, P.F., Yang, M., Shi, Y.B.: A novel algorithm of image fusion using shearlets. Opt. Commun. **284**, 1540–1547 (2011)
10. Bhatnagar, G., Wu, Q.M.J., Liu, Z.: Directive contrast based multimodal medical image fusion in NSCT domain. IEEE Trans. Multimed. **15**, 1014–1024 (2013)
11. Easley, G., Labate, D., Lim, W.Q.: Sparse directional image representation using the discrete shearlet transforms. Appl. Comput. Harmon. Anal. **25**, 25–46 (2008)
12. Burt, P.J., Kolcznski, R.J.: Enhanced image capture through fusion. In: Proceedings Conference on Computer Vision, vol. 1, pp. 173–182 (1993)
13. Palsson, F., Sveinsson, J.R., Ulfarsson, M.O., Benediktsson, J.A.: Model-based fusion of multi- and hyperspectral images using PCA and wavelets. IEEE Trans. Geosci. Remote Sens. **53**, 2652–2663 (2015)
14. Mitianoudis, N., Stathaki, T.: Optimal contrast correction for ICA-based fusion of multimodal images. IEEE Sens. J. **8**, 2016–2026 (2008)
15. Broussard, R.P., Rogers, S.K., Oxley, M.E., Tarr, G.L.: Physiologically motivated image fusion for object detection using a pulse coupled neural network. IEEE Trans. Neural Netw. **10**, 554–563 (1999)
16. Kinser, J.M.: Simplified pulse-coupled neural network. In: Proceedings of Conference on Applied Artificial Neural Network, vol. 1, pp. 563–567 (1996)
17. Abdullah, A., Omar, A.J., Inad, A.A.: Image mosaicing using binary edge detection algorithm in a cloud-computing environment. Int. J. Inf. Technol. Web Eng. **11**, 1–14 (2016)
18. Sathiyamoorthi, V.: A novel cache replacement policy for web proxy caching system using web usage mining. Int. J. Inf. Technol. Web Eng. **11**, 1–13 (2016)
19. Sylvaine, C., Insaf, K.: Reputation, image, and social media as determinants of e-reputation: the case of digital natives and luxury brands. Int. J. Technol. Hum. Interact. **12**, 48–64 (2016)
20. Wu, Z.M., Lin, T., Tang, N.J.: Explore the use of handwriting information and machine learning techniques in evaluating mental workload. Int. J. Technol. Hum. Interact. **12**, 18–32 (2016)
21. Kong, W.W., Lei, Y., Ren, M.M.: Fusion method for infrared and visible images based on improved quantum theory model. Neurocomputing **212**, 12–21 (2016)
22. Kong, W.W., Wang, B.H., Lei, Y.: Technique for infrared and visible image fusion based on non-subsampled shearlet transform and spiking cortical model. Infrared Phys. Technol. **71**, 87–98 (2015)
23. Kong, W.W., Lei, Y., Zhao, H.X.: Adaptive fusion method of visible light and infrared images based on non-subsampled shearlet transform and fast non-negative matrix factorization. Infrared Phys. Technol. **67**, 161–172 (2014)

24. Kong, W.W., Liu, J.P.: Technique for image fusion based on NSST domain improved fast non-classical RF. Infrared Phys. Technol. **61**, 27–36 (2013)
25. Kong, W.W., Lei, Y.J., Lei, Y., Zhang, J.: Technique for image fusion based on non-subsampled contourlet transform domain improved NMF. Sci. China Ser. F-Inf. Sci. **53**, 2429–2440 (2010)
26. Kong, W.W., Lei, Y., Ma, J.: Virtual machine resource scheduling algorithm for cloud computing based on auction mechanism. Optik **127**, 5099–5104 (2016)
27. Kong, W.W., Lei, Y., Zhao, R.: Fusion technique for multi-focus images based on NSCT-ISCM. Optik **126**, 3185–3192 (2015)
28. Kong, W.W.: Technique for image fusion based on NSST domain INMF. Optik **125**, 2716–2722 (2014)
29. Kong, W.W., Lei, Y.: Technique for image fusion between gray-scale visual light and infrared images based on NSST and improved RF. Optik **124**, 6423–6431 (2013)
30. Kong, W.W., Lei, Y.: Multi-focus image fusion using biochemical ion exchange model. Appl. Soft Comput. **51**, 314–327 (2017)
31. Hinton, G.E., Salakhutdinov, R.R.: Reducing the dimensionality of data with neural networks. Science **313**, 504–507 (2006)
32. Gu, X.D., Zhang, L.M., Yu, D.H.: General design approach to unit-linking PCNN for image processing. In: Proceedings Conference on Neural Networks, vol. 1, pp. 1836–1842 (2005)
33. Liu, Q., Ma, Y.D.: A new algorithm for noise reducing of image based on PCNN time matrix. J. Electron. Inf. Technol. **30**, 1869–1873 (2008)
34. Guillamet, D., Vitria, J., Scheile, B.: Introducing a weighted non-negative matrix factorization for image classification. Pattern Recogn. Lett. **24**, 2447–2454 (2003)
35. Li, S.Z., Hou, X.W., Zhang, H.J.: Learning spatially localized, parts-based representation. In: Proceedings International Conference on Computer Vision Pattern Recognition, vol. 1, pp. 207–212 (2001)
36. Konsstantinides, K., Yao, K.: Statistical analysis of effective singular values in matrix rank determination. IEEE Trans. Acoust. Speech Signal Process. **36**, 757–763 (1988)

CNN-Based Age Classification via Transfer Learning

Jian Lin[✉], Tianyue Zheng, Yanbing Liao, and Weihong Deng

Beijing University of Posts and Telecommunications, Beijing, China
hellolinjian@163.com

Abstract. Age estimation has always hit people's eyes. While most previous works have focused on constrained images taken under lab condition, which is far from real-world age estimation. The benchmark we used in this paper is the unconstrained Adience [3], which is believed to better reflect the traits of age in the wild condition. In this paper, we adapted contrastive loss to fine-tune the pre-trained VGG-16 over FG-NET to get a better start point and proposed AvgOut-FC Layer to enhance the performance of the models over Audience. We have achieved better results over the Adience benchmark than previous works, which demonstrated the effectiveness of our methods.

Keywords: Unconstrained age estimation · Convolutional Neural Network · Transform learning

1 Introduction

As an important personal trait, human age has become a particularly prevalent topic recently. There are a lot of applications based on age estimations such as video surveillance, access control, and demography [10,11]. Though People have the ability to easily distinguish faces in different age ranges, it is still a challenge for a machine or computer to do that. There are many factors that contribute to the difficulty of automatic age estimation. Age estimation is sensitive to intrinsic factors, such as identity, ethnicity and so on, as well as extrinsic factors, for instance, pose, illumination, and expression. All of these have posed great challenges to modeling the age pattern.

As one of the most powerful tools of machine learning, Convolutional Neural Network has developed rapidly in recent years. Lots of complex problems such as human pose estimation [19], face parsing [18] has been solved used this architecture. However, for some computer vision tasks, a large amount of annotated data may not be available for deep CNN training, especially in our age estimation problem, whose quantities of training data are usually very limited and the face image of same people only covered a narrow range of ages. Such insufficient data is hard to exploiting the most general features of the age so that the models tend to suffer from overfitting.

© Springer International Publishing AG 2017
Y. Sun et al. (Eds.): IScIDE 2017, LNCS 10559, pp. 161–168, 2017.
DOI: 10.1007/978-3-319-67777-4_14

Still, recent studies [7,9] have shown that a pre-trained deep CNN model can be transferred and used for another image classification problem. The idea of transferring pre-trained deep CNN models is also very practical, since training a deep CNN model from scratch requires significant computational resources and time. Besides, fine-tuning pre-trained models is not so strict with train data size that our age estimation problem can benefit a lot because the limit of the exists age data size.

In this paper, we used the pre-trained deep VGG-16, which is trained on the large ImageNet [16] dataset for image classification, as our start model. We firstly fine-tuned the pre-trained model with small FG-NET benckmark [6] to adapt the base model to the age estimation problem. Fine-tuning the CNNs on face images with age annotations is a necessary step for superior performance, as the CNN can be adapted to best fit to the particular data distribution and target of age estimation. Then, we fine-tuned the model over the recently released Adience benchmark [3].

We proposed two approaches to enhance the performance of our models. The first one is to expand the training data by selecting pairs from FG-NET dataset and creatively take pair loss to fine-tune the pre-trained model. This can solve the insufficiency problem of data. The second one is an ensemble way by adapting proposed AvgOut-FC Layer, which is believed to enlarge the capacity of the networks.

The remainder of this paper is organized as follows. In Sect. 2, we introduce the datasets and the CNN architecture we used in this paper. Our methods to fine-tune are detailed in Sect. 3. The details of training are discussed in Sect. 4, experiments results and analysis are also given in Sect. 4.

2   Related Works

2.1   Age Benchmarks

We evaluated our methods in the newly released benchmark Adience [3] for wild age estimations, which collects images from Flicker.com albums and were labeled manually with attributes such as age or gender. This posted benchmark contains about 26,580 images of 2,284 subjects and poses great variability in terms of poses and viewing conditions of the photos as true as the real-world challenges, which can be seen as Fig. 1 shows. Be worth mentioning, we only used the age information that is labeled into 8 groups (0–2, 4–6, 8–13, 15–20, 25–32, 38–43, 48–53, 60– years).

Another benchmark we take is FG-NET to fine-tune the pre-trained VGG-16 model to best fit to the particular data distribution and target of age estimation. FG-NET age set is possibly the most well-used benchmark for age estimations. It contains about 1000 images of 82 subjects, labeled for accurate age and these photos are collected under experiment condition and contains usually frontal faces with little variability in poses and viewing condition.

Fig. 1. Images from the Adience benchmark. These photos are selected randomly and to some extend represent the challenges, for example, low-resolution, out-of-plane pose variations, expressions, occlusions and so on, of the real-world, unconstrained images.

2.2 VGG-16 Architecture

The VGG-16 [17] model was built upon ImageNet ILSVRC challenge by training on 1.3M images of 1000 class objects. In VGG-16 model, there are five convolutional layer blocks. For the first two blocks, one convolution layer is followed by another one and the output of each layer activated by a ReLU. At the end of the block, a max pooling layer is set to reduce the size. For the remaining three blocks, three convolutional layer followed one by one and each layer in the block is again activated by a ReLU, then followed by a max pooling layer. The last convolution layer block is followed by three fully connected layers (FC6, FC7, FC8). FC6 and FC7 have an output size of 4096 with a dropout ratio of 0.5 while FC8 is responsible for classification with a size of 1000. The model was fed with a three channel input images with a size of 224 * 224.

When fine-tuning the pre-trained VGG-16 in particular tasks, we should change the number of output neuron according to the dataset. We usually change number of output neuron to 1 in regression problems, to classification numbers in classification problems.

3 Proposed Methods

3.1 Contrastive Loss in Fine-Tune

To fine-tune a model, the previous works usually use softmax loss or square loss to supervise the training process. It works well but if the dataset is very small, such as FG-NET, it is possible that the networks are trained inadequately. To enlarge our small dataset, we constructed image pairs to increase the data size from N to $\frac{1}{2}N(N-1)$ and assigned label 1 to the pairs whose two images have the same label and assigned label 0 to pairs in which the images have different labels. The previous methods to train models over FG-NET regards it as a regression

problem and adopt square loss define as Eq. 1 where $f(x_n)$ is the output of the
model when fed image x_n and y_n is the corresponding label.

$$loss_{square} = \frac{1}{2N} \sum_{n=1}^{N} (f(x_n) - y_n)^2 \tag{1}$$

To use image pairs as the input of the network, we adapt contrastive loss [4]
as our loss function as defined in Eq. 2.

$$loss_{contrastive} = \frac{1}{2N} \sum_{n=1}^{N} y_n * dis_n^2 + (1 - y_n) * max(margin - dis_n, 0)^2 \tag{2}$$

where dis is the Euclidean distance of the model outputs of two images in the
same pair, defined as Eq. 3, and $margin$ is a constant indicated the tolerance of
the distance between two different classes.

$$dis = \frac{1}{2}(f(x1) - f(x2))^2 \tag{3}$$

In the above equation $f(x1)$ is the model output of first image of the pair
while $f(x2)$ is the model output of second of the same pair.

3.2 AvgOut-FC Layer

To enhance the performance of the models, we proposed AvgOut-FC Layer,
which is inspired by the definition of Maxout [1]. Given an input x, which had
been flattened into 1D if it was a 2D feature map, with dimension d, then the
jth output neuron of AvgOut-FC Layer is defined as follows:

$$h_j(x) = \frac{1}{k} \sum_{i=1}^{k} \sum_{t=1}^{d} x_t w_{jt}^i \tag{4}$$

where the parameter k is a hyperparameter of AvgOut-FC layer, x_t is the tth
feature of the input x, and w_{jt}^i is the jth row, tth col of ith filter weight matrix.
In short, we defined a group of FC layers, they act on the input vector (or matrix
if the batch size is not 1) independently, and we then computed the average of
the corresponding nodes of these layers to get the output. By adding AvgOut
mechanism, the model can be more robust in the fine-tuning process, due to
the fact that it can avoid bad initialization of the FC layer to some extent. The
architecture of AvgOut-FC Layer was as the Fig. 2 shows. The parameter k of
the shown AvgOut Layer was set to 3.

4 Training

Our training process can be divided into two parts. The first stage is to fine-
tune the pre-trained VGG-16 over FG-NET database, while the second stage is
to fine-tune the intermediate pre-trained models over the Adience benchmark
and evaluate the effectiveness of our proposed methods.

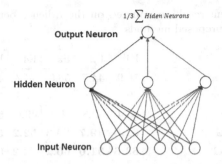

Fig. 2. The architecture of AvgOut-FC layer. The input neurons is the input of the AvgOut Layer, the output neuron is the output of the AvgOut Layer. The k in this figure was set to 3 so that we have 3 hidden neurons. The value of the output neuron is the average of the 3 hidden neurons.

4.1 Fine-Tuning with Contrastive Loss

We trained two models in this stage: Ft-Pairs-VGG and Ft-Single-VGG, which, by definition, are fine-tuned with contrastive loss and square loss respectively. To train the Ft-Pairs-VGG, we selected 7683 training pairs and 557 validation pairs from FG-NET, compared with the Ft-VGG training process, which using all 1008 images of FG-NET and obeys LOPO validation protocol. The two models start with the pre-trained ImageNet weights from [17]. The fine-tuning strategy is to change the number of outputs in FC8 to 1 as we treated the age estimation of FG-NET as a regression problem. The learning rate for all layers except the last layer is set to 0.0001. As we have changed the number of output neurons, the weights of the last layer were initialized randomly. To allow quick adjustment of those new weights, we set the learning rate for the output layer to 0.001. We trained with a momentum of 0.9 and a weight decay of 0.0005. The learning rate has reduced a factor of 10 according to the loss of validation data.

For the reason that the loss definition and the data organization of the two models are far from each other, we can't compare these two models directly. We took ways by fine-tuning these two models over the Adience benchmark and compared the results of the Adience benchmark under the same judgment rule. We fine-tuned two models with the same network configuration: a momentum of 0.9, a weight decay of 0.0005, a gamma of 0.1 with an initial learning rate of 0.01 with 50000 iterations. The learning rate reduced every 10 passes through the entire data by a factor of 10.

The fine-tuned models over Adience named by Ft-Con-VGG and Ft-Squ-VGG respectively according to their start pre-trained models: Ft-Pairs-VGG and Ft-Single-VGG.

We evaluated our methods with the exact age-group classification. The exact age-group classification can be calculated by:

$$exact = \frac{100}{N} \sum_{n=1}^{N} 1\{y_n = \hat{y_n}\} \tag{5}$$

Table 1. The experiment results conducted on the Adience benchmark. In bold are results achieved by our proposed methods.

Method	F1	F2	F3	F4	F5	Avg.
GilNet	56.9	44.7	56.6	46.7	48.8	50.7
Ft-AlexNet-like	59.8	46.6	56.5	47.2	51.7	52.3
Ft-VGG-Face	65.4	52.8	59.1	49.9	59.0	57.2
Ft-Squ-VGG	**61.0**	**49.7**	**59.3**	**52.2**	**55.1**	**55.4**
Ft-Con-VGG	**63.6**	**51.0**	**60.2**	**52.2**	**57.5**	**56.9**
Ft-Squ-VGG (AvgOut)	**66.4**	**53.9**	**62.1**	**50.3**	**57.6**	**58.0**
Ft-Con-VGG (AvgOut)	**68.7**	**52.0**	**61.7**	**54.2**	**59.5**	**59.2**

where y_n denotes the ground truth label while $\hat{y}_n$ denotes the predicted label. $1\{statement\}$ equal to 1 if *statement* is true while equal to 0 else. Here, we referred to *exact* classification as the mean accuracy, across all age groups, of predicting the true age label. Testing for the age classification is performed using a stander five-fold, subject-exclusive cross-validation protocol as defined in [3].

The results are reported in Table 1. The accuracy of the model fine-tuned with the contrastive loss, named by Ft-Con-VGG, has a significant margin to the Ft-Squ-VGG fine-tuned with the square loss. The reported results convinced us that with contrastive loss fine-tuning, the model gets a more optimal start point than the square loss.

4.2   Fine-Tuning with AvgOut-FC Layer

In this part, we fine-tuned the intermediate models with AvgOut-FC Layer take different parameter k over the Adience benchmark. We still used the *exact* evaluation metric to judge our models, a stander five-fold, subject-exclusive cross-validation protocol is maintained.

We replaced the FC8 of the Ft-Pairs-VGG and Ft-Single-VGG by the AvgOut-FC layer and designed our experiments by comparing models which take different hyperparameter k of the AvgOut-FC Layer. The remainder of the two pre-trained models except the last FC layer kept the same and assigned with a learning rate of 0.0001. We defined k filters with shape 4096 * 8 and initialized randomly, the learning rate of the k filters was set to 0.01 at the beginning, with a decay factor of 10 every 10 epochs. Be worth mentioning, if the parameter k was set to 1, the two models with AvgOut-FC Layer are equal to the previous Ft-Con-VGG and Ft-Squ-VGG models.

The comparison of models with different hyperparameter k was showed in Fig. 3. The blue bars are the test results of age estimators in five folds with a mean accuracy percentage of models fine-tuned from Ft-Pairs-VGG with different k while the red bars are the test results of models fine-tuned from Ft-Single-VGG.

From Fig. 3 we can make some conclusions. The first is that the pre-trained model used contrastive loss guaranteed a better start point for the following

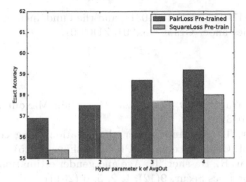

Fig. 3. The comparison of models with different AvgOut-FC parameters. The blue bars and the red bars represent the mean accuracy of the models fine-tuned from Ft-Pairs-VGG and Ft-single-VGG respectively. (Color figure online)

fine-tuning as in the same k all blue bars are higher than red bars. The second is the effectiveness of our proposed AvgOut-FC as we can see that the same color bars are getting higher along with the increasements of k.

We also compared our results with previous works, which can be seen in Table 1. The GilNet [2] is the first approach used CNN over the Adience, provided a remarkable baseline with mean accuracy 50.4 for deep-learning-based methods. All of our results outperform GilNet by substantial margins. The Ft-AlexNet-like [5] and the Ft-VGG-FACE [5] is the first two models that fine-tune the pre-trained models as far as we are concerned. The Ft-AlexNet-like is fine-tuned in the AlexNet-like pre-trained model [15]. The Ft-VGG-Face is fine-tuned on the VGG-Face CNN model, which was trained on 2.6M facial images of 2622 identities. The reported results of the two models are 52.3 and 57.2 respectively, from which we can draw a conclusion that depth matters when comparing the two models. Our models without AvgOut-FC Layer reaches comparable results with them by 55.4 and 56.9. When adding AvgOut-FC Layer, our best model shows encouraging result by about 2% better than Ft-VGG-Face. Those facts have demonstrated the effectiveness of our proposed methods.

5   Conclusion

In this paper, we adapted transform learning in the unconstrained age estimation. The start model we used is the VGG-16 built upon ImageNet ILSVRC challenge by training on 1.3M images of 1000 class objects. To adapt the pre-trained model to best fit to the particular data distribution and target of age, we firstly fine-tuned the model over FG-NET pairs with the contrastive loss. Then we fine-tuned the intermediate models used the AvgOut-FC Layer over the Adience. The results have proven the effectiveness of our proposed methods.

Acknowledgements. This work was partially supported by Beijing Nova Program under Grant No. Z161100004916088, the National Natural Science Foundation of China

(Project 61573068, 61471048, and 61375031), and the Fundamental Research Funds for the Central Universities under Grant No. 2014ZD03-01.

References

1. Goodfellow, I.J., Wardefarley, D., Mirza, M., et al.: Maxout networks. Comput. Sci. 1319–1327 (2013)
2. Levi, G., Hassner, T.: Age and gender classification using convolutional neural networks. In: IEEE Conference on CVPR Workshops (2015)
3. Eidinger, E., Enbar, R., Hassner, T.: Age and gender estimation of unfiltered faces. Trans. Inform. Forensics Secur. 9(12), 1, 2, 5, 6 (2014)
4. Chopra, S., Hadsell, R., LeCun, Y.: Learning a similarity metric discriminatively, with application to face verification. In: IEEE Computer Society Conference on Computer Vision and Pattern Recognition, CVPR 2005, vol. 1. IEEE (2005)
5. Ozbulak, G., Aytar, Y., Ekenel, H.K.: How transferable are CNN-based features for age and gender classification? In: International Conference of the Biometrics Special Interest Group, pp. 1–6 (2016)
6. Lanitis, A.: The FG-NET Aging Database (2002). www-prima.inrialpes.fr/FGnet/html/benchmarks.html
7. Krizhevsky, A., Sutskever, I., Hinton, G.E.: ImageNet classification with deep convolutional neural networks. In: NIPS, pp. S. 1097–1105 (2012)
8. Yosinski, J., et al.: How transferable are features in deep neural networks? In: Advances in Neural Information Processing Systems (2014)
9. Oquab, M., et al.: Learning and transferring mid-level image representations using convolutional neural networks. In: CVPR. IEEE (2014)
10. Fu, Y., Guo, G., Huang, T.: Age synthesis and estimation via faces: a survey. IEEE Trans. Pattern Anal. Mach. Intell. (TPAMI) 32(11), 1955–1976 (2010)
11. Han, H., Otto, C., Liu, X., Jain, A.: Demographic estimation from face images: human vs. machine performance. IEEE Trans. Pattern Anal. Mach. Intell. (TPAMI) 37(6), 1148–1161 (2015)
12. Jia, Y., et al.: Caffe: convolutional architecture for fast feature embedding. In: International Conference on Multimedia, pp. S. 675–678. ACM (2014)
13. Russakovsky, O., Deng, J., Su, H., Krause, J., Satheesh, S., Ma, S., Huang, Z., Karpathy, A., Khosla, A., Bernstein, M., Berg, A.C., Fei-Fei, L.: Imagenet large scale visual recognition challenge. Int. J. Comput. Vis. (IJCV) 115(3), 211–252 (2015)
14. Simonyan, K., Zisserman, A.: Very deep convolutional networks for large-scale image recognition. CoRRabs/1409.1556 (2014)
15. Luo, P., Wang, X., Tang, X.: Hierarchical face parsing via deep learning. In: Proceedings of Conference on Computer Vision Pattern Recognition, vol. 3, pp. 2480–2487. IEEE (2012)
16. Toshev, A., Szegedy, C.: Deeppose: human pose estimation via deep neural networks. In: Proceedings of Conference on Computer Vision Pattern Recognition, vol. 3, pp. 1653–1660. IEEE (2014)

Deep Attentive Structured Language Model Based on LSTM

Di Cao and Kai Yu[✉]

Key Lab. of Shanghai Education Commission for Intelligent Interaction and
Cognitive Engineering SpeechLab, Department of Computer Science and Engineering,
Brain Science and Technology Research Center, Shanghai Jiao Tong University,
Shanghai, China
{caodi0207,kai.yu}@sjtu.edu.cn

Abstract. Language model (LM) plays an essential role in natural language processing tasks. Given the context, the language model can predict the next word. However, when the history becomes longer, the single hidden vector may be not big enough to store the entire information. In this paper, we propose a deep attentive structured language model (DAS LM), which extends the Long Short-Term Memory (LSTM) neural network with the attention mechanism. With the alternative input of part of speech (POS) tags, the language model is capable of extracting relations between a word and its context. Our model is evaluated on Penn Treebank, Chinese short message and Swb-Fisher corpora. The experiments in language modeling show that our model achieves significant improvements compared to the conventional LSTM language model.

Keywords: Language model · Long Short-Term Memory · Attention mechanism · Part of speech

1 Introduction

Language model plays an essential role in natural language processing tasks. Given the word sequence, the language model can compute the probability distribution of the next word. N-gram language model used to be the most popular language model. However, the prediction is only related to the closest several words, the N-gram language model cannot model the long-term dependencies. When the N becomes larger, the sample space will increase exponentially, which induces the problem of data sparseness.

Feedforward Neural Network (FNN) was first introduced into language modeling in 2003 [1]. The neural network based language model has solved some problems such as data sparseness, sample space, and word similarity, but the problem of long-term dependencies still remained unsettled. In 2010, recurrent neural network (RNN) was applied to establish language models, which solved the long-term dependencies problem to a great extent [2,3]. Currently, Long Short-Term Memory language model (LSTM LM) [4,5] is widely used for its good performance.

© Springer International Publishing AG 2017
Y. Sun et al. (Eds.): IScIDE 2017, LNCS 10559, pp. 169–180, 2017.
DOI: 10.1007/978-3-319-67777-4_15

However, there still remains limitations for these RNN based language models. The next word is predicted based on its history, which is compressed and blended into a single dense vector [6], conditionally independent of other states. The relation between the current word and its context is also not known. And, although word embeddings may carry abundant information, they are mainly produced according to the relative positions of words, which may ignore the linguistical information of the sentences.

In this paper, a deep attentive structured language model is proposed. The model is capable of extracting the relation between the hidden state and its input history, applied with the attention module. Using the attention mechanism, we can compute a context input to enhance the language model. Introducing linguistical information to the language model proves to be effective [7]. So, POS tags are applied as an alternative input of our attention based model. The experiments on Penn Treebank, Chinese short messages (SMS) and Swb-Fisher show that our model gains obvious improvements on perplexity (PPL). Interpolated with conventional LSTM LM, it gains significant improvements.

The rest of the paper is organized as follows, Sect. 2 is the background. Section 3 explains the deep attentive structured language model and Sect. 4 shows the experimental setup and results. Finally, conclusion will be given in Sect. 5.

2   Background

2.1   Long Short-Term Memory

RNN is a neural network used to deal with sequential data with its circle structure. Encoding the contextual information into the history hidden, RNN can deal with data of unlimited history length. However, conventional RNN is usually faced with the situations of vanishing or exploding gradients [8], which make it difficult to learn long-term dependency.

LSTM is raised for solving the vanishing and exploding gradient problems. Using three gates, input gate, forget gate and output gate, which control the data flow, LSTM can achieve better performances compared to conventional RNN, and it is widely used in natural language processing.

2.2   External Memory and Attention Mechanism

RNN based models compress and blend history information into a single dense vector [6], conditionally independent of other states. However, when the history becomes longer, the single vector may be not big enough to store the entire information. Using external memory to store the whole history states is one solution [9]. However, approaches to taking full advantage of the external memory remains to study. Attention is a mechanism for reasoning the relation between two vectors [6]. In external memory structure, attention mechanism can also be applied to generate vectors representing the history information [10–12].

Bahdanau et al. proposed a general attention mechanism [13]. The main job of the attention model is to score each context vector with respect to the current hidden of the decoder. [11]

2.3 Long Short-Term Memory-Network

Jianpeng et al. proposed a Long Short-Term Memory-Network (LSTMN) [6], which is a reading simulator that can be used for sequence processing tasks. The LSTMN stores the contextual states in a memory slot and is capable of reasoning about relations between tokens with a neural attention layer [6]. The hidden states and memory states of the LSTM are stored in two tapes, H_{t-1} and C_{t-1}. The attention layer computes the relation between x_t and h_i according to the formulas below.

$$e_i^t = v^T Tanh(W_h h_i + W_x x_t + W_{\tilde{h}}\tilde{h}_{t-1}) \tag{1}$$

$$a_i^t = Softmax(e_i^t) \tag{2}$$

where W_* are weight parameters, and v is a vector parameter. a_i^t is a probability distribution over the hidden state vectors of previous tokens [6].

$$\begin{bmatrix} \tilde{h}_t \\ \tilde{c}_t \end{bmatrix} = \sum_{i=1}^{t-1} a_i^t \cdot \begin{bmatrix} h_i \\ c_i \end{bmatrix} \tag{3}$$

$\tilde{h}_t$ and $\tilde{c}_t$ can be computed respectively by the weighted sum of h_i and c_i. Replacing the h_t, c_t by $\tilde{h}_t$, $\tilde{c}_t$, the next hidden vector h_{t+1} and memory vector c_{t+1} can be computed by the LSTM cell.

$$\begin{bmatrix} h_{t+1} \\ c_{t+1} \end{bmatrix} = LSTM(x_t, \begin{bmatrix} \tilde{h}_t \\ \tilde{c}_t \end{bmatrix}) \tag{4}$$

3   Deep Attentive Structured Language Model

3.1 Attention Mechanism

Conventional RNN based language model predicts the next word based on the context information stored in the hidden layer, which is a single dense vector. When the history becomes longer, the single vector may be not big enough to store the entire information. In the paper, a deep attentive structured language model (DAS LM) is proposed which considers long-term contexts. The model utilizes external memory to store entire inputs. An attention mechanism is applied to produce the context vector as the additional input, to enhance the LSTM. The network of the DAS LM is shown in Fig. 1.

In the DAS language model, the module to compute the context vector w_t' is called the controller module. According to the input history w_i and the previous hidden state h_{t-1}, the controller module is able to extract their relation and compute the corresponding context vector w_t'.

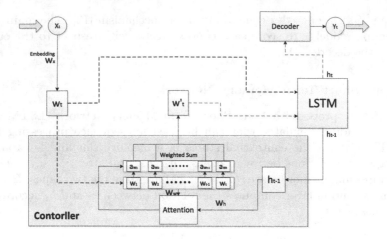

Fig. 1. The network of the deep attentive structured language model. The dashed arrows indicate that the connections are applied with dropout operation.

The relation between the previous hidden state h_{t-1} and the input history $w_1, w_2 \ldots w_{t-1}, w_t$ can be scored by the following formulas.

$$w_t = W_x x_t \tag{5}$$

$$e_i^t = v^T Tanh(W_w w_i + W_h h_{t-1} + b_{att}) + b_v \tag{6}$$

where W_x, W_w, W_h are weight parameters, v^T is a vector parameter, b_{att}, b_v are bias parameters, and e_i^t denotes the unnormalized relation weight between h_{t-1} and w_i.

Softmax operation is applied to normalize the weights. The values of the weights indicate the relation between the previous hidden state and the input history. Then we can obtain the context vector w_t' by calculating the weighted sum of the input history.

$$a_i^t = \frac{exp(e_i^t)}{\sum_{j=1}^t exp(e_j^t)} \tag{7}$$

$$w_t' = \sum_{i=1}^t a_i^t \cdot w_i \tag{8}$$

Feeding w_t and w_t' into the LSTM cell, we can compute the hidden and the memory vectors h_t, c_t.

$$f_t = \sigma(W_f[h_{t-1}, w_t, w_t'] + b_f) \tag{9}$$

$$i_t = \sigma(W_i[h_{t-1}, w_t, w_t'] + b_i) \tag{10}$$

$$o_t = \sigma(W_o[h_{t-1}, w_t, w_t'] + b_o) \tag{11}$$

$$\hat{c}_t = tanh(W_{\hat{c}}[h_{t-1}, w_t, w'_t] + b_{\hat{c}}) \tag{12}$$

$$c_t = f_t * c_{t-1} + i_t * \hat{c}_t \tag{13}$$

$$h_t = o_t * tanh(c_t) \tag{14}$$

Finally, the probability distribution of the predicted word y_t can be calculated.

$$y_t = Softmax(W_y h_t + b_y) \tag{15}$$

where W_* are weight parameters, b_* are bias parameters and σ is the sigmoid function.

3.2   Attention Mechanism with a POS Tagging Model

Although word embeddings carry abundant information, they are mainly produced according to the relative positions of words, which may ignore the linguistical information of the sentences. In our model, linguistical information such as part of speech (POS) can be applied as an alternative input to promote the language model.

POS Tagging Model. In traditional grammar, a part of speech is a category of words (or, more generally, of lexical items) which have similar grammatical properties [14]. POS tagging is a classic classification task. Given the word sequence w_1, w_2, ..., w_n, the tagging model predicts the aligned tag sequence $y_1, y_2, \ldots, y_n$. Mainstream tagging models [15] utilize the contextual information of the input word sequence, which means the former tags may be decided by latter words. However, for most language modeling task, future information shouldn't be provided, which means bidirectional tagging models may not be taken into consideration. In this case, an unidirectional LSTM based model is established for POS tagging.

Combination with a POS Tagging Model. The network of the DAS LM with a POS Tagging model is shown in Fig. 2. x_t and l_t denote the one-hot representations of the word and its POS tag at time step t. They are embedded separately and passed to controller modules.

$$w_t = W_x x_t \tag{16}$$

$$p_t = W_l l_t \tag{17}$$

where W_x and W_l are weight parameters.

Applying the attention mechanism in Sect. 3.1, we can calculate both the context vectors w'_t and p'_t of the word history and the tag history respectively. Concatenating the current inputs w_t and p_t, w'_t and p'_t respectively, we can obtain the combined input m_t and context vector m'_t.

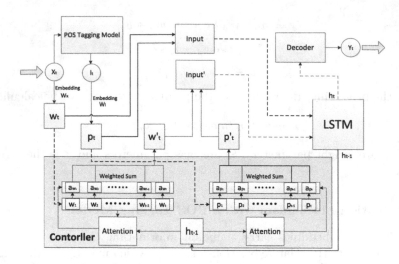

Fig. 2. The network of the deep attentive structured language model with a POS tagging model. The dashed arrows indicate that the connections are applied with dropout operation.

$$m_t = [w_t, p_t] \tag{18}$$
$$m'_t = [w'_t, p'_t] \tag{19}$$

Feeding m_t and m'_t into the LSTM cell instead of w_t and w'_t in Eqs. 9 to 12, we can compute the hidden and the memory vectors h_t, c_t.

3.3   Integration with LSTMN

LSTMN model [6] stores the entire hidden vectors, and is able to select useful content with the attention layer. In this section, we introduce a simple method to integrate our model with the LSTMN model.

We can score the relation between the current input and the hidden history according to the Eq. 1, just replacing x_t by m_t if the POS tagging model is applied, where m_t is the concatenation input of the word and its POS tag. Thus, both $\tilde{h}_{t-1}$ and $\tilde{c}_{t-1}$ can be computed. Using $\tilde{h}_{t-1}, \tilde{c}_{t-1}$ to update the LSTM cell, we can further improve the performances of the language model. The integration model is called DAS-LSTMN in the paper. The results of the experiments comparing these attention based models are shown in the Sect. 4.2.

4   Experiments

4.1   Experimental Setup

Our model is evaluated on the English Penn TreeBank (PTB) [16], Chinese short message (SMS) and Swb-Fisher corpora. The Penn TreeBank corpus is a famous English dataset, with a vocabulary size of 10K and 4.8% words out of vocabulary (OOV), which is widely used to evaluate the performance of a language model. The training set contains approximately 42K sentences with 887K words. The Chinese SMS corpus is collected from short messages. The corpus has a vocabulary size of about 30K. The training set contains 380K sentences with 1931K words. The Swb-Fisher is an English corpus containing over 26 million words and 2 million sentences. The corpus has a vocabulary size of about 63K. The training set consists of approximately 2.4M sentences with 26.5M words.

For the motivation to improve word-level language model, we only considered the word-level tags such as part of speech (POS). Stanford NLP tools [17] have been utilized to generate the POS tags.

In language modeling, we also use dropout technology [18] for regularizing the networks. In the following experiments, different attention based models are evaluated. We apply different tagging models to the baseline LSTM language model respectively to observe the performances in language modeling. Experiments of our attention based language model combined with the POS tagging model are also carried out. Finally, experiments in language modeling are conducted respectively on PTB, SMS and Swb-Fisher corpora.

4.2 Experiments on the Attention Mechanisms

As mentioned in the Sect. 3.3, experiments on the attention based model are conducted. In the experiments, we compare the LSTMN[1] [6], the DAS model, and the DAS-LSTMN model without the POS tagging model. The hidden size of all the models are set to 300. We use stochastic gradient descent (SGD) for optimization. Batch size is set to 128 and dropout rate is set to 0.5 at the stage of training.

The Table 1 shows the DAS model outperforms the baseline LSTM and performs as well as the LSTMN. The DAS-LSTMN model gains the best PPL score on PTB corpus, which indicates the integration is effective.

Table 1. Performances of different attention based models without the POS tagging model

Model	PPL		PPL (dropout)	
	Valid	Test	Valid	Test
LSTM	120.0	114.9	91.3	88.1
LSTMN [6]	114.9	110.4	90.9	87.0
DAS LM	114.7	111.2	90.5	86.6
DAS-LSTMN	**113.6**	**109.1**	**88.8**	**85.3**

---

[1] The LSTMN language model is implemented by us. The PPL score in the original paper is 108.

4.3   Experiments on Different POS Tagging Models

We have trained two unidirectional tagging models of different network sizes, using the POS tags generated by Stanford bidirectional tagging tools. The model(a) contains about 3.7M parameters while the model(b) contains about 1.1M parameters. The Table 2 shows the tagging performances on PTB corpus.

Table 2. Tagging performances of the unidirectional POS tagging models on PTB corpus

Model	Size	Accuracy (%)
Model(a)	3.7M	96.4
Model(b)	1.1M	92.4

We apply the Stanford tagging tools and these two unidirectional POS tagging models to the baseline LSTM LM respectively to observe the improvements contributed by POS tags. We concatenate the word embedding and the POS tag embedding as the combined input for the baseline LM. The baseline language model is a 1-layer LSTM LM with a hidden size of 300. The Table 3 shows the performances of the baseline language model with different POS tagging models.

Table 3. Performances of the baseline LSTM LM with different tagging models on PTB corpus

Model	PPL		PPL (dropout)	
	Valid	Test	Valid	Test
Baseline LSTM	120.0	114.9	91.3	88.1
LSTM + Stanford tools	**110.6**	**107.1**	**85.8**	**83.2**
LSTM + tagging model(a)	115.0	111.7	90.9	87.9
LSTM + tagging model(b)	116.2	112.9	91.3	88.9

The Table 3 shows the POS tags do help to promote the performances in language modeling. The language model with the Stanford tools, which are based on bidirectional models, achieves the best performances. However, the generated POS tags carry future information, which is not suitable for most language modeling tasks. Tags of this kind may be useful in particular language modeling tasks such as rescoring in speech recognition. The language model with the unidirectional tagging model doesn't utilize future information, but still gains obvious improvement in the experiments. The language model with the tagging model(a) gains the higher PPL score than the model(b), which also achieves the better accuracy score in POS tagging task. In the following experiments, we utilize model(a) as the tagging model.

4.4 Experiments on the Attention Based Models Combined with the POS Tagging Model

POS tags and attention mechanisms prove to be effective in promoting performances in language modeling according to the experiments above. We conduct the experiments of combining these two methods in language modeling.

Table 4. Performances of the attention based models with the POS tagging model on PTB corpus

Model	PPL		PPL (dropout)	
	Valid	Test	Valid	Test
Baseline LSTM	120.0	114.9	91.3	88.1
DAS LM	114.7	111.2	90.5	86.6
DAS LM + tagging model	113.3	110.3	89.4	86.4
DAS-LSTMN	113.6	109.1	88.8	85.3
DAS-LSTMN + tagging model	**110.9**	**108.3**	**88.3**	**85.0**

The Table 4 shows the attention based language model still achieves minor improvements when combined with our tagging model. The contribution of POS tags to the language model relates highly to the performance of the tagging model. The promotion may be obvious if a better POS tagging model can be applied.

4.5 Experiments in Language Modeling on Different Corpora

Our models are evaluated on PTB, SMS and Swb-Fisher corpora respectively. The compared models include the 4-gram language model with kneser-Ney smoothing (KN4), the baseline LSTM LM, DAS-LSTMN with the POS tagging model and it averagely interpolated with the LSTM LM. For the fair comparison, the hidden size of all the models is set to 300. We use SGD for optimization. Batch size is set to 128 and dropout rate is set to 0.5 at the stage of training. Table 5 shows the performances of different LMs on the three corpora.

The Table 5 shows that the DAS-LSTMN model achieves the best PPL scores among all the single models. Averagely interpolated with the conventional LSTM LM, the model achieves consistent and significant improvements on both validation and test sets on all the corpora. The improvements on SMS corpus is not so obvious as the others, for the reason that the sentences of SMS corpus may be too short. The PPL scores are reduced by about 8.4%, 4.8% and 9.4% compared to the conventional LSTM model on the three corpora respectively, which also indicates our model behaves well on various kinds of datasets.

We also take an example from the PTB corpus to observe the relation between the words extracted by the DAS-LSTMN model.

Table 5. Performances of different LMs on PTB, SMS and Swb-Fisher corpora

Corpus	Model	PPL		PPL (dropout)	
		Valid	Test	Valid	Test
PTB	KN4	151.1	143.6	–	–
	LSTM	120.0	114.9	91.3	88.1
	DAS-LSTMN	110.9	108.3	88.3	85.0
	DAS-LSTMN + LSTM	**97.4**	**94.5**	**84.2**	**80.7**
SMS	KN4	123.1	106.8	–	–
	LSTM	112.5	102.2	102.4	95.3
	DAS-LSTMN	109.4	100.0	101.2	93.0
	DAS-LSTMN + LSTM	**101.2**	**92.3**	**98.2**	**90.8**
Swb-Fisher	KN4	80.9	83.4	–	–
	LSTM	45.8	58.3	**50.2**	63.1
	DAS-LSTMN	49.5	56.5	78.1	65.3
	DAS-LSTMN + LSTM	**43.0**	**52.8**	55.5	**62.4**

Figure 3 shows an example sentence from the PTB test set. The arrows indicate high attention scores between the predicted word and its history. We can see most words are highly related to the previous word, which agrees with the linguistic intuition that long-term dependencies are relatively rare [6].

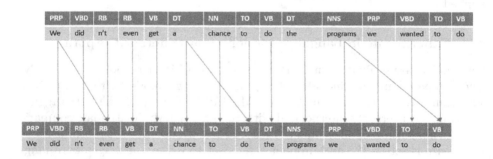

Fig. 3. An example from the PTB test set. The arrows indicate the attention scores higher than 0.2.

5   Conclusion

For the limitation of fixed memory space for conventional RNN based language model, we propose a deep attentive structured language model in this paper. The model is capable of extracting the relation between the hidden state and its input history, applied with the attention module. POS tags are applied as an alternative input of our attention based model. The experiments show that the attention mechanisms do help to achieve improvements in language modeling and POS tags can further promote the language model. The experiments in language modeling indicate that our model outperforms other language models. Interpolated with the LSTM LM, the DAS-LSTMN model achieves significant improvements on the PPL scores. The PPL scores are reduced by 8.4%, 4.8% and 9.4% respectively compared to the baseline LSTM. The results of the experiments on PTB, SMS and Swb-Fisher corpora also indicate our model behaves well on various kinds of datasets.

Acknowledgement. This work was supported by the Shanghai Sailing Program No. 16YF1405300, the China NSFC projects (Nos. 61573241 and 61603252) and the Interdisciplinary Program (14JCZ03) of Shanghai Jiao Tong University in China. Experiments have been carried out on the PI supercomputer at Shanghai Jiao Tong University.

References

1. Bengio, Y., et al.: A neural probabilistic language model. J. Mach. Learn. Res. **3**(February), 1137–1155 (2003)
2. Mikolov, T., et al.: Recurrent neural network based language model. In: Interspeech, vol. 2 (2010)
3. Mikolov, T., et al.: Extensions of recurrent neural network language model. In: 2011 IEEE International Conference on Acoustics, Speech and Signal Processing (ICASSP). IEEE (2011)
4. Hochreiter, S., Schmidhuber, J.: Long short-term memory. Neural Comput. **9**(8), 1735–1780 (1997)
5. Sundermeyer, M., Schlter, R., Ney, H.: LSTM neural networks for language modeling. In: Interspeech (2012)
6. Cheng, J., Dong, L., Lapata, M.: Long short-term memory-networks for machine reading. arXiv preprint arXiv:1601.06733 (2016)
7. Chelba, C., Jelinek, F.: Structured language modeling for speech recognition. arXiv preprint cs/0001023 (2000)
8. Hochreiter, S., et al.: Gradient flow in recurrent nets: the difficulty of learning long-term dependencies (2001)
9. Graves, A., Wayne, G., Danihelka, I.: Neural turing machines. arXiv preprint arXiv:1410.5401 (2014)
10. Sukhbaatar, S., Weston, J., Fergus, R.: End-to-end memory networks. In: Advances in Neural Information Processing Systems (2015)
11. Cho, K., Courville, A., Bengio, Y.: Describing multimedia content using attention-based encoder-decoder networks. IEEE Trans. Multimedia **17**(11), 1875–1886 (2015)
12. Feng, S., Liu, S., Li, M., et al.: Implicit distortion and fertility models for attention-based encoder-decoder NMT model. arXiv preprint arXiv:1601.03317 (2016)
13. Bahdanau, D., Cho, K., Bengio, Y.: Neural machine translation by jointly learning to align and translate. arXiv preprint arXiv:1409.0473 (2014)
14. Part of speech. https://en.wikipedia.org/wiki/Part_of_speech
15. Wang, P., et al.: A unified tagging solution: bidirectional LSTM recurrent neural network with word embedding. arXiv preprint arXiv:1511.00215 (2015)

16. Marcus, M.P., Marcinkiewicz, M.A., Santorini, B.: Building a large annotated corpus of English: the Penn Treebank. Comput. Linguist. **19**(2), 313–330 (1993)
17. The Stanford Natural Language Processing Group. https://nlp.stanford.edu/software/stanford-dependencies.shtml
18. Zaremba, W., Sutskever, I., Vinyals, O.: Recurrent neural network regularization. arXiv preprint arXiv:1409.2329 (2014)

Classification of Motor Imagery EEG Signals with Deep Learning Models

Yurun Shen[1(✉)], Hongtao Lu[1], and Jie Jia[2]

[1] Shanghai Jiao Tong University, 800 Dongchuan Road, Shanghai 200240,
People's Republic of China
shenyurun@gmail.com, htlu@sjtu.edu.cn
[2] Huashan Hospital of the Shanghai FuDan University Medical College,
Shanghai 200040, People's Republic of China
shannonjj@126.com

Abstract. Motor imagery (MI) is a mental process of a motor action including preparation for movement, passive observations of action and mental operations of motor representations. Brain computer interfaces can discriminate different status of individuals according to their EEG signals during imagery tasks. Power spectral density and common spatial patterns are both feature extraction methods that are commonly used to in the classification tasks of EEG series. In this paper, we combine recurrent neural networks and convolutional neural networks inspired by speech recognition and natural language processing. Furthermore, we apply deep models consist of stacking random forests to enhance the ability of feature representation and classification abilities for motor imagery EEG signals. Compared with traditional feature extraction methods, our approaches achieve significant improvements both in the MI-EEG dataset of BCI competitions with healthy individuals and the dataset collected from stroke patients.

Keywords: Motor imagery EEG · Recurrent convolutional neural network · Deep forest

1 Introduction

Brain computer interface is a system that establishes a direct communication pathway between human brain and an electrical device. One motivation of developing brain computer interfaces is to help stroke people to recover from neuromuscular disabilities or to create a new communication channel for those disabled people. Experiments have demonstrated that motor imaginary based brain computer interfaces can improve the recovery training of stroke patients using BCI based stroke rehabilitation systems. Most of these systems use electroencephalography (EEG) signals and feed back detection results through visual display or mechanical movements. Therefore, it is useful to enhance the accuracy of real-time detection in BCI systems. Generally, invasive EEG data are more accurate but they may cause damage to subjects who take the experiments. Non-invasive BCIs are more competitive due to low set-up cost and fine temporal resolution potential to accomplish tasks close to invasive BCIs. Neurophysiological fundamentals are event related desynchronization (ERD), event related

© Springer International Publishing AG 2017
Y. Sun et al. (Eds.): IScIDE 2017, LNCS 10559, pp. 181–190, 2017.
DOI: 10.1007/978-3-319-67777-4_16

synchronization (ERS) and sensorimotor rhythms (SMR). Both movement and imagery are associated with desynchronization of μ and β rhythms. The main difficulties of BCI-FES systems include that stroke patients have damaged or displayed the motor cortical area which commands the movements of its affected limbs, which leads to almost no evidence of an ERD common pattern. EEG signals have a low signal-to-noise ratio and usually have large individual differences. Also, it only reflects summation of excitatory and inhibitory postsynaptic potentials of neurons in the more superficial layers of the cortex. In this paper, we applied deep learning methods to the classification tasks of motor imagery EEG signals to extract features automatically.

2 Related Work

Recently, machine learning techniques are becoming more and more popular in BCI and other EEG-based systems. Linear discriminant analysis are used in classifications of motor imagery tasks by Aldea and Fira [1]. Common Spatial Pattern (CSP) is frequently used in motor imagery based BCI [2]. Filter Bank Common Spatial Pattern (FBCSP) is an improved method based on original CSP method [3]. Leamy et al. use FBCSP to explore EEG features during recovery following stroke [6]. Results show that stroke-affected EEG dataset has a lower classification accuracy than healthy ones. Deep learning is a hot branch of machine learning that has achieved remarkable results in many applications. Zheng used Deep Belief Networks in EEG-based emotion classification [4]. Recurrent Neural Networks and Convolutional Neural Networks are combined together to learn the features after converting EEG time series to 2D images using spectral topography maps [5]. Since EEG signals often have more dimensions besides time and space, tensor decomposition is often applied to represent EEG signals. A. Cichocki derived tensor decomposition for signal processing applications from two-way to multiway component analysis [7]. Zhao proposed a novel decomposition model of slice oriented decomposition to extract slice features from multichannel time-frequency representation of EEG signals for motor imagery tasks [8]. Herein, we explore the feature representation capabilities of deep neural networks for modelling motor imagery EEG signals. More specifically, we use convolutional neural networks and LSTM networks to extract spatial and temporal patterns from EEG data. This architecture has been successfully applied in text classification [13] and end-to-end speech recognition [14]. Furthermore, we apply deep forest models proposed by Zhou and Feng [9] by stacking random forests to obtain a stronger classifier. Two MI-EEG dataset are used to evaluate the performance of these deep models.

3 Data Analysis

3.1 Preprocessing

Raw EEG signals are composed of voltages across temporal dimension in each channel. Time window intercepting and band-pass filtering are often applied to focus on short temporal features and time frequency statistics of EEG series. For the dataset that collected by ourselves, we also need to perform Independent Component Analysis (ICA) to remove artifacts and improve the low signal-noise ratio before these pre-processing methods.

3.2 Feature Extraction

In order to apply the EEG series to classification, we need to transform each EEG data sample into a feature vector. Power spectral density and common spatial pattern algorithm are commonly used in analyzing EEG features.

Power Spectral Density. Power spectral density (PSD) denotes the energy of specific frequency. PSD of EEG series can be computed directly by Fast Fourier Transformation or Wavelet Transformation. It shows the strength of variations as a function of frequency bands. PSD is often used to evaluation the power of brain activity when subjects are conducting different tasks. It has been successfully applied in many classification tasks base on EEG signals, such as emotion recognition and fatigue detection. Motor imagery EEG analysis can also be realized based on features extracted by power spectral density function.

Common Spatial Pattern. Common spatial pattern (CSP) is a kind of principle component analysis (PCA) algorithm successfully applied in brain-computer interfaces to select channels and extract features of EEG signals. For a two-class classification problem, EEG data will be linearly transformed into a subspace with low dimension to achieve maximum differences in variance between these two classes. It performs well based on the phenomenon of ERD/ERS. The details of CSP algorithm are described as follows with the example of classifying EEG data of imaging left and right hand movement.

Given the input data X_L and X_R denoting the preprocessed motor imagery EEG signals. Both the samples of X_L and X_R have dimensions $c \times t$, where c is the number of channels and t is the number of samples across the time domain of each trial. We denote the CSP filter by

$$Z = W^T X \tag{1}$$

where $W \in \mathbb{R}^{c \times d}$ is the spatial filter matrix, $Z \in \mathbb{R}^{d \times t}$ is the filtered matrix and d is the dimension of EEG data after spatial filtering. The normalized covariance matrices of the two classes can be computed as

$$\Sigma_L = \frac{X_L X_L^T}{trace(X_L X_L^T)} \tag{2}$$

$$\Sigma_R = \frac{X_R X_R^T}{trace(X_R X_R^T)} \tag{3}$$

The criterion of CSP can be described as

$$\text{maximize} \quad tr(W^T \Sigma_L W) \tag{4}$$

$$\text{subject to} \quad W^T(\Sigma_L + \Sigma_R)W = I \tag{5}$$

First we can factorize the composite covariance

$$\Sigma_L + \Sigma_R = UDU^T \tag{6}$$

where U is a matrix of eigenvectors and D is a diagonal matrix of eigenvalues. Then the whitening transformation matrix can be computed as

$$P = D^{-\frac{1}{2}}U^T \tag{7}$$

and it can be applied to transform Σ_L and Σ_R into

$$\hat{\Sigma}_L = P\Sigma_L P^T \tag{8}$$

$$\hat{\Sigma}_R = P\Sigma_R P^T \tag{9}$$

The sum of the two matrices is always an identity matrix

$$\hat{\Sigma}_L + \hat{\Sigma}_R = I \tag{10}$$

which means any orthonormal matrix V satisfies that

$$V^T(\hat{\Sigma}_L + \hat{\Sigma}_R)V = I \tag{11}$$

Then we can decompose $\hat{\Sigma}_L = V\Lambda V^T$, where V is the matrix of eigenvectors for $\hat{\Sigma}_L$ and Λ is a diagonal matrix of corresponding eigenvalues. The spatial filter matrix can be computed finally

$$W_{all} = P^T V \tag{12}$$

Specifically, the first column of W_{all} corresponds to the largest eigenvalue of Σ_L and the last column of W_{all} corresponds to the largest eigenvalues of Σ_R since the sum of their eigenvalues are always equal to one. So both of these two vectors are the most significant spatial filters that maximize the variance of two classes. If we select the first and last m columns in W_{all}, we can get the final spatial filter $W \in \mathbb{R}^{c \times 2m}$. The feature vector $x = [x_1, x_2, \ldots, x_d]$ can be calculated by

$$x_i = \log\left(\frac{var[z_i]}{\sum_{i=1}^{d} var[z_i]}\right) \tag{13}$$

3.3 Deep Models

Deep learning focuses on the distributed representations of data by constructing hierarchical models layer by layer. In this section we will focus on several deep models that are commonly used in feature learning.

Recurrent Convolutional Neural Network. Convolutional neural networks and recurrent neural networks both achieve great success in many applications such as image recognition, video analysis and natural language processing. CNNs usually have fewer parameters than other deep neural networks since they share weights of the filters. For time series modelling, 1-D convolution layer is often used to capture the local patterns of data. Since EEG signals represent the temporal evolution of brain activities, convolutional layers can be used to extract spectral invariant representations from each frame of data. Time delay neural networks based on 1-D convolution with more flexibility in window size and weight sharing perform well in phoneme recognition [10]. CNNs also need little pre-processing of training data and easy to train.

Given a sequence of input data $x = (x_1, x_2, \ldots, x_\tau)$ on each channel and a filter $W_f \in \mathbb{R}^{h \times m}$ of window size h, the output of convolution operation by this filter starting from time step t is

$$o = \text{ReLU}(\sum_{t=1}^{\tau-h+1} (W_f \cdot x_{t:t+h-1} + b_f)) \tag{14}$$

Performing the above convolution operations with n filters on sequences from c channels, we will finally get the feature matrix $O \in \mathbb{R}^{c \times n}$.

Recurrent neural networks are another reasonable choice for modelling temporal sequences which share parameters in a different way from CNNs. Relevant past context can be represented by the hidden units and then be used to compute the states of hidden units at the next time step, and the computations of all time steps share same parameters. One problem of RNNs is that they are expensive to train in both perspectives of time and memory. In order to learn long-term dependencies, we use Long Short-Term Memory networks (LSTMs) which are more advanced architectures than vanilla RNNs. Suppose we have a sequence of input $x = (x_1, x_2, \ldots, x_\tau)$ and the corresponding hidden units are $h = (h_1, h_2, \ldots, h_\tau)$, then the output sequence is $y = (y_1, y_2, \ldots, y_\tau)$. We have the following computational formulas for a tradition LSTM unit iterating from $t = 1$ to τ proposed by Hochreiter in 1997 [11]:

$$f_t = \sigma(W_f x_t + U_f h_{t-1} + b_f) \tag{15}$$

$$i_t = \sigma(W_i x_t + U_i h_{t-1} + b_i) \tag{16}$$

$$\tilde{C}_t = \tanh(W_C x_t + U_C h_{t-1} + b_C) \tag{17}$$

$$C_t = f_t \odot C_{t-1} + i_t \odot \tilde{C}_t \tag{18}$$

$$o_t = \sigma(W_o x_t + U_o h_{t-1} + b_o) \tag{19}$$

$$h_t = o_t \odot \tanh(C_t) \tag{20}$$

where $\sigma(\cdot)$ and $\tanh(\cdot)$ refer to the element-wise logistic sigmoid function and hyperbolic tangent function, $\odot$ refer to the element-wise multiplication and i, f, o and C denote the input gate, forget gate, output gate and cell activation vectors respectively. In the training process, we need to update the parameters of each component by backpropagation through time algorithm. By adding layers of convolutional layers and recurrent layers together, we obtain a deep recurrent-convolutional neural network model to learn high-level features of EEG data instead of using traditional feature extraction methods such as power spectral density or common spatial pattern filters.

Deep Forests. Deep forest is a novel decision tree ensemble method with a cascade structure which enables the model to do representation learning [9]. Each level contains multiple decision tree forests including random forests and complete-random forests. Random forests are an ensemble learning method that is commonly used in many classification and regression tasks. The output of each cascade is the average estimation of class distribution computed by each forest. It will be concatenated with original feature vector as the input of next level of cascade. Multi-grained scanning is also applied to further enhance the model's representational learning ability, especially when the inputs have high dimensionality. Sliding windows are used to handle local feature relationships. According to the results experimented on different tasks such as image categorization, face recognition, music classification, hand movement recognition and sentiment classification, deep forest models are highly competitive to deep neural networks. Besides the impressive performance, deep forest models are much easier to train compared to deep neural networks which require great effort in hyper-parameter tuning. They also work well when the given training data are much smaller.

4 Experiments

4.1 Training Data

We use the BCI Competition III Dataset V [12] to evaluate the deep models. The dataset contains EEG data experimented on 3 healthy subjects during 4 non-feedback sessions. There are 3 tasks in these sessions: imagination of left hand movements, imagination of right hand movements and generation of words beginning with the same random letter. EEG signals were firstly spatially filtered by means of a surface Laplacian. Then the power spectral density in the frequency band 8–30 Hz were computed over the last second of data with a frequency resolution of 2 Hz for the 8 centro-parietal channels C3, Cz, C4, CP1, CP2, P3, Pz and P4. Feature vector of each sample contains 96 dimensions. For deep models, we use raw data only with simple band-pass filtering and channel selection to eliminate the electrodes influenced by large artifacts. We only use 1 s window for every 0.5 s to output a label in order to realize fast interactions in BCI systems. In out deep learning models, we use 8 overlapping frames of EEG series and an additional max pooling layer to get the final output label for each 0.5 s. The diagram of the proposed deep model is presented in Fig. 1.

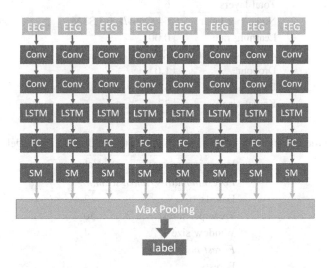

Fig. 1. Diagram of recurrent convolutional neural networks for multi-frame EEG series. The notation here is as follow. Conv: 1-D convolution layer across temporal series; LSTM: LSTM recurrent layer; FC: fully-connected layer; SM: softmax layer. Max Pooling: max pooling layer across different frames of EEG series.

Besides the BCI competition dataset, we also use the motor imagery EEG data collected from 5 post-stroke patients using the BCI-FES rehabilitation system. There are 2 tasks in the training sessions: imagination left hand movements and right hand movements. The training was conducted during 2 months during the medical treatments of these patients. The sampling frequency of the experiments is 256 Hz and the total channels are 16 including the reference channel using Ag/AgCl electrodes (FC3, FCz, FC5, C1-C6, Cz, CP3, CPZ, CP4, P3, Pz and P4). EEG signals can be divided into 4-s windows. Before data preprocessing and feature learning, we have already eliminated some samples with extremely low accuracy in the training process. Also, CSP algorithm is used to eliminate some channels with large artifacts.

4.2 Configuration

In this section we compare different deep models and traditional classification algorithms on different EEG data. The main hyper-parameters of the deep neural networks and the deep forest models for the BCI competition dataset are mainly summarized in Tables 1 and 2. As for the dataset of stroke patients, we use only 1 convolutional layer to reduce the number of parameters since the total samples for each subject are limited, which only amount to less than 1,000 in average. Also, we can add more forests in the deep forest model which can improve the classification accuracy further, but it will cost too much computation time without adequate computing resources.

Table 1. Summary of main hyper-parameters of recurrent convolutional neural network.

Hyper-parameters	Value
Total layers	6
Nodes per layer	64 Conv/100 LSTM
Learning rate	0.001
Activation functions	ReLU
Dropout	1.0
Batch size	32
Kernel size	32

Table 2. Summary of main hyper-parameters of deep forest model.

Hyper-parameters	Value
Forest in multi-grained scanning:	
Forests	2
Trees per forest	50
Window size	4
Forest in cascade:	
Forests	4
Trees per forest	100

4.3   Results

The performance of deep models and traditional classifiers on the test samples of BCI competition III dataset V is shown in Table 3. On the BCI competition dataset, recurrent-convolutional neural network performs better than traditional classifiers with extracted power spectral density features to 68.51% which is close to the best result in the competition. Deep forest model is less powerful in accuracy and also involves large computation given large training data, which is inappropriate in online BCIs.

The classification accuracies of deep models and traditional feature extraction methods on the MI-EEG dataset of the 5 post-stroke patients are shown in Table 4. One intractable problem is that the number of samples are much smaller than data dimension. Deep models like CNNs and RNNs are easy to overfit on small scale dataset. So

Table 3. The classification accuracies of deep models and traditional classifiers on BCI competition III dataset V.

Method	Subject 1	Subject 2	Subject 3	Average
SVM	72.46%	62.13%	48.88%	61.16%
Naive Bayes	63.50%	54.00%	44.67%	54.06%
MLP	74.60%	60.51%	46.59%	60.57%
Random forest	76.60%	63.51%	50.17%	63.43%
Deep forest	77.17%	55.30%	49.31%	60.60%
1D-Conv + LSTM	82.17%	68.42%	54.93%	**68.51%**

Table 4. The classification accuracies of deep models and traditional feature extraction methods on the MI-EEG dataset of stroke patients.

Method	Subject 1	Subject 2	Subject 3	Subject 4	Subject 5	Average
PSD	48.35%	60.05%	45.83%	50.11%	56.92%	52.25%
CSP	59.70%	57.60%	57.20%	53.48%	57.90%	57.18%
1D-Conv + LSTM	56.96%	54.63%	54.02%	51.75%	53.85%	54.24%
Deep forest	60.00%	60.00%	71.06%	54.35%	63.03%	**61.69%**

deep forest is a better choice in this situation. Besides, brain activities are physiologically influenced for stroke patients and it can be harder for them to concentrate on imagery tasks. So the accuracy on this dataset remains to be improved by collecting more data from stricter experiments and use more sophisticated models.

5   Conclusion

In this paper we apply some deep learning models in the domain of EEG analysis, mainly focusing on the classification problem of motor imagery EEG data. Deep learning methods can be used to learn distributed representations of EEG signals automatically across temporal dimension among different channels. Results verify that deep models outperform classifiers with traditional feature extraction methods in motor imagery classification tasks with both healthy individuals and stroke patients.

Acknowledgement. This paper is supported by the 863 National High Technology Research and Development Program of China (SS2015AA020501), the Basic Research Project of "Innovation Action Plan" (16JC1402800), the Major Basic Research Program (15JC1400103) of Shanghai Science and Technology Committee and the interdisciplinary Program of Shanghai Jiao Tong University (YG2015MS43).

References

1. Classifications of Motor Imagery Tasks in Brain Computer Interface Using Linear Discriminant Analysis
2. Wang, Y., Gao, S., Gao, X.: Common spatial pattern method for channel selection in motor imagery based brain-computer interface. In: 27th Annual International Conference of the Engineering in Medicine and Biology Society, IEEE-EMBS 2005, pp. 5392–5395. IEEE (2006)
3. Ang, K.K., Chin, Z.Y., Zhang, H., et al.: Filter bank common spatial pattern (FBCSP) in brain-computer interface. In: IEEE International Joint Conference on Neural Networks, IJCNN 2008 (IEEE World Congress on Computational Intelligence), pp. 2390–2397. IEEE (2008)
4. Zheng, W.L., Zhu, J.Y., Peng, Y., et al.: EEG-based emotion classification using deep belief networks. In: IEEE International Conference on Multimedia and Expo (ICME), pp. 1–6. IEEE (2014)

5. Bashivan, P., Rish, I., Yeasin, M., et al.: Learning representations from EEG with deep recurrent-convolutional neural networks. arXiv preprint arXiv:1511.06448 (2015)
6. Leamy, D.J., Kocijan, J., Domijan, K., et al.: An exploration of EEG features during recovery following stroke–implications for BCI-mediated neurorehabilitation therapy. J. Neuroeng. Rehabil. 11(1), 9 (2014)
7. Cichocki, A., Mandic, D., De Lathauwer, L., et al.: Tensor decompositions for signal processing applications: from two-way to multiway component analysis. IEEE Sig. Process. Mag. 32(2), 145–163 (2015)
8. Zhao, Q., Caiafa, C.F., Cichocki, A., Zhang, L., Phan, A.H.: Slice oriented tensor decomposition of EEG data for feature extraction in space, frequency and time domains. In: Leung, C.S., Lee, M., Chan, J.H. (eds.) ICONIP 2009. LNCS, vol. 5863, pp. 221–228. Springer, Heidelberg (2009). doi:10.1007/978-3-642-10677-4_25
9. Zhou, Z.H., Feng, J.: Deep forest: towards an alternative to deep neural networks. arXiv preprint arXiv:1702.08835 (2017)
10. Waibel, A., Hanazawa, T., Hinton, G., et al.: Phoneme recognition using time-delay neural networks. IEEE Trans. Acoust. Speech Sig. Process. 37(3), 328–339 (1989)
11. Hochreiter, S., Schmidhuber, J.: Long short-term memory. Neural Comput. 9(8), 1735–1780 (1997)
12. Millán, J. del R.: On the need for on-line learning in brain-computer interfaces. In: Proceedings of Intrelational Joint Conference on Neural Networks (2004)
13. Lai, S., Xu, L., Liu, K., et al.: Recurrent convolutional neural networks for text classification. In: AAAI, vol. 333, pp. 2267–2273 (2015)
14. Amodei, D., Anubhai, R., Battenberg, E., et al.: Deep speech 2: end-to-end speech recognition in English and mandarin. In: International Conference on Machine Learning, pp. 173–182 (2016)

Jointly Using Deep Model Learned Features and Traditional Visual Features in a Stacked SVM for Medical Subfigure Classification

Hongyu Wang[1], Jianpeng Zhang[1], and Yong Xia[1,2(✉)]

[1] Shaanxi Key Lab of Speech and Image Information Processing (SAIIP),
School of Computer Science and Engineering,
Northwestern Polytechnical University,
Xi'an 710072, People's Republic of China
yxia@nwpu.edu.cn
[2] Centre for Multidisciplinary Convergence Computing (CMCC),
School of Computer Science and Technology,
Northwestern Polytechnical University,
Xi'an 710072, People's Republic of China

Abstract. Classification of diagnose images and illustrations in the literature is a major challenge towards automated literature review and retrieval. Although being widely recognized as the most successful image classification technique, deep learning models, however, may need to be complemented by traditional visual features to solve this problem, in which there are intra-class variation, inter-class similarity and a small training dataset. In this paper, we propose an approach to classifying diagnose images and biomedical publication illustrations. This algorithm jointly uses the image representations learned by three pre-trained deep convolutional neural network models and ten types of traditional visual features in a stacked support vector machine (SVM) classifier. We have evaluated this algorithm on the ImageCLEF 2016 Subfigure Classification dataset and achieved an accuracy of 85.62%, which is higher than the top performance of purely visual approaches in this challenge.

Keywords: Medical image classification · Feature extraction · Deep convolutional neural network · Stacked support vector machine

1 Introduction

Medical imaging plays a pivotal role in modern healthcare, which has resulted in the proliferation of digital images in all types of electronic biomedical publications. Such proliferation poses major issues for retrieval, review and in assimilating these data for clinical practices and research settings. Hence there has been considerable research directed at classifying the diagnose images and illustrations in the literature to improve data mining in this area. The ImageCLEF Subfigure Classification Challenge [1] recognized the increasing significance and complexity of this problem and provided a benchmark dataset with 30 heterogeneous classes of data for comparing automated image classification techniques.

© Springer International Publishing AG 2017
Y. Sun et al. (Eds.): IScIDE 2017, LNCS 10559, pp. 191–199, 2017.
DOI: 10.1007/978-3-319-67777-4_17

A number of solutions to medical image classification have been published and most consist of two steps: extracting visual features to characterize images and using those features to train a classifier. Commonly used visual features include the descriptors for color, texture, edges and combined descriptors. Cirujeda and Binefa [2] extracted the first and second order color features at the image level and built a distinctive signature of an image in the form of a covariance matrix based descriptor. Lazebnik et al. [3] proposed an opponent SIFT based on a modification to pyramid match kernels. Kitanovski et al. [4] employed different feature fusion techniques to combine multiple color and textual features. Abedini et al. [5] implemented the fusion of a set of low level global and local visual features extracted at different spatial granularities. Unay and Ekin [6] extracted intensity and texture descriptors and complemented them with spatial context. Valavanis et al. [7] attempted various visual features, including the bag-of-visual-words and quad-tree bag-of-colors. Once features are extracted, most classification technique, such as support vector machine (SVM) [8], back-propagation neural network (BPNN) [9] and random forest [10], can be applied to solve this problem.

Recently, deep learning, particularly the deep convolutional neural network (DCNN), has become the most successful image classification technique, as it provides a unified framework for joint feature extraction and classification. Kumar et al. [11] created an optimized feature extractor by fine-tuning a DCNN that has been pre-trained on general image data for subfigure modality classification. Lopez et al. [12] adopted the VGG model to solve the problem of image classification. Koitka and Friedrich [13] used the ResNet-152 model trained on the ImageNet dataset to win the ImageCLEF 2016 competition.

Despite these achievements, classifying diagnose images according to the modalities by which they were produced and labelling biomedical publication illustrations according to their production attributes remain a major challenge. The first difficulty is the intra-class variation and inter-class similarity [14], which is particularly acute in classifying diagnosis images based on their modalities. For instance, a brain CT image and a pleural CT image look dissimilar due to viewing different anatomical structures although they belong to the same category; whereas a brain CT image and a brain MR image look very similar but belonging to different categories. Although deep neural networks have enough capacity to forcibly remember all training samples [15], the ambiguity produced by intra-class variation and inter-class similarity may tease a neural network and make it fall into confusion. Another difficulty is the lack of sufficient training data. DCNN models generally involve a tremendous number of parameters, and hence require a huge amount of labeled data for supervised training [16], which is not available for this image classification problem, since collection and annotation of diagnosis images and illustrations are time consuming, expensive and tedious. Therefore, when applied to this problem, DCNN models may suffer from over-fitting and are not able to produce satisfying classification.

To overcome both difficulties, the high-level and abstract image representations learned by deep learning models must be complemented by the low-level image characteristics described by traditional visual features [17]. In this paper, we propose a novel approach to classifying diagnose images and biomedical publication illustrations, in which the image representations learned by three pre-trained deep convolutional neural network models and ten types of traditional visual features are jointly used in a stacked support vector machine (SVM) classifier. We have evaluated the proposed algorithm against three best-performed purely visual solutions in the ImageCLEF 2016 Subfigure Classification Challenge on the challenge data [18].

2 Method

As shown in Fig. 1, the proposed algorithm consists of three major steps: feature extraction, feature dimensionality reduction and training a stacked SVM classifier. During feature extraction, both traditional visual features and deep model learned image representations are calculated.

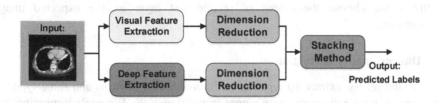

Fig. 1. Flowchart of the proposed algorithm

2.1 Feature Extraction

The traditional visual features [19] used in the proposed algorithm include

(1) a 500-dimensional bag of features (BoF) with a vocabulary size of 500 [20],
(2) a 256-dimensional color histogram (CH) [21],
(3) angular second moment, entropy, inverse different moment, contrast and correlation on four gray-level co-occurrence matrices (GLCM) counted at and [22],
(4) a 58-dimensional local binary patterns (LBP) [23],
(5) a 160-dimensional Gabor features created by a Gabor filter bank with five scales and eight orientations [24],
(6) a 256-dimensional CENsus TRansform hISTogram (CENTRIST) with the first and last components dropped [25],
(7) a 850-dimensional pyramid histogram of oriented gradients (PHOG) computed over 3 pyramid levels with 10 bins per histogram [26],
(8) a 64-dimensional edge histogram descriptor (EHD) [27],
(9) a 256-dimensional auto color correlogram (ACC) calculated with 64 quantized colors and four different distances [28], and
(10) a 144-dimensional color and edge directivity descriptor (CEDD) with 24 color bins and 6 texture descriptors [29].

To take the advantage of deep learning in effective image representation, we adopt three DCNN models for feature extraction, including VGG-16 [30], VGG-19 [30] and ResNet-50 [31], which contains 16, 19 and 50 learnable layers, respectively. These deep models were previously trained on the ImageNet training set, which is a 1000-category large-scale natural image database. Each model takes an input image of size and generates a prediction vector of 1000 dimension. To adapt these models to our 30-category image classification problem, we first randomly select 30 neurons in the last fully connected layer and remove other output neurons and the weights attached to them. Then, we resize each of our training images to the size of using the bicubic interpolation [32] and apply it to each model to fine tune model parameters. We fix the maximum iteration number to 200, choose the min-batch stochastic gradient decent with a batch size of 50, set the learning rate as small as 0.0001 and further reduce the learning rate by one-tenth every 20 iterations. To implement the early stop strategy, we randomly choose 20% of training images to form a validation set. The training of each model will be early terminated, if the error on the other 80% of training images continues to decrease but the error on the validation set stops to decrease. Once fine-tuned, we choose the output of second last layer as the expected image representation.

2.2 Dimensionality Reduction

For each image, we extract 10 types of traditional visual features and three types of deep features, whose dimensionality ranges from 20 to 4096. We apply dimensionality reduction to these features on a group-by-group basis before using them to train a classifier. Since manifold leaning performs well on the low-dimensional data and the principal components analysis (PCA) [33] performs better on high-dimensional data [34], we apply the locality preserving projections (LPP) [35] to each type of traditional visual features and PCA to each type of deep features.

In LPP, the reduced dimension is determined by the maximum likelihood estimation of intrinsic dimensionality [36]; whereas, in PCA, each type of features is mapped into a low-dimensional space spanned by the first eigenvectors of the covariance matrix, which give an accumulative eigenvalues above 97% of the sum of all eigenvalues. The dimensionality of each type of features before and after reduction is shown in Table 1.

Table 1. Original and reduced dimension of each type of features

Dimension reduction method	Features	Original dimension	Reduced dimension
LPP	BOF	500	23
	CH	256	8
	LBP	58	10
	GLCM	20	5
	Gabor	160	31
	CETRIST	254	23
	PHOG	850	30
	EHD	64	22
	ACC	256	18
	CEDD	144	7
PCA	VGG16	512	235
	VGG19	4096	1223
	Resnet50	2048	1144

2.3   Stacking-Based Ensemble Learning

Since the dimensionality of traditional visual features is much lower than that of deep features, we concatenate all of them into one feature group. Thus, we design a two-layer stacked SVM classifier, as shown in Fig. 2 to jointly use those four groups of features. In the first layer, each group of features is used to train a SVM classifier with linear kernel. After performing the 10-fold cross-validation on the training dataset, we obtain a N × 435 result matrix for each SVM classifier, where N is the number of training samples and 435 is the number of class pairs among 30 classes. Then we use the weighted sum of those four result matrixes as the input to train the SVM classifier in the second layer, where the weight of each result matrix is proportional to the classification accuracy of the corresponding SVM classifier in the cross-validation. Finally, the SVM classifier in the second layer gives the predicted class label of each input subfigure.

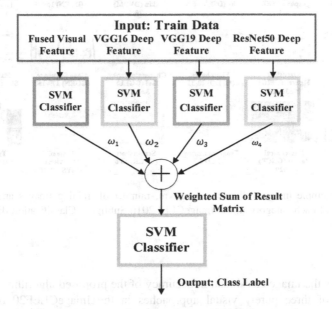

Fig. 2. Diagram of our two-layer stacking framework

3   Dataset

The dataset used for this study are from the subfigure classification task of the ImageCLEF2016 Medical Image Classification Challenge. It is composed of 6,776 training subfigures and 4,166 test subfigures, ranging from 18 categories of diagnose images acquired from different modalities, such as radiology, microscopy, CT and MRI, to 12 categories of biomedical publication illustrations. These subfigures are from the biomedical literature and distributed by the PubMed Central in their original form. One example from each category, together with the full name, acronym, number of training images and number of testing images of the category, is shown in Fig. 3.

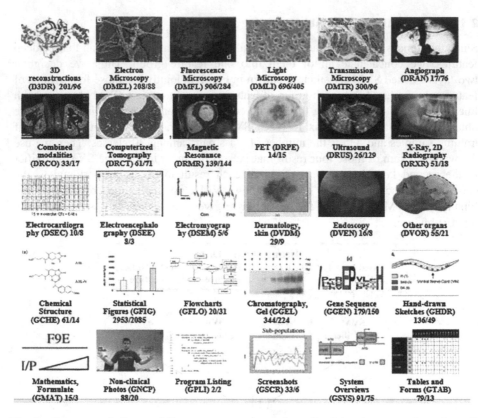

Fig. 3. One example image, full name, acronym, number of training images and number of testing images of each category in the ImageCLEF2016 Subfigure Classification dataset

4   Results

Table 2 shows the image classification accuracy of the proposed algorithm and the top performance of three purely visual approaches in the ImageCLEF2016 Subfigure Classification Challenge. Due to not using any textural information, we compared our proposed algorithm with only visual-based solutions, among which Koitka's algorithm employs a pre-trained 152-layer ResNet and Kumar's algorithm uses 11 types of traditional visual descriptors. It shows that our proposed algorithm can achieve even higher classification accuracy by jointly using DCNN models with less than 50 layers and traditional visual features.

Table 2. Accuracy of three methods in subfigure classification

Method	Feature type	Accuracy
Proposed	Visual and deep	**85.62%**
Koitka and Friedrich [13]	Visual and deep	85.35%
Kumar et al. [11]	Visual	84.01%
Valavanis's [7]	Deep	77.55%

5 Discussion on Computational Complexity

As a deep learning-based approach, the proposed algorithm is very time consuming during training. At the feature extraction step, calculating traditional visual features is not that complex, but fine-tuning each DCNN model may cost about 10 h. At the classifier training step, it takes less than twenty minutes to train the stacked SVM classifier. Nevertheless, applying the trained model to medical image classification is relatively efficient, as it costs around 6 s to predict the class label of one test image.

6 Conclusion

This paper proposes a novel method to classify diagnose images and illustrations in the biomedical literature, which jointly uses the image representations learned by three pre-trained DCNN models and ten types of traditional visual features in a stacked SVM classifier. Our experimental results suggest that the proposed method can produce satisfying classification accuracy, which is higher than the top performance of three purely visual approaches in the ImageCLEF2016 Subfigure Classification Challenge. Our future work will focus on developing more effective ensemble learning techniques to further improve the image classification performance.

Acknowledgment. This work was supported in part by the National Natural Science Foundation of China under Grants 61471297 and 61771397, and in part by the Seed Foundation of Innovation and Creation for Graduate Students in Northwestern Polytechnical University. We appreciate the efforts devoted by the organizers of the ImageCLEF2016 Medical Image Classification Challenge to collect and share the data for comparing algorithms of classifying diagnose images and illustrations in the biomedical literature.

References

1. Liverman, C.T., Fulco, C.E., Kipen, H.M.: Internet access to the national library of medicine's toxicology and environmental health databases. National Academies Press (US), Washington, D.C (1998). doi:10.17226/6327
2. Cirujeda, P., Binefa, X.: Medical image classification via 2D color feature based covariance descriptors. In: CLEF (Working Notes) (2015)
3. Lazebnik, S., Schmid, C., Ponce, J.: Beyond bags of features: spatial pyramid matching for recognizing natural scene categories. In: 2006 IEEE Computer Society Conference Computer Vision and Pattern Recognition, pp. 2169–2178 (2006)
4. Kitanovski, I., Dimitrovski, I., Loskovska, S.: FCSE at medical tasks of ImageCLEF 2013. In: CLEF (Working Notes) (2013)
5. Abedini, M., Cao, L., Codella, N., et al.: IBM research at ImageCLEF 2013 medical tasks. In: American Medical Informatics Association (AMIA) ImageCLEF, Medical Image Retrieval Workshop (2013)
6. Unay, D., Ekin, A.: Intensity versus texture for medical image search and retrieval. In: IEEE International Symposium on Biomedical Imaging: From Nano to Macro (2008)

7. Valavanis, L., Stathpoulos, S., Kalamboukis, T.: IPL at CLEF 2016 medical task. In: CLEF (Working Notes) (2016)
8. Chang, C., Lin, C.: LIBSVM: a library for support vector machines. ACM Trans. Intell. Syst. Technol. (TIST) 2(3), 27 (2011)
9. Dandil, E., et al.: Artificial neural network-based classification system for lung nodules on computed tomography scans. In: 2014 6th International Conference of Soft Computing and Pattern Recognition (2014)
10. Breiman, L.: Random forests. Mach. Learn. 45(1), 5–32 (2001)
11. Kumar, A., et al.: Subfigure and multi–label classification using a fine–tuned convolutional neural network. In: CLEF2016 Working Notes, CEUR Workshop Proceedings. CEUR-WS. org, Évora (2016)
12. Lopez, A.R., et al.: Skin lesion classification from dermoscopic images using deep learning techniques. In: 2017 13th IASTED International Conference on Biomedical Engineering, pp. 40–54 (2017)
13. Koitka, S., Friedrich, C.M.: Traditional feature engineering and deep learning approaches at medical classification task of ImageCLEF 2016. In: CLEF2016 Working Notes, CEUR Workshop Proceedings. CEUR-WS.org, Évora (2016)
14. Song, Y., et al.: Large margin local estimate with applications to medical image classification. IEEE Trans. Med. Imaging 34(6), 1362–1377 (2015)
15. Zhang, C., Bengio, S., et al.: Understanding deep learning requires rethinking generalization. In: International Conference on Learning Representations, Toulon France (2017)
16. Donahue, J., Jia, Y., Vinyals, O., et al.: DeCAF: a deep convolutional activation feature for generic visual recognition. Int. Conf. Mach. Learn. 50, 815–830 (2014)
17. Müller, H., Kalpathy-cramer, J., Demner-Fushman, D., et al.: Creating a classification of image types in the medical literature for visual categorization. In: SPIE Medical Imaging. International Society for Optics and Photonics, pp. 75–84 (2012)
18. Gilbert, A., Piras, L., Wang, J., et al.: Overview of the ImageCLEF 2016 scalable concept image annotation task. In: Conference and Labs of the Evaluation Forums (2016)
19. Chen, C.: Computer Vision in Medical Imaging, vol. 2. World Scientific, Singapore (2014)
20. Cula, O.G., Dana, K.J.: Compact representation of bidirectional texture functions. In: IEEE Computer Society Conference on Computer Vision and Pattern Recognition, vol. 1, I-1041–I-1047 (2001)
21. Novak, C.L., Shafer, S.A.: Anatomy of a color histogram. In: 1992 IEEE Computer Society Conference on Computer Vision and Pattern Recognition, pp. 599–605 (2002)
22. Guo, D., Song, Z.: A study on texture image classifying based on gray-level co-occurrence matrix. For. Mach. Woodworking Equip. 7, 007 (2005)
23. Ojala, T., Pietikalnen, M., et al.: Multiresolution gray-scale and rotation invariant texture classification with local binary patterns. IEEE Trans. Pattern Anal. Mach. Intell. 24(7), 971–987 (2002)
24. Fogel, I., Sagi, D.: Gabor filters as texture discriminator. Biol. Cybern. 61(2), 103–113 (1989)
25. Wu, J., Rehg, J.M.: CENTRIST: a visual descriptor for scene categorization. IEEE Trans. Pattern Anal. Mach. Intell. 33(8), 1489–1501 (2011)
26. Bosch, A., Zisserman, A., Munoz, X.: Representing shape with a spatial pyramid kernel. In: ACM International Conference on Image and Video Retrieval, pp. 401–408 (2007). ACM
27. Sikora, T.: The MPEG-7 visual standard for content description-an overview. IEEE Trans. Circ. Syst. Video Technol. 11(6), 696–702 (2001)
28. Huang, J., Kumar, S.R., Mitra, M., et al.: Image indexing using color correlograms. In: IEEE Computer Society Conference on Computer Vision and Pattern Recognition, p. 762 (1997)

29. Chatzichristofis, S.A., Boutalis, Y.S.: CEDD: color and edge directivity descriptor: a compact descriptor for image indexing and retrieval. In: Gasteratos, A., Vincze, M., Tsotsos, J.K. (eds.) ICVS 2008. LNCS, vol. 5008, pp. 312–322. Springer, Heidelberg (2008). doi:10. 1007/978-3-540-79547-6_30

30. Simonyan, K., Zisserman, A.: Very deep convolutional networks for large-scale image recognition. arXiv preprint arXiv:1409.1556 (2014)

31. He, K., et al.: Deep residual learning for image recognition. In: Proceedings of the IEEE Conference on Computer Vision and Pattern Recognition (2016)

32. Keys, R.G.: Cubic convolution interpolation for digital image processing. IEEE Trans. Acoust. Speech Sig. Process. **29**(6), 1153–1160 (1981)

33. Dunteman, G.H.: Principal Components Analysis. Sage, Thousand Oaks (1989)

34. van der Maaten, L., Postma, E., van den Herik, J.: Dimensionality reduction: a comparative review. J. Mach. Learn. Res. **10**(1), 66–71 (2007)

35. He, X., Niyogi, P.: Locality preserving projections. In: Advances in Neural Information Processing Systems (2004)

36. Levina, E., Bickel, P.J.: Maximum likelihood estimation of intrinsic dimension. Adv. Neural. Inf. Process. Syst. **17**, 777–784 (2004)

Faces and People

Facial Emotion Recognition via Discrete Wavelet Transform, Principal Component Analysis, and Cat Swarm Optimization

Shui-Hua Wang[1,2], Wankou Yang[3], Zhengchao Dong[4],
Preetha Phillips[5], and Yu-Dong Zhang[1,6(✉)]

[1] School of Computer Science and Technology, Nanjing Normal University,
Nanjing, Jiangsu 210023, China
yudongzhang@ieee.org
[2] Department of Electrical Engineering, The City College of New York, CUNY,
New York, NY 10031, USA
[3] Laboratory of Measurement and Control of Complex Systems of Engineering,
Ministry of Education, Southeast University, Nanjing 210096, Jiangsu, China
[4] Translational Imaging Division and MRI Unit, Columbia University and New
York State Psychiatric Institute, New York, NY 10032, USA
[5] West Virginia School of Osteopathic Medicine, Lewisburg, WV 24901, USA
[6] Department of Informatics, University of Leicester, Leicester LE1 7RH, UK

Abstract. Facial emotion recognition is important in many academic and industrial applications. In this paper, our team proposed a novel facial emotion recognition method. First, we used discrete wavelet transform to extract wavelet coefficients from facial images. Second, principal component analysis was utilized to reduce the features. Third, a single-hidden-layer neural network was used as the classifier. Finally and most importantly, we introduced the cat swarm optimization to train the weights and biases of the classifier. The ten-fold stratified cross validation showed cat swarm optimization method achieved an overall accuracy of $89.49 \pm 0.76\%$. It was better than genetic algorithm, particle swarm optimization, and time-varying-acceleration-coefficient particle swarm optimization. Besides, our facial emotion recognition system was better than two state-of-the-art approaches.

Keywords: Facial emotion recognition · Discrete wavelet transform · Principal component analysis · Cat swarm optimization · Genetic algorithm · Particle swarm optimization

1 Introduction

Facial emotion recognition (FER) studies the identification of human facial emotions [1], which academically means the positions and motions of facial muscles below the face skins. The facial emotion conveys non-verbal communications to the observers. Currently, FER has been employed in many applications, such as Parkinson's disease [2], schizophrenia [3], speech emotion recognition [4], autism spectrum disorder [5], bipolar disorder [6], social influence [7], etc.

© Springer International Publishing AG 2017
Y. Sun et al. (Eds.): IScIDE 2017, LNCS 10559, pp. 203–214, 2017.
DOI: 10.1007/978-3-319-67777-4_18

Many existing approaches were proposed to solve FER. For instance, Drume and Jalal [8] combined principal component analysis (PCA) and support vector machine (SVM). Later, Ali et al. [9] employed higher-order spectral (HOS), radon transform (RT), and support vector machine (SVM). Boubenna and Lee [10] suggested to use pyramid histogram of oriented gradient (PHOG) and genetic algorithm (GA). Finally, linear discriminant analysis (LDA) was used. Lu [11] employed biorthogonal wavelet entropy (BWE).

Nevertheless, above approaches suffer from following shortcomings: (i) Their algorithm can not detect "disgust" emotion efficiently. (ii) The computation time cost too long. (iii) Their classification accuracy is low. After analyzing above shortcomings, we believe they are due to the low-efficiency features and classifier. In this paper, we proposed a novel system based on wavelet transform, principal component analysis, and single-hidden-layer neural network (SNN). Besides, we trained the SNN via cat swarm optimization (CSO) method.

The structure of this paper is organized as follows: Sect. 2 gives the subjects and methodology used in this paper. Section 2 gives the methodology. Section 3 offers the experiments, results, and discussions. Finally, Sect. 4 presents the concluding remarks.

2 Methodology

2.1 Subjects

The dataset used is from reference [11]. Here we have 700 images in total, and 100 image for each emotion

- Class 1: happy
- Class 2: sadness
- Class 3: surprise
- Class 4: anger
- Class 5: disgust
- Class 6: fear
- Class 7: neutral

Figures 1 and 2 offers seven emotion classes of a male and a female faces, respectively.

Fig. 1. Seven emotions of a male face

Fig. 2. Seven emotions of a female face

2.2 Feature Extraction and Reduction

Figure 3 shows three generations of signal processing: The 1st generation is Fourier transform (FT), the 2nd generation is short-time Fourier transform (ST-FT), and the 3rd generation is wavelet transform (WT). The discrete version of WT is called discrete wavelet transform (DWT).

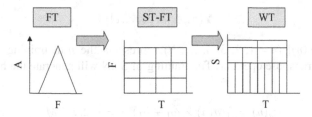

Fig. 3. Signal processing development (A = Amplitude, F = Frequency, T = Time, S = Scale)

Take one-dimensional signal as an example, suppose $x(n)$ represents a time-domain signal, $g(n)$ and $h(n)$ are the impulse response of a low-pass filter and a high-pass filter, we have the approximation coefficient $a_1(n)$ and detail coefficient $d_1(n)$ of first level as below:

$$a_1 = (x * g) \downarrow 2 \tag{1}$$

$$d_1 = (x * h) \downarrow 2 \tag{2}$$

where * denotes the convolution operation, and the downwards arrow $\downarrow$ denotes the undersampling. Then, the 2-level approximation coefficient a_2 and detail coefficient d_2 are obtained by

$$a_2 = (a_1 * g) \downarrow 2 \tag{3}$$

$$d_2 = (a_1 * h) \downarrow 2 \tag{4}$$

This cascading procedure iterates until the given decomposition level j. There are other important feature selection methods, for example, the contourlet transform, the curvelet transform, joint filter, etc. The wavelet coefficients are reduced by the canonical principal component analysis [12].

2.3   Classifier

Scholars have proposed many classifiers, for example, decision tree, logistic regression [13], naive Bayes classifier, support vector machine, extreme learning machine, and artificial neural network (ANN). All of them achieved great success in both academic and industrial fields. The ANN is the most popular, since it can approximate to any function at any degree, provided by universal approximation theorem.

The single-hidden-layer neural network (SNN) is one type of ANN. It maps given input feature to target labels, by generating a fully connected feedforward neural network, which merely contains one hidden layer. Suppose c represents the number of classes and D the dimension of input features, we have:

$$\mathbf{d}(n) = [d_1(n), d_2(n), \ldots, d_D(n)]^T \tag{5}$$

$$\mathbf{Y}(n) \in [1, 2, \ldots, c] \tag{6}$$

where $[\mathbf{d}(n), \mathbf{Y}(n)]$ $(n = 1, 2, 3, 4, \ldots, N)$ represents the n-th training data, and $()^T$ represents the transpose operator. The training of SNN will generate the hidden neuron $z_j(n)$ as

$$z_j(n) = f_1(d(n) \times \omega_1 + b_1), j = 1, 2, \ldots, M \tag{7}$$

where ω_1 and b_1 are the weight and bias matrix between input and hidden layers. The f_1 is the activation function between input and hidden layers. M is the number of hidden neurons.

The result of output neuron $\mathbf{y}(n)$ is obtained as

$$y_k(n) = f_2(z(n) \times \omega_2 + b_2) \tag{8}$$

where ω_2 and b_2 are the weight and bias matrix between hidden and output layers. The f_2 is the activation function between hidden and output layers.

Finally, the predicted label is

$$\mathbf{y}(n) = \mathrm{argmax}\left([y_1(n), y_2(n), \ldots, y_c(n)]^T\right) \tag{9}$$

In a word, the aim of all training algorithms is to minimize the sum of mean-squared error (MSE) between the target label $\mathbf{Y}(n)$ and predicted label $\mathbf{y}(n)$ as

$$\min \sum_{n=1}^{N} (\mathbf{y}(n) - \mathbf{Y}(n))^2 \tag{10}$$

Figure 4 gives the structure of SNN pictorially.

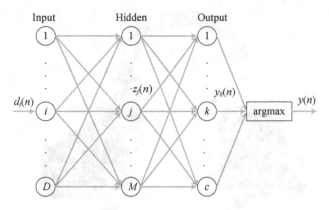

Fig. 4. Structure of SNN

2.4 Cat Swarm Optimization

To train the weights and biases of SNN, traditional methods used gradient descent method and its variants. Nevertheless, traditional methods may trap into local optimal point. Hence, bio-inspired methods are proposed. For example, particle swarm optimization (PSO) [14] is currently widely used to train SNN. Besides, Jotheeswaran and Koteeswaran [15] used genetic algorithm (GA) to train SNN. Yang and Sun [16] demonstrated that time-varying acceleration coefficient PSO (TPSO) can achieve better performance than PSO. Sun [17] demonstrated quantum-behaved PSO obtained better performance than PSO. Yang [18] and Wu [19] both used biogeography-based optimization to train SNN. Wei [20] used an improved artificial bee colony algorithm to train SNN. Cattani and Rao [21] employed Jaya algorithm to train SNN.

To further improve the performance, we introduced the cat swarm optimization (CSO) method in this study. Chu et al. [22] proposed the CSO by mimicking the behaviors of cat. In their opinion, two modes exist for cats: seeking mode and tracing mode, as shown in Fig. 5. The two modes are introduced below:

(i) Seeking Mode: The cats rest most of the time when they are awake. Their motions are slowly.
(ii) Tracing Mode: The cats running after the targets.

In summary, the pseudocode of PSO can be portrayed in Table 1.

2.5 Model Validation

A ten-fold stratified cross validation (TFSCV) was used to compare our method with other methods. Figure 6 shows the setting of TFSCV, where 8 folds are utilized for training, 1 fold for validation, and the final fold for test. The stratification guarantees that each fold contains the same distribution.

The confusion matrixes of 10 trials are summed. In this way, we can calculate the performance of TFSCV. Its no-error cost matrix $\mathcal{F}$ is presented as:

Fig. 5. Two modes in CSO algorithm

Table 1. Pseudocode of CSO

Step	Implementation
1	Initiate the cat swarm
2	Randomly sprinkle the cats into the solution space, and assign random values to them
3	Haphazardly pick the tracing mode cats, and the others are set to seeking mode. This ratio is called mixture ratio (MR)
4	Evaluate the fitness value of each cat, keep the best cat in memory
5	Move the cats according to their flags. If a cat is in seeking mode, apply the cat to seeking process, otherwise, apply it to tracing process. The description of the two processes are in reference [22]
6	Divide the swarm into seeking mode cat and tracing mode again
7	Check the termination condition. If satisfied, terminate the program and output the position of best cat; otherwise repeat from Step 4 to 6

Facial Expression Image Dataset

	I	II	III	IV	V	VI	VII	VIII	IX	X
Trial 1	I	II	III	IV	V	VI	VII	VIII	IX	X
Trial 2	I	II	III	IV	V	VI	VII	VIII	IX	X
Trial 3	I	II	III	IV	V	VI	VII	VIII	IX	X
Trial 9	I	II	III	IV	V	VI	VII	VIII	IX	X
Trial 10	I	II	III	IV	V	VI	VII	VIII	IX	X

Training
Validation
Test

Fig. 6. Setting of TFSCV

$$\mathcal{F} = \begin{bmatrix} 100 & 0 & 0 & 0 & 0 & 0 & 0 \\ 0 & 100 & 0 & 0 & 0 & 0 & 0 \\ 0 & 0 & 100 & 0 & 0 & 0 & 0 \\ 0 & 0 & 0 & 100 & 0 & 0 & 0 \\ 0 & 0 & 0 & 0 & 100 & 0 & 0 \\ 0 & 0 & 0 & 0 & 0 & 100 & 0 \\ 0 & 0 & 0 & 0 & 0 & 0 & 100 \end{bmatrix} \tag{11}$$

which means all seven classes are identified correctly. In real condition, the classifier without fail make mistakes; therefore, Suppose

$$\mathcal{F} = [F_{ij}], i \& j \in [1, 2, \ldots, 7] \tag{12}$$

we can measure the classifier by two types of indicators:

(i) the sensitivity (S_k) of each class k. We defined S_k as

$$S_k = \frac{F_{kk}}{\sum_j F_{kj}} \tag{13}$$

(ii) overall accuracy (O) is defined as

$$O = \frac{\sum_i F_{ii}}{\sum_j \sum_i F_{ij}} \tag{14}$$

We carried out the TFSCV 10 runs. Each run, the ten-fold segmentation is performed haphazardly. Therefore, we report the mean and standard deviation of S_k and O.

3 Experiments and Results

We developed the program in-house. The program was run on the platform of Dell desktop with i5-3470 3.20 GHz CPU and 4 GB Ram. Matlab 2015a is our development platform. The optimal parameters of our algorithm are obtained by trial-and-error methods.

3.1 Statistical Results of Our Method

Table 2 shows the sensitivities of each class by 10 times of TFSCV. Here the 1st class corresponds to Anger, 2nd class to Disgust, 3rd class to Fear, 4th class to Happy, 5th class to Neutral, 6th class to Sadness, and 7th class to Surprise. Table 3 shows the overall accuracy of our classifier.

Table 2. Sensitivities (S_k) of each class by CSO (Unit: %)

Run	S_1	S_2	S_3	S_4	S_5	S_6	S_7
1	89	91	87	87	89	87	92
2	88	92	89	86	90	86	91
3	91	91	86	88	89	86	91
4	92	93	88	88	90	89	92
5	90	93	87	88	91	88	87
6	91	92	86	88	93	87	93
7	90	91	88	90	91	89	92
8	90	94	88	90	93	90	91
9	89	90	88	88	91	86	92
10	87	90	86	88	89	90	91
Average	89.70 ± 1.49	91.70 ± 1.34	87.30 ± 1.06	88.10 ± 1.20	90.60 ± 1.51	87.80 ± 1.62	91.20 ± 1.62

Table 3. Overall accuracies (O) by CSO (Unit: %)

Run	O
1	88.86
2	88.86
3	88.86
4	90.29
5	89.14
6	90.00
7	90.14
8	90.86
9	89.14
10	88.71
Average	89.49 ± 0.76

3.2   Training Algorithm Comparison

In the second experiment, we compared the CSO with traditional genetic algorithm (GA) [15], particle swarm optimization (PSO) [14], time-varying acceleration coefficient PSO (TPSO) [16]. The input features are the reduced features from this study. All the other settings are the same. The comparison results are shown in Table 4.

Table 4. Training algorithm comparison (Unit: %)

Approach	O
GA [15]	73.44 ± 1.21
PSO [14]	82.37 ± 0.64
TPSO [16]	85.64 ± 0.45
CSO (Our)	89.49 ± 0.76

The sensitivities of GA [15], PSO [14], and TPSO [16] are shown in Tables 5, 6, and 7, respectively. Their overall accuracies are listed in Table 8.

Table 5. Sensitivities (S_k) of each class by GA [15] (Unit: %)

Run	S_1	S_2	S_3	S_4	S_5	S_6	S_7
1	75	77	71	77	70	73	76
2	65	76	68	76	72	68	74
3	75	76	73	74	70	73	73
4	76	76	72	77	72	71	76
5	76	75	72	78	72	74	73
6	75	77	71	77	71	68	73
7	73	75	73	75	71	73	74
8	75	73	69	77	71	70	76
9	72	75	67	76	70	69	74
10	76	78	71	79	73	74	75
Average	73.80 ± 3.36	75.80 ± 1.40	70.70 ± 2.06	76.60 ± 1.43	71.20 ± 1.03	71.60 ± 2.55	74.40 ± 1.26

Table 6. Sensitivities (S_k) of each class by PSO [14] (Unit: %)

Run	S_1	S_2	S_3	S_4	S_5	S_6	S_7
1	85	83	83	82	78	81	83
2	83	87	80	82	79	81	84
3	83	85	82	81	81	81	83
4	85	86	78	81	81	81	83
5	84	87	83	84	82	83	83
6	85	82	83	81	78	80	80
7	84	80	81	83	80	82	83
8	84	85	80	83	80	82	84
9	85	86	79	84	78	82	85
10	83	86	82	82	80	81	85
Average	84.10 ± 0.88	84.70 ± 2.31	81.10 ± 1.79	82.30 ± 1.16	79.70 ± 1.42	81.40 ± 0.84	83.30 ± 1.42

Table 7. Sensitivities (S_k) of each class by TPSO [16] (Unit: %)

Run	S_1	S_2	S_3	S_4	S_5	S_6	S_7
1	87	86	85	82	85	87	89
2	87	87	84	83	82	86	91
3	85	90	86	82	84	83	90
4	88	90	81	83	85	84	91
5	85	90	86	82	83	85	90
6	84	90	78	85	82	84	88
7	85	90	84	82	84	85	91
8	88	88	85	83	81	86	90
9	86	90	84	84	84	82	90
10	87	90	85	84	84	85	86
Average	86.20 ± 1.40	89.10 ± 1.52	83.80 ± 2.49	83.00 ± 1.05	83.10 ± 1.52	84.70 ± 1.49	89.60 ± 1.58

3.3 Comparison to State-of-the-Art Approaches

We compared our proposed "DWT + PCA + SNN + CSO" method with two state-of-the-art methods. They are: (i) PCA-SVM [8], and (ii) HOS-RT-SVM [9]. The comparison results wither terms of overall accuracy (O) are listed in Table 9.

Table 8. Overall accuracies (*O*) by GA, PSO, and TPSO (Unit: %)

Run	GA [15]	PSO [14]	TPSO [16]
1	74.14	82.14	85.86
2	71.29	82.29	85.71
3	73.43	82.29	85.71
4	74.71	82.14	86.00
5	74.29	83.71	85.86
6	73.14	81.29	84.43
7	73.43	81.86	85.86
8	73.00	82.57	85.86
9	71.86	82.71	85.71
10	75.14	82.71	85.43
Average	73.44 ± 1.21	82.37 ± 0.64	85.64 ± 0.45

Table 9 shows the HOS-RT-SVM [9] is the worst algorithm. It yields an overall accuracy of 83.43 ± 2.15%. PCA-SVM [8] ranks the second, and it yields an overall accuracy of 89.14 ± 2.91%. Nevertheless, our method yields better overall accuracy than those two methods, viz., our method obtains an overall accuracy of 89.49 ± 0.76%.

Table 9. Comparison to state-of-the-art methods

Approach	*O* (Unit: %)	Rank
PCA-SVM [8]	89.14 ± 2.91	2
HOS-RT-SVM [9]	83.43 ± 2.15	3
DWT + PCA + SNN + CSO (Our)	89.49 ± 0.76	1

A shortcoming of our method is that we did not estimate the light change. In the future, we shall try to remove the effect of light variation. Besides, we shall try to test other advanced classifiers, such as extreme learning machine, twin support vector machine, fuzzy support vector machine, and generalized eigenvalue proximal SVM [23]. Deep learning approaches, such as convolution neural network [24] and autoencoder [25].

4   Conclusion

Our team put forward a novel facial emotion recognition system, which combines discrete wavelet transform, principal component analysis, single-hidden-layer neural network, and cat swarm optimization. The ten-fold stratified cross validation results demonstrated that CSO was better than GA, PSO, and TPSO. Besides, our system was better than two recent methods: "PCA-SVM" and "HOS-RT-SVM".

Acknowledgment. This paper is supported by Natural Science Foundation of China (61602250), Natural Science Foundation of Jiangsu Province (BK20150983), Program of Natural Science Research of Jiangsu Higher Education Institutions (16KJB520025, 15KJB470010), Key Laboratory of Measurement and Control of Complex Systems of Engineering, Southeast University, Ministry of Education (MCCSE2017A02).

References

1. Lee, S.H., Ro, Y.M.: Partial matching of facial expression sequence using over-complete transition dictionary for emotion recognition. IEEE Trans. Affect. Comput. **7**, 389–408 (2016)
2. Argaud, S., et al.: Does facial amimia impact the recognition of facial emotions? An EMG study in Parkinson's disease. PLoS One **11**, Article ID: e0160329 (2016)
3. Hargreaves, A., et al.: Detecting facial emotion recognition deficits in schizophrenia using dynamic stimuli of varying intensities. Neurosci. Lett. **633**, 47–54 (2016)
4. Lalitha, S., et al.: Speech emotion recognition using DWT. In: International Conference on Computational Intelligence And Computing Research, pp. 20–23. IEEE (2015)
5. Garman, H.D., et al.: Wanting it too much: an inverse relation between social motivation and facial emotion recognition in autism spectrum disorder. Child Psychiatry Hum. Dev. **47**, 890–902 (2016)
6. Martino, D.J., et al.: Stability of facial emotion recognition performance in bipolar disorder. Psychiatry Res. **243**, 182–184 (2016)
7. Mishra, P., Hadfi, R., Ito, T.: Multiagent social influence detection based on facial emotion recognition. In: Bajo, J., et al. (eds.) PAAMS 2016. CCIS, vol. 616, pp. 148–160. Springer, Cham (2016). doi:10.1007/978-3-319-39387-2_13
8. Drume, D., Jalal, A.S.: A Multi-level classification approach for facial emotion recognition. In: International Conference on Computational Intelligence And Computing Research, pp. 288–292. IEEE (2012)
9. Ali, H., et al.: Facial emotion recognition based on higher-order spectra using support vector machines. J. Med. Imaging Health Inform. **5**, 1272–1277 (2015)
10. Boubenna, H., Lee, D.: Feature selection for facial emotion recognition based on genetic algorithm. In: 12th International Conference on Natural Computation, Fuzzy Systems and Knowledge Discovery (ICNC-FSKD), pp. 511–517. IEEE (2016)
11. Lu, H.M.: Facial emotion recognition based on biorthogonal wavelet entropy, fuzzy support vector machine, and stratified cross validation. IEEE Access **4**, 8375–8385 (2016)
12. Chen, Y., Chen, X.-Q.: Sensorineural hearing loss detection via discrete wavelet transform and principal component analysis combined with generalized eigenvalue proximal support vector machine and Tikhonov regularization. Multimedia Tools Appl. (2016). doi:10.1007/s11042-016-4087-6
13. Zhan, T.M., Chen, Y.: Multiple sclerosis detection based on biorthogonal wavelet transform, RBF kernel principal component analysis, and logistic regression. IEEE Access **4**, 7567–7576 (2016)
14. Ji, G.: A comprehensive survey on particle swarm optimization algorithm and its applications. Math. Probl. Eng. **2015**, Article ID: 931256 (2015)
15. Jotheeswaran, J., Koteeswaran, S.: Mining medical opinions using hybrid genetic algorithm-neural network. J. Med. Imaging Health Inform. **6**, 1925–1928 (2016)

16. Yang, J.F., Sun, P.: Magnetic resonance brain classification by a novel binary particle swarm optimization with mutation and time-varying acceleration coefficients. Biomed. Eng.-Biomed. Tech. **61**, 431–441 (2016)
17. Sun, P.: Preliminary research on abnormal brain detection by wavelet-energy and quantum-behaved PSO. Technol. Health Care **24**, S641–S649 (2016)
18. Yang, G.: Automated classification of brain images using wavelet-energy and biogeography-based optimization. Multimed Tools Appl. **75**, 15601–15617 (2016)
19. Wu, J.: Fruit classification by biogeography-based optimization and feedforward neural network. Expert Syst. **33**, 239–253 (2016)
20. Wei, L.: Fruit classification by wavelet-entropy and feedforward neural network trained by fitness-scaled chaotic ABC and biogeography-based optimization. Entropy **17**, 5711–5728 (2015)
21. Cattani, C., Rao, R.: Tea category identification using a novel fractional fourier entropy and Jaya algorithm. Entropy **18**, Article ID: 77 (2016)
22. Chu, S.-C., Tsai, P.-W., Pan, J.-S.: Cat swarm optimization. In: Yang, Q., Webb, G. (eds.) PRICAI 2006. LNCS, vol. 4099, pp. 854–858. Springer, Heidelberg (2006). doi:10.1007/978-3-540-36668-3_94
23. Yang, J.: Preclinical diagnosis of magnetic resonance (MR) brain images via discrete wavelet packet transform with Tsallis entropy and generalized eigenvalue proximal support vector machine (GEPSVM). Entropy **17**, 1795–1813 (2015)
24. Ghamisi, P., et al.: A self-improving convolution neural network for the classification of hyperspectral data. IEEE Geosci. Remote Sens. Lett. **13**, 1537–1541 (2016)
25. Hou, X.-X., Chen, H.: Sparse autoencoder based deep neural network for voxelwise detection of cerebral microbleed. In: 22nd International Conference on Parallel and Distributed Systems, pp. 1229–1232. IEEE (2016)

Using LFDA to Learn Subset-Haar-Like Intermediate Feature Weights for Pedestrian Detection

Kai Zang[1,2], Jifeng Shen[3], and Wankou Yang[1,2(✉)]

[1] School of Automation, Southeast University, Nanjing 210096, China
wkyang@seu.edu.cn
[2] Key Lab of Measurement and Control of Complex Systems of Engineering,
Ministry of Education, Nanjing 210096, China
[3] School of Electrical and Information Engineering, Jiangsu University,
Zhenjiang 212013, China

Abstract. In this paper, we propose a pedestrian detection method by using LFDA to learn the weights of Subset-Haar-like intermediate features. The Subset-Haar-like intermediate features are generated by using Subset-Haar-like template filtering low-level Aggregated Channel Features (ACF). The target features are constituted by the weighted sum of these Subset-Haar-like intermediate features, where Local Fisher Discriminant Analysis (LFDA) algorithm is used to learn weighting coefficients vector. Our detector is modelled by the target features in combination with a cascade adaboost classifier using depth-2 decision trees. In order to evaluate our proposed method, we have conducted several experiments on two pedestrian datasets: INRIA and Caltech dataset. The experimental results show that our proposed method can achieve lower miss rate and higher precision, compared with other state-of-arts pedestrian detection methods.

Keywords: Pedestrian detection · Subset-Haar-like intermediate features · ACF · Local Fisher Discriminant Analysis · Adaboost

1 Introduction

Pedestrian detection has attracted much attention in recent years because of the development in computer vision field. It is widely used in visual surveillance, behavior analysis, and automated personal assistance field. Pedestrian detection is a challenging task in practical applications, which is mainly affected by illumination change, occlusion, a wide range of poses and human deformation.

Based on traditional pattern recognition theory, pedestrian detection task can be regarded as a binary classification problem. That is to say judging whether the input image is pedestrian or not. Therefore, the performance of pedestrian detection is determined by feature extraction and classifier design parts.

When it comes to the selection of features descriptors, HOG (Histograms of Oriented Gradient) descriptors [1] and Haar-like feature [2] are typical features in pedestrian detection. HOG descriptors have unique advantages for pedestrian detection,

© Springer International Publishing AG 2017
Y. Sun et al. (Eds.): IScIDE 2017, LNCS 10559, pp. 215–230, 2017.
DOI: 10.1007/978-3-319-67777-4_19

compared with other existing features including wavelets [3, 4]. However, HOG descriptors for pedestrian detection show poor performance in the case of severe occlusion and complex background. Haar-like feature stands out for its calculation speed, which is due to the use of integral images [5]. However, Haar-like feature exhibits poor performance in classification ability because of its simple structure. Piotr proposed ACF (Aggregate Channel Features) [6] which is easy to compute, but still face the bottleneck of overload of computation, such as multiple scales of image features and image resizing; However, the multi-scale feature approximation [7] can handle this problem. In this paper, we propose the improved ACF (combining ACF with Haar-like feature), which learns the optimized weights of Haar-like intermediate features by LFDA (Local Fisher Discriminant Analysis) [8] algorithm. The improved feature brings out lower miss rate and higher precision.

This paper mainly focuses on the feature representation rather than classifier design. The effective feature extraction of pedestrian is the prerequisite for the design of high-precision pedestrian detection system. The classifier learning process adopts the framework in reference [6], which utilizes the soft-cascade [9] adaboost [10] classifiers and the weak classifier is depth-2 decision tree. Because of the convergence of ACF and Haar features, and the efficient combination of original features through LFDA, Our proposed weighted Subset-Haar-like intermediate features learnt by LFDA, along with the soft-cascade adaboost classifiers framework, exploit competitive performance against state-of-the-art algorithms in both miss rate and precision.

The remaining parts of this paper are organized as follows. Section 2 discusses the related work. Section 3 introduces our proposed Subset-Haar-like intermediate feature and intermediate feature weights learnt by LFDA in detail. Section 4 gives some experimental results and compares our method with previous algorithms for pedestrian detection. Section 5 concludes the paper.

2 Aggregated Channel Features (ACF)

Piotr first proposed the ICF [11], which models the feature C of image I as a channel generation function Ω, so the feature of image I can be represent as $C_i = \Omega_i(I)$, where $\Omega = \{\Omega_i\}$, $i = 1, 2, \ldots, n$. C_i is the ith channel feature for image I, Ω_i is the ith channel generation function. The channel generation function can be linear (such as gray-level image of I) or nonlinear (such as gradient image). Each channel represents a different feature space which is derived from original image. Figure 1 shows the different channels of the original image. ICF comprises of ten channels including three color channels (LUV), gradient magnitude channel and gradient histograms channels (six different directions).

After the ICF, Piotr proposed a more efficient feature ACF, whose detection framework is conceptually straightforward (Fig. 2). ACF uses the same channels as ICF (ten channels): normalized gradient magnitude, histogram of oriented gradients (six channels), and LUV color channels. Original image I (Fig. 2(a)) is smoothed with a [1 2 1]/4 filter and then compute ten channels features, which is shown as (Fig. 2(b)). Then channels are divided into 4×4 blocks and pixels in each block are summed (Fig. 2(c)). Afterwards, the channels are smoothed, again with a [1 2 1]/4 filter. Finally,

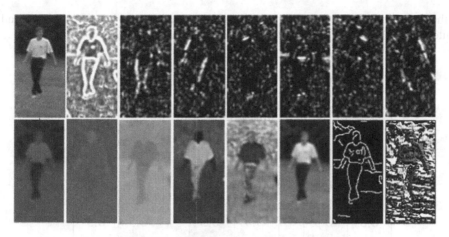

Fig. 1. Different channels of the input image (First row corresponds to the original, magnitude, gradient angle (*1–6*); Second row corresponds to the LUV, HSV, canny image and LBP)

all features in ten channels are integrated into a column vector as (Fig. 2(d)). Adaboost classifiers of the soft-cascade structure are used to train and combine decision trees over these features (pixels) to distinguish object from the background and a multi-scale sliding-window approach is employed.

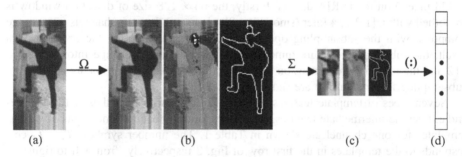

Fig. 2. Generation flow chart of ACF ((a), (b), (c), (d) correspond to the original image, computing channels, aggregating and vectorization operation respectively)

3 Subset-Haar-Like Intermediate Features Based on ACF

3.1 Subset-Haar-Like Intermediate Feature

Since the pedestrian not only has strong edge characteristics but also has obvious structural features, in this work, seven types of Subset-Haar-like template features are chosen, which are used to filter low-level ACF and extract the intermediate features. The chosen seven types of Subset-Haar-like template features are shown in Fig. 3. Every cell of templates is used to matched with each feature (pixel) in ACF, the number in the cell represents the corresponding weight of each cell, the sum of weights in all white cells is 1 and the yellow cells is -1 in each template.

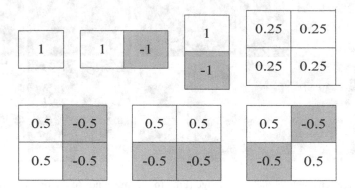

Fig. 3. Seven types of Subset-Haar-like template features

All pedestrian and non-pedestrian samples are resized to the size of detection windows and then be used to train our detector. Larger window size includes more pixels in feature pool and thus may improve the pedestrian detection performance. On the other hand, too large window will miss some small pedestrians and diminish the detection efficiency. In this work, we adopt 64×128 size (width is 64 pixels and height is 128 pixels) on the INRIA dataset. Firstly, the 64×128 size of detection window is smoothed with a [1 2 1]/4 filter (smoothing filtering) and then ten channels features are computed with the subsampling operation followed (subsampling factor is 4). As a result, the subsampling feature dimensions of every channel change into $16 \times 32 = 512$ and the total feature dimension of ten channels is 5120. Experiments on different subsampling factors (shrink) are shown in Fig. 9.

Seven types of template features in Fig. 3 are chosen to filter these 5120 features and extract the intermediate features. The corresponding feature dimensions of various templates for one channel are shown in Table 1. The number symbols $T_1 - T_4$ corresponds to the templates in the first row of Fig. 3 respectively (from left to right), as

Table 1. Template features (filter features) description

Template	Feature distribution	Feature dimension
T_1	16×32	512
T_2	15×32	480
T_3	16×31	496
T_4	15×31	465
T_5	15×31	465
T_6	15×31	465
T_7	15×31	465
Total		3348

the same way, symbols $T_5 - T_7$ are related to the templates in the second row. The sliding stride of every template in each channel is one pixel both in the horizontal and vertical directions.

As can be seen from Table 1, total Subset-Haar-like intermediate feature dimensions of ten channels are $3348 \times 10 = 33480$.

The Subset-Haar-like intermediate features generation of template T_4 in channel one is shown in Fig. 4. Though the dimensions of intermediate features have risen compared with ACF, the size and all possible positions of these templates in the feature map can be calculated in advance. Once ACF is calculated, the calculation of each intermediate feature only needs to using these templates scan the ACF feature map. The calculation of intermediate feature is quick and without significantly affects the speed of feature extraction.

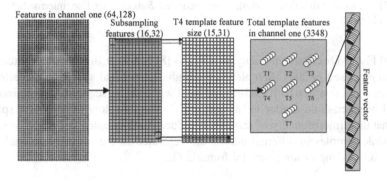

Fig. 4. The feature extraction work-flow of Subset-Haar-like intermediate feature

3.2 Weighted Subset-Haar-Like Intermediate Feature Learnt by LFDA

It can be seen from Fig. 8, the detector based on Subset-Haar-like intermediate features is able to improve the performance of pedestrian detection; However, the structure of Subset-Haar-like intermediate feature is relatively simple and some candidate features only reflect the local information of the pedestrian. In view of this, further research work is carried out. The single candidate feature selected randomly from the Subset-Haar-like intermediate features is combined linearly by the weighting coefficients vector, which is learnt using Local Fisher Discriminant Analysis (LFDA).

After Subset-Haar-like intermediate features extracted, n (experiments on different values of n are shown in Fig. 12, which reveals that $n = 3$ has the lowest miss rate, so that we set $n = 3$ in this paper) random candidate features are selected from 3348 features in one channel. These n candidate features are combined with a weighting coefficients vector W, which is learnt by LFDA and n random candidate features change into one new weighted feature (the target feature). In this way, 3348 target features are generated in one channel in total. Similarly, 33480 features can be obtained in all ten channels (the same method is adopted in all channels). The feature extraction work-flow of weighted Subset-Haar-like intermediate feature learnt by LFDA is shown in Fig. 5.

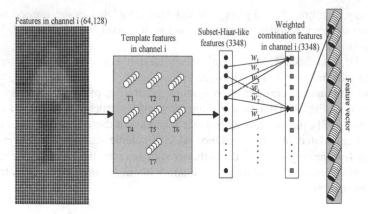

Fig. 5. The feature extraction work-flow of weighted Subset-Haar-like intermediate feature learnt by LFDA

Local Fisher Discriminant Analysis (LFDA) [8] is a liner dimensionality reduction algorithm proposed in 2007. Samples in the high-dimensional space are projected into low-dimensional space, which leads to the local within-class scatter is 'minimized' and the local between-class scatter is 'maximized' in the new low-dimensional space. It means that the dimensionality of samples is reduced and samples in the same class are closer, while samples in different classes are separated in the low-dimensional space. The embedded samples are given by formula (1).

$$z_i = T^T x_i. \tag{1}$$

where $x_i \in R^d (i = 1, 2, \ldots, m)$ represents m d-dimensional samples, T is a $d \times r$ transformation (projection) matrix and $z_i(i = 1, 2, \ldots, m) \in R^r (1 \leq r \leq d)$ is the embedded samples after projection, where r is the reduced dimension (the dimension of the embedding space). Effectively we consider d to be large and r to be small, but not limited to such cases.

In order to seek the transformation matrix T, fisher discriminant criterion is introduced as follow.

$$T_{LFDA} = \arg\max_T \frac{T^T S_b T}{T^T S_w T}. \tag{2}$$

where S_b and S_W are the local between-class scatter matrix and the local within-class scatter matrix respectively, which are defined by formula (3) and (4).

$$S_b = \frac{1}{2} \sum_{i,j=1}^m W_{i,j}^b (x_i - x_j)(x_i - x_j)^T. \tag{3}$$

$$S_w = \frac{1}{2}\sum_{i,j=1}^{m} W_{i,j}^{w}(x_i - x_j)(x_i - x_j)^{T}. \tag{4}$$

where

$$W_{i,j}^{b} = \begin{cases} A_{i,j}(1/m - 1/m_l), & y_i = y_j = l \\ 1/m, & y_i \neq y_j \end{cases}. \tag{5}$$

$$W_{i,j}^{w} = \begin{cases} A_{i,j}/m_l, & y_i = y_j = l \\ 0, & y_i \neq y_j \end{cases}. \tag{6}$$

where $y_i \in \{1, 2, \ldots, c\}$ is the associated class label, m is the number of samples and c is the number of classes. m_l is the number of samples in class l:

$$\sum_{l=1}^{c} m_l = m. \tag{7}$$

where A is an affinity matrix, that is, the m-dimensional matrix with the (i,j)-th element $A_{i,j}$ being the affinity between x_i and x_j. We assume that $A_{i,j} \in [0, 1]$; $A_{i,j}$ is large if x_i and x_j are 'close', $A_{i,j}$ is small if x_i and x_j are 'far part'. The affinity matrix A is defined as follow.

$$A_{i,j} = \exp(-\frac{\|x_i - x_j\|^2}{\sigma_i \sigma_j}). \tag{8}$$

where $\sigma_i(\sigma_j)$ represents the local scaling of the data samples around $x_i(x_j)$, which is determined by formula (9) and (10) respectively.

$$\sigma_i = \left\| x_i - x_i^{(K)} \right\|. \tag{9}$$

$$\sigma_j = \left\| x_j - x_j^{(K)} \right\|. \tag{10}$$

where $x_i^{(K)}(x_j^{(K)})$ is the K-th nearest neighbor of $x_i(x_j)$. The parameter K is a tuning parameter. In this work, K is 801, which works well on the whole.

As can be seen from Fig. 5, three Subset-Haar-like intermediate features are converted to one feature, which is equivalent to change the 3-dimensional feature to a 1-dimensional feature. Therefore, the key point of this work is to seek the weighting coefficients matrix W, which reduces the dimension of the original feature and makes the combined feature more distinguished. The whole process is consistent with the ideal of LFDA. So, in this paper, LFDA is applied to learn the weighting coefficients matrix W.

Based on formula (2), T_{LFDA} is the optimum projection matrix when the function $\frac{T^T S_b T}{T^T S_w T}$ takes the maximum value. When it comes to the solution of the maximum value problem, lagrange multiplier method is introduced. In this work, T is a 1-dimensional column vector, so both $T^T S_b T$ and $T^T S_w T$ are scalars. Supposing $T^T S_w T = 1$, and lagrange constructor is given as follow.

$$L(T) = T^T S_b \text{T} - \lambda(T^T S_w T - 1)$$
$$\Rightarrow \frac{dL}{dT} = 2S_b T - 2\lambda S_w T = 0$$
$$\Rightarrow S_b T = \lambda S_w T \tag{11}$$
$$\Rightarrow S_w^{-1} S_b T = \lambda T.$$

Based on formula (11), we can conclude that:

$$\frac{T^T S_b T}{T^T S_w T} = \frac{|T^T S_b T|}{|T^T S_w T|} = \frac{|T^T \lambda S_w T|}{|T^T S_w T|} = \frac{\lambda |T^T S_w T|}{|T^T S_w T|} = \lambda. \tag{12}$$

$$T_{LFDA} = \arg\max_T \frac{T^T S_b T}{T^T S_w T} \Rightarrow \max\lambda. \tag{13}$$

As can be seen from formula (13), $\arg\max_T \frac{T^T S_b T}{T^T S_w T}$ is equivalent to the maximum λ, and the optimum projection vector T_{LFDA} is equivalent to the eigenvector, which is the $\lambda_{\max}$ of the matrix $S_w^{-1} S_b$ corresponded to. Therefore, T_{LFDA} can be obtained when the eigenvector that $\lambda_{\max}$ of the matrix $S_w^{-1} S_b$ corresponded to is calculated. The weighting coefficients matrix T_{LFDA} is applicable to all ten channels.

The workflow of using LFDA to learn the weight coefficients matrix T is shown in Fig. 6 and the procedure is shown as follow (on the INRIA dataset).

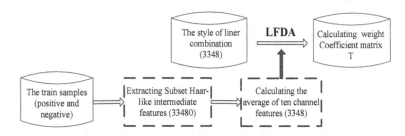

Fig. 6. The workflow of using LFDA to learn the weighting coefficients matrix T

(1) Preparing the positive and negative training samples.
(2) Extracting Subset-Haar-like intermediate features (3348) of all training samples.
(3) Calculating the average of ten channel features and obtaining 3348 Subset-Haar-like intermediate features.
(4) Selecting three random intermediate features from 3348 and combining three into one target feature. 3348 types of combinations are chosen in total.

(5) Applying LFDA to learn the weighting coefficients vector corresponding to 3348 types of combinations.

(6) Obtaining the final weighting coefficients vector T_{LFDA}.

4 Experiments

4.1 Experiment Setting

In order to validate the effectiveness of our method, we carry out extensive experiments on two public datasets for pedestrian detection, which briefly described in Table 2.

Table 2. Datasets for pedestrian detection

Dataset name	Description
INRIA	http://pascal.inrialpes.fr/data/human/ The training data: 2416 pedestrian annotations in 614 and 1218 non-pedestrian images The testing data: 1132 pedestrian annotations in 288 and 453 non- pedestrian images
Caltech-USA	http://www.vision.caltech.edu/Image_Datasets/CaltechPedestrians/ The training data: 6325 pedestrian annotations in 4250 images (set00–set05) The testing data: 5051 pedestrian annotations in 4024 images (set06–set10)

The detailed experiment setting for pedestrian detection is described as follows. The size of the pedestrian window is set to 128×64, and each positive sample is cropped from the annotated image. Each annotation of pedestrian is jittered to mitigate the misalignment problem. The total number of positive samples used to train is 2474 for the INRIA dataset and 3262 for the Caltech dataset. The acquisition method of these training samples is described in reference [6]. The pooling template size is of 4×4 pixels, which shrinks the original channel maps (size $128 \times 64 \times 10$) into pooled channel maps (size $32 \times 16 \times 10$). We make use of adaboost algorithm to per- form feature selection, with depth-2 decision trees as the weak classifiers. The adaboost training is conducted by four rounds (32, 128, 512, 2048), in the first round, 5000 negative samples are selected from original 1218 non-pedestrian images randomly, which is the same as the reference [6]. After the first training round, 1218 original non-pedestrian images are tested by the model obtained in the first training round, and false detections of the pedestrian are marked as 'hard negative sample', 5000 'hard negative sample' are chosen to add to the negative training samples in the next training round, which means the number of negative training samples in the next training round is 10000. Similarly, the acquisition method of negative training samples is applied to the third and forth training round (10000 negative samples are adopted and positive samples remain unchanged). We use public available Piotr's toolbox [12] to calculate the channel features and use the evaluation code [12] to evaluate the detector.

Piotr proposed an evaluation standard for pedestrian detection, which is FPPI (False Positives Per Image) [13]. In recent years, the basic evaluation criteria are all based on the miss rate-FPPI curve. In this work, two types of evaluation standards are introduced:

(1) Miss rate-FPPI Curve, which is one of the most important evaluation standard. The x coordinate FPPI is the false detection rate of each image, $FPPI = \frac{FP}{M}$; The y coordinate miss rate is the miss detection rate of positives, $missrate = \frac{FN}{P}$. where FP is False Positive, M is the number of testing images, FN is False Negative, P is the number of positive.

(2) Precision-Recall Curve. The x coordinate is Recall rate, $Recall = \frac{TP}{P}$; The y coordinate $Precision = \frac{TP}{TP+FP}$. where TP is True Positive.

4.2   Weighted Subset-Haar-Like Intermediate Feature Learnt by LFDA

In order to compare the discriminative ability between ACF and our proposed Subset-Haar-like intermediate feature, we have shown the miss rate-FPPI and precision-recall curve which are shown in Fig. 7.

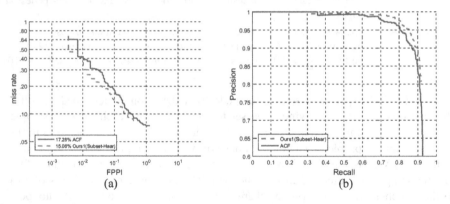

Fig. 7. The performance comparison between ACF and Subset-Haar like on the INRIA dataset

As can be seen from Fig. 7(a), the miss rate of the Subset-Haar-like intermediate feature (Ours1) equals 15.08% when FPPI is 0.1, which is lower than ACF (17.28%); In addition, from the trend of curve, when FPPI equals other different values, miss rate of Subset-Haar-like intermediate feature is always less than ACF, which illustrates that the performance of Subset-Haar-like intermediate feature is superior to ACF. The reason is that Subset-Haar-like intermediate feature gathers more first-order statistical information, which is more discriminative than the traditional zero-order feature ACF. From Fig. 7(b), the Subset-Haar-like intermediate feature has a higher precision and recall rate than ACF, which also reveals that the Subset-Haar-like intermediate feature has a better performance than ACF.

In order to validate the effectiveness of our Subset-Haar intermediate feature, we also compare our methods with other state-of-art algorithms with the miss rate-FPPI and precision-recall curve, the result is shown in Fig. 8. Other six state-of-art algorithms are VJ [14], HOG [1], LatSvm-V1 [15], HogLBP [16], ICF [11] and LatSvm-V2 [17].

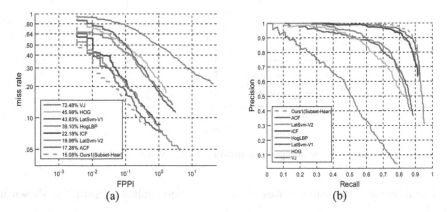

(a)   (b)

Fig. 8. The performance comparison between other state-of-art algorithms and Subset-Haar-like on the INRIA dataset

4.3 Discussion on the Value of Factor for Subsampling

As can be seen from Fig. 4, there is a down-sampling stage in the process of features extraction for pedestrian detection. In this work, the factor of down-sampling (shrink) is selected as four. Therefore, the feature dimension is reduced to 1/16 of the original pedestrian detection window (size 128×64). The factor for subsampling can be regarded as the perceptive scale for that it controls the scale at which the aggregation is done. Changing the factor from large to small leads to the feature representation shifting from coarse to fine, meanwhile, the detection time changes from less to more. As a result, the selection of down-sampling factor is discussed in experiment on the INRIA dataset, the result is shown in Table 3 and Fig. 9. The experiment is done on DELL computer (CPU i5-4590, 3.3 GHZ, 8G, win 64) with matlab 2015a.

It is easy to see from Table 3, as the down-sampling factor (shrink) increases, the detection speed becomes faster.

Table 3. The influence of shrink on detection speed (the size of testing image is 1060×605)

Shrink	Time (s)
2	0.12
4	0.08
6	0.02
8	0.014

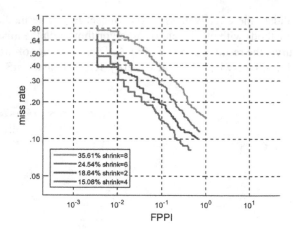

Fig. 9. The comparison among different values of shrink

The miss rate-FPPI curve corresponding to various values of shrink is shown in Fig. 9.

As can be seen from Fig. 9, the miss rate of shrink = 2 is less than shrink = 4 when FPPI is lower than 0.01, however, the miss rate is too high at this time, which is an unreasonable range of values in practice. When FPPI is higher than 0.01, shrink = 4 has the lowest miss rate, and the average miss rate of shrink = 4 is 15.08% which is lowest.

Due to the insignificant difference in detection time among various values of shrink, as a result, shrink set to 4 may have the optimal performance on the whole.

4.4   Comparing Subset-Haar Learnt by LFDA with ACF and Other State-of-Art Algorithms

In order to validate the effectiveness of Subset-Haar-like intermediate feature learnt by LFDA, we compare our method with other state-of-art algorithms with the miss rate-FPPI and precision-recall curve, the result is shown in Fig. 10 and the comparison results (miss rate) on different datasets (the INRIA and Caltech dataset) are shown in Table 4.

As can be seen from Fig. 10(a), our proposed method Subset-Haar-like intermediate feature learnt by LFDA achieves state-of-the-art result at a miss rate approximately 4.16% lower than ACF and 1.96% lower than Subset-Haar-like intermediate feature. We think the reason behind this phenomenon is that single Subset-Haar-like intermediate feature only reflect the local information of pedestrians, which can be compensated by combining these single features with LFDA. From Fig. 10(b), Subset-Haar-like intermediate feature learnt by LFDA outperforms other algorithms mentioned above both in the precision and recall rate, which substantially suggests that our proposed Subset-Haar-like intermediate feature learnt by LFDA has a superior performance on the INRIA dataset.

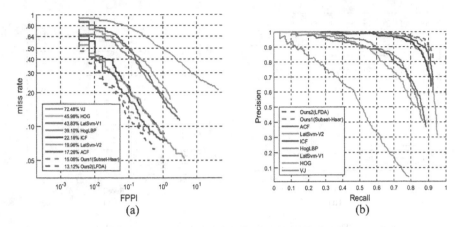

Fig. 10. The performance comparison between other state-of-art algorithms and weighted Subset-Haar-like features learnt by LFDA on the INRIA dataset

Table 4. The detection results comparison on different datasets (miss rate/FPPI (0.1))

Algorithm	Dataset	
	INRIA	Caltech
VJ	72.48%	94.73%
HOG	45.98%	68.46%
LatSvm-V1	43.83%	79.78%
HogLBP	39.10%	67.77%
ICF	22.18%	56.34%
LatSvm-V2	19.96%	63.26%
ACF	17.28%	51.36%
Ours1 (Subset-Haar)	15.08%	45.55%
Ours2(LFDA)	13.12%	42.39%

From Table 4, we can see that Ours1 (Subset-Haar) achieves a lower miss rate compared with other seven types of state-of-art algorithms, and it is approximately 2.2% and 5.81% lower than ACF on the INRIA and Caltech dataset respectively; In particular, our proposed method Ours2 (LFDA) significantly outperforms ACF by reporting a 4.16% and 8.97% lower average miss rate on the INRIA and Caltech dataset respectively. As a result, the miss rate in both the INRIA and Caltech dataset can effectively reflect that our proposed method is superior.

In the above experiment, the weighting coefficients matrix T_{LFDA} of the Subset-Haar-like intermediate feature is learnt by LFDA. In order to validate the effectiveness of adopting LFDA, we also compare the matrix T generated randomly with T_{LFDA} learnt by LFDA, the result is shown in Fig. 11.

As can be seen from Fig. 11(a), our method significantly outperforms the method of generating matrix T randomly by reporting a 2.46% lower average miss rate. In

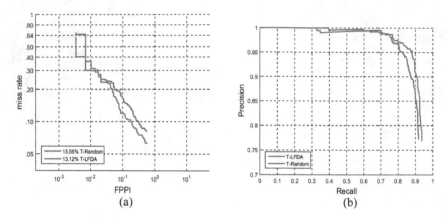

Fig. 11. The performance comparison between T_{Random} and T_{LFDA}

addition to the miss rate, our method is also advantageous both in precision and recall rate, which is shown in Fig. 11(b).

The parameter n is the number of candidate features selected from 3348 Subset-Haar-like intermediate features randomly, and the value of n is set to three in this paper. The value of n has a direct impact on the detection speed. Larger n means more complex feature calculations and thus may lead to slower detection speed. In order to derive the optimal parameter n, we have conducted a set of experiments on the INRIA dataset, which is shown in Table 5 and Fig. 12.

Table 5. The influence of n on detection speed (the size of testing image is 1060×605)

n	Time (s)
2	0.16
3	0.24
4	0.30
5	0.34

As can be seen from Table 5, the greater n, the slower the detection speed; however, the difference in the detection time is insignificant. Figure 12 provides the experimental results of different n on the miss rate on the INRIA dataset. It demonstrates that $n = 3$ considerably outperforms other values (4.43% lower than $n = 2$, 3.4% lower than $n = 4$ and 3.83% lower than $n = 5$). Overall, $n = 3$ is the optimal parameter configuration when both the detection time and miss rate are taken into consideration on the INRIA dataset.

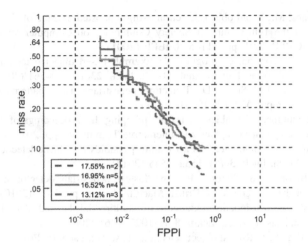

Fig. 12. The comparison among different values of n

5 Conclusions

In this paper, we propose a Subset-Haar-like intermediate feature learnt by LFDA method, which generated using Subset-Haar-like template filtering low-level Aggregated Channel Features (ACF) and constituted by the weighted sum of these Subset-Haar-like intermediate features, LFDA algorithm used to learn weighting coefficients vector. The application of weighting coefficients vector learnt by LFDA significantly contributes to the performance gains.

Experimental results for pedestrian detection based on the INRIA and Caltech datasets show that our method can achieve state-of-the-art results both in the miss rate and in precision, compared with other handcrafted features. The reason behind this is that LFDA combines single candidate features of the original Subset-Haar-like features by weighting, which makes the target features after combination more distinguishable.

In this paper, seven types of Subset-Haar-like template features are chosen, we can see that it only use simple weights $(1, -1, 0.5, -0.5, 0.25)$ which may not be discriminative enough. Therefore, how to learn an optimized weight for these Subset-Haar-like template features and select more complex templates is our future work.

Acknowledgments. This work is supported by NSF of China (61473086), partly supported by the Fundamental Research Funds for the Central Universities (2242017K40124).

References

1. Dalal, N., Triggs, B.: Histograms of oriented gradients for human detection. In: Proceedings of the 23rd Computer Society Conference on Computer Vision and Pattern Recognition, CVPR 2005, pp. 886–893. IEEE (2005)

2. Viola, P., Jones, M.: Rapid object detection using a boosted cascade of simple features. In: Proceedings of the 19th Computer Society Conference on Computer Vision and Pattern Recognition, CVPR 2001, pp. 511–518. IEEE (2001)
3. Mohan, A., Papageorgiou, C., Poggio, T.: Example-based object detection in images by components. IEEE Trans. Pattern Anal. March. Intell. **23**(4), 349–361 (2001)
4. Viola, P., Jones, M.J., Snow, D.: Detecting pedestrians using patterns of motion and appearance. J. Comput. Vis. **63**, 153–161 (2005)
5. Franklin, C.: Summed-area tables for texture mapping. In: Proceedings of the 11th Annual Conference on Computer Graphics and Interactive Techniques, pp. 207–212 (1984)
6. Dollár, P., Appel, R., Belongie, S.: Fast feature pyramids for object detection. IEEE Trans. Pattern Anal. Mach. Intell. **36**(8), 1532–1545 (2014)
7. Dollár, P., Belongie, S., Perona, P.: The fastest pedestrian detector in the west. In: Proceedings of the Conference on British Machine Vision, pp. 7–17 (2010)
8. Sugiyama, M.: Dimensionality reduction of multimodal labeled data by local fisher discriminant analysis. J. Mach. Learn. **8**(5), 1027–1061 (2007)
9. Bourdev, L., Brandt, J.: Robust object detection via soft cascade. In: Proceedings of the 23rd Computer Society Conference on Computer Vision and Pattern Recognition, CVPR 2005, pp. 236–243. IEEE (2005)
10. Freund, Y., Schapire, R.E.: A desicion-theoretic generalization of on-line learning and an application to boosting. In: Vitányi, P. (ed.) EuroCOLT 1995. LNCS, vol. 904, pp. 23–37. Springer, Heidelberg (1995). doi:10.1007/3-540-59119-2_166
11. Dollár, P., Perona, P., Belongie, S.: Integral Channel Features. In: Proceedings of the Conference on British Machine Vision, pp. 38–48 (2009)
12. Yang, B., Yan, J., Lei, Z., Li, S.Z.: Aggregate channel features for multi-view face detection. In: Proceedings of the International Joint Conference on Biometrics, pp. 1–8. IEEE (2014)
13. Dollár, P., Wojek, C., Schiele, B., Perona, P.: Pedestrian detection. In: Proceedings of the 27th Computer Society Conference on Computer Vision and Pattern Recognition, CVPR 2009, pp. 304–311. IEEE (2009)
14. Viola, P., Jones, M.J.: Robust real-time face detection. J. Com. Vis. **57**, 137–154 (2004)
15. Felzenszwalb, P., McAllester, D., Ramanan, D.: A discriminatively trained, multiscale, deformable part model. In: Proceedings of the 26th Computer Society Conference on Computer Vision and Pattern Recognition, CVPR 2008, pp. 1–8. IEEE (2008)
16. Wang, X., Han, T.X., Yan, S.: An HOG-LBP human detector with partial occlusion handling. In: Proceedings of the 12th International Conference on Computer Vision, pp. 32–39. IEEE (2009)
17. Felzenszwalb, P., Girshick, R., McAllester, D.: Object detection with discriminatively trained part-based models. IEEE Trans. Pattern Anal. March. Intell. **32**(9), 1627–1645 (2010)

Using Original Face Image and Its Virtual Image for Face Recognition

Jianguo Wang[✉] and Shucai Fu

Tangshan College, Tangshan 063000, People's Republic of China
wjgfwjg@163.com

Abstract. In this paper, we propose an approach to produce the half image of the face and integrate the original face image and its half image for representation-based face recognition. Because we take the geometrical nature of faces into full account, the proposed approach not only is quite simple but also can well represent possible changes in illuminations, facial expressions and poses of faces. Extensive experiments conducted on face databases (Yale, FERET face databases) show that the proposed one can greatly improve the accuracy of the representation-based classification approaches, particularly in small sample sizes.

Keywords: Face recognition · Pattern recognition · Half image · Sparse representation

1 Introduction

Face recognition (FR) [1–4] has attracted many researchers in the field such as pattern recognition, computer vision, machine learning, and etc. In the past decades, various face recognition approaches have been proposed, such as principle component analysis (PCA) [5], independent component analysis (ICA) [6], linear discriminant analysis (LDA) [7], and its kernel-based approaches [8–10]. However, up to now, a face recognition approach always suffers from various changes owing to its varying appearance such as illuminations, facial expressions and poses. As we know, in real world face recognition applications, there are always only a small number of training samples of a face; a limited number of training samples cannot comprehensively convey many changes of the face [11–14].

In order to improve the accuracy of face recognition, previous studies [11, 15–18] have proposed some approaches to generate new samples from true face images, which can enlarge the size of the set of the training samples.

Xu et al. exploited the symmetry of the face to generate virtual images, and proposes symmetrical faces. Their extensive experiments convincingly proved that symmetrical faces are good alternative representations of faces and the combination of symmetrical faces and original face images are very useful for improving the performance of face recognition [14]. From the literatures [18, 21], we also see that the incorporation of RBCM with original face images and mirror face images can bring surprising accuracy improvement.

© Springer International Publishing AG 2017
Y. Sun et al. (Eds.): IScIDE 2017, LNCS 10559, pp. 231–238, 2017.
DOI: 10.1007/978-3-319-67777-4_20

To improve the accuracy of face recognition, a novel approach takes advantages of the score level fusion was proposed, which has proven to be very competent and is usually better than the decision level and feature level fusion.

This paper has the following main contributions. First, it takes into account the fact that the face itself has a symmetrical structure and the face image is usually not a symmetrical image. Based on this fact, the paper very reasonably uses the mirror image of the face image to reflect some possible poses and illuminations change of the original face image. Second, it proposes the approach to generate left half face images of the original face images and its mirror images and for the first time integrates the left half face images with CRC for face recognition. In addition, though this study is performed after we are inspired by the idea of [14, 18, 21], our scheme seems to be very different from the previous ones. In [14, 18, 21], the virtual face image has the same size as the original face image. However, in this our work, the size of the virtual face image is just one second of the size of the original face image. As a result, the virtual face image generated using our approach needs much less memory.

The rest of the paper is organized as follows. Section 2 presents related works. Section 3 describes our proposed approach. Section 4 shows the experimental results and Sect. 5 offers the conclusion.

2   Related Works

We assume that there are c classes and each class has n training samples in the form of column vectors. Let $x_1, \ldots, x_N$ represent all the N training samples in the form of column vectors ($N = nc$). Column vector $x_{(i-1)n+k}(k = 1, \ldots, n)$ stands for the k-th training sample of the i-th subject, and $i = 1, 2, \ldots, c$. Let column vector z stands for the test sample.

2.1   Linear Regression Classification (LRC) [19]

LRC formulates the face identification task as a problem of linear regression, which uses a fundamental concept that patterns from a single-object class lie on a linear subspace. LRC establishes an equation for each class. If z belongs to the i-th class, it should be represented as a linear combination of the training samples from the same class (lying in the same subspace), i.e., the equation of the i-th class is:

$$z = X_i A_i, \ i = 1, \ldots, c \tag{1}$$

where $A_i = [a_1^i \ldots a_n^i]$, $X_i = [x_{(i-1)*n+1} \ldots x_{i*n}]$.

The solution of Eq. (1) can be obtained using least-squares estimation as follows:

$$\tilde{A}_i = (X_i^T X_i)^{-1} X_i^T z \tag{2}$$

The estimated vector of parameters $\tilde{A}_i$, along with the predictors X_i are used to predict the response vector for each class i.

$$\tilde{z}_i = X_i \tilde{A}_i, \quad i = 1, \ldots, c \tag{3}$$

The deviation between the predicted response vector $\tilde{z}_i$ and the test sample z is defined as $d_i = \|z - \tilde{z}_i\|$. If $k = arg\,min_i d_i$, then the test sample is assigned to the k-th class.

2.2 Collaborative Representation Classification (CRC) [20]

CRC uses the training samples from all classes to represent the test sample z, i.e., collaborative representation, which assumes that Eq. (4) is approximately satisfied:

$$z = XB \tag{4}$$

where $B = [b_1 \ldots b_N]^T$, $X = [x_1 \ldots x_N]$.

The solution of Eq. (4) can be easily and analytically derived as:

$$\tilde{B} = (X^T X + \mu I)^{-1} X^T z \tag{5}$$

where μ is a small positive constant, I is the identity matrix and $\tilde{B} = [\tilde{b}_1 \ldots \tilde{b}_N]^T$. Let $P = (X^T X + \mu I)^{-1} X^T$. It is obvious that P is independent of z, so, it can be pre-calculated as a projection matrix. Once given a test sample z, we can just simply project onto P via Pz; this makes CR very computationally efficient.

Of course, if $X^T X$ is not singular, the solution of Eq. (4) can be also obtained using:

$$\tilde{B} = (X^T X)^{-1} X^T z \tag{6}$$

CRC calculates the residual of the test sample z with respect to the i-th class using $r_i = \|z - X_i \tilde{B}_i\|$ where $X_i = [x_{(i-1)*n+1} \ldots x_{i*n}]$ and $\tilde{B}_i = [\tilde{b}_{(i-1)*n+1} \ldots \tilde{b}_{i*n}]^T$. If $k = arg\,min_i r_i$, then CRC assigns the test sample z to the k-th class.

The main difference between LRC and CRC is that CRC first exploits a linear combination of all the training samples to represent the test sample.

3 Description of the Proposed Approach

In this section, we present the main steps of the proposed approach in detail. Suppose that there are c classes and each class has n training samples. Let $x_1, \ldots, x_N$ represent all the N training samples ($N = nc$).

The proposed approach includes the following main steps. The first step generates mirror face training samples. The second step generates left half face samples of the original and mirror face training samples, respectively. The third step uses left half face samples of the original and mirror face training samples to perform CRC face recognition. The fourth step uses the original and mirror face training samples to perform LRC face recognition. The fifth step combines the scores obtained using the third and fourth steps to conduct weighted score level fusion, getting the ultimate classification result. Finally, the algorithm of the proposed approach is presented as follows:

Step 1. Use every original training sample to generate mirror face training samples. Let $x_i \in R^{p \times q}$ be the i-th training sample in the form of image matrix, the mirror image of an original face image has the same size. For the i-th training sample, mirror image $m^0_{(i-1)n+k}$, $(k = 1, \ldots, n)$ is generated using $m^0_{(i-1)n+k}$ $(p, q) = x^0_{(i-1)n+k} (p, Q - q + 1)$, $p = 1, \ldots, P$, $q = 1, \ldots, Q$, P and Q stand for the numbers of the rows and columns of the face image matrix, respectively; $x^0_{(i-1)n+k}(p, q)$ and $m^0_{(i-1)n+k}(p, q)$ denote the pixels located in the p-th row and q-th column of $x^0_{(i-1)n+k}$ and $m^0_{(i-1)n+k}$, respectively. $m^0_{(i-1)n+k}$ is then converted into a column vector and is denoted by $m_{(i-1)n+k}$. As all $x_{(i-1)n+k}$ and $m_{(i-1)n+k}$ act as training samples, such that, there are $2cn$ training samples in total.

Step 2. Use the original and mirror face training samples to generate left half face samples. Let $x_j \in R^{p \times q}$ be the j-th training sample in the form of image matrix, the left half face image $(h^0_{(j-1)2n+k}, k = 1, \ldots, 2n)$ of the original and mirror face images is generated using $h^0_{(j-1)2n+k}(u, v) = x^0_{(j-1)2n+k}(u, v)$, $u = 1, \ldots, P$, $v = 1, \ldots, Q/2$, P and Q stand for the numbers of the rows and columns of the face image matrix, respectively. $h^0_{(j-1)2n+k}$ is then converted into a column vector and is denoted by $h_{(j-1)2n+k}$. So, there are $2cn$ left half face training samples in total.

Step 3. Use the original and mirror face training samples to perform LRC face recognition. Let s_{j1} denote the score of test sample z with respect to the j-th class. For the algorithm, please see Sect. 2.1.

Step 4. Use the left half face training samples to perform CRC face recognition. Let s_{j2} denote the score of test sample z with respect to the j-th class. For the algorithm, please see Sect. 2.2.

Step 5. Combine the scores obtained using the third and fourth steps to conduct weighted score level fusion. For test sample z, we use $s_j = w_1 s_{j1} + w_2 s_{j2}$ to calculate the ultimate score with respect to the j-th class, w_1 and w_2 are the weights. Let $w_1 + w_2 = 1$, our approach assigns a larger value to w_1 in comparison with w_2. If $k = arg\,min_j s_j$, then test sample z is assigned to the k-th class.

4 Experimental Results

In all the experiments, w_1 and w_2 satisfy the condition $w_1 + w_2 = 1$, and only the value of w_1 is given.

4.1 Experiment on the FERET Database

The proposed approach was first tested on the FERET database. This subset includes 1400 images of 200 individuals (each individual has seven images), which involves variations in facial expression, illumination, and pose. In our experiment, the facial portion of each original face image is cropped automatically based on the location of eyes and resized to 40 × 40 pixels and without histogram equalization. Some facial portion images of one individual are shown in Fig. 1.

Fig. 1. Seven original samples of one individual from the FERET face database as well as the mirror images of these face samples; the left half face images of the samples and mirror images, respectively. The first and second rows show the samples and its mirror images, respectively. The third row shows the left half face images of the face samples, and the fourth row shows the left half face images of the mirror images.

In the experiment, we use the first one, two, three and four face images of each individual as training samples, respectively; and the remaining images are taken as testing samples. w_1 in our approach was set to 0.85, and the regularization parameter λ in SRC method was set to 0.01. The experimental results show that λ has only little influence on the classification performance if it is set to a value near to 0.01. Table 1 shows the classification errors. This table shows that our approach classifies much more accurately than all the other approaches. For example, when the first two face images of each individual and the rest face images were respectively used as original training samples and test samples, the classification errors (%) of the proposed approach with $w_1 = 0.85$, LRC, CRC and SRC [22] are 19.0%, 35.9%, 41.6%, 41.9%, respectively. The face classification errors of the well-known SRC is 22.9% higher than that of the proposed approach demonstrated that the proposed one can perform very well in recognizing the face.

Table 1. The rates of classification errors (%) of different approaches on the FERET database.

Training samples per class	1	2	3	4
The proposed approach	46.67	19.0	19.25	14.17
LRC	55.08	35.9	40.13	21.50
CRC	55.67	41.6	55.63	44.67
SRC	58.58	41.9	49.25	36.67

4.2   Experiments on the Yale Database

In the first experiment conducted on the Yale database, the first one, two, three, four and five face images of each individual are used as training samples, respectively; and the remaining images are taken as testing samples. w_1 in our approach was set to 0.85, and λ was set to 0.01, respectively. The experimental results were shown in Table 2. It shows again that the proposed one is able to perform well than all the other approaches.

Table 2. The rates of classification errors (%) of different approaches on the Yale database.

Training samples per class	1	2	3	4	5
The proposed approach	22.67	6.67	3.33	1.90	1.11
LRC	32.00	19.26	11.67	8.57	6.67
CRC	25.33	17.78	9.17	3.81	4.44
SRC	26.00	17.78	9.17	2.86	4.44

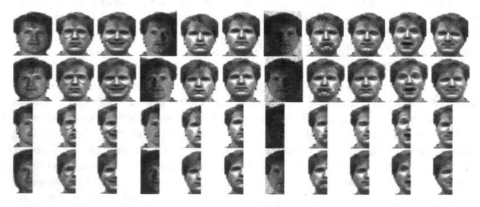

Fig. 2. Eleven original samples of one individual from the Yale face database as well as the mirror images of these face samples; the left half face images of the samples and mirror images, respectively. The first and second rows show the samples and its mirror images, respectively. The third row shows the left half face images of the face samples, and the fourth row shows the left half face images of the mirror images.

To evaluate the effect of the parameter w_1 of the proposed approach, another experiment was conducted on the Yale database. In this experiment, w_1 was set to 0.95,

0.85, 0.75, 0.65, and 0.55, respectively. The experimental results were shown in Table 3. From Table 3, we can find that the parameter w_1 has effect on the recognition rates.

Table 3. The rates of classification errors (%) of the proposed approach with different weighted fusion scheme (w_1, w_2 and $w_1 + w_2 = 1$) on the Yale database.

Training samples per class	1	2	3	4	5
$w_1 = 0.95$, $w_2 = 0.05$	22.67	6.67	5.00	0.95	1.11
$w_1 = 0.85$, $w_2 = 0.15$	22.67	6.67	3.33	1.90	1.11
$w_1 = 0.75$, $w_2 = 0.25$	21.33	5.93	3.33	1.90	2.22
$w_1 = 0.65$, $w_2 = 0.35$	20.67	5.93	2.50	1.90	2.22
$w_1 = 0.55$, $w_2 = 0.45$	20.67	4.44	2.50	1.90	2.22

5 Conclusions

In this paper, we proposed a very promising approach to exploit limited training samples for face recognition. The proposed one first generates the mirror image of the original face image and then applies LRC approach to both the original face image and its mirror image to perform face recognition. The real face image is usually not an axial symmetry image owing to the non-frontal poses, so to use the left half image of the original and mirror face images to produce new samples. This is very helpful for overcoming the drawback of limited training samples in the real world face recognition applications. And then applies CRC approach to left half face image generated by original and mirror images. Finally, the weighted fusion scheme in the proposed approach was used to combine the scores obtained from the LRC approach and CRC approach and the combined score ultimately was used to classify the test sample. The experiments are conducted on three benchmark data sets (Yale database, FERET database), and the experimental results indicate that the proposed one has better performance.

Acknowledgements. This work was supported by the Research Program of Hebei Municipal Science & Technology Department under Grant No. 10213551.

References

1. Zhao, W., Chellappa, R., Phillips, P.J., Rosenfeld, A.: Face recognition: a literature survey. ACM Comput. Surv. **35**(4), 399–458 (2003)
2. Yang, J., Zhang, D., Xu, Y., Yang, J.Y.: Two-dimensional discriminant transform for face recognition. Pattern Recogn. **38**(7), 1125–1129 (2005)
3. He, X.F., Yan, S.C., Hu, Y.X., Zhang, H.J.: Face recognition using laplacian faces. IEEE Trans. Pattern Anal. Mach. Intell. **27**(3), 328–340 (2005)
4. Xu, Y., Zhang, D., Yang, J., Yang, J.Y.: A two-phase test sample sparse representation method for use with face recognition. IEEE Trans. Circuits Syst. Video Technol. **21**(9), 1255–1262 (2011)

5. Turk, M., Pentland, A.: Eigenfaces for recognition. J. Cogn. Neurosci. **3**(1), 71–86 (1991)
6. Comon, P.: Independent component analysis - a new concept? Sig. Process. **36**, 287–314 (1994)
7. Belhumeur, P.N., Hespanha, J.P., Kriengman, D.J.: Eigenfaces versus Fisherfaces: recognition using class specific linear projection. IEEE Trans. Pattern Anal. Mach. Intell. **19**(7), 711–720 (1997)
8. Schölkopf, B., Smola, A., Müller, K.-R.: Kernel principal component analysis. In: Gerstner, W., Germond, A., Hasler, M., Nicoud, J.-D. (eds.) ICANN 1997. LNCS, vol. 1327, pp. 583–588. Springer, Heidelberg (1997). doi:10.1007/BFb0020217
9. Mika, S., Rätsch, G., Weston, J., Schölkopf, B., Müller, K.R.: Fisher discriminant analysis with kernels. In: Proceedings of IEEE International Workshop Neural Networks for Signal Processing, Madison, WI, pp. 41–48 (1999)
10. Bach, F.R., Jordan, M.I.: Kernel independent component analysis. J. Mach. Learn. Research **3**, 1–48 (2002)
11. Tan, X., Chen, S., Zhou, Z.-H., Zhang, F.: Face recognition from a single image per person: a survey. Pattern Recogn. **39**(9), 1725–1745 (2006)
12. Qiao, L., Chen, S., Tan, X.: Sparsity preserving discriminant analysis for single training image face recognition. Pattern Recogn. Lett. **31**(5), 422–429 (2010)
13. Naseem, I., Togneri, R., Bennamoun, M.: Robust regression for face recognition. Pattern Recogn. **45**(1), 104–118 (2012)
14. Xu, Y., Zhu, X., Li, Z., Liu, G., Lu, Y., Liu, H.: Using the original and 'symmetrical face' training samples to perform representation based two-step face recognition. Pattern Recogn. **46**(4), 1151–1158 (2013)
15. Ryu, Y.S., Oh, S.Y.: Simple hybrid classifier for face recognition with adaptively generated virtual data. Pattern Recogn. Lett. **23**(7), 833–841 (2002)
16. Vetter, T.: Synthesis of novel views from a single face image. Int. J. Comput. Vision **28**(2), 102–116 (1998)
17. Sharma, A., Dubey, A., Tripathi, P., Kumar, V.: Pose invariant virtual classifiers from single training image using novel hybrid-eigenfaces. Neurocomputing **73**(10–12), 1868–1880 (2010)
18. Xu, Y., Li, X., Yang, J., Zhang, D.: Integrate the original face image and its mirror image for face recognition. Neurocomputing **131**(9), 191–199 (2014)
19. Naseem, I., Togneri, R., Bennamoun, M.: Linear regression for face recognition. IEEE Trans Pattern Anal. Mach. Intell. **32**(11), 2106–2112 (2010)
20. Zhang, L., Yang, M., Feng, X.: Sparse representation or collaborative representation: which helps face recognition. In: Proceedings of IEEE International Conference on Computer Vision, pp. 471–478 (2011)
21. Xu, Y., Li, X., Yang, J., Lai, Z., Zhang, D.: Integrating conventional and inverse representation for face recognition. IEEE Trans. Cybern. **44**(10), 1738–1746 (2014)
22. Koh, K., Kim, S.J., Boyd, S.: l1 ls: a MATLAB solver for large-scale l1-regularized least squares problems. http://www.stanford.edu/boyd/l1_ls/

A Novel Representation for Abnormal Crowd Motion Detection

Songbo Liu$^{(\boxtimes)}$, Ye Jin, Ye Tao, and Xianglong Tang

School of Computer Science and Technology, Harbin Institute of Technology,
Mailbox 352, 92 West Dazhi Street, Nan Gang District, Harbin, China
sbliu@hit.edu.cn

Abstract. A lot of methods of abnormal crowd motion detection in videos have been proposed in recent years. Most of them are still based on low semantic features, such gray value, velocity and gradient. Usually, the low-level features cannot represent discriminative information of the scene. In addition, former representations often ignore information in time or space dimension. Thus, it is necessary to establish representations with discriminative features and spatio-temporal information. In this work, a crowd abnormal detection framework is proposed. Slow feature analysis (SFA), which can provide high semantic inherent features, is adopted in representation. Besides, the effect of spatio-temporal information is added into the representation. We conduct extensive experiments on two datasets to demonstrate the effectiveness of proposed method. Experimental results suggest that our method improves the detection performance.

Keywords: Video surveillance · Crowd analysis · Abnormal events · Slow feature analysis

1 Introduction

There exist many human disasters in crowded scenes in recent years. If these abnormal events could be detected at the beginning, the time for rescue could be reduced and crowd disasters and deadly stampedes may even be completely avoided.

Crowded scenes usually contain two different events, normal events and abnormal events. Usually, the unusual or irregular behaviors of the whole scene (Fig. 1(a) and (b)) and the appearances of special local objects such as cars, bicycles, and skaters (Fig. 1(c) and (d)) are considered to be abnormal events. Previous researches has given many definitions of abnormal events. The most widely accepted definition tends to define the abnormal events as events of low probability to be normal behaviors [1].

The central task of abnormal detection in videos is to find new patterns (abnormal events) that significantly deviate from the expected patterns (normal events). Therefore, patterns/events representation is crucial and may significantly affect the efficiency or accuracy of an abnormal events detection system.

Previous researches proposed many representation methods, most of which are based on low semantic features. However, low semantic features often ignore discriminative information of the scenes. In addition, most of representations ignore the spatio-temporal information of the scene.

© Springer International Publishing AG 2017
Y. Sun et al. (Eds.): IScIDE 2017, LNCS 10559, pp. 239–248, 2017.
DOI: 10.1007/978-3-319-67777-4_21

Fig. 1. Examples of abnormal events. (a) and (b) Show irregular behaviors. (c) and (d) Show bicycle and car crossing the road

We propose a framework for abnormal events detection in video surveillances. The framework contains an event representing step and an events detection step. To obtain high semantic feature, slow feature analysis (SFA) is applied to learn a few temporally invariant and slowly varying features from low sematic input. Usually, the gray pixel values of a CCD camera varies quickly within a short period of time, while high level responses in a human brain tend to vary slowly for a long time [2]. The gray pixel values of scenes vary quickly while the macro-actions of crowd vary slowly. With SFA, the pixel-level features of a crowded scene can be transformed into high semantic features of macro-actions. In order to remain the space-temporal information, the effect of a cuboid's neighbors in the dimensions of space and time are modeled and added into our representation. In events detection step, sparse reconstruction method [3] is adopted to separate the abnormal samples from normal ones. Figure 2 shows the overall pipeline of our method.

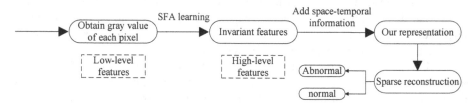

Fig. 2. Framework for anomaly detection

The rest of this paper is organized as follows: in Sect. 2, the previous work of crowd representation is listed. Section 3 is the theory of SFA. The adding of space-temporal information is introduced in Sect. 4. In Sect. 5, the detection method is given. Experiments and comparisons are presented in Sect. 6. Conclusion is given at last.

2   Relate Work

An efficient abnormal event detection system relies on a robust representation. The previous representations can be roughly grouped into three categories.

- Trajectory based representation
  This representation usually needs a tracking method to extract trajectories of each individual [4, 5]. The tracking methods tend to miss the targets when the scene is over-crowded. Therefore, accurate trajectories is still unreachable.
- Foreground-based representation
  This representation usually extract objects from foreground blobs obtained by a background subtraction or pedestrian detection method [6–9]. Mixture Gaussian model [10], which often achieves good counting and detections, is widely used in foreground-based representations. Unfortunately, the foreground cannot be extracted accurately in complex background.
- Spatio-temporal cuboid based representation
  The representation usually divides the scenes into many equal sized 2-D patches or 3-D cuboids without relying on tracking methods and foreground extraction. Therefore, it is the most widely used representation in crowd analysis. Spatio-temporal gradients [11, 12] and motion histogram [13] are typical features adopted in the representations. However, they are time consuming. To reduce the dimension of the feature, Jodoin et al. [13] proposed orientation distribution function (ODF) feature, which contains no magnitude information. Without magnitude information, the feature cannot describe the crowd motion completely. Different from ODF, multiscale histogram of optical flow (MHOF) [3] contains small amounts of important magnitude information. With reasonable dimension and more magnitude information, MHOF features are more widely accepted. In addition, though these features are extracted from spatio-temporal cuboids, the spatio-temporal relations between cuboids are often ignored.

3   Slow Feature Analysis

In this work, low semantic features are expressed as input signals. SFA learns a family of slowness functions and then use them to extract slow features from the input signals. The output signals, which contain the information of slowly varying macro-actions, can provide inherent features for events representation.

Given a L-dimensional input signal $\mathbf{x}(t) = [x_1(t), \cdots, x_L(t)]^T$ with $t \in [t_0, t_I]$ indicating time, a set of slowness functions $\mathbf{g}(\mathbf{x}) = [g_1(\mathbf{x}(t)), \cdots, g_I(\mathbf{x}(t))]^T$ should be found to transform $\mathbf{x}(t)$ into slowly varying output, $\mathbf{y}(t) = [y_1(t), \cdots, y_J(t)]^T$, which vary as slowly as possible but carry inherent information. Here, $y_j(t) = g_j(\mathbf{x}(t))$, $j \in \{1, \cdots, J\}$. Therefore, the following optimization problem should be worked out.

$$\Delta_j = \Delta(y_j) = \left\langle \dot{y}_j^2 \right\rangle_t \text{ is minimal,} \tag{1}$$

subject to

$$\langle y_j \rangle_t = 0 \text{ zero mean;} \tag{2}$$

$$\left\langle y_j^2 \right\rangle_t = 1 \text{ unit variance;} \tag{3}$$

$$\text{And } \forall j' < j : \langle y_{j'} y_j \rangle_t = 0 \text{ decorrelation,} \tag{4}$$

where $\dot{y}$ denotes the operator of computing the first order derivative of y and $\langle y \rangle_t$ is the mean of signal y over time. Equation (1) expresses the primary objective of minimizing the temporal variation of the output signal. Constraints (2) and (3) help avoiding the trivial solution $y_j(t) = const$. Constraint (4) guarantees that different output signals carry different information. It also induces an order in output signals, so that $y_1(t)$ is the optimal output signal, while $y_2(t)$ is a less optimal one. Through the optimization problem, SFA can achieve the purpose of transforming input signals into a group of slowly varying uncorrelated signals. Given a video sequence, we spatio-temporally divide it into a set of small cuboids. We concatenate the gray values of all pixels in the cuboid into a long vector $\mathbf{m}$, the output feature of slow feature analysis can be shown as $y = g(\mathbf{m})$.

4   Representation with Spatio-Temporal Information

In previous work which adopted SFA in abnormal detection, the external information between cuboids in the dimension of space has been taken into account. However, the contact between adjacent cuboids in time dimension were ignored [15]. Besides, they clustered the cuboids of the same time into several clusters, and only kept the cluster with largest number of cuboids as the main motion. Some motion patterns in other clusters, which contains important local information, are removed.

In this section, a new method to learning spatio-temporal information for events representation is proposed. The effect of cuboids located in nearby spatio-temporal positions is considered as the spatio-temporal information of a certain cuboid. We do not delete features of any cuboid, as these features may reflect meaningful information. The representation of a certain cuboid is shown as

Proposed representation = SFA feature + effect of nearby cuboids

The SFA features is the output signal $\mathbf{y}(t)$, which can be obtained by the process introduced in Sect. 3. The 26 nearby cuboids (see Fig. 3) of a certain cuboid are chosen to estimate the effect of spatio-temporal information.

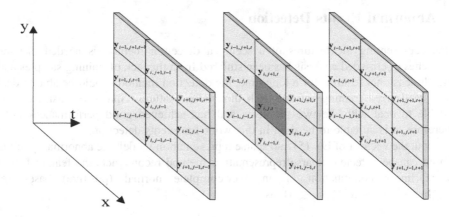

Fig. 3. Spatio-temporal positions of nearby cuboids. (The index i, j, t denotes the spatio-temporal position of the cuboid. The target cuboid is highlighted in red). (Color figure online)

Figure 3 shows that a cuboid has 26 surround cuboids. The effect of each cuboid is measure by $\omega_k \mathbf{y}_k$, where ω_k is the weight of kth cuboid. Thus, the representation can be expressed as

$$\mathbf{f} = \mathbf{y}(t) + \frac{1}{n}\sum_{k=1}^{n} \omega_k \mathbf{y}_k(t), \quad n = 26 \tag{5}$$

The weight ω_k, which denotes the rate of effect of k-th cuboid, is measured by correlation coefficient. The correlation coefficient shows the correlation between the target cuboid and the k-th cuboid. In this work, the effect is proportional to the correlation coefficient, and is given by

$$\omega_k = \frac{\sum_{i=1}^{m}(y_i - \langle y\rangle_t)(y_{ki} - \langle y_k\rangle_t)}{\sqrt{\sum_{i=1}^{m}(y_i - \langle y\rangle_t)^2}\sqrt{\sum_{i=1}^{m}(y_{ki} - \langle y_k\rangle_t)^2}} \tag{6}$$

where m is the dimension of SFA features. $\sqrt{\sum_{i=1}^{m}(y_i - \langle y\rangle_t)^2}$ and $\sqrt{\sum_{i=1}^{m}(y_{ki} - \langle y_k\rangle_t)^2}$ are the standard deviations.

The range of ω_k is $[-1, 1]$. If $\omega_k > 0$, a positive correlation can be found between the two cuboids. If $\omega_k < 0$, a negative correlation exists between the two cuboids.

Finally, the representation in Eq. (5) includes high semantic features and spatio-temporal information of the scenes.

5 Abnormal Events Detection

After representing the features of a scene, a detection method is needed. As the researches of crowded analysis has been suffered from the lack of training samples and high dimensional features for a long time, the detection method should be able to deal with small sized training set and high dimension features. Sparse reconstruction is suitable to deal with the two problems and has achieved good performances [14]. Therefore, it is suitable to be used in this work for events detection.

Input the vector $\mathbf{f}$ of Eq. (5) into sparse representation model, the abnormality of the cuboid can be detected. Sparse representation method reconstructs the feature $\mathbf{f}$ by a sparse linear combination of an over-complete normal (positive) bases set $\Phi = [\mathbf{b}_1, \cdots, \mathbf{b}_n] \in R^{m \times n}(m \ll n)$, as

$$\mathbf{a}^* = \arg\min_{\mathbf{a}} \frac{1}{2}\|\mathbf{f} - \Phi\mathbf{a}\|_2^2 + \lambda\|\mathbf{a}\|_1, \tag{7}$$

where $\mathbf{a}$ is the reconstruction coefficient vector. $\|\mathbf{f} - \Phi\mathbf{a}\|_2^2$ denotes the distance between $\mathbf{f}$ and $\Phi\mathbf{a}$. Minimizing $\|\mathbf{f} - \Phi\mathbf{a}\|_2^2$ can make $\Phi\mathbf{a}$ equals to $\mathbf{f}$. Minimizing $\|\mathbf{a}\|_1$ makes the reconstruction coefficient contains least number of nonzero-elements. It means that least bases (normal samples) are needed. Thus, sparse representation can achieve the goal of detecting abnormal events with limited number of training samples. $\mathbf{a}^*$ is the vector of optimized coefficient. A normal testing sample is likely to generate a coefficient vector with small number of nonzero-elements, while an abnormal sample generates a coefficient vector which has large number of nonzero-elements.

The sparse reconstruction cost (SRC) is proposed to quantify the normalness.

$$SRC = \frac{1}{2}\|\mathbf{f} - \Phi\mathbf{a}^*\|_2^2 + \lambda\|\mathbf{a}^*\|_1. \tag{8}$$

A normal sample has a small SRC, while an abnormal one usually generates a large SRC. Therefore, the SRC is adopted as an anomaly measurement for the one-class classification problem in this work.

6 Experimental Results

Different detection methods are selected to compare with proposed method. First, the chaotic invariant-based method is chosen, for the chaotic invariant tends to extract high semantic inherent feature like the SFA. Second, sparse representation method [3] is chosen. Multi histogram optical flow (MHOF), which has been widely used for events representation, is adopted in sparse representation method. The comparison with the two methods can verify the advantages of high semantic features. Third, heterogeneous representation [15], which also use SFA features, is brought to the comparison. The method ignore the information in time dimension, while the proposed method does not. Thus, the comparison can verify whether it is necessary to add time information. In addition, several state-of-the-art methods, e.g., social force [16], Mixture dynamic texture (MDT) [17], and energy model [18] are compared, too.

UMN dataset [19] is used in global abnormal events detection. The dataset is consisted of 3 different crowd scenes of escape events, and the total frame number is 7740 (1450, 4415 and 2145 for scenes 1−3, respectively) with a 320 × 240 resolution.

The UCSD dataset [20] is used for local abnormal events detection. It contains local abnormal events such as bikers, skaters, small cars, etc. The data was split into 2 subsets, each corresponding to a kind of human scenes. The UCSD Ped1 has 44 short clips, and the UCSD Ped2 has 46 short clips. Each clip has 180 frames, with a 158 × 238 resolution.

6.1   Global Abnormal Events Detection

Table 1 provides the equal error rate (EER) of these methods. The proposed method has the smallest EER. Equal error rate is the mean of a method's error rates in all the video clips in a dataset.

Table 1. Equal error rate of each method

Method	Equal error rate (EER) (%)
Proposed model	4.3
Sparse representation [3]	10
Chaotic invariant [21]	5.5
Heterogeneous representation [15]	8.3
Social force [16]	12.6
Mixture dynamic texture [22]	15
Energy model [18]	28

As shown in Fig. 4 and Table 1, proposed method outperform heterogeneous representation [15], chaotic invariant [21], and sparse representation [3]. The phenomenon shows a few advantages of the proposed method.

- The adding of information in time dimension makes the representation more discriminative.
- SFA can provide better inherent features than chaotic invariant.
- The high semantic discriminative representation obtained by our method outperforms traditional low-level representations, such as MHOF, which is used in sparse representation.

In addition, social force [16], mixture dynamic texture [22], and energy model [18] are based on optical flow. The performances of the three methods are not good enough. It is obviously that the low semantic feature based on velocities has low discriminability.

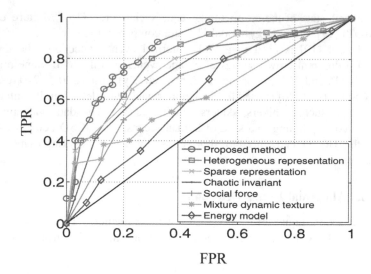

Fig. 4. ROC curves of abnormal events detection in UMN and UCSD dataset

6.2 Local Abnormal Events Detection

In Table 2, we compare our method with the other state-of-the-art methods through Equal error Rate (EER) and Detection Rate (DR). Dividing the number of detected abnormal events by the number of true abnormal events, the detection rate (DR) can be obtained. The results of other state-of-the-art methods are obtained from their respective references [3, 17, 23, 24]. Some methods are not suitable to detect abnormal events in subset Ped2, thus, the authors do not provide the results in their papers. Then, some results in Table 2 are not listed in this section. The lower the EER value, the better the performance that can be achieved, while DR value is just the opposite. For Ped1 subset, the EER value of proposed method is 18% which is lower than the rest; the DR value is 53% which is higher than all the other state-of-the-art approaches. For Ped2 subset, the EER value of our approach is 20% which is lower than other approaches, and the DR value is 55% which is the highest value. The comparison of the results demonstrates that our approach outperforms the state-of-the-art methods on the UCSD dataset.

6.3 Local Abnormal Events Detection

Figure 4 shows the ROC curves of the above methods in both UMN and UCSD dataset. The result of Fig. 4 shows that proposed method is better than the rest.

The proposed method takes into account the spatio-temporal information and all the local features. In addition, SFA provides high discriminative features. Thus, the proposed method can achieve highest detection accuracy.

Table 2. Summary of quantitative performance and comparison with state-of-the-art approaches on Ped1 and Ped2 subsets.

Method	Ped1		Ped2	
	EER (%)	DR (%)	EER (%)	DR (%)
Proposed method	18	53	20	55
Sparse representation [3]	19	46	–	–
Chaotic invariant [21]	22	41	26	35
Heterogeneous representation [15]	23	47	–	–
Social force [16]	25	45	25	–
Mixture dynamic texture [22]	38.9	32.6	45.8	22.4
Energy model [18]	36.5	40.9	35.0	27.6

7 Conclusion

In this work, slow feature analysis is used to extract the inherent features. It provides more discriminative representation than low semantic representations. Besides, spatio-temporal information is also added. Our representation can significantly improve the detection accuracy compared with other approaches. Experimental results show that proposed approach can accurately detect different types of anomalies and outperforms the state-of-art methods.

References

1. Li, W., Mahadevan, V., Vasconcelos, N.: Anomaly detection and localization in crowded scenes. IEEE Trans. Pattern Anal. Mach. Intell. **36**(1), 18–32 (2014)
2. Zhang, Z., Tao, D.: Slow feature analysis for human action recognition. IEEE Trans. Pattern Anal. Mach. Intell. **34**(3), 436–450 (2012)
3. Cong, Y., Yuan, J., Liu, J.: Abnormal event detection in crowded scenes using sparse representation. Pattern Recogn. **46**(7), 1851–1864 (2013)
4. Jiang, F., Wu, Y., Katsaggelos, A.K.: A dynamic hierarchical clustering method for trajectory-based unusual video event detection. IEEE Trans. Image Process. **18**(4), 907–913 (2009)
5. Piciarelli, C., Micheloni, C., Foresti, G.L.: Trajectory-based anomalous event detection. IEEE Trans. Circuits Syst. Video Technol. **18**(11), 1544–1554 (2008)
6. Benezeth, Y., et al.: Abnormal events detection based on spatio-temporal co-occurences. In: IEEE Conference on Computer Vision and Pattern Recognition, CVPR 2009. IEEE (2009)
7. Reddy, V., Sanderson, C., Lovell, B.C.: Improved anomaly detection in crowded scenes via cell-based analysis of foreground speed, size and texture. In: 2011 IEEE Computer Society Conference on Computer Vision and Pattern Recognition Workshops (CVPRW). IEEE (2011)
8. Xiang, T., Gong, S.G.: Video behavior profiling for anomaly detection. IEEE Trans. Pattern Anal. Mach. Intell. **30**(5), 893–908 (2008)
9. Yuan, Y., Fang, J.W., Wang, Q.: Online anomaly detection in crowd scenes via structure analysis. IEEE Trans. Cybern. **45**(3), 562–575 (2015)

10. Fernando, B., et al.: Supervised learning of Gaussian mixture models for visual vocabulary generation. Pattern Recogn. **45**(2), 897–907 (2012)
11. Kratz, L., Nishino, K.: Anomaly detection in extremely crowded scenes using spatio-temporal motion pattern models. In: CVPR: 2009 IEEE Conference on Computer Vision and Pattern Recognition, vol. 1–4, pp. 1446–1453 (2009)
12. Kratz, L., Nishino, K.: Tracking pedestrians using local spatio-temporal motion patterns in extremely crowded scenes. IEEE Trans. Pattern Anal. Mach. Intell. **34**(5), 987–1002 (2012)
13. Jodoin, P.-M., Benezeth, Y., Wang, Y.: Meta-tracking for video scene understanding. In: 10th IEEE International Conference on Advanced Video Signal Based Surveillance, pp. 1–6 (2013)
14. Cong, Y., Yuan, J., Liu, J.: Sparse reconstruction cost for abnormal event detection. In: 2011 IEEE Conference on Computer Vision and Pattern Recognition (CVPR). IEEE (2011)
15. Hu, X., et al.: Robust and efficient anomaly detection using heterogeneous representations. J. Electron. Imaging **24**(3), 033021–033022 (2015)
16. Raghavendra, R., Del Bue, A., Cristani, M., Murino, V.: Abnormal crowd behavior detection by social force optimization. In: Salah, A.A., Lepri, B. (eds.) HBU 2011. LNCS, vol. 7065, pp. 134–145. Springer, Heidelberg (2011). doi:10.1007/978-3-642-25446-8_15
17. Mahadevan, V., et al.: Anomaly detection in crowded scenes. In: 2010 IEEE Conference on Computer Vision and Pattern Recognition (CVPR). IEEE (2010)
18. Xiong, G., et al.: Abnormal crowd behavior detection based on the energy model. In: 2011 IEEE International Conference on Information and Automation (ICIA). IEEE (2011)
19. Mehran, R., Oyama, A., Shah, M.: Abnormal crowd behavior detection using social force model. In: IEEE Conference on Computer Vision and Pattern Recognition, CVPR 2009. IEEE (2009)
20. Stroppi, L.J.R., Chiotti, O., Villarreal, P.D.: Defining the resource perspective in the development of processes-aware information systems. Inf. Softw. Technol. **59**, 86–108 (2015)
21. Wu, S., Moore, B.E., Shah, M.: Chaotic invariants of lagrangian particle trajectories for anomaly detection in crowded scenes. In: 2010 IEEE Conference on Computer Vision and Pattern Recognition (CVPR). IEEE (2010)
22. Mahadevan, V., et al.: Anomaly detection in crowded scenes (2010)
23. Cong, Y., Yuan, J.S., Tang, Y.D.: Video anomaly search in crowded scenes via spatio-temporal motion context. IEEE Trans. Inf. Forensics Secur. **8**(10), 1590–1599 (2013)
24. Adam, A., et al.: Robust real-time unusual event detection using multiple fixed-location monitors. IEEE Trans. Pattern Anal. Mach. Intell. **30**(3), 555–560 (2008)

Robust Face Hallucination via Locality-Constrained Nuclear Norm Regularized Regression

Guangwei Gao[1,2,5(✉)], Jian Yang[2,3], Pu Huang[4], Zuoyong Li[5],
and Dong Yue[1]

[1] Institute of Advanced Technology, Nanjing University of Posts
and Telecommunications, Nanjing, People's Republic of China
{csgwgao,yued}@njupt.edu.cn
[2] Key Laboratory of Intelligent Perception and Systems for High-Dimensional
Information of Ministry of Education, Nanjing University of Science
and Technology, Nanjing, People's Republic of China
csjyang@njust.edu.cn
[3] School of Computer Science and Engineering, Nanjing University of Science
and Technology, Nanjing, People's Republic of China
[4] Jiangsu Key Laboratory of Big Data Security and Intelligent Processing,
Nanjing University of Posts and Telecommunications,
Nanjing, People's Republic of China
huangpu@njupt.edu.cn
[5] Fujian Provincial Key Laboratory of Information Processing and Intelligent
Control, Minjiang University, Fuzhou, People's Republic of China
fzulzytdq@126.com

Abstract. The performance of traditional face recognition approaches is sharply reduced when facing a low-resolution (LR) probe face image. Various face hallucination methods have been proposed in the past decade to obtain much more facial details. The basic idea of face hallucination is to desire a high-resolution (HR) face image from an observed LR one with the help of a set of training examples. In this paper, we propose a locality-constrained nuclear norm regularized regression (LCNNR) model for face hallucination task and use the alternating direction method of multipliers to solve it. LCNNR attempts to directly use the image matrix to compute the representation coefficients to maintain the essential structural information. Moreover, a locality constraint is also enforced to preserve the locality and the sparsity simultaneously. Experiments carried out on the benchmark FEI face database show that LCNNR outperforms some state-of-the-art algorithms.

Keywords: Face hallucination · Locality constrained · Position-patch · Nuclear norm

1 Introduction

Face hallucination, or face super-resolution, is a technology to obtain high-resolution (HR) face images from observed low-resolution (LR) inputs, thus providing more facial details for the following recognition process [1–3]. Inspired by the pioneering work of

© Springer International Publishing AG 2017
Y. Sun et al. (Eds.): IScIDE 2017, LNCS 10559, pp. 249–258, 2017.
DOI: 10.1007/978-3-319-67777-4_22

Barker and Kanade [4], various learning-based face hallucination methods have been presented in the past few decades. Wang and Tang [5] represented the input LR image as a linear combination over the LR training samples by principal component analysis (PCA). An [6] used canonical correlation analysis (CCA) to find a coherent subspace that maximizes the correlation between the PCA coefficients of corresponding LR and HR images. Chang et al. [7] presented a neighbor embedding (NE) based super-resolution method by the well-known locally linear embedding. Different from those approaches using a fixed number of neighbors for reconstruction, Yang et al. [8] and Gao and Yang [9] introduced sparse coding technique that adaptively selects the most relevant neighbors for reconstruction.

Recognizing that human face is a highly structured object, the position prior can be fully incorporated into face hallucination procedure. Ma et al. [10] proposed a position-patch based face hallucination method using all patches from the same position in a dictionary, and used a least square representation (LSR) to obtain the optimal reconstruction weights. However, if the number of the training samples is much larger than the patch dimension, the solutions to least square become unstable. To tackle this problem, Jung et al. [11] employed sparsity prior to improve the reconstruction result with several principal training patches. Recently, Wang et al. [12] further proposed a weighted adaptive sparse regularization (WASR) method to super-resolve face images. Jiang et al. [13] incorporated a locality constraint into the least square inverse problem to maintain locality and sparsity simultaneously. The local manifold distance is used to determine weights on the representation coefficients by following the observation that nearer neighborhoods make greater contributions to the final reconstruction.

The aforementioned patch based methods are all vector-based methods. In other words, before calculating the representation coefficients, we must convert the patch matrices into vectors beforehand. However, the converting step neglects the whole structure of the error image. In addition, the l_2 and l_1 norm are pixel-based error model, which assumes that errors of pixels are independent. This assumption does not hold in the case of contiguous error, where the corruptions are spatially correlated. Fortunately, nuclear norm not only can alleviate these correlations via the involved singular value decomposition, but also directly characterizes the holistically structure of error image. In this paper, we propose to add a nuclear norm of the representation residual image into the regression model to compute the representation coefficients straightforward (without the matrix-to-vector conversion). In addition, we also incorporate a locality constraint into the objective function to reach sparsity and locality simultaneously. The locality constraint can capture fundamental similarities between neighbor patches and derives an analytical solution to the constrained problem. The whole model can be solved via the alternating direction method of multipliers. Performance comparison with the state-of-the-art algorithms on the benchmark FEI face database shows the effectiveness of the proposed method for a face image super-resolution in general.

The rest of the paper is organized as follows. In Sect. 2, we present our face hallucination method based on locality-constrained nuclear norm regularized regression. Section 3 evaluates the performance of the proposed methods on commonly used face hallucination databases. Section 4 concludes this paper.

2 Locality-Constrained Nuclear Norm Regularized Regression for Face Hallucination

2.1 Problem Formulation

In this subsection, all the face image patches are denoted in the matrix form. In other words, the input patch and training face image patches can be denoted as $y(i, j) \in \Re^{d \times d}$ and $A^m(i, j) \in \Re^{d \times d}$ $(m = 1,\ldots,N)$, respectively. For the convenience of expression, we omit the position index (i,j) in the following text. Then, the problem can be formulated as follows:

$$\min_{w} \; \|A(w) - y\|_F^2 + \alpha \|A(w) - y\|_*, \tag{1}$$

where $A(w) = w_1 A^1 + w_2 A^2 + \ldots + w_N A^N$, $w \in \Re^{N \times 1}$ is the coefficient, $\|\cdot\|_*$ is the nuclear norm (i.e. the sum of the singular values) of a matrix, $\|\cdot\|_F$ is the Frobenius norm, α is balancing parameters for controlling contributions from the low-rank property. As in [13], we also introduce a local manifold constraint via a similarity metric between the input patch and dictionary atoms to reveal prior information from nearest atoms. The locality-constrained nuclear norm regularized regression can be formulated as follows:

$$\min_{w} \; \|A(w) - y\|_F^2 + \alpha \|A(w) - y\|_* + \beta \|d \otimes w\|_2^2, \tag{2}$$

where $\otimes$ denotes the element wise product, β is balancing parameters for controlling contributions from the locality constraints, and $d_i = \|y - A^i\|_*$ is the locality metric to measure the distance between the input and dictionary atoms.

2.2 Optimization via ADMM

For the convenience of expression, we rewrite the optimization problem of (2) as:

$$\min_{w,E} \; \|E\|_F^2 + \alpha \|E\|_* + \beta \|d \otimes w\|_2^2$$
$$s.t. \quad A(w) - y = E. \tag{3}$$

The above problem can be solved via the alternating direction method of multipliers (ADMM) [14, 15] with the following augmented Lagrangian function:

$$L_\mu(w, E) = \|E\|_F^2 + \alpha \|E\|_* + \beta \|d \otimes w\|_2^2$$
$$+ Tr\left(Z^T(A(w) - y - E)\right) + \frac{\mu}{2} \|A(w) - y - E\|_F^2, \tag{4}$$

where Z is the Lagrange multiplier, μ is a penalty parameter. Following some simple algebraic steps, problem (4) can be rewritten as

$$L_\mu(w, E) = \|E\|_F^2 + \alpha\|E\|_* + \beta\|d \otimes w\|_2^2$$
$$+ \frac{\mu}{2}\left\|A(w) - y - E + \frac{1}{\mu}Z\right\|_F^2 - \frac{1}{2\mu}\|Z\|_F^2. \tag{5}$$

Updating w

Given E, the optimization problem can be reformulated as

$$L_\mu(w) = \|d \otimes w\|_2^2 + \frac{\mu}{2\beta}\left\|y + E - \frac{1}{\mu}Z - A(w)\right\|_F^2. \tag{6}$$

Following [13], the solution of problem (6) can be derived analytically as

$$w^{k+1} = (G + \tau D^2)\backslash ones(N, 1), \tag{7}$$

where $ones(N,1)$ is a $N \times 1$ column vector of ones, the operator "\" denotes the left matrix division operation, $\tau = 2\beta/\mu$, D is a $N \times N$ diagonal matrix with entries $D_{mm} = d_m$, and G is the covariance matrix $G = C^T C$ with

$$C = \left(y + E - \frac{1}{\mu}Z\right)ones(N, 1)^T - H, \tag{8}$$

where $H = [Vec(A^1), Vec(A^2), \ldots, Vec(A^N)]$ and $Vec(\cdot)$ denotes the vectorization operator. The final optimal solution is obtained by rescaling to satisfy $\sum_{m=1}^{N} w_m = 1$.

Updating E

Given x, the optimization problem can be rewritten as

$$\begin{aligned}
L_\mu(E) &= \|E\|_F^2 + \alpha\|E\|_* - Tr(Z^T E) + \frac{\mu}{2}\|A(w) - y - E\|_F^2 \\
&= \|E\|_F^2 + \alpha\|E\|_* - Tr(Z^T E) + \frac{\mu}{2}Tr(((A(w) - y)^T - E^T)(A(w) - y - E)) \\
&= \alpha\|E\|_* + \frac{\mu}{2}Tr\left(\left(\frac{\mu}{2} + 1\right)E^T E - 2\left(A(w)^T - y^T + \frac{1}{\mu}Z^T\right)E\right) + const1 \\
&= \alpha\|E\|_* + \frac{\mu + 2}{2}\left\|E - \frac{\mu}{\mu + 2}\left(A(w)^T - y^T + \frac{1}{\mu}Z^T\right)^T\right\|_F^2 + const2,
\end{aligned} \tag{9}$$

where $const1$ and $const2$ are constant terms, which are independent of the variable E. The optimal solution can be obtained by

$$E^{k+1} = \arg\min_E\left(\frac{\alpha}{\mu + 2}\|E\|_* + \frac{1}{2}\left\|E - \frac{\mu}{\mu + 2}\left(A(w) - y + \frac{1}{\mu}Z\right)\right\|_F^2\right). \tag{10}$$

Its solution is [16]

$$E^{k+1} = UT_{\frac{\alpha}{\mu+2}}[S]V, \tag{11}$$

where $(U, S, V^T) = svd\left(\frac{\mu}{\mu+2}\left(A(w) - y + \frac{1}{\mu}Z\right)\right)$.

The singular value thresholding operator $T_\tau(\cdot)$ is defined as

$$T_{\frac{\alpha}{\mu+2}}[S] = \text{diag}\left(\left\{\max\left(0, s_{j,j} - \frac{\alpha}{\mu+2}\right)\right\}_{1 \le j \le r}\right), \tag{12}$$

where r is the rank of S.

The detailed algorithm via ADMM to solve problem (6) is summarized in Algorithm 1.

2.3 Face Hallucination via LCNNR

As for face hallucination, the training set consists of HR and LR face image pairs. A_H^m denote the HR face images, while $A_H^m (m = 1, \ldots, N)$ denote their LR counterparts. The face hallucination task aims to acquire the HR face image X from its LR observation y.

Firstly, we divide each training faces and the LR input into overlapped patch matrices and denote them as $A_L^m(i,j), A_H^m(i,j), y(i,j)$. For each LR input image patch matrix $y(i,j)$, it is represented as a linear combination over the LR training patch matrices $A_L^m(i, j)$ $(m = 1, \ldots, N)$ using LCNNR. By keeping the combination coefficients and replacing the LR training patch matrices with the corresponding HR counterparts, the desired HR patch matrix at the same position can be synthesized. By concatenating all the HR patch matrices to their relevant positions and averaging values in the overlapping regions, an estimated HR target face image can be obtained. We summarize the whole face hallucination algorithm in Algorithm 2.

Algorithm 1. ADMM algorithm for solving LCNNR

Input: Training patch matrices $A^1, \ldots, A^N$ and test patch matrix y, parameters α and β, the termination condition parameter ε.

Initialize: $w = 0, E = 0, Z = 0$

1: Fix the others and update w according to (7);
2: Fix the others and update E according to (11);
3: Update the multiplies:
   $$Z \leftarrow Z + \mu(A(w) - y - E);$$
4: Check for convergence
   $$\|A(w) - y - E_2\|_\infty > \varepsilon.$$
   Go to 1;

5. **Output:** Optimal coding vector w

Algorithm 2. Face hallucination via LCNRR

Input: HR training images $A_H^1, \ldots, A_H^N$, corresponding LR training images $A_L^1, \ldots, A_L^N$, input LR images y.

1:    **For** each input patch matrix in y:

    a) Compute the distance between the LR input $y(i,j)$ and each of the LR training patch matrices $A_L^m(i,j)$ ($m=1, \ldots, N$):

$$d_m(i,j) = \left\| y(i,j) - A_L^m(i,j) \right\|_2^2, m = 1, \ldots, N$$

    b) Calculate the optimal weights $w^*(i,j)$ with regard to the LR input $y(i,j)$ using **Algorithm 1**;

    c) Construct the desired HR patch by

$$x(i,j) = \sum_{m=1}^N A_H^m(i,j) w_m^*(i,j)$$

2:    **End for**

3:    The target HR image X can be obtained by integrating all the reconstructed HR patch matrices.

Output: The hallucinated HR face image X.

---

3   Experimental Results and Discussions

3.1   Datasets Description and Experimental Setting

We perform experiments on the FEI database [17] to demonstrate the effectiveness of the proposed algorithm. All the face images are manually aligned using the locations of three points: centers of left and right eyeballs and center of the mouth (some examples are shown in Fig. 1). We crop the region of the faces and normalize the HR images to the size of 120 × 100. The LR images are obtained by smoothing (an averaging filter of size 4 × 4) and down-sampling (the down-sampling factor is 4) the corresponding HR images, thus the LR images have size 30 × 25. 400 images from 200 subjects are selected and each subject has two frontal images, one with a neutral expression and the other with a smiling facial expression.

In this part, we compare our proposed methods with Chang's neighbor embedding (NE) method [7], Ma's least square representation (LSR) method [10], Jung's sparse representation (SR) method [11] and Jiang's locality-constrained representation (LcR) method [13]. We randomly choose 250 images for training, and 40 images for

Fig. 1. Some sample images from the FEI face database.

testing. For all comparative methods, we tune their parameters to achieve the best possible results. Specifically, for Chang's method, the number of neighbors is set to 50. As for other patch-based methods, we suggest using the size 3×3 pixels for LR patch and the overlaps between its neighbor patches are 3×1 pixels, while the corresponding HR patch size is 12×12 pixels with an overlap of 12×4 pixels.

3.2 Hallucination Comparisons on FEI Database

In this subsection, we conduct hallucination experiments on FEI database with both noise and noise free two different configurations. As usual, we adopt the same assessment methods (PSNR and SSIM) to measure the objective and subjective qualities of the reconstructed images with other algorithms.

(1) Experimental results without noise: Some representative hallucinated results generated by different methods are listed in Fig. 2. We can find that Jiang's LcR method and our LCNNR method hallucinate competitive face images with more details in the eye, mouth and face contour than other methods and meanwhile the results of LCNNR are much more similar to the original HR ones. Chang's NE method tends to generate blurring effects on the hallucinated face images especially on locations around eyes. Ma's LSR method induces some smoothness effects on the eyes and mouth. Jung's SR method enhances the edges on the mouth to some extent, while the ringing effects also take place around the face contour. We also

Fig. 2. Comparisons of hallucinated results based on different methods. From left to right are the Input LR image, the results of NE [7], LSR [10], SR [11], LcR [13], proposed LCNNR and the original HR image.

provide the average PSNR and SSIM values of different methods in Table 1. It can be seen that the proposed method has better performance than state-of-the-arts with noise free scenario. From the above analysis, we can have the following observations: (i) by taking consideration of the face position semantics priors, position-patch based methods (LSR, SR, LcR and our method) perform better than NE method; (ii) by incorporating the nuclear norm regularized regression and the local manifold constraint, our method can achieve the best performances in both of visual quality and objective assessment.

Table 1. The average PSNR and SSIM values on the FEI database.

Methods	NE [7]	LSR [10]	SR [11]	LcR [13]	LCNNR
PSNR (dB)	31.1680	31.8206	31.8860	32.2757	**32.4377**
SSIM	0.8920	0.9013	0.9087	0.9114	**0.9189**

(2) Experimental results with noise: We test the robustness of the proposed method in this part. Each input is corrupted by a randomly located square block whose elements are random numbers between 0 and 255. For visual comparison of different methods, we list some representative face images in Fig. 3 to show the subjective reconstruction image equality. From Fig. 3, we can observe that outputs

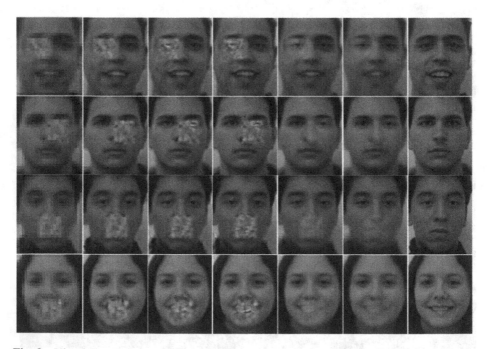

Fig. 3. Visual Comparisons of different hallucination methods with noise. From left to right are the inputs with noise, the results of NE [7], LSR [10], SR [11], LcR [13], proposed LCNNR and the original HR image.

of NE, LSR and SR look similar to input noise images. Results from LCR seem more reliable than previous methods with less noise. However, block effects are still evident due to biased solution from input noise. Compared with other methods, our approach render smoother images with less block effect and more details with the help of structure noise depiction of nuclear norm. The average evaluation measures of different methods are tabulated in Table 2. The reconstruction qualities of all the super-resolution methods reduce in noise situation. While our method can still obtains the best performance in terms of PSNR and SSIM.

Table 2. The average PSNR and SSIM values on the FEI database.

Methods	NE [7]	LSR [10]	SR [11]	LcR [13]	LCNNR
PSNR (dB)	23.6105	23.6790	23.7460	25.3105	**25.7632**
SSIM	0.7946	0.7995	0.8042	0.8121	**0.8270**

4 Conclusions

In this paper, we present a novel locality-constrained nuclear norm regularized regression (LCNNR) model and apply the alternating direction method of multipliers to solve it. LCNNR can directly compute the combination coefficients of input patch matrix without matrix-to-vector conversion by taking advantage of the structure characteristics of the representation error. Furthermore, LCNNR also enforces the locality constraint which can capture fundamental similarities between neighbor patches and derives an analytical solution to the constrained problem. Experiments conducted on the benchmark FEI face database clearly validate the advantages of our proposed LCNNR over the state-of-the-art methods in face hallucination both objectively and subjectively.

5 Acknowledgement

This work was partially supported by the National Natural Science Foundation of China under Grant no. 61502245, the China Postdoctoral Science Foundation under Grant no. 2016M600433, the Natural Science Foundation of Jiangsu Province under Grant no. BK20150849, Open Fund Project of Fujian Provincial Key Laboratory of Information Processing and Intelligent Control (Minjiang University) (No. MJUKF 201717), Open Fund Project of Key Laboratory of Intelligent Perception and Systems for High-Dimensional Information of Ministry of Education (Nanjing University of Science and Technology) (No. JYB201709).

References

1. Jain, A.K., Ross, A., Prabhakar, S.: An introduction to biometric recognition. IEEE Trans. Circuits Syst. Video Technol. **14**(1), 4–20 (2004)
2. Gao, G., Yang, J., Wu, S., Jing, X., Yue, D.: Bayesian sample steered discriminative regression for biometric image classification. Appl. Soft Comput. **37**, 48–59 (2015)
3. Jing, X., Wu, F., Zhu, X., Dong, X., Ma, F., Li, Z.: Nulti-spectral low-rank structured dictionary learning for face recognition. Pattern Recogn. **59**, 14–25 (2016)
4. Baker, S., Kanade, T.: Limits on super-resolution and how to break them. IEEE Trans. Pattern Anal. Mach. Intell. **24**(9), 1167–1183 (2002)
5. Wang, X., Tang, X.: Hallucinating face by eigen transformation. IEEE Trans. Syst. Man Cybern. Part C Appl. Rev. **35**(3), 425–434 (2005)
6. An, L., Bhanu, B.: Face image super-resolution using 2D CCA. Sig. Process. **103**, 184–194 (2014)
7. Chang, H., Yeung, D.-Y., Xiong, Y.: Super-resolution through neighbor embedding. In: IEEE Computer Society Conference on Computer Vision and Pattern Recognition (CVPR), pp. 1275–1282 (2004)
8. Yang, J., Wright, J., Huang, T.S., Ma, Y.: Image super-resolution via sparse representation. IEEE Trans. Image Process. **19**(11), 2861–2873 (2010)
9. Gao, G., Yang, J.: A novel sparse representation based framework for face image super-resolution. Neurocomputing **134**, 92–99 (2014)
10. Ma, X., Zhang, J., Qi, C.: Hallucinating face by position-patch. Pattern Recogn. **43**(6), 2224–2236 (2010)
11. Jung, C., Jiao, L., Liu, B., Gong, M.: Position-patch based face hallucination using convex optimization. IEEE Sig. Process. Lett. **18**(6), 367–370 (2011)
12. Wang, Z., Hu, R., Wang, S., Jiang, J.: Face hallucination via weighted adaptive sparse regularization. IEEE Trans. Circuits Syst. Video Technol. **24**(5), 802–813 (2014)
13. Jiang, J., Hu, R., Wang, Z., Han, Z.: Noise robust face hallucination via locality-constrained representation. IEEE Trans. Multimedia **16**(5), 1268–1281 (2014)
14. Yang, J., Luo, L., Qian, J., Tai, Y., Zhang, F., Xu, Y.: Nuclear norm based matrix regression with applications to face recognition with occlusion and illumination changes. IEEE Trans. Pattern Anal. Mach. Intell. **39**(1), 156–171 (2017)
15. Gao, G., Yang, J., Jing, X.-Y., Shen, F., Yang, W., Yue, D.: Learning robust and discriminative low-rank representations for face recognition with occlusion. Pattern Recogn. **66**, 129–143 (2017)
16. Cai, J.F., Candes, E.J., Shen, Z.W.: A singular value thresholding algorithm for matrix completion. SIAM J. Optim. **20**(4), 1956–1982 (2010)
17. Thomaz, C.E., Giraldi, G.A.: A new ranking method for principal components analysis and its application to face image analysis. Image Vis. Comput. **28**(6), 902–913 (2010)

Multi-task Learning for Person Re-identification

Hua Gao[1](✉), Lingyan Yu[2], Yujiao Huang[1], Yiwei Dong[1],
and Sixian Chan[1]

[1] College of Computer Science, Zhejiang University of Technology,
288 Liuhe Road, West Lake District, 310023 Hangzhou City, China
{ghua, hyj0507}@zjut.edu.cn,
melinni0108@163.com, sxchan@163.com
[2] School of Electronic Engineering and Optoelectronic Technology,
Nanjing University of Science and Technology Zijin College, 89 Wenlan Road,
Qixia District, 210023 Nanjing City, China
106695110@qq.com

Abstract. Person re-identification is a hot topic due to its huge application potentials. Siamese network is a good method to learn feature representation in verification tasks and has been used in previous person re-identification research, but hard to convergence during training process. This paper presents a multi-task learning pipeline including Siamese loss for learning deep feature representations of people appearance. Firstly, we point out the defects of training a convolutional neural network (CNN) only with Siamese loss which is usually used for person re-identification. Secondly, a multi-task CNN for person re-identification combing the Softmax loss with Siameses loss is proposed. Finally, some experiments are carried out to test the performance of proposed multi-task person appearance learning pipeline. Experiments on various pedestrian dataset shows the effectiveness of our pipeline. Our method outperforms state-of-the-art person re-identification methods in some public datasets.

Keywords: Person re-identification · Siamese network · Multi-task learning · Convolutional neural network

1 Introduction

Person re-identification [1–17, 22, 27, 30, 33] is a hot topic and has attracted a great deal of research interests due to its huge application potentials. It is generally carried out by representing the appearances of person, giving a probe and finding a most similar person from gallery sets according to the distance of appearance representations. One of the most important task is to learning generic and robust feature representations of pedestrian. It is challenging because of complex variations of poses, viewpoints, lightings, blurring effects, image resolutions, camera settings, occlusions and background clutter across camera views.

Siamese network [28, 29, 31] is a kind of network framework with a Siamese loss, works well in verification tasks with few samples per class in train set. It is a method for training a similarity metric from data, which is to learn a function that maps input

© Springer International Publishing AG 2017
Y. Sun et al. (Eds.): IScIDE 2017, LNCS 10559, pp. 259–268, 2017.
DOI: 10.1007/978-3-319-67777-4_23

patterns into a target space such that the feature presentation in the target space approximates the "semantic" distance in the input space. It has been applied in face verification, image comparison and person re-identification [4, 5, 8, 13]. Yi et al. [8] use a Siamese deep neural network to learn a similarity metric from pedestrian images, the proposed method can jointly learn the color feature, texture feature and metric. To deal with the big variations of person images, they use binomial deviance to evaluate the cost between similarities and labels. Article [13] trains a feature extractor including convolutional network, recurrent layer and temporal pooling layer for video-based re-identification using a Siamese network architecture.

However, Siamese networks including no other kind of loss are hard to converge because of the unbalance of positive pair (image pair with same identification) and negative pairs (image pair with different identification). In this paper, we analyze the insufficient of training a CNN with a single Siamese network and present a multi-task learning method including Siamese loss for person re-identification. The remainder of this paper is organized as follows. Section 2 discusses related work, focusing on deep learning methods of person re-identification. Section 3 details the Siamese loss and proposed multi-task CNN. Section 4 presents the experimental methodology and results on a suite of standard datasets. We conclude with a summary and ideas for future work.

2   Related Work

Recent person re-identification methods focus on learning generic and robust feature representations of pedestrian. Feature representations learned by CNNs have shown their effectiveness in a wide range of person re-identification researches. In the person re-identification results [6] collected by smart surveillance interest group, and Market-1501 competition [7], many methods from top five in different dataset are CNNs based. This paper focus on deep learning methods of person re-identification.

The most challenging issues of person re-identification tasks are complex variations of poses, viewpoints and lightings. Article [2] proposes a filter pairing neural network (FPNN) to jointly handle misalignment, photometric and geometric transforms, occlusions and background clutter. All the key components are jointly optimized to maximize the strength of each component when cooperating with others. The learned filter pairs encode photometric transforms. Its deep architecture makes it possible to model a mixture of complex photometric and geometric transforms.

Article [3] proposes a method for simultaneously learning features and a corresponding similarity metric for person re-identification. The architecture include a layer that computes cross-input neighborhood differences, which capture local relationships between the two input images based on midlevel features from each input image. A high-level summary of the outputs of this layer is computed by a layer of patch summary features, which are then spatially integrated in subsequent layers. They also demonstrate that by initially training on an unrelated large data set before fine-tuning on a small target data set. It is worth mentioning that it works well in VIPeR dataset which is challenging.

Article [8] uses a Siamese deep neural network to learn a similarity metric from pedestrian images, the proposed method jointly learn the color feature, texture feature and metric in a unified framework. To deal with the big variations of person images, binomial deviance is used to evaluate the cost between similarities and labels.

[9] formulates a unified deep ranking framework that jointly tackles both of these key components to maximize their strengths. An effective learning-to-rank algorithm is proposed to minimize the cost corresponding to the ranking disorders of the gallery. The ranking model is solved with a deep CNN that builds the relation between input image pairs and their similarity scores through joint representation learning directly from raw image pixels. The proposed framework allows us to get rid of feature engineering and does not rely on any assumption. Additionally, our approach has better ability to generalize across datasets without fine-tuning.

Xiao et al. [10] present a pipeline for learning deep feature representations called domain guided dropout algorithm from multiple domains with CNNs, low level representations shared across several domains, while some others are effective only for a specific one.

Wu et al. [11] uses hand-crafted histogram features and deep features to represent person features, they propose a novel feature extraction model called feature fusion net, in which back propagation makes CNN features constrained by the handcrafted features. They find that hand-crafted features, such as color histogram features and multi-scale and multi-orientation Gabor features texture features are good complementary to deep features.

[12] presents fine-tuned CNN features for person re-identification. They improve the CNN features by conducting a fine-tuning on a pedestrian attribute dataset. In addition to the classification loss for multiple pedestrian attribute labels, they propose new labels by combining different attribute labels and use them for an additional classification loss function. The combination attribute loss forces CNN to distinguish more person specific information, yielding more discriminative features. After extracting features from the learned CNN, they apply conventional metric learning on a target re-identification dataset for further increasing discriminative power.

Article [13] proposes a recurrent neural network architecture for video-based person re-identification. Given the video sequence of a person, features are extracted from each frame using a CNN that incorporates a recurrent final layer, which allows information to flow between time-steps. The features from all timesteps are then combined using temporal pooling to give an overall appearance feature for the complete sequence. The convolutional network, recurrent layer, and temporal pooling layer, are jointly trained to act as a feature extractor for video-based re-identification using a Siamese network architecture. Their approach makes use of color and optical flow information in order to capture appearance and motion information which is useful for video re-identification.

[14] presents a novel multi-channel parts-based CNN model under the triplet framework for person re-identification. Specifically, the proposed CNN model consists of multiple channels to jointly learn both the global full-body and local body-parts features of the input persons. The CNN model is trained by an improved triplet loss function that serves to pull the instances of the same person closer, and at the same time push the instances belonging to different persons farther from each other in the learned feature space.

The triplet loss pays main attentions on obtaining correct orders on the training set. It still suffers from a weaker generalization capability from the training set to the testing set, thus resulting in inferior performance. [15] design a quadruplet loss, which can lead to the model output with a larger inter-class variation and a smaller intra-class variation compared to the triplet loss. As a result, our model has a better generalization ability and can achieve a higher performance on the testing set. In particular, a quadruplet deep network using a margin-based online hard negative mining is proposed based on the quadruplet loss for the person ReID. In extensive experiments, the proposed network outperforms most of the state-of-the-art algorithms on representative datasets which clearly demonstrates the effectiveness of our proposed method.

In order to improve person re-identification for real-world scenarios, [16] propose a new deep learning framework for person search, jointly handle pedestrian detection and person re-identification in a single CNN. An online instance matching loss function is proposed to train the network effectively, which is scalable to datasets with numerous identities.

3   The Proposed Approach

3.1   Siamese Loss

The function of Siamese loss [28, 29, 31] is to pull the samples with same labels and push away samples with different labels. Given an image patch x, the descriptor of x is a nonlinear mapping $D(x)$ that is expected to be discriminative, i.e. descriptors for image patches corresponding to the same point should be similar, and dissimilar otherwise. Therefore, taking the $L2$ norm as a similarity metric between descriptors, for an ideal descriptor we would wish that

$$d_D(x_1, x_2) = \|D(x_1) - D(x_2)\|_2 = \begin{cases} 0 & \text{if } p_1 = p_2 \\ \infty & \text{if } p_1 \neq p_2 \end{cases} \tag{1}$$

where p_1 and p_2 are labels of image patch x_1 and x_2 respectively.

In order to facilitate the solution of (1), formulate the Siamese loss as follows,

$$l_{verif}(x_1, x_2) = \delta \cdot d_D(x_1, x_2) + (1 - \delta) \cdot \max(T - d_D(x_1, x_2), 0). \tag{2}$$

where δ is the indicator function, which is 1 if $p_1 = p_2$, and 0 otherwise, T is a positive number. When performing back-propagation, the gradients are independently accumulated for both descriptors, but jointly applied to the weights, as they are shared.

3.2   Risk of Siamese Loss

It is hard to converge when training a CNN with a single Siamese loss because of the unbalance of positive pair (image pair with same identification) and negative pair (image pair with different identification). Given a train set which has N labels and L samples per label averagely. Then for a example, get a sample randomly from the remainders of the train set, the probability of same label is about $1/N$ while of different label is about $(N -1)/N$. However, N may very big in deep learning tasks based classification, the probability of negative pair is much bigger than positive pair. As a result, the Siamese loss will tend to increase the distances of samples with different labels.

Figure 1 illustrates the risk of non-convergence in a two-dimensional coordinate system. For a selected sample A1, samples A2, B1, C1 and D1 are the alternative points which can be choose for a pair with A1. A1 indexed the same label with A2 sample, and with different labels with B1, C1 and D1, illustrated with different colors and shapes in Fig. 1. Figure 1(a) illustrates gradient directions of different pairs relative to A1 sample with arrow lines, and the solid line circle illustrate the distance range of two samples with same label A1 and A2. In this case, only A2 sample pull A1, all the other three samples push A1 away. As a result, after the back propagation, A1 will away from samples B1, C1 and D1, but also away from A2, as illustrated in Fig. 1(b). In Fig. 1(b), all samples move a little, the solid line illustrates the original distance range of two samples with same label A1 and A2 as illustrated in Fig. 1(a), and the dotted line circle illustrates the new distance range of A1 and A2, the radius change bigger which means the distance of A1 and A2 is bigger, as the result of the unbalance of positive pair and negative pair (three samples with different labels and only one sample with same label, so the ratio is 3:1). In practice, this unbalance is even more serious because the more labels.

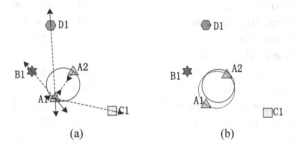

(a)   (b)

Fig. 1. Risk of non-convergence with Siamese loss. (a) gradient directions of different pairs relative to A1 sample with dotted lines, three samples with different labels and only one sample with same label relative to A1, the solid line circle illustrate the distance range of two samples with same label A1 and A2; (b) the new position of these 5 samples after the back propagation, the solid line illustrates the original distance range of two samples with same label A1 and A2 as illustrated in (a), and the dotted line circle illustrates the new distance range of A1 and A2.

3.3   Multi-task CNN with Siamese Loss

When training a Siamese network, a global constraint is help to decrease the risk of non-convergence. Softmax loss is used in classification task frequently which can be used all alone. Softmax loss is denoted as

$$l_{ident}(x_i) = p_i \log \hat{p}_i. \tag{3}$$

where x_i is person image patch and its feature vector represented as $D(x_i)$, p_i is the target probability distribution, where $p_i = 0$ for all i except $p_i = 1$ for the target class.

We combine Siamese loss and Softmax loss as the target of our CNN. The loss is as follows,

$$loss = \alpha \Sigma_{i,j \in N} l_{verif}\left(x_i, x_j\right) + (1 - \alpha) \sum\nolimits_{i \in N} l_{ident}(x_i) = p_i \log \hat{p}_i. \qquad (4)$$

where $\alpha \in (0, 1)$ is a trade-off between Siamese loss and Softmax loss, we suggest that $\alpha \leq 0.5$.

Since pedestrian images are usually quite small, we modify the network in [10] with three preceding 3×3 convolutional layers followed by six inception v3 modules [34] and one fully connected layers. In the final, add a Softmax loss and Siamese loss layer after FC1 layer. The Batch Normalization (BN) [32] layers are employed before each ReLU [35] layer, which accelerate the convergence process and avoid manually tweaking the initialization of weights and biases. Table 1 illustrate the outline of the network architecture.

Table 1. The outline of the network architecture

Type	Patch size/stride or remarks	Input size
Conv1	$3 \times 3/1$	$3 \times 144 \times 56$
Conv2	$3 \times 3/1$	$32 \times 144 \times 56$
Conv3	$3 \times 3/1$	$32 \times 144 \times 56$
MaxPool1	$2 \times 2/2$	$32 \times 144 \times 56$
Inception1a	Inception a	$32 \times 72 \times 28$
Inception1b	Inception b	$256 \times 72 \times 28$
Inception2a	Inception a	$384 \times 36 \times 14$
Inception2b	Inception b	$512 \times 36 \times 14$
Inception3a	Inception a	$768 \times 18 \times 7$
Inception3b	Inception b	$1024 \times 18 \times 7$
AvePool1	$9 \times 4/1$	$1536 \times 9 \times 4$
FC1		$1536 \times 1 \times 1$
Softmax	Classification	$256 \times 1 \times 1$
Siamese	Verification	$2 \times 256 \times 1 \times 1$

Inception structures a and b are illustrated in Fig. 2.

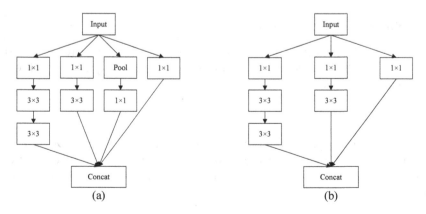

(a)   (b)

Fig. 2. Inception structures used in this paper. (a) Inception a structure; (b) Inception b structure

4   Experiments

4.1   Datasets and Protocols

We conducted experiments on several popular pedestrian datasets, including CUHK01 [18], CUHK03 [2], PRID [19], VIPeR [20], 3DPeS [21], Market-1501, ETH123 [23] and i-LIDS [36]. There exist many challenging person re-identification benchmark datasets. In our experiments, we chose seven of them followed [10] and [17] for training. A difference is that we replace Shinpuhkan dataset with ETH123 dataset. Table 2 illustrates the statistics of the training datasets and evaluation protocols.

Table 2. Statistics of the training datasets and evaluation protocols

Dataset	#ID	#Training images	#Training images	#Prob ID	#Galory ID
CUHK03	1467	21012	21012	100	100
CUHK01	971	1552	1552	485	485
PRID	385	2997	2997	100	649
VIPeR	632	506	506	316	316
3DPeS	193	420	420	96	96
i-LIDS	119	194	194	60	60
ETH123	146	6870	6870	0	0

During training, we set α in Eq. (4) 0.2, and add Siamese loss after 100 epochs with Softmax loss lonely when its loss is stable.

4.2   Comparison with State-of-the-Art Methods

We compare the results of our approach with state-of-the-art methods on all the six test datasets from the survey of Person Re-identification Results [6] collected by smart surveillance interest group and Market-1501 competition [7], which are VIPeR, PRID, CUHK01, CUHK03, i-LIDS and Market-1501 dataset. For the six test datasets, the methods with the highest accuracies are [10, 13, 24–26] respectively.

In order to validate our approach, we first obtain a baseline by training the CNN in training set with Softmax and Siamese loss respectively. Then we train our multi-task

Table 3. CMC top-1 accuracies of different methods

Method	CUHK01	CUHK03	PRID	VIPeR	3DPeS	Market-1501	i-LIDS
Xiao [10]	66.6	75.3	64.0	38.6	56.0	–	64.6
McLaughlin [13]	–	–	70.0	17.0	–	–	58.0
Chen [24]	–	–	–	**53.54**	57.29	51.9	–
Raphael [25]	61.2	–	68.8	51.60	–	–	–
Hermans [26]	–	–	–	–	–	**86.67**	–
Softmax	81.0	99.8	99.9	10.8	74.2	54.4	99.9
Multi-task	**87.6**	**99.9**	**99.9**	15.2	**82.6**	62.8	**99.9**

framework. The results are summarized in Table 3. Experiments on CUHK01, CUHK03, PRID, 3DPeS and i-LIDS pedestrian datasets show the effectiveness of our pipeline. Accuracy in VIPeR dataset is much lower than the best method. There are only 2 images per ID in VIPeR dataset, and great pose changes in those 2 images for every people. We do not do any data balance for the training set. And the multi-task method outperforms the same CNN but with only Siamese loss and Softmax loss in those 8 data set (Table 2).

5   Conclusion

In this paper, we raise the question of training Siamese network in person re-identification. The risk of non-convergence and its reason is analyzed, and an effective multi-task pipeline with Siamese loss is presented to improve the feature learning process. We conduct extensive experiments on multiple person re-identification datasets to validate our method. Moreover, our results outperform state-of-the-art ones by large margin on most of the datasets, which demonstrates the effectiveness of the proposed method.

Acknowledgements. A project supported by Scientific Research Fund of Zhejiang Provincial Education Department (Grant No. Y201534841); Supported by National Natural Science Foundation of China (Grant No. 61503338) and Natural Science Foundation of Zhejiang Province (Grant No. LQ15F030005).

References

1. Alexander, H., Lucas, B., Bastian, L.: In defense of the triplet loss for person re-identification. arXiv preprint arXiv:1703.07737 (2017)
2. Li, W., Zhao, R., Xiao, T., Wang, X.: DeepReID: deep filter pairing neural network for person re-identification. In: 2014 IEEE Conference on Computer Vision and Pattern Recognition (CVPR), pp. 152–159. IEEE Press, Columbus (2014)
3. Ahmed, E., Jones, M., Marks, T.K.: An improved deep learning architecture for person re-identification. In: 2015 IEEE Conference on Computer Vision and Pattern Recognition (CVPR), pp. 3908–3916. IEEE Press, Boston (2015)
4. Zagoruyko, S., Komodakis, N.: Learning to compare image patches via convolutional neural networks. 2015 In: IEEE Conference on Computer Vision and Pattern Recognition (CVPR), pp. 4353–4361. IEEE Press, Boston (2015)
5. Sergey, Z., Nikos, K.: MatchNet: unifying feature and metric learning for patch-based matching. In: 2015 IEEE Conference on Computer Vision and Pattern Recognition (CVPR), pp. 3279–3286. IEEE Press, Boston (2015)
6. Person Re-identification Results. http://www.ssig.dcc.ufmg.br/reid-results/
7. State of the art on the Market-1501 dataset. http://liangzheng.com.cn/Project/state_of_the_art_market1501.html
8. Yi, D., Lei, Z., Liao, S.C., Li, S.Z.L.: Deep metric learning for practical person re-identification. In: 22nd International Conference on Pattern Recognition, Stockholm, pp. 34–39 (2014)

9. Chen, S.Z., Guo, C.C., Lai, J.H.: Deep ranking for person re-identification via joint representation learning. Trans. Image Process. **25**, 2353–2366 (2016)
10. Xiao, T., Li, H., Ouyang, W., Wang, X.: Learning deep feature representations with domain guided dropout for person re-identification. In: 2016 IEEE Conference on Computer Vision and Pattern Recognition (CVPR), pp. 1249–1258. IEEE Press, Las Vegas (2016)
11. Wu, S., Chen, Y.C., Li, X., Wu, A.C., You, J.J., Zheng, W.S.: An enhanced deep feature representation for person re-identification. In: 2016 IEEE Winter Conference on Applications of Computer Vision, pp. 1–8. IEEE Press, Waikoloa (2016)
12. Tetsu, M., Einoshin, S.: Person re-identification using CNN features learned from combination of attributes. In: 23rd International Conference on Pattern Recognition (ICPR), pp. 2428–2433. IEEE Press, Cancun (2016)
13. McLaughlin, N., Martinez del Rincon, J., Miller, P.: Recurrent convolutional network for video-based person re-identification. In: 2016 IEEE Conference on Computer Vision and Pattern Recognition (CVPR), pp. 1325–1334. IEEE Press, Las Vegas (2016)
14. Cheng, D., Gong, Y., Zhou, S.: Person re-identification by multi-channel parts-based CNN with improved triplet loss function. In: 2016 IEEE Conference on Computer Vision and Pattern Recognition (CVPR), pp. 1335–1344. IEEE Press, Las Vegas (2016)
15. Chen, W.H., Chen, X.T., Zhang, J.G., Huang, K.Q.: Beyond triplet loss: a deep quadruplet network for person re-identification. CoRR abs/1704.01719 (2017)
16. Xiao, T., Li, S., Wang, B.C., Lin, L., Wang, X.G.: Joint detection and identification feature learning for person search. In: IEEE Conference on Computer Vision and Pattern Recognition (CVPR). IEEE Press, Salt Lake City (2017)
17. Paisitkriangkrai, S., Shen, C., Hengel, A.V.D.: Learning to rank in person re-identification with metric ensembles. In: 2015 IEEE Conference on Computer Vision and Pattern Recognition (CVPR), pp. 1846–1855. IEEE Press, Boston (2015)
18. Li, W., Wang, X.: Locally aligned feature transforms across views. In: 2013 IEEE Conference on Computer Vision and Pattern Recognition, pp. 3594–3601. IEEE Press, Portland (2013)
19. Hirzer, M., Beleznai, C., Roth, P.M., Bischof, H.: Person re-identification by descriptive and discriminative classification. In: Heyden, A., Kahl, F. (eds.) SCIA 2011. LNCS, vol. 6688, pp. 91–102. Springer, Heidelberg (2011). doi:10.1007/978-3-642-21227-7_9
20. Gray, D., Brennan, S., Tao, H.: Evaluating appearance models for recognition, reacquisition, and tracking. In: 10th IEEE International Workshop on Performance Evaluation of Tracking and Surveillance (PETS), pp. 71–91 (2007)
21. Baltieri, D., Vezzani, R., Cucchiara, R..: 3DPeS: 3D people dataset for surveillance and forensics. In: 11th Proceedings of 2011 Joint ACM Workshop on Human Gesture and Behavior Understanding, pp. 59–64. ACM Press, New York (2011)
22. Zheng, W.S., Gong, S., Xiang, T.: Associating groups of people. In: British Machine Vision Conference (BMVC 2009), London, pp. 1–11 (2009)
23. Schwartz, W.R., Davis, L.S.: Learning discriminative appearance-based models using partial least squares. In: XXII Brazilian Symposium on Computer Graphics and Image Processing (SIBGRAPI 2009), Rio de Janeiro, pp. 322–329 (2009)
24. Chen, D.P., Yuan, Z.J., Chen, B.D., Zheng, N.N.: Similarity learning with spatial constraints for person re-identification. In: 2016 IEEE Conference on Computer Vision and Pattern Recognition (CVPR), pp. 1268–1277. IEEE Press, Las Vegas (2016)
25. Raphael, P., William, R.S.: Kernel cross-view collaborative representation based classification for person re-identification. In: 2016 IEEE Conference on Computer Vision and Pattern Recognition (CVPR). IEEE press, Las Vegas (2016)
26. Hermans, A., Beyer, L., Leibe, B.: In defense of the triplet loss for person re-identification. Arxiv (2017)

27. Zheng, J.W., Yang, P., Chen, S.Y., Shen, G.J., Wang, W.L.: Iterative re-constrained group sparse face recognition with adaptive weights learning. IEEE Trans. Image Process. **26**, 2408–2423 (2017)
28. Chopra, S., Hadsell, R., Lecun, Y.: Learning a similarity metric discriminatively, with application to face verification. In: 2005 IEEE Computer Society Conference on Computer Vision & Pattern Recognition, vol. 1, pp. 539–546 (2005)
29. Norouzi, M., Fleet, D.J., Salakhutdinov, R.: Hamming distance metric learning. In: 2013 Neural Information Processing Systems, vol. 2, pp. 1061–1069 (2013)
30. Shen, F.M., Shen, C.H., Zhou, X., Yang, Y., Shen, H.T.: Face image classification by pooling raw features. Pattern Recogn. (PR) **54**, 94–103 (2016)
31. Sun, Y., Chen, Y.H., Wang, X.G., Tang, X.O.: Deep learning face representation by joint identification-verification. In: NIPS 2014, pp. 1988–1996 (2014)
32. Ioffe, S., Szegedy, C.: Batch normalization: accelerating deep network training by reducing internal covariate shift. In: Computer Science 2015 (2015)
33. Kumar, V., Namboodiri, A.M., Paluri, M., Jawahar, C.V.: Pose-aware person recognition. CoRR (2017)
34. Szegedy, C., Vanhoucke, V., Ioffe, S., Shlens, J., Wojna, Z.: Rethinking the inception architecture for computer vision. In: 2016 IEEE Conference on Computer Vision and Pattern Recognition (CVPR), pp. 2818–2826. IEEE press, Las Vegas (2016)
35. Krizhevsky, A., Sutskever, I., Hinton, G.E.: ImageNet classification with deep convolutional neural networks. In: 2012 International Conference on Neural Information Processing Systems, vol. 25, pp. 1097–1105 (2012)
36. Ding, S., Lin, L., Wang, G., Chao, H.: Deep feature learning with relative distance comparison for person re-identification. Pattern Recogn. **48**(10), 2993–3003 (2015)

Database Search Algorithm Based on Track Predicting in Fingerprinting Localization

Deyue Zou$^{(\boxtimes)}$, Yuqun Guo, and Xin Liu

Dalian University of Technology, Linggong Street 2, Dalian 116024, China
{zoudeyue, liuxinstar1984}@dlut.edu.cn,
guoyiqun1996@outlook.com

Abstract. Fingerprint positioning is a common technique in the field of indoor positioning. As a result of avoiding and utilizing the complex indoor structure to block and reflect the signal effectively, it has the most necessary room-level positioning accuracy. Due to the large amount of data in the radio-map, the clustering algorithm has become a commonly used method to reduce the workload of the search. But the artificial clustering and automatic clustering have their own limitations. This paper proposes a predictive radio-map search strategy to accelerate the database search, which means taking the positioning results, predicted by the filtering algorithm, to be the priori information to accelerate the next positioning. The simulation results show that the algorithm is superior to the traditional clustering - localization strategy in positioning accuracy.

Keywords: Fingerprint location algorithm · Positioning accuracy · Clustering · Priori information

1 Introduction

Fingerprint positioning is often faced with the problem of the large radio-map, resulting in a very large search volume. Positioning algorithm cannot extract valid information quickly from the fingerprint database, making the real-time of positioning poor. The conventional solution is to cluster the radio-map to reduce the search workload of the fingerprints. Before the specific positioning, the user calculates the Euclidean distance from cluster heads (Which can be regarded as a signal feature vector that characterizes the entire cluster) to find their own cluster and run a deep traversal search in its radio-map database to locate. In this way, we can grade the search work of the radio-map to reduce the search volume.

The clustering algorithms are also being improved continuously. For example, in the literature [1], the HIT team compared three different clustering algorithms to strike a balance between positioning accuracy and algorithm complexity. The literature [2], published in the IWCMC in 2013, used a clustering algorithm to facilitate the fingerprint positioning in the outdoor environment applications. In 2016, Zhang [3] proposed domain clustering method to improve the positioning accuracy. It is well known that the compression sensing algorithm is widely used to reduce the computational complexity of the system. In literature [4], the clustering algorithm is used to

© Springer International Publishing AG 2017
Y. Sun et al. (Eds.): IScIDE 2017, LNCS 10559, pp. 269–276, 2017.
DOI: 10.1007/978-3-319-67777-4_24

assist the compression sensing algorithm. So we can see that the clustering algorithm is widely used. The clustering in the PMLC algorithm proposed in literature [5] is also a very important part. The literature [6] presented that the cluster recognition process itself can be seen as a positioning process in some applications where the accuracy is not high. The literature [7] proposed that floor resolution itself is a low-precision positioning process, so the clustering process is even more important. Therefore, reducing the computational cost of the clustering algorithm is also a direction of research in literature [8]. In the literature [9] published in 2010, RSS-based clustering is also used to reduce the amount of computing system. Lee proposed a new SVM algorithm for automatic clustering of fingerprints in literature [10]. The dynamic clustering idea proposed in literature [11] makes the application of fingerprints in non-calibration environments a reality. The grid estimation algorithm proposed in literature [12] further improved the reliability of the clustering process.

However, there are inherent contradictions in clustering. If the radio-map is arranged in too many clusters, the benefits of clustering will be lost. While, if radio-map is arranged in too few clusters, number of RP in each cluster is still a lot. On the other hand, it will introduce positioning error when the user's location is at the edge of the cluster, because all the RPs are on one side of the user's location. This cannot be completely avoided even if the overlapping area is preliminarily drawn while clustering. In addition, if the cluster head recognition error occurs, it will lead to a larger positioning error. The current clustering means include manual clustering which is based on the subjective cognition of the designer and automatic clustering which is based on mathematical algorithm. The former may cause the cluster signal feature domain ununiformed, and the latter may cause the location space domain ununiformed. Therefore, it is possible to introduce errors in the cluster head recognition, which will reduce the positioning accuracy.

To solve the problems above, this paper proposes a predictive search algorithm to provide a more efficient radio-map search method to improve the accuracy of positioning by taking priori information, which is predicted by filtering algorithm, into the search.

The arrangement of this article is as follows: The second part briefly introduces the background of the algorithm, and then the basic introduction of the algorithm. The fourth part is the simulation of the algorithm, and finally the conclusion and the acknowledgment.

2   Relevant Background

Fingerprint positioning technology refers to establishing a mapping relationship between the physical space position and the signal space feature by means of point-by-point measurement in the region, and allowing the user to estimate their position by comparing the mapping relationship. Fingerprint positioning is divided into offline process and online process, and the offline process is responsible for establishing the database. The relationship between the physical spatial location and the signal space feature is often stored in the database, and the database is called radio-map. The radio-map is composed of several reference points (RP), each of which stores its physical location and the signal feature vector.

Users complete the positioning in the online process. In the positioning process, the user calculates the Euclidean distance between the signal feature of test point (TP) and all the RPs in the radio-map to calculate the similarity between them, and records the K most similar RPs. The location of the K RPs can estimate the location of the user, and that is the KNN algorithm.

Considering the facts that the user uses positioning service in movement generally and the user is likely to have left the location after the positioning algorithm is completed, the positioning process needs to predict the user's location.

Position prediction is done by filtering algorithms, and classical filtering algorithms include Kalman filtering and particle filtering. The filtering algorithm predicts the next positioning by the result of the previous positioning of the user, thereby compensating for the delay of the positioning result due to the existence of the solution time.

3 Proposed Radio-Map Searching Strategy

The algorithm uses the filtering algorithm to predict the location information which is used to be the priori information, and then centers them and searches the surroundings in a circle. During the search, the user constantly updates the K selected RPs with the closest Euclidean distance between the signal feature (TP) and all the RPs.

When we search continuously for L times and K RPs does not change, we identify the K RPs the final RPs for the position estimation and the K RPs will be taken into the KNN algorithm to solve. The workflow of the algorithm is shown in Fig. 1.

Here, we take RP (i, j) to presents the reference point of row i in column j of the radio-map. The algorithm predicts the user's current position (X_i, Y_i). Then, it calculates RP (a, b), which is closest to (X_i, Y_i), and calculate the Euclidean distance between the RP (a, b) and TP in the signal feature space. Take RP (a, b) as the center, search the existence of the surrounding RP, and calculate the Euclidean distance in the signal feature space. Record the K RPs with the closest Euclidean distance in the signal feature space and set the search counter Nc to zero. Extend the search range outward a circle, calculate the RP on the extended circle and calculate the Euclidean distance in the signal feature space. Find the K RPs with the closest Euclidean distance from TP in the signal feature space. Check if the selected RP has changed. If it is, set the search counter Nc to zero; if not, then $Nc = Nc + 1$. Expand the search range outward and determine if Nc is equal to the default search limit L. If it is reached, put the K RPs into the KNN algorithm to locate the result (X_i, Y_i). If not, expand the search range.

The search range extends in the shape of a circle. Assuming that the search center point is recorded as RP (a, b), the total number of RPs is

$$N(f) = (2f - 3)^2 - 1 \tag{1}$$

when the search range is expanded to the f-th circle. And the number of RPs on the circle f is $N(f) - N(f - 1)$. Because the search process is based on the square-shaped diffusion, the search RP is composed of four parts, corresponding to the four sides of

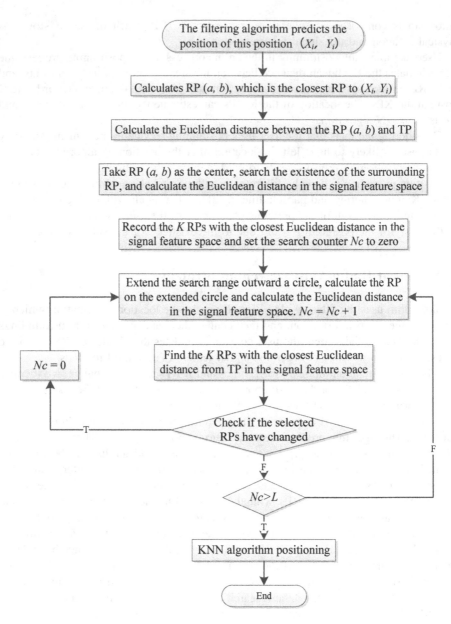

Fig. 1. The flow chart of the algorithm

the box. Take one of the most right sides of the edge for example, the RP number of the search can be calculated by the following way. Firstly, it can be expressed as $RP\left(a + f, b'_1 \sim b'_{2f-2}\right)$, where b'_i can be calculated as $b'_i = b - (i - f)$, where $i = 1 \sim f$. In the same way, we can get the calculation method of the other three side's search order.

By taking this gradual diffusion of the search order, we can effectively avoid users' jamming positioning error when crossing the cluster and improve positioning accuracy. At the same time, the algorithm also eliminates the need for hierarchical processes, which reduce the complexity of the system.

4 Simulation Verification

In order to verify the effectiveness of the algorithm, the research team conducted a simulation experiment. The simulation environment is based on the cost 231 model, and the indoor environment is constructed by MATLAB software.

The environment consists of 6 rooms and 1 corridor, as shown in Fig. 2. In order to contrast with the traditional cluster-locating scheme, we also cluster the radio-map in the simulation process. Clustering is based on the structure of the room, represented by the Arabic numerals in the figure. And in the process of localization of this algorithm, the system uses uncluttered radio-map.

The simulation of the positioning accuracy is shown in Fig. 3. It can be seen that the positioning accuracy of the system is smaller than that without clustering. After the application of the algorithm, we can see that the positioning accuracy is equal to the algorithm without clustering when $L = 10$, and the positioning accuracy is higher than

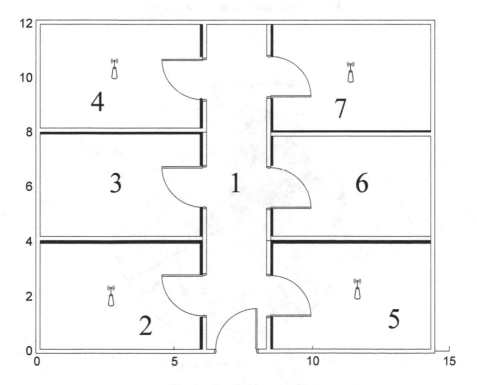

Fig. 2. Simulation scenario

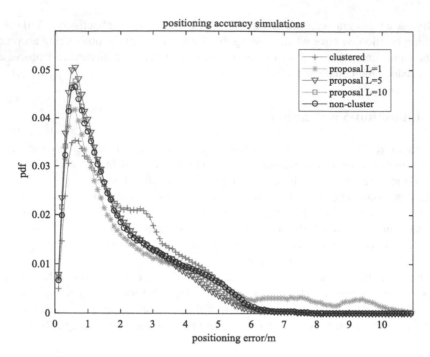

Fig. 3. Positioning accuracy simulations

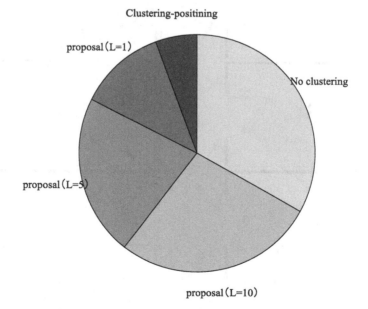

Fig. 4. Search speed comparison

that without clustering when $L = 5$. The reason is that when the L value is small, the system can avoid searching the remote RP. Because the person position changes continuously, the far RP will only be interference for positioning. Therefore, the L value constraints the search range dynamically and reasonably to avoid the introduction of interference.

Figure 4 shows the difference in the positioning speed of the centralized algorithm. The proportion indicates the average number of searched RPs to complete positioning. It can be seen that the lowest precision clustering is the fastest, and the algorithm slows down with the increase of the L value, but it is still better than the non-clustering system.

5 Conclusion

In this paper, a radio-map searching algorithm is proposed to eliminate the shortcomings of traditional clustering algorithm. The algorithm centers the last position and search the radio-map on the center in a diffuse way. The algorithm can improve the speed of radio-map search, increase the positioning accuracy, and avoid the phenomenon of jamming and so on.

Acknowledgment. This research is supported by the Fundamental Research Funds for the Central Universities DUT16RC (3)100.

References

1. Zou, G., Ma, L., Zhang, Z., Mo, Y.: An indoor positioning algorithm using joint information entropy based on WLAN fingerprint. In: Fifth International Conference on Computing, Communications and Networking Technologies (ICCCNT), Hefei, pp. 1–6 (2014)
2. Li, K., Bigham, J., Tokarchuk, L., Bodanese, E.L.: A probabilistic approach to outdoor localization using clustering and principal component transformations. In: 2013 9th International Wireless Communications and Mobile Computing Conference (IWCMC), Sardinia, pp. 1418–1423 (2013)
3. Zhang, W., Hua, X., Yu, K., Qiu, W., Zhang, S.: Domain clustering based WiFi indoor positioning algorithm. In: 2016 International Conference on Indoor Positioning and Indoor Navigation (IPIN), Alcala de Henares, pp. 1–5 (2016)
4. Feng, C., Au, W.S.A., Valaee, S., Tan, Z.: Received-signal-strength-based indoor positioning using compressive sensing. IEEE Trans. Mob. Comput. **11**(12), 1983–1993 (2012)
5. Premchaisawatt, S., Ruangchaijatupon, N.: Enhancing indoor positioning based on partitioning cascade machine learning models. In: 2014 11th International Conference on Electrical Engineering/Electronics, Computer, Telecommunications and Information Technology (ECTI-CON), Nakhon Ratchasima, pp. 1–5 (2014)
6. Dousse, O., Eberle, J., Mertens, M.: Place learning via direct WiFi fingerprint clustering. In: 2012 IEEE 13th International Conference on Mobile Data Management, Bengaluru, Karnataka, pp. 282–287 (2012)

7. Zhong, W., Yu, J.: WLAN floor location method based on hierarchical clustering. In: 2015 3rd International Conference on Computer and Computing Science (COMCOMS), Hanoi, pp. 41–44 (2015)
8. Lin, H., Chen, L.: An optimized fingerprint positioning algorithm for underground garage environment. In: 2016 International Conference on Information Networking (ICOIN), Kota Kinabalu, pp. 291–296 (2016)
9. Liu, X.-C., Zhang, S., Zhao, Q.-Y., Lin, X.-K.: A real-time algorithm for fingerprint localization based on clustering and spatial diversity. In: International Congress on Ultra-Modern Telecommunications and Control Systems, Moscow, pp. 74–81 (2010)
10. Lee, C.W., Lin, T.N., Fang, S.H., Chou, Y.C.: A novel clustering-based approach of indoor location fingerprinting. In: 2013 IEEE 24th Annual International Symposium on Personal, Indoor, and Mobile Radio Communications (PIMRC), London, pp. 3191–3196 (2013)
11. Lin, Y.T., Yang, Y.H., Fang, S.H.: A case study of indoor positioning in an unmodified factory environment. In: 2014 International Conference on Indoor Positioning and Indoor Navigation (IPIN), Busan, pp. 721–722 (2014)
12. Cai, D.: A retail application based on indoor location with grid estimations. In: 2014 International Conference on Computer, Information and Telecommunication Systems (CITS), Jeju, pp. 1–4 (2014)

Cascade Error-Correction Mechanism for Human Pose Estimation in Videos

Huibing Dai, Lihuo He$^{(\boxtimes)}$, Xinbo Gao, Zhaoqi Guo, and Wen Lu

School of Electronic Engineering, Xidian University, Xi'an 710071, China
hbdai@stu.xidian.edu.cn,
{lhhe,xbgao,luwen}@mail.xidian.edu.cn,
zhaoqi.aiden.guo@gmail.com

Abstract. This paper aims to estimate constantly changing human poses in videos. Traditional methods fail to locate wrists accurately, which is a tremendously challenging task. We propose a three-stage framework for human pose estimation, emphasizing on the improvement of wrist location accuracy. The first stage applies the pictorial structure model to localize the positions of all joints in each frame and calculate the posterior edge distribution probability of wrists. In the second stage, a visual tracking based method is fused into the posterior edge distribution probability of wrists to obtain the wrist location. Instead of directly predicting the wrist location, the third stage designs a novel cascade error-correction mechanism (CECM) to correct the predicted results. In addition, a skin-based proposal and multifarious reinitializing modes are also involved in CECM. Experiments are conducted on the two public datasets, and results demonstrate the superiority of the proposed algorithm compared to state-of-the-art methods.

Keywords: Human pose estimation · Pictorial structure · Visual tracking · Skin-based proposal · Cascade error-correction mechanism

1 Introduction

Human pose estimation is the process of inferring the 2D or 3D human body joint positions from still images or videos. In this paper, we focus our attention on human pose estimation in 2D videos. Human pose estimation is a fundamental problem in computer vision, because it is widely used in many real-world applications, such as video surveillance, human-computer interaction, digital entertainment, medical imaging and sports scenes. It is also a very challenging problem, since there are many sources of uncertainty for human pose estimation, e.g., variation of pose, cluster background, complex scene and severe target occlusion.

To address these issues, a large number of methods are proposed to estimate human poses in still images or videos. Most methods in still images can be divided into two groups. The first is based on the pictorial structure (PS) models [1, 2] that represent the body configuration as a collection of rigid parts and a set of pair part connections. Ramanan et al. [1] propose an improved pictorial structure model called a flexible mixture-of-parts model, whose key idea is the "mini part" model can approximate

Y. Sun et al. (Eds.): IScIDE 2017, LNCS 10559, pp. 277–289, 2017.
DOI: 10.1007/978-3-319-67777-4_25

deformations. In [2], another improved pictorial structure model with poselet presentations attempts to capture important dependencies between non-adjacent body parts. The second group focuses on training ConvNets [3–6] as powerful feature generators for the task of human pose estimation. Toshev *et al.* [3] employ a multi-stage convolutional network to regress joint coordinates. Fan *et al.* [4] utilize a multi-task network to extract local and global features for predicting joint heatmaps. Wei *et al.* [5] design a pose machine that provides a sequential prediction framework for learning spatial relationships. Chu *et al.* [6] propose an end-to-end framework with a multi-context attention mechanism for human pose estimation. Compared to still images, temporal information is another cue in videos, which promotes more accurate results for human pose estimation in videos. Most methods for human pose estimation in videos are based on the tracking-by-detection framework [7–10]. Ramanan *et al.* [7] select the optimal pose from generated pose candidates in each frame according the consistency and continuity of the video. Ramakrishna *et al.* [8] propose an algorithm to model the symmetric structures of human body parts. Tokola *et al.* [9] focus on solving the graph optimization problem by introducing the tracking-by-detection framework. In [10], Cherian *et al.* design a strategy to decompose poses into limbs and recombine limbs to obtain the optimal pose in videos. Although existing methods have make process in human pose estimation, they are still unable to overcome complete difficulties, particularly for the low accuracy of the joint wrist.

In this work, we divide the task of human pose estimation in videos into the following three stages. In the first stage, we infer all body joint positions and calculate the posterior edge distribution probability of wrists in each frame by using the pictorial structure model. In the second stage, a method based visual tracking is fused into the posterior edge distribution probability of wrists from the first stage, which aims to obtain the wrist location. Instead of directly predicting the wrist, a novel strategy for correcting the predicted wrist positions is designed in the third stage, in which a skin-based proposal and multifarious reinitializing modes are involved, in a process we call Cascade Error-Correction Mechanism (CECM).

The rest of the paper is organized as follows. Section 2 introduces the proposed human pose estimation method. Then the experimental results and analysis are presented in Section 3. Finally, Section 4 concludes the paper.

2   Proposed Human Pose Estimation Method

Figure 1 shows an overview of the proposed algorithm. In the first stage, we infer all body joint positions and calculate the posterior edge distribution probability of wrists in each frame by using the pictorial structure model. In the second stage, a method based visual tracking is fused the posterior edge distribution probability of wrists from the first stage, which aims to obtain the wrist location. Instead of directly predicting the wrist, the cascade error-correction mechanism for correcting the predicted wrist positions is designed in the third stage. Finally, we can obtain the optimal pose in each frame.

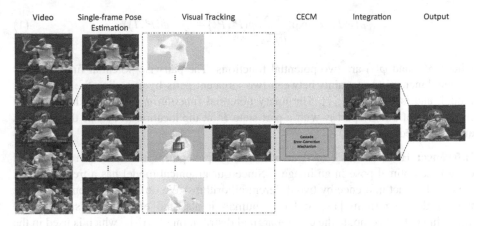

Fig. 1. Framework of the proposed human pose estimation method.

2.1 Human Pose Estimation in Each Frame

The tasks of this section are all joint locations of the upper body in each frame and the inference of the posterior edge distribution probability of wrists. Here, we employ the pictorial structure (PS) model in [11] as our basic presentation and two inference algorithms [12] to estimate the human pose in each frame.

PS Model: The pictorial structure model is shown in Fig. 2. It represents the human body as an undirected graph $G(E, V)$, in which the graph vertex $v_i \in V$ corresponds to a body part and the graph edge E models the kinematic constraints between two adjacent parts v_i and v_j. For instance, the lower limb of the left arm is physically constrained to the upper limb of the left arm.

The body pose is defined by the body configuration $P = \{P^1, \cdots, P^i, \cdots P^n\}$. A body part state P^i is defined as a vector $P^i = (x^i, y^i, \theta^i)$, in which (x^i, y^i) and θ^i denote the central coordinates and the direction of a body part rectangular box, respectively. Given an image I, the posterior of the body configuration P is defined as

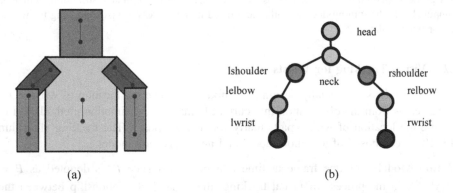

Fig. 2. The graphical model. (a) the human upper body structure, (b) the pictorial structure model of the upper body.

$$p(P|I) \propto \exp(\sum_{(i,j)\in E} \varphi(P^i, P^j) + \sum_{i\in V} \phi(P^i, I)) \qquad (1)$$

where $\varphi(\cdot)$ and $\phi(\cdot)$ are two potential functions. The pairwise potential function $\varphi(\cdot)$ models kinematic constraints between two adjacent parts by encoding the body prior model, and is defined as [1]. The unary potential function $\phi(\cdot)$ reflects the detection confidence of a body part v_i with a part state P^i at the location (x^i, y^i) of the image I, and is defined as [1].

Inference: Here, we seek to deal with the graph optimization problem in (1) and obtain the optimal pose in an image I. Since our graphical model has a tree structure, we can do exact inference by two different algorithms. We do MAP inference with the max-product algorithm [11] to locate human joints, and employ the sum-product algorithm [11] to compute the exact marginal distribution of wrists, which is used in the process of visual tracking for wrists. The results of inference by two different algorithms are shown in Fig. 3.

(a)   (b)

Fig. 3. Results of inference. (a) the globally optimal body configuration obtained by the max-product algorithm, (b) green areas are the exact wrist marginal distribution, which is computed with the sum-product algorithm and fused in the process of visual tracking for wrists. (Color figure online)

2.2   Visual Tracking for Wrists

This section aims to specially design a method to estimate the location of wrists. Our idea streams from the observation that current human pose estimation methods fail to capture the location of wrists, particularly for unusual poses. The tracking algorithm [13–18] is composed of a motion model and an observation model.

Motion Model: For one frame at time t, a body part state P^i is denoted as $P^i = (x^i, y^i, \theta^i)$. In the process of visual tracking, the dynamical relationship between the states P^i_t and P^i_{t-1} obeys Brownian motion. The affine parameters of P^i_t is closely linked with those of P^i_{t-1}, which can be modeled by a Gaussian distribution $\mathcal{N}(\cdot)$. Consequently, $p(P^i_t|P^i_{t-1})$ becomes

$$p(P^i_t|P^i_{t-1}) = \mathcal{N}(P^i_t; P^i_{t-1}, \theta^i) \tag{2}$$

where θ^i is the diagonal covariance of each state variable P^i_t.

Observation Model: In the process of visual tracking, we can model the image observation using a probabilistic interpretation of principal component analysis in [13, 19]. The observation model $p(I_t|P^i_t)$ presents the probability that an image patch I_t is generated from the subspace, and it is codetermined by the distance-to-subspace d_t, and the distance-within-subspace d_w [20]. Consequently, $p(I_t|P^i_t)$ becomes

$$p(I_t|P^i_t) = p_{d_t}(I_t|P^i_t)p_{d_w}(I_t|P^i_t) = \mathcal{N}(I_t; \mu_i, U_iU^T_i + \varepsilon E)\mathcal{N}(I_t; \mu_i, U_i\sum_i^{-2} U^T_i) \tag{3}$$

where μ_i is the mean of the subspace U_i, εE is the additive Gaussian noise related to the identity matrix E, and $\sum_i^{-2} U^T_i$ is the matrix of singular values of the subspace U_i.

The process of visual tracking for wrists is shown in Fig. 4. In this section, we apply the tracking algorithm based particle filtering on optical flow images to generate the wrist candidates, and finally obtain the location of wrists by the fusion of the wrist candidates generated by visual tracking and the exact marginal distribution of wrists M_t obtained from the first stage of our framework. This fusion can guarantee that the wrist location results in the second stage are constrained by the human body structure. After the fusion, the observation model $p_{in}(I_t|P_t)$ becomes

$$p_{in}(I_t|P_t) = \lambda_{vt}p_{vt}(I_t|P_t) + \lambda_{pe}p_{pe}(M_t|P_t) \tag{4}$$

where $\lambda_{vt}, \lambda_{pe} \in \{\lambda_{vt}, \lambda_{pe}|\lambda_{vt} + \lambda_{pe} = 1\}$ are two parameters to balance the fusion; $p_{vt}(\cdot)$ is the observation model before the fusion, and $p_{pe}(\cdot)$ reflects the exact marginal distribution of wrists.

(a)   (b)   (c)

Fig. 4. A process of visual tracking for wrists. (a) the right wrist candidates are generated by visual tracking, (b) the exact marginal distribution of the right wrist obtained from the first stage of our framework, (c) the right wrist location in the image I by combining (a) and (b).

2.3   Cascade Error-Correction Mechanism for Wrists

In Sect. 2.2, a tracking algorithm is applied on optical flow images to estimate the location of wrists. However, we can find a fact that the method above can't effortlessly provide the exact location of wrists for each frame of a video. Particularly when the wrist keeps still in adjacent frames, the wrong wrist location happens because of fuzzy wrist HOG feature extracted from optical images. In addition, the elbow and the wrist have the similar HOG feature in optical flow images so that we can't distinguish between the elbow and the wrist easily.

To optimize the wrist location obtained from visual tracking, we propose a cascade error-correction mechanism (CECM). This is done in three steps: (1) given the optical flow response from visual tracking, it is compared with the setting thresholds to judge whether the location of wrists by visual tracking is correct; (2) when the wrist error location happens, the pictorial structure model is used to relocate the wrist positions; (3) a skin-based proposal is then proposed to judge whether the location of wrists by the pictorial structure model is correct. Considering the skin-based proposal, the location of wrists in the current frame is selectively replaced by the location of wrists in the last frame. Below we discuss these steps in detail.

Step 1: Utilizing the Optical Flow Response as a Proposal. For a given optical flow image I, the elbow is easily mistaken as the wrist because of their similar HOG feature in I, which produces a higher proportion of wrist error location in visual tracking. Therefore, the task of this step is to distinguish between the wrist and the elbow. We find a fact that the optical flow response from visual tracking changes in different tendency for wrists and elbows. Therefore, we can regard the optical flow response as a proposal of distinguishing between the wrist and the elbow. Given the optical flow response B, the rule that judges whether the location of wrists by visual tracking is correct becomes

$$\begin{cases} B(j_1) < \delta_1 \quad or \quad B(j_2) < \delta_2 \quad v_i = elbow \\ B(j_1) > \delta_1 \quad and \quad B(j_2) > \delta_2 \quad v_i = wrist \end{cases} \tag{5}$$

where $B(j_1)$ and $B(j_2)$ is respectively the j_1 column of and the j_2 column of the matrix; δ_1 and δ_2 are two threshold parameters. Here, we set $j_1 = 1$, $j_2 = 200$, $\delta_1 = 1.5$ and $\delta_2 = 1$.

Step 2: Relocation. Following Step 1, if the predicted joint is judged to be an elbow, we discard the tracking result and relocate the wrist by the location of wrists from Sect. 2.1, which is obtained by the pictorial structure model.

Step 3: Skin-Based Proposal. Since the pictorial structure model has limitation on effective expression for human poses in real world, we propose a skin-based proposal to examine if the relocation result of wrists is acceptable.

The skin-based proposal proceeds in two steps. Firstly, we design an adaptive skin detection model. Our idea streams from the observation that skin is one of classical descriptors for bare wrists and the face is almost the same with the wrist in color for one person. The adaptive skin detection model consists of a face detector and a skin detector. A face detector is used to determine the skin color of one person, which enhances the skin detection result when unusual lighting happens. A global skin detector is applied to detect the skin regions within the square boxes of locating wrists. The proportion γ of skin regions within a square box is

$$\gamma = \frac{n}{w \times l} \tag{6}$$

where n is the number of pixels assumed to be skin regions. w is the width of a square boxing. l is the length of a square boxing.

Secondly, we separately compute the proportion of skin regions within the square box of locating wrists in the last frame γ_1 and the proportion of skin regions within the square box of locating wrists in the current frame γ_2. Considering γ_1 and γ_2, the rule that judges if the relocation result of wrists in Step 3 is acceptable becomes

$$\begin{cases} \gamma_1 > \gamma_2 & and \quad \gamma_2 < \delta_3 \quad f = false \\ \gamma_1 \leq \gamma_2 & or \quad \gamma_2 \geq \delta_3 \quad f = true \end{cases} \tag{7}$$

where δ_3 is a threshold parameter and is taken as 0.05. f presents the judgement result for the relocation result of wrists in Step 3. When the value of f is false, it can judge that the wrist relocation in Step 3 performs badly and we discard the wrist relocation result once again. Inspired by detection-by-tracking [21], the positions of wrist in the current and last frames have close relationships, particularly when the wrist keeps still in adjacent frames of a video. When relocation fails, we assume the extreme case that the wrist keep still in adjacent frames and replace the location of wrist in the current frame by the location of wrist in the last frame to modify the inaccurate relocation result.

3 Experiments

In this work, we focus on estimating the upper human body pose, which is determined by six parts (head, torso, left/right upper arms, left/right lower arms). Our work is particularly dedicated to improving the lower arm location performance, since state of the art methods perform badly on to the lower arm location, which plays a key role on obtaining the optical human pose.

We first describe the datasets and the evaluation measure; then present analysis of gains from each stage in our algorithm; and finally compare our results to state of the art.

3.1 Datasets and Evaluation

Our experiments are conducted on two video datasets, which contain VideoPose2.0 [22] and VIPS-VideoPose [23].

VideoPose2.0. This dataset is composed of 44 video clips with a total 1286 frames, which is taken from the TV shows Friends and Lost. The length of each clip ranges from 2 s to 3 s. Referring to [22], we select 26 short clips as training data and 18 short clips as testing data from 44 video clips of VideoPose2.0. The dataset covers varied human poses particularly with foreshortened lower arms, which hindering the process of human pose estimation.

VIPS-VideoPose. This dataset is composed of 4 video clips with a total 414 frames, which is taken from the TV series Buffy the Vampire Slayer and the Wimbledon 2012 Mens Final. The dataset covers everyday actions and tennis-playing actions. Referring to [22], we employ 26 short clips as training data selected from VideoPose2.0 and use the whole clips on the dataset VIPS-VideoPose as testing data for doing the cross-dataset testing to verify the rationality of our method.

Evaluation Measure. The performance of the pose estimator is evaluated using the evaluation measure the same as [22]. A joint location is considered true or false according to its located distance to the ground truth. A joint location is acceptable if its located distance to the ground truth is within a given matching distance. A recall curve for each joint is obtained when a given matching distance ranges from 15 pixels to 40 pixels.

3.2 Component Evaluation

In this work, we propose a three-stage framework for human pose estimation. Here we respectively evaluate the stages of our algorithm on two dataset VideoPose2.0 and VIPS-VideoPose. Firstly, we remove the third stage called the CECM from our framework and get the location results of all joints by the pictorial structure model and the visual tracking. Secondly, we remove the second stage called visual tracking once again so that only the pictorial structure acts in the process of human pose estimation and obtain the location results of all joints without using visual tracking. By making comparison for them, we can make the conclusion that each stage of our algorithm greatly impacts on the location accuracy of all human body joints, which is shown in Fig. 5.

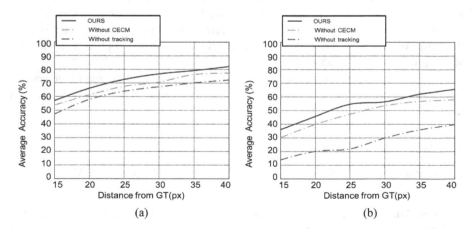

Fig. 5. Component evaluation. (a) the average location accuracy of all joints on VideoPose2.0, (b) the average location accuracy of all joints on VIPS-VideoPose.

3.3 Comparison to State-of-the-Art

On two datasets of VideoPose2.0 and VIPS-VideoPose, we compare our algorithm with four baseline pose estimators, including Ferrari et al.'s PS implementation [24], Park and Ramanan's Nbest model [7], Sapp et al.'s stretchable model [22] and Yang and Ramanan's flexible mixtures of parts model [1]. The training data for all baseline methods are the same as training data of our algorithm with 26 clips of VideoPose2.0, which guarantees the experiment fairness.

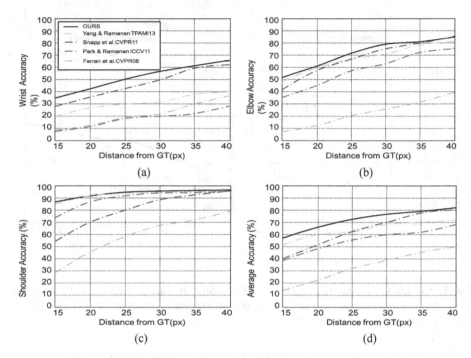

Fig. 6. Comparison to state-of-the-arts on VideoPose2.0. (a) the location accuracy of the wrist, (b) the location accuracy of the elbow, (c) the location accuracy of the shoulder, (d) the location average accuracy of all joints.

Since the accuracy of head and torso is nearly 100%, we respectively present recall curves for 6 joints (left/right shoulders, left/right elbows, left/right wrists) and an average recall curves for all 6 joints on VideoPose2.0 in Fig. 6 and on VIPS-VideoPose in Fig. 7. From two figures, we can see the superiority of our algorithm in joint location accuracy, particularly for the wrist joints. For example, giving 30 pixels as a matching

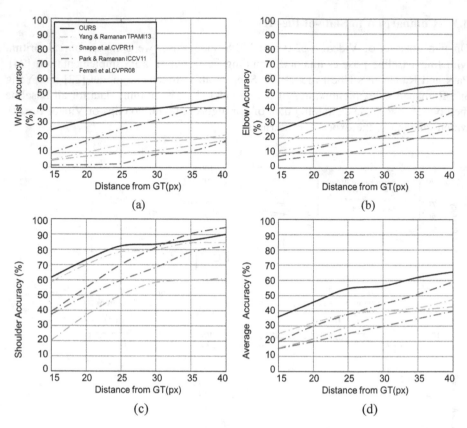

Fig. 7. Comparison to state-of-the-arts on VIPS-VideoPose. (a) the location accuracy of the wrist, (b) the location accuracy of the elbow, (c) the location accuracy of the shoulder, (d) the location average accuracy of all joints.

distance, our algorithm obtains 57.3% location accuracy for the joint wrist on VideoPose2.0, greatly boosting over the accuracy of state of the art, including 22.1% of [24], 20.1% of [7], 50.2% of [22] and 32.1% of [1]. In addition, example results of human pose estimation on the two datasets are shown in Fig. 8.

Fig. 8. Example predictions on poses on VideoPose2.0 and VIPS-VideoPose

4   Conclusion

We propose a three-stage framework for human pose estimation and have shown that our algorithm achieves greater improvement over the results of state-of-the-arts on the joint location. In the future, we can extend our algorithm to realize multiple human pose estimation. Furthermore, we would add fast methods to three stages of our framework in order to reduce the time for human pose estimation.

Acknowledgments. This work was supported in part by the National Natural Science Foundation of China under Grant 61432014, 61501349 and U1605252, in part by the National Key Research and Development Program of China under Grant 2016QY01W0204, in part by Key Industrial Innovation Chain in Industrial Domain under Grant 2016KTZDGY-02, in part by the Fundamental Research Funds for the Central Universities under Grant XJS17074 and JBX170218, in part by National High-Level Talents Special Support Program of China under Grant CS31117200001, in part by the Natural Science Basic Research Plan in Shaanxi Province of China under Grant 2017JM6050.

References

1. Yang, Y., Ramanan, D.: Articulated Human Detection with Flexible Mixtures of Parts. IEEE Trans. Software Eng. **35**(12), 2878–2890 (2013)
2. Pishchulin, L., Andriluka, M., Gehler, P.: Poselet Conditioned Pictorial Structures. IEEE Conference on Computer Vision and Pattern Recognition, pp. 588–595, Portland (2013)
3. Toshev, A., Szegedy, C.: Deeppose: human pose estimation via deep neural networks. IEEE Conference on Computer Vision and Pattern Recognition, pp. 1653–1660, Portland (2013)
4. Fan, X., Zheng, K., Lin, Y., Wang, S.: Combining local appearance and holistic view: Dual-Source Deep Neural Networks for human pose estimation. IEEE Conference on Computer Vision and Pattern Recognition, pp. 1347–1355, Boston (2015)
5. Wei, S. E., Ramakrishna, V., Kanade, T., Sheikh, Y.: Convolutional pose machines. IEEE Conference on Computer Vision and Pattern Recognition, pp. 4724–4732, Las Vegas (2016)
6. Chu, X., Yang, W., Ouyang, W., Ma, C., Yuille, A. L., Wang, X.: Multi-context attention for human pose estimation. IEEE Conference on Computer Vision and Pattern Recognition, Honolulu (2017)
7. Park, D., & Ramanan, D..: N-best maximal decoders for part models. IEEE International Conference on Computer Vision, pp. 2627–2634, Spain (2011)
8. Ramakrishna, V., Kanade, T., Sheikh, Y.: Tracking Human Pose by Tracking Symmetric Parts. IEEE Conference on Computer Vision and Pattern Recognition, Portland, pp. 3728–3735 (2013)
9. Tokola, R., Choi, W., Savarese, S.: Breaking the chain: liberation from the temporal Markov assumption for tracking human poses. In: IEEE Conference on Computer Vision and Pattern Recognition, Portland, pp. 2424–2431 (2013)
10. Cherian, A., Mairal, J., Alahari, K., Schmid, C.: Mixing body-part sequences for human pose estimation. In: IEEE Conference on Computer Vision and Pattern Recognition, Columbus, pp. 2353–2360 (2014)
11. Felzenszwalb, P.F., Huttenlocher, D.P.: Pictorial structures for object recognition. Int. J. Comput. Vis. **61**(1), 55–79 (2005)
12. Bishop, C.M.: Pattern Recognition and Machine Learning (Information Science and Statistics). Springer, New York (2006)
13. Li, X., Hu, W., Zhang, Z., Zhang, X., Luo, G.: Robust visual tracking based on incremental tensor subspace learning. In: IEEE Conference on Computer Vision and Pattern Recognition, Minneapolis, pp. 1–8 (2007)
14. Shao, L., Jones, S., Li, X.: Efficient search and localization of human actions in video databases. IEEE Trans. Circ. Syst. Video Technol. **24**(3), 504–512 (2014)
15. Saegusa, R., Metta, G., Sandini, G., Natale, L.: Developmental perception of the self and action. IEEE Trans. Neural Netw. Learn. Syst. **25**(1), 183 (2014)
16. Bousmalis, K., Zafeiriou, S., Morency, L.P., Pantic, M.: Infinite hidden conditional random fields for human behavior analysis. IEEE Trans. Neural Netw. Learn. Syst. **24**(1), 170 (2013)
17. Tao, D., Jin, L., Wang, Y., Li, X.: Rank preserving discriminant analysis for human behavior recognition on wireless sensor networks. IEEE Trans. Ind. Inform. **10**(1), 813–823 (2014)
18. Ding, C., Xu, C., Tao, D.: Multi-task pose-invariant face recognition. IEEE Trans. Image Process. **24**(3), 980 (2015)
19. Zhen, X., Shao, L., Li, X.: Action recognition by spatio-temporal oriented energies. Inf. Sci. **281**, 295–309 (2014)
20. Ross, D.A., Lim, J., Lin, R.S., Yang, M.H.: Incremental learning for robust visual tracking. Int. J. Comput. Vis. **77**(1), 12–141 (2008)

21. Andriluka, M., Roth, S., Schiele, B.: People-tracking-by-detection and people-detection-by-tracking. In: IEEE Conference on Computer Vision and Pattern Recognition, Anchorage, pp. 1–8 (2008)
22. Sapp, B., Weiss, D., Taskar, B.: Parsing human motion with stretchable models. In: IEEE International Conference on Computer Vision, Spain, pp. 1281–1288 (2011)
23. Zhao, L., Gao, X., Tao, D., Li, X.: Learning a tracking and estimation integrated graphical model for human pose tracking. IEEE Trans. Neural Netw. Learn. Syst. 26(12), 3176–3186 (2015)
24. Ferrari, V., Marinjimenez, M., Zisserman, A.: Progressive search space reduction for human pose estimation. In: IEEE International Conference on Computer Vision, Anchorage, pp. 1–8 (2008)

Objects

Index Tracking by Using Sparse Support Vector Regression

Yue Teng[1], Li Yang[2], Kunpeng Yuan[2], and Bo Yu[1(✉)]

[1] School of Mathematical Sciences, Dalian University of Technology,
Dalian 116024, People's Republic of China
yubo@dlut.edu.cn
[2] School of Science, Dalian University of Technology,
Panjin 124221, People's Republic of China

Abstract. In this paper a sparse support vector regression (SVR) model and its solution method are considered for the index tracking problem. The sparse SVR model is structured by adding a cardinality constraint in a ε-SVR model and the piecewise linear functions are used to simplify the model. In addition, for simplifying the parameter selection of the model a sparse variation of the ν-SVR model is considered too. The two models are solved by utilizing the penalty proximal alternating linearized minimization (PALM) method and the structures of the two models satisfy the convergence conditions of the penalty PALM method. The numerical results with practical data sets demonstrate that for the fewer sample data the sparse SVR models have better generalization ability and stability especially for the large-scale problems.

Keywords: Index tracking · Sparse support vector regression · Proximal alternating linearized minimization method · Cardinality constraints

1 Introduction

Index tracking is a popular passive investment strategy in stock market. The goal of index tracking is to reproduce the performance of a given market index, that is, given n constituent stocks of a market index, one selects some stocks and their weights to construct a tracking portfolio and uses this portfolio to represent the index as closely as possible. A market index can be considered as a combination of all the constituent stocks contained in it, thus a natural way to track the index is full replication, where all of the constituent stocks are purchased according to their weights in the index. However, in some indexes many constituent stocks may take up very small weights. In the practical market environment, fully replicating these indexes will cause high administration and transaction costs. Therefore, portfolio managers are often required to limit the number of constituent stocks in their tracking portfolios.

The index tracking problem has received a great deal of attention in the literature. Two comprehensive reviews can be found in [1, 2]. The classical approaches

Y. Sun et al. (Eds.): IScIDE 2017, LNCS 10559, pp. 293–315, 2017.
DOI: 10.1007/978-3-319-67777-4_26

for the index tracking problem include integer programming methods and heuristic methods. Canakgoz and Beasley [2] proposed two mixed integer linear programming formulations by adopting the regression viewpoint for index tracking and enhanced indexation problem, respectively. Konno and Wijayanayake [3] considered the minimal cost index tracking problem under non-linear and non-convex transaction costs and proposed a branch-and-bound algorithm to solve this class of problems. By using the constraint aggregation method, Okay and Akman [4] changed the formulations based on [1] to a mixed integer nonlinear programming problem and improved the calculation efficiency of the model. Most formulations of the index tracking problem are NP-hard, thus finding an exact solution of the problem will usually cost a large amount of computation, especially for large-scale problems. As a consequence, several heuristic methods had been proposed. Beasley et al. [1] proposed a generalized formulation of enhanced index tracking problem, where the index tracking return and excess return can be weighed by the decision maker, and used an evolutionary method to solve it. Gilli and Kellezi [5] investigated the performance of the threshold accepting heuristic method for the index tracking problem. Derigs and Nickel [6] considered using a multi-factor model to structure the stock returns and covariances and used a simulated annealing method to solve the problem.

The index tracking problem also can be described as a sparse optimization problem. In last decade, sparse optimization has obtained widespread development and application, many sparse optimization approaches has been applied to the index tracking problem. Jansen and van Dijk [7] considered minimizing a weighted objective function which comprised tracking error and a continuous function of the number of stocks in the tracking portfolio. Maringer and Oyewumi [8] considered the index tracking problem with a cardinality constraint and lower and upper limits on the portfolio weights. Fastrich et al. [9] relaxed the cardinality constraint in the index tracking problem by using a l_p-norm ($0 < p < 1$) constraint. Brodie et al. [10] proposed using l_1-norm regularization method to structure sparse and stable portfolios. Chen et al. [11] proposed some new l_p-norm regularized sparse portfolio models and used an interior-point algorithm [13] to seek the near-optimal sparse portfolios. And Xu et al. [12] considered a special case with $p = 1/2$ and proposed a half thresholding algorithm for solving the $l_{1/2}$-regularized model. Recently, Xu et al. [14] proposed a nonmonotone projected gradient method to solve a sparse least squares (LS) model for the index tracking problems. In addition, the methods [15–18] for solving general sparse optimization problems can also be used to solve the index tracking problems.

In machine learning, support vector machine (SVM) and support vector regression (SVR) are data classification and regression models, which base on the structural risk minimization principle and have good generalization ability on a limited number of learning patterns. The SVM and SVR have been widely applied in the text categorization, image processing, biology and other sciences. Several comprehensive reviews for SVM and SVR can be found in Steinwart and Christmann [19], Basak et al. [20] and Smola and Schölkopf [21]. For the index tracking problem, Takeda et al. [22] proved the equivalence between SVR and conditional value-at-risk minimization and applied the SVR model to the index tracking problem, but the proposed model did not limit the number of assets in the portfolio. De Leone [23] considered a 0–1 mixed integer programming formulation for the index tracking problem based on the SVR model and feature selection, however, that formulation was still hard to solve.

For taking the advantage of the SVR model and limiting the number of assets in the tracking portfolio, in this paper, under the cardinality constraint we consider two kinds of sparse SVR models for index tracking problem. In the sparse SVR models, a parametric tracking error measure needs to be minimized in the objective function. There is a fixed tolerable error bound in one kind of sparse SVR model and this kind of model can be viewed as a sparse variation of the ε-SVR model [24]. The other kind of sparse SVR model uses an extra parameter to tradeoff the parametric tracking error and the tolerable error bound and can be viewed as a sparse variation of the ν-SVR model [25]. In addition, for satisfying the budget and diversification request, we also add the budget constraint and upper bounds of the weights of assets in the models. The optimization problems corresponding to our models are general cardinality constrained minimization problems. Inspired by [27], we use the penalty proximal alternating linearized minimization (PALM) method and sparse projection techniques to solve these problems. Under some suitable assumptions, any accumulation point of the sequence generated by penalty PALM method satisfies the first-order optimality conditions of the problems. Furthermore, under another suitable assumption such an accumulation point is also a local minimizer of the problems. Finally, we test the performance of our models by comparing them with a sparse LS model and a heuristic model. The computational results demonstrate that when the training data is relatively fewer, the solutions of sparse SVR models have better generalization ability and stability.

The rest of this paper is organized as follows. In Sect. 2, we present two cardinality constrained SVR models for the index tracking problem and introduce the penalty PALM method for solving the two models. In Sect. 3, we establish the first-order optimality conditions for optimization problems corresponding to our models and present the convergence result of the penalty PALM method for the problems. The numerical experiments for testing the performance of our models are summarized in Sect. 4. Finally, we give some concluding remarks in Sect. 5.

Notation. In this paper, the symbol $\mathbb{R}^n$ denotes the n-dimensional Euclidean space. For any real vector, $\| \cdot \|_0$ and $\| \cdot \|$ denote the l_0-norm (i.e., the cardinality or the number of the nonzero entries) and the Euclidean norm, respectively. Given an index set $I \subseteq \{1, \ldots, n\}$, $|I|$ denotes the size of I, and the elements of I are always arranged in ascending order. x_I denotes the subvector formed by the entries of x indexed by I. For any two sets $\mathcal{A}$ and $\mathcal{B}$, the subtraction of $\mathcal{A}$ and $\mathcal{B}$ is given by $\mathcal{A} \setminus \mathcal{B} = \{x \in \mathcal{A} : x \notin \mathcal{B}\}$. And for a point x and a closed set $\mathcal{A}$, $\mathrm{dist}(x, \mathcal{A})$ denotes the Euclidean distance from x to $\mathcal{A}$.

2   Sparse SVR Model and Penalty PALM Method

In this section we propose two kinds of sparse SVR models for the index tracking problem and introduce the penalty PALM scheme for solving these models.

The index tracking problem can be considered as finding an optimal or suitable tracking portfolio for a time period $[1, \ldots, T]$ and ensuring that this portfolio is also close to optimal in the period $[T + 1, \ldots, T + L]$. The evolution of the market index and its constituent stocks in the period $[1, \ldots, T]$ is known, however, the evolution of the market index and its constituent stocks in the period $[T + 1, \ldots, T + L]$ is not known. A common strategy that can be used to track a market index is to determine an investment weight on each constituent stock in the beginning and maintain the weights constant throughout the tracking period. Let $x \in \mathrm{R}^n$ be the weight vector of n investable index constituent stocks, then x represents a tracking portfolio. Let $r_t \in \mathrm{R}$ and $R_t \in \mathrm{R}^n$ be the observed historical return of the market index and its n constituents at time t, where $t = 1, \ldots, T$, respectively. Then, the tracking error during the period $[1, \ldots, T]$ can be simply defined by using the linear least squares approach. Further, by adding an upper bound on the number of constituent stocks in the portfolio and other constraints, a classical sparse LS model for index tracking problem can be defined as

$$\min_{x} \quad \frac{1}{2} \sum_{t=1}^{T} (R_t^\top x - r_t)^2$$

$$\text{s.t.} \quad e^\top x = 1, \tag{1}$$

$$0 \le x \le u,$$

$$\|x\|_0 \le K,$$

where e is a n-dimensional vector with all components one, $u \in \mathrm{R}^n$ is a vector of the upper bounds on investment weights, $\|x\|_0$ represents the number of the nonzero entries of x, K is a positive integer and denotes the upper limit of the number of assets which are invested to track the index and the constraint $\|x\|_0 \le K$ is called cardinality constraint.

When the training data is relatively fewer, the generalization and stability of the SVR model are usually better than that of the LS model. Hence, in this paper, we consider the following sparse SVR model for the index tracking problem:

$$\min_{x, \xi^u, \xi^l} \quad \frac{1}{2} \|x\|^2 + C_1 \sum_{t=1}^{T} (c(\xi_t^u) + c(\xi_t^l))$$

$$\text{s.t.} \quad R_t^\top x - r_t \le \varepsilon + \xi_t^u,$$

$$r_t - R_t^\top x \le \varepsilon + \xi_t^l,$$

$$\xi_t^u, \xi_t^l \ge 0, \quad t = 1, \ldots, T, \tag{2}$$

$$e^\top x = 1,$$

$$0 \le x \le u,$$

$$\|x\|_0 \le K,$$

where $x \in R^n$, $\xi^u, \xi^l \in R^T$, constant $C_1 > 0$ is the weight of the deviations which are larger than ε, $c(\xi)$ is the loss function of deviation $\xi \in R_+$ and constant $\varepsilon \geq 0$ is the maximum deviation tolerance. Problem (2) is a ε-SVR model [24] with a soft margin loss function [26] and a cardinality constraint. For problem (2), our goal is to find a weight vector x^* such that the sum of deviations $|R_t^\top x^* - r_t|$ which are more than ε is as small as possible, and at the same time we need the cardinality of x^* is no more than K. In other words, we do not care about the tracking errors $|R_t^\top x^* - r_t|$ as long as they are less than ε, but we need minimize any deviation larger than ε. Since we need to find a sparse portfolio to track the underlying index, the excellent tracking is difficult. Therefore, it may be more suitable if we try to control the tracking errors in an acceptable level.

The classical SVR problem is solved by turning it into its dual problem which is more easily to be solved than the primal problem. However, it is difficult to set up a dual problem for the SVR problem with a cardinality constraint and it may be more difficult to solve the dual problem than to solve the primal problem. Hence, we need to simplify the primal problem and solve the primal problem by using the sparse optimization approach.

For the convenience of solving the problem, we choose the Gaussian loss function as the loss function in the model, i.e., $c(\xi) = (1/2)\xi^2$, and problem (2) can be changed to the following optimization problem:

$$\min_{x} \quad \frac{1}{2}\|x\|^2 + C_1 \sum_{t=1}^{T}[(R_t^\top x - r_t - \varepsilon)_+^2 + (r_t - R_t^\top x - \varepsilon)_+^2]$$

$$\text{s.t.} \quad e^\top x = 1, \tag{3}$$

$$0 \leq x \leq u,$$

$$\|x\|_0 \leq K,$$

where, for simplicity, we substitute C_1 for $C_1/2$ in the objective function.

In the rest of the paper, we consider problem (3) instead of problem (2). In problem (3) maximum deviation tolerance ε is a prior parameter. If there does not have a good estimation of ε, inspired by the ν-SVR model [25], we can set ε to be a variable and construct a new model as following:

$$\min_{x,\varepsilon} \quad \frac{1}{2}\|x\|^2 + C_1 \sum_{t=1}^{T}[(R_t^\top x - r_t - \varepsilon)_+^2 + (r_t - R_t^\top x - \varepsilon)_+^2] + C_2\varepsilon$$

$$\text{s.t.} \quad \varepsilon \geq 0,$$

$$e^\top x = 1, \tag{4}$$

$$0 \leq x \leq u,$$

$$\|x\|_0 \leq K,$$

where the constants C_1, C_2 determine the tradeoff of the deviations which are larger than ε, the flatness of the regression function and the maximum deviation tolerance ε.

Problems (3) and (4) are general cardinality constrained optimization problems. We use the penalty PALM scheme in [27] to solve these two problems. For convenience, we set $C_1 = \{x \in \mathbb{R}^n : e^\top x = 1\}$, $C_2 = \{(x, \varepsilon) \in \mathbb{R}^{n+1} : \varepsilon \geq 0, e^\top x = 1\}$, $\mathcal{D} = \{x \in \mathbb{R}^n : \|x\|_0 \leq K, 0 \leq x \leq u\}$ and denote the objective functions of problems (3) and (4) by $f_1(x)$ and $f_2(x, \varepsilon)$, respectively. Then we rewrite problems (3) and (4) as

$$\min_x \{f_1(x) : x \in C_1 \cap \mathcal{D}\},$$
$$\min_{x, \varepsilon} \{f_2(x, \varepsilon) : (x, \varepsilon) \in C_2, x \in \mathcal{D}\}.$$

Similarly with the penalty PALM method in [27], we reformulate problem (3) by adding an artificial variable y:

$$\min_{x, y} \{f_1(x) : x = y, x \in C_1, y \in \mathcal{D}\}. \tag{5}$$

For any given penalty parameter $\rho > 0$, the quadratic penalty subproblem of (5) can be defined as following:

$$\min_{x, y} \{f_1(x) + \frac{\rho}{2}\|x - y\|^2 : x \in C_1, y \in \mathcal{D}\}. \tag{6}$$

Like the classical quadratic penalty function method, at each outer iteration of the penalty PALM method we approximately solve problem (3) by solving a penalty subproblem in the form of (6). The penalty subproblem (6) can be further rewritten as an unconstrained minimization problem:

$$\min_{x, y} \Psi_{1,\rho}(x, y) = h_1(x) + q_{1,\rho}(x, y) + g_1(y) \text{ over all } (x, y) \in \mathbb{R}^n \times \mathbb{R}^n, \tag{7}$$

by setting

$$\begin{cases} h_1(x) = I_{C_1}(x); \\ q_{1,\rho}(x, y) = f_1(x) + \frac{\rho}{2}\|x - y\|^2; \\ g_1(y) = I_{\mathcal{D}}(y), \end{cases} \tag{8}$$

where $I_{C_1}(x)$ denotes the indicator function of C_1 that satisfies $I_{C_1}(x) = 0$ if $x \in C_1$ and $+\infty$ otherwise as well as $I_{\mathcal{D}}(x)$.

Problem (7) is a nonconvex and nonsmooth minimization problem. Bolte et al. have developed a novel PALM method [18] which builds on the Kurdyka-Łojasiewicz (KL) property for solving a broad class of nonconvex and nonsmooth minimization problems. The objective function of problem (7) is a proper and lower semicontinuous function. From Fermat's rule, if a point is a local minimizer of a proper and lower semicontinuous function, then this point is a critical point of the function. It is shown in [18] that for any proper, lower semicontinuous and bounded below functions $h_1(x)$, $g_1(y)$ and smooth function $q_{1,\rho}(x, y)$, if these functions satisfy Assumption A and B in [18] and the objective function $\Psi_{1,\rho}(x, y)$ is a KL function then each bounded sequence generated by the PALM method globally converges to a critical point of problem (7).

From the definition of $h_1(x)$, $g_1(y)$ and $q_{1,\rho}(x,y)$, it is not hard to observe that for fixed ρ they satisfy Assumption A and B in [18]. And when ρ is fixed there exists $L_1 = 1 + 2C_1 \sum_{t=1}^{T} \|R_t\|^2 + \rho$ such that for any fixed y the partial gradient $\nabla_x q_{1,\rho}(x,y)$ is globally Lipschitz with moduli L_1, that is

$$\|\nabla_x q_{1,\rho}(x_1, y) - \nabla_x q_{1,\rho}(x_2, y)\| \leq L_1 \|x_1 - x_2\|, \quad \forall x_1, x_2 \in \mathbf{R}^n. \tag{9}$$

Likewise, there exists $L_2 = \rho$ such that for any fixed x the partial gradient $\nabla_y q_{1,\rho}(x,y)$ is globally Lipschitz with moduli L_2. The definition of the critical point of a proper and lower semicontinuous function and the KL function will be given in Sect. 4 as well as the proof of the function $\Psi_{1,\rho}(x,y)$ is a KL function. Next, we present the PALM method for solving the unconstrained problem (7) which is equivalent to the penalty subproblem (6).

Algorithm 1. PALM method for (7)

Choose an initial point $(x^0, y^0) \in \mathbf{R}^n \times \mathbf{R}^n$. Set $l = 0$.

(1) Take $\gamma_1 > 1$, $t_1 = \gamma_1 L_1$ and compute

$$x^{l+1} \in P_{\mathcal{C}_1}(x^l - (1/t_1)\nabla_x q_{1,\rho}(x^l, y^l)).$$

(2) Take $\gamma_2 > 1$, $t_2 = \gamma_2 L_2$ and compute

$$y^{l+1} \in P_{\mathcal{D}}(y^l - (1/t_2)\nabla_y q_{1,\rho}(x^{l+1}, y^l)).$$

(3) Set $l \leftarrow l + 1$ and go to step (1).

The PALM method solves the problem (7) by alternately solving two proximal linearized minimization problems:

$$x^{l+1} \in \arg\min_x \left\{ h_1(x) + \langle x - x^l, \nabla_x q_{1,\rho}(x^l, y^l)\rangle + \frac{t_1}{2}\|x - x^l\|^2 \right\}, \tag{10}$$

$$y^{l+1} \in \arg\min_y \left\{ g_1(y) + \langle y - y^l, \nabla_y q_{1,\rho}(x^{l+1}, y^l)\rangle + \frac{t_2}{2}\|y - y^l\|^2 \right\}, \tag{11}$$

where $t_1 = \gamma_1 L_1$, $t_2 = \gamma_2 L_2$, $\gamma_1, \gamma_2 > 1$ and t_1, t_2 are two appropriately chosen step sizes.

Note that $h_1(x) = I_{\mathcal{C}_1}(x)$, $g_1(y) = I_{\mathcal{D}}(y)$ are indicator functions of nonempty and closed sets. The proximal linearized minimization problems (10) and (11) reduce to the projection problems onto $\mathcal{C}_1$ and $\mathcal{D}$, defined by

$$x^{l+1} \in P_{\mathcal{C}_1}(w_1) := \arg\min_x \{\|x - w_1\| : x \in \mathcal{C}_1\}, \tag{12}$$

$$y^{l+1} \in P_{\mathcal{D}}(w_2) := \arg\min_y \{\|y - w_2\| : y \in \mathcal{D}\}, \tag{13}$$

where $w_1 = x^l - (1/t_1)\nabla_x q_{1,\rho}(x^l, y^l)$, $w_2 = y^l - (1/t_2)\nabla_y q_{1,\rho}(x^{l+1}, y^l)$.

$\mathcal{C}_1 = \{x \in \mathbb{R}^n : e^\top x = 1\}$ is a nonempty closed convex set, hence projection problem $P_{\mathcal{C}_1}(w_1)$ has a unique closed form solution:

$$P_{\mathcal{C}_1}(w_1) = w_1 + \frac{1 - e^\top w_1}{n} e. \tag{14}$$

Similarly with [17,27], $\mathcal{D} = \{y \in \mathbb{R}^n : \|y\|_0 \leq K, 0 \leq y \leq u\}$ is a nonempty closed set, $P_{\mathcal{D}}(w_2)$ is a projection onto the sparse feasible set and projection map $P_{\mathcal{D}}$ is a general multi-valued map.

Suppose $\bar{y}^*$ is a point of set $P_{\mathcal{D}}(w_2)$. It follows from Proposition 3.1 of [17] that $\bar{y}^*$ can be computed as follows:

$$\bar{y}_i^* = \begin{cases} \bar{y}_i, & \text{if } i \in S^*; \\ 0, & \text{otherwise,} \end{cases} \quad i = 1, \ldots, n,$$

where $\bar{y}_i \in \arg\min_{y_i}\{(y_i - (w_2)_i)^2 : 0 \leq y_i \leq u_i\}$ and $S^* \subseteq \{1, \ldots, n\}$ be the index set corresponding to K largest values of $\{((w_2)_i)^2 - (\bar{y}_i - (w_2)_i)^2\}_{i=1}^n$. Therefore, problem (7) can be solved by alternately solving two projection problems which have closed form solutions. We rewrite the PALM method for solving problem (7) in Algorithm 1.

Now, we propose the penalty PALM method for solving problem (3) (or equivalently problem (5)) in Algorithm 2.

Algorithm 2. Penalty PALM method for problem (3)

Let $\{\varepsilon_k\}$ be a positive decreasing sequence. Let $\rho_0 > 0$, $\sigma > 1$, $\gamma_1 > 1$, $\gamma_2 > 1$ be given. Choose an arbitrary point $(x_0^0, y_0^0) \in \mathbb{R}^n \times \mathbb{R}^n$. Set $k = 0$.

(1) Set $l = 0$, $L_1^k = 1 + 2C_1 \sum_{t=1}^T \|R_t\|^2 + \rho_k$, $L_2^k = \rho_k$, $t_1^k = \gamma_1 L_1^k$, $t_2^k = \gamma_2 L_2^k$ and apply the PALM method to find an approximate critical point $(x^k, y^k) \in \mathcal{C}_1 \times \mathcal{D}$ of the penalty subproblem

$$\min_{x,y} \Psi_{1,\rho_k}(x,y) = I_{\mathcal{C}_1}(x) + q_{1,\rho_k}(x,y) + I_{\mathcal{D}}(y), \quad (x,y) \in \mathbb{R}^n \times \mathbb{R}^n \tag{15}$$

by performing steps (1.1)-(1.4):
(1.1) Compute $x_{l+1}^k \in P_{\mathcal{C}_1}(x_l^k - (1/t_1^k)\nabla_x q_{1,\rho_k}(x_l^k, y_l^k))$.
(1.2) Compute $y_{l+1}^k \in P_{\mathcal{D}}(y_l^k - (1/t_2^k)\nabla_y q_{1,\rho_k}(x_{l+1}^k, y_l^k))$.
(1.3) If (x_{l+1}^k, y_{l+1}^k) satisfies

$$\text{dist}(0, \partial\Psi_{1,\rho_k}(x_{l+1}^k, y_{l+1}^k)) < \varepsilon_k, \tag{16}$$

set $(x^k, y^k) := (x_{l+1}^k, y_{l+1}^k)$ and go to step (2).
(1.4) Set $l \leftarrow l + 1$ and go to step (1.1).
(2) Set $\rho_{k+1} = \sigma\rho_k$, and $(x_0^{k+1}, y_0^{k+1}) := (x^k, y^k)$.
(3) Set $k \leftarrow k + 1$ and go to step (1).

Remark 1. The condition (16) is used to guarantee the global convergence of the penalty PALM method. But in real computation, just like [17,27], this condition can be replaced by another practical termination condition which is based on the relative change of the sequence $\{(x_l^k, y_l^k)\}$, that is,

$$\max\left\{\frac{\|x_{l+1}^k - x_l^k\|_\infty}{\max(\|x_l^k\|_\infty, 1)}, \frac{\|y_{l+1}^k - y_l^k\|_\infty}{\max(\|y_l^k\|_\infty, 1)}\right\} \leq \varepsilon_I \qquad (17)$$

for some $\varepsilon_I > 0$. In addition, we can terminate the out iterations of the penalty PALM method by condition

$$\|x^k - y^k\|_\infty \leq \varepsilon_O \quad \text{or} \quad \frac{\|(x^{k+1}, y^{k+1}) - (x^k, y^k)\|_\infty}{\max(\|(x^k, y^k)\|_\infty, 1)} \leq \varepsilon_O \qquad (18)$$

for some $\varepsilon_O > 0$. For enhancing the performance and convergence of the penalty PALM method, we suggest to check that whether the set C_1 and the sparse subset of $\mathcal{D}$ which contains y^k have an intersection and recompute the penalty subproblem multiple times by restarting the PALM method from a suitable perturbation of the current best approximate solution.

We now consider the penalty PALM method for solving problem (4). Similarly, for any given penalty parameter $\rho > 0$, we can define the quadratic penalty subproblem and rewrite it as a unconstrained minimization problem as following:

$$\min_{x,\varepsilon,y} \Psi_{2,\rho}(x, \varepsilon, y) = h_2(x, \varepsilon) + q_{2,\rho}(x, \varepsilon, y) + g_2(y) \quad \text{over all} \quad (x, \varepsilon, y) \in \mathbb{R}^{n+1} \times \mathbb{R}^n,$$

where

$$\begin{cases} h_2(x) = I_{C_2}(x, \varepsilon); \\ q_{2,\rho}(x, \varepsilon, y) = f_2(x, \varepsilon) + \dfrac{\rho}{2}\|x - y\|^2; \\ g_2(y) = I_{\mathcal{D}}(y). \end{cases} \qquad (19)$$

Likewise, for fixed ρ, $h_2(x)$, $g_2(y)$ and $q_{2,\rho}(x, \varepsilon, y)$ satisfy Assumption A and B in [18]. Set $\bar{R}_t^\top = (R_t^\top, 1)$, for fixed ρ there exists $\bar{L}_1 = 1 + 2C_1 \sum_{t=1}^T \|\bar{R}_t\|^2 + \rho$, $\bar{L}_2 = \rho$. Choose $\bar{\gamma}_1, \bar{\gamma}_2 > 1$ and set $\bar{t}_1 = \bar{\gamma}_1 \bar{L}_1$, $\bar{t}_2 = \bar{\gamma}_2 \bar{L}_2$, the proximal linearized minimization problems reduce to the projection problems onto sets C_2 and $\mathcal{D}$, defined by

$$(x^{l+1}, \varepsilon^{l+1}) \in P_{C_2}(\bar{w}_1) := \arg\min_{x,\varepsilon} \{\|(x, \varepsilon) - \bar{w}_1\| : (x, \varepsilon) \in C_2\}, \qquad (20)$$

$$y^{l+1} \in P_{\mathcal{D}}(\bar{w}_2) := \arg\min_y \{\|y - \bar{w}_2\| : y \in \mathcal{D}\}, \qquad (21)$$

where $\bar{w}_1 = (x^l, \varepsilon^l) - (1/\bar{t}_1)\nabla_{x,\varepsilon} q_{2,\rho}(x^l, \varepsilon^l, y^l)$, $\bar{w}_2 = y^l - (1/\bar{t}_2)\nabla_y q_{2,\rho}(x^{l+1}, \varepsilon^{l+1}, y^l)$. The projection problem $P_{\mathcal{D}}(\bar{w}_2)$ is same to the projection problem in the PALM method for problem (7). And the projection problem $P_{C_2}(\bar{w}_1)$ also has a unique closed form solution: set $\bar{w}_1 = (\bar{w}_{1,a}^\top, \bar{w}_{1,b})^\top$, where $\bar{w}_{1,a} \in \mathbb{R}^n$ and $\bar{w}_{1,b} \in \mathbb{R}$, then it has

$$P_{C_2}(\bar{w}_1) = ((\bar{w}_{1,a} + \frac{1 - e^\top \bar{w}_{1,a}}{n} e)^\top, (\bar{w}_{1,b})_+)^\top.$$

Similarly, we can use the penalty PALM method to solve problem (4).

3   Convergence Analysis

In this section, for the sake of brevity, we give the first-order optimality conditions for problem (3), which are special cases of the optimality conditions for
general cardinality constrained optimization problems in [17], and establish the
global convergence of the penalty PALM method for problem (3). The optimality conditions and the convergence of the penalty PALM method for problem
(4) can be obtained by imitating the relevant results of problem (3).

Next, we give the first-order necessary optimality conditions for problem (3).

Theorem 1. *Assume that x^* is a local minimizer of problem (3). Let $I^* \subseteq$
$\{1, \cdots, n\}$ be an index set with $|I^*| = K$ such that $x_i^* = 0$ for all $i \in \bar{I}^*$, where
$\bar{I}^* = \{1, \cdots, n\} \setminus I^*$. Suppose that at x^* the following Mangasarian-Fromovitz
constraint qualification (MFCQ) condition holds, that is, there exists a point
$x \in \mathbb{R}^n$ such that*

$$
\begin{aligned}
e_i^\top (x - x^*) &< 0, \text{ if } x_i^* = u_i, \\
-e_i^\top (x - x^*) &< 0, \text{ if } x_i^* = 0, i \in I^*, \\
e_i^\top (x - x^*) &= 0, \text{ if } i \in \bar{I}^*, \\
e^\top (x - x^*) &= 0,
\end{aligned}
\tag{22}
$$

*are satisfied, where e_i is an n-dimensional vector of which the i-th compo-
nent is equal to 1 and other components are equal to 0. Then, there exists
$(\lambda_1^*, \lambda_2^*, \mu^*, z^*) \in \mathbb{R}^{3n+1}$ together with x^* satisfying*

$$
\begin{aligned}
\nabla f_1(x^*) + \lambda_1^* - \lambda_2^* + \mu^* e + z^* &= 0, \\
(\lambda_1^*)_i (x_i^* - u_i) = 0, \ (\lambda_2^*)_i x_i^* &= 0, \ i = 1, \cdots, n, \\
e^\top x^* = 1, \ 0 \leq x^* \leq u, \ \lambda_1^*, \lambda_2^* &\geq 0, \\
(\lambda_2^*)_i = 0, \ i \in \bar{I}^*; \ z_i^* = 0, \ i &\in I^*.
\end{aligned}
\tag{23}
$$

Proof. It is obvious that if x^* is a local minimizer of problem (3), x^* is also a
minimizer of the problem:

$$
\min_x \left\{ f_1(x) : \ e^\top x = 1, \ 0 \leq x_{I^*} \leq u, \ x_{\bar{I}^*} = 0 \right\}.
\tag{24}
$$

Together with the MFCQ condition and Theorem 3.25 of [29], the conclusion
holds.

Similarly with the problem in [27], the cardinality constraint is the only
nonconvex part of problem (3). Using this observation and the conclusion in [17]
we can establish the first-order sufficient optimality conditions for problem (3).

Theorem 2. *For problem (3), the objective function is convex and the constraints except the cardinality constraint are affine. Let x^* be a feasible point of
the problem, and let $\mathcal{I}^* = \{I^* \subseteq \{1, \cdots, n\} : |I^*| = K, \ x_i^* = 0, \ \forall i \in \{1, \cdots, n\} \setminus
I^*\}$. Suppose that for any $I^* \in \mathcal{I}^*$, there exists some $(\lambda_1^*, \lambda_2^*, \mu^*, z^*) \in \mathbb{R}^{3n+1}$ sat-
isfying (23). Then, x^* is a local minimizer of problem (3).*

Proof. By the assumptions and Theorem 3.27 of [29], it has that x^* is a minimizer of problem (24) for all $\bar{I}^* \in \{\{1, \cdots, n\} \setminus I^* : I^* \in \mathcal{I}^*\}$. Then there exists $\delta > 0$ such that $f_1(x) \geq f_1(x^*)$ for all $x \in \cup_{I^* \in \mathcal{I}^*} \mathcal{O}_{I^*}(x^*; \delta)$ where

$$\mathcal{O}_{I^*}(x^*; \delta) = \{x \in \mathbb{R}^n : e^\top x = 1, 0 \leq x_{I^*} \leq u, x_{\bar{I}^*} = 0, \|x - x^*\| < \delta\}$$

with $\bar{I}^* = \{1, \cdots, n\} \setminus I^*$. We can observe from problem (3) that for any $x \in \mathcal{O}(x^*; \delta)$, where

$$\mathcal{O}(x^*; \delta) = \{x \in \mathbb{R}^n : e^\top x = 1, 0 \leq x \leq u, \|x\|_0 \leq K, \|x - x^*\| < \delta\},$$

there exists $I^* \in \mathcal{I}^*$ such that $x \in \mathcal{O}_{I^*}(x^*; \delta)$ and hence $f_1(x) \geq f_1(x^*)$. It then implies that x^* is a local minimizer of (3).

From Theorem 2 we observe that for any point x if $\|x\|_0 = K$ and x satisfies the first-order necessary conditions of problem (3), then x is a local minimizer of the problem. Next, we prove the global convergence of the penalty PALM method for problem (3). Before proving the convergence of the penalty PALM method, we need to introduce the definition of the critical point of a nonconvex and nonsmooth function and establish the convergence of the PALM method for the penalty subproblem (7).

Definition 1 *(Subdifferentials and critical points). Let $\sigma : \mathbb{R}^n \to (-\infty, +\infty]$ be a proper and lower semicontinuous function and $\mathrm{dom}\,\sigma = \{x \in \mathbb{R}^n : \sigma(x) < +\infty\}$.*

(i) For a given $x \in \mathrm{dom}\,\sigma$, the Fréchet subdifferential of σ at x is defined as

$$\widehat{\partial}\sigma(x) = \left\{u : \liminf_{y \to x, y \neq x} \frac{\sigma(y) - \sigma(x) - \langle u, y - x \rangle}{\|y - x\|} \geq 0\right\}$$

and $\widehat{\partial}\sigma(x) = \emptyset$ if $x \notin \mathrm{dom}\,\sigma$.
(ii) The limiting subdifferential, or simply the subdifferential, of σ at $x \in \mathbb{R}^n$, written $\partial\sigma(x)$, is defined as

$$\partial\sigma(x) = \left\{u \in \mathbb{R}^n : \exists x^k \to x, \sigma(x^k) \to \sigma(x) \text{ and } u^k \in \widehat{\partial}\sigma(x^k) \to u \text{ as } k \to \infty\right\}.$$

(iii) The point x is a critical point of σ if $0 \in \partial\sigma(x)$.

For a general proper and lower semicontinuous function σ, if a point x is a local minimizer of function σ, then x is a critical point of σ. Bolte et al. have showed in [18] that for any nonconvex and nonsmooth function, if it satisfies Assumptions A and B in [18] and is a KL function, then any bounded sequence generated by the PALM method globally converges to a critical point of this function. From the description in Sect. 2, we have that $\Psi_{1,\rho}(x, y)$ is a proper and lower semicontinuous function and satisfies the assumptions in [18]. In addition, the level set of function $\Psi_{1,\rho}(x, y)$ is bounded, and from the conclusions of [18] the function value sequence $\{\Psi_{1,\rho}(x^l, y^l)\}$ is nonincreasing. It then follows that the iterative sequence $\{(x^l, y^l)\}$ generated by the PALM method is bounded. Therefore, to prove the convergence of the PALM method, we only need to prove function $\Psi_{1,\rho}(x, y)$ is a KL function. If a proper and lower semicontinuous function σ satisfies the KL property at each point of dom $\partial\sigma$ then σ is called a KL function. The KL property plays a central role in the convergence analysis of the PALM method. We can see the definition in [18] for more details of the KL property. From [30] we know that a semi-algebraic function is a KL function. Next, we give the definitions of the semi-algebraic set and function, and show that function $\Psi_{1,\rho}(x, y)$ is a semi-algebraic function.

Definition 2 *(Semi-algebraic sets and functions).*

(i) *A set $S \subseteq \mathrm{R}^n$ is a real semi-algebraic set if there exists a finite number of real polynomial functions g_{ij}, $h_{ij} : \mathrm{R}^n \to \mathrm{R}$ such that*

$$S = \bigcup_j \bigcap_i \{u \in \mathrm{R}^n : g_{ij}(u) = 0 \text{ and } h_{ij}(u) < 0\}.$$

(ii) *A function $h : \mathrm{R}^n \to (-\infty, +\infty]$ is called semi-algebraic if its graph*

$$\{(u, t) \in \mathrm{R}^{n+1} : h(u) = t\}$$

is a semi-algebraic subset of R^{n+1}.

From the definition above and the basic properties of the semi-algebraic sets and functions in [18, 28], it can be observed that function $q_{1,\rho}(x, y)$ is a semi-algebraic function and function $\Psi_{1,\rho}(x, y)$ is also a semi-algebraic function. Hence, $\Psi_{1,\rho}(x, y)$ is a KL function. Under this condition and the conclusions of [18], we give the convergence theorem which ensures the global convergence of the iterative sequences generated in solving the penalty subproblems (7).

Theorem 3. *The sequence $\{(x^l, y^l)\}$ generated by the PALM method in solving the penalty subproblem (7) converges to a critical point of (7). And for any given $\varepsilon > 0$ there exists a constant $L > 0$ such that for all $l > L$ it has*

$$dist(0, \partial\Psi_{1,\rho}(x^l, y^l)) < \varepsilon.$$

Proof. It can be observed that function $\Psi_{1,\rho}(x, y)$ satisfy Assumption A and B in [18] and the level set of the function $\Psi_{1,\rho}(x, y)$ is bounded. And it is shown in [18] that the function value sequence $\{\Psi_{1,\rho}(x^l, y^l)\}$ is nonincreasing. Therefore the sequence $\{(x^l, y^l)\}$ is bounded. Let $z^l = (x^l, y^l)$. It follows from Lemma 3.4 of [18] that there exists a constant $M > 0$ and $(A_x^l, A_y^l) \in \partial\Psi_{1,\rho}(x^l, y^l)$ such that

$$dist(0, \partial\Psi_{1,\rho}(x^l, y^l)) \leq \|(A_x^l, A_y^l)\| \leq M\|z^l - z^{l-1}\|.$$

And from Lemma 3.3 and Theorem 3.1 of [18], it has $\|z^l - z^{l-1}\| \to 0$ as $l \to \infty$ and the sequence $\{z^l\}$ converges to a critical point of problem (7). Then using the conditions above, we can see that the conclusion holds.

Now we show that under a suitable assumption, any accumulation point of the sequence generated by the penalty PALM method for problem (3) satisfies the first-order necessary optimality conditions of problem (3). Moreover, we show that if the cardinality of the accumulation point is equal to K, the accumulation point is also a local minimizer of problem (3).

Theorem 4. *Assume that* $\{\varepsilon_k\} \rightarrow 0$. *Let* $\{(x^k, y^k)\}$ *be the sequence generated by the penalty PALM method for problem (3),* $I_k = \{i_1^k, \ldots, i_K^k\}$ *be a set of* K *distinct indices in* $\{1, \cdots, n\}$ *such that* $(y^k)_i = 0$ *for any* $i \notin I_k$. *Then the following statements hold:*

(i) *The sequence* $\{(x^k, y^k)\}$ *is bounded.*

(ii) *Suppose* (x^*, y^*) *is an accumulation point of* $\{(x^k, y^k)\}$. *Then* $x^* = y^*$ *and* x^* *is a feasible point of problem (3). Moreover, there exists a subsequence* J *such that* $\{(x^k, y^k)\}_{k \in J} \rightarrow (x^*, y^*)$, $I_k = I^*$ *for some index set* $I^* \subseteq \{1, \cdots, n\}$ *when* $k \in J$ *is sufficiently large.*

(iii) *Let* x^*, J *and* I^* *be defined as above, and let* $\bar{I}^* = \{1, \cdots, n\} \setminus I^*$. *Suppose that the MFCQ condition (22) holds at* x^* *for* I^* *and* $\bar{I}^*$. *Then* x^* *satisfies the first-order optimality conditions (23). Moreover, if* $\|x^*\|_0 = K$, x^* *is a local minimizer of (3).*

Proof. The proof of this theorem is similar to the proof of Theorem 4.3 in [27].

The convergence of the penalty PALM method for problem (4) is similar to the convergence of the penalty PALM method for problem (3). One can obtain the convergence conclusions of problem (4) by imitating the relevant results of problem (3).

4   Numerical Experiments

In this section, we compare the performance of the sparse SVR models (3) and (4) with that of the sparse LS model (1) and a heuristic model for some testing problems from the OR-Library [31]. The OR-library is a publicly available collection of testing data sets for a variety of Operations Research problems. For the index tracking problem it contains the weekly data of the indexes and the constituent stocks included in major world markets from 1992 to 1997, such as Hang Seng (Hong Kong), DAX (Germany), FTSE (Great Britain), Standard and Poor's 100 (USA), Nikkei (Japan), Standard and Poor's 500 (USA), Russell 2000 (USA) and Russell 3000 (USA).

The sparse LS model (1) has been introduced in Sect. 2. The heuristic model which is compared with sparse SVR models in this section is also a common stock picking strategies. The idea of this model is to solve a LS problem without cardinality constraint directly and use the variables with the K largest values of the solution to construct a sparse tracking portfolio by solving a new low dimensional LS problem. It is reasonable to assume that the statistical properties of the asset returns in the near future are similar to those in the recent past. Hence, if the time period $[1, \ldots, T]$ is long enough and horizon $T + L$ is not too far in the future, the portfolios found by the sparse LS model and the heuristic model will perform well in period $[T+1, \ldots, T+L]$. But, if the period $[1, \ldots, T]$ is not long enough and $T + L$ is far in the future, the sparse LS model and the heuristic model may tend to find over-optimized portfolios. Comparing with the LS model, the SVR model has better generalization ability and stability in general when the training data is limited. For comparing the performance of the sparse SVR models with that of the sparse LS model and the heuristic model, we divide the above data sets into two groups: small data sets including Hang Seng, DAX, FTSE, Standard and Poor's 100, and large data sets including Nikkei, Standard and Poor's 500, Russell 2000, Russell 3000. And each data set is partitioned into three subsets: one training set and two testing sets. The training set contains the first 25% of the data and is used to find the appropriate index tracking portfolios. The testing sets are used to test the performance of the index tracking portfolios found by the models. The first testing set contains the second 25% of the data, and excepting for the first 25% of the data the rest 75% of the data is contained in the second testing set. For a portfolio x, the training tracking error or testing tracking error is defined by the mean squared error $\sum_{t=1}^{m}(R_t^\top x - r_t)^2/m$ where R_t, $t = 1, \ldots, m$ are fetched from the training set or the testing set and m is the length of the corresponding subsets.

For the small data sets, we select small numbers as the upper bounds K of the number of constituent stocks in the portfolio, and for the large data sets the span of the upper bounds become wider. For convenience, the upper bound u of the investment weight is set to positive infinity for each testing problem. The sparse LS model (1) is solved by the CPLEX(12.6) solver which uses the default parameters of the software. Although it is easier to solve the sparse LS model than to solve the sparse SVR models the optimization problem corresponding to the sparse LS model is still generally NP-hard and will cost a large amount of computation to find a good solution. Hence, for each testing problem we limit the computation time of the CPLEX to no more than 3600 s. If the computation time is more than one hour we use the best feasible solution found by the CPLEX in one hour as the solution of the model. And in the heuristic model we use the CPLEX with default parameters to solve the two LS problems. For comparison, the parameters of two sparse SVR models are chosen by using fourfold cross validation and two sparse SVR models are solved by the penalty PALM method which is coded in Matlab (R2014b). For the penalty PALM method, we choose $\rho_0 = 0.2$, $\sigma = 1.1$, $\gamma_1 = 1.0001$, $\gamma_2 = 1.0001$, L_1, L_2, $\bar{L}_1$ and $\bar{L}_2$ are chosen as described in Sect. 2. And (17) and (18) are used as the inner and outer termination criteria with $\varepsilon_I = 10^{-5}$, $\varepsilon_O = 10^{-6}$, respectively. The penalty PALM method can find a local minimizer or a stationary point of the optimization problems corresponding to the sparse SVR models. In order to compare with the sparse LS model and the heuristic model, in each testing problem, we choose the variables with the $K+m$ largest values of the solution of the LS model without the cardinality constraint to randomly generate ten groups of different initial points of the penalty PALM method, where m is from -10 to 20. Then we solve the sparse SVR models by using the penalty PALM method ten times with these initial points and choose the solution with the minimum objective value as the solution of the sparse SVR models to compare with the solutions of the sparse LS model and the heuristic model. In the following, we display the numerical results of these models on the two groups of the data sets. All the computations described above are performed on a PC (Intel core i7-4790CPU, 3.6 GHz, 16 GB RAM).

The numerical results on the small data sets are presented in Tables 1 and 2. In Table 1, we report the training errors and the testing errors on the two testing sets of the solutions of the sparse SVR model (3), the sparse LS model (1) and the heuristic model. And we also report the total computation time of these three models in Table 1. The training errors and two testing errors of the solutions of the sparse SVR model (4) and the total computation time are reported in Table 2. The numerical results on the large data sets are presented in Tables 3 and 4. The training errors and testing errors of the solutions of the sparse SVR model (3), the sparse LS model (1) and the heuristic model and the total computation time are reported in Table 3. The training errors and testing errors of the solutions of the sparse SVR model (4) and the total computation time are reported in Table 4. In addition, in Table 5, we report the influences of the hyper-parameters for the sparse SVR models and the standard deviations of the testing errors of the ten solutions which are found by the penalty PALM method in the solving processes of each sparse SVR model on the two testing sets for a example of the small data set and a example of the large data set. In Tables 1, 2, 3, 4 and 5, "Training" stands for the training errors; "Testing1" stands for the testing errors on the first testing set; "Testing2" stands for the testing errors on the second testing set; "Time" (in seconds) stands for the computation time; "Std1" and "Std2" are the standard deviations of the testing errors on two testing sets, respectively.

When the number of the constituent stocks is relatively fewer, the sparse LS model and the heuristic model can find tracking portfolios with good performance. From Tables 1 and 2, we can make the observations that the training errors and testing errors of the solutions of the sparse LS model, the heuristic model and the sparse SVR models decrease when the upper bound of the number of constituent stocks in the portfolio increases. The training errors of the solutions of the sparse SVR models are bigger than that of the sparse LS model and the heuristic model but the testing errors of the solutions of the sparse SVR models are smaller than that of the sparse LS model and the heuristic model or similar to that of the heuristic model. From the numerical results on the large data sets, we observe that the training errors and testing errors of the solutions of the sparse LS model, the heuristic model and the sparse SVR models also decrease when the upper bound of the number of constituent stocks in the portfolio increases. Like the numerical results on the small data sets, the training errors of the solutions of the sparse SVR models are bigger than that of the sparse LS model and the heuristic model, while the testing errors of the solutions of the sparse SVR models are smaller than that of the sparse LS model and the heuristic model, especially for the large-scale problems. And from Tables 1,

Table 1. Comparison of sparse SVR model (3) and sparse LS model (1) for small data sets.

Index	K	Sparse SVR (3)				Sparse LS				Heuristic			
		Training	Testing1	Testing2	Time	Training	Testing1	Testing2	Time	Training	Testing1	Testing2	Time
Hang Seng $n=31$	5	6.0883e−05	7.1746e−05	6.2428e−05	1.8860	3.1427e−05	7.3302e−05	7.1771e−05	0.2800	3.9569e−05	6.8228e−05	5.8850e−05	0.4709
	6	3.8258e−05	6.2825e−05	4.7627e−05	2.1264	1.9976e−05	6.8521e−05	5.9251e−05	0.1090	2.9133e−05	6.3557e−05	4.5037e−05	0.5037
	8	2.2460e−05	3.8214e−05	3.2442e−05	1.7926	1.2411e−05	4.5449e−05	4.3035e−05	0.1720	1.9648e−05	4.4815e−05	3.3523e−05	0.4571
	10	1.1720e−05	3.6434e−05	2.4904e−05	1.5887	8.7445e−06	3.1256e−05	2.6270e−05	0.1250	1.0095e−05	2.2926e−05	2.5909e−05	0.4634
	12	7.8314e−06	2.3843e−05	2.1333e−05	1.5465	6.3867e−06	2.6477e−05	2.1917e−05	0.3280	8.2188e−06	2.6509e−05	2.1458e−05	0.4687
	15	4.6566e−06	2.1063e−05	1.8070e−05	1.8861	4.0062e−06	2.4699e−05	2.0468e−05	0.1150	4.3942e−06	1.9022e−05	1.6739e−05	0.4687
DAX $n=85$	5	1.8954e−05	3.6832e−05	8.2162e−05	5.5044	1.4491e−05	3.3504e−05	8.9039e−05	13.1500	1.7727e−05	3.9671e−05	8.7634e−05	0.6614
	6	1.5951e−05	3.3500e−05	6.8575e−05	5.0448	1.1412e−05	3.4524e−05	7.8923e−05	49.8120	1.5218e−05	3.3585e−05	7.1391e−05	0.5183
	8	1.4842e−05	2.9256e−05	6.4651e−05	4.1295	6.9822e−06	3.2873e−05	7.9047e−05	515.5800	1.2165e−05	2.8022e−05	6.4841e−05	0.5437
	10	1.4422e−05	2.7887e−05	6.3764e−05	3.6102	4.9584e−06	3.7318e−05	7.7635e−05	3600.0820	9.7680e−06	2.3228e−05	5.9822e−05	0.5063
	12	1.0807e−05	2.1216e−05	5.5092e−05	3.5482	3.1175e−06	2.6772e−05	6.5091e−05	3600.0810	8.2334e−06	2.1940e−05	5.6970e−05	0.5396
	15	9.5125e−06	1.8622e−05	4.5304e−05	3.2372	1.9992e−06	2.2043e−05	5.8290e−05	3600.0970	6.0994e−06	1.4425e−05	4.7657e−05	0.5195
FTSE $n=89$	5	1.7551e−04	1.1453e−04	1.3634e−04	5.1423	4.9141e−05	1.2868e−04	1.3097e−04	190.4460	8.8901e−05	1.7186e−04	1.8413e−04	0.5016
	6	1.3770e−04	9.9365e−05	1.1695e−04	5.5644	3.6637e−05	1.4311e−04	1.3492e−04	3023.8920	8.4644e−05	1.6743e−04	1.6834e−04	0.5081
	8	8.4103e−05	7.6889e−05	7.9848e−05	4.4761	2.0746e−05	9.6738e−05	8.8616e−05	3600.0809	4.1124e−05	1.1638e−04	1.2794e−04	0.5384
	10	7.2054e−05	3.7681e−05	4.2864e−05	4.4102	1.3580e−05	6.8823e−05	6.2224e−05	3600.0910	2.9696e−05	8.5164e−05	1.0649e−04	0.5242
	12	6.9257e−05	3.4260e−05	3.5341e−05	3.6051	8.9293e−06	6.7023e−05	7.1684e−05	3600.1129	1.3759e−05	7.0126e−05	7.0451e−05	0.5057
	15	4.2216e−05	2.2973e−05	3.0581e−05	3.6985	5.0194e−06	4.4741e−05	5.2095e−05	3600.0960	9.0756e−06	5.3376e−05	6.0770e−05	0.5358
S&P $n=98$	5	1.2109e−04	9.7138e−05	1.0598e−04	2.8651	3.4736e−05	1.0966e−04	1.0508e−04	213.8620	7.7836e−05	9.6925e−05	1.4518e−04	0.5207
	6	4.8413e−05	6.0370e−05	8.2641e−05	2.4533	2.6004e−05	7.7941e−05	8.1164e−05	3092.8290	6.6548e−05	7.6908e−05	1.0036e−04	0.5215
	8	2.6211e−05	3.7982e−05	5.8424e−05	2.0214	1.5385e−05	3.7360e−05	6.8300e−05	3600.0820	2.8781e−05	4.5365e−05	6.0093e−05	0.5029
	10	2.3757e−05	3.0113e−05	4.7547e−05	1.8610	9.9834e−06	4.1410e−05	1.0336e−04	3600.0810	1.9609e−05	3.5791e−05	4.4009e−05	0.5822
	12	1.7233e−05	2.5221e−05	3.9609e−05	1.6491	5.7967e−06	3.8950e−05	4.0797e−05	3600.0659	1.7353e−05	2.9804e−05	3.6192e−05	0.5287
	15	1.3748e−05	2.1972e−05	3.1634e−05	1.3839	3.3068e−06	2.4011e−05	3.1356e−05	3600.0980	9.1530e−06	2.0885e−05	2.7415e−05	0.5277

Table 2. Comparison of sparse SVR model (4) and sparse LS model (1) for small data sets.

Index	K	Sparse SVR (4)				Sparse LS				Heuristic			
		Training	Testing1	Testing2	Time	Training	Testing1	Testing2	Time	Training	Testing1	Testing2	Time
Hang Seng $n=31$	5	3.9229e−05	5.2441e−05	6.9021e−05	8.6636	3.1427e−05	7.3302e−05	7.1771e−05	0.2800	3.9569e−05	6.8228e−05	5.8850e−05	0.4709
	6	2.7626e−05	5.6101e−05	4.2854e−05	7.4741	1.9976e−05	6.8521e−05	5.9251e−05	0.1090	2.9133e−05	6.3557e−05	4.5037e−05	0.5037
	8	1.5384e−05	3.4998e−05	3.0622e−05	6.0910	1.2411e−05	4.5449e−05	4.3035e−05	0.1720	1.9648e−05	4.4815e−05	3.3523e−05	0.4571
	10	9.6631e−06	3.1102e−05	2.5423e−05	4.6692	8.7445e−06	3.1256e−06	2.6270e−05	0.1250	1.0095e−05	2.2926e−05	2.5909e−05	0.4634
	12	7.3456e−06	2.3498e−05	1.9740e−05	3.7962	6.3867e−06	2.6477e−05	2.1917e−05	0.3280	8.2188e−06	2.6509e−05	2.1458e−05	0.4687
	15	5.1030e−06	2.1435e−05	1.8785e−05	3.1055	4.0062e−06	2.4699e−05	2.0468e−05	0.1150	4.3942e−06	1.9022e−05	1.6739e−05	0.4687
DAX $n=85$	5	2.2754e−05	4.7893e−05	7.9656e−05	14.7706	1.4491e−05	3.3504e−05	8.9039e−05	13.1500	1.7727e−05	3.9671e−05	8.7631e−05	0.6614
	6	1.7711e−05	3.0653e−05	7.2177e−05	13.1490	1.1412e−05	3.4524e−05	7.8923e−05	49.8120	1.5218e−05	3.3585e−05	7.1391e−05	0.5183
	8	1.1243e−05	2.5892e−05	6.0742e−05	11.2832	6.9822e−06	3.2873e−05	7.9047e−05	515.8800	1.2165e−05	2.8022e−05	6.4841e−05	0.5437
	10	8.4239e−06	1.8625e−05	5.6353e−05	8.8283	4.9584e−06	3.7318e−05	7.7635e−05	3600.0820	9.7680e−06	2.3228e−05	5.9822e−05	0.5063
	12	7.2310e−06	1.6273e−05	4.8907e−05	8.0473	3.1175e−06	2.6772e−05	6.5091e−05	3600.0810	8.2334e−06	2.1940e−05	5.6970e−05	0.5396
	15	3.6702e−06	1.4383e−05	5.0457e−05	7.2526	1.9992e−06	2.2043e−05	5.8290e−05	3600.0970	6.0994e−06	1.4425e−05	4.765*e−05	0.5195
FTSE $n=89$	5	7.8661e−05	1.4051e−04	1.2838e−04	11.0463	4.9141e−05	1.2868e−04	1.3097e−04	190.4460	8.8901e−05	1.7186e−04	1.8413e−04	0.5016
	6	5.5307e−05	1.1403e−04	1.1435e−04	9.4984	3.6637e−05	1.4311e−04	1.3492e−04	3023.8920	8.4644e−05	1.6743e−04	1.683*e−04	0.5081
	8	3.0488e−05	7.3401e−05	6.6979e−05	7.5342	2.0746e−05	9.6738e−05	8.8616e−05	3600.0809	4.1124e−05	1.1638e−04	1.2793e−04	0.5384
	10	1.5397e−05	6.1084e−05	5.3919e−05	6.5903	1.3580e−05	6.8823e−05	6.2224e−05	3600.0910	2.9696e−05	8.5164e−05	1.0643e−04	0.5242
	12	1.2023e−05	4.4835e−05	3.7667e−05	5.8964	8.9293e−06	6.7023e−05	7.1684e−05	3600.1129	1.3759e−05	7.0126e−05	7.0451e−05	0.5057
	15	1.0310e−05	2.7380e−05	3.0437e−05	5.3471	5.0194e−06	4.4741e−05	5.2095e−05	3600.0960	9.0756e−06	5.3376e−05	6.0770e−05	0.5358
S&P $n=98$	5	7.1653e−05	9.5159e−05	1.0688e−04	14.5384	3.4736e−05	1.0966e−04	1.0508e−04	213.8620	7.7836e−05	9.6925e−05	1.4518e−04	0.5207
	6	4.5976e−05	7.5869e−05	8.0982e−05	12.8893	2.6004e−05	7.7941e−05	8.1164e−05	3092.8290	6.6548e−05	7.6908e−05	1.0036e−04	0.5215
	8	3.2999e−05	5.6537e−05	5.5807e−05	11.0334	1.5385e−05	3.7360e−05	6.8300e−05	3600.0820	2.8781e−05	4.5365e−05	6.0093e−05	0.5029
	10	1.9926e−05	4.7511e−05	4.6874e−05	9.6983	9.9834e−06	4.1410e−05	1.0336e−04	3600.0810	1.9609e−05	3.5791e−05	4.4009e−05	0.5822
	12	1.1320e−05	3.1943e−05	3.4280e−05	9.0605	5.7967e−06	3.8950e−05	4.0797e−05	3600.0659	1.9609e−05	3.5791e−05	4.4009e−05	0.5822
	15	7.3819e−06	2.7084e−05	2.9543e−05	7.4877	3.3068e−06	2.4011e−05	3.1356e−05	3600.0980	9.1530e−06	2.0885e−05	2.7415e−05	0.5277

Table 3. Comparison of sparse SVR model (3) and sparse LS model (1) for large data sets.

Index	K	Sparse SVR (3)				Sparse LS				Heuristic			
		Training	Testing1	Testing2	Time	Training	Testing1	Testing2	Time	Training	Testing1	Testing2	Time
Nikkei $n = 225$	5	9.0230e−05	9.8139e−05	1.4916e−04	52.6221	3.7815e−05	1.2190e−04	2.6233e−05	3600.0120	1.6073e−04	1.2844e−04	1.4134e−04	0.7094
	10	3.6770e−05	5.7376e−05	7.3665e−05	42.1410	1.2211e−05	7.0131e−05	8.5029e−05	3600.0190	6.0255e−05	8.4927e−05	9.8321e−05	0.6662
	15	3.3050e−05	4.4422e−05	4.8400e−05	31.0362	3.3498e−06	3.2352e−05	6.7353e−05	3600.0199	2.4790e−05	3.7251e−05	7.3229e−05	0.6475
	20	2.2408e−05	3.0128e−05	3.8044e−05	25.3132	2.0282e−06	3.7968e−05	6.2176e−05	3600.0200	1.9086e−05	2.9917e−05	7.5137e−05	0.6508
	30	1.3257e−05	2.0019e−05	2.8926e−05	26.4403	4.2168e−07	2.6284e−05	4.4453e−05	3600.0040	3.5080e−06	2.6508e−05	6.3774e−05	0.6675
	40	1.1677e−05	1.4039e−05	2.4966e−05	22.4751	1.0215e−07	1.8742e−05	3.4370e−05	3600.0120	1.3601e−06	1.6135e−05	3.9807e−05	0.6979
S&P $n = 457$	5	2.5315e−04	2.6388e−04	3.6871e−04	61.0291	4.6219e−05	3.5415e−04	4.0505e−04	3600.0130	1.4324e−04	3.0686e−04	3.7971e−05	0.8684
	10	1.2465e−04	8.4853e−05	1.6172e−04	52.8294	1.4839e−04	2.5976e−04	3.3918e−04	3600.0110	6.4128e−05	1.5375e−04	2.7304e−04	0.9242
	15	8.0322e−05	6.4121e−05	1.8542e−04	46.3220	3.9548e−06	2.1221e−04	2.3723e−04	3600.0120	2.4947e−05	1.3688e−04	1.7001e−04	0.8627
	20	5.3527e−05	5.7316e−05	1.2050e−04	41.6894	2.3441e−06	1.0524e−04	1.7456e−04	3600.0122	1.9236e−05	1.2520e−04	1.6439e−04	0.8661
	30	3.2516e−05	5.3925e−05	1.1879e−04	37.3706	5.2750e−07	8.9958e−05	1.2720e−04	3600.0120	1.1640e−05	1.1020e−04	1.6308e−04	0.9192
	40	2.7066e−05	4.5264e−05	1.0869e−04	30.5452	1.4163e−07	8.7162e−05	1.1832e−04	3600.0121	7.8044e−04	1.0749e−04	1.4419e−04	0.9032
Russell 2000 $n = 1318$	5	1.3900e−04	4.9989e−04	6.2571e−04	48.1692	8.0169e−05	5.0185e−04	7.4603e−04	3600.0190	6.4431e−04	0.0011	0.0015	4.4572
	10	9.2235e−05	4.2306e−04	4.9297e−04	44.8831	5.5245e−05	5.1924e−04	5.8017e−04	3600.0191	3.3580e−04	6.8992e−04	7.2279e−04	4.6606
	15	4.6152e−05	2.8178e−04	3.5483e−04	37.4462	3.9936e−05	4.7989e−04	4.9004e−04	3600.0209	2.4159e−04	4.9619e−04	4.8351e−04	4.4189
	20	4.0950e−05	1.9363e−04	2.7980e−04	32.9616	2.2457e−05	3.2694e−04	3.7387e−04	3600.0210	1.5540e−04	5.0399e−04	6.0774e−04	5.2620
	30	2.6952e−05	1.7744e−04	2.7448e−04	26.4051	4.8834e−06	3.3511e−04	3.0519e−04	3600.0201	1.1712e−04	4.1284e−04	5.1932e−04	5.2437
	40	1.5942e−05	1.6869e−04	2.0010e−04	22.8590	3.5761e−06	2.3555e−04	3.8944e−04	3600.0201	9.5211e−05	3.5578e−04	4.2894e−04	4.5839
Russell 3000 $n = 2151$	5	1.2022e−04	2.7650e−04	4.2662e−04	58.5171	9.1549e−05	6.2532e−04	8.8021e−04	3600.0211	2.3190e−04	3.9944e−04	7.6840e−04	9.4637
	10	6.9992e−05	2.1837e−04	3.6897e−04	57.4182	5.3522e−05	4.4538e−04	5.2766e−04	3600.0212	1.2274e−04	2.6383e−04	5.4300e−04	9.5291
	15	3.9153e−05	1.6381e−04	3.0252e−04	53.8821	2.8373e−05	2.2925e−04	3.9686e−04	3600.0209	9.7655e−05	2.2624e−04	3.8350e−04	9.7157
	20	3.6539e−05	1.1547e−04	2.7974e−04	51.1527	2.9011e−05	1.8164e−04	3.2314e−04	3600.0210	7.6021e−05	1.5168e−04	2.9136e−04	10.3651
	30	1.4975e−05	1.1871e−04	1.5790e−04	42.9381	7.5081e−06	1.4996e−04	2.6408e−04	3600.0210	4.8177e−05	1.5733e−04	2.6873e−04	9.2701
	40	1.3780e−05	8.5298e−05	1.4887e−04	40.3707	7.6230e−06	1.2628e−04	2.1566e−04	3600.0200	4.0169e−05	1.6253e−04	2.4485e−04	10.1545

Table 4. Comparison of sparse SVR model (4) and sparse LS model (1) for large data sets.

Index	K	Sparse SVR (4)				Sparse LS				Heuristic			
		Training	Testing1	Testing2	Time	Training	Testing1	Testing2	Time	Training	Testing1	Testing2	Time
Nikkei $n=225$	5	7.6001e−05	9.8110e−05	1.4121e−04	29.7992	3.7815e−05	1.2190e−04	2.6233e−04	3600.0120	1.6073e−04	1.2844e−04	1.4134e−04	0.7094
	10	2.0632e−05	3.9455e−05	7.5790e−05	19.5535	1.2211e−05	7.0131e−05	8.5029e−05	3600.0190	6.0255e−05	8.4927e−05	9.8321e−05	0.6662
	15	2.0436e−05	2.6656e−05	5.2378e−05	15.8314	3.3498e−06	3.2352e−05	6.7353e−05	3600.0199	2.4790e−05	3.7251e−05	7.3229e−05	0.6475
	20	1.2904e−05	2.9703e−05	4.7959e−05	13.7981	2.0282e−06	3.7968e−05	6.2176e−05	3600.0200	1.9086e−05	2.9917e−05	7.513″e−05	0.6508
	30	8.2579e−06	2.2601e−05	2.9380e−05	11.3166	4.2168e−07	2.6284e−05	4.4453e−05	3600.0040	3.5080e−06	2.6508e−05	6.3774e−05	0.6675
	40	5.4580e−06	1.3146e−05	2.2092e−05	10.8292	1.0215e−07	1.8742e−05	3.4370e−05	3600.0120	1.3601e−06	1.6135e−05	3.980″e−05	0.6979
S&P $n=457$	5	7.8835e−05	1.8714e−04	3.3008e−04	47.7442	4.6219e−05	3.5415e−04	4.0505e−04	3600.0130	1.4324e−04	3.0686e−04	3.7971e−05	0.8684
	10	3.0257e−05	1.4820e−04	2.2965e−04	31.1626	1.4839e−05	2.5976e−04	3.3918e−04	3600.0110	6.4128e−05	1.5375e−04	2.7304e−04	0.9242
	15	2.3723e−05	1.1620e−04	1.8358e−04	25.2646	3.9548e−06	2.1221e−04	2.3723e−04	3600.0120	2.4947e−05	1.3688e−04	1.7001e−04	0.8627
	20	1.3661e−05	8.3195e−05	1.4574e−04	22.0631	2.3441e−06	1.0524e−04	1.7456e−04	3600.0122	1.9236e−05	1.2520e−04	1.6435e−04	0.8661
	30	1.1638e−05	7.3939e−05	1.2851e−04	18.5386	5.2750e−07	8.9955e−05	1.2720e−04	3600.0120	1.1640e−05	1.1020e−04	1.6305e−04	0.9192
	40	5.8876e−06	5.7782e−05	1.1687e−04	17.0115	1.4163e−07	8.7162e−05	1.1832e−04	3600.0121	7.8044e−06	1.0749e−04	1.4419e−04	0.9032
Russell 2000 $n=1318$	5	1.3879e−04	4.9972e−04	6.4374e−04	146.7276	8.0169e−05	5.0185e−04	7.4603e−04	3600.0190	6.4431e−04	0.0011	0.0015	4.4572
	10	6.5636e−05	4.1504e−04	4.9714e−04	102.7041	5.5245e−04	5.1924e−04	5.8017e−04	3600.0191	3.3580e−04	6.8992e−04	7.2279e−04	4.6606
	15	5.9517e−05	3.4606e−04	3.7841e−04	90.4402	3.9936e−05	4.7989e−04	4.9004e−04	3600.0209	2.4159e−04	4.9619e−04	4.8351e−04	4.4189
	20	4.2270e−05	3.0145e−04	3.0325e−04	74.3545	2.2457e−05	3.2694e−04	3.7387e−04	3600.0210	1.5540e−04	5.0399e−04	6.0774e−04	5.2620
	30	4.0200e−05	1.9799e−04	2.9711e−04	64.9030	4.8834e−06	3.3511e−04	3.0519e−04	3600.0201	1.1712e−04	4.1284e−04	5.1932e−04	5.2437
	40	1.3132e−05	1.7533e−04	2.4468e−04	57.1566	3.5761e−06	2.3555e−04	3.8944e−04	3600.0201	9.5211e−05	3.5578e−04	4.2894e−04	4.5839
Russell 3000 $n=2151$	5	1.2077e−04	3.4966e−04	5.3782e−04	416.2771	9.1549e−05	6.2532e−04	8.8021e−04	3600.0211	2.3190e−04	3.9944e−04	7.6840e−04	9.4637
	10	5.9355e−05	1.6546e−04	3.4083e−04	304.0703	5.3522e−05	4.4538e−04	5.2766e−04	3600.0212	1.2274e−04	2.6383e−04	5.4300e−04	9.5291
	15	4.1252e−05	1.2495e−04	2.9550e−04	261.5331	2.8373e−05	2.2925e−04	3.9686e−04	3600.0209	9.7655e−05	2.2624e−04	3.8350e−04	9.7157
	20	2.6621e−05	1.4513e−04	2.6988e−04	234.2974	2.9011e−05	1.8164e−04	3.2314e−04	3600.0210	7.6021e−05	1.5168e−04	2.9136e−04	10.3651
	30	1.5412e−05	1.4243e−04	1.9433e−04	189.1744	7.5081e−06	1.4996e−04	2.6408e−04	3600.0210	4.8177e−05	1.5733e−04	2.6873e−04	9.2701
	40	1.7554e−05	7.8449e−05	1.6050e−04	169.7401	7.6230e−06	1.2628e−04	2.1566e−04	3600.0200	4.0169e−05	1.6253e−04	2.4485e−04	10.1545

Table 5. The influences of the hyper-parameters for the sparse SVR models.

C_1	ε	Training	Testing1	Testing2	Std1	Std2	C_1	C_2	Training	Testing1	Testing2	Std1	Std2
Model (3) for Hang Seng $K = 15$							Model (4) for Hang Seng $K = 15$						
50	1e−03	6.2858e−06	2.7807e−05	2.1100e−04	4.3984e−07	1.1255e−06	50	50	5.4983e−06	2.4719e−05	2.0821e−05	6.7332e−06	3.6311e−06
	1e−04	6.0192e−06	2.6671e−05	1.9919e−05	4.5435e−07	5.2915e−07		500	5.3649e−06	2.4443e−05	1.9753e−05	6.1687e−06	4.3054e−06
	1e−05	4.6537e−06	2.1064e−05	1.8066e−05	1.7380e−06	1.6083e−06		5e03	4.6530e−06	2.1064e−05	1.8065e−05	9.3540e−06	5.0202e−06
	1e−06	5.3142e−06	2.4895e−05	2.0313e−05	1.1560e−05	1.6623e−06		5e04	4.6530e−06	2.1064e−05	1.8065e−05	6.5568e−06	4.3254e−06
5	1e−05	9.7871e−06	3.2513e−05	2.4499e−05	5.7346e−06	3.6181e−06	5	5e03	1.1249e−05	4.3891e−05	3.0170e−05	5.2958e−06	3.6972e−06
50		4.6537e−06	2.1064e−05	1.8066e−05	1.7380e−06	1.6083e−06	50		4.6530e−06	2.1064e−05	1.8065e−05	9.3504e−06	5.0202e−06
500		4.3555e−06	1.9128e−06	1.6819e−05	1.0139e−06	5.5459e−07	500		4.5151e−06	1.9699e−05	1.6491e−05	7.6516e−06	4.5922e−06
5e03		4.3493e−06	1.9036e−06	1.6750e−05	0	0	5e03		4.3493e−06	1.9037e−05	1.6745e−05	9.5061e−06	6.2158e−06
Model (3) for Nikkei $K = 40$							Model (4) for Nikkei $K = 40$						
5	5e−03	1.3798e−05	2.5069e−05	3.7748e−05	3.4659e−05	6.2939e−06	5	50	5.5041e−06	1.8556e−06	2.7172e−05	2.7990e−06	5.1551e−06
	5e−04	1.3401e−05	1.2488e−05	2.6398e−05	2.5190e−05	5.5478e−06		500	6.3253e−06	9.3235e−06	2.2639e−05	3.8165e−06	6.0460e−06
	5e−05	1.3695e−05	1.4779e−05	2.2765e−05	2.3979e−06	4.6675e−06		5e3	4.4037e−06	1.3056e−05	2.1249e−05	2.6990e−06	8.7752e−06
	5e−06	1.3483e−05	1.2104e−05	2.3688e−05	2.9105e−06	2.7909e−06		5e4	5.2191e−06	1.7046e−05	3.6524e−05	3.0832e−06	7.1352e−06
0.5	5e−05	3.5865e−05	1.3705e−05	1.8082e−05	2.0021e−06	2.9118e−06	0.5	5e03	4.0081e−05	1.1304e−05	1.7631e−05	3.5270e−06	4.1607e−06
5		1.3695e−05	1.4779e−05	2.2765e−05	2.3979e−06	4.6675e−06	5		4.4037e−06	1.3056e−05	2.1249e−05	2.6990e−06	8.7752e−06
50		2.3294e−06	1.8049e−05	3.6287e−05	3.6620e−06	6.0469e−06	50		3.2256e−06	9.2742e−06	3.0214e−05	4.7656e−06	1.0570e−05
500		1.8116e−06	2.8260e−05	5.4165e−05	4.3550e−06	6.6983e−06	500		1.6458e−06	2.1577e−05	3.2509e−05	3.9804e−05	8.4200e−06

2, 3 and 4, we can observe that although we need find the solution of the sparse SVR model by using the penalty PALM method to solve the sparse SVR model ten times from different initial points, the computational efficiency of the sparse SVR model is better than that of the sparse LS model. Since the heuristic model only contains two LS problem, the total computation time of the heuristic model is less than the other models, but the heuristic model can not guarantee to find a good tracking portfolio excepting some small-scale problems. In addition, from Table 5 we also observe that, for the sparse SVR models, the parameters selection of model (4) is easier than that of model (3) but the stability of the performance of the solutions of model (3) is better than that of model (4).

5   Conclusion

In this paper we consider two sparse SVR models (3) and (4) for the index tracking problem and the penalty PALM method for solving these models. For structuring the sparse SVR model, we add a cardinality constraint in a commonly used ε-SVR model and use the piecewise linear functions to reduce the number of constraints in the model. In addition, if there does not have a good estimation of the maximum deviation tolerance ε, we consider another new sparse SVR model which makes a tradeoff between the tracking error larger than the maximum deviation tolerance and the maximum deviation tolerance itself. The optimization problems (3) and (4) corresponding to two sparse SVR models are generally NP-hard. We choose the penalty PALM method to solve these problems. Under some suitable assumptions, we can establish that any accumulation point of the sequence generated by the penalty PALM method for problem (3) or (4) satisfies the first-order necessary optimality conditions of the problem. Furthermore, under another assumption, such an accumulation point is also a local minimizer of problem (3) or (4). The proposed models are tested and compared with a common sparse LS model and a common heuristic model. The computational results demonstrate that when the training data is relatively fewer, the sparse SVR models have better generalization ability and stability especially for the large-scale problems. And we also observe from the numerical experiment that, for the sparse SVR models, the parameters selection of model (4) is easier than that of model (3) and the stability of the performance of the solutions of model (3) is better than that of model (4).

Acknowledgments. The authors are grateful to anonymous reviewers for many helpful suggestions.

Funding. This work was supported by the National Natural Science Foundation of China (11571061, 11301050) and the Fundamental Research Funds for the Central Universities (DUT15RC(3)037).

References

1. Beasley, J.E., Meade, N., Chang, T.-J.: An evolutionary heuristic for the index tracking problem. Eur. J. Oper. Res. **148**, 621–643 (2009)

2. Canagkoz, N.A., Beasley, J.E.: Mixed-integer programming approaches for index tracking and enhanced indexation. Eur. J. Oper. Res. **196**, 384–399 (2009)
3. Konno, H., Wijayanayake, A.: Minimal cost index tracking under nonlinear transaction costs and minimal transaction unit constraints. Int. J. Theoret. Appl. Financ. **4**, 939–958 (2001)
4. Okay, N., Akman, U.: Index tracking with constraint aggregation. Appl. Econ. Lett. **10**, 913–916 (2003)
5. Gilli, M., Kellezi, E.: The threshold accepting heuristic for index tracking. In: Financial Engineering, Ecommerce and Supply Chain, Applied Optimization, vol. 70, pp. 1–18. Kluwer Academic Publishers, Dordrecht (2002)
6. Derigs, U., Nickel, N.-H.: Meta-heuristic based decision support for portfolio optimisation with a case study on tracking error minimization in passive portfolio management. OR Spectr. **25**, 345–378 (2003)
7. Jansen, R., van Dijk, R.: Optimal benchmark tracking with small portfolios. J. Portf. Manag. **28**, 33–39 (2002)
8. Maringer, D., Oyewumi, O.: Index tracking with constrained portfolios. Intell. Syst. Account. Financ. Manag. **15**, 57–71 (2007)
9. Fastrich, B., Paterlini, S., Winker, P.: Cardinality versus q-norm constraints for index tracking. Quant. Financ. 1–14 (2012, ahead-of-print)
10. Brodie, J., Daubechies, I., De Mol, C., Giannone, D., Loris, D.: Sparse and stable Markowitz portfolios. Proc. Natl. Acad. Sci. **106**(30), 12267–12272 (2009)
11. Chen, C., Li, X., Tolman, C., Wang, S., Ye, Y.: Sparse portfolio selection via quasi-norm regularization. arXiv preprint arXiv:1312.6350 (2013)
12. Xu, F., Wang, G., Gao, Y.: Nonconvex L1/2 Regularization For Sparse Portfolio Selection (2014)
13. Bian, W., Chen, X., Ye, Y.: Complexity analysis of interior point algorithms for non-Lipschitz and nonconvex minimization. Math. Program. **149**, 301–327 (2015)
14. Xu, F., Lu, Z., Xu, Z.: An efficient optimization approach for a cardinality-constrained index tracking problem. Optimization Methods and Software **31**(2), 258–271 (2016)
15. Zheng, X., Sun, X., Li, D., Sun, J.: Successive convex approximations to cardinality-constrained convex programs: a piecewise-linear DC approach. Comput. Optim. Appl. **59**(1–2), 379–397 (2014)
16. Burdakov, O., Kanzow, C., Schwartz, A.: Mathematical programs with cardinality constraints: reformulation by complementarity-type constraints and a regularization method. SIAM J. Optim. **26**(1), 397–425 (2016)
17. Lu, Z., Zhang, Y.: Sparse approximation via penalty decomposition methods. SIAM J. Optim. **23**(4), 2448–2478 (2013)
18. Bolte, J., Sabach, S., Teboulle, M.: Proximal alternating linearized minimization for nonconvex and nonsmooth problems. Math. Program. **146**(1–2), 459–494 (2014)
19. Steinwart, I., Christmann, A.: Support Vector Machines. Springer Science Business Media, New York (2008)
20. Basak, D., Pal, S., Patranabis, D.C.: Support vector regression. Neural Inf. Process.-Lett. Rev. **11**(10), 203–224 (2007)
21. Smola, A.J., Schölkopf, B.: A tutorial on support vector regression. Stat. Comput. **14**(3), 199–222 (2004)
22. Takeda, A., Gotoh, J., Sugiama, M.: Support vector regression as conditional value-at-risk minimization with application to financial time-series analysis. In: IEEE International Workshop on Machine Learning for Signal Processing (MLSP), pp. 118–123. IEEE (2010)

23. De Leone, R.: Support vector regression for time series analysis. In: Hu, B., Morasch, K., Pickl, S., Siegle, M. (eds.) Operations Research Proceedings 2010, pp. 33–38. Springer, Heidelberg (2010)
24. Vapnik, V.: The Nature of Statistical Learning Theory. Springer, New York (2013)
25. Schölkopf, B., Smola, A., Williamson, R.C., Bartlett, P.L.: New support vector algorithms. Neural Comput. **12**, 1207–1245 (2000)
26. Bennett, K.P., Mangasarian, O.L.: Robust linear programming discrimination of two linearly inseparable sets. Optim. Methods Softw. **1**, 23–34 (1992)
27. Teng, Y., Yang, L., Yu, B., Song, X.: A penalty PALM method for sparse portfolio selection problems. Optim. Methods Softw. **32**(1), 126–147 (2017)
28. Attouch, H., Bolte, J., Redont, P., Soubeyran, A.: Proximal alternating minimization and projection methods for nonconvex problems: an approach based on the Kurdyka-Łojasiewicz inequality. Math. Oper. Res. **35**, 438–457 (2010)
29. Ruszczyński, A.P.: Nonlinear Optimization, vol. 13. Princeton University Press, Princeton (2006)
30. Bolte, J., Daniilidis, A., Lewis, A.: The Lojasiewicz inequality for nonsmooth subanalytic functions with applications to subgradient dynamical systems. SIAM J. Optim. **17**(4), 1205–1223 (2007)
31. Beasley, J.E.: OR-library: distributing test problems by electronic mail. J. Oper. Res. Soc. **41**(11), 1069–1072 (1990)

Two-Stage Transfer Learning of End-to-End Convolutional Neural Networks for Webpage Saliency Prediction

Wei Shan, Guangling Sun$^{(\boxtimes)}$, Xiaofei Zhou, and Zhi Liu

School of Communication and Information Engineering,
Shanghai University, Shanghai, China
shanwei1993@126.com, sunguangling@shu.edu.cn,
zxforchid@163.com, liuzhisjtu@163.com

Abstract. With the great success of convolutional neural networks (CNN) achieved on various computer vision tasks in recent years, CNN has also been applied in natural image saliency prediction. As a specific visual stimuli, webpages exhibit evident similarities whereas also significant differences from natural image. Consequently, the learned CNN for natural image saliency prediction cannot be directly used to predict webpage saliency. Only a few researches on webpage saliency prediction have been developed till now. In this paper, we propose a simple yet effective scheme of two-stage transfer learning of end-to-end CNN to predict the webpage saliency. In the first stage, the output layer of two typical CNN architectures with instances of AlexNet and VGGNet are reconstructed, and the parameters between the fully connected layers are relearned from a large natural image database for image saliency prediction. In the second stage, the parameters between the fully connected layers are relearned from a scarce webpage database for webpage saliency prediction. In fact, the two-stage transfer learning can be regarded as a task transfer in the first stage and a domain transfer in the second stage, respectively. The experimental results indicate that the proposed two-stage transfer learning of end-to-end CNN can obtain a substantial performance improvement for webpage saliency prediction.

Keywords: Convolutional neural networks · End-to-End · Webpage saliency prediction · Two-stage transfer learning

1 Introduction

Pervasive usage of social network, search engine and online commerce have been undergoing a dominant development around the world. According to the report by International Telecommunication Union (ITU), the number of Internet users has reached 3.4 billion in 2016 [1]. Consequently, as a base platform of Internet information, webpage has become more important than ever before. Accordingly, automatically predicting fixations when browsing webpage has significant commercial value for web based applications and academic value for computer vision community.

A closely related issue is the natural image saliency prediction and the pioneer work came from Itti *et al.* [2], in which saliency is predicted by calculating the

© Springer International Publishing AG 2017
Y. Sun et al. (Eds.): IScIDE 2017, LNCS 10559, pp. 316–324, 2017.
DOI: 10.1007/978-3-319-67777-4_27

center-surrounded contrast of colors, intensity, and orientations. Since then, plenty of saliency models based on various frameworks have been proposed [3]. In recent years, Convolutional Neural Networks (CNN), as one of the high-performance deep learning neural networks, has obtained encouraging results in many computer vision tasks including image classification, object detection and segmentation. In particular, it also has been successfully applied to predict visual saliency on natural images. An early attempt of predicting saliency model with a CNN was the ensembles of Deep Networks (eDN) [4], which proposed an optimal blend of features from three different CNN layers. Then the features were combined with a simple linear classifier trained with positive (salient) or negative (non-salient) local regions. Inspired by this work, DeepGaze [5] was proposed to combine features from different layers. It was worth pointing out that DeepGaze exploited the pre-trained AlexNet [6], which was originally trained for image classification.

Compared to natural image, webpage has different characteristics: webpage is full of salient stimuli, such as text, logos and pictures [7]. In addition, human's web-viewing patterns on webpages are different from that on natural images, such as F-bias to scan top-left region at the start of web-viewing [8, 9]. These aspects determine that the model used for natural image saliency prediction cannot be directly applied to effectively predict webpage saliency. A handful of approaches are developed for predicting webpage saliency. In [10], low-level, middle-level and high-level features are extracted from AlextNet and then fed into SVM classifier to determine the attention value for each pixel, yielding the final saliency map. In [11], a machine learning framework with multiple features is employed to be adaptive to various types of webpage saliency prediction tasks.

In this work, we explore to utilize transfer learning to train an end-to-end CNN with the goal of predicting webpage saliency. Transfer learning aims at transferring knowledge between source and target task or source and target domain [12]. Transfer learning is critically meaningful when training data is scarce or distribution discrepancy exists between training data and testing data. In our case, the amount of available webpage with human fixation data is quite limited so that it is unfeasible to train a complex and large-scale deep network in a traditional way. We exploit the idea of transfer learning to cope with this challenge. Specifically, we design a two-stage transfer learning framework composed of task transfer and domain transfer, in which we adopt an end-to-end architecture to directly generate the saliency map.

The rest of the paper is organized as follows. The proposed model is detailed in Sect. 2. Experimental results and analysis are presented in Sect. 3, and conclusions are given in Sect. 4.

2 Proposed Method

2.1 Overview

Since the public available webpage data with human fixations is lacking, the usual training scheme is unfeasible for the deep networks. Therefore, we propose a two-stage transfer learning strategy. During the first-stage transfer learning, the output node number is firstly set to 4096 and reshaped to a 64 * 64 saliency map. To obtain the same resolution as the input image, the output saliency map is interpolated to obtain the same size as the input image. Secondly, the parameters of CNN originally for image classification are transferred by using pre-trained CNN for initialization and the network is re-trained from a large natural image data labeled with human fixations. Since the output of the network has been changed from classification result to saliency map prediction, the first-sage transfer learning can be regarded as task transfer; During the second-stage transfer learning, the parameters of the network are further fine-tuned with the webpage dataset. Since the final output of the network is webpage saliency map, the second-stage transfer learning can be regarded as domain transfer. The procedure is illustrated in Fig. 1. To make the proposed model more convictive, we explore two widely used CNN architectures including AlexNet and VGGNet [13] as network instances.

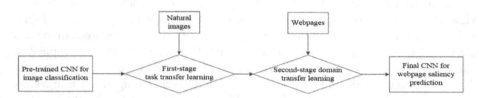

Fig. 1. An overview of two-stage transfer learning for webpage saliency prediction.

2.2   CNN for Image Classification

The structure of AlexNet and VGGNet are shown in Fig. 2(a) and (b). Red box denotes convolutional layer, blue box denotes pooling layer and green box denotes fully-connected layer. Both AlexNet and VGGNet take the image with fixed dimension of 224 × 224 × 3 pixels as the input layer.

(a) *AlexNet:* The AlexNet architecture was published in [6], which achieved the state-of-the-art performance on ImageNet Large Scale Visual Recognition Challenge (ILSVRC) 2012. ImageNet is a large database consisting of 1.2 million images belonging to 1000 categories. AlexNet is composed of five convolution layers, three pooling layers, and three fully-connected layers with an approximation of 60 million free parameters.

(b) *VGGNet:* The VGGNet architecture is introduced in [13], which is designed to significantly increase the depth with 16 or 19 layers compared to AlexNet. To reduce the number of parameters in the deeper networks, small 3 × 3 convolutional filters are adopted in all convolution layers with a convolutional stride of size 1. Since VGGNet is much deeper than AlexNet, VGGNet is more susceptible to the vanishing gradient problem, and training VGGNet requires far more memory and computation time than AlexNet. In this work, we use the VGGNet with 16 layers.

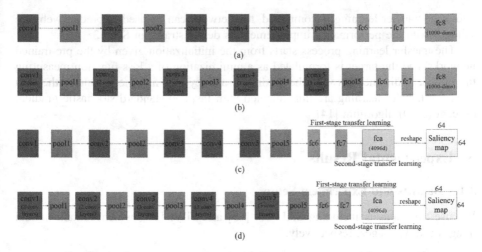

Fig. 2. Illustration of two-stage transfer learning. (a) AlexNet; (b) VGGNet; (c) two-stage transfer learning with AlexNet as the pre-trained network; (d) two-stage transfer learning with VGGNet as the pre-trained network. (Color figure online)

2.3   Two-Stage Transfer Learning of End-to-End CNN

a. *First-stage Task Transfer Learning*

For saliency prediction task, we use the dimension of $224 \times 224 \times 3$ as the input layer size. However, we change the dimension of the last layer (denoted with yellow box "fca" in Fig. 2(c) and (d)) from 1000 to 4096, followed by a 64×64 reshaping operation. The matrix 64×64 is the output saliency map and then is interpolated to obtain the final saliency map with the same resolution as the original image. The dimension of 4096 is set from the perspective of balance between efficiency and performance. The parameters of all the other layers, except for the last fully-connected layer, are initialized with that of pre-trained AlexNet or VGGNet, and the parameters of the last fully-connected layer are initialized with Gaussian distribution of mean of 0 and standard deviation of 1. Only the parameters of layers "fc6" and "fc7" (denoted with green box in Fig. 2(c) and (d)) are modified and the parameters of layer "fca" are learned during the first-stage transfer learning, while the parameters of previous convolutional layers remain fixed. Once the first-stage transfer learning is completed, the CNN could be used to perform saliency prediction on natural images.

b. *Second-stage Domain Transfer Learning*

As aforementioned, the natural image exhibits different characteristics from webpage so that the trained CNN via the first-stage transfer learning cannot be used to perform saliency prediction on webpages yet. Fortunately, transfer learning is expected to further train the CNN using webpage dataset labeled with human fixations. Thus, the second-stage transfer learning is a domain transfer learning where only the parameters of the layer "fca" (denoted with yellow box in Fig. 2(c) and (d)) are updated. Once the two-stage transfer learning is completed, the network can be used to predict webpage saliency with the performance improvement as demonstrated in Sect. 3.

The transfer learning process starts from the initialization given by the pre-trained networks, and the target is formulated as a minimization of a loss function measuring the Euclidean distance between the predicted saliency map and the ground truth. Both stages of transfer learning are implemented with fast and standard stochastic gradient descent (SGD) algorithm [14].

3   Experimental Results

3.1   Datasets and Experiment Setup

We use two datasets including SALION [15] and FiWI [16], which contain natural images and webpages, respectively.

SALICON: It is the largest dataset available for natural image saliency prediction and is used in the first-stage transfer learning. The images of SALICON are from *Microsoft CoCo* dataset [17]. However, different from most popular datasets for saliency prediction, the human fixations as the ground truth in SALICON were not obtained with eye-trackers but with mouse clicks captured in a crowd-sourcing platform. Some images of the dataset are shown in Fig. 3(a).

FiWI: In our work, we exploit FiWI, which is the only public webpage dataset with human fixation data, in the second-stage transfer learning and as well as performance evaluation. It consists of 149 webpages and corresponding human fixation data. The ground truth, *i.e.* fixation density map, was generated by convolving a 2D Gaussian filter on all human fixation data collected from 11 subjects. According to the content, the webpages are categorized as pictorial, text and mixed. Each category contains around 50 images. Examples of webpage in each category are shown in Fig. 3(b).

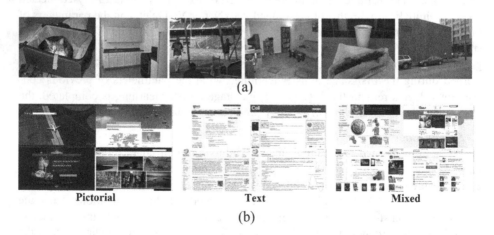

(a)

Pictorial   Text   Mixed

(b)

Fig. 3. (a) Samples in SALICON; (b) Samples in FiWI.

The proposed model is implemented using Caffe [14] and all experiments were performed on a NVIDIA GPU GeForce GTX Titan X with 12 GB RAM. The parameters used in SGD are set as follows: the learning rate is 0.01, the momentum coefficient is 0.9, the weight decay is 0.0005, and the batch size is 10.

3.2 Quantitative Assessment

To quantitatively assess the performance, we calculate the metric scores including linear Correlation Coefficient (CC), Normalized Scanpath Saliency (NSS), and shuffled Area Under Curve (sAUC). CC measures the linear correlations between the ground truth and the predicted saliency map. The closer CC to 1, the better the performance. NSS measures the average response values at fixation locations along the scanpath in the normalized saliency map. sAUC is calculated as the area under a Receiver Operating Characteristics (ROC) curve using fixations of other images in the same dataset as negatives. sAUC can eliminate the effect of center-bias. For the sAUC score, 1 means perfect prediction while 0.5 indicates chance level.

(1) Roles of two-stage transfer learning

To validate the roles of the two-stage transfer learning, we compare with the other two scenarios: transfer learning only with natural images and transfer learning only with webpages. For brevity, the three scenarios are denoted with Network-NW (both natural images and webpages), Network-N (only natural images) and Network-W (only webpages), respectively. Network represents AlexNet or VGGNet. Network-N and Network-NW are trained with 10000 samples of SALICON, and Network-W and Network-NW are trained and tested with webpages by cross validation. Specifically, FiWI is randomly separated into 119 training samples and 30 testing samples with 5 rounds. The final result is the average of 5 metric scores on the 5 random webpage testing sets.

As shown in Table 1, since the amount of webpage samples is very low, the performance of "AlexNet-W" and "VGGNet-W" is low. On the other hand, "AlexNet-N" and "VGGNet-N" achieve the better performance due that the large dataset SALICON is used to retrain the pre-trained network. Furthermore, "AlexNet-NW" and "VGGNet-NW" achieve the best performance because both SALICON and FiWI are used to retrain the pre-trained network in two stages respectively. These results demonstrate the effectiveness of the two-stage transfer learning framework. In addition, it is observed

Table 1. Webpage saliency prediction performances of networks based on different training sets.

Network and training set	CC	sAUC	NSS
AlexNet-W	0.2185	0.5233	0.6349
AlexNet-N	0.3658	0.7536	0.9062
AlexNet-NW	**0.3872**	**0.7658**	**0.9386**
VGGNet-W	0.2207	0.5128	0.6337
VGGNet-N	0.3786	0.7623	0.9275
VGGNet-NW	**0.4013**	**0.7765**	**0.9641**

that the performances of VGGNet are better than AlexNet under the condition of the same training setting and evaluation metric. It may attribute to the deeper architecture of VGGNet. However, the fact that the performance of AlexNet-NW is slightly better than VGGNet-N indicates that the second-stage transfer learning, *i.e.* domain transfer learning, has the vital role for improving webpage saliency prediction performance. Since VGGNet-NW achieves the best performance, we only test VGGNet of two-stage transfer learning in the following experiments.

(2) **Performance of three specific webpage categories**

In this experiment, we tested VGGNet-NW on three specific categories of webpage: text, pictorial and mixed. The results show the trivial performance differences among the three webpage categories. Different from the results in Table 2, the performance on "pictorial" is much better than "text" as reported in [16]. This is because no appropriate features for "text" webpage saliency prediction are defined in [16] while VGGNet can automatically learn powerful features for all three webpage categories.

Table 2. The performances of VGGNet on three specific webpage categories.

Webpage categories	CC	sAUC	NSS
Text	0.4001	0.7761	0.9613
Pictorial	0.4015	0.7775	0.9682
Mixed	0.4023	0.7759	0.9628

(3) **Comparison with other models**

We compare the proposed model with other state-of-the-art saliency models on the webpage saliency dataset. These saliency models include Itti [2], Li [11], Shen [16], and AWS [18]. The results in Table 3 show that the proposed model outperforms all the other compared models in terms of sAUC and NSS, while in terms of CC, our model achieves the second best performance.

Table 3. The performance comparison of the proposed model with other models.

Models	CC	sAUC	NSS
Proposed	0.4013	**0.7765**	**0.9641**
Li [11]	**0.4412**	0.7316	0.9048
Shen [16]	0.3876	0.7485	0.9379
AWS [18]	0.2628	0.6613	0.8227
Itti [2]	0.1835	0.5122	0.4268

3.3 Qualitative Assessment

For a qualitative assessment, we also visualize the fixation density maps (ground truths) and saliency maps generated using the proposed model and the compared saliency models for several images from the webpage dataset FiWI in Fig. 4. We can see from

Fig. 4 that the proposed model better highlights the salient objects such as face, text and logo. This demonstrates the advantage of the proposed model for the better saliency prediction.

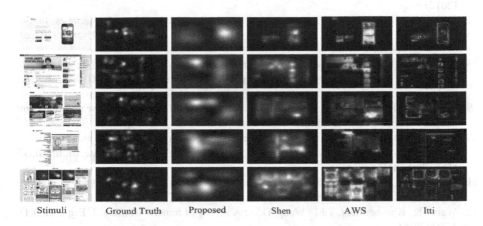

| Stimuli | Ground Truth | Proposed | Shen | AWS | Itti |

Fig. 4. Qualitative assessment on webpage samples.

4   Conclusions

Despite the large amount of existing saliency models that predict where humans look at in natural images, there are few studies on saliency prediction of webpages. In this work, we propose the two-stage transfer learning of end-to-end CNN for predicting webpage saliency. Specifically, the CNN originally used for image classification is transferred to serve for webpage saliency prediction through the task transfer in the first stage and the domain transfer in the second stage. Experimental results show that the proposed model has indeed improved the performance of webpage saliency prediction.

Acknowledgments. This work was supported by Shanghai Municipal Natural Science Foundation under Grant No. 16ZR1411100 and National Natural Science Foundation of China under Grant No. 61471230.

References

1. http://www.internetlivestats.com/internet-users/
2. Itti, L., Koch, C., Niebur, E.: A model of saliency-based visual attention for rapid scene analysis. IEEE Trans. Pattern Anal. Mach. Intell. **20**(11), 1254–1259 (1998)
3. Borji, A., Itti, L.: State-of-the-art in visual attention modeling. IEEE Trans. Pattern Anal. Mach. Intell. **35**(1), 185–207 (2013)
4. Vig, E., Dorr, M., Cox, D.: Large-scale optimization of hierarchical features for saliency prediction in natural images. In: Proceedings of the IEEE Conference on Computer Vision and Pattern Recognition, pp. 2798–2805 (2014)

5. Kümmerer, M., Theis, L., Bethge, M.: Deep gaze I: boosting saliency prediction with feature maps trained on imagenet (2014). arXiv preprint arXiv:1411.1045
6. Krizhevsky, A., Sutskever, I., Hinton, G.E.: Imagenet classification with deep convolutional neural networks. In: Advances in Neural Information Processing Systems, pp. 1097–1105 (2012)
7. Still, J.D., Masciocchi, C.M.: A saliency model predicts fixations in web interfaces. In: 5th International Workshop on Model Driven Development of Advanced User Interfaces, pp. 25–28 (2010)
8. Buscher, G., Cutrell, E., Morris, M.R.: What do you see when you're surfing?: using eye tracking to predict salient regions of web pages. In: Proceedings of the SIGCHI Conference on Human Factors in Computing Systems, pp. 21–30. ACM (2009)
9. Nielsen, J.: F-shaped pattern for reading web content (2006). http://www.nngroup.com/articles/f-shaped-pattern-reading-web-content
10. Shen, C., Huang, X., Zhao, Q.: Predicting eye fixations on webpage with an ensemble of early features and high-level representations from deep network. IEEE Trans. Multimed. **17** (11), 2084–2093 (2015)
11. Li, J., Su, L., Wu, B., et al.: Webpage saliency prediction with multi-features fusion. In: 2016 IEEE International Conference on Image Processing (ICIP), pp. 674–678 (2016)
12. Weiss, K., Khoshgoftaar, T.M., Wang, D.D.: A survey of transfer learning. J. Big Data **3**(1), 1–40 (2016)
13. Simonyan, K., Zisserman, A.: Very deep convolutional networks for large-scale image recognition (2014). arXiv preprint arXiv:1409.1556
14. Jia, Y., Shelhamer, E., Donahue, J., et al.: Caffe: convolutional architecture for fast feature embedding. In: Proceedings of the 22nd ACM International Conference on Multimedia, pp. 675–678. ACM (2014)
15. Jiang, M., Huang, S., Duan, J., et al.: SALICON: saliency in context. In: 2015 IEEE Conference on Computer Vision and Pattern Recognition (CVPR), pp. 1072–1080 (2015)
16. Shen, C., Zhao, Q.: Webpage saliency. In: Fleet, D., Pajdla, T., Schiele, B., Tuytelaars, T. (eds.) ECCV 2014. LNCS, vol. 8695, pp. 33–46. Springer, Cham (2014). doi:10.1007/978-3-319-10584-0_3
17. Lin, T.-Y., Maire, M., Belongie, S., Hays, J., Perona, P., Ramanan, D., Dollár, P., Zitnick, C. L.: Microsoft COCO: common objects in context. In: Fleet, D., Pajdla, T., Schiele, B., Tuytelaars, T. (eds.) ECCV 2014. LNCS, vol. 8693, pp. 740–755. Springer, Cham (2014). doi:10.1007/978-3-319-10602-1_48
18. Garcia-Diaz, A., Leboran, V., Fdez-Vidal, X.R., et al.: On the relationship between optical variability, visual saliency, and eye fixations: a computational approach. J. Vis. **12**(6), 17 (2012)

Saliency Detection via Combining Global Shape and Local Cue Estimation

Qiang Qi[1,2], Muwei Jian[1,2(✉)], Yilong Yin[1], Junyu Dong[2],
Wenyin Zhang[3], and Hui Yu[4]

[1] School of Computer Science and Technology,
Shandong University of Finance and Economics, Jinan, China
jianmuweihk@163.com
[2] Department of Computer Science and Technology,
Ocean University of China, Qingdao, China
[3] School of Information Science and Engineering, Linyi University, Linyi, China
[4] School of Creative Technologies, University of Portsmouth, Portsmouth, UK

Abstract. Recently, saliency detection has become a hot issue in computer vision. In this paper, a novel framework for image saliency detection is introduced by modeling global shape and local cue estimation simultaneously. Firstly, Quaternionic Distance Based Weber Descriptor (QDWD), which was initially designed for detecting outliers in color images, is used to model the salient object shape in an image. Secondly, we detect local saliency based on the reconstruction error by using a locality-constrained linear coding algorithm. Finally, by integrating global shape with local cue, a reliable saliency map can be computed and estimated. Experimental results, based on two widely used and openly available databases, show that the proposed method can produce reliable and promising results, compared to other state-of-the-art saliency-detection algorithms.

Keywords: Saliency detection · QDWD · Locality-constrained linear coding · Local cue

1 Introduction

Visual saliency detection is used to predict the most interesting regions or objects in an image, which is extremely valuable in many various computer-vision applications, such as video surveillance [1], vision tracking [2, 3], image classification/retrieval [4], image retargeting [5], image segmentation [6], etc.

Saliency detection models can be categorized from different perspectives. Basically, most approaches adopt a bottom-up approach via low-level features while a few works incorporate a top-down solution by task driven. In the pioneering work [7] by Itti et al. a center-surround contrast method, based on the multi-scale image features, was presented for saliency detection. Later, a large number of methods were proposed, including the graph-based saliency model, proposed by Harel et al. [8] and the fuzzy growing contrast method, proposed by Ma and Zhang [9]. In [10], a salient object segmentation method by using a saliency measure and a conditional random field

© Springer International Publishing AG 2017
Y. Sun et al. (Eds.): IScIDE 2017, LNCS 10559, pp. 325–334, 2017.
DOI: 10.1007/978-3-319-67777-4_28

(CRF) model was proposed. The works in [11], a new saliency model, which is based on the image regions and context-aware, was presented for saliency detection. Borji et al. [12] made an exhaustive comparison of thirty-five saliency models, based on fifty-four challenging synthetic patterns, three image datasets, and two video datasets. In [13], the final saliency maps are obtained by computing the covariance matrices of image features. Yang et al. [14] used the contrast, center and smoothness priors to estimate the saliency maps. The works in [15], a salient object detection algorithm was designed in multiple scales, based on the superpixels. Cheng et al. [16] proposed a novel saliency framework based on regional contrast (RC) to compute the global-contrast differences and spatial-weighted-coherence scores. In [17], a novel model based on Cellular Automata (CA) was proposed to compute the saliency of the objects. Jian et al. [18] proposed a visual-attention-aware model to mimic the human visual system (HVS) for salient-object detection. In [19], a bottom-up saliency-detection method was proposed by integrating directional, center and color cues. In [20], a novel self-paced multiple-instance learning (SP-MIL) framework, which considers both multiple instance learning (MIL) and self-paced learning (SPL), was proposed for co-saliency detection. The works in [21], a multistage saliency detection framework based on multilayer cellular automata (MCA) was proposed. Lu et al. [22] proposed a saliency detection framework, which exploring the fusion of various saliency models in a manner of bootstrap learning. In [23], a bottom-up saliency model that both consider the background and foreground cues was proposed.

The top-down methods need to consider both visual information and prior knowledge, thus these models usually contain complex learning process. In [24], a top-down saliency model, which using the global scene configuration, was proposed for saliency detection. Cholakkal et al. [25] adopted three locality constraints to compute the final saliency maps. In [26], a novel top-down saliency model that jointly learns a Conditional Random Field (CRF) and a visual dictionary was proposed. The works in [27] used a few exemplars and deep association to estimate the saliency of the image. In [28], a top-down contextual weighting saliency model, which incorporates high-level knowledge of the gist context of images was proposed.

In this paper, a novel saliency-detection method by modeling global shape and local estimation, is proposed for saliency detection. Different from existing saliency detection models, our proposed method integrates both global information and local features into a unified framework. We first computed the QDWD, which was initially designed for detecting outliers in color images, to represent the salient object shape in an image. Then, we incorporated an improved locality-constrained linear coding algorithm (LLC) to detect the local saliency. Finally, the two maps are combined to represent the most important saliency. In order to evaluate the performance of our proposed method, we carried out some experiments on two datasets and the comparison results with other state-of-the-art saliency-detection algorithms show that our approach is effective and efficient for saliency detection.

The rest of the paper is organized as follows. In Sect. 2, we introduce the proposed saliency-detection algorithm in detail. In Sect. 3, we demonstrate our experimental results based on two image datasets and compared the results with other eight saliency-detection methods. The paper closes with a conclusion in Sect. 4.

2 The Proposed Saliency Detection Model

In this section, we present the proposed method, which incorporates the global shape of the salient object by using QDWD and local cue by utilizing a locality-constrained method, to simulate saliency detection. QDWD will first be described, followed by the locality-constrained method. All these different types of information are fused to form a saliency map, which indicates the effective fusion of global and local information as well.

2.1 Quaternionic Distance Based Weber Descriptor

The global methods [16, 29, 30], which are characterized by holistic rarity and uniqueness, take the entire image information into consideration to detect the salient regions, and thus detect large objects and uniformly assign saliency values to the contained regions. Unlike local methods, which are sensitive to high frequency image contents like edges and noise, global methods are less effective when the textured regions of salient objects are similar to the background. In order to reduce the effect of background and generate saliency maps with little noise, QDWD, which was initially proposed to detect the outliers and edges in an image [31], is utilized to represent the global shape information for the HVS to detect saliency [19, 35].

A quaternion $\dot{q}$ is made up of one real part and three imaginary parts, as follows:

$$\dot{q} = a + ib + jc + kd, \tag{1}$$

where $a, b, c, d \in \Re$; i, j, k are complex operators; a is the real part; and $\{ib, jc, kd\}$ are the imaginary parts. By considering two pixels in a color image, $\dot{q}_m = r_m i + g_m j + b_m k$ and $\dot{q}_n = r_n i + g_n j + b_n k$, different types of distances can be defined to measure the distance of quaternions. Let $D_t(\dot{q}_m, \dot{q}_n)$ represents the tth type Quaternionic distances (QD) of quaternions $\dot{q}_m$ and $\dot{q}_n$, we compute the different types of quaternions distances $D_t(\dot{q}_m, \dot{q}_n)$ using the method in [31]. The computed quaternionic distances, which can be defined as the increments between two quaternions, can also be used to measure the similarity between different quaternions from different perspective viewpoints. That is to say, the quaternionic increment between quaternions can be computed by using quaternionic distances.

Assume that $\dot{q}_c$ denotes the center quaternion in a local patch, and $\dot{q}_l$ ($l \in L$, where L is the index set) represents the residual quaternions in the patch. Thus, the total quaternionic increment in a local patch can be written as $\sum_{l=0}^{l=7} D_t(\dot{q}_c, \dot{q}_l)$. With the aid of $|\dot{q}_c|$ as the quaternionic intensity, the differential features of QDWD defined by D_t can be represented as follows:

$$\Phi_2^t(\dot{q}_c) = \arctan\left(\frac{\sum_{l \in L} D_t(\dot{q}_c, \dot{q}_l)}{|\dot{q}_c|}\right). \tag{2}$$

The nonlinear mapping, arctan($\cdot$), aims to enhance $\Phi_2^t(\dot{q}_c)$ to become more robust.

In order to achieve a better performance, we normalize these differential features achieved from Eq. (2) and add them to form an integrated global shape map. Thus, the global saliency can be defined as follows:

$$G = \frac{1}{N} \sum_{t=1}^{N} \Phi_2^t(\dot{q}_c) \tag{3}$$

where $\Phi_2^t(\dot{q}_c)$ is the differential features defined in Eq. (2), and $N = 6$ in our implement.

Figure 1(a) shows an example of the QDWD features for an input color image. Figure 1(b)–(f) are the differential feature maps $\Phi_2^t(\dot{q}_c)$ of QDWD, obtained by using $D_t(t = 1, 2, 3, 4, 6)$. The image generated by the 5th QD is not given, since it is equivalent to the 3rd QD. As shown in Fig. 1(b)–(f), these QDWD features can be utilized to reflect the global shape information of the salient object. Then, we normalize these global shape maps and add them to form an integrated holistic global shape map G. Figures 1(g) and 2(b) show the fusion of the different feature maps to form an integrated global shape map, which can be utilized for saliency detection. More details about QDWD features for saliency detection, please refer to [19].

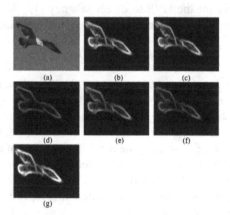

Fig. 1. An example of the QDWD features. (a) an input image, (b)–(f) the differential feature maps of QDWD produced by different QDs, and (g) the integrated global shape map.

2.2   Locality-Constrained Method

The motivation of local estimation is the local outliers, which are standing out from their neighbors with different colors or textures and tend to attract human attention. In order to detect local outliers and get acceptable performance, local coordinate coding method, which described the locality is more essential than sparsity, has been used in saliency detection. Furthermore, the proposed QDWD model in Sect. 2.1 only takes the global shape information into consideration, which highlight the edges of salient object but miss some local cues. Thus, we employ an approximated algorithm based on locality-constrained linear coding (LLC) [33] to estimate the local saliency.

(a)   (b)   (c)   (d)

Fig. 2. The proposed scheme for saliency detection. (a) input images, (b) global shape saliency maps, (c) local saliency maps, (d) the final saliency maps.

For a given image, we first over-segmented the image into N regions, $\{r_i\}, i = 1, 2, \ldots, N$. For each region r_i, let X be a set of 64-dimensions local descriptors extracted from the image, and $X = [x_i^0, x_i^1, \ldots, x_i^{63}]^T, i = 1, 2, \ldots N$. Therefore, the original function of LLC method is written as follows [33]:

$$\min_{B}\left\{ \sum_{i=1}^{N} \left(||x_i - Db_i||^2 + \omega||dr_i.^*b_i||^2 \right) \right\}, \quad s.t. \ \mathbf{1}^T b_i = 1, \ \forall i, \quad (4)$$

where $B = [b_1, b_2, \ldots, b_N]$ is the set of codes for X, $D = [d_1, d_2, \ldots, d_M]$ is the codebook with M entries, and the parameter ω is used to balance the weight between the penalty term and regularization term. The constraint $\mathbf{1}^T b_i = 1$ follows the shift-invariant requirements of the LLC coding and $.^*$ denotes an element-wise multiplication. Here, the dr_i is the locality adaptor that gives different freedom for each codebook vector based on its similarity to the input descriptor x_i, and is defined:

$$dr_i = \exp\left(\frac{dist(x_i, D)}{\lambda} \right), \quad (5)$$

where $dist(x_i, D) = [dist(x_i, d_1), \ dist(x_i, d_2), \ \ldots, \ dist(x_i, d_M)]^T$, $dist(x_i, d_i)$ denotes the Euclidean distance between x_i and the codebook vector d_i, λ is used to adjust the weight decay speed for the locality adaptor, and M is the number of elements in the codebook. More details about LLC can be referred to [32, 33].

In this paper, we adopted an approximated LLC algorithm [32] to detect the local cue. We consider the K nearest neighbors in spatial as the local basis D_i owing to the vector b_i in Eq. (4) with a few non-zero values, which means that it is sparse in some extent. It should be noted that the K is smaller than the size of the original codebook M. Thus, the Eq. (4) can be rewritten as follows:

$$\min_{B} \left(\sum_{i=1}^{N} ||x_i - D_i b_i||^2 \right), \quad s.t. \ \mathbf{1}^T b_i = 1, \ \forall_i, \tag{6}$$

where D_i denotes the new codebook for each region r_i, $i = 1, 2, \ldots, N$ and K is the size of the new codebook and is empirically set at $K = 2M/3$.

Unlike the traditional LLC algorithm, solving the approximated LLC algorithm [32] is simple and the solution can be derived analytically by

$$b_i = 1/(C_i + \omega \times dig(C_i)), \tag{7}$$

$$\tilde{b}_i = b_i/\mathbf{1}^T b_i, \tag{8}$$

where $C_i = (D_i - 1x_i^T)(D_i - 1x_i^T)^T$ represents the covariance matrix of the feature and ω is a regularization parameter, which is set to be 0.1 in the proposed algorithm. As the solution of the improved LLC method is simple and fast, therefore, we defined the local saliency value of the region r_i as follows [32]:

$$S(r_i) = ||x_i - D_i \tilde{b}_i||^2, \tag{9}$$

where $\tilde{b}_i$ is the solution of Eq. (6), which is achieved by Eqs. (7) and (8).

2.3 Final Saliency Fusion

We have already generated global shape saliency map G and local saliency map S for an input image. As is shown in Fig. 2(b), global shape maps detect salient object accurately with more complete shape due to the QDWD, which highlight the edges of salient object but miss some inside object cues. While the local saliency maps achieve more reliable local details owing to the locality-constrained coding model. By integrating the two kind of maps, both global and local information are take into account simultaneously. Therefore, the final saliency map can be defined as follows:

$$Sal = G \odot S, \tag{10}$$

where $\odot$ denotes the addition operator in Eq. (10). It should be noted that we explore a lot of methods for integration of the global shape maps and local saliency maps, and we find that combining them straightforwardly can obtain the best saliency maps. Although the fusion algorithm seems simple, it achieves satisfactory performance as shown in Fig. 2(d).

3 Experimental Results

We present some experimental results based on two widely used datasets: MSRA10K [16] and THUR15K [34], and compare our method with other eight state-of-the-art methods including the Itti [7], Graph-Based (GB) [8], Segmenting-E. Rahtu-J. Kannala

(SEG) [10], Context-aware (CA) [11], Nonlinearly Covariance (NC) [13], Graph-regularized (GR) [14], Multi-Scale Superpixel (MSS) [15] and Background-Single Cellular Automata (BSCA) [17]. In order to quantitatively compare the state-of-the-art saliency-detection methods, the average precision, recall, and F-measure are utilized to measure the quality of the saliency maps.

3.1 MSRA10K

We first evaluate the proposed algorithm on the MSRA10K database. It contains 10,000 images with their pixel-wise ground truth, which are randomly chosen from the MSRA dataset.

Figure 3 shows some saliency detection results of different methods based on MSRA10K database. It can be seen that the proposed algorithm consistently generates saliency maps closer to the ground truth. We also compared the performance of the proposed method with other eight state-of-the-art saliency detection methods. Figure 4 shows the precision, recall and the F-measure values of all the different methods. From the comparison, we can see that our proposed method outperforms the other state-of-the-art methods. Meanwhile, our proposed method outperforms the GR [8] method which only used the local contrast priors.

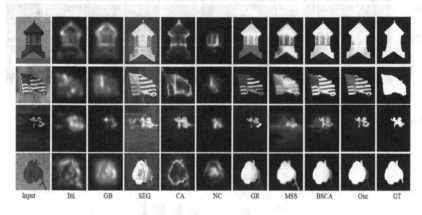

Input   Itti   GB   SEG   CA   NC   GR   MSS   BSCA   Our   GT

Fig. 3. Saliency detection results of different methods based on MSRA10K database.

3.2 THUR15K

We also test the proposed model on the THUR15K database. Figure 5 shows some results of saliency maps generated by nine different algorithms including our method. As shown in Fig. 5, the proposed model produces saliency maps with better outlines than other eight methods attributing to the contribution of QDWD, which is used to detect the object's global shape information.

For objective evaluation, we also used the precision, recall and the F-measure to assess the performance of the nine different methods. Figure 6 shows the comparisons of different methods according to different evaluation criterions. From Fig. 6, the

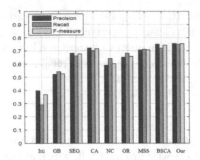

Fig. 4. Comparison of different saliency-detection methods in terms of average precision, recall, and F-measure based on the MSRA10K database.

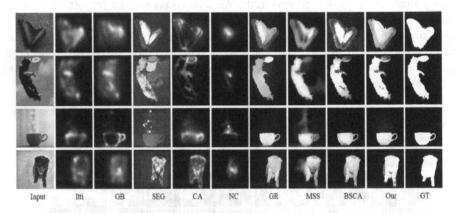

Fig. 5. Qualitative comparisons of different approaches based on THUR15K database.

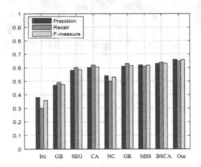

Fig. 6. Comparison of different saliency-detection methods in terms of average precision, recall, and F-measure based on THUR15K database.

proposed method achieves the best performance on the THUR15K database in term of both precision and F-measure values.

4 Conclusion and Discussion

This paper proposes a novel bottom-up method, which incorporates the global shape and local estimation into a unified framework for saliency detection. The global shape maps are obtained by using QDWD rather than directly computing global contrast. The local saliency cues are achieved by utilizing a locality-constrained method. The final saliency maps are generated by integrating the global with the local saliency maps, which can take both global and local cues into account simultaneously. Experiments show that the proposed algorithm can produce satisfactory results.

Acknowledgments. This work was supported by National Natural Science Foundation of China (NSFC) (61601427); Natural Science Foundation of Shandong Province (ZR2015FQ011); Applied Basic Research Project of Qingdao (16-5-1-4-jch); China Postdoctoral Science Foundation funded project (2016M590659); Postdoctoral Science Foundation of Shandong Province (201603045); Qingdao Postdoctoral Science Foundation funded project (861605040008) and The Fundamental Research Funds for the Central Universities (201511008, 30020084851).

References

1. Huang, K., et al.: Biologically inspired features for scene classification in video surveillance. IEEE Trans. Syst. Man Cybern. B Cybern. **41**(1), 307–313 (2010)
2. Zhang, K., Zhang, L., Yang, M.H.: Fast compressive tracking. IEEE Trans. PAMI **36**(10), 2002–2015 (2014)
3. Zhang, K., Liu, Q., Wu, Y.: Robust visual tracking via convolutional networks without training. IEEE TIP **25**(4), 1779–1792 (2016)
4. Jian, M.W., Dong, J.Y., Ma, J.: Image retrieval using wavelet-based salient regions. Imaging Sci. J. **59**(4), 219–231 (2011)
5. Zhu, G., Wang, Q., Yuan, Y., Yan, P.: Learning saliency by MRF and differential threshold. IEEE Trans. Cybern. **43**(6), 2032–2043 (2013)
6. Hsu, C.Y., Ding, J.J.: Efficient image segmentation algorithm using SLIC superpixels and boundary-focused region merging. In: Information, Communications and Signal Processing, pp. 1–5 (2013)
7. Itti, L., Koch, C., Niebur, E.: A model of saliency based visual attention for rapid scene analysis. IEEE Trans. PAMI **20**(11), 1254–1259 (1998)
8. Harel, J., Koch, C., Perona, P.: Graph-based visual saliency. In: Advances in NIPS, pp. 545–552 (2006)
9. Ma, Y.F., Zhang, H.J.: Contrast-based image attention analysis by using fuzzy growing. In: ACM Conference on Multimedia, pp. 374–381 (2003)
10. Rahtu, E., Kannala, J., Salo, M., Heikkilä, J.: Segmenting salient objects from images and videos. In: Proceedings of 11th ECCV, pp. 366–379 (2010)
11. Goferman, S., Zelnik-Manor, L., Tal, A.: Context-aware saliency detection. IEEE Trans. PAMI **34**(10), 1915–1926 (2012)
12. Borji, A., Sihite, D.N., Itti, L.: Quantitative analysis of human-model agreement in visual saliency modeling: a comparative study. IEEE TIP **22**(1), 55–69 (2013)
13. Erdem, E., Erdem, A.: Visual saliency estimation by nonlinearly integrating features using region covariances. J. Vis. **13**(4), 11 (2013)

14. Yang, C., Zhang, L., Lu, H.: Graph-regularized saliency detection with convex-hull-based center prior. IEEE Sig. Process. Lett. **20**(7), 637–640 (2013)
15. Tong, N., Lu, H., Zhang, L., et al.: Saliency detection with multi-scale superpixels. IEEE Sig. Process. Lett. **21**(9), 1035–1039 (2014)
16. Cheng, M.M., Mitra, N.J., Huang, X., et al.: Global contrast based salient region detection. IEEE Trans. PAMI **37**(3), 569–582 (2015)
17. Qin, Y., Lu, H., Xu, Y., et al.: Saliency detection via cellular automata. In: IEEE CVPR, pp. 110–119 (2015)
18. Jian, M., Lam, K.M., Dong, J.J., Shen, L.L.: Visual-patch-attention-aware saliency detection. IEEE T. Cybernetics **45**(8), 1575–1586 (2015)
19. Jian, M.M., Qi, Q., Sun, Y., Lam, K.M., et al: Saliency detection using quaternionic distance based weber descriptor and object cues. In: APSIPA ASC 2016, Korean (2016)
20. Zhang, D., Meng, D., Han, J.: Co-saliency detection via a self-paced multiple-instance learning framework. IEEE Trans. Pattern Anal. Mach. Intell. (2016)
21. Wang, A., Wang, M.: RGB-D salient object detection via minimum barrier distance transform and saliency fusion. IEEE Sig. Process. Lett. **24**(5), 663–667 (2017)
22. Lu, H., Zhang, X., Qi, J., et al.: Co-Bootstrapping saliency. IEEE Trans. Image Process. **26** (1), 414–425 (2017)
23. Lin, X., Yan, Z., Jiang, L.: Saliency detection via foreground and background seeds. In: Kim, K., Joukov, N. (eds.) ICISA 2017. LNEE, vol. 424, pp. 145–154. Springer, Singapore (2017). doi:10.1007/978-981-10-4154-9_18
24. Oliva, A., Torralba, A., Castelhano, M.S., et al.: Top-down control of visual attention in object detection. In: IEEE ICIP, vol. 1 (2003)
25. Cholakkal, H., Rajan, D., Johnson, J.: Top-down saliency with locality-constrained contextual sparse coding. In: BMVC (2015)
26. Yang, J., Yang, M.H.: Top-down visual saliency via joint CRF and dictionary learning. IEEE Trans. PAMI **39**(3), 576–588 (2017)
27. He, S., Lau, R.W.H., Yang, Q.: Exemplar-driven top-down saliency detection via deep association. In: IEEE CVPR, pp. 5723–5732 (2016)
28. Rahman, I., Hollitt, C., Zhang, M.: Contextual-based top-down saliency feature weighting for target detection. Mach. Vis. Appl. **27**(6), 893–914 (2016)
29. Achanta, R., Hemami, S., Estrada, F., Susstrunk, S.: Frequency-tuned salient region detection. In: IEEE CVPR, pp. 1597–1604 (2009)
30. Perazzi, F., Krahenbuhl, P., Pritch, Y., Hornung, A.: Saliency filters: contrast based filtering for salient region detection. In: IEEE CVPR, pp. 733–740 (2012)
31. Lan, R.S., Zhou, Y.C., Tang, Y.: Quaternionic weber local descriptor of color images. IEEE Trans. Circuits Syst. Video Technol. (2015)
32. Tong, N., Lu, H., Zhang, Y., et al.: Salient object detection via global and local cues. Pattern Recogn. **48**(10), 3258–3267 (2015)
33. Wang, J., Yang, J., Yu, K., Lv, F., Huang, T., Gong, Y.: Locality-constrained linear coding for image classification. In: IEEE CVPR, pp. 3360–3367 (2010)
34. Cheng, M.M., Mitra, N.J., Huang, X.: Salient shape: group saliency in image collections. Vis. Comput. **30**(4), 443–453 (2014)
35. Jian, M., Qi, Q., et al.: Saliency detection using quaternionic distance based weber local descriptor and level priors. Multimedia Tools Appl. (2017). doi:10.1007/s11042-017-5032-z

Online Vehicle Tracking in Aerial Imagery

Zihao Liu[1], Zhihui Wang[1], Huimin Lu[2], and Dong Wang[1(✉)]

[1] School of Information and Communication Engineering,
Dalian University of Technology, Dalian, China
wdice@dlut.edu.cn
[2] Department of Electrical and Electronic Engineering,
Kyushu Institute of Technology, Kitakyushu, Japan

Abstract. Compared with the traditional visual tracking, online vehicle tracking in aerial imagery brings lots of unique challenges including low frame rate sampling, small tracked targets, to name a few. As we know, color information provides rich information of the tracked objects especially when the texture features of small objects are not easily observed. Thus, in this work, we attempt to combine different color models within the correlation-filter-based tracking framework for tracking vehicles in aerial images. First, we exploit a set of color models to describe the appearance of the tracked object based on correlation filters. Second, the confidence maps generated by these correlation filters are selected based on the variance rule and combined using an adaptive fusion method. Finally, we conduct numerous experiments on the KIT_IPF aerial dataset to compare the proposed tracker with other competing methods and analyze the effects of different components. The experimental results not only demonstrate that our tracker performs better than other state-of-the-art algorithms but also show that the adopted feature selection and fusion schemes could facilitate improving the tracking performance.

Keywords: Areal imagery · Object tracking · Color models · Feature selection · Feature fusion

1 Introduction

Visual tracking is a technique to estimate the appearance, scales, locations of the tracked objects by detecting, extracting and identifying video image sequences. As demonstrated in [18], visual tracking includes lots of challenges, such as illumination variation, scale variation, occlusion, deformation, motion blur, background clutters and so on. Besides, online vehicle tracking in aerial imagery poses a number of additional difficulties [3,14], including low frame rate sampling, low resolution, limited contrast, geometric occlusion, to name a few. All these factors make visual tracking in aerial imagery more challenging than the traditional tracking problem. As shown in Fig. 1, there are many clutter targets and the same object moves fast due to low frame rate sampling.

Z. Liu and Z. Wang—Contributed equally to this work.

© Springer International Publishing AG 2017
Y. Sun et al. (Eds.): IScIDE 2017, LNCS 10559, pp. 335–345, 2017.
DOI: 10.1007/978-3-319-67777-4_29

Fig. 1. An sample from KIT_IPF arial imagery datasets.

From the perspective of object searching, most of existing trackers can be categorized into methods based on random sampling or dense sampling. The popular trackers from the former one include incremental visual tracking [15], ℓ_1-based tracking [13], online sparse prototypes tracking [17], and so on. While the popular trackers from the latter one contain multiple instance learning tracking [1], structured output tracking [8], compressive tracking [20], to name a few. Recently, the trackers using dense sampling can be efficiently speeded up by using the correlation filter technique. The convolution theorem indicates that the convolution of two image patches corresponds to a element-wise product in Fourier domain, which makes the convolution operator could be efficiently calculated by fast Fourier transform (FFT). Hence, the tracker based on correlation filter calculates the correlation between candidates and templates via FFT, and then the correlation response can be transformed back into the spatial domain using inverse FFT (IFFT). The MOSSE (minimum error sum of squares filter) method [2] is the first attempt to exploit the convolution theorem for visual tracking. By introducing the concept of the circulant matrix, the CSK (circulant structure with kernels) tracker [9] promotes the efficiency of correlation filter. After that, the KCF (kernelized correlation filter) method [10] extends CSK to multi-channel with HOG (histogram of oriented gradient) features.

Multiple features are usually used in visual tracking to establish appearance models of the tracked objects (such as color, texture and so on). However, color information is often ignored in visual tracking although it could provide many valuable cues. In [6], an effective color-based tracking method is proposed, in which the traditional RGB color model is converted into a 11-dimensional color description. After that, Liang *et al.* [11] comprehensively evaluate the contribution of 10 color models in 16 different trackers, which demonstrates that the color information facilitate designing a robust tracker in realistic scenes. In addition, feature selection and fusion are usually exploited to improve the tracking performance in terms of both accuracy and speed. In [21], Zhong *et al.* use a sparse-representation-based classifier to select discriminative image patches, thereby effectively reducing the number of features and improving the tracking accuracy. Wang *et al.* [16] design a tracking method based on multi-cues spatial pyramid matching, where feature models are fused with adaptive linear weights and therefore achieve robust performance. In [19], a multiple kernel boosting method is proposed to choose and combine different kernel functions.

In this paper, we use different color models to generate different confidence maps and construct a set of correlation filters to conduct dense sampling by circulant structures. In addition, we combine different selection and fusion schemes to combine different confidence maps and fuse them to determine the final location of the tracked object. The contributions of this work include: (1) we apply color models to online vehicle tracking in aerial imagery; (2) we investigate the effectiveness of feature selection and fusion schemes; (3) we conduct a detailed evaluation to compare the proposed tracker with other state-of-the-art methods and analyze the effects of different components.

2   Proposed Tracking Method

In this work, we develop a tracking framework to conduct online vehicle tracking in aerial imagery. The overall pipeline is illustrated in Fig. 2, the main components of which include color models, correlation filter, feature selection and feature fusion. The detailed information is presented in the following subsections.

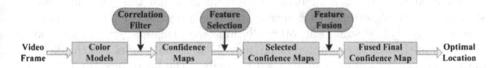

Fig. 2. The overall pipeline of the proposed tracking method. (Color figure online)

2.1   Correlation Filter

In this work, we construct our tracker based on the KCF framework [10], which is introduced in the following contexts. The image patch $\mathbf{x}$ of size $M \times N$ is a training patch centered around the target and padded with appropriate proportion. Thanks to the advantages of cyclic properties, the tracker exploits dense sampling based on the base samples. All shifted samples of $\mathbf{x}$, $\mathbf{x}_{m,n}$, $(m, n) \in \{0, \ldots, M - 1\} \times \{0, \ldots, N - 1\}$, constitute the examples for training the filter parameters. We also introduce the symbol $\mathbf{y}$ to denote the regression label following a Gaussian distribution, which takes the maximum value for a centered target, and smoothly decays to 0 for any other shifts, i.e., $\mathbf{y}(m,n)$ is the label of $\mathbf{x}_{m,n}$.

The basic idea of KCF is to learn a function $f(\mathbf{x}) = \mathbf{w}^\top \mathbf{x}$ to minimize the mean square error between training samples $\mathbf{x}_{m,n}$ and their labels $\mathbf{y}(m, n)$,

$$\min_{m} \sum_{m,n} |\langle \phi(\mathbf{x}_{m,n}), \mathbf{w} \rangle - \mathbf{y}(m, n)|^2 + \lambda \|\mathbf{w}\|^2, \tag{1}$$

where ϕ is the mapping of $\mathbf{x}_{m,n}$ in Hilbert space related with a kernel function $k(\langle\phi(\mathbf{x}),\phi(\bar{\mathbf{x}})\rangle = k(\mathbf{x},\bar{\mathbf{x}}))$, and λ is a regularization constant to avoid over-fitting. The solution of the objective function (1) can be derived as $\mathbf{w} = \sum_{m,n} \alpha(m,n)\phi(\mathbf{x}_{m,n})$, and the coefficient α can be calculated as,

$$\alpha = \mathcal{F}^{-1}\left(\frac{\mathcal{F}(\mathbf{y})}{\mathcal{F}(\langle\phi(\mathbf{x}),\phi(\mathbf{x})\rangle) + \lambda}\right), \tag{2}$$

where $\mathcal{F}$ and $\mathcal{F}^{-1}$ denote the Fourier transform and its inverse, respectively.

Given the learned coefficient α and appearance model $\bar{x}$[1], the tracking task can be carried out on an image patch z in the new frame with the same size of $\bar{x}$ by computing the response map as

$$\bar{\mathbf{y}} = \mathcal{F}^{-1}(\mathcal{F}(\alpha) \odot \mathcal{F}(\langle\phi(\mathbf{z}),\phi(\bar{\mathbf{x}})\rangle)). \tag{3}$$

where $\odot$ is the Hadamard product and $\bar{\mathbf{y}}$ denotes the response map. Then, the target position can be detected based on the maximal value in the response map.

2.2 Color Models

Illumination variation is a challenging problem in visual tracking, thus, most state-of-the-art algorithms that merely use the gray model are sensitive to shadow, highlight, shading and other illumination elements. [7] evaluates lots of color models used in computer vision, and the diagonal model and the dichromatic reflection model are usually used to analyze the invariance properties of different photometric representations. In [11], Liang et al. list invariance properties of nine color models and demonstrate their improvements on state-of-the-art algorithms by embedding different color models respectively in existing algorithms. These nine color models include RGB, TRGB, OPP, C-OPP, N-OPP, NRGB, HSV, Hue and LAB. The detailed descriptions and evaluations can be found in [11]. In this work, we adopt ten color models comprised of the gray model and the above-mentioned nine models.

2.3 Feature Selection with Variance

Feature selection is a crucial step in our tracking framework, which will remove some bad features and keep good ones for the further combination. An effective selection scheme could not only select good candidates but also save computational load. In this work, we exploit the variance rule to conduct feature evaluation and selection. The confidence map generated by the correlation between the candidate patches and target usually provides rich information for feature selection. In frame t, we extract feature map $\mathbf{x}_t$, then correlate $\mathbf{x}_t$ with the template, and the position of object on current frame will be obvious with the help of the maximum of the confidence map $\mathbf{C}_t$. Confidence map is wildly used in feature

---

[1] The target appearance model $\bar{x}$ is learned over time.

selection as it is able to show the distribution of foreground and background clearly. The variance rule can be used to measure the distribution of a given confidence map. The variance of confidence map is high when foreground and background are easily to be distinguished. But when they are similar, the variance will be lower. Thus, we can use the variances of different confidence maps to rank them according to their discriminations. As shown in Fig. 3, the first row demonstrates the good color models that distinguish the tracked objects from their cluttered backgrounds and the second row shows some bad color models with lower variances.

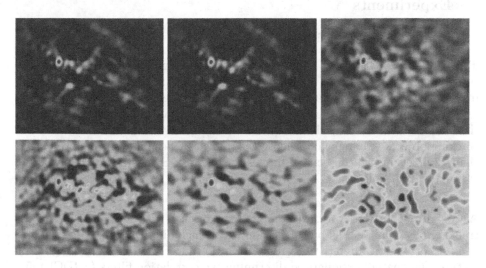

Fig. 3. An example of confidence maps of the 4-th object in MunichCrossroad01. The first line denotes the color models with high discrimination (left to right: GRAY, TRGB, N-OPP), and the second row shows bad color models (left to right: HSV, NRGB, Hue). (Color figure online)

2.4   Feature Fusion

After feature selection, it requires to combine these selected confidence maps into a final confidence map for determining the location of the tracked object. Lots of fusion approaches are discussed in [4,14]. In this work, we investigate three fusion manners: an additive fusion manner, a multiplicative fusion manner, and an adaptive additive fusion manner. We adopt the method presented in [16] to determine optimal weights by fitting a pre-defined Gaussian function, $\rho(x, x^*) = \exp\left(-\| x - x^* \|^2 / 2\sigma^2\right)$ (σ is the variance of the Gaussian function, x^* denotes the optimal position and x stands for a candidate position). Then the optimal weights can be obtained by solving the optimization problem (4),

$$w^* = \arg\min \sum_{m=1}^{M} c(\varepsilon_m), \text{ subject to}: \sum_{i=1}^{N} w_i = 1, \ w_i \geq 0, \qquad (4)$$

where

$$\frac{\sum_{i=1}^{N} w_i C_i (x_m)}{\sum_{i=1}^{N} w_i C_i (x^*)} = \rho (x_m, x^*) + \varepsilon_m, \forall x_m \in N (x^*), \tag{5}$$

where ε_m is a slack variable, which determines the discrimination ability of $\rho(x, x^*)$. $c(\varepsilon_m)$ is a cost function to penalize $\varepsilon_m > 0$, i.e., $c(\varepsilon) = max(0, \varepsilon)$. C_i denotes the i-th pixel in the confidence map. N is the number of candidate features and M is the number of candidate locations.

3 Experiments

3.1 Dataset and Algorithms

The KIT_IPF Dataset: KIT_IPF is an aerial image sequences dataset captured from a plane flying at an altitude of about 1500 m, which can be download from http://www.ipf.kit.edu/code.php#tracking_vehicles. Three scenes are contained in the image sequences, crossroad, street and autobahn. There are more than one vehicle in each video clip, and all of the vehicles are considered as tracking objects (we manually choose the vehicles whose movements are bigger than 12 pixels in two consecutive frames for testing). Some detailed information is provided in Table 1.

Algorithms: In this work, we compare our tracker with four state-of-the-art tracking algorithms based on correlation filters including trackers based on kernelized correlation filter (KCF) [10], correlation filter with color naming (CN) features [6], spatially regularized discriminative correlation filters (SRDCF) [5], and hierarchical convolutional features (HCF) [12]. For fair evaluation, we use the precision score of one-pass evaluation (OPE) [18] at 20 pixels, which is the number of object whose Euclidean distance between the estimate position with the ground truth of the target is less than 20 pixels, to evaluate all algorithms.

3.2 Parameter Setting

According to our statistics, the size of most target objects is less than 60×50 pixels. To deal with low frame rate sampling, we use the padding method to extend the search region. The value of padding is chosen as 2.5, which means the size of search region is initially set to 3.5 times of the size of target. For the basic correlation filter model, the spatial bandwidth of label $\mathbf{y}$ is set to $\frac{1}{16}$, and the regularization parameter is chosen as 0.01. The learning rate is $\gamma = 0.075$ for online updating the correlation filter. For the adaptive fusion scheme, the initial weights of different confidence map are equal and the $\sigma = 6$ is used to generate a Gaussian weight distribution. For the other competing methods, the parameters are all set as their original version except the padding value (=2.5).

3.3   Quantitative Comparison

Table 1 demonstrates a detailed quantitative comparison of different trackers on each video clip in the KIT_IPT dataset. It can be seen from this table that our method achieves the best overall performance (0.703 OPE value in average) compared with other competing ones. The CN tracker achieves the second best overall performance. The CN method is also designed based on color information, however, it only refines RGB to 11 colors without using special techniques to adapt the illumination variance. The MunichCrossroad02 video clip includes 32 objects with different sizes, complex geographical conditions and uneven illumination conditions. The proposed method provides the most robust tracking result on this video. The KCF method extracts gray and HOG to express the appearance model of targets, and therefore is also robust to illumination variation. But when the size of target is even less than 60×50 pixels, the image patch will provide very limited edge information. Thus, it cannot achieve good performance in this dataset. For the similar reason, the HCF method with deep features also achieves unsatisfactory performance. The average OPE plots of different trackers are illustrated in Fig. 4, which also show that our tracker performs best.

Table 1. The detailed comparisons of different trackers.

Video name	#Frames	#Objects	KCF [10]	CN [6]	HCF [12]	SRDCF [5]	Ours
MunichAutobahn01	16	7	0.098	0.256	0.232	0.143	**0.351**
MunichCrossroad01	20	5	0.460	0.771	0.500	0.490	**0.771**
MunichCrossroad02	45	32	0.557	0.854	0.385	0.597	**0.871**
MunichStreet01	25	5	0.400	0.432	0.400	**0.552**	0.480
MunichStreet02	20	7	0.650	**0.871**	0.650	0.6857	0.871
MunichStreet04	29	27	0.538	0.790	0.447	0.659	**0.820**
StuttgartAutobahn01	23	7	0.205	**0.902**	0.236	0.3975	0.883
StuttgartCrossroad01	14	22	0.240	0.352	0.321	0.283	**0.352**
Average	-	-	0.432	0.682	0.503	0.391	**0.703**

Qualitative Comparison: Figure 5 illustrates some screenshots of the proposed method and four competing trackers. The images in the fist two rows are captured from MunichCrossroad02, and the other three rows are selected from StuttgartCrossroad01, StuttgartAutobahn01, MunichStreet02 video clips. From this figure, we can see that the proposed method achieves more robust performance to heavy illumination variations since it could exploit the photometric invariance of combined color models. The SRDCF method puts the maximum weights to the center of feature map and therefore learns convictive appearance models. When the object is very small as shown in the second row and has big in-plane rotation, the other three algorithms fail (except the SRDCF and our

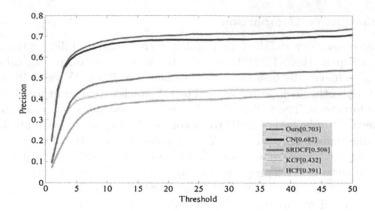

Fig. 4. The average OPE plots of different trackers. (Color figure online)

trackers). In addition, for the flat objects with less texture information (the 4-th and 5-th rows), only our tracker successfully achieves long-term tracking.

3.4 Effects of Different Components

In this subsection, we investigate the effects of feature selection and fusion schemes and report the detailed comparisons in Table 2. From this table, we can see that the better performance could be achieved when the selected number of features is 3. If this number is too small (such as 1), the tracker cannot exploit the rich color information. However, the tracker will be easy to be disturbed when this number is too large (like 10). In addition, the adaptive weight fusion scheme usually performs better than the traditional additive and multiplicative fusion algorithms.

Table 2. The effects of selected and fusion methods. SN: selected feature number, FM: fusion method, ADD: additive; MUL: multiplicative; and AWF: adaptive weight.

FM	SFN			
	1	3	5	10
ADD	0.597	0.698	0.690	0.678
MUL	0.597	0.699	0.691	0.612
AWF	0.597	0.703	0.683	0.690

4   Conclusions

In this paper, we design a tracking framework for online vehicle tracking in aerial imagery, including different color models, correlation filters, feature selection and

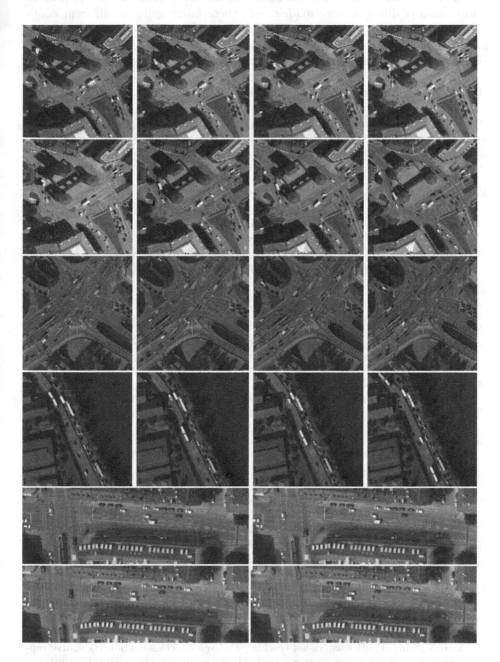

Fig. 5. Qualitative comparisons with different tracking algorithms. Ours: red; CN: green; KCF: yellow; SRDCF: blue; and HCF: cyan. (Color figure online)

feature fusion. Different correlation filters are adopted to exploit the appearance information of different color models, and these filters generate different confidence maps to depict different aspects regarding the tracked object. Then, some good confidence maps are selected from all candidate confidence maps. Finally, the selected confidence maps are fused using an adaptive weight scheme. The experimental results on many challenging aerial image sequences demonstrate that our tracker performs better than other competing methods.

Acknowledgements. This work was supported by the Undergraduate Innovation and Entrepreneurship Training Program (No. 2017101410201010981) and Fundamental Research Funds for the Central Universities (No. DUT16RC(4)16).

References

1. Babenko, B., Yang, M., Belongie, S.J.: Robust object tracking with online multiple instance learning. IEEE Trans. Pattern Anal. Mach. Intell. **33**(8), 1619–1632 (2011)
2. Bolme, D.S., Beveridge, J.R., Draper, B.A., Lui, Y.M.: Visual object tracking using adaptive correlation filters. In: CVPR, pp. 2544–2550 (2010)
3. Candemir, S., Palaniappan, K., Bunyak, F., Seetharaman, G.: Feature fusion using ranking for object tracking in aerial imagery. Geosp. Infofus. II **8396**, 839604-1–839604-9 (2012)
4. Collins, R.T., Liu, Y., Leordeanu, M.: Online selection of discriminative tracking features. IEEE Trans. Pattern Anal. Mach. Intell. **27**(10), 1631–1643 (2005)
5. Danelljan, M., Häger, G., Khan, F.S., Felsberg, M.: Learning spatially regularized correlation filters for visual tracking. In: ICCV, pp. 4310–4318 (2015)
6. Danelljan, M., Khan, F.S., Felsberg, M., van de Weijer, J.: Adaptive color attributes for real-time visual tracking. In: CVPR, pp. 1090–1097 (2014)
7. Everts, I., van Gemert, J.C., Gevers, T.: Evaluation of color STIPs for human action recognition. In: CVPR, pp. 2850–2857 (2013)
8. Hare, S., Golodetz, S., Saffari, A., Vineet, V., Cheng, M., Hicks, S.L., Torr, P.H.S.: Struck: structured output tracking with kernels. IEEE Trans. Pattern Anal. Mach. Intell. **38**(10), 2096–2109 (2016)
9. Henriques, J.F., Caseiro, R., Martins, P., Batista, J.: Exploiting the circulant structure of tracking-by-detection with kernels. In: Fitzgibbon, A., Lazebnik, S., Perona, P., Sato, Y., Schmid, C. (eds.) ECCV 2012. LNCS, vol. 7575, pp. 702–715. Springer, Heidelberg (2012). doi:10.1007/978-3-642-33765-9_50
10. Henriques, J.F., Caseiro, R., Martins, P., Batista, J.: High-speed tracking with kernelized correlation filters. IEEE Trans. Pattern Anal. Mach. Intell. **37**(3), 583–596 (2015)
11. Liang, P., Blasch, E., Ling, H.: Encoding color information for visual tracking: algorithms and benchmark. IEEE Trans. Image Process. **24**(12), 5630–5644 (2015)
12. Ma, C., Huang, J., Yang, X., Yang, M.: Hierarchical convolutional features for visual tracking. In: ICCV, pp. 3074–3082 (2015)
13. Mei, X., Ling, H.: Robust visual tracking and vehicle classification via sparse representation. IEEE Trans. Pattern Anal. Mach. Intell. **33**(11), 2259–2272 (2011)
14. Palaniappan, K., Bunyak, F., Kumar, P., Ersoy, I., Jäger, S., Ganguli, K., Haridas, A., Fraser, J., Rao, R.M., Seetharaman, G.: Efficient feature extraction and likelihood fusion for vehicle tracking in low frame rate airborne video. In: FUSION, pp. 1–8 (2010)

15. Ross, D.A., Lim, J., Lin, R., Yang, M.: Incremental learning for robust visual tracking. Int. J. Comput. Vis. **77**(1–3), 125–141 (2008)
16. Wang, D., Lu, H., Chen, Y.: Object tracking by multi-cues spatial pyramid matching. In: ICIP, pp. 3957–3960 (2010)
17. Wang, D., Lu, H., Yang, M.: Online object tracking with sparse prototypes. IEEE Trans. Image Process. **22**(1), 314–325 (2013)
18. Wu, Y., Lim, J., Yang, M.: Online object tracking: a benchmark. In: CVPR, pp. 2411–2418 (2013)
19. Yang, F., Lu, H., Chen, Y.-W.: Human tracking by multiple kernel boosting with locality affinity constraints. In: Kimmel, R., Klette, R., Sugimoto, A. (eds.) ACCV 2010. LNCS, vol. 6495, pp. 39–50. Springer, Heidelberg (2011). doi:10.1007/978-3-642-19282-1_4
20. Zhang, K., Zhang, L., Yang, M.: Fast compressive tracking. IEEE Trans. Pattern Anal. Mach. Intell. **36**(10), 2002–2015 (2014)
21. Zhong, W., Lu, H., Yang, M.: Robust object tracking via sparse collaborative appearance model. IEEE Trans. Image Process. **23**(5), 2356–2368 (2014)

Autonomous Object Segmentation in Cluttered Environment Through Interactive Perception

Rui Wu[✉], Dongfang Zhao, Jiafeng Liu, Xianglong Tang,
and Qingcheng Huang

Research Center of Pattern Recognition and Intelligent System,
Harbin Institute of Technology, Harbin, China
simple@hit.edu.cn

Abstract. This paper investigates the problem of object segmentation in cluttered environment. This problem enables a large variety of exciting and important applications. An interactive perception method is proposed to segment scene into constituent objects based on principal angle. Trajectory data of feature points reflecting the essence of scene structure changes is extracted by a robot arm to calculate least stable regions for the interactive task. The segmentation task is achieved by the principal angle of stable regions because the principal angle essentially estimates the similarity between two regions. In contrast to probability based approach, our method performs well on efficiency of segmentation as it works without estimating parameters of the system. Experimental results on real world scene confirm the effectiveness of our method.

Keywords: Object segmentation · Interactive perception · Trajectory · Principal angle

1   Introduction

Object segmentation is a traditional and important problem in computer vision. It serves as a prerequisite in many exciting applications such as scene understanding [1] and intelligence manipulation [2]. Object segmentation is to represent a scene into a state that is easier to analyze by partitioning a cluttered environment into multiple segments.

Methods that can segment cluttered environment into its constituent objects are required as starting point for many significant applications. Such problem is not as easy task as it seems because visual boundary does not always correspond to authentic object boundaries especially meet with cluttered environment. The difficulties mainly come from the following two aspects of object segmentation. First, environment is cluttered, for instance, multiple objects crowded and overlapped. Second, the environment is dynamic, even multiple objects may be moved together. These environmental complexities result in the segmentation ambiguities which would confuse segmentation system and lead to segmentation error.

Interactive perception is an effective method in solving segmentation ambiguities. It autonomously obtains object segmentation of cluttered environment

© Springer International Publishing AG 2017
Y. Sun et al. (Eds.): IScIDE 2017, LNCS 10559, pp. 346–355, 2017.
DOI: 10.1007/978-3-319-67777-4_30

with a sequence of images. Physical interaction with a cluttered environment provides a sensor information that reflects the essence of environment structure. Object segmentation approach using this strategy usually obtains segmentation and interacts with environment at the same time [3–6]. Sensor information is contained in a sequence of images which are visual observation of interaction. Thus interactive perception is introduced to object segmentation approaches for better segmentation quality. State-of-the-art object segmentation approaches using this strategy tend to employ a lot of heuristic knowledge provided by human teacher or learned from interaction. This prior knowledge ranges from information of the clustered environment to necessary parameters of segmentation system. It might be set by human teacher or learned by the segmentation system itself. However, this will lead to high computational complexity and unnecessary interactions, especially for the simple environment. To deal with this problem, we propose a trajectory-based non-parameter method to segment scene into its constituent objects by using principal angle as segmentation measurement criterion.

In contrast to traditional approaches for interactive object segmentation problem, we initially over-segmented the scene into several small regions. Then we use interactive information to merge these regions until all objects are segmented. Neither learning parameters nor prior knowledge about the cluttered environment is used to this system which will decrease the computational complexity. These advantages fundamentally increase the autonomy of the system and segmentation efficiency.

2   Approach

In cluttered environment, the number, location, and model of the scene into its constituent objects are initially unknown. In this section, we describe an interactive perception framework for object segmentation. Probabilistic reasoning method mentioned in [5] take movement model as input. System have to learn four unknown parameters and hidden variables at the first few steps. This self-adaptive process results in system overhead especially when the scene includes few objects. Our region-based principal angle segmentation approach overcomes the defect and enhance system performance. In Subsect. 2.1, we introduce the overview of our system. Feature points, movement model and region space are defined in Subsect. 2.2. The principal angle acquiring method is described in Subsect. 2.3.

2.1   System Overview

Our algorithm consists of six components. In scene observation step, a color image was observed by CMOS camera. The scene was over segmented into amount of small regions. Each region contains a feature points which consists candidate interactive point set. We select a prefer point and execute interactive manipulation. A new scene was observed. Subsequently, feature points was tracked in the new scene and former scenes for trajectory date using KLT algo-rithm. A set of trajectory data constitutes movement space. And after obtaining this movement space of the scene, we compute region space through selecting the feature points located in local region. Movement space can be divided into sev-eral region space. In object segmentation step, we compute the principal angle matrix between each pair of region space. This principal angle matrix describes essentially the distance of two region spaces. And it implies whether two parts located on the same object in physical level. Then we use clustering algorithm to segment the principal angle data. And this result achieves a segmentation of feature points. At last, we use region growing algorithm to obtain object segmentation. Figure 1 illustrates the entire system processes.

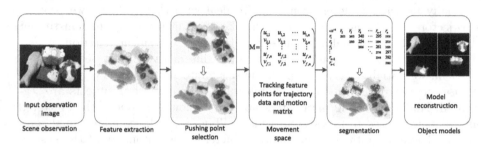

Fig. 1. The entire process of our non-parameter interactive object segmentation app-roach. Some scene and brief intermediate result was presented.

Our movement space which constitutes by trajectory data utilize a sequence of frames provided by former interaction. Moreover, our principal angle consid-ers correlation between each pair of over segmented regions. As the constraint of distance between two movement subspaces reflects the principal of object segmentation essentially. The method we proposed can segment cluttered envi-ronment into consist objects efficiently. In general, our non-parameter interactive segmentation algorithm does not need hypothesis parameters, so it can improve computation speed obviously.

2.2   Scene Description and Movement Space

In this subsection, we use feature points and movement space to describe scene. Then we compute region spaces as input of next subsection. Scene description process is an abstract expression of scene. As our scene description process allows segmentation system to estimate at object level instead of pixel level. It can help enhancing computation speed obviously.

– Feature points extraction

In our experiment, we utilize 2D grid to cover the initial observed cluttered scene. The intersections of grid are defined as feature points. This method made the average feature points similar each other as the size of object is roughly the same. The decision made for the size of grid should be optimal. A too large size leads to bad segmentation. Meanwhile if the size is too small, it leads to unnecessary computation. With the robot arm continuously interacts with scene. Some feature points previously unseen can became visible. Meanwhile the points previously visible may be lose. These points are ignored as they are disadvantageous to our algorithm.

- Tracking and movement space

We employ a robot arm to interact with objects and select a feature point as pushing point randomly. Interacting with the environment results in movement space. We observe the result. And record it with a sequence of images. A Kanade-Lucas-Tomasi feature tracker is employed to track feature points. And this algorithm helps to construct movement matrix. Each pushing interaction process generates a single image and we track the feature points to obtain trajectory data. Let $\mathbf{M} = \{\mathbf{t_1}, \mathbf{t_2}, \cdots, \mathbf{t_n}\}$ be the movement space of n feature points. The element t_k is the trajectory data of feature point k. A trajectory data contains coordinates of feature points essentially. The trajectory data of a feature point:

$$\mathbf{t}_k = \begin{pmatrix} u_{1,k} \\ v_{1,k} \\ \vdots \\ u_{f,k} \\ v_{f,k} \end{pmatrix} \tag{1}$$

The set of trajectory data or movement space:

$$\mathbf{M}_{2f \times n} = \begin{pmatrix} u_{1,1} & u_{1,2} & \cdots & u_{1,n} \\ v_{1,1} & v_{1,2} & \cdots & v_{1,n} \\ \vdots & \vdots & \vdots & \vdots \\ u_{f,n} & u_{f,2} & \cdots & u_{f,n} \\ v_{f,1} & v_{f,2} & \cdots & v_{f,n} \end{pmatrix} \tag{2}$$

Each column m_k is a trajectory data obtained by tracking feature point k. The coordinates $(u_{i,j}, v_{i,j})$ represent the location of feature point j in interaction i. The movement space is a matrix with size of $2f \times n$, f is interaction steps.

- Region space

Region spaces represent over-segmentation of a scene which consists of several trajectories of feature points. It is a partitioning of movement space $\mathbf{M}$. We represent a scene's segmentation using region spaces: $\{\mathbf{R_1}, \mathbf{R_2}, \cdots, \mathbf{R_p}\}$.

Before first interaction, the number of region spaces equals to n (number of feature points). For initial observation, We use famous mean shift algorithm to obtain initial over-segmentation. The initial movement matrix is shown in Eq. 3. We define the trajectory of feature points which located in EDISON region as a initial region space. This initial procedure outputs over-segmentation of initial scene which presented by region spaces. The movement space after one interaction:

$$M_{2 \times n} = \begin{pmatrix} u_{1,1} & u_{1,2} & \cdots & u_{1,n} \\ v_{1,1} & v_{1,2} & \cdots & v_{1,n} \end{pmatrix} \tag{3}$$

The region spaces $\{R_1, R_2, \cdots, R_p\}$ after initial over-segmentation can be present:

$$[R_1 | R_2 | \cdots | R_p] = M_{2 \times n} \tag{4}$$

where $p < n$.

Subsequently, we interact with the environment repeatedly and a series of images are obtained. We use KLT tracker to track feature points and then segmentation algorithm are used to update region spaces. As interaction goes on. At last, the number of region spaces equals to the number of objects in this environment.

For the same reason, the region spaces $\{R_1, R_2, \cdots, R_q\}$ after entire interactive segmentation process can be present:

$$[R_1 | R_2 | \cdots | R_q] = M_{2f \times n} \tag{5}$$

where $q \ll n$ and q equals to the number of objects in the ideal situation. For convenience of calculation, region spaces have the same dimension.

One thing to note here is that the feature point which compose region space should be on a single object. To ensure this, we restrain the rank of each region space less than or equals to 4 in the experiment. This principal has been proved in [7].

Lastly, we decompose region spaces $\{R_1, R_2, \cdots, R_q\}$ using SVD.

$$R_{2f \times m} = U_{2f \times k} \Sigma_{k \times k} V_{m \times k}^T \tag{6}$$

where $k = rank(R_{2f \times m})$, subspace $V_{m \times k}$ becomes new representation of region space, its dimension is m. And v_i $(i = 1...m) \in V_{m \times k}$ are new representation of feature points' trajectory. In next subsection, we will use subspace $V_{m \times k}$ to compute principal angles and merge region spaces.

2.3   Object Segmentation

In this subsection, we merge region spaces according to the distance between them. Principal angle is used to measure the distance between subspace of two region space. The number of region spaces decrease as interactions are processed.

Firstly, we define subspaces of two different region spaces using symbol F and G. And compute principal angles between them. F and G are computed by Eq. 6. Principal angle is a basic concept in matrix computation. It measures the distance between a pair of subspaces [8]. The idea of principal angle has been used in computer vision for motion segmentation. Yan et al. apply this concept to learn the structure of articulated object [9]. We review the definition of principal angle briefly, and analysis the rationality of using it to solve our object segmentation problem.

In matrix computations [8], Golub and van Loan determine the principal angles between a pair of given subspaces $\mathbf{F}$ and $\mathbf{G}$. Their dimensions satisfy

$$\begin{aligned} dim(\mathbf{F}) &= p \geq 1 \\ dim(\mathbf{G}) &= q \geq 1 \end{aligned} \tag{7}$$

The principal angles $\{\theta_{i,j}\} \in [0, \pi/2]$ between two subspaces $\mathbf{F}$ and $\mathbf{G}$ are defined by

$$\cos(\theta_{i,j}) = \mathbf{f}_i^T \mathbf{g}_j \ (i = 1...p, j = 1...q) \tag{8}$$

In most cases, p and q are different in our problem. The idea of using SVD to compute the principal angles is due to Björck and Golub [10]. Principal angle essentially estimates the distance between two multi-dimensional vector space. In physical level, the value of principal angle reflects whether the two subspace have similar trajectory or not. In our algorithm, the region space is a representation of object region motion in physical level. So the principal angle we computed is essential constraint for segmentation.

The key of our segmentation algorithm is that the trajectory of two feature points located on a same object should be similar. In our algorithm, region space contains trajectory data essentially. And its subspace obtained by SVD is a representation of trajectory. Thus, estimation on subspaces and trajectory are equivalent. We use principal angle to estimate the similarity of two subspaces. This estimates the similarity of trajectories in disguise.

For a single object, its trajectory subspaces are orthogonal. In this case, for two region spaces located on the same object, their trajectory should be similar, and their subspaces obtained by SVD are intersecting. The max principal angles are close to $\pi/2$. In the ideal situation, if two region spaces belong to the same object, their subspaces are orthogonal. The max principal angles are exact $\pi/2$. In practice, they will not be exact orthogonal so a threshold is required.

Secondly, the distance between two subspaces is represented by $dist(\mathbf{F}, \mathbf{G})$ which ranges from 0 to 1. We define the function as below.

$$dist(\mathbf{F}, \mathbf{G}) = 1 - e^{-\sum_{i=1,...,p,j=1,...,q} \cos^2(\theta_{i,j})} \tag{9}$$

where $\theta_{i,j}$ are the principal angles between subspaces $\mathbf{F}$ and $\mathbf{G}$.

For two regions which have the same motion, their principal angle equals to $\pi/2$. Their dist equals to 0. This means that the two regions belong to the same object. However, not all this region belongs to a single object as the existence of adhesion. Fortunately, we own multi-images by interactive perception to solve this problem. Each interaction results in an observation image and intermediate segmentation result.

Lastly, we use clustering algorithm to realize object segmentation. Consider the problem of clustering p samples into c categories. As the object number c is unknown, we use bottom-up hierarchical clustering algorithm to merge p region spaces $\{\mathbf{R}_1, \mathbf{R}_2, \cdots, \mathbf{R}_p\}$ into c objects. The key of hierarchical clustering is distance metric. Distance metric defines similarity of samples. In our problem, distance matric of each hierarchical is shown in Eq. 10.

$$d_{\min}(\mathbf{R}_i, \mathbf{R}_j) = \min_{i,j=1...p} dist(\mathbf{V}_i, \mathbf{V}_j) \tag{10}$$

where $\mathbf{V}_i, \mathbf{V}_j$ are subspaces of $\mathbf{R}_i, \mathbf{R}_j$ respectively. p is the number of segmentation. A cut-off distance metric of 0.4 works well in our experiments. $dist(\mathbf{V}_i, \mathbf{V}_j)$ is computed by Eq. 9.

2.4   Action Planning Logic

When a robot use interactive perception strategy to obtain object segmentation of a cluttered environment, action planning logic is a basic problem. When we get a set of constituent interactive points. Many different next actions are possible, not all of these actions are equally helpful. Therefore, it will be necessary to select a prefer interactive pushing point from constitute set.

In this paper, we use a heuristic action planning method to implement this functionality. The key of our action planning algorithm is center point selection. When over-segmentation $\{\mathbf{R}_1, \mathbf{R}_2, \cdots, \mathbf{R}_p\}$ is obtained, we initially mark these regions as 0 and compute the center point of each region $\mathbf{C} = \{\mathbf{c}_1, \mathbf{c}_2, \cdots, \mathbf{c}_p\}$ as candidate interaction point. In order to maximally change the structure of environment, we compute the center point of entire regions. And then we compute the center of entire regions. Consider this center may not happened to be at a object. So we select a point $\mathbf{c}_k$ from $\mathbf{C}$ that is closest to it. $\mathbf{c}_k$ is the interaction point of this round. Then interaction action can be processed.

Subsequently, we update region markers according to the result of clustering algorithm (presented in Subsect. 2.3). For two merged regions, if they have same markers, the marker of new region equals to original marker plus 1. Otherwise, the two merged regions have different markers, the marker of new region equals to the bigger one. Then repeat the entire process until all the regions' marker becoming 2.

3   Experimental Results

We evaluate our system on real-world scenes with a 6 degrees of freedom robot arm which equipped with a camera and a holder. Each manipulation of pushing action results in one image. The holder is responsible for executing action. Our camera is a MV-VD200SC 1200×1600 high resolution CCD camera. In each experiment, the robot was presented with a clutter scene of novel objects. A scene may contain different numbers of objects according to experimental needs. The average size of the object used in our experiment is 10 cm which limited by 83 cm length of our robot arm.

Following [11], we estimate the segmentation quality using the measurement presented by Fowlkes and Mallows in [12].

$$Q = \frac{|H \cap P|}{\sqrt{|H|\,|P|}} \tag{11}$$

where H and P denotes the set of feature points belong to the same object according to real segmentation from human annotation or according to system result respectively. The operator $|*|$ denotes the cardinality of a set, it is a measure of the number of elements of the set. The segmentation quality Q ranges from 0 to 1. When it comes to 0, none of feature points correspond to the real segmentation. And in turn, Q equals to 1 if all the feature points system obtained correspond to human annotation. We manually mark the real segmentation in our experiment. Instead of marking every image, we only need to annotate the first image and KLT tracker described in Sect. 2.2 can track them subsequently. Hence, we can get human annotation of every observation easily.

Next, we estimate the availability of our interactive segmentation system on two scenes. All of the four objects are rigid body in scene 1. Interactive manipulation can not change their shape. By comparison, four objects are elastic and rotatable in scene 2. It means that manipulation can change object representation. Fortunately, shape would recover once manipulation disappeared. This property ensure that observation we obtained can reflect actual object appearance. We use a famous mean shift algorithm system—the EDISON system to obtain initial over-segmentation. EDISON system is a mean shift image segmentation software. Segmentation results are shown in lower row. It is intuitively that all objects can be mostly segmented.

Figure 2(a) and (b) shows the dist matching matrix in step 15. This Matching matrix shows the distance of a pair of region spaces. Detail description of algorithm are shown in Sect. 2.3. The row and column indexes represent region numbers in the 15 interaction. The values are computed by Eq. 9 which represent similarity of two regions.

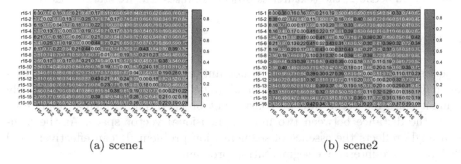

(a) scene1   (b) scene2

Fig. 2. Dist matching matrix of two scenes in step 15. The bigger the value, the larger probability two regions belong to the same object.

The segmentation qualities of two scenes are shown in Fig. 3. In the task of segmenting an cluttered scene into its constituent object. We use principal angle-based approach to estimate the distance between two multi-dimensional vector spaces instead of clustering with Euler distance [13]. In physical level, the value of principal angle reflects whether the two subspace have similar trajectory or not. Thus, the region space is a representation of object region in physical level. So the non-parameter principal angle based interactive segmentation approach can obtain object segmentation effectively.

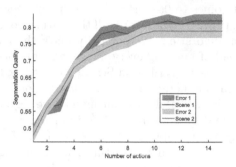

Fig. 3. Segmentation quality of two scenes using our non-parameter principal angle-based approach. The blue and red lines correspond to scene1 and scene2 respectively. The areas with low transparence show their segmentation standard error (Color figure online)

Results in Fig. 3 show that our approach is effective to rotation. However, the segmentation quality is slightly lower than scene 1. This is mainly caused by missing of feature points. In the experiment, 500 feature points are set in the initial scene. Unfortunately, about 13% are lost in the tracking step especially 15 steps later. Occasionally the feature points are even tracked mistakenly. Moreover, in scene 2, the textureless object confuses feature tracker badly.

4 Conclusions

Object segmentation problem is an important problem in computer vision. This paper propose a non-parameter principal angle based approach to interactively segment cluttered scene into its constituent objects. The principal angle estimates the distance between two multi-dimensional vector space. As the principal angle reflects the essence of segmentation problem. It can effectively and efficiently solve interactive segmentation problem.

References

1. Schiebener, D., Ude, A., Asfour, T.: Physical interaction for segmentation of unknown textured and non-textured rigid objects. In: 2014 IEEE International Conference on Robotics and Automation (ICRA), pp. 4959–4966. IEEE (2014)
2. Lyubova, N., Ivaldi, S., Filliat, D.: From passive to interactive object learning and recognition through self-identification on a humanoid robot. Auton. Robot. **40**(1), 33–57 (2016)

3. Hausman, K., Niekum, S., Osentoski, S., Sukhatme, G.S.: Active articulation model estimation through interactive perception. In: 2015 IEEE International Conference on Robotics and Automation (ICRA), pp. 3305–3312, May 2015

4. Chang, L., Smith, J.R., Fox, D.: Interactive singulation of objects from a pile. In: 2012 IEEE International Conference on Robotics and Automation (ICRA), pp. 3875–3882, May 2012

5. van Hoof, H., Kroemer, O., Peters, J.: Probabilistic segmentation and targeted exploration of objects in cluttered environments. IEEE Trans. Robot. **30**(5), 1198–1209 (2014)

6. Kenney, J., Buckley, T., Brock, O.: Interactive segmentation for manipulation in unstructured environments. In: 2009 IEEE International Conference on Robotics and Automation, ICRA 2009, pp. 1377–1382, May 2009

7. Yan, J., Pollefeys, M.: A factorization-based approach to articulated motion recovery. In: 2005 IEEE Computer Society Conference on Computer Vision and Pattern Recognition (CVPR 2005), vol. 2, pp. 815–821, June 2005

8. Golub, G.H., Van Loan, C.F.: Matrix Computations, vol. 3. JHU Press, Baltimore (2012)

9. Yan, J., Pollefeys, M.: A general framework for motion segmentation: independent, articulated, rigid, non-rigid, degenerate and non-degenerate. In: Leonardis, A., Bischof, H., Pinz, A. (eds.) ECCV 2006. LNCS, vol. 3954, pp. 94–106. Springer, Heidelberg (2006). doi:10.1007/11744085_8

10. Björck, A., Golub, G.H.: Numerical methods for computing angles between linear subspaces. Math. Comput. **27**(123), 579–594 (1973)

11. van Hoof, H., Kroemer, O., Amor, H.B., Peters, J.: Maximally informative interaction learning for scene exploration. In: 2012 IEEE/RSJ International Conference on Intelligent Robots and Systems, pp. 5152–5158, October 2012

12. Fowlkes, E.B., Mallows, C.L.: A method for comparing two hierarchical clusterings. J. Am. Stat. Assoc. **78**(383), 553–569 (1983)

13. Hausman, K., Balint-Benczedi, F., Pangercic, D., Marton, Z.-C., Ueda, R., Okada, K., Beetz, M.: Tracking-based interactive segmentation of textureless objects. In: 2013 IEEE International Conference on Robotics and Automation (ICRA), pp. 1122–1129. IEEE (2013)

Robust Variational Auto-Encoder for Radar HRRP Target Recognition

Ying Zhai[✉], Bo Chen, Hao Zhang, and Zhengjue Wang

National Laboratory of Radar Signal Processing, Xidian University, Xi'an, China
Yingzhai_rainie@163.com, bchen@mail.xidian.edu.cn,
zhanghao_xidian@163.com, zhengjuewang@163.com

Abstract. Traditional deep networks used for radar High-Resolution Range Profile (HRRP) target recognition usually ignore the inherent characteristics of the target, which result in the limited capability to learn effective features for classification task. To address this issue, a novel nonlinear feature learning method, called Robust Variational Auto-Encoder model (RVAE) is proposed. According to the stable physical properties of the average profile in each HRRP frame without migration through resolution cell, RVAE is developed based on variational auto-encoder, and such model is able to not only explore the latent representations of HRRP but reserve structure characteristics of the HRRP frame. We use the measured HRRP data to show the effectiveness and efficiency of our algorithm.

Keywords: Radar Automatic Target Recognition (RATR) · High-Resolution Range Profile (HRRP) · Feature Extraction · Robust Variational Auto-Encoder (RVAE)

1 Introduction

A high-resolution range profile (HRRP) is the amplitude of the coherent summations of the complex time returns from target scatterers in each range cell, which represents the projection of the complex returned echoes from the target scattering centers onto the radar line-of-sight (LOS). It contains abundant target structure signatures, such as target size, scatterer distribution, etc. Therefore, radar HRRP target recognition has received intensive attention from the radar automatic target recognition (RATR) community [1–17].

For most applications of machine learning, such as HRRP-based RATR, feature extraction is extremely crucial for them. For that reason, feature extraction captures many researchers' attention and pushes them to pay a lot of effort to seek various methods. Extracting the bispectra feature from real HRRP data to restrain noise is investigated in [5]. The recognition performance of power spectrum, FFT-magnitude feature [3, 7], and various high-order spectra features is analyzed in [6]. Such features are useful but rely on researchers' perception of data and experience. The PCA-based feature subspace is constructed to minimize reconstruction error for RATR in [2]. Feng et al. [7] utilize dictionary learning to represent the favor sparse overcomplete features of HRRP data, and obtain better generalization ability. Those models work well in

© Springer International Publishing AG 2017
Y. Sun et al. (Eds.): IScIDE 2017, LNCS 10559, pp. 356–367, 2017.
DOI: 10.1007/978-3-319-67777-4_31

practice, however, because of shallow linear architecture they cannot obtain hierarchical features to effectively represent HRRP.

Deep networks such as Deep Belief Networks (DBN) [19] and Stacked Denoising Autoencoders (SDAE) [20] have been widely applied to various machine learning tasks. Those models are distinguished from the conventional multi-layer perceptron (MLP) and other shallow models by their hierarchical features and the unique layer-wise unsupervised greedy learning method [21].

As for HRRP target recognition, the properties of data such as 'speckle effect' [2, 8, 15] can affect the models' final recognition performance, which also limits the feature learning capability of those traditional deep networks. In this paper, we first analyze these properties and then by coupling them with the deep generative model, we propose a novel nonlinear architecture named Robust Variational Auto-Encoder (RVAE) for HRRP target recognition. RVAE employs the average profile, having stable physical properties in each HRRP frame without migration through resolution cell, as the constraint to learn more effective and robust features of the single HRRP sample, which is expected to not only represent the sample, but also reflect the stable property of the corresponding average profile. Moreover, as an extension of Variational Auto-Encoder (VAE), RVAE not only can be used for target recognition but also has the ability to generate data. Both of them are examined and analyzed by detailed experiments.

The reminder of the paper is organized as follows. A brief description of Auto-Encoder and Variational Auto-Encoder is provided in Sect. 2. Then we introduce HRRP and propose a novel nonlinear architecture called RVAE for HRRP target recognition in Sect. 3. The detailed experiments based on measured HRRP data are provided in Sect. 4, followed by conclusions and future work in Sect. 5.

2 Preliminaries

In this section, we will briefly review the concepts of Auto-encoder (AE) [20] and Variational Auto-Encoder (VAE) [22].

2.1 Auto-Encoder

We briefly recall the traditional Auto-Encoder (AE) framework and its terminology here.

Encoder: The deterministic mapping that transforms an input vector x into hidden representation $h \in R^K$ is called the encoder. Its typical form is an affine mapping followed by a nonlinearity:

$$h = \text{sigm}(W^T x + b). \tag{1}$$

where W is a $D \times K$ weight matrix and b is an offset vector of dimensionality K.

Decoder: The resulting hidden representation h is then mapped back to a reconstructed D-dimensional vector $\hat{x}$ in input space. This mapping is called the decoder. Its typical form is again an affine mapping optionally followed by a squashing non-linearity, that is, either $\hat{x} = Wh + c$ or

$$\hat{x} = \text{sigm}(Wh + c). \tag{2}$$

here W is a tied weight matrix, i.e., using the same weights for encoding the input and decoding the hidden representation.

The training criterion of AE is to minimize the reconstruction error, and the process is simple and easy to implement. However, AE is lack of describing the underlying probabilistic distribution of data.

2.2 Variational Auto-Encoder

Variational Auto-Encoder (VAE) utilizes reparameterization trick to combine the probabilistic model and neural network, which takes advantage of the backpropagation and stochastic gradient learning to scale up the generative models.

VAE is shown in Fig. 1, solid lines denote the generative model $p_\theta(z)p_\theta(x|z)$, dashed lines denote the variational approximation $q_\phi(z|x)$ to the intractable posterior $p_\theta(z|x)$. The variational parameters ϕ are learned jointly with the generative model parameters θ.

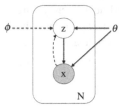

Fig. 1. Variational Auto-Encoder (VAE) architecture.

VAE contains inference and generative process, and the model's objective function is to maximize the variational lower bound [22]. Since true posterior density $p_\theta(z|x)$ is intractable, VAE constructs a variational approximation $q_\phi(z|x)$ to $p_\theta(z|x)$. What's more, it applies the reparameterization trick to obtain a differentiable estimator of the variational lower bound, and updates model parameters by the stochastic gradient. Moreover, VAE is a probabilistic model, which has the ability to describe the distribution of data and provides the uncertainty of parameters.

3   Robust Variational Auto-Encoder for Radar HRRP Target Recognition

Conventional deep networks used for radar HRRP target recognition usually ignore the inherent characteristics of the target. To resolve this issue, we develop Robust Variational Auto-Encoder (RVAE), which aims at extracting more robust and useful features, which can not only represent each sample, but also reflect the structure characteristics of the HRRP frame.

3.1   Preprocessing and the Average Profile

Due to the characteristics of HRRP, we cannot straightforwardly utilize the original data as the input for networks, where there are some issues should be considered. First, we adopt centroid alignment [17] as the time-shift compensation technique in this paper since it is simple and easy use. Then we use l_2 normalization method to eliminate the amplitude scale sensitivity of HRRP.

From the literatures [2, 3, 6], a small target-aspect sector without scatterers' motion through range cells (MTRC) is defined as an HRRP frame. According to [2, 6, 9, 10, 15], the definition of the average profile is:

$$\mathbf{m} \triangleq \left[\frac{1}{N} \sum_{n=1}^{N} |x_{n1}|, \frac{1}{N} \sum_{n=1}^{N} |x_{n2}|, \ldots, \frac{1}{N} \sum_{n=1}^{N} |x_{nD}| \right]^T = \frac{1}{N} \sum_{n=1}^{N} |x_n|. \tag{3}$$

where $X = \{x_n\}_{n=1}^{N}$ is an HRRP frame, with the n th HRRP sample $x_n = [x_{n1}, x_{n2}, \ldots, x_{nD}]^T$. T denotes the transpose operation.

Figure 2 illustrates a single sample of Yak-42 and its corresponding average profile. It shows that the average profile has a smoother and concise signal form than the single HRRP, and can better reflect the scattering property of the target in the given aspect-frame [18]. What's more, the average profile of each HRRP frame can not only enhance echo signal-to-noise ratio (SNR) but also efficiently depress the amplitude fluctuation property, the speckle effect and outliers.

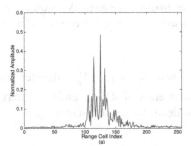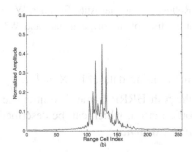

Fig. 2. The single HRRP and the corresponding average profile of Yak-42. (a) the single HRRP. (b) the average profile.

In a word, the average profile in a small target-aspect sector has better generalization performance than the single HRRP [2, 6, 9, 10, 15]. In the following part, we will construct a network embedded with average processing for HRRP-based RATR.

3.2   Robust Variational Auto-Encoder

For HRRP target recognition, a good feature is one that is useful for mapping the original signal to a discriminative subspace and preserving the structure information of targets. Therefore, we propose Robust Variational Auto-Encoder (RVAE) to learn such features. RVAE's training process contains two parts: feature extraction and data generation. Moreover, data generation is divided into two procedures, one is to learn a nonlinear network to generate the reconstruction of the original data based on the extracted features, the other is to learn a new nonlinear network generating the corresponding average profile with the same features, both of which are expected to close to their goals as much as possible. We utilize the average profile as a constraint to help us to learn effective and robust features of the single HRRP sample, which involve detailed information of the sample and reflect the stable property of the corresponding average profile. RVAE incorporates the HRRP frame and average processing into one variational auto-encoder. Such architecture helps us to learn stable and useful features for RATR. The architecture of RVAE is shown in Fig. 3. In addition, the number of model layers depends on the number of latent variables.

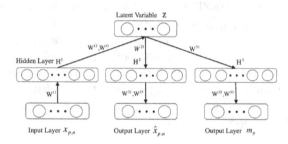

Fig. 3. Robust Variational Auto-Encoder. W^1 is used to extract features, W^2 is used to generate the reconstruction of the original data and W^3 is used to generate the corresponding average profile.

Let the training dataset be $X = \{X_p\}_{p=1}^P$, where $X_p = \{x_{p,n}\}_{n=1}^N$ denotes the samples in the p th HRRP frame. We infer the parameters of RVAE under the maximum likelihood criterion, which can be described in detail by several stages.

Feature extraction:

$$h_{p,n}^1 = f(\alpha_1) = Relu(x_{p,n}W^{11} + b^{11}). \tag{4}$$

$$\begin{cases} \mu\_z_{p,n} = g(\alpha_2) = h_{p,n}^1 W^{12} + b^{12} \\ \log \sigma\_z_{p,n} = g(\alpha_3) = h_{p,n}^1 W^{13} + b^{13} \end{cases} \tag{5}$$

$$z_{p,n} = \mu\_z_{p,n} + \sigma\_z_{p,n} \cdot \varepsilon. \tag{6}$$

where $\alpha_1 = x_{p,n}W^{11} + b^{11}$, $\alpha_2 = h^1_{p,n}W^{12} + b^{12}$, $\alpha_3 = h^1_{p,n}W^{13} + b^{13}$, $h^1_{p,n}$ is the element of the first hidden layer, $z_{p,n} \sim N(\mu\_z_{p,n}, \sigma^2\_z_{p,n})$, $z_{p,n}$ is the element of the latent variable, $\varepsilon \sim N(0,1)$.

Reconstruction of the single HRRP:

$$h^2_{p,n} - f(\alpha_4) = Relu(z_{p,n}W^{21} + b^{21}). \tag{7}$$

$$\begin{cases} \mu\_\hat{x}_{p,n} = g(\alpha_5) = h^2_{p,n}W^{22} + b^{22} \\ \log \sigma\_\hat{x}_{p,n} = g(\alpha_6) = h^2_{p,n}W^{23} + b^{23} \end{cases} \tag{8}$$

where $\alpha_4 = z_{p,n}W^{21} + b^{21}$, $\alpha_5 = h^2_{p,n}W^{22} + b^{22}$, $\alpha_6 = h^2_{p,n}W^{23} + b^{23}$, $h^2_{p,n}$ is the element of the second hidden layer, $\hat{x}_{p,n} \sim N(\mu\_\hat{x}_{p,n}, \sigma^2\_\hat{x}_{p,n})$, $\hat{x}_{p,n}$ is the reconstruction of $x_{p,n}$.

Generation of the average profile:

$$h^3_{p,n} = f(\alpha_7) = Relu(z_{p,n}W^{31} + b^{31}). \tag{9}$$

$$\begin{cases} \mu\_m_p = g(\alpha_8) = h^3_{p,n}W^{32} + b^{32} \\ \log \sigma\_m_p = g(\alpha_9) = h^3_{p,n}W^{33} + b^{33} \end{cases} \tag{10}$$

where $\alpha_7 = z_{p,n}W^{31} + b^{31}$, $\alpha_8 = h^3_{p,n}W^{32} + b^{32}$, $\alpha_9 = h^3_{p,n}W^{33} + b^{33}$, $h^3_{p,n}$ is the element of the third hidden layer, $m_p \sim N(\mu\_m_p, \sigma^2\_m_p)$, m_p denotes the average profile in the pth HRRP frame. The objective function of RVAE:

$$L(\hat{X}, M) = E_{q(z|X)}\{\log p(\hat{X}|z)\} + E_{q(z|X)}\{\log p(M|z)\} - KL(q(z|X) \| p(z)). \tag{11}$$

where

$$KL(q(z|X)\| p(z)) = E_{q(z|X)} \log \frac{q(z|X)}{p(z)}. \tag{12}$$

with $M = [m_1, m_1, \ldots, m_1, m_2, \ldots, m_p, \ldots, m_P, m_P]$. The first term of the objective function expects the latent variable to match the distribution characteristics of the original data and makes the extracted features containing the detailed information of the sample. The second term compels the latent variable to fit the corresponding average profile in order to make the learned features having the structure information of the HRRP frame. The KL-divergence term makes the difference between $q(z|X)$ and $p(z)$ as small as possible.

Based on the above parameter inference, we summarize the procedure of RVAE as the following:

Step 1: suppose the training dataset $X = \{X_p\}_{p=1}^{P}$ and the corresponding average profile $M = \{M_p\}_{p=1}^{P}$, and then initialize the parameter set θ_{RVAE} in random way.

Step 2: according to the formula (4)–(10), build network based on the Python platform. Under the maximum likelihood criterion, update model parameters using gradients based on BP algorithm [23, 24].

Step 3: repeat step 2 until the algorithm converges.

The features extracted by RVAE can be used in recognition tasks. Typical radar HRRP target recognition framework based on RVAE is illustrated in Fig. 4.

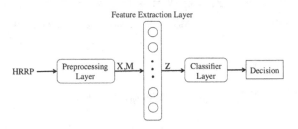

Fig. 4. Typical radar HRRP target recognition framework based on RVAE.

4 Experimental Results and Analysis

4.1 Data Description

The results presented in this section are based on measured HRRP data from three real airplanes, which are extensively used in [1–3, 5–7, 13–17]. The specific parameters of the targets and radar are shown in Table 1 and the projections of target trajectories onto the ground plane are shown in Fig. 5.

Table 1. Parameters of planes and radar in the ISAR experiment

Radar parameters	Center frequency	5520 MHz	
	Bandwidth	400 MHz	
Aircraft	Length (m)	Width (m)	Height (m)
Yark-42	36.38	34.88	9.83
Cessna Citation S/II	14.40	15.90	4.57
An-26	23.80	29.20	9.83

There are two points should be considered when choosing training and test datasets [1–3, 13]: (a) the training dataset should cover all possible target-aspect theoretically, (b) the elevation angles of the test dataset are different from those of the training dataset. Therefore, we take the second and the fifth segments of Yark-42, the sixth and the seventh segments of Cessna Citation S/II and the fifth and the sixth segments of

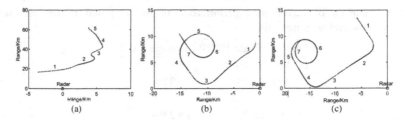

Fig. 5. Projections of target trajectories onto the ground plane. (a) Yak-42. (b) Cessna Citation S/II. (c) An-26.

An-26 as the training samples, and take other data segments as test samples. The training dataset has 2800 HRRP frames and each of them has 50 adjacent HRRP samples. The average profile can be calculated by the mean of the aligned HRRPs in each frame. Finally, we have 700 average profiles for Yak-42, 1050 for Cessna, and 1050 for An-26. The size of the test dataset is 5200. Each sample is a 256-dimensional vector, i.e., D = 256. The number of target classes is C = 3.

4.2   Recognition Performance

According to the algorithm in Sect. 3.2, we initialize the model parameters at first. Let $W \sim N(0, 1)$ and $b = 0$, where $W = \{W^{11}, W^{12}, W^{13}, W^{21}, W^{22}, W^{23}, W^{31}, W^{32}, W^{33}\}$, $b = \{b^{11}, b^{12}, b^{13}, b^{21}, b^{22}, b^{23}, b^{31}, b^{32}, b^{33}\}$. The size of minibatch is 200 and the learning rate is 0.001. According to the above, we set the number of hidden units in each layer is $K^1 = K^2 = K^3 = 400$ and the dimension of the latent variable is 20.

We compare the recognition performance of RVAE with three shallow models, including Linear discriminant analysis (LDA), K-SVD and Principal component analysis (PCA), and three deep networks, i.e., DBN, SDAE and Stacked Corrective Autoencoders (SCAE) [18], the results are shown in Table 2. All the deep networks have the same framework, and the size of deep architecture is 256-1500-500-50. The feature dimension of LDA is C-1 according to the literature [25]. The feature dimension of K-SVD is set to 1500 and the number of nonzero elements is less than 50. We use these models to extract features, and then taking these features as the input to the classifier. LSVM here only serves as a simple baseline, so it does not employ any feature extraction.

From Table 2, the recognition rate of PCA taking single HRRP as the input is 83.81%, which is better than other shallow non-statistical models but still less than the accuracy of the deep networks about 6–8%. Therefore, compared with the shallow architecture, the nonlinear hierarchical structure of the deep networks is helpful to learn more expressive features, which improves the recognition performance. Meanwhile, among those deep networks, SCAE has the best recognition performance but its accuracy is still less than RVAE, which can be attributed to that RVAE has the ability to extract the robust and effective discriminative features. Moreover, it is worth of noting that RVAE only learned 20 latent features as the final input for LSVM to reach

Table 2. Comparison of classification performance of the proposed model with several traditional shallow and deep methods.

Input	Single HRRP	Average profile
LDA	81.3%	73%
K-SVD	74.7%	70.83%
PCA	83.81%	83.66%
LSVM	86.7%	88.28%
DBN	90.29%	90.64%
SDAE	90.42%	91.20%
SCAE	92.03%	
RVAE	*92.12%*	

the best performance. It proves that RVAE is also good at dimension reduction, which will be discussed in detail in the following experiment.

4.3   Generalization Performance

Figure 6 shows the visualizations of the original HRRP and the features extracted by RVAE respectively, via using two-dimensional PCA. Compared with the original HRRP, RVAE can produce more discriminative feature space, although RVAE, without using the class information, is an unsupervised learning model, where the learning of the model is independent of the final recognition task.

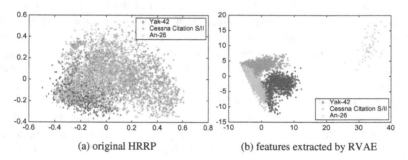

(a) original HRRP   (b) features extracted by RVAE

Fig. 6. Visualizations of the test HRRP samples and their corresponding features, via using two-dimensional PCA.

In practice, the training data is limited and the test data is huge. Under this circumstance, models may be trained to overfitting and tend to mismatch the unseen samples which can influence the models' generalization performance. In this experiment, we present the reconstruction performance of RVAE to evaluate its generalization ability. There are three types of the reconstruction: the single training sample, the corresponding average profile and the single test sample. Note that the last one is unseen in the training phase.

Figure 7 shows that RVAE is able to retain the main details of the single sample as well as the structure information of the HRRP frame, and does not have the mismatch phenomenon. It indicates that RVAE has a good generalization capability to describe and capture the underlying characteristics of HRRP.

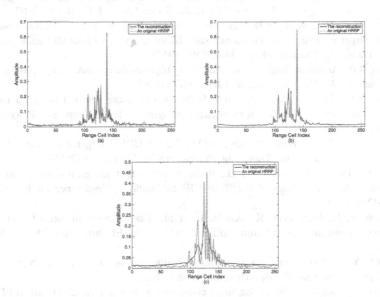

Fig. 7. The reconstruction performance of RVAE. (a) reconstruction of the single training sample. (b) reconstruction of the corresponding average profile. (c) reconstruction of the single test sample.

5 Conclusions and Future Work

In this paper, considering the inherent characteristic of HRRP, we propose a novel nonlinear feature learning method, robust variational auto-encoder (RVAE), for the task of radar HRRP target recognition. The proposed model employs the average profile as a constraint to learn more effective and robust features of the single HRRP sample, which involve detailed information of the sample and reflect the stable property of the corresponding average profile. The experimental results on measured HRRP data show that our model achieves superior recognition performance compared with several traditional shallow and deep methods based on inferred representations, meanwhile, the detailed analysis about generalization ability is also provided.

The model proposed in this paper with only one latent variable can be seen as a shallow network. In order to further improve the performance of the model and get hierarchical recognition features, we will develop a deep model based on the existing network. Unfortunately, the work about the deep model is still at the research stage.

References

1. Chen, B., Liu, H.W., Chai, J., Bao, Z.: Large margin feature weighting method via linear programming. IEEE Trans. Knowl. Data Eng. **21**(10), 1475–1488 (2009)
2. Du, L., Liu, H.W., Bao, Z., Zhang, J.Y.: Radar automatic target recognition using complex high-resolution range profiles. Proc. IET Radar Sonar Navig. **1**(1), 18–26 (2007)
3. Du, L., Liu, H.W., Wang, P.H., Feng, B., Pan, M., Chen, F., Bao, Z.: Noise robust radar HRRP target recognition based on multitask factor analysis with small training data size. IEEE Trans. Sig. Process. **60**(7), 3546–3559 (2012)
4. Jacobs, S.P.: Automatic target recognition using high-resolution radar range profiles. (Ph.D. dissertation), Washington Univ., St. Louis, MO (1999)
5. Zhang, X.D., Shi, Y., Bao, Z.: A new feature vector using selected bispectra for signal classification with application in radar target recognition. IEEE Trans. Sig. Process. **49**(9), 1875–1885 (2001)
6. Du, L., Liu, H.W., Bao, Z., Xing, M.D.: Radar HRRP target recognition based on higher-order spectra. IEEE Trans. Sig. Process. **53**(7), 2359–2368 (2005)
7. Feng, B., Du, L., Liu, H.W., et al.: Radar HRRP target recognition based on K-SVD algorithm. In: Proceedings of the IEEE CIE International Conference on Radar, China, pp. 642–645 (2011)
8. Molchanov, P., Egiazarian, K., Astola, J., et al.: Classification of aircraft using micro-Doppler bicoherence-based features. IEEE Trans. Aerosp. Electron. Syst. **50**(2), 1455–1467 (2014)
9. Li, H.J., Yang, S.H.: Using range profiles as feature vectors to identify aerospace objects. IEEE Trans. Antennas Propag. **41**(3), 261–268 (1993)
10. Zyweck, A., Bogner, R.E.: Radar target classification of commercial aircraft. IEEE Trans. Aerosp. Electron. Syst. **32**(2), 598–606 (1996)
11. Liao, X.J., Runkle, P., Carin, L.: Identification of ground targets from sequential high-range-resolution radar signatures. IEEE Trans. Aerosp. Electron. Syst. **38**(4), 1230–1242 (2002)
12. Zhu, F., Zhang, X.D., Hu, Y.F., Xie, D.: Nonstationary hidden Markov models for multiaspect discriminative feature extraction from radar targets. IEEE Trans. Sig. Process. **55**(5), 2203–2213 (2007)
13. Du, L., Liu, H.W., Bao, Z.: Radar HRRP statistical recognition: parametric model and model selection. IEEE Trans. Sig. Process. **56**(5), 1931–1944 (2008)
14. Du, L., Wang, P.H., Liu, H.W., Pan, M., Chen, F., Bao, Z.: Bayesian spatiotemporal multitask learning for radar HRRP target recognition. IEEE Trans. Sig. Process. **59**(7), 3182–3196 (2011)
15. Xing, M.D., Bao, Z., Pei, B.: The properties of high-resolution range profiles. Opt. Eng. **41**(2), 493–504 (2002)
16. Shi, L., Wang, P.H., Liu, H.W., et al.: Radar HRRP statistical recognition with local factor analysis by automatic Bayesian Ying-Yang harmony learning. IEEE Trans. Sig. Process. **59**(2), 610–617 (2011)
17. Chen, B., Liu, H.W., Bao, Z.: Analysis of three kinds of classification based on different absolute alignment methods. Mod. Radar **28**(3), 58–62 (2006)
18. Feng, B., Chen, B., Liu, H.: Radar HRRP target recognition with deep networks. Pattern Recogn. **61**, 379–393 (2017)
19. Hinton, G., Salakhutdinov, R.: Reducing the dimensionality of data with neural networks. Science **313**(5786), 504–507 (2006)

20. Vincent, P., Larochelle, H., Lajoie, I., Bengio, Y., Manzagol, P.A.: Stacked denoising autoencoders: Learning useful representations in a deep network with a local denoising criterion. J. Mach. Learn. Res. **11**, 3371–3408 (2010)
21. Bengio, Y., Courville, A., Vincent, P.: Representation learning: a review and new perspectives. IEEE Trans. Pattern Anal. Mach. Intell. **35**(8), 1798–1828 (2013)
22. Kingma, D.P., Welling, M.: Auto-encoding variational bayes. In: Proceedings of the International Conference on Learning Representations (ICLR) (2014)
23. Bengio, Y., Lamblin, P., Popovici, D., Larochelle, H.: Greedy layer-wise training of deep networks. In: Proceedings of Neural Information and Processing Systems (2007)
24. Nair, V., Hinton, G.E.: 3D object recognition with deep belief nets. In: Proceedings of Neural Information and Processing Systems (2009)
25. Yu, H., Yang, J.: A direct LDA algorithm for high-dimensional data with application to face recognition. Pattern Recogn. **34**(10), 2067–2070 (2001)

Probabilistic Hypergraph Optimization for Salient Object Detection

Jinxia Zhang[1,2], Shixiong Fang[1], Krista A. Ehinger[3], Weili Guo[1],
Wankou Yang[1], and Haikun Wei[1(✉)]

[1] Key Laboratory of Measurement and Control of CSE,
Ministry of Education, School of Automation, Southeast University,
2 Sipailou Street, Nanjing 210096, Jiangsu, China
hkwei@seu.edu.cn
[2] Key Laboratory of Intelligent Perception and Systems for High-Dimensional
Information of Ministry of Education, Nanjing University of Science and Technology,
Nanjing 210094, China
[3] Centre for Vision Research, York University, Toronto, Canada

Abstract. In recent years, many graph-based methods have been introduced to detect saliency. These methods represent image regions and their similarity as vertices and edges in a graph. However, since they only represent pairwise relations between vertices, they give an incomplete representation of the relationships between regions. In this work, we propose a hypergraph based optimization framework for salient object detection to include not only the pairwise but also the higher-order relations among two or more vertices. In this framework, besides the relations among vertices, both the foreground and the background queries are explicitly exploited to uniformly highlight the salient objects and suppress the background. Furthermore, a probabilistic hypergraph is constructed based on local spatial correlation, global spatial correlation, and color correlation to represent the relations among vertices from different views. Extensive experiments demonstrate the effectiveness of the proposed method.

Keywords: Hypergraph · Optimization framework · Saliency detection

1 Introduction

Since Itti and Koch et al. [1] introduced the first computational model of visual saliency detection, a large amount of approaches have been proposed to detect salient objects in images [2–12]. Visual saliency detection can be applied to many computer vision related applications, including image segmentation, image compression, image quality assessment, object detection, and object recognition.

One series of salient object detection approaches are unsupervised and graph-based [6–10], which model each image as a graph and propagate saliency information via weighted edges connecting image parts. These approaches do not need manually annotated samples and overcome a major drawback of classical,

© Springer International Publishing AG 2017
Y. Sun et al. (Eds.): IScIDE 2017, LNCS 10559, pp. 368–378, 2017.
DOI: 10.1007/978-3-319-67777-4_32

pixel-level bottom-up saliency methods: namely, they tend to highlight object edges more than homogeneous object interiors. Graph-based approaches [6–10] use a pairwise simple graph to model the relations between two image regions. In a simple graph, image regions are vertices; two vertices are connected by an edge and the edge weight represents the similarity between regions.

However, representing the relations among image regions in a pairwise simple graph gives an incomplete representation of the image. It is important to consider not only the pairwise relation between two vertices but also local grouping information among two or more vertices [13]. These higher-order relations can be represented by a hypergraph, a generalization of a simple graph [14]. In a hypergraph, hyperedges are used to describe complex relations among any number of vertices. Hypergraphs have been proved to be useful in many applications, such as image retrieval [13] and image segmentation [15]. In the salient object detection field, very few works have made use of the hypergraph [16]. In the work of Li et al. [16], a binary hypergraph is constructed using a clustering algorithm at multiple image scales, which describes the binary relations between vertices and hyperedges, i.e. whether a vertex belongs to a hyperedge. A hypergraph modeling approach is then proposed to capture the contextual saliency information on image superpixels and the final hypergraph-based saliency map is essentially the average of saliency detection results from different scales.

In this work, we introduce a new hypergraph-based method for salient object detection, which has two main contributions. Firstly, a novel hypergraph based optimization framework is proposed. Salient object detection aims to separate the salient foreground from the non-salient background. Thus, we explicitly exploit both the foreground queries and the background queries in this optimization framework to uniformly highlight all the salient regions and suppress the background. These two kinds of queries are generated based on the boundary prior of visual saliency and the local grouping information among different vertices.

Our second contribution is that a probabilistic hypergraph is constructed to better represent the local grouping information, which captures not only the binary relations but also the probability that a vertex belongs to a hyperedge. We fuse local spatial correlation, global spatial correlation, and color correlation among multiple vertices into a probabilistic hypergraph by adding different types of hyperedges. Each vertex in the hypergraph serves as the centroid for a local spatial hyperedge which connects a centroid vertex to its local spatial neighbors, a global spatial hyperedge which connects a centroid vertex to the global border vertices of the image, and a color hyperedge which connects a centroid vertex to its neighbors in the color space. We further define the probability that a vertex belongs to a hyperedge to be the similarity between this vertex and the corresponding centroid vertex of the hyperedge. We then define the weight of each hyperedge as the quadratic sum of the probabilities that different vertices belong to this hyperedge.

We test our probabilistic hypergraph based optimization framework on three databases and demonstrate that our method outperforms eleven state-of-the-art models. The remainder of this paper is organized as follows: The hypergraph based optimization framework is introduced in Sect. 2; Sect. 3 describes how we construct a probabilistic hypergraph; Our testing procedure and experimental comparison are described in Sect. 4; Sect. 5 concludes the paper.

2   Hypergraph Based Optimization Framework

Let V denote a finite set of vertices and E denote a series of hyperedges, which is a family of subsets of V such that $\bigcup_{e \in E} = V$. A hypergraph can be represented by a $|V| \times |E|$ incident matrix $H(v_i, e_j)$ where i is an index of a vertex and j is an index of a hyperedge.

$$H(v_i, e_j) = \begin{cases} p(v_i|e_j) & \text{if } v_i \in e_j \\ 0 & \text{otherwise} \end{cases} \tag{1}$$

If a vertex v_i is contained in a hyperedge e_j, the corresponding value $H(v_i, e_j)$ equals to the probability that v_i belongs to e_j; otherwise, the value equals to 0. Thus, each column of the incident matrix H represents the composition information of a hyperedge in the hypergraph. Let a diagonal matrix W record the hyperedge weights, where each hyperedge e is assigned a positive weight $W(e)$. Based on the incident matrix H and the weight matrix W, a diagonal matrix D_v which records the degree of each vertex $D_v(v_i)$ can be defined as $D_v(v_i) = \sum_{e_j \in E} W(e_j)H(v_i, e_j)$, and a diagonal matrix D_e which records the degree of each hyperedge $D_e(e_j)$ can be defined as $D_e(e_j) = \sum_{v_i \in V} H(v_i, e_j)$.

Based on hypergraph, a ranking algorithm has been proposed for classification [14]. In this algorithm, a smoothness term is used to include the relations among different vertices and a fitting term is used to constrain the final result to be close to the original queries. The hypergraph based ranking algorithm is suitable for the situation where one kind of specific queries has been given, e.g. image retrieval [13]. The goal of salient object detection is to separate the salient foreground from the non-salient background. Thus, both the foreground queries and the background queries are important to detect saliency in images. In this work, we propose a novel hypergraph-based optimization framework which includes both the foreground queries and the background queries for saliency detection. The proposed hypergraph-based optimization framework aims to compute a vector $S : V \Rightarrow \mathbb{R}$, which assigns a saliency value $S(v_i)$ to each vertex v_i.

$$S^* = \arg\min_S(\frac{1}{2}\Omega + \lambda_f \Psi + \lambda_b \Phi), where$$

$$\Omega = \sum_{e_k \in E} \sum_{v_i, v_j \in e_k} \frac{W(e_k)}{D_e(e_k)} H(v_i, e_k)H(v_j, e_k)(S(v_i) - S(v_j))^2$$

$$\Psi = \sum_{i=1}^{n} Q_f(v_i) \|S(v_i) - L_f\|^2 \tag{2}$$

$$\Phi = \sum_{i=1}^{n} Q_b(v_i) \|S(v_i) - L_b\|^2$$

In the term Ω, $S(v_i)$ and $S(v_j)$ are the saliency values of the vertices v_i and v_j separately. $H(v_i, e_k)$ and $H(v_j, e_k)$ indicate the probabilities that v_i and v_j belong to the hyperedge e_k. $W(e_k)/D_e(e_k)$ can be understood as the normalized weight of the hyperedge e_k. From the formula, we can see that Ω indicates the smoothness constraint defined in a hypergraph: If two vertices v_i and v_j have high probabilities to belong to the same hyperedge frequently and these hyperedges have high weights, the saliency values of these vertices should be close.

Among the term Ψ, L_f is the label of the salient foreground and is set as 1 in our paper. $Q_f(v_i)$ indicates whether the vertex v_i is a foreground query: If v_i is a foreground query, $Q_f(v_i) = 1$; otherwise, $Q_f(v_i) = 0$. Thus, Ψ can be understood as a foreground fitting term, which encourages the saliency value of a vertex v_i to be close to the label of the salient foreground if this vertex is a foreground query.

Similarly, the term Φ can be understood as a background fitting term. L_b is the label of the non-salient background and is set as -1. $Q_b(v_i)$ indicates whether the vertex v_i is a background query: If v_i is a background query, $Q_b(v_i) = 1$; otherwise, $Q_b(v_i) = 0$. This term Φ encourages a vertex v_i which is a background query to take a saliency value close to the background label.

In Formula 2, λ_f and λ_b are the weighting parameters, which specify the relative balance of three terms. By setting the derivative of the function in Formula 2 to be zero, the saliency values of different vertices can be computed as:

$$S^* = (D_v - HWD_e^{-1}H^T + \lambda_f Q_f + \lambda_b Q_b)^{-1}(\lambda_f Q_f - \lambda_b Q_b) \quad (3)$$

In this work, we first choose initial foreground queries Q_f and initial background queries Q_b based on the boundary prior of visual saliency. According to the boundary prior, the vertices on the four sides of the image are more likely to be background. We then use the vertices which are not on the image borders as the initial foreground queries and set each of the four image borders as the initial background queries separately to get four maps based on Formula 3. These maps are then normalized to be in $[0, 1]$ and multiplied to get a temporal result T. The final foreground and background queries are acquired based on this result T using different thresholds th_f and th_b: If a vertex v_i has a value $T(v_i) > th_f$, then this vertex is a foreground query; If a vertex v_i has a value $T(v_i) < th_b$, this vertex is a background query.

According to Formula 3, besides the foreground and background queries, the hypergraph represented by H and W is also critical to compute the saliency value of each vertex. In Sect. 3, we describe how we construct the hypergraph for salient object detection.

3   Probabilistic Hypergraph Construction

We first employ the SLIC algorithm [17] to segment an input image I into n homogeneous superpixels and then define these superpixels as the vertices of our hypergraph. The number of the homogeneous superpixels in each input image is set to 300 in this work. For each superpixel, we compute its mean in the CIELab color space and the image spatial coordinates as the color and spatial features of this superpixel.

3.1 Hyperedges and Weights

To get a probabilistic hypergraph, we need to construct the incident matrix H, i.e. choose different hyperedges and compute the probabilities that different vertices belongs to these hyperedges. In this work, given a set of vertices V, we choose different hyperedges based on three kinds of correlations: local spatial correlation, global spatial correlation and color correlation. Thus, for each vertex v_i, three hyperedges are chosen: a local spatial hyperedge, which contains v_i itself as a centroid vertex and its immediate spatial neighbors that share a boundary with the centroid vertex in the input image; a global spatial hyperedge, which contains v_i as a centroid vertex and the global boundary vertices on the four borders of the image; and a color hyperedge which contains v_i as a centroid vertex and its neighbors in the CIELab color space. We further compute the probability that a contained vertex belongs to a hyperedge as the similarity between this vertex and the centroid vertex of the hyperedge. Our method to compute the similarity between two vertices are introduced in Sect. 3.2.

Besides the incident matrix H, the diagonal hyperedge weight matrix W is also important for salient object detection. In our work, we define the weight of each hyperedge $w(e_j)$ as the sum of the squares of the probability that each contained vertex belongs to the hyperedge e_j.

$$w(e_j) = \sum_{v_i \in e_j} H(v_i, e_j)^2 \tag{4}$$

If all vertices belonging to a hyperedge e_j have a higher probability to be part of the hyperedge, this hyperedge has higher inner group similarity and thus should be assigned a higher weight. Otherwise, this hyperedge should be assigned a lower weight.

3.2 Similarity Matrix

Let a $n * n$ matrix SIM contain the similarity between each pair of the vertices. In this work, we compute the similarity between two vertices using exponential function based on both the color distance and the spatial distance.

$$SIM(v_i, v_j) = exp(-\frac{D_c(v_i, v_j) + D_s(v_i, v_j)}{2\sigma}) \tag{5}$$

A scale parameter σ is used to control the strength of the color and spatial distance. If σ has a small value, only vertices with close color features and spatial positions would make a contribution. If σ has a large value, the vertices with large distances would also have big influence on each other. In this work, σ is set to 0.1 for all experiments. $D_c(v_i, v_j)$ and $D_s(v_i, v_j)$ respectively represents the distance of the color features and the spatial positions between the vertices v_i and

v_j. We use the Euclidean distance to compute the color distance $D_c(v_i, v_j)$ and employ the sine spatial distance proposed in our previous work [18] to compute the spatial distance $D_s(v_i, v_j)$, as shown in Formula 6.

$$D_s(v_i, v_j) = \sqrt{(\sin(\pi \cdot |x_i - x_j|))^2 + (\sin(\pi \cdot |y_i - y_j|))^2} \tag{6}$$

In this formula, x_i and y_i represent the horizontal and vertical coordinates of a vertex i in the image plane, which have been normalized to be in $[0, 1]$. Experimental comparison in the work [18] has demonstrated that the sine spatial distance helps to uniformly suppress the non-salient background.

4   Experimental Comparison

We compare with eleven state-of-the-art saliency detection methods: GB [2], FT [3], RC [4], CB [5], HM [16], GR [6], BD [7], CL [8], GP [9], PM [10] and MST [11]. Following [3,4], we chose different methods based on various principles: recency (PM, MST, CL and GP) and variety (FT is frequency tuned; RC uses regional contrast; CB is based on shape and context knowledge; GB, GR, CL, GP and PM are graph based; HM is hypergraph based; and MST employs minimum spanning tree).

We test the performances of different methods on three standard salient object databases: the 10000-image MSRA10K database [4], the 643-image iCoseg database [19] and the 1000-image Extended Complex Scene Saliency Database (ECSSD) [20]. The MSRA10K database contains many images with one salient object. The iCoseg database contains a lot of images with multiple salient foreground objects. The ECSSD database contains many natural images with complex foreground and background patterns.

The most common evaluation metric for salient object detection, i.e. precision-recall curves (PR curves) is used in this work for experimental comparison. To get PR curves, the saliency map is binarized at each threshold in the range $[0:1:255]$ and the precision and recall values are computed at each threshold by comparing the binary masks against the ground truth.

There are four main parameters in the proposed probabilistic hypergraph-based optimization framework: the balance weights λ_f and λ_b in Formula 3; the thresholds th_f and th_b to get the final foreground and background queries. These parameters are empirically chosen, $\lambda_f = 0.05$, $\lambda_b = 0.1$, $th_f = 0.05$ and $th_b = 0.3$, for all the experiments.

Quantitative Comparison: The quantitative comparison between our method and eleven state-of-the-art methods based on PR curves is shown in Fig. 1.

Figure 1(a) shows PR curves of different methods in images which have one salient object in the content. Previous graph-based methods, e.g. CL, GP and PM already have competitive performances. Our proposed method outperforms these methods, demonstrating the effectiveness of our method. HM and our method are both hypergraph based models, but the algorithms proposed by

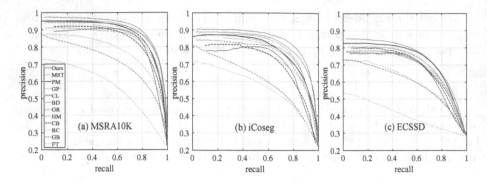

Fig. 1. The quantitative comparison of different methods on MSRA10K, iCoseg and ECSSD databases. Our method gives the best performance on all three databases.

these two models are totally different. Firstly, HM constructs a binary hypergraph based on a clustering algorithm while our method constructs a probabilistic hypergraph based on local spatial correlation, global spatial correlation and color correlation. Secondly, HM detects salient objects based on a hypergraph modeling algorithm which result is essentially the average of saliency detection maps from different scales, while in this work we propose a novel hypergraph-based optimization framework which make use of both the foreground queries and the background queries. The quantitative comparison demonstrates that our method outperforms HM by a large margin.

We further evaluate our method on the iCoseg database which contains a lot of images with multiple foreground objects, shown in Fig. 1(b). We note that the PR curves of different saliency detection methods drop a little in comparison to the PR curves on the MSRA10K database. This phenomenon indicates that it's harder to detect all of the foreground regions when there are multiple salient objects in an image. From the comparison, it can be seen that the proposed method outperforms other saliency detection methods. It demonstrates that the construction of the probabilistic hypergraph and the proposed optimization framework can help to detect multiple salient objects in an image.

The ECSSD database contains 1000 complex natural images. The performance comparison in Fig. 1(c) demonstrates that the proposed probabilistic hypergraph based optimization framework can better detect salient foreground objects in complex images than other state-of-the-art methods.

We further compare the performances of different hypergraphs, which use all or part of local spatial correlation, global spatial correlation and color correlation on the ECSSD database, shown in Fig. 2. The hypergraphs with only one correlation have the worst performances. The hypergraphs with two of the correlations outperforms the hypergraphs considering only one correlation. And our method using all three correlations has the best performance.

Visual Comparison: Different evaluated saliency detection methods can be visually compared in Fig. 3, which displays representative images to highlight

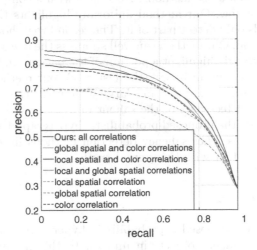

Fig. 2. The PR curves on ECSSD database with different hypergraphs. (Color figure online)

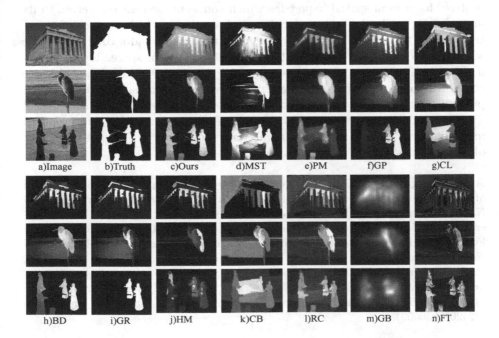

Fig. 3. Comparison of different salient object detection methods. The first column is the original image, the second is the ground truth. The third row is our saliency detection result, and the remaining columns are results of other evaluated methods.

the differences between these methods. An image with a complex salient object is shown in the first row. Our method uniformly highlights the building while most other methods only detect part of it. The second row shows an image with a complex background. Our method can well suppress the background while a lot of other methods wrongly highlight a portion of the background. When there are multiple salient objects in an image (the last row), our method can uniformly highlight all of the salient objects and largely suppress the background while other methods only detect part of the salient objects or wrongly highlight the background. In short, the proposed probabilistic hypergraph based optimization framework outperforms other saliency detection methods in generating results which are more consistent with the ground truth.

5   Conclusion

In this work we have proposed a probabilistic hypergraph based optimization framework to detect salient objects in images. In this framework, both foreground queries and background queries are explicitly exploited to better detect salient objects. To construct a probabilistic hypergraph, we use three different correlations in natural images, including local spatial correlation, global spatial correlation, and color correlation. Each vertex in the hypergraph serves as the centroid for a local spatial hyperedge which connects the centroid vertex to its immediate spatial neighbors, a global spatial hyperedge which connects the centroid vertex to the vertices on the borders of the image, and a color hyperedge which connect the centroid vertex to its neighbors in color space. The probability that a vertex belongs to a hyperedge and the weight of each hyperedge are computed based on a similarity matrix, which encodes the similarities between pairs of vertices and is computed based on both color distance and sine spatial distance. Extensive experimental comparisons demonstrate that the proposed method outperforms state-of-the-art methods on a variety of images for salient object detection.

Acknowledgments. This work was supported by the National Natural Science Fund of China (Grant numbers 61233011, 61374006, 61473086, 61703100); Major Program of National Natural Science Foundation of China (Grant number 11190015); Natural Science Foundation of Jiangsu (Grant number BK20131300, BK20170692); the Innovation Fund of Key Laboratory of Intelligent Perception and Systems for High-Dimensional Information of Ministry of Education (Nanjing University of Science and Technology, Grant number JYB201601); the Innovation Fund of Key Laboratory of Measurement and Control of Complex Systems of Engineering (Southeast University, Grant number MCCSE2017B01); and the Fundamental Research Funds for the Central Universities (2242016k30009).

References

1. Itti, L., Koch, C., Niebur, E.: A model of saliency-based visual attention for rapid scene analysis. IEEE Trans. Pattern Anal. Mach. Intell. **20**(11), 1254–1259 (1998)

2. Harel, J., Koch, C., Perona, P.: Graph-based visual saliency. In: Advances in Neural Information Processing Systems, pp. 545–552 (2006)
3. Achanta, R., Hemami, S., Estrada, F., et al.: Frequency-tuned salient region detection. In: IEEE Conference on Computer Vision and Pattern Recognition, pp. 1597–1604 (2009)
4. Cheng, M.-M., Zhang, G.-X., Mitra, N.J., et al.: Global contrast based salient region detection. In: IEEE Conference on Computer Vision and Pattern Recognition, pp. 409–416 (2011)
5. Jiang, H., Wang, J., Yuan, Z., et al.: Automatic salient object segmentation based on context and shape prior. In: British Machine Vision Conference, vol. 6, p. 7 (2011)
6. Yang, C., Zhang, L., Lu, H., et al.: Saliency detection via graph-based manifold ranking. In: Conference on Computer Vision and Pattern Recognition, pp. 3166–3173 (2013)
7. Zhu, W., Liang, S., Wei, Y., et al.: Saliency optimization from robust background detection. In: IEEE Conference on Computer Vision and Pattern Recognition, pp. 2814–2821 (2014)
8. Gong, C., Tao, D., Liu, W., et al.: Saliency propagation from simple to difficult. In: IEEE Conference on Computer Vision and Pattern Recognition, pp. 2531–2539 (2015)
9. Jiang, P., Vasconcelos, N., Peng, J.: Generic promotion of diffusion-based salient object detection. In: IEEE International Conference on Computer Vision, pp. 217–225 (2015)
10. Kong, Y., Wang, L., Liu, X., Lu, H., Ruan, X.: Pattern mining saliency. In: Leibe, B., Matas, J., Sebe, N., Welling, M. (eds.) ECCV 2016. LNCS, vol. 9910, pp. 583–598. Springer, Cham (2016). doi:10.1007/978-3-319-46466-4_35
11. Tu, W.-C., He, S., Yang, Q., et al.: Real-time salient object detection with a minimum spanning tree. In: IEEE Conference on Computer Vision and Pattern Recognition, pp. 2334–2342 (2016)
12. Zhang, J., Ehinger, K.A., Wei, H., et al.: A novel graph-based optimization framework for salient object detection. Pattern Recogn. **64**, 39–50 (2017)
13. Huang, Y., Liu, Q., Zhang, S., et al.: Image retrieval via probabilistic hypergraph ranking. In: IEEE Conference on Computer Vision and Pattern Recognition, pp. 3376–3383 (2010)
14. Zhou, D., Huang, J., Schölkopf, B.: Learning with hypergraphs: clustering, classification, and embedding. In: Advances in Neural Information Processing Systems, pp. 1601–1608 (2006)
15. Kim, S., Nowozin, S., Kohli, P., et al.: Higher-order correlation clustering for image segmentation. In: Advances in Neural Information Processing Systems, pp. 1530–1538 (2011)
16. Li, X., Li, Y., Shen, C., et al.: Contextual hypergraph modeling for salient object detection. In: IEEE International Conference on Computer Vision, pp. 3328–3335 (2013)
17. Achanta, R., Shaji, A., Smith, K., et al.: SLIC superpixels compared to state-of-the-art superpixel methods. IEEE Trans. Pattern Anal. Mach. Intell. **34**(11), 2274–2282 (2012)
18. Zhang, J., Ehinger, K.A., Ding, J., et al.: A prior-based graph for salient object detection. In: IEEE International Conference on Image Processing, pp. 1175–1178 (2014)

19. Batra, D., Kowdle, A., Parikh, D., et al.: iCoseg: interactive co-segmentation with intelligent scribble guidance. In: IEEE Conference on Computer Vision and Pattern Recognition, pp. 3169–3176 (2010)
20. Yan, Q., Xu, L., Shi, J., et al.: Hierarchical saliency detection. In: IEEE Conference on Computer Vision and Pattern Recognition, pp. 1155–1162 (2013)

A Cascaded Segmentation Method Based on Region Merging to Change Detection in Remote Sensing Images

Ning Lv[(⊠)] and Xinbo Gao

School of Electronic Engineering/VIP Lab., Xidian University,
Taibai South Road 2, Xi'an, Shaanxi, China
nlv@mail.xidian.edu.cn

Abstract. Change detection based on image superpixels can extract more geomorphologic information among multitemporal remote sensing images than methods based on pixel difference. In this paper, we presented a cascaded segmentation method to extract clear change region boundry with noise supperssion. First, Simple linear iterative clustering (SLIC) is used to generate super pixels which adhere difference image boundries tightly for purpose of searching change regions. Second, one Statistical Region Merging (SRM) with dynamic sorting algorithm is modified to merge those homogeneous super pixels. After the candidate change regions established, classified change map are remerged by using simplified SRM. Finally, the proposed method are compared with methods based on PCA and MRF. Experimental results shows our method restrain the over segmentation and obtain better performance of change detection than conventional SRM algorithms.

Keywords: Change detection · Remote sensing · Image segment · SLIC · SRM

1 Introduction

Change detection of the remote sensing image is a process of identifying differences in the state of an object or phenomenon by observing it at different times. [1] Therefore, remote sensing image change detection mainly focus on disaster (floods, fires and earthquakes) area location, urban regional change analysis and other similar applications. Because of fixed running cycle of the space borne, Synthetic Aperture Radar (SAR) imaging has advantages in change detection task than other imaging sensors, especially when the ground region is observed in cloudy, rainy or haze weather. However, SAR imaging is susceptible to speckle noise, and its denoising [2, 3] or noise suppression in change detection are very important.

Recently, the methods based on difference image are often adopted in change detection tasks. These methods extract the difference image directly by comparing the

---

N. Lv—Foundation item: National Natural Science Foundation of China (61571347, 61201293).

Y. Sun et al. (Eds.): IScIDE 2017, LNCS 10559, pp. 379–389, 2017.
DOI: 10.1007/978-3-319-67777-4_33

temporal images, then the difference image is divided into the change class and un-change class. For example, pixel-based methods [3, 4] generally use differentiated pixel grayscale features to classify and use clustering methods to obtain as good a possible classification as possible. CVA (Change Vector Analysis) [5] using the change feature of the difference image to extract the change region. Celik [6] using PCA dimensionality reduction and K-means clustering to carry out the change detection process. Gong et al. [7] combines Markov random field and the fuzzy clustering with proposed energy function to improve the change detection performance. However, there is a *semantic gap* problem that the pixel feature does not include the semantic features. And these binary classification result cannot offer the detailed change information which are more significant in practice application.

Hence, image segmentation technology arouses more interesting in how to acquire heterogeneous regions by extracting semi-semantic information for change detection. Image segmentation methods mainly include spectral clustering segmentation [8, 9], active contour segmentation [10–12], MRF segmentation [13, 14], SRM [15, 16] and SLIC [17]. Hichri et al. [18] introduced Markov Random Filed (MRF) and level set segmentation method into change detection task, and use Supporting Vector Machine (SVM) to train the classifier, which is a kind of supervised learning method needs a certain amount of training data. Honglin et al. [19] introduced interactive segmentation and decision fusion into SAR image change detection, but this method is very sensitive to seed point selection. The method of Meanshift [20] is also used to obtain high-resolution remote sensing images for multi-scale segmentation. Wang et al. [21] extract the Scale-invariant Feature Transform (SIFT) points and use the Otsu method to segment by these feature points, its effect depends on the effective extraction of the SIFT point. Most of the above methods require training data to obtain stable supervised segmentation. But in Change detection task, change of geomorphic itself is variety and unpredictable, so labeling and training data often is difficult and impractical for supervised segmentation.

The *Statistical Region Merging* (SRM) is a fast region generation segmentation algorithm. With the establishment of the statistical model of the image, the statistical reasoning is used to control the merging process of pixels and regions, and the homogeneous region decision rule is set to obtain accurate segmentation results. The SRM method does not depend on the probability distribution assumptions and have a good anti-noise ability. It is important to note that SRM algorithm need a good initialization of adjacent image regions to start merging. *Simple Linear Iterative Clustering* (SLIC) algorithm is one approach fit for preprocessing segmentation as its fast speed. Boundaries of super pixels segmented by SLIC adhere well to the edge of the difference image. SLIC can offer good difference image super pixels for SRM to merge those to change regions.

Therefore, In this paper, an cascade segmentation framework is designed by using SLIC and DSSRM segmentation model. As Fig. 1 shown, firstly three criterions is used to get the difference data set of multiple channels sources for segmentation. Next, SLIC is used to map image pixels to super pixels. After that, a new DSSRM algorithm based on multi-feature Markov distance criterion and dynamic sorting mode is proposed for the candidate change region segmentation. Finally, one simplified SRM method are modified to output the change map of the difference data set.

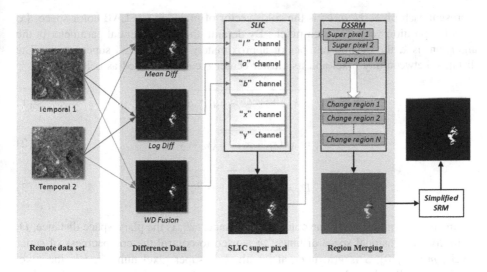

Fig. 1. Framework of cascade segmentation to change detection

The rest of this paper is organized as follow. Section 2 introduces the modified SLIC-SRM algorithm, to presents the framework of the cascade segmentation, including the establishment of the difference data set. The Sect. 3 discusses the simulation experiment results comparing with state of the art methods. In the Sect. 4 the advantage and disadvantage of the cascade segmentation method proposed in this paper are concluded.

2   Cascade Segmentation Based on SRM

Change detection algorithm on SAR images often degrade because of speckle noise and vague imaging quality. In this section, we present a cascade segmentation method to locate detailed change in homogeneous change region. SLIC algorithm is used as the first segmentation to generate super pixels which approximate image boundaries because of its fast speed and good edges adherence ability. And then, a novel SRM algorithm is modified to merge homogeneous super pixels to form candidate change regions. In these change regions, a simplified SRM is implemented to compute the fine change map. Besides, we constructed different data sets based on the three difference type images to suppress the affection of the SAR image noise.

2.1   Simple Linear Iterative Clustering (SLIC) and Distance Metric

Simple linear iterative clustering (SLIC) algorithm is faster than existing methods, more memory efficient, exhibits state-of-the-art boundary adherence, and improve the performance segmentation algorithm, and it is like the approach used as preprocessing step for depth estimation described in [23], which was not fully explored in the context of super pixel generation. In SLIC, *"labxy"* five-dimensional space are used to represent each pixel, $\{l, a, b\}$ is the color vector of pixels in CIELAB color space, $\{x, y\}$ is the position vector of each pixels. By default, the only essential parameter in the algorithm is k, the desired number of approximately equally sized super pixels. The distance between any pixel and its cluster center c can be defined as below:

$$d_{lab} = \sqrt{(l_c - l_i)^2 + (a_c - a_i)^2 + (b_c - b_i)^2} \tag{1}$$

$$d_{xy} = \sqrt{(x_c - x_i)^2 + (y_c - y_i)^2} \tag{2}$$

$$D_s = \sqrt{(\frac{d_{lab}}{m})^2 + (\frac{d_{xy}}{S})^2} \tag{3}$$

In above formula, d_{lab} is the color space distance, d_{xy} is the plane space distance, D_s is the five-dimensional space distance; m is used to control the compactness of super pixels, $m = [1,40]$; S is grid interval length, k is super pixel number, i is the pixel number. Usually, the color space distance and the plane space distance can be calculated by the same distance metric.

It is need to mention that SLIC algorithm take color image as its input. And there is only one single intensity data channel for SAR image. Furthermore, the difference data set of remote sensing image must be constructed to improve the quality of change detection and bonding capacity of merging of SRM algorithm. So, the two-input temporal image I_1, I_2 are used to build three channel of difference image: log ratio difference image D_1, mean ratio difference image D_2, Gauss ratio difference image D_3, μ_1 and μ_2 represents the neighborhood gray mean of I_1 and I_2, g_1 and g_2 represents the neighborhood Gauss filter value. These three difference images are defined as formula (4) shown.

$$\begin{cases} D_1 = \left| \log \frac{I_1}{I_2} \right| \\ D_2 = 1 - MIN\left(\frac{\mu_1}{\mu_1}, \frac{\mu_2}{\mu_1}\right) \\ D_3 = \log\left[\max\left(\frac{g_1}{g_2}, \frac{g_2}{g_1}\right) \right] \end{cases} \tag{4}$$

Two temporal image of Berne data set, log ratio difference image, mean ratio difference image and Gauss ratio difference image as the following Fig. 2(a), (b), (c), (d), (e).

2.2   Statistical Region Merging (SRM) with Dynamic Sorting

Statistical Region Merging (SRM). Classical SRM was proposed as a kind of image statistical model in 2004 by Richard Nock and Frank Nielsen. Algorithm use homogeneity of image regions to seek the optimal merging regions in the probability space. It can fast percept and capture the primary structure of the image, and has strong noise suppression performance and ability to carry out multi-scale segmentation with the

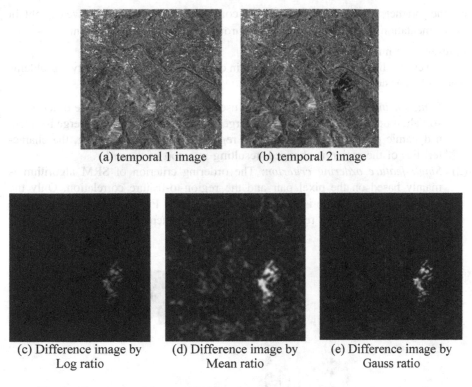

(a) temporal 1 image   (b) temporal 2 image

(c) Difference image by   (d) Difference image by   (e) Difference image by
        Log ratio   Mean ratio   Gauss ratio

Fig. 2. Two temporal images of Berne dataset and three difference image results

different complexity of contents. So, in this paper SRM is consider to execute change region merging in change detection task.

As the same as SLIC, SRM is also designed for RGB color image, suppose the image including N pixels, each pixel includes (R, G, B) three color channels, the range of each channel is $\{1, 2, \ldots, g\}, g = 2^q$, q is the quantization order of each channel. Each color channel is represented by the independent random variables which number is Q and the range is $[0, g / Q]$. Image I is an observation of the real scene I^*, each statistical region in I^* represents a real object, therefore, the statistical region should satisfy the homogeneity. And the consolidation rule of unconsolidated paired regions (R, R'):

$$P(R, R') = \begin{cases} \text{true} & \forall a \in (R, G, B), \max_{a \in \{R,G,B\}} \left| \overline{R'_a} - \overline{R_a} \right| \leq \sqrt{b^2(R) - b^2(R')} \\ \text{false} & others \end{cases} \tag{5}$$

$$b(R) = g\sqrt{\frac{1}{2Q|R|} \ln \frac{\left\| R_{|R|} \right\|}{\delta}} \tag{6}$$

$\overline{R'_a}$ and $\overline{R_a}$ represent the mean observation values of statistical regions R' and R in the channel of a. $R_{|R|}$ represents the region set with $|R|$ pixels, $\left\| R_{|R|} \right\| \leq (n+1)^{\min(|R|, g)}$, and the parameter Q represent the statistical complexity of the real scene, we can obtain the segmentation results of different scales through adjusting Q and constant $\delta = \frac{1}{6|I|^2}$, $|I|$ represents the number of pixels.

However, when SRM algorithm is used in change detection, there are two problems need to be considered:

(1) *Static sorting method*: SRM algorithm uses a static sorting mode. The order of the sort alignment is not updated in the merger process. But the region merge is in fact a dynamic process. Each time a new region is formed by merging, the characteristics of the region is changed, resulting in resort of the order.

(2) *Single-feature ordering criterion*: The ordering criterion of SRM algorithm is mainly based on the pixel pair and the region-to-feature correlation. Only the gray-scale mean feature is considered. As shown in Fig. 3, the circle part of the Berne data is corroded, resulting in missed alarm increased in change map.

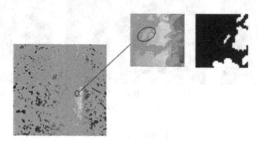

Fig. 3. Local over-segmentation of SRM algorithm on Berne difference image

To improve the efficiency of static sorting in SRM, a dynamic sorting method DSRM [24] was proposed to use the non-similarity of pixels as the sorting rule. But in change detection task, the change region in difference image may contains various intensity value, and over-segmentation are also inevitable.

Hence, to modified classical SRM to fit change detection task, a dynamic multiple features sorting algorithm of SRM is proposed, which considers the regional pixels mean value, the area and the histogram of the regional intensity value scale, and constructs the sorting function by dimensionless Mahalanobis distance. Define the best merging region pair (R, R'):

$$(R, R') = \arg \max_{a \in \{D_1, \ldots, D_n\}} f(R_a, R'_a) \tag{7}$$

$f(R, R')$ is a sorting function, D_1, D_n represent n data channel, $n = 3$ in nature image for RGB channels. Sorting function $f(R, R')$ is defined as formula (8):

$$\begin{cases} f(R,R') = \sqrt{[H_R - H_{R'}] \times \mathbf{S}^{-1} \times [H_R - H_{R'}]^T} \\ r(R,R') = 1 \\ \max[|\frac{A_R}{A_{R'}}|, |\frac{A_{R'}}{A_R}|] < \delta \end{cases} \tag{8}$$

Which, H_R and $H_{R'}$ respectively represent the corresponding eigenvector in the region R and R', $H_{R'} = \{\overline{R'}, A(R'), |R'|\}$, $H_R = \{\overline{R}, A(R), |R|\}$, the gray scale mean $\overline{R}$, the statistical histogram $A(R)$ and the area of the region $|R|$ are combined to produce the eigenvector. $\overline{R}$ expresses the regional mean gray similarity, $A(R)$ expresses the similarity of the regional gray probability distribution, and $|R|$ shows the similarity degree of the region size. Dimensionless *Mahalanobis* distance $\sqrt{[H_R - H_{R'}] \times \mathbf{S}^{-1} \times [H_R - H_{R'}]^T}$ is used to define the similarity of the feature vector H_R and $H_{R'}$ in regions to be merged. $\mathbf{S}$ is the covariance matrix of all the feature vectors to be merged, and $\mathbf{S}^{-1}$ is its inverse matrix.

The function $r(R,R')$ represents the adjacent relation of the region pair (R,R'), 1 is adjacent, 0 is not adjacent, and the region pair satisfying the adjacent relation is determined according to the adjacency matrix element value. The adjacency matrix is dynamically updated with the merge. A_R indicates the number of pixels in the area, $\max[|\frac{A_R}{A_{R'}}|, |\frac{A_{R'}}{A_R}|] < \delta$ indicating that the area ratio of the area to be combined can not be too large to prevent loss of image detail. δ is an empirical constant.

Finally, to adapt to the dynamic characteristics of the regional characteristics in the process of regional merging, we use the dynamic sorting method to recalculate the feature vector of the new region every time it is merged, update the adjacent matrix, recalculate $f(R,R')$ and update the sorted list, to keep the next merging region is most similar. From the above, the algorithm is summarized as shown below:

Step1: traverse the adjacent regions in image I which to be segmented, get the number of adjacent region pairs M, generate the adjacent matrix B;
Step2: compute the similarity of each adjacent region pair $f(R',R)$; according to the formula (7) (8), get the candidate region pair (R,R');
Step3: decide (R,R') to be merged or not by formula (5) (6); if merge decided, update B and M, go to step2; if not, continue traversing region to be merged; otherwise, go to start of step3.
End.

2.3 Cascade Segmentation Framework Based on SLIC and SRM

As previously mentioned, the remote sensing image suffers more noise than nature optical image. It is difficult to accurately locate the difference between the two temporal images and restrain the effect of the speckle noise at the same time. So we present the cascade segmentation framework based on SLIC and SRM. At preprocess stage, a non-local mean filter is first used to remove the speckle noise as usual [1].

First Stage in framework, SLIC algorithm is used to map the difference image from pixel feature space to super pixel feature space. In the difference image, the SLIC super pixels are segmented with the boundary of latent change region. And by adjusting the number of clusters in SLIC, these boundary are guaranteed to be consistency as possible. In second stage, SRM uses the multi-feature combination to segment based on the super-pixel initial segmentation of the difference image previously output by SLIC stage. When using classical SRM to merge the super pixels by sorting once, over-segmentation is often inevitable. In particular, the over-segmentation of the SAR difference image is more serious, so the SRM with dynamic sorting method is used to make the segmentation result more stable.

After that, one simplified SRM method is designed to loosens the condition that the merged regions must be adjacent in the original SRM algorithm. The intensity value of each region in the second stage merge result are sorted in ascending order and merged according to the merge rule until the image is divided into the change area and Non-changing two regions. Finally, the change map of the two temporal images is built after these merging stages, which is what we call the cascade segmentation framework. The results of the merging in different stage on the Berne dataset are shown in Fig. 4.

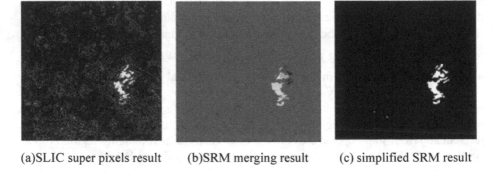

(a)SLIC super pixels result   (b)SRM merging result   (c) simplified SRM result

Fig. 4. The results of the merging of the cascade segmentation

3   Experimental Results and Analysis

To perform the framework based on cascade segmentation in the SAR image change detection, we compared with performance indicators in SAR image change detection of four methods, including the SRM + K-means method, [6] PCA+K-means method, Fuzzy + MRF [7], and the cascading segmentation method based on the dual channel difference map referred to as SLIC-SRM. The simulation data set is based on the Berne data set. The data set has a resolution of: 301×301.

Four methods used in Berne data set have results shown in Fig. 5(a), (b), (c), (d), as shown in Table 1 there are measuring performance indicators of four methods in three data sets, including False alarm pixel F_p, missing detection pixel F_n, pixel correct classified probability Pcc and $Kappa$ consistency coefficient. The best detection indicators are also shown in bold.

As shown in Table 1, compared with other methods SLIC-SRM obtain the better results, especially on the number of false alarm error and won a better control on the

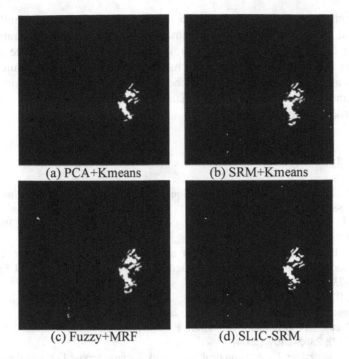

(a) PCA+Kmeans   (b) SRM+Kmeans

(c) Fuzzy+MRF   (d) SLIC-SRM

Fig. 5. Change detection results comparison of four algorithms

Table 1. Performance comparison for four algorithms

Data set	Methods	F_p	F_n	Kappa	Pcc
Berne	PCA+Kmeans	258	146	0.8437	0.9955
	SRM	287	114	0.8363	0.9956
	FCM+MRF	364	**47**	0.8413	0.9955
	SLIC-SRM	**208**	117	**0.8585**	**0.9963**

total number of errors, making the false alarm and missing detection be relatively balanced. Therefore, the Kappa coefficient also has a good Performance indicators. Because the initial segmentation by SLIC converts pixel space into super pixel space, the pixel in difference image has more change boundary information. And then, the region merging segmentation leverages regional features to allow the most similar areas to merge first, avoiding over-segmentation, and the multiple difference image channel data also enhances the effectiveness of the merge.

4   Conclusions

As the experimental results shown, the false alarm number and total of error number by the proposed method are decreased thereby the performance of *Kappa* can get higher than methods based on PCA and MRF. Our method is based on dynamic sorting algorithm with Manhattan distance of multi-feature of regions, it makes similar regions to be merged firstly. Experiments on construction of multi-channel illustrates that the more is the difference between channels the better is the performance of change detection. At the same time, the proposed method improved the performance of SRM algorithm to avoid the over-segmentation phenomenon. Comparison experiments show that this method can obtain better performance of change detection than conventional SRM and other state-of-art algorithms.

References

1. Singh, A.: Digital change detection techniques using remotely-sensed data. Int. J. Remote Sens. (1988)
2. Salmon, J.: On two parameters for denoising with non-local means. IEEE Sig. Process. Lett. **17**(3), 269–272 (2010)
3. Soni, V., Bhandari, A.K., Kumar, A., et al.: Improved sub-band adaptive thresholding function for denoising of satellite image based on evolutionary algorithms. IET Sig. Proc. **10**(4), 720–730 (2013)
4. Coppin, B.P., Jonckheere, I., Nachaerts, K.: Digital change detection in ecosystem monitoring: a review. Int. J. Remote Sens. **25**(9), 1565–1596 (2004)
5. Lu, D., Mausel, P., Brondízio, E., Moran, E.: Change detection techniques. Int. J. Remote Sens. **25**(12), 2365–2407 (2004)
6. Bovolo, F., Bruzzone, L.: A theoretical framework for unsupervised change detection based on change vector analysis in the polar domain. IEEE Trans. Geosci. Remote Sens. **45**(1), 218–236 (2007)
7. Celik, T.: Unsupervised change detection in satellite images using principal component analysis and k-means clustering. IEEE Geosci. Remote Sens. Lett. **6**(4), 772–776 (2009)
8. Gong, M., Su, L., Jia, M.: Fuzzy clustering with a modified MRF energy function for change detection in synthetic aperture radar images. IEEE Trans. Fuzzy Syst. **22**(1), 98–109 (2014)
9. Jia, J., Jiao, L.: Image segmentation by spectral clustering algorithm with spatial coherence constraints. J. Infrared Millim. Waves **29**(1), 69–74 (2010)
10. Gou, S., Zhuang, X., Zhu, H.: Parallel sparse specral clustering for SAR image segmentation. IEEE J. Sel. Top. Appl. Earth Observ. Remote Sens. **6**(4), 1949–1963 (2013)
11. Feng, J., Cao, Z., Pi, Y.: Multiphase SAR image segmentation with G0-statistical-model-based active. IEEE Trans. Geosci. Remote Sens. **51**(7), 4190–4199 (2013)
12. Gao, X., Wang, B., Tao, D., Li, X.: A relay level set method for automatic image segmentation. IEEE Trans. Syst. Man Cybern. B Cybern. **41**(2), 518–525 (2011)
13. Wang, B., Gao, X., Tao, D., Li, X.: A nonlinear adaptive level set for image segmentation. IEEE Trans. Cybern. **44**(3), 418–428 (2014)
14. Song, X., Wang, S., Liu, F.: SAR image segmentation using markov random field based on regions and Bayes belief propagation. Acta Electronica Sinica **38**(12), 2810–2816 (2010)
15. Xiong, B., Chen, Q., Jiang, Y.: A threshold selection method using two SAR change detection measures based on the Markov random field model. IEEE Geosci. Remote Sens. Lett. **9**(2), 287–291 (2012)
16. Nock, R., Nielsen, F.: Statistical region merging. IEEE Trans. Pattern Anal. Mach. Intell. **26**(11), 1452–1458 (2004)
17. Achanta, R., et al.: SLIC superpixels compared to state-of-the-art superpixel methods. IEEE Trans. Pattern Anal. Mach. Intell. **34**(11), 2274–2282 (2012)

18. Lang, F., Yang, J., Li, D.: Polarimetric SAR image segmentation using statistical region merging. IEEE Geosci. Remote Sens. Lett. **11**(2), 509–513 (2014)
19. Hichri, H.: Interactive segmentation for change detection in multispectral remote-sensing images. IEEE Geosci. Remote Sens. Lett. **10**(2), 298–302 (2013)
20. Dian, Y.Y., Fang, S.H., Yao, C.H.: Change detection for high-resolution images using multilevel segment method. J. Remote Sens. **20**(1), 129–137 (2016)
21. Wan, H., Jiao, L., Xin, F.: Interative segmentation technique and decision-level fusion based change detection for SAR image. Acta Geodaet. Cartograhpica Sinica. **41**(12), 74–80 (2012)
22. Wang, Y., Lan, D., Dai, H.: Unsupervised SAR image change detection based on SIFT keypoints and region information. IEEE Geosci. Remote Sens. Lett. **13**(7), 931–935 (2016)
23. Smith, T.F., Waterman, M.S.: Identification of common molecular subsequences. J. Mol. Biol. **147**, 195–197 (1981)
24. Huang, Z., Zhang, J., Li, X., Zhang, H.: Remote sensing image segmentation based on dynamic statistical region merging. Int. J. Light Electron Opt. **125**(2), 870–875 (2014)

Saliency Detection by Unifying Regression and Propagation

Jianwu Ai, Lihe Zhang$^{(\boxtimes)}$, and Xiukui Li

School of Information and Communication Engineering,
Dalian University of Technology, Dalian 116024, China
aijianwu@mail.dlut.edu.cn, {zhanglihe,xli}@dlut.edu.cn

Abstract. This paper introduces a novel saliency detection method by incorporating logistic regression into the label propagation framework, along with a principled weight computation for saliency fusion. First, the initial map is generated by computing objectness and backgroundness. Second, we unify logistic regression and label propagation to predict saliency labels. Last, we fuse the predicted result and initial map and further refine the fused map across multiple scales. Moreover, we use clustering random forest to learn the pairwise affinities between superpixels for backgroundness computation and saliency prediction. Extensive experiments on three large benchmark datasets demonstrate the proposed algorithm performs well against the state-of-the-art methods.

Keywords: Saliency detection · Logistic regression · Label propagation · Affinity learning

1 Introduction

Saliency detection has been an important preprocessing step to tackle the information overload problem in computer vision, which makes certain regions in a scene stand out from their neighbors and catch human visual attention. Saliency detection has been widely applied to numerous vision tasks, such as object segmentation, object recognition and image retrieval. Although numerous saliency methods have been proposed, it remains a challenging problem.

Saliency detection in general can be categorized by top-down or bottom-up models. The former is knowledge-driven, which incorporates task demands and high-level knowledge learned from a large set of training images with ground-truth labeling to distinguish salient regions. In contrast, the latter is data-driven, which models saliency by exploiting various visual cues of foreground objects or background, such as color contrast and boundary connectivity.

The learning-based approaches train a classifier or regressor using a large number of human-annotated labels to predict saliency for image elements. The propagation-based methods transmit saliency information from a few labeled elements to the unlabeled ones in a semi-supervised manner. In this work, motivated by [1], we incorporate a logistic function into the propagation model to

© Springer International Publishing AG 2017
Y. Sun et al. (Eds.): IScIDE 2017, LNCS 10559, pp. 390–399, 2017.
DOI: 10.1007/978-3-319-67777-4_34

constrain the label inference. We learn the logistic regression classifier and label propagation using a unified framework, and then employ the logistic function to predict the saliency labels. The proposed model can more accurately predict the labels than traditional label propagation. And, the model uses both the labeled and unlabeled samples to train the classifiers, which possibly avoids the over-fitting problem in logistic regression when only a limited labeled data are available.

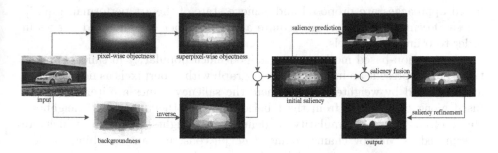

Fig. 1. Pipeline of the proposed algorithm.

The overview of the proposed algorithm is shown in Fig. 1. We first over-segment an image into superpixels. Based on boundary prior, we calculate the backgroundness for each superpixel. Along with the objectness, which can accurately indicate the location of object, we compute the initial saliency map, thereby obtaining the foreground and background seeds. Then, we predict the saliency labels by using logistic regression combined with the graph-based propagation, which transmits the saliency information from the labeled superpixels to the unlabeled ones. Next, we fuse the predicted result and the initial map. In order to further uniformly and completely highlight the salient object, we finally refine the fused result using the multi-scale saliency cues. The contributions of this paper are summarized as follows:

- We present a novel saliency prediction approach by incorporating logistic regression and label propagation into a unified framework.
- We learn pairwise affinities to depict the grouping cues of superpixels for saliency prediction.

2   Related Work

In the past decades, numerous approaches have been proposed to detect salient objects or predict human fixations. We refer the readers to recent survey papers [2] for detailed discussions and quantitative analysis. In this section, we mainly focus on the computational approaches on salient object detection.

Based on the early cognitive studies of visual search [3], Itti *et al.* [4] calculate the center-surround contrast over a set of low-level image features to define image saliency. While Goferman *et al.* [5] combine the feature difference at multiple scales to comprehensively evaluate the local and global context of the dominant objects. In a number of literature, sparse reconstruction and low-rank matrix decomposition are used to depict pattern rarity of salient regions. They consider that the regions with high encoding residual may stand out from image backgrounds.

Supervised learning is also applied in saliency detection, which learns to distinguish salient regions from the background. Recently, many deep learning based approaches are proposed and achieve state-of-the-art performance [6,7]. These aforementioned methods require a large number of annotated images in order to train their models.

Propagation-based methods have gained much popularity in saliency detection. They represent an input image as a graph with superpixels as nodes, which are connected by weighted edges. Then, the saliency values are iteratively diffused along these edges from the labeled nodes to their unlabeled neighbors. Harel *et al.* [8] use dissimilarity to define edge weights on graphs which are interpreted as Markov chains. Jiang *et al.* [9] construct an absorbing Markov chain on the image graph model and use the absorbed time to measure the saliency. While some other approaches exploit the graphical model to integrate multiple saliency cues.

In this work, we incorporate the regression learning and label propagation into a unified framework. We exploit the semi-supervised learning mechanism of the propagation model to train a logistic regressor for saliency prediction.

3   The Proposed Algorithm

In this section, we detail the proposed method, which mainly includes affinity learning, seed selection, saliency prediction and saliency refinement.

3.1   Affinity Learning

A good affinity metric in feature space is crucial. Motivated by [10], we introduce unsupervised clustering random forest to infer robust pairwise similarity, thereby better capturing the underlying semantic structure among superpixels. Its main idea is to exploit the path sharing mechanism of sample pairs in the hierarchical structure of the forest to define similarity. A sample pair is considered dissimilar if they are split at the very beginning, and greatly similar if they travel together passing the same set of internal nodes till the identical leaf node.

Specifically, given a clustering tree, assume sample i goes through path $\mathcal{P}^i = \{\gamma, s_1^i, ..., s_k^i, ..., \ell^i\}$ from root node γ to leaf node ℓ^i. s_k^i denotes the k-th internal node traveled by node i, and its corresponding weight is w_k. This weight depicts the importance of node s_k. Thus, the similarity between sample pair (i, j) is defined as

$$a_{ij}^t = \frac{\sum_{k=1}^{\lambda} w_k}{\sum_{k=1}^{K} w_k}, \tag{1}$$

where $K = max(|\mathcal{P}^i|, |\mathcal{P}^i|) - 1$, and λ is the length of the common path shared by node i and j in the t-th clustering tree. We combine the pairwise similarity from multiple decision trees in the forest to get the final affinity as

$$a_{ij} = \frac{1}{T} \sum_{t=1}^{T} a_{ij}^t. \tag{2}$$

Please refer [10] for more details. The affinity matrix implicitly contains the notion of global distance metric and more accurately depicts the grouping cues of nodes.

It is well known that object regions usually present spatial compactness and appearance homogeneousness, while background regions are the opposite. This kind of visual organization cues are easy to be captured through a local-to-global propagation mechanism. Therefore, we further refine the resulting affinity model to highlight the interaction of neighboring nodes. Specifically, we only reserve the weights among the K-regular neighbors in the coordinate space and K nearest neighbors in the feature space, while the other weights are enforced to be zero.

3.2 Seed Selection

As a preliminary criterion of saliency detection, the quality of seeds directly influences the performance of seed-based solutions. Several seed mechanisms have been proposed in the literature. The image elements along the image border often are selected as background seeds as used in [11,12]. In this work, we use a purely computational method to determine the foreground seeds. We respectively compute the objectness and backgroundness prior maps, and combine them to obtain an initial saliency map. Then we binarize the initial map to get the foreground seeds.

Objectness. Alexe et al. [13] propose a generic objectness measure to quantify the likelihood of an image window containing an object. We extend the window-level objectness to the superpixel-level objectness. Assume that the input image is over-segmented into N superpixels $S = \{s_1, ..., s_N\}$. Let $p(w)$ denote the probability of window w containing an object, the objctness score of superpixel s_i is defined as:

$$O(s_i) = \frac{1}{|s_i|} \sum_{j \in s_i} p(j), \tag{3}$$

where $|s_i|$ is the number of the pixels that superpixel s_i contains, and $p(j) = \sum_{\{w|j \in w\}} p(w)$, which indicates the probability of pixel j belonging to an object.

Backgroundness. The backgroundness describes the possibility of an image element belonging to the image background. We take the boundary superpixels as the background seeds, and then iteratively compute the similarity with them for each superpixel. Let $B(s_i)$ denote the score of superpixel s_i, defined as:

$$B_{k+1}(s_i) = \sum_{j=1}^{N} a_{ij} B_k(s_j) \tag{4}$$

where a_{ij} denotes the learnt pairwise affinity computed by Eq. 2. and the subscript $k \in [0, K-1]$ indexes the number of iterations. Initially, $B_0(s_i) = 1$ if s_i is a background seed, and 0 otherwise. In order to strengthen the effect of the boundary superpixels, we enforce their scores to be 1 after each iteration. Finally, we normalize the vector B_K by min-max normalization.

Seed. We compute the initial saliency score for each superpixel as follows:

$$S_{in}(s) = O(s) \circ (1 - \overline{B}(s)), \tag{5}$$

where $\circ$ denotes the product operator, and $\overline{B}$ is the normalized backgroundness score. Then, we select the superpixels which scores are larger than a fixed threshold as the foreground seeds. While the boundary superpixels are still taken as the background seeds, similar to previous methods [9].

3.3   Saliency Prediction

We represent the image as a graph $G = (\mathcal{V}, \mathcal{E})$ with superpixels as the nodes $\mathcal{V}$, and the undirected edges $\mathcal{E}$ are weighted by the above-defined affinity a_{ij}. We combine logistic function with quadratic energy function to establish a unified model for salient object detection, which objective function is written as

$$\begin{aligned}
J(\boldsymbol{w}, b) = \sum_{i,j \in \mathcal{V}} a_{ij}(\hat{y}_i - \hat{y}_j)^2 + \kappa \sum_{i \in \mathcal{L}}(\hat{y}_i - y_i)^2 \\
- \lambda \sum_{i \in \mathcal{L}} \{y_i log(\hat{y}_i) + (1 - y_i)log[(1 - \hat{y}_i)]\},
\end{aligned} \tag{6}$$

where κ and λ are two balancing parameters. $\mathcal{L} \subset \mathcal{V}$ denotes the set of the labeled nodes (i.e. seeds). y_i is the initial label of node i, in which $y_i = 1$ if node i is foreground and $y_i = 0$ if it is background. $\hat{y}_i \in [0, 1]$ denotes the predicted label, which is regarded as posterior probability for the foreground class and computed by using logistic regression as

$$\hat{y}_i = \frac{exp(\boldsymbol{w}'\boldsymbol{x}_i + b)}{1 + exp(\boldsymbol{w}'\boldsymbol{x}_i + b)}, \tag{7}$$

where $\boldsymbol{x}_i$ denotes the feature vector of node i. $\boldsymbol{w}$ and b are the parameters of the logistic function.

In Eq. 6, the first term encodes the pairwise smoothness constraint, i.e., the labellings of nearby nodes should not change too much. The second term encodes the unary fitting constraint, i.e., the labellings of seed nodes should be similar to the initial assignments. The third term measures the classification deviations across the labeled nodes by negative log-likelihood as in logistic regression. We minimize the objective to obtain the parameters $\boldsymbol{w}$ and b for logistic regression, and then employ Eq. 7 to predict the labels of all nodes. Compared with logistic regression which separately estimates the label of each node without considering the intrinsic geometric structure of nodes, the unified model can prompt similar nodes on the image manifold have similar labels by using the smoothness constraint. The smoothness property is very important and essential in saliency detection, since we expect that salient object can be uniformly highlighted from the background. Moreover, by incorporating logistic regression classifier, we more accurately predict the labels than directly using quadratic energy function. We use the gradient descent method to minimize the cost function in Eq. 6, which derivative with respect to the parameters w and b are respectively written as, for all $j \in V$,

$$
\begin{aligned}
\frac{\partial J}{\partial w} = &[2\sum_{j \in V}(D_{ij} - a_{ij})\hat{y}_j - 2\kappa\sum_{j \in L}\hat{y}_j]\hat{y}_j(1 - \hat{y}_j)x_j \\
&- \lambda\sum_{j \in L}[y_j(1 - \hat{y}_j) + \hat{y}_j(1 - y_j)]x_j
\end{aligned}
\tag{8}
$$

and

$$
\begin{aligned}
\frac{\partial J}{\partial b} = &[2\sum_{j \in V}(D_{ij} - a_{ij})\hat{y}_j - 2\kappa\sum_{j \in L}\hat{y}_j]\hat{y}_j(1 - \hat{y}_j) \\
&- \lambda\sum_{j \in L}[y_j(1 - \hat{y}_j) + \hat{y}_j(1 - y_j)]
\end{aligned}
\tag{9}
$$

Since the cost function in Eq. 6 is not necessarily convex, we determine the initial w and b based on the solution by linear regression.

Fig. 2. Predicted saliency results by the regression-propagation model.

The predicted labels ranged from 0 to 1 are taken as the saliency values of superpixels. Some examples are shown in Fig. 2, from which we can see that although most of the background regions are adequately suppressed, the salient objects aren't uniformly highlighted for some images. That is, the logistic diffusion method tends to generate a precision-preference result. In contrast, the initial map is likely to coarsely detect salient objects, but lose fine details and global shapes. Hence, we further fuse the two results by multiplication. Some visual examples are shown in Fig. 3.

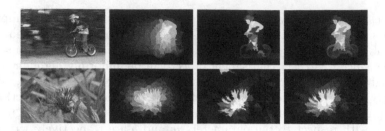

Fig. 3. Saliency fusion. From left to right: input, initial maps, predicted results by the regression-propagation model and fused results.

3.4 Multi-scale Saliency Refinement

As salient objects are likely to appear at different scales consisting of numerous perceptually heterogeneous regions in a scene, single quantization is difficult to obtain stable and reliable detection results. To this end, we further refine the saliency results by classifying the superpixels at multiple scales. Specifically, we first over-segment the input image into tri-level superpixels, and compute the average saliency value for each superpixel. Second, we set two thresholds to select the pseudo positive and negative samples to train a SVM classifier. Third, we use the learnt classifier to predict the saliency labels for all superpixels and respectively obtain a saliency map at each scale. Last, the three maps are combined by multiplication.

4   Experimental Results

We evaluate the proposed algorithm on several public datasets, including the ECSSD [14], PASCAL-S [15] and THUS [16] datasets. We compare the performance with sixteen state-of-the-art methods, including the MST [17], BL [18], RR [19], wCO [20], RCJ [16], DSR [21], GMR [22], HS [14], AMC [9], GC [23], UFO [24], GS [12], LR [25], CB [26], SVO [27], and GB [8] methods.

We set the number of superpixel nodes $N = 200$ in all the experiments. There are several parameters in the proposed algorithm: the threshold for selecting foreground seeds and the number of iteration steps for backgroundness computation. They are respectively set to be 0.8 and 30. In the multi-scale refinement stage, the numbers of the tri-level superpixels are set to be 150, 250 and 300, respectively. The parameters are empirically chosen based on the observation of experimental results.

The precision-recall curves of all the methods on the three benchmark datasets are demonstrated in Fig. 4. We can see that the proposed method consistently performs better than the other methods on the ECSSD and THUS datasets. Table 1 demonstrates the quantitative comparisons in terms of the F-measure and MAE scores. On the ECSSD and THUS datasets, the proposed algorithm consistently performs better than these competitors in terms of F-measure and MAE scores, and averagely achieve 2.4% and 5.5% improvement

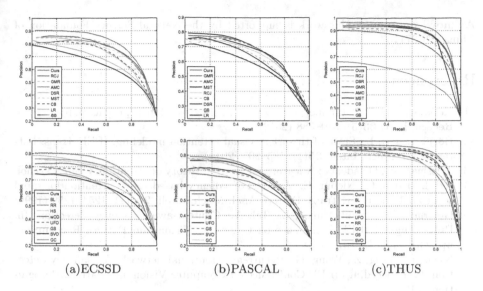

(a)ECSSD   (b)PASCAL   (c)THUS

Fig. 4. Precision-recall curves of different methods.

Table 1. Quantitative comparisons of different methods.

Dataset	Metric	Methods									
		CB	GS	UFO	RCJ	HS	BL	GMR	AMC	RR	Ours
ECSSD	F-m	0.627	0.606	0.643	0.681	0.636	0.683	0.693	**0.699**	0.698	0.724
	MAE	0.241	0.208	0.205	0.187	0.228	0.217	0.188	0.204	**0.185**	0.172
PASCAL	F-m	0.517	**0.622**	0.550	0.603	0.609	0.571	0.630	0.599	0.587	0.618
	MAE	0.270	0.224	0.232	**0.223**	0.264	0.249	0.232	0.232	0.231	0.215
THUS	F-m	0.762	0.785	0.813	0.816	0.804	0.822	0.834	**0.836**	0.818	0.850
	MAE	0.178	0.139	0.147	0.137	0.149	0.158	**0.126**	0.145	0.131	0.119

over the second best methods, respectively. On the PASCAL-S dataset, the proposed method performs best in terms of MAE score.

5   Conclusions

In this paper, we unify logistic regression and graph-based label inference to detect salient regions in images. And, we combine the precision-preference saliency map generated by the regression-propagation model and the recall-preference initial map. In order to further address the scale problem of salient objects, we train a classifier across multiple scales to refine the detection result. Extensive experiments on three large benchmark datasets demonstrate that the proposed method performs favorably against the state-of-the-art methods in terms of accuracy.

Acknowledgement. This work is supported by the Natural Science Foundation of China #61371157.

References

1. Kobayashi, T., Watanabe, K., Otsu, N.: Logistic label propagation. Pattern Recogn. Lett. **33**(5), 580–588 (2012)
2. Borji, A., Itti, L.: State-of-the-art in visual attention modeling. IEEE Trans. Pattern Anal. Mach. Intell. **35**(1), 185–207 (2013)
3. Triesman, A., Gelade, G.: A feature-integration theory of attention. Cogn. Psychol
4. Itti, L., Koch, C., Niebur, E.: A model of saliency-based visual attention for rapid scene analysis. IEEE Trans. Pattern Anal. Mach. Intell. **20**(11), 1254–1259 (1998)
5. Goferman, S., Zelnik-Manor, L., Tal, A.: Context-aware saliency detection. In: Proceedings of IEEE Conference on Computer Vision and Pattern Recognition, pp. 2376–2383 (2010)
6. Kuen, J., Wang, Z., Wang, G.: Recurrent attentional networks for saliency detection. In: Proceedings IEEE Conference on Computer Vision and Pattern Recognition (2016)
7. Kim, J., Pavlovic, V.: A shape-based approach for salient object detection using deep learning. In: Proceedings of European Conference on Computer Vision, pp. 455–470 (2016)
8. Harel, J., Koch, C., Perona, P.: Graph-based visual saliency. In: Proceedings of Neural Information Processing Systems, pp. 545–552 (2006)
9. Jiang, B., Zhang, L., Lu, H., Yang, C., Yang, M.-H.: Saliency detection via absorbing Markov chain. In: Proceedings of IEEE International Conference on Computer Vision, pp. 1665–1672 (2013)
10. Zhu, X., Loy, C.C., Gong, S.: Constructing robust affinity graphs for spectral clustering. In: Proceedings IEEE Conference on Computer Vision and Pattern Recognition, pp. 1450–1457 (2014)
11. Grady, L., Jolly, M., Seitz, A.: Segmentation from a box. In: Proceedings of IEEE International Conference on Computer Vision, pp. 367–374 (2011)
12. Wei, Y., Wen, F., Zhu, W., Sun, J.: Geodesic saliency using background priors. In: Proceedings of European Conference on Computer (2012)
13. Alexe, B., Deselaers, T., Ferrari, V.: Measuring the objectness of image windows. IEEE Trans. Pattern Anal. Mach. Intell. **34**(11), 2189–2202 (2012)
14. Yan, Q., Xu, L., Shi, J., Jia, J.: Hierarchical saliency detection. In: Proceedings IEEE Conference on Computer Vision and Pattern Recognition, pp. 1155–1162 (2013)
15. Li, Y., Hou, X., Koch, C., Rehg, J.M., Yuille, A.L.: The secrets of salient object segmentation. In: Proceedings of IEEE Conference on Computer Vision and Pattern Recognition, pp. 280–287 (2014)
16. Cheng, M., Mitra, N., Huang, X., Torr, P., Hu, S.-M.: Global contrast based salient region detection. IEEE Trans. Pattern Anal. Mach. Intell. **37**(3), 569–582 (2015)
17. Tu, W.-C., He, S., Yang, Q., Chien, S.-Y.: Real-time salient object detection with a minimum spanning tree. In: Proceedings of IEEE Conference on Computer Vision and Pattern Recognition (2016)
18. Tong, N., Lu, H., Ruan, X., Yang, M.-H.: Salient object detection via bootstrap learning. In: Proceedings of IEEE Conference on Computer Vision and Pattern Recognition, pp. 1884–1892 (2015)

19. Li, C., Yuan, Y., Cai, W., Xia, Y., Feng, D.D.: Robust saliency detection via regularized random walks ranking. In: Proceedings IEEE Conference on Computer Vision and Pattern Recognition, pp. 2710–2717 (2015)
20. Zhu, W., Liang, S., Wei, Y., Sun, J.: Saliency optimization from robust background detection. In: Proceedings of IEEE Conference on Computer Vision and Pattern Recognition, pp. 2814–2821 (2014)
21. Li, X., Lu, H., Zhang, L., Ruan, X., Yang, M.-H.: Saliency detection via dense and sparse reconstruction. In: Proceedings of IEEE International Conference on Computer Vision, pp. 2976–2983 (2013)
22. Yang, C., Zhang, L., Lu, H., Ruan, X., Yang, M.-H.: Saliency detection via graph-based manifold ranking. In: Proceedings of IEEE Conference on Computer Vision and Pattern Recognition, pp. 3166–3173 (2013)
23. Cheng, M., Warrell, J., Lin, W., Zheng, S., Vineet, V., Crook, N.: Efficient salient region detection with soft image abstraction. In: Proceedings of IEEE International Conference on Computer Vision, pp. 1529–1536 (2013)
24. Jiang, P., Ling, H., Yu, J., Peng, J.: Salient region detection by UFO: uniqueness, focusness and objectness. In: Proceedings of IEEE International Conference on Computer Vision, pp. 1976–1983 (2013)
25. Shen, X., Wu, Y.: A unified approach to salient object detection via low rank matrix recovery. In: Proceedings of IEEE Conference on Computer Vision and Pattern Recognition, pp. 853–860 (2012)
26. Jiang, H., Wang, J., Yuan, Z., Liu, T., Zheng, N., Li, S.: Automatic salient object segmentation based on context and shape prior. In: Proceedings of British Machine Vision Conference (2011)
27. Chang, K.-Y., Liu, T.-L., Chen, H.-T., Lai, S.-H.: Fusing generic objectness and visual saliency for salient object detection. In: Proceedings of IEEE International Conference on Computer Vision, pp. 914–921 (2011)

Classification and Clustering

Classification and Clustering

Classification and Clustering via Structure-enforced Matrix Factorization

Lijun Xu[1]([✉]), Yijia Zhou[2], and Bo Yu[1]

[1] Dalian University of Technology, Dalian 116024, Liaoning,
People's Republic of China
xulijundlut@gmail.com, yubo@dlut.edu.cn
[2] Dalian Neusoft University of Information, Dalian 116023, Liaoning,
People's Republic of China
zhouyijia@neusoft.edu.cn

Abstract. In this paper, we present a new classification and clustering framework via structure-enforced matrix factorization which represents a large class of mathematical models appearing in many applications. We are factorizing the data as representations of several blocks, one block for each cluster or class. The signals which are best reconstructed by the same block are clustered together. The proposed framework additionally imposes incoherence in the blocks since an incoherence promoting term encourages blocks associated to different classes to be as independent as possible. In addition, the new framework is applicable both to supervised and unsupervised learning. We first illustrate the proposed framework for the MNIST digit dataset in the supervised and unsupervised way, respectively, obtaining results comparable to the state-of-the-art. We then present experiments for fully unsupervised clustering on simple texture images for instance, also yielding excellent performance.

Keywords: Matrix factorization · Alternating direction method · Classification · Unsupervised clustering · Dictionary learning

1 Introduction

Classification and clustering are main task of exploratory data mining, and common techniques for statistical data analysis, used in many fields, including machine learning, pattern recognition, image analysis, information retrieval, bioinformatics, data compression, and computer graphics. In machine learning and statistics, classification is the problem of identifying to which of a set of categories a new observation belongs, on the basis of a training set of data containing observations (or instances) whose category membership is known. Classification is also terminologically considered an instance of supervised learning in machine learning, i.e. learning where a training set of correctly identified observations is available. The corresponding unsupervised procedure is known as clustering, and involves grouping data into categories based on some measure of inherent similarity or distance. Here, we do not discuss and review various models

© Springer International Publishing AG 2017
Y. Sun et al. (Eds.): IScIDE 2017, LNCS 10559, pp. 403–411, 2017.
DOI: 10.1007/978-3-319-67777-4_35

and algorithms for classification and clustering in numerous applications. In this paper, motivated by the framework via dictionary learning with structured incoherence and shared features in [1], we focus on how to utilize the technique of matrix factorization with imposing structures to do classification and clustering.

In the work [2], the structure-enforced matrix factorization (SeMF) model is studied which is a general and unifying mathematical model encompassing numerous problem classes arising from diverse application backgrounds. For instance, dictionary learning for sparse representation can be formulated as a SeMF problem. Thus, the clustering framework via dictionary learning in [1] could be extended to structure-enforced matrix factorization. In [1], given K clusters, ones learn K dictionaries for representing the data, and then associate each signal to the dictionary for which the best sparse decomposition is obtained. A single dictionary (block) is selected per data point, and the point is sparsely represented (subspace) with atoms only from this dictionary. Specifically, the dictionary learning model in [1] is

$$\min_{D_i, C_i} \sum_{i=1}^{K} \sum_{x_j \in C_i} \mathcal{R}(x_j, D_i) + \eta \sum_{i \neq j} \mathcal{Q}(D_i, D_j), \tag{1}$$

where $D_i = [d_1 \ d_2 \cdots d_{k_i}] \in \mathbb{R}^{n \times k_i}$ is a dictionary of k_i atoms associated with the class C_i, $x_j \in \mathbb{R}^n$ are the data vectors, $\mathcal{R}$ is a function that measures how good the sparse decomposition for the signal x_j under the dictionary D_i is, and $\mathcal{Q}$ is a block (dictionary) incoherence term that promotes incoherence between the different blocks (dictionaries). The measurement $\mathcal{R}$ naturally takes into account both the reconstruction error and the sparseness of the representation on the corresponding learned dictionary. Such measurement can be applied to image patches directly or to image features and this measurement has shown enormous discrimination power in practice (see [3–10] for example).

Inspired in part by their work in [1,11], we rebuild the clustering model as the structure-enforced matrix factorization problem. We concatenate the K dictionaries of K clusters as a whole large dictionary matrix $D = [D_1 \ D_2 \cdots D_K]$ for representing the data, and allow each data point to be "best" sparsely represented only by one block (dictionary). Let $X = [x_1 \ x_2 \cdots x_m] \in \mathbb{R}^{n \times m}$ be the data matrix, each column of which is the signal vector, and $A = [a_1 \ a_2 \cdots a_m]$ be the coefficient matrix representing signals over the whole dictionary D. Refer to the earlier work on SeMF in [2], the original SeMF model is

$$\min_{D, A} \|X - DA\|_F^2 \text{ s.t. } D \in \mathcal{D}, A \in \mathcal{A}. \tag{2}$$

where $\| \cdot \|_F$ is Frobenius norm, $\mathcal{D}$ and $\mathcal{A}$ are structured subsets of $\mathbb{R}^{n \times p}$ and $\mathbb{R}^{p \times m}$, respectively. Common structures imposed on $\mathcal{D}$ include but not limited to normalization, nonnegativity, orthogonality, etc. And $\mathcal{A}$ usually can be enforced some specific sparsity (i.e. group sparsity) and nonnegativity and so on. In this paper, we consider to impose incoherence in the blocks of $\mathcal{D}$ to obtain an extended model based on (2). Although we use the same incoherence promoting term (that is, $\mathcal{Q}(D_i, D_j) = \|D_i^T D_j\|_F^2$) as one in [1], the novel model and algorithm we proposed are different from that in [1]. We then extend the classic alternating direction method of multipliers (ADMM or simply ADM) to the variant of SeMF and utilize adaptive penalty parameter updating strategy to enhance stability and accelerationbe when solving the proposed model. In addition, similar to dictionary model in [1], our proposed model can also be rendered to the case of unsupervised or semi-supervised classification and clustering since imposed structures are independent of the data and intrinsic to the factorized matrices.

In Sect. 2 we introduce the structure-enforced matrix factorization designed for classification and clustering and propose an ADMM-based algorithm for the new model. In Sect. 3, we implement our model on supervised and unsupervised classification for MNIST digit dataset and present experiments on fully unsupervised clustering for several simple texture images. Finally, we conclude this paper in Sect. 4.

2 Structure-enforced Matrix Factorization for Classification and Clustering

2.1 Extended SeMF Model for Classification and Clustering

The general structured-enforced matrix factorization (SeMF) model (2) is firstly proposed in the earlier work in [2], which is considered as an approximate and structured matrix factorization. In model (2), the objective function measures data fidelity in a least-square sense, which is the most popular metric for data fidelity although other measures are frequently used as well. In this model, prior knowledge are explicitly enforced to two constraint sets $\mathcal{D}$ and $\mathcal{A}$ whose members possess desirable matrix structures. The most useful structures of this kind include nonnegativity and diverse sparsity. We note that in the literature unconstrained optimization models are widely used where prior knowledge are handled through penalty or regularization functions added to the data fidelity term so that a weighted sum of the two is to be simultaneously minimized. In unconstrained optimization models, it is generally the case that desired structures are encouraged or promoted, but not exactly enforced to the constraint. Obviously, both types of formulations have their distinct advantages and disadvantages under different circumstances.

As claimed in [2], the original SeMF model is applicable to a range of constraint sets that are *easily projectable*. In many practically relevant applications constraint sets are indeed easily projectable. However, there still exist sets with some structures that do not possess *easily projectable* but are really intrinsic and desired for the data. Therefore, we combine SeMF with the approach promoting desired structures in unconstrained optimization. The obtained variant is more rich and versatile to model practical problems. For classification and clustering, we encourage the incoherence in the blocks D_i of $D \in \mathcal{D}$, which is difficult to do projection. Here we add the term $\mathcal{Q}(D_i, D_j) = \|D_i^T D_j\|_F^2$ to the classic fidelity referring to [1]. Thereby the extended SeMF model becomes

$$\min_{D,A} \|X - DA\|_F^2 + \eta \sum_{i \neq j} \|D_i^T D_j\|_F^2 \text{ s.t. } D \in \mathcal{D}, A \in \mathcal{A}, \tag{3}$$

where $D = [D_1\ D_2 \cdots D_K]$ is K dictionaries (or blocks) corresponding to K clusters. This model can lead to the learning of blocks which are optimized to represent the corresponding classes (or clusters) properly.

2.2 Alternating Direction Algorithm for the Proposed SeMF

As introduced in the work [2], to facilitate an efficient use of alternating minimization, we introduce two auxiliary variables U and V and consider the following model equivalent to (3),

$$\min_{D,A,U,V} \frac{1}{2} \|X - DA\|_F^2 + \frac{\eta}{2} \sum_{i \neq j} \|U_i^T U_j\|_F^2 \tag{4}$$

$$\text{s.t. } D - U = 0, A - V = 0, U \in \mathcal{D}, V \in \mathcal{A},$$

where $U = [U_1\ U_2 \cdots U_K] \in \mathbb{R}^{n \times p}$ and $V \in \mathbb{R}^{p \times m}$. Obviously, U possesses the same group information with D. The augmented Lagrangian function of (4) is

$$\mathcal{L}_A(D, A, U, V, \Lambda, \Pi) = \frac{1}{2} \|X - DA\|_F^2 + \frac{\eta}{2} \sum_{i \neq j} \|U_i^T U_j\|_F^2$$

$$+ \Lambda \bullet (D - U) + \Pi \bullet (A - V) \tag{5}$$

$$+ \frac{\alpha}{2} \|D - U\|_F^2 + \frac{\beta}{2} \|A - V\|_F^2,$$

where $\Lambda \in \mathbb{R}^{n \times p}$, $\Pi \in \mathbb{R}^{p \times m}$ are Lagrangian multipliers and $\alpha, \beta > 0$ are penalty parameters for the constraints $D - U = 0$ and $A - V = 0$, respectively. Note that the scalar product "$\bullet$" of two equal-size matrices Y and Z is the sum of all element-wise products, i.e., $Y \bullet Z = \sum_{i,j} y_{ij} z_{ij}$.

The alternating direction method of multiplier (ADMM) [12,13] for (4) is derived by successively minimizing the augmented Lagrangian function $\mathcal{L}_A$ with respect to D, A and (U, V), one at a time while fixing others at their most recent values, and then updating the multipliers after each sweep of such alternating minimization. The introduction of the two auxiliary variables U and V makes it easy to carry out each of the alternating minimization steps. Specifically, these steps can be written in the following forms,

$$D_+ \approx \underset{D}{\text{argmin}}\, \mathcal{L}_A(D, A, U, V, \Lambda, \Pi), \tag{6a}$$

$$A_+ \approx \underset{A}{\text{argmin}}\, \mathcal{L}_A(D_+, A, U, V, \Lambda, \Pi), \tag{6b}$$

$$U_+ \approx \underset{U \in \mathcal{D}}{\text{argmin}}\, \mathcal{L}_A(D_+, A_+, U, V, \Lambda, \Pi), \tag{6c}$$

$$V_+ \approx \operatorname*{argmin}_{V \in \mathcal{A}} \mathcal{L}_A(D_+, A_+, U_+, V, \Lambda, \Pi), \tag{6d}$$

$$\Lambda_+ = \Lambda + \alpha(D_+ - U_+), \tag{6e}$$

$$\Pi_+ = \Pi + \beta(A_+ - V_+), \tag{6f}$$

where the subscript "+" is used to denote iterative values at the new iteration. Actually, we can write (6a) (6b) and (6d) exactly in closed forms,

$$D_+ = (XA^T + \alpha U - \Lambda)(AA^T + \alpha I)^{-1}, \tag{7a}$$

$$A_+ = (D_+^T D_+ + \beta I)^{-1}(D_+^T X + \beta V - \Pi). \tag{7b}$$

$$V_+ = \mathcal{P}_\mathcal{A}(A_+ + \Pi/\beta), \tag{7c}$$

where $\mathcal{P}_\mathcal{A}$ stands for the projection onto the set $\mathcal{A}$ in Frobenius norm. Since the augmented Lagrangian function (5) is block-wise coupled over U in terms of $\|U_i^T U_j\|_F^2$, the U-updating (6c) can be approximately obtained by successively updating each block U_i, $i = 1, ..., K$. That is,

$$
\begin{aligned}
U_{i+} &\approx \operatorname*{argmin}_{U_i \in \mathcal{D}_i} \mathcal{L}_A(D_+, A_+, U, V, \Lambda, \Pi) \\
&= \operatorname*{argmin}_{U_i \in \mathcal{D}_i} \frac{\eta}{2} \sum_{j \neq i} \|U_i^T U_j\|_F^2 + \Lambda_i \bullet (D_{i+} - U_i) + \frac{\alpha}{2} \|D_{i+} - U_i\|_F^2 \\
&\approx \mathcal{P}_\mathcal{D}\left((I + \eta/\alpha \sum_{j \neq i} U_j U_j^T)^{-1}(D_{i+} + \Lambda_{i+}/\alpha) \right),
\end{aligned}
\tag{8}
$$

where $\mathcal{P}_\mathcal{D}$ stands for the projection onto the set $\mathcal{D}$ in Frobenius norm. Based on the formulas in (7) and (8), we can implement ADMM algorithmic framework to solve our model. Moreover, we can adopt adaptive penalty parameter updating rule to accelerate the algorithm and improve the quality of solution. Here we omit this part since the algorithms in [2] can be directly applied to our proposed framework.

3   Experimental Results

This section contains two sets of numerical experiments. In Sect. 3.1, we apply the proposed model to MNIST dataset in supervised and unsupervised learning settings, respectively, and illustrate the comparable results compared to the state-of-the-art. In Sect. 3.2, we apply our model for the texture segmentation problem of which the goal is to assign each pixel on an objective image to one of K possible textures and obtain very low rates of miss-clustered pixels.

3.1   Classification for MNIST Datasets

In this section, a handwritten digit database MNIST is used to evaluate the classification performance. The MNIST is one of the widely used database in machine learning community which consists of 10 classes ("0–9") with 60000 dataset for training and 10000 dataset for testing, respectively. The dimension

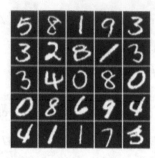

Fig. 1. An example of MNIST handwritten digit dataset.

of MNIST digit image is centered in a fixed-size 28×28 grayscale image field as an example illustrated in Fig. 1.

Since the image size of the MNIST is 28×28 pixels, the input vector dimension is $\mathbb{R}^{784}$. From Fig. 1, it's natural to consider that digit images from one class are more similar compared with ones from other classes and possess their own representing basis. Thus we can factorize the dataset using the proposed model to learn a block-wise basis D for 10 classed digits, one block for each digital class.

Firstly, we test our algorithm for supervised classification. It is easy to implement our model (3) to the labeled training dataset since the known labels of digit images impose explicit structures on A that obviously permit *easy projections*. Actually, each column of A is sparse over one block D_i associated with the label of the corresponding digit image. After we obtain the block-wise matrix D learned from the training dataset, we can classify the testing dataset using our model with fixed D and U which is reduced to a general sparse coding problem. In this experiment, we use penalty parameter $\eta = 0.1$ in incoherence promoting term, set $k_i = 100$ as the size of each dictionary (or block) and follow other settings for algorithms in [2]. Table 1 shows the misclassification rate of proposed method and lists results of several other classification algorithms. Although it does not lead to the best classification rate, the result obtained by our model is comparable and desirable. In addition, it should be evidence that imposing incoherence term in our model can lead to a more discriminative classifier.

Table 1. Error rate (in percentage) of supervised classification by different algorithms

Dataset	SeMF ($\eta = 0.1$)	SeMF ($\eta = 0$)	[1]	[14]	[4]	SVM	k-NN
MNIST	1.09	1.21	1.26	3.41	1.05	1.4	5.0

Next, we apply our algorithm directly to the testing data in an unsupervised way. We opt to cluster the digits from 0 to 5 ($K = 6$) which follows the experiment in [1]. For our model, we choose the size of basis matrix are $p = 150$, in the other word, the block representing each cluster is of size $k_i = 25$. The structure enforced on A follows that in supervised case, that is, each column of A is sparse and nonzeros must be mainly located in one block corresponding certain D_i. Then, it is straightforward to label each test image according to block sparsity property in the obtained representation A. As can be see in Table 2, we obtain comparable classification rate compared with method in [1]. Note that our model takes almost less than half time since the size of initial dictionary used in [1] is 300 and in their model it needs to do sparse coding on every dictionaries (blocks) for all testing images every time before refining the set of dictionaries (10 times refinements is set in [1]).

Table 2. Error rate (in percentage) and running time (in second) for unsupervised methods on the MNIST test digits from 0 to 5.

	SeMF ($\eta \neq 0$)	SeMF ($\eta = 0$)	[1] ($\eta \neq 0$)	[1] ($\eta = 0$)
Error rate	3.2	7.2	3.0	6.9
Time	738.21	540.95	1447.59	1236.36

Through testing our algorithm to the supervised and unsupervised classification for MNIST dataset, we can conclude that encouraging incoherence in the dictionaries (or blocks) is of paramount importance. As claimed in [2], our SeMF model is versatile in terms of various enforced structures associated with specific applications. Next we illustrate our algorithm for the texture segmentation problem.

3.2   Unsupervised Clustering for Texture Segmentation

Texture segmentation deals with identification of regions where distinct textures exist. In other words, the goal is to assign each pixel or patch on an objective image to one of K possible textures. It is a natural application of our framework, since it can be formulated as a local feature extraction and patch classification process. We choose to evaluate our method on the Brodatz dataset introduced in [15]. Overlapped 16×16 patches were extracted from the original images and used as input signals for our model. After decomposing the input signal matrix using our SeMF algorithm, we cluster each patch according to the representation matrix A. Then we employ the techniques in [1,4,11] to obtain the pixel-wise classification.

In Fig. 2, we show some of the results. The comparable rates of misclassified pixels are obtained by our method. For example, we obtain 0.31% (top row) and 1.55% (bottom row) competitive with 0.25% and 1.75% in [1] which is one of the state-of-the-art unsupervised clustering algorithms.

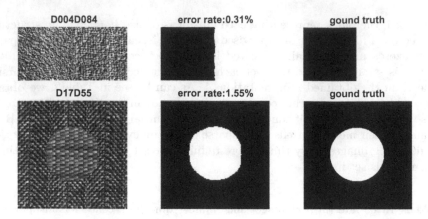

Fig. 2. Texture segmentation results on two sets (D004D084 and D17D55) in the Brodatz database.

4   Conclusion and Future

We have introduced a novel framework for classification and clustering based on the frame of matrix factorization. The proposed framework impose incoherence between atoms from different classes and enforce the special sparsity to coefficient matrix. These structures actually come from intrinsic properties of the problems. The obtained model can be equivalently transformed to an ADMM-applicable model which is easy implemented. We further speed up and stabilize our algorithm referring to the adaptive penalty parameter updating rule in [2]. Numerical experiments also show the applicable of our model for clustering problems.

Although structured matrix factorization problems are generally highly non-convex, they widely and variously exist in real-world applications. Many problems could be considered as a structured-enforced matrix factorization or its variants, such as wave separation, face clustering, principal component analysis. In future, we will explore more experiments in related fields.

References

1. Ramirez, I., Sprechmann, P., Sapiro, G.: Classification and clustering via dictionary learning with structured incoherence and shared features. In: 2010 IEEE Computer Society Conference on Computer Vision and Pattern Recognition, San Francisco, CA, pp. 3501–3508 (2010)
2. Xu, L., Yu, B., Zhang, Y.: An alternating direction and projection algorithm for structure-enforced matrix factorization. Comput. Optim. Appl. (2017). doi:10.1007/s10589-017-9913-x
3. Lee, H., Battle, A., Raina, R., Ng, A.Y.: Efficient sparse coding algorithms. In: NIPS, vol. 19 (2007)
4. Mairal, J., Bach, F., Ponce, J., Sapiro, G., Zisserman, A.: Discriminative learned dictionaries for local image analysis. In: CVPR (2008)

5. Peyré, G.: Sparse modeling of textures. J. Math. Imaging Vis. **34**, 17–31 (2009)
6. Yang, J., Yu, K., Gong, Y., Huang, T.: Linear spatial pyramid matching using sparse coding for image classification. In: CVPR (2009)
7. Yuan, J., Wang, D., Cheriyadat, A.M.: Factorization-based texture segmentation. IEEE Trans. Image Process. **24**(11), 3488–3497 (2015)
8. Zdunek, R., Phan, A.H., Cichocki, A.: Image classification with nonnegative matrix factorization based on spectral projected gradient. In: Koprinkova-Hristova, P., Mladenov, V., Kasabov, N.K. (eds.) Artificial Neural Networks: Methods and Applications in Bio-/Neuroinformatics. SSB, vol. 4, pp. 31–50. Springer, Cham (2015). doi:10.1007/978-3-319-09903-3_2
9. Bampis, C.G., Maragos, P., Bovik, A.C.: Projective non-negative matrix factorization for unsupervised graph clustering. In: 2016 IEEE International Conference on Image Processing (ICIP), Phoenix, AZ, pp. 1255–1258 (2016)
10. Wang, Y., Wu, S., Mao, B., Zhang, X., Luo, Z.: Correntropy induced metric based graph regularized non-negative matrix factorization. Neurocomputing **204**, 172–182 (2016)
11. Sprechmann, P., Sapiro, G.: Dictionary learning and sparse coding for unsupervised clustering. In: Proceedings of ICASSP (2010)
12. Glowinski, R., Marroco, A.: Sur lapproximation, par elements finis dordre un, et la resolution, par penalisation-dualite dune classe de problemes de dirichlet non lineaires. Revue francaise dautomatique, informatique, recherche operationnelle. Analyse numerique **9**(2), 41–76 (1975)
13. Gabay, D., Mercier, B.: A dual algorithm for the solution of nonlinear variational problems via finite element approximation. Comput. Math. Appl. **2**(1), 17–40 (1976)
14. Mairal, J., Bach, F., Ponce, J., Sapiro, G., Zisserman, A.: Supervised dictionary learning. In: NIPS, vol. 21, pp. 1033–1040 (2009)
15. Randen, T., Husøy, J.H.: Filtering for texture classification: a comparative study. IEEE Trans. Pattern Anal. Mach. Intell. **21**(4), 291–310 (1999)

Reweighted Sparse Subspace Clustering Based on Fractional-Order Function

Yiqiang Zhai and Zexuan Ji[(⊠)]

School of Computer Science and Engineering,
Nanjing University of Science and Technology, Nanjing 210094, China
jizexuan@njust.edu.cn

Abstract. Sparse subspace clustering (SSC) achieves state-of-the-art clustering performance via solving a ℓ_1 minimization problem which is a convex relaxation of ℓ_0 minimization. In this paper, we propose a unified fractional-order function based reweighted ℓ_1 minimization framework, which can approximate ℓ_0 norm better than ℓ_1 norm and reweighted ℓ_1 minimization framework. Based on the unified framework, a fractional-order function is introduced to reweight the sparse subspace clustering algorithm (FRSSC) to further improve the sparsity representation of data. By imposing constraints on coefficient matrix, the proposed weights are embedded into the sparse formulation to obtain the sparsest representation in each iteration. Experimental results demonstrate the advantage of FRSSC over state-of-the-art methods.

Keywords: Sparse subspace clustering · Fractional-order function · Reweighted ℓ_1 minimization · Spectral clustering

1 Introduction

As the basic assumption of manifold learning, the high-dimensional data usually distributes near low-dimensional manifolds. Subspace clustering [1] refers to the problem of clustering a set of high-dimensional data samples drawn from multiple linear subspaces to their respective subspaces while recognizing possible noises and outliers simultaneously [2], which has been widely utilized in machine learning, data mining, image processing and computer vision, including image segmentation [3], motion segmentation and face clustering. According to the mathematical framework, existing subspace clustering algorithms can be mainly divided into four categories: iterative, algebraic, statistical, and spectral clustering-based methods [2, 4–6].

As an important approach of spectral clustering based method, sparse subspace clustering (SSC), proposed by Elhamifar [2], solves the clustering problem by seeking a sparse representation of data points. Later, Elhamifar proposed a robust version of SSC [4] to deal with noises and corruptions or missing entries. The sparse representation problem uses ℓ_0 norm to capture the subspace, which is, however, a NP-hard problem. SSC uses the sparsest representation produced by ℓ_1 minimization to construct an affinity matrix of an undirected graph. Then subspace clustering is performed by spectral clustering algorithms such as the Normalized Cuts (Ncut) [3]. However, the affinity matrix obtained by SSC is not sparse enough. In addition, the ℓ_1 norm is based

© Springer International Publishing AG 2017
Y. Sun et al. (Eds.): IScIDE 2017, LNCS 10559, pp. 412–422, 2017.
DOI: 10.1007/978-3-319-67777-4_36

on the one-dimensional sparsity of vectors to represent the data, so SSC finds the sparsest representation of each data vector individually and there is no global constraint on its solution.

Liu et al. [5] proposed low rank representation (LRR) based on the two-dimensional sparsity of matrices, which is another recently proposed spectral clustering based method for subspace clustering. LRR seeks a low-rank representation of the data over the data itself, which can capture the global structure of the data. Essentially, instead of using the ℓ_1-norm, Liu et al. [6] used the nuclear norm to serve as a convex relaxation of the rank function. Lu et al. proposed a method based on least squares regression (LSR) [7] that took the advantage of data correlation and tended to group highly correlated data together. They analyzed the grouping effect of representation deeply and proposed the smooth representation (SMR) [8]. By combining SSC with LRR, Wang et al. [9] proposed Low-Rank Sparse Subspace Clustering (LRSSC), which takes the sparsity and the global low-rank structure of the data into consideration simultaneously. Recently, in order to obtain a sparser representation, Xu et al. [10] proposed a reweighted sparse subspace clustering (RSSC) by introducing a reweighted ℓ_1 minimization framework into SSC to replace the original ℓ_1 minimization problem.

SSC algorithm uses ℓ_1 norm to approximate ℓ_0 norm to represent the data, so the representation is not sparse enough. RSSC seek for a sparser representation by introducing a reweighted ℓ_1 minimization framework. However, the log-sum penalty used in RSSC cannot approximate ℓ_0 norm well. Inspired by the above observations, our goal is to seek for a sparser representation than RSSC. In our work, we firstly propose a fractional-order function based reweighted ℓ_1 minimization framework, which can approximate ℓ_0 norm better. Next, we propose a novel method, called the fractional-order based reweighted sparse subspace clustering (FRSSC) to address the subspace clustering problem. Given a set of data vectors drawn from a union of multiple subspaces, FRSSC finds the sparsest representation of data. The sparsest representation can be used to define an affinity matrix of an undirected graph, and then the final clustering results can be obtained by spectral clustering. FRSSC can be applied into many subspace clustering problems, such as motion segmentation and face clustering. We finally demonstrate our FRSSC in three real-world datasets: Hopkins155 dataset [11], the Freiburg-Berkeley Motion Segmentation dataset [12, 13] and the Extended Yale B dataset [14].

2 Reweighted Sparse Subspace Clustering Based on Fractional-Order Function

2.1 Reweighted l_1 Minimization Framework Based on Fractional-Order Function

Majorization-Minimization (MM) algorithms are more general than EM algorithms and work by iteratively minimizing a simple surrogate function majorizing a given objective function. Candès [15] gave the general form of MM algorithms:

$$\min_{v} g(v) \, s.t. \, v \in C, \tag{1}$$

where C is a convex set. One can improve on a guess v at the solution by minimizing a linearization of g around v. Starting with $v^{(0)} \in C$, we can define the following MM algorithm [1]:

$$v^{(\ell+1)} = \text{argmin} \, g\left(v^{(\ell)}\right) + \nabla g\left(v^{(\ell)}\right) \cdot \left(v - v^{(\ell)}\right) s.t. \, v \in C. \tag{2}$$

Each iterate is now the solution to a convex optimization problem. Candès [15] proposed a reweighted ℓ_1 minimization framework based on log-sum function, the problem can be defined as follows:

$$\min_{x \in R^n} \sum_{i=1}^{n} \log(|x_i + \varepsilon|) \, s.t. \, y = \Phi x, \tag{3}$$

which actually equals to

$$\min_{x,u \in R^n} \sum_{i=1}^{n} \log(u_i + \varepsilon) \, s.t. \, y = \Phi x, |x_i| \leq u_i, \quad i = 1, \cdots, n. \tag{4}$$

In other words, if x^* is a solution to Eq. (3), then (x^*, u^*) is a solution to (4), and vice versa. The surrogate function $\sum_{i=1}^{n} \log(u_i + \varepsilon)$ is concave, and is smooth on any convex constraints. So the problem can be locally minimized via an iterative linearization method. The problem in Eq. (4) satisfies the general problem in Eq. (1) and it can be derived by Eq. (2):

$$\sum_{i=1}^{n} \log\left(u_i^{(\ell+1)} + \varepsilon\right) \approx \sum_{i=1}^{n} \log\left(u_i^{(\ell)} + \varepsilon\right) + \sum_{i=1}^{n} \frac{1}{u_i^{(\ell)} + \varepsilon}\left(u_i - u_i^{(\ell)}\right). \tag{5}$$

After the deduction, we can obtain

$$\left(x^{(\ell+1)}, u^{(\ell+1)}\right) = \text{argmin} \sum_{i=1}^{n} \frac{u_i}{u_i^{(\ell)} + \varepsilon} s.t. \, y = \Phi x, |x_i| \leq u_i, i = 1, \cdots, n, \tag{6}$$

which equals to

$$x^{(\ell+1)} = \text{argmin} \sum_{i=1}^{n} \frac{|x_i|}{\left|x_i^{(\ell)}\right| + \varepsilon} s.t. \, y = \Phi x. \tag{7}$$

Note that in each iteration, we solve a weighting ℓ_1 minimization problem and hence we can implement our solution via existing tools. Specifically, if we choose $x^0 = [1, \cdots, 1]$, the solution of x^1 is the minimizer of $\sum_{i=1}^{n} |x_i|$. Based on Eq. (7), we can obtain the weights:

$$w_i^{(\ell+1)} = \frac{1}{\left(x_i^{(\ell)} + \varepsilon\right)}. \tag{8}$$

Besides the log-sum penalty function, Candès [15] proposed an alternative arc-tangent function as follows:

$$g(x) = \sum_{i=1}^{n} arctan(|x_i|/\varepsilon). \tag{9}$$

With the similar deductions of Eq. (4), we can obtain the weights

$$w_i^{(\ell+1)} = \frac{1}{\left(x_i^{2(\ell)} + \varepsilon^2\right)}. \tag{10}$$

Considering the $f_{log}(x)$ and $f_{arctan}(x)$ penalty function, we cannot make sure which is better because the results can be influenced by the value of ε and the magnitude of the datasets. For one dataset, we need to compare the two penalty function based methods simultaneously and then pick up the better. Compared the two weights in Eqs. (8) and (10), we can find that they actually are similar with each other. The difference mainly lies in the index value of the data at denominator. Therefore, we next construct a unified framework to include both, which is inspired by the concept of fractional-order. Fractional-order function is widely used in pattern recognition, which solves the problem of limiting the index value to some constant value. Therefore, we define the penalty function as

$$g(x) = x^{\alpha}. \tag{11}$$

After the deduction of Eq. (4), we can obtain the weights

$$w_i^{(\ell+1)} = \frac{\alpha}{x_i^{1-\alpha(\ell)}}. \tag{12}$$

To maintain the stability and flexibility of weights, we introduce ε_1 and β. In order to fit datasets and obtain the best results, we introduce ε_2 to balance the magnitude of datasets and transform Eq. (12) into the following:

$$w_i^{(\ell+1)} = \frac{\varepsilon_1}{\left(x_i^{\beta(\ell)} + \varepsilon_2\right)}. \tag{13}$$

Considering the weights in Eq. (13): when $\beta = 1$, it resembles weights in Eq. (8) produced by log-sum penalty function and when $\beta = 2$, it resembles weights in Eq. (10) produced by arctangent penalty function.

Next, we compared the ℓ_1 norm, ℓ_0 norm, log-sum penalty function, arc-tangent penalty function and our fractional-order penalty function. To visually show the performance of each penalty function, we draw five penalty functions for scalar magnitudes x (see Fig. 1). The first (ℓ_0-like) penalty function $f_0(x)$ has infinite slope at $x = 0$ (positive infinity as $x \to 0^+$, negative infinity as $x \to 0^-$), while its convex (ℓ_1-like) relaxation $f_1(x)$ has unit slope at the origin (1 as $x \to 0^+$, -1 as $x \to 0^-$). On the other hand, the slope of $f_0(x)$ is always 0 as $x \neq 0$, while $f_1(x)$ has slope of 1 as $x \geq 0$ (-1 as $x < 0$). We know the ℓ_0 minimization problem is NP-hard, so we use ℓ_1 norm as its relaxation. The concave $f_{log}(x)$ and $f_{arctan}(x)$ has slope of positive infinity at $x \to 0^+$ (negative infinity as $x \to 0^-$) as $\varepsilon \to 0$ and when $|x|$ grows to positive infinity, the scope approaches to 0. So $f_{log}(x)$ and $f_{arctan}(x)$ better approximate $f_0(x)$ (see Fig. 1). Like the ℓ_0 norm, this allows a relatively large penalty to be placed on small nonzero coefficients and encourages them to be set to 0. In fact, $f_{log}(x)$ and $f_{arctan}(x)$ tends to $f_0(x)$ as $\varepsilon \to 0$. Following this argument, it would appear that ε should be set arbitrarily small, to most closely make the power penalty resemble the ℓ_0 norm. Our fractional-order penalty function ranges widely and approximates the ℓ_0 norm highly. It should be noted that the penalty function of the reweighted ℓ_1 minimization algorithms can only produce a local minimum.

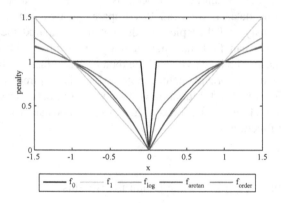

Fig. 1. Illumination of penalty functions for scalar magnitudes x. At the origin, the canonical ℓ_0 sparsity count $f_0(x)$ is better approximated by the log-sum penalty function $f_{log}(x)$ and the arc-tangent penalty function $f_{arctan}(x)$ than by the traditional convex ℓ_1 relaxation $f_1(x)$

2.2 Fractional-Order Function Based Reweighted Sparse Subspace Clustering

Considering noises and sparse outliers in affine space, SSC solves the following ℓ_1 minimization problem

$$\min_C \|C\|_1 + \lambda_e \|E\|_1 + \tfrac{\lambda_z}{2} \|Z\|_F^2$$
$$s.t.\, Y = YC + E + Z,\, A^T 1 = 1,\, A = C - \mathrm{diag}(C),$$
(14)

where C corresponds to a sparse coefficient matrix, E corresponds to a matrix of sparse outlying entries, and Z is a noise matrix, $A \in R^{N \times N}$ is an auxiliary matrix that helps to obtain faster updates through hard-threshold on the optimization variables. The two parameters $\lambda_e > 0$ and $\lambda_z > 0$ are used to balance the three terms in the object functions.

Inspired by SSC [4], our goal is to obtain sparser representation of the data. In Sect. 2.1, we have analyzed that reweighted ℓ_1 framework based on fractional-order function approximates the ℓ_0 norm than the ℓ_1 norm. By introducing the reweighted ℓ_1 framework based on fractional-order function into SSC, we can formulate fractional-order based reweighted sparse subspace clustering (FRSSC) model as follow:

$$\min_{C} \|W \odot C\|_1 + \lambda_e \|E\|_1 + \frac{\lambda_z}{2} \|Y - YA - E\|_F^2$$
$$s.t. A^T 1 = 1, A = C - \text{diag}(C), \tag{15}$$

where $\odot$ denotes the element-wise product between two matrices. Thus, Eq. (13) is a simple ℓ_1 minimization problem, which can be solved by Alternating Direction Method of Multipliers (ADMM) algorithm [17].

Finally, the detailed optimization procedure of FRSSC can be summarized as follows:

Algorithm 1: FRSSC solved via ADMM

Input: A set of points Y lying in a union of n linear subspaces $\{S_i\}_{i=1}^n$
Initialization: maxIter $= 10^4$, $C^{(0)}, A^{(0)}, E^{(0)}, \delta^{(0)}$ and $\Delta^{(0)}$ to zero, $W^{(0)} = I, k = 0$.
   While not converged do

1.   fix the others and update $A^{(k+1)}$ by

$$(\lambda_e Y^T Y + \rho I + \rho 11^T) A^{(k+1)} = \lambda_z Y^T (Y - E^{(k)}) + \rho(11^T + C^{(k)}) - 1\delta^{(k)^T} - \Delta^{(k)}$$

2.   fix the others and update $C^{(k+1)}$ by

$$C^{(k+1)} = J - \text{diag}(J), J = \Theta_{\frac{1}{\rho}}^{W^{(k)}} (A^{(k+1)} + \Delta^{(k)}/\rho),$$

$\Theta_\eta^W = \max \ (|X| - \eta W, 0) \odot sgn(X)$ is the shrinkage-thresholding operator [16].

3.   fix the others and update $W^{(k+1)}$ by $W^{(k+1)} = \frac{\varepsilon_2}{\left|C_{ij}^{(k+1)}\right|^\beta + \varepsilon_1}$

4.   fix the others and update $E^{(k+1)}$ by $E^{(k+1)} = \Theta_{\frac{\lambda_e}{\lambda_z}}(Y - YA^{(k+1)})$

5.   fix the others and update $\delta^{(k+1)}$ by $\delta^{(k+1)} = \delta^{(k)} + \rho\left(A^{(k+1)^T} 1 - 1\right)$

6.   fix the others and update $\Delta^{(k+1)}$ by $\Delta^{(k+1)} = \Delta^{(k)} + \rho\left(A^{(k+1)} - C^{(k+1)}\right)$

7.   check the convergence conditions:

$$\left\|A^{(k)^T} 1 - 1\right\|_\infty \leq \tau \ \left\|A^{(k)} - C^{(k)}\right\|_\infty \leq \tau \ \left\|A^{(k)} - A^{(k-1)}\right\|_\infty \leq \tau$$
$$\left\|E^{(k)} - E^{(k-1)}\right\|_\infty \leq \tau \ (k \geq maxIter)$$

   end while

8.   Optimal sparse coefficient matrix : $C^* = C^{(k)}$
9.   Normalize the columns of C as $c_i \leftarrow \frac{c_i}{\|c_i\|_\infty}$.
10. Construct a similarity matrix: $W = |C| + |C|^T$.
11. Apply spectral clustering [46] to the similarity graph.
Output: Segmentation of the data: $Y_1, \cdots, Y_n$.

---

3 Experimental Results and Analysis

We tested our FRSSC and other methods, including LSR [7], SSC [4], LRR [5, 6], RSSC [10], LRSSC [9] on motion segmentation and face clustering. In motion segmentation experiment, we utilized the Hopkins155 dataset [11]. However, the Hopkins155 dataset has few no-rigid motion objects, we further used the Freiburg-Berkeley dataset [12, 13], most sequences of which are no-rigid. Besides, for the segmentation of human faces, we evaluate more algorithms, such as SMR [8]. In face clustering, we used the Extended Yale B face dataset [14]. These datasets have different noise levels, therefore suitable for testing the influence of noises and corruptions on the performance.

- The Hopkins155 dataset [11] contains 156 sequences, each of which is composed of 39–550 data points drawn from two or three motion objects. Each sequence is independent, so there are 156 clustering task in total. The dataset has 120 two motion objects sequences and 35 three motion objects sequences.
- Freiburg-Berkeley dataset [12, 13] contains 137 motion sequences, 22 of which are rigid motion sequences and 115 of which are non-rigid motion sequences.
- The Extended Yale B dataset [14] consists of 192×168 pixel cropped face images of 38 individuals. For each individual, there are 64 frontal face images in different illuminations. Thus each individual corresponds to a subspace. More than half of the face images are corrupted with shadows noise. So the dataset can be regarded as heavily corrupted. To reduce the computational cost, we downsampled the images to 48×42 pixels and treated each 2016-dimensional vectorized image as a sample.

We evaluated our method on three datasets. Figure 2 presents the clustering error rates varying according to β on three datasets. On Hopkins155 dataset, we fixed $\varepsilon_1 = 0.001$, $\varepsilon_2 = 0.02$ and we set $\beta = 1.007$ when the best results obtained. On the FBMS137 dataset, we fixed $\varepsilon_1 = 0.0002$, $\varepsilon_2 = 0.0014$ and we set $\beta = 1.001$ when the best results occurred. On the Extended Yale B dataset, we fixed $\varepsilon_1 = 0.009$, $\varepsilon_2 = 0.00027$ and we obtained the best results of 2/3/5/8/10 subjects at $\beta = 3, 2.1, 2.1, 1.9, 1.3$.

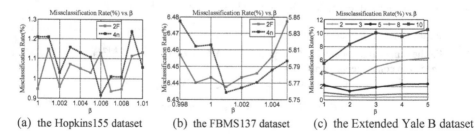

(a) the Hopkins155 dataset   (b) the FBMS137 dataset   (c) the Extended Yale B dataset

Fig. 2. Clustering error (%) of FRSSC on three datasets as a function of β. (a): clustering error (%) of FRSSC on Hopkins155 dataset as a function of β for the two cases of clustering of 2F-dimensional data and 4n-dimensional data obtained by PCA. (b): clustering error (%) of FRSSC on Freiburg-Berkeley dataset as a function of β for the two cases of clustering of 2F-dimensional data and 4n-dimensional data obtained by PCA. (c): clustering error (%) of FRSSC on Extended Yale B dataset as a function of β for the cases of clustering of different numbers of subject.

Table 1 tabulates the results of the original 2F-dimensional feature trajectories. In Table 2, the results are obtained when we project the original 2F-dimensional feature trajectories onto a 4n-dimensional subspace via PCA where n is the number of moving objects. It shows that our FRSSC outperforms the other methods on 3-motion objects and all-motion objects. Actually, the differences among all methods are limited. This is mainly because the noises and corruptions in the Hopkins155 dataset are small.

Table 1. Segmentation errors (%) on the Hopkins155 dataset with the 2F-dimensional data points. For our FRSSC, the parameter β is set as 1.0007. The parameters of other methods are also tuned to be the best.

No. of motion	Algorithm	LSR1	LSR2	SSC	LRR	LRSSC	RSSC	FRSSC
2 Motions	Mean	1.86	2.07	1.53	2.13	1.32	**0.64**	0.75
120 seqs	Median	0.22	0.21	0.00	0.00	0.00	0.00	**0.00**
3 Motions	Mean	4.14	4.87	4.4	4.03	3.79	2.01	**1.59**
35 seqs	Median	1.60	1.63	0.56	1.43	0.59	0.56	**0.31**
All	Mean	2.38	2.70	2.18	2.56	1.87	0.95	**0.94**
155 seqs	Median	0.30	0.28	0.00	0.00	0.19	0.00	**0.00**

Table 2. Segmentation errors (%) on the Hopkins155 dataset with the 4n-dimensional data points. For our FRSSC, the parameter β is set as 1.0007. The parameters of other methods are also tuned to be the best.

No. of motion	Algorithm	LSR1	LSR2	SSC	LRR	LRSSC	RSSC	FRSSC
2 Motions	Mean	2.18	2.45	1.83	3.39	1.42	**0.83**	0.85
120 seqs	Median	0.26	0.21	0.00	0.00	0.00	0.00	**0.00**
3 Motions	Mean	4.50	4.98	4.40	4.86	3.88	2.50	**1.55**
35 seqs	Median	1.60	1.83	0.56	1.47	0.59	**0.37**	0.41
All	Mean	2.70	3.02	2.41	3.73	1.98	1.21	**1.00**
155 seqs	Median	0.38	0.31	0.00	0.00	0.18	0.00	**0.00**

Table 3 tabulates the results of the original 2F-dimensional feature trajectories and Table 4 tabulates the results of the 4n-dimensional feature trajectories. In general, each feature trajectories in the same subspace can be represented as a linear combination of all other points in the same subspace, but the non-rigid motions are not lying in a linear subspace. Considering the property of the dataset, we compare our FRSSC algorithm with other algorithms on non-rigid motions and all motions particularly. We find that our FRSSC algorithm performs better than other algorithms both on 2F-dimensional data and 4n-dimensional data. Especially, only our algorithm achieves clustering error rate on 4n-dimensional data less than 5.8%.

Table 5 presents the clustering error rates of the six methods and our FRSSC algorithm with different numbers of individuals. It shows that in this test our FRSSC significantly has a clear advantage over other methods. Owing to the noises and outliers in the dataset, the robustness of our RWSSC algorithm is testified. Compared with

Table 3. Segmentation errors (%) on the FBMS137 dataset with the 2F-dimensional data points. For our FRSSC, the parameter β is set as 1.0001.The parameters of other methods are also tuned to be the best.

Type of motion	Algorithm	LSR1	LSR2	SSC	LRR	LRSSC	RSSC	FRSSC
Rigid	Mean	**0.59**	**0.59**	1.31	3.14	0.40	1.20	1.23
(22 seqs)	Median	**0.00**	**0.00**	0.00	0.00	0.00	0.00	**0.00**
Non-rigid	Mean	7.56	7.56	8.63	10.97	8.33	7.45	**7.43**
(115 seqs)	Median	0.38	0.38	0.00	0.00	0.25	0.00	**0.00**
All	Mean	6.44	6.44	7.45	9.72	7.05	6.44	**6.43**
(137 seqs)	Median	0.00	0.00	0.00	0.00	0.00	0.00	**0.00**

Table 4. Segmentation errors (%) on the FBMS137 dataset with the 4n-dimensional data points. For our FRSSC, the parameter β is set as 1.0001. The parameters of other methods are also tuned to be the best.

Type of motion	Algorithm	LSR1	LSR2	SSC	LRR	LRSSC	RSSC	FRSSC
Rigid	Mean	0.57	0.57	1.31	3.09	**0.40**	1.24	1.23
(22 seqs)	Median	0.00	0.00	0.00	0.00	**0.00**	0.00	0.00
Non-rigid	Mean	7.28	7.28	8.78	10.53	7.69	6.68	**6.63**
(115 seqs)	Median	0.50	0.50	0.00	0.25	0.25	0.00	**0.00**
All	Mean	6.20	6.20	7.58	9.34	6.52	5.81	**5.76**
(137 seqs)	Median	0.00	0.00	0.00	0.00	0.00	0.00	**0.00**

Table 5. Clustering error (%) of our algorithm and some state-of-art algorithms on the Extended Yale B dataset. The highest mean and median clustering accuracies are highlighted bold.

No. of subject	Algorithm	LSR1	LSR2	SMR	LRR	SSC	RSSC	FRSSC
2 Subjects	Mean	7.34	7.35	1.75	2.13	1.85	0.57	**0.38**
(163)	Median	7.03	7.03	0.78	0.78	0.00	0.00	**0.00**
3 Subjects	Mean	10.06	9.92	3.03	3.50	3.30	1.09	**0.70**
(416)	Median	10.42	10.42	2.08	2.08	1.04	0.00	**0.00**
5 Subjects	Mean	17.99	17.57	3.91	5.91	4.33	2.21	**1.30**
(812)	Median	18.13	17.81	2.50	5.00	2.81	0.62	**0.62**
8 Subjects	Mean	28.91	27.52	6.83	11.05	5.88	3.97	**2.46**
(136)	Median	29.69	27.83	3.71	7.42	4.49	1.86	**1.37**
10 Subjects	Mean	36.46	33.49	7.81	16.93	7.24	4.79	**3.59**
(3)	Median	36.09	32.81	7.03	18.91	5.47	3.28	**1.09**

RSSC algorithm, our FRSSC algorithm achieves lower clustering error rates in all the combinations of subjects, which demonstrate the unified reweighted ℓ_1 minimization framework based on fractional-order function framework efficiently and evaluate our RWSSC algorithm sufficiently.

4 Conclusion

In this paper, we analyzed the mechanism of the reweighted ℓ_1 minimization problem and proposed a unified fractional-order function based reweighted ℓ_1 minimization framework, which can approximate the ℓ_0 norm better than original ℓ_1 norm. Based on the framework, we propose a fractional-order function based reweighted sparse subspace clustering algorithm (FRSSC). Experimental results on three real-world dataset illustrate the effectiveness of FRSSC. However, our FRSSC depends on the settings of β and we seek for the best β on unsupervised learning. Our future work aims at seeking for the best β by 10-fold cross-validation and obtaining the best results of different datasets automatically.

Acknowledgments. This work was supported in part by the National Natural Science Foundation of China under Grant 61401209, in part by the Natural Science Foundation of Jiangsu Province, China under Grant BK20140790, in part by the Fundamental Research Funds for the Central Universities under Grant 30916011324, and in part by China Postdoctoral Science Foundation under Grants 2014T70525 & 2013M531364.

References

1. Vidal, R.: Subspace clustering. IEEE Sig. Process. Mag. **28**(2), 52–68 (2011)
2. Elhamifar, E., Vidal, R.: Sparse subspace clustering. In: CVPR, pp. 2790–2797 (2009)
3. Shi, J., Malik, J.: Normalized cuts and image segmentation. IEEE Trans. Pattern Anal. Mach. Intell. **22**(8), 888–905 (2000)
4. Elhamifar, E., Vidal, R.: Sparse subspace clustering: algorithm, theory, and applications. IEEE Trans. Pattern Anal. **35**(11), 2765–2781 (2013)
5. Liu, G., Lin, Z., Yu, Y.: Robust subspace segmentation by low-rank representation. In: Proceedings of the 27th International Conference on Machine Learning, Haifa, Israel (2010)
6. Liu, G., Lin, Z., Yan, S., Sun, J., Yu, Y., Ma, Y.: Robust recovery of subspace structures by low-rank representation. IEEE Trans. Pattern Anal. **35**(1), 171–184 (2013)
7. Lu, C.-Y., Min, H., Zhao, Z.-Q., Zhu, L., Huang, D.-S., Yan, S.: Robust and efficient subspace segmentation via least squares regression. In: Fitzgibbon, A., Lazebnik, S., Perona, P., Sato, Y., Schmid, C. (eds.) ECCV 2012. LNCS, vol. 7578, pp. 347–360. Springer, Heidelberg (2012). doi:10.1007/978-3-642-33786-4_26
8. Hu, H., Lin, Z., Feng, J., Zhou, J.: Smooth representation clustering. In: 2014 IEEE Conference on Computer Vision and Pattern Recognition (CVPR), pp. 3834–3841 (2014)
9. Wang, Y.X., Xu, H., Chen, L.L.: Provable subspace clustering: when LRR meets SSC. In: Advances in Neural Information Processing Systems (NIPS-13), pp. 64–72 (2013)
10. Xu, J., Xu, K., Chen, K., Ruan, J.: Reweighted sparse subspace clustering. Comput. Vis. Image Underst. **138**, 25–37 (2015)
11. Tron, R., Vidal, R.: A Benchmark for the comparison of 3-d motion segmentation algorithms. In: CVPR (2007)
12. Brox, T., Malik, J.: Object segmentation by long term analysis of point trajectories. In: Daniilidis, K., Maragos, P., Paragios, N. (eds.) ECCV 2010. LNCS, vol. 6315, pp. 282–295. Springer, Heidelberg (2010). doi:10.1007/978-3-642-15555-0_21
13. Ochs, P., Malik, J., Brox, T.: Segmentation of moving objects by long term video analysis. IEEE Trans. Pattern Anal. Mach. Intell. **36**(6), 1187–1200 (2013)

14. Georghiades, A.S., Belhumeur, P.N., Kriegman, D.J.: From few to many: illumination cone models for face recognition under variable lighting and pose. IEEE Trans. Pattern Anal. Mach. Intell. **23**(6), 643–660 (2001)
15. Candès, E.J., Michael, W.B., Boyd, S.P.: Enhancing sparsity by reweighted ℓ_1 minimization. J. Fourier Anal. Appl. **14**(5), 877–905 (2008)
16. Candès, E., Li, X., Ma, Y., Wright, J.: Robust principal component analysis? J. ACM **58**(3), 1–37 (2011)
17. Boyd, S., Parikh, N., Chu, E., Peleato, B., Eckstein, J.: Distributed optimization and statistical learning via the alternating direction method of multipliers. Found. Trends Mach. Learn. **3**(1), 1–122 (2011)

Ensemble Re-clustering: Refinement of Hard Clustering by Three-Way Strategy

Pingxin Wang[1,4]($\boxtimes$), Qiang Liu[2], Xibei Yang[2,4], and Fasheng Xu[3,4]

[1] School of Science, Jiangsu University of Science and Technology,
Zhenjiang 212003, China
pingxin_wang@hotmail.com
[2] School of Computer Science, Jiangsu University of Science and Technology,
Zhenjiang 212003, China
[3] School of Mathematical Sciences, University of Jinan, Jinan 250022, China
[4] Department of Computer Science, University of Regina,
Regina, SK S4S 0A2, Canada

Abstract. In this paper, we propose a three-way ensemble re-clustering method based on ideas of cluster ensemble and three-way decision. In the proposed method, we use hard clustering methods to produce different clustering results and cluster labels matching to align each clustering results to a given order. The intersection of the clusters with same labels are regarded as the core region and the difference between the union and the intersection of the clusters with same labels are regarded as the fringe region of the specific cluster. Therefore, a three-way result of the cluster is naturally formed. The results on UCI data sets show that such strategy is effective in improving the structure of clustering results and F_1 values.

Keywords: Three-way decisions · Three-way clustering · Cluster ensemble · Label matching

1 Introduction

Clustering is one of the most significant unsupervised learning problems which has been used in diverse areas like machine vision and pattern recognition as well as in medical applications. The fundamental objective of data clustering is to group similar objects in one cluster and divide dissimilar objects into different clusters. Research on clustering algorithm has received much attention and a number of clustering methods have been developed over the past decades.

The various methods for clustering can be divided into two categories: hard clustering and soft clustering. Hard clustering methods, such as C-means [11], spectral clustering [4], are based on an assumption that a cluster is represented by a set with a crisp boundary. That is, a data point is either in or not in a specific cluster. The requirement of a sharp boundary leads to easy analytical results, but may not adequately show the fact that a cluster may not have a well-defined cluster boundary. Furthermore, It is not the best way to absolutely

© Springer International Publishing AG 2017
Y. Sun et al. (Eds.): IScIDE 2017, LNCS 10559, pp. 423–430, 2017.
DOI: 10.1007/978-3-319-67777-4_37

divide the boundary objects into one cluster. In such cases, an object in the boundary should belong to more than one cluster.

In order to relax the constraint of hard clustering methods, many soft clustering methods were proposed for different application backgrounds. Fuzzy sets are a well known generalization of crisp sets, first introduced by Zadeh [23]. Incorporating fuzzy sets into C-means clustering, Bezdek [2] proposed Fuzzy C-Means (FCM), which is assumed that a cluster is represented by a fuzzy set that models a gradually changing boundary. Another effective tool for uncertain data analysis is rough set theory [15], which use a pair of exact concepts, called the lower and upper approximations, to approximate a rough (imprecise) concept. Based on the rough set theory, Lingras and West [10] introduced Rough C-Means (RCM) clustering, which describes each cluster not only by a center, but also with a pair of lower and upper bounds. Incorporating membership in the RCM framework, Mitra et al. [13] put forward a Rough-Fuzzy C-Means (RFCM) clustering method. Shadowed set, proposed by Pedrycz [16], provides an alternate mechanism for handling uncertainty. As a conceptual and algorithmic bridge between rough sets and fuzzy sets, shadowed set has been successfully used for clustering analysis, resulting in Shadowed C-Means (SCM) [14]. A brief summary of existing clustering methods can be shown in Fig. 1.

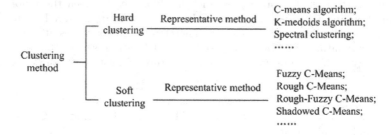

Fig. 1. Classification diagram of existing clustering methods

Although there are a lot of clustering methods, the performance of many clustering algorithms is critically dependent on the characteristics of the data set and the input parameters. Improper input parameters may lead to clusters that deviate from those in the data set. There is not one clustering method that can identify any form of data structure distribution. In order to solve this problem, Strehl and Ghosh [17] proposed cluster Ensemble algorithm, which combines multiple clusterings of a set of objects into one clustering result without accessing the original features of the objects. It has been shown that cluster ensembles are useful in many applications, such as knowledge-reuse [6], multi-view clustering [8], distributed computing [12] and in improving the quality and robustness of clustering results [3,5,7,9].

Recently, three-way decisions for problem solving was proposed by Yao [18, 19], which is an extension of the commonly used binary-decision model by adding a third option. The approach of three-way decisions divides the universe into the Positive, Negative and Boundary regions which denote the regions of acceptance, rejection and non-commitment for ternary classifications. Specifically, for the objects partially satisfy the classification criteria, it is difficult to directly identify them without uncertainty. Instead of making a binary decision, we use thresholds on the degrees of satisfiability to make one of three decisions: accept, reject, non-commitment. The third option may also be referred to as a deferment decision that requires further judgments. Three-way decisions have been proved to build on solid cognitive foundations and are a class of effective ways commonly used in human problem solving and information processing [20]. Many soft computing models for leaning uncertain concepts, such as interval sets, rough sets, fuzzy sets and shadowed sets, have the tri-partitioning properties and can be reinvestigated within the framework of three-way decisions [19].

Motivated by the three-way strategy, Yu [21,22] proposed a new soft clustering framework, three-way clustering, which uses two regions to represent a cluster, i.e., core region (Co) and fringe region (Fr) rather than one set. The core region is an area where the elements are highly concentrated of a cluster and fringe region is an area where the elements are loosely concentrated. There are maybe common elements in the fringe region among different clusters.

This paper aims at presenting a three-way clustering method by using cluster ensemble and three-way decisions based on the results of hard clustering. In the proposed method, hard clustering methods are used to produce different clustering results and cluster labels matching are used to align each clustering results to a given order. The three-way ensemble re-clustering results are obtained by the following strategy. The intersection of the clusters with same order are regarded as the core region and the difference between the union and the intersection of the clusters with same order are regarded as the fringe region of the specific cluster.

The study is organized into five sections. We start with a briefly introduction of the background knowledge in Sect. 2 and in Sect. 3 we present the process of Ensemble Re-clustering by two main steps. Experiment results are reported in Sect. 4.

2   A Three-Way Ensemble Re-clustering

By following ideas of cluster ensemble and three-way decisions, we present a three-way ensemble re-clustering algorithm. In this section, we assume that the universal has been divided into k disjoint sets m times by existing hard clustering algorithms. We discuss how to design a valid consensus function to obtain a three-way clustering based on the hard clustering results.

We begin our discussion by introducing some notations. We suppose that $V = \{v_1, \cdots, v_n\}$ is a set of n objects and $\mathbb{C}_i, (i = 1, \cdots m)$, denotes i-th clustering of V, where $\mathbb{C}_i = \{C_{i1}, \cdots, C_{ik}\}$ is a hard clustering results of V. Although we have obtained the the clustering results of V, $\mathbb{C}_i$ can not be directly used for the conclusion of the next stage due to the lack of priori category information.

As an example, consider the dataset $V = (v_1, v_2, v_3, v_4, v_5, v_6)$ that consists of six objects, and let $\mathbb{C}_1, \mathbb{C}_2$ and $\mathbb{C}_3$ be three clusterings of V which are shown in Table 1. Although they are expressed in different orders, they represent the same clustering result, so in order to combine the clustering results, the cluster labels must be matched to establish the correspondence between each other.

Table 1. Different ways of the same clustering results

	$\mathbb{C}_1$	$\mathbb{C}_2$	$\mathbb{C}_3$
v_1	1	2	3
v_2	1	2	3
v_3	2	3	2
v_4	2	3	2
v_5	3	1	1
v_6	3	1	1

In general, the number of identical objects covered by the corresponding cluster labels should be the largest, so the cluster labels can be registered based on this heuristic. Assuming that there are two clustering results $\mathbb{C}_1$ and $\mathbb{C}_2$. Each divides the dataset into k clusters, respectively, denoted by $\{C_{11}, \cdots, C_{1k}\}$ and $\{C_{21}, \cdots, C_{2k}\}$. First, the numbers of identical objects covered by each pair of cluster labels C_{1i} and C_{2j} in the two clusters are recorded in the overlap matrix of $k \times k$. And then select the cluster label that covers the largest number of identical objects to establish the correspondence and remove the result from the overlap matrix. Repeat the above process until all the cluster labels have established the corresponding relationship.

When there are $m(m > 2)$ clustering results, we can randomly select one as the matching criterion and match the other clustering results with the selected results. The matching algorithm only needs to check the $m - 1$ clustering results and store the overlap matrix with the storage space of $(m - 1) \times k^2$. The whole matching process is fast and efficient.

After all clustering labels match, all objects of V can be divided into three types for a given label j based on the results of labels matching:

$$\text{Type I} = \{v \mid \forall i = 1, \cdots, m, v \in C_{ij}\},$$
$$\text{Type II} = \{v \mid \exists i \neq p, i, p = 1, \cdots, m, v \in C_{ij} \wedge v \notin C_{pj}\},$$
$$\text{Type III} = \{v \mid \forall i = 1, \cdots, m, v \notin C_{ij}\},$$

From the above classifications, it can be seen that the objects in Type I are assigned to j-th cluster in all clustering results. The objects of Type II are assigned to j-th cluster in part of clustering results. The objects in Type III have not intersection with j-th in each clustering results. Based on the ideas of three-way decisions and three-way clustering, The elements in Type I are clearly attributable to the j-th cluster. And should be assigned to core region of j-th cluster. The elements in Type II should be assigned to fringe region of j-th cluster and all the elements in Type III should be assigned to trivial region of j-th cluster. From the above discussion, we get the following strategy to obtain a three-way clustering by cluster ensemble.

$$\mathrm{Co}(C_j) = \{v \mid \forall i = 1, \cdots, m, v \in C_{ij}\} = \bigcap_{i=1}^{m} C_{ij},$$

$$\mathrm{Fr}(C_j) = \{v \mid \exists i \neq p, i, p = 1, \cdots, m, v \in C_{ij} \wedge v \notin C_{pj}\} = \bigcup_{i=1}^{m} C_{ij} - \bigcap_{i=1}^{m} C_{ij}$$

The above clustering method are called a three-way ensemble re-clustering. The procedure of three-way ensemble re-clustering consists mainly of three steps.

1. Obtain a group of hard clustering results $\mathbb{C}_i, (i = 1, \cdots m)$ by using existing methods.
2. Randomly select one clustering result in step 1 as the matching criterion and match the other clustering results with the selected results
3. Compute the intersection of the clusters with same labels and the difference between the union and the intersection of the clusters with same labels.

The above procedure can be depicted by Fig. 2. Finally, we present Algorithm 1, which describes the proposed three-way ensemble re-clustering based on hard clustering results. In Algorithm 1, we choose the first clustering results $\mathbb{C}_1$ as matching criterion and match the other clustering results with $\mathbb{C}_1$ during labels matching.

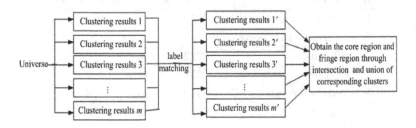

Fig. 2. Procedure diagram of three-way ensemble re-clustering

3   Experimental Illustration

To illustrate the effectiveness of Algorithm 1, some experiments on UCI [1] data sets are employed in this section. Before running Algorithm 1 on date sets, we need to obtain m clustering results. We use NJW spectral clustering with different scale parameter to get 10 clustering results in our experiments. All codes

Algorithm 1. Three-way ensemble re-clustering

1: Input: m clustering results $\mathbb{C}_1, \cdots, \mathbb{C}_m$
2: Output: Three-way ensemble re-clustering result
     $\mathbb{C} = \{(Co(C_1), Fr(C_1)), (Co(C_2), Fr(C_2)), \cdots, (Co(C_k), Fr(C_k))\}$
3: **for** each $\mathbb{C}_i$ in $\{\mathbb{C}_i\}, i = 2, \cdots, m$ **do**
4:   **for** $j = 1$ to k, $p = 1$ to k **do**
5:     overlap (j, p)=Count (C_{ij}, C_{1p});
       //overlap is a $k \times k$ matrix;
       //Count (C_{ij}, C_{1p}) counts the number of same elements of C_{ij} and C_{1p}
6:   **end for**
7:   $\Gamma = \phi$
8:   **while** $\Gamma \neq \{C_{i1}, C_{i2}, \cdots, C_{ik}\}$ **do**
9:     (u, v)=argmax(overlap(j, p)) //(u, v) is the biggest element
10:    $C_{iu} = C_{iv}$ //align C_{iu} to C_{1v}
11:    Delete overlap$(u, *)$
12:    Delete overlap$(*, v)$
13:    $\Gamma = \Gamma \cup \{C_{iu}\}$
14: **end while**
15: **end for**
16: **for** $j = 1$ to k **do**
17:   Calculate $Co(C_j) = \bigcap_{i=1}^{m} C_{ij}$
18:   Calculate $Fr(C_j) = \bigcup_{i=1}^{m} C_{ij} - \bigcap_{i=1}^{m} C_{ij}$
19: **end for**
20: **return** $\mathbb{C} = \{(Co(C_1), Fr(C_1)), (Co(C_2), Fr(C_2)), \cdots, (Co(C_k), Fr(C_k))\}$

Table 2. A description of 5 data sets

ID	Data sets	Samples	Attributes	Classes
1	Banknote authentication	1372	4	2
2	Congressional voting	435	16	2
3	Hill valley	1212	100	2
4	Ionosphere	351	34	2
5	Vertebral column	310	6	2

are run in Matlab R2013b on a personal computer. The details of these data are shown in Table 2.

In order to measure the tests accuracy, we use Macro F_1 values and Micro F_1 values [24] of cluster results as the evaluation indicator, which are two commonly used methods of testing the effect of classification. The results of above 5 UCI data sets by spectral clustering and three-way ensemble re-clustering are computed respectively. We list the experimental results in Table 3.

With a deep investigation of Table 3, it is not difficult to observe that both the Macro F_1 values and the Micro F_1 values of three-way ensemble re-clustering are

Table 3. Comparisons of clustering results

Data sets	Spectral Micro F_1	Clustering Macro F_1	Ensemble Micro F_1	Re-clustering Macro F_1
Banknote authentication	0.6150	0.6131	**0.6346**	**0.638**
Congressional voting	0.8639	0.8608	**0.9640**	**0.9633**
Hill valley	0.6359	0.6358	**0.6390**	**0.6389**
Ionosphere	0.7012	0.6927	**0.7280**	**0.7550**
Vertebral column	0.6785	0.6690	**0.7487**	**0.7425**

higher than those based on spectral clustering. From Table 3, we can conclude that three-way ensemble re-clustering can significantly improve the structure of classification results by comparing with the traditional spectral clustering algorithm.

4 Concluding Remarks

In this paper, we developed a three-way ensemble re-clustering method by employing the ideas of three-way decisions and cluster ensemble. Hard clustering methods are used to produce different clustering results and cluster labels matching is used to align each clustering results to a given order. The intersection of the clusters with same labels are regarded as the core region and the difference between the union and the intersection of the clusters with same labels are regarded as the fringe region of the specific cluster. Based on the above strategy, a three-way explanation of the cluster is naturally formed and experimental results demonstrate that the new algorithm can significantly improve the structure of classification results by comparing with the traditional clustering algorithm. The present study is the first step for the research of three-way clustering. How to determine the number of clusters is an interesting topic to be addressed for further research.

Acknowledgements. This work was supported in part by National Natural Science Foundation of China (Nos. 61503160 and 61572242), and Natural Science Foundation of the Jiangsu Higher Education Institutions of China (No. 15KJB110004).

References

1. UCI Machine Learning Repository (2005). http://www.ics.uci.edu/mlearn/MLRepository.html
2. Bezdek, J.: Pattern Recognition with Fuzzy Objective Function Algorithms. Plenum Press, New York (1981)
3. Fern, X.Z., Brodley, C.E.: Random projection for high dimensional clustering: A cluster ensemble approach. In: Proceedings of the Twentieth International Conference on Machine Learning (2003)

4. Fiedler, M.: Algebraic connectivity of graphs. ICzechoslovak Math. J. **23**, 298–305 (1973)
5. Fred, A., Jain, A.K.: Data clustering using evidence accumulation. In: Proceedings of the Sixteenth International Conference on Pattern Recognition (ICPR), pp. 276–280 (2002)
6. Ghosh, J., Strehl, A., Merugu, S.: A consensus framework for integrating distributed clusterings under limited knowledge sharing. In. In Proceedings of NSF Workshop on Next Generation Data Mining, pp. 99–108 (2002)
7. Hadjitodorov, S., Kuncheva, L., Todorova, L.: Moderate diversity for better cluster ensembles. Inf. Fusion **7**, 264–275 (2006)
8. Kreiger, A.M., Green, P.: A generalized rand-index method for consensus clustering of separate partitions of the same data base. J. Classif. **16**, 63–89 (1999)
9. Kuncheva, L., Hadjitodorov, S.: Using diversity in cluster ensembles. In: Proceedings of IEEE International Conference on Systems, Man and Cybernetics, pp. 1214–1219 (2004)
10. Lingras, P., West, C.: Interval set clustering of web users with rough k-means. J. Intell. Inf. Syst. **23**, 5–16 (2004)
11. Macqueen, J.: Some methods for classification and analysis of multivariate observations. In: Proceedings of 5-th Berkeley Symposium on Mathematical Statistics and Probability, vol. 1, pp. 281–297. University of California Press, Berkeley (1967)
12. Merugu, S., Ghosh, J.: Privacy-preserving distributed clustering using generative models. In: Proceedings of The Third IEEE International Conference on Data Mining (ICDM), pp. 211–218 (2003)
13. Mitra, S., Banka, H., Pedrycz, W.: Rough-fuzzy collaborative clustering. IEEE Trans. Syst. Man Cybern. (Part B) **36**, 795–805 (2006)
14. Mitra, S., Pedrycz, W., Barman, B.: Shadowed c-means: integrating fuzzy and rough clustering. Pattern Recogn. **43**, 1282–1291 (2010)
15. Pawlak, Z.: Rough sets. Int. J. Comput. Inf. Sci. **11**, 314–356 (1982)
16. Pedrycz, W.: Shadowed sets: representing and processing fuzzy sets. IEEE Trans. Syst. Man Cybern. (Part B) **28**, 103–109 (1998)
17. Strehl, A., Ghosh, J.: Cluster ensembles – a knowledge reuse framework for combining multiple partitions. J. Mach. Learn. Res. **3**, 583–617 (2002)
18. Yao, Y.: Three-way decisions with probabilistic rough sets. Inf. Sci. **180**, 341–353 (2010)
19. Yao, Y.: An outline of a theory of three-way decisions. In: Yao, J.T., Yang, Y., Słowiński, R., Greco, S., Li, H., Mitra, S., Polkowski, L. (eds.) RSCTC 2012. LNCS (LNAI), vol. 7413, pp. 1–17. Springer, Heidelberg (2012). doi:10.1007/978-3-642-32115-3_1
20. Yao, Y.: Three-way decisions and cognitive computing. Cogn. Comput. **8**, 543–554 (2016)
21. Yu, H., Jiao, P., Yao, Y., Wang, G.: Detecting and refining overlapping regions in complex networks with three-way decisions. Inf. Sci. **373**, 21–41 (2016)
22. Yu, H., Liu, Z., Wang, G.: A tree-based incremental overlapping clustering method using the three-way decision theory. Knowl.-Based Syst. **91**, 189–203 (2016)
23. Zadeh, L.: Fuzzy sets. Inform. Control **8**, 338–353 (1965)
24. Zhou, Z.: Machine Learning. Tsinghua University Press, Beijing (2016)

Motor Imagery EEG Classification Based on Multi-scale Time Windows

Jun Jiang[✉], Boxin Zhao, Peng Zhang, and Yang Yu

Aeronautical and Astronautical Engineering College,
Air Force Engineering University, Xi'an, China
jiang_mail@163.com

Abstract. To increase the performance Motor imagery (MI) EEG classification, a multi-scale window length algorithm was proposed in this paper. Under this algorithm, eight different time windows with the length of 600–2000 ms were selected to classify the MI data. The confidence level of the classification results from the eight windows were calculated simultaneously, to determine the weights for the outputs of each pre-classifier. The final classification results of MI tasks were generated by the weighted results from the eight windows. Using the proposed algorithm, the response time of MI classification was decreased by 11.4%, comparing to the single window method. On the same level of response time, the classification accuracy was increased by 2.8% by the multi-scale window algorithm. These results show that the proposed algorithm can speed up the detection of MI tasks with reliable classification accuracy.

Keywords: Brain-computer interface · Motor imagery · Electroencephalograph · Multi-scale windows

1 Introduction

Brain-computer interface (BCI) provides a new communication and control channel between brain and computers, without the participation of peripheral nerves and muscles [1,2]. A BCI can be used as an assistive technology for disabled individuals, helping to maintain or restore their lost communicate or motor functions [3–5]. The spontaneous BCI based on motor imagery (MI) has been widely applied on assistant devices control, due to its high response speed and autonomy [6–8]. By imaging the movement of human bodies (e.g., left and right hands), the characteristic of sensorimotor rhythms (SMRs, μ (8–14 Hz) and β (18–25 Hz)) in electroencephalograph (EEG) signals can be changed as effective features for classification [9,10]. The strength of these SMRs decreases when subjects voluntarily imagine kinesthetic movements (event-related desynchronization, or ERD), and increases after the imagined movements cease (event-related synchronization, or ERS) [11].

Improving the detection speed of MI task is an important research direction in BCI field, which can make the control intention of the subject respond in time

© Springer International Publishing AG 2017
Y. Sun et al. (Eds.): IScIDE 2017, LNCS 10559, pp. 431–439, 2017.
DOI: 10.1007/978-3-319-67777-4_38

and therefore improve the BCI system's controllability [12]. At the same time, reducing the delay of MI detection can also increase system stability, improving the security and reliability of BCI system. Gao *et al.* proposed a fast classification algorithm based on the fusion of ERD/ERS and Bereitschaftspontential (BP) feaures [13]. BP, also named as readiness potential or movement-related cortical potentials (MRCPs), is induced before the real or imagined movement onset [14]. Niazi and Bai *et al.* also designed a new algorithm with the combining of ERD/ERS and MRCPs features [15,16]. The above studies showed that MI tasks classification could be speeded up by adding new sensorimotor features of MRCPs. However, MRCPs is one kind of temporal potential (like P300), which is difficult to detect from single-trial EEG signals. In Niazi and Bai's study, the accuracy of MRCPs detection was below 60%.

Optimization on the length of sliding window during processing of EEG signals is another way to speed up the MI tasks classification. The sliding window is usually applied to intercept the latest EEG signal with the same length of the window. The ERD/ERS, or other MI features extracted from this window are classified to determine the current brain state. The selection of the sliding window length has a great influence on the MI classification result. The reliability of ERD/ERS feature may decrease in a sliding window with short length, while the response time may lengthen if the sliding window is too long.

In this paper, a novel signal processing algorithm based on multi-length sliding windows was proposed to enhance the classification performance of MI tasks. The classification results from different sliding windows were calculated simultaneously, to determine the final result of the MI tasks. The algorithm was tested on the EEG dataset from our previous study [17]. The MI samples in the dataset consisted of the left or right hand MI tasks, and the sequential MI (sMI) samples. The sMI samples were constructed by alternately imagining movements of the left or right hand.

2 Materials and Methods

2.1 Experimental Dataset

The EEG dataset in this experiment was collected from our previous study [17]. The MI samples were acquired from four healthy, right-handed subjects. In the experiments, the subjects were seated in a comfortable armchair facing a computer screen. The detailed information about the experimental procedures were told to the subject before the experiments. The EEG signals from C3, C4, F3, F4, FC5, FC1, FC2, FC6, T7, Cz, T8, CP5, CP1, CP2, CP6, P3, P4 were recorded (referenced to P8 and grounded to FPz), in accordance with the international 10–20 system. The impedances of all the electrodes were maintained below 10 kΩ. EEG Data were acquired with a BrainAmp DC Amplifier (BrainProducts GmbH, Germany) at a sampling rate of 250 Hz.

The EEG dataset was collected as the following procedure. A trial for one sMI task lasted 15 s, as illustrated in Fig. 1. For the first 6 s, "START" was displayed on the screen, and the subjects were asked to relax. After the rest

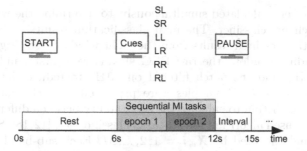

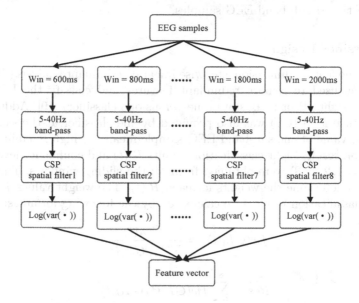

Fig. 1. The experimental procedure to collect the sMI samples in our previous study.

Fig. 2. The feature extraction algorithm based on multi-length sliding windows.

period, one of the four literal cues ("LR", "RL", "SL" or "SR") was displayed on the screen for 6 s in random order, and the subjects attempted to execute the corresponding task. "PAUSE" was then displayed for a 3 s interval before the next trial started. Four sMI tasks were executed in the experiment, name as "LR", "RL", "SL" and "SR". "LR" contained one left hand MI and one right hand MI subsequently. "RL" contained one right hand MI and one left hand MI subsequently. "SL" and "SR" contained only one left or right hand MI.

2.2 Feature Extraction

Under the multi-length sliding windows algorithm, eight different sliding windows with the length from 600 to 2000 ms (200 ms overlap) were selected to classify the MI data. The confidence level of the classification results from the eight windows were calculated simultaneously, to determine the weights for the outputs of each pre-classifier. The final classification results of MI tasks were generated by the weighted results from the eight windows (see Fig. 2).

In each sliding window, the raw EEG signals were first band-pass filtered between 5–40 Hz, and the notch filtered on 50 Hz to reduce the signal noise. Then, each sub-band EEG samples were processed with the common spatial pattern (CSP) algorithm to enhance the separability between different MI tasks. CSP is a widely used algorithm for MI tasks classification [12, 18]. Suppose there are M sub-bands denoted by $X_i, i = 1, 2, ..., M$. Each sub-band EEG sample produce a CSP filter W_i in a $C \times C$ matrix, where C is the channel number of EEG signals. The final feature vector was constructed by combining the CSP features of the M sub-band EEG samples.

2.3   Classifier Design

To reduce the feature dimension, stepwise linear discriminant analysis (SWLDA) method was used to remove redundant features and classify the EEG samples to left, right MI or relax state (one-versus-one classifier) [19]. Additionally, the confidence level (CL) was also provided by the classifier at the same time. A large CL value of one sub-band EEG sample mean the high reliability of the classification result. Eight classification results $R_i(t)$ and confidence level values $CL_i(t)$ $(i = 1, 2, ..., 8)$ were obtained from all the sliding windows. The $CL_i(t)$ was used to determine the weights for each $R_i(t)$. The weight values $fw(\cdot)$ and the final classification result were calculated by the following formulas:

$$fw(CL_i(t)) = \frac{CL_i(t)}{\sum_{j=1}^{8} CL_j(t)} \qquad (1)$$

$$Res = \sum_{i=1}^{8} fw(CL_i(t)) \cdot R_i(t) \qquad (2)$$

The value of $CL_i(t)$ was determined by the location of the corresponding feature in the feature space. If one feature located far away from the hyperplane in the feature space, it has high confidence level of the classification result.

3   Results

Table 1 showed the classification accuracy of the sMI tasks from the four subjects. The average accuracy of the four subjects was 90.2%. Subject A got the highest accuracy of 93.8%, while subject D had the lowest accuracy of 87.0%.

Table 2 showed the response time of the sMI detection. The Δt value of the four subjects was 1.72 s. Subject A got the fastest response speed with the Δ of 1.54, while subject C and D had the largest Δt value of 1.85 s.

Figure 3 showed the continuous classification results of "LR" and "RL" sMI tasks from subject B. It could be found that the results from the proposed multi-length sliding window algorithm (red curve) had faster response speed than the

Table 1. sMI classification accuracy based on the multi-length sliding windows.

Subject	sMI tasks [%]				
	SL	SR	LR	RL	Mean
A	93.8	95.8	93.8	91.7	93.8
B	91.7	89.6	89.6	93.8	91.2
C	89.6	91.8	87.5	85.4	88.6
D	85.4	89.6	87.5	85.4	87.0
AVG	90.1	91.7	89.6	89.1	90.2

Table 2. Response time of the sMI tasks detection based on the multi-length sliding windows [s].

Subject	Δt of the sMI task				
	SL	SR	LR	RL	Mean
A	1.40	1.20	1.95	1.60	1.54
B	1.50	1.40	1.90	1.70	1.63
C	1.67	1.76	2.01	1.95	1.85
D	1.70	1.67	1.91	2.10	1.85
AVG	1.57	1.51	1.94	1.84	1.72

results from long length (e.g., 1600–2000 ms) single sliding window (blue curves). Furthermore, although the curves of short length single sliding window (below 1200 ms) had shorter response time, their classification accuracies is lower than the multi-length sliding window results.

To compare the classification performance of multi-length sliding window algorithm with the traditional single-length sliding window method, one sliding window win_acc with the closest accuracy to the one of multi-length sliding window algorithm was selected. The response times Δt of the win_acc condition and the proposed algorithm in this paper was compared in Fig. 4. It could be found that Δt value of the multi-length sliding windows was 1.71 s, while for win_acc was the 1.93 s. The win_acc of the four subject A, B, C, D were 1600 ms, 2000 ms, 1800 ms, and 2000 ms respectively.

Figure 5 showed the classification accuracy comparison of the four subjects in the same response speed level. The other sliding window win_RT with the closest Δt value to the one of multi-length sliding window algorithm was selected. The classification accuracy on the win_RT condition and the one of the proposed algorithm were presented on this figure. The average accuracy based on the our method was 90.2%, while the accuracy of single-length sliding window was 87.4%. On the single-length method, the win_RT of subject A, B, C, D were 1000 ms, 1400 ms, 1800 ms, and 1800 ms respectively, which had the same level of the response speed with the multi-length sliding windows algorithm.

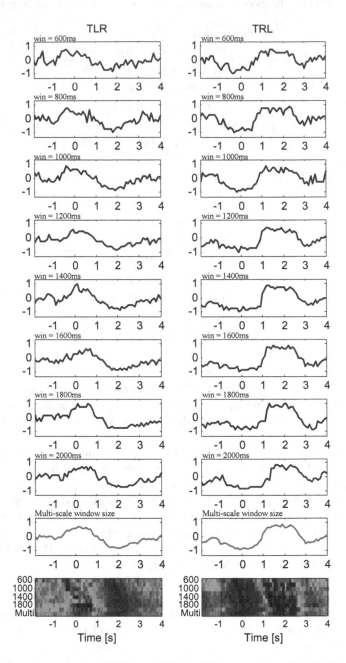

Fig. 3. The response speed of "LR" and "RL" sMI from subject B. The blue curves is the continuous classification results using single sliding window, and the red curve is the results from the proposed multi-length sliding windows. (Color figure online)

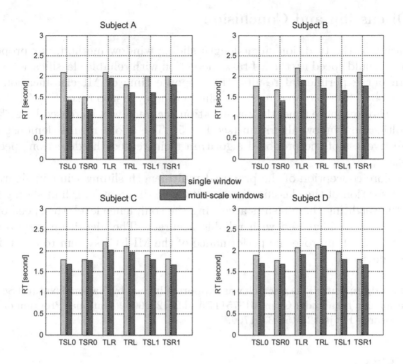

Fig. 4. The response time comparison of the four subject in the same accuracy level.

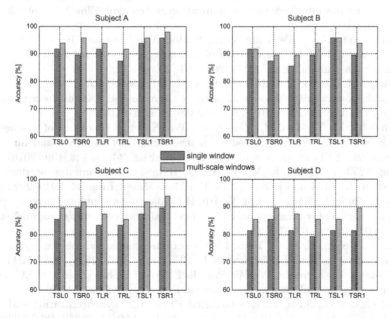

Fig. 5. The classification accuracy comparison of the four subject in the same response time level.

4   Discussion and Conclusion

Comparing to the tradition single-length sliding window method, our proposed algorithm could speed up the MI tasks detection with reliable classification accuracy. In our experimental results, the response time of MI classification was decreased by 11.4%, comparing to the single window method (see Fig. 4). On the same level of response time, the classification accuracy was increased by 2.8% by the multi-scale window algorithm (see Fig. 5). The above results demonstrated the effectiveness of the proposed algorithm to increased the detection speed of MI tasks.

The core conception of the proposed multi-length sliding window algorithm is synthesization of the classification results from different length of sliding windows. By weighting these results according the confidence level, the speed of MI detection can be increased with reliable accuracy. The algorithm proposed in this paper could improve the performance of the MI-BCI system to control the external devices.

Acknowledgement. This research was supported in part by National Natural Science Foundation of China under Grant 61375117 and 9142030002; National Program on Key Basic Research Project 2015CB351706.

References

1. Wolpaw, J.R., Birbaumer, N., McFarland, D.J., Pfurtscheller, G., Vaughan, T.M.: Brain-computer interfaces for communication and control. Clin. Neurophysiol. **113**, 767–791 (2002)
2. Pfurtscheller, G., Müller-Putz, G.R., Schlöl, A., Graimann, B., et al.: 15 years of BCI research at Graz university of technology: current projects. IEEE Trans. Neural Syst. Rehabil. Eng. **14**(2), 205–210 (2006)
3. Millán, J.R., Rupp, R., Müller-Putz, G.R., Murray-Smith, R., et al.: Combining brain-computer interfaces and assistive technologies: state-of-the-art and challenges. Front. Neurosci. (2010). doi:10.3389/fnins.2010.00161
4. Zhang, R., Li, Y., Yan, Y., Zhang, H., Wu, S., et al.: Control of a wheelchair in an indoor environment based on a brain-computer interface and automated navigation. IEEE Trans. Neural Syst. Rehabil. Eng. **24**(1), 128–139 (2016)
5. Jiang, J., Zhou, Z., Yin, E., Yu, Y., Hu, D.: Hybrid brain-computer interface (BCI) based on the EEG and EOG signals. Bio-Med. Mater. Eng. **24**, 2919–2925 (2014)
6. Yue, J., Zhou, Z., Jiang, J., Liu, Y., Hu, D.: Balancing a simulated inverted pendulum through motor imagery: an EEG-based real-time control paradigm. Neurosci. Lett. **101**(51), 96–100 (2012)
7. Chae, Y., Jeong, J., Jo, S.: Toward brain-actuated humanoid robots: asynchronous direct control using an EEG-based BCI. IEEE Trans. Rob. **28**(5), 1031–1044 (2012)
8. McFarland, D., Sarnacki, W., Wolpaw, J.: Electroencephalographic (EEG) control of three dimensional movement. J. Neural Eng. **7**(3), 036007 (2010)
9. Pfurtscheller, G., Silva, F.: Event-related EEG/MEG synchronization and desynchronization: basic principles. Clin. Neurophysiol. **110**(11), 1842–1857 (1999)
10. Pfurtscheller, G., Neuper, C., Brunner, C., Silva, F.: Beta rebound after different types of motor imagery in man. Neurosci. Lett. **378**(3), 156–159 (2005)

11. Quandt, F., Reichert, C., Hinrichs, H., Knight, R., Rieger, J.: Single trial discrimination of individual finger movements on one hand: a combined MEG and EEG study. NeuroImage **59**(4), 3316–3324 (2012)
12. Lotte, F., Congedo, M., Lécuyer, A., Lamarche, F., Arnaldi, B.: A review of classification algorithms for EEG-based brain-computer interfaces. J. Neural Eng. (2007). doi:10.1088/1741-2560/4/R01
13. Yijun, W., Zhiguang, Z., Yong, L., et al.: BCI competition 2003-Data Set IV: an algorithm based on CSSD and FDA for classifying single-trial EEG. IEEE Trans. Biomed. Eng. **51**(6), 2004–2009 (2003)
14. Shibasaki, H., Hallett, M.: What is the Bereitschaftspotential? Clin. Neurophysiol. **117**, 2341–2356 (2006)
15. Bai, O., Rathi, V., Lin, P., et al.: Prediction of human voluntary movement before it occurs. Clin. Neurophysiol. **122**, 364–372 (2011)
16. Niazi, I.K., Jing, N., Tiberghien, O., et al.: Detection of movement intention from single-trial movement-related cortical potentials. J. Neural Eng. **8**(6), 066009 (2011)
17. Jiang, J., Zhou, Z., Yin, E., Yu, Y., Liu, Y., Hu, D.: A novel Morse code-inspired method for multiclass motor imagery brain-computer (BCI) design. Comput. Biol. Med. **66**, 11–19 (2015)
18. Ramoser, H., Muller-Gerking, J., Pfurtscheller, G.: Optimal spatial filtering of single trial EEG during imagined hand movement. IEEE Trans. Neural Syst. Rehabil. Eng. **8**(4), 441–446 (2000)
19. Krusienski, D.J., Sellers, E.W., McFarland, D.J., et al.: Toward enhanced P300 speller performance. J. Neurosci. Methods **167**(1), 15–21 (2008)

Hyperspectral Image Classification Based on Empirical Mode Decomposition and Local Binary Pattern

Changli Li[1(✉)], Hang Zuo[1], Xin Wang[1], Aiye Shi[1],
and Tanghuai Fan[2]

[1] College of Computer and Information,
Hohai University, Nanjing 211100, China
charlee@hhu.edu.cn
[2] School of Information Engineering, Nanchang Institute of Technology,
Nanchang 330099, China

Abstract. Traditional hyperspectral image classification methods always focused on spectral information, and lots of spatial information was neglected. Therefore, this paper introduces the spatial texture information in the process of hyperspectral image classification, and focuses on how to deeply combine the texture information and the spectral information. Based on empirical mode decomposition and local binary pattern, the method of support vector machine is used to classify hyperspectral image, in order to improve the image classification accuracy.

Keywords: Hyperspectral image classification · Empirical mode decomposition · Local binary patterns · Support vector machine

1 Introduction

In the late 20th century, the hyperspectral remote sensing technology is a major breakthrough in the field of remote sensing science and technology, and the high spatial resolution and abundant spectral information of hyperspectral remote sensing images have been applied in geological survey and military applications m [1]. At present, so many scholars have studied the classification methods of hyperspectral images, such as Bayesian mode, feature extraction and random forest, neural network, fuzzy clustering and support vector machine and other methods [2–4]. Except rich spectrum information and spatial information, hyperspectral images also have affluent texture information. Texture information can fully reflect the texture distribution in remote sensing images, it could be a good way to suppress different body with same spectrum" or "same body with different spectrum" phenomenon. Frequently-used methods are gray level co-occurrence matrix, wavelet transform [5]. In 1996, Professor Ojala proposed local binary pattern [6, 7], which is based on the statistical texture description operator, and have the advantages of easy understanding and fast running. In 2009, the application of LBP operator in hyperspectral images was first reported by Masood et al., then the domestic and foreign scholars carried out the application of LBP operator based on

© Springer International Publishing AG 2017
Y. Sun et al. (Eds.): IScIDE 2017, LNCS 10559, pp. 440–449, 2017.
DOI: 10.1007/978-3-319-67777-4_39

spatial spectrum combining algorithm in hyperspectral image processing [8]. Empirical mode decomposition was proposed by Huang with others in 1998, which have a very desirable method of dealing nonstationary and non-linear signal [9], it could adaptively decompose the signal into numbers of intrinsic mode functions with a residual signal. Empirical mode decomposition could adaptively obtain the signal characteristics, making it have obvious advantages in the extraction of hyperspectral data characteristics, empirical mode decomposition has also already been applied in the classification of hyperspectral images.

Previous classification methods have many shortcomings, either didn't considered to extract the spatial information, or classification methods were too cumbersome, needed to set so many parameters. Based on the above problems, this paper proposed a method which based on empirical mode decomposition combine the spatial and spectral information of hyperspectral images, it could calculate the spatial and spectral information with high classification accuracy at the same time, meanwhile, there are not have many parameters need be set.

2   Empirical Mode Decomposition

Empirical mode decomposition is a new adaptive signal time-frequency approach proposed by Huang with others in NASA in 1998 [9]. Its self-adaptive characteristics makes it different from wavelet transform, which need to set the basic function, because of this, theoretically, any type of signal can be decomposed by empirical mode decomposition, so empirical mode decomposition has been proposed in different engineering areas application. Basic principle of empirical mode decomposition is to find intrinsic mode functions through sifting progress, and decompose the original signal into a combination of intrinsic mode functions and a residual.

IMF must satisfy the following conditions: (a) The number of local extremal points (including the maximum and the minimum) of the function must be equal to the number of zero crossing points and the error less than two; (b) The mean of lower envelope lines and upper envelope lines always be zero over the entire signal time domain.

EMD must follow these conditions: (a) The signal have at least two extreme points, a maximum value and a minimum value; (b) The time scale between the extreme points uniquely determines the local time domain characteristics of the data. (c) If the signal data have only inflection points without extreme points, differentia signal data once or several times to obtain extreme values, and then reverse the process, obtained the results. This type of decomposition process could be vividly call "filtering".

Now we assume original image is $X(m, n)$, and the ith IMF referred to as IMF_i, the following are the concrete steps of two-dimensional empirical mode decomposition:

(1) Initialize the image $x_{ij}(m, n) = X(m, n)$, $i = 1, j = 1$, and $x_{ij}(m, n)$ will be used as the original image.
(2) All local extreme points are obtained by morphological method. The radial basis function is used to fit the upper and lower envelope surfaces, then obtained mean function.

$$e_{ij}(m,n) = \frac{e_{max}(m,n) + e_{min}(m,n)}{2} \tag{1}$$

(3) Original image minus the mean function:

$$h_{ij}(m,n) = x_{ij}(m,n) - e_{ij}(m,n) \tag{2}$$

(4) Judging if the IMF satisfies the above-mentioned conditions. if not, than $j = j + 1$, and return to step 2, and $h_{ij}(m,n)$ will be the original signal, the second time sifting:

$$x_{ij}(m,n) = h_{ij}(m,n) \tag{3}$$

we will get the first IMF:

$$IMF_1 = c_1(m,n) = h_{ij}(m,n) \tag{4}$$

(5) Original signal minus the first IMF will get the first residual:

$$r_1(m,n) = x_{ij}(m,n) - c_1(m,n) \tag{5}$$

(6) The residual is re-enter as an input signal and cycle the steps 1 to 5 to obtain the second IMF and a new residual. Judging whether the first residual is monotonic, if not, then $i = i + 1$, $j = 1$, $x_{ij}(m,n) = r_i(m,n)$, and return to step 2. So when we cycle p times will get:

$$\begin{aligned} IMF_2 &= c_2(m,n) \\ r_2(m,n) &= r_1(m,n) - c_2(m,n) \\ IMF_3 &= c_3(m,n) \\ r_3(m,n) &= r_2(m,n) - c_3(m,n) \\ &\vdots \\ IMF_p &= c_p(m,n) \\ r_p(m,n) &= r_{p-1}(m,n) - c_p(m,n) \end{aligned} \tag{6}$$

When the residual becomes monotonous, that means the IMF couldn't be decomposed, so the entire two-dimensional empirical mode decomposition of the signal process is finish.

(7) The original signal can be composed of IMFs with a residual:

$$X(m,n) = \sum_{k=1}^{p} c_k(m,n) + r_p(m,n) \tag{7}$$

In step 4, Huang proposed the SD criterion, which γ requires the final selection of an appropriate value by multiple tests. Normally $\gamma \in (0.2, 0.3)$, in this paper we choose $\gamma = 0.2$.

3   Local Binary Pattern

The local binary pattern is used to compare the relationship between a pixel and its surrounding pixels. The original LBP operator acts on a range of 3×3 pixels neighborhood blocks, each pixel of the neighborhood block is binarized with the center pixel as the threshold. In the binarization processing, if the surrounding pixel gray value is less than the gray value of the center pixel, the pixel position is marked as 0, otherwise marked as 1; the value after the threshold (0 or 1) is multiplied by the weight of the corresponding position pixel, the sum of the products is the LBP value of the neighborhood.

The local binary pattern is very sensitive to the direction of the image. If the image changes in angle, the LBP texture coding will completely change, so Mäenpää et al. extended the LBP algorithm [10] and proposed the rotation invariant LBP for:

$$LBP_{P,R}^{ri} = \min\{ROR(LBP_{P,R}, i)| i = 0, 1, \cdots, P - 1\} \tag{8}$$

Figure 1 is definition of rotation LBP operators. In a circular area with radius is R, the center of the circular area is the central pixel, and the center pixel is evenly distributed around P neighborhood pixels.

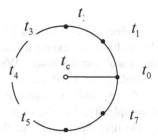

Fig. 1. Definition of rotation LBP operators

Professor Ojala through careful observation and extensive experiments found that the probability of a small number of LBP patterns appeared were very high, and sometimes even more than 90%, and the leaving majority of the LBP patterns were so low, even sometimes can be ignored. Thus Professor Ojala considered to name the pattern of parts with a high probability as uniform pattern and recorded as $LBP_{P,R}^{u2}$. The formula as follows:

$$U = |s(t_{P-1} - t_c) - s(t_0 - t_c)| + \sum_{p=1}^{P-1} |s(t_p - t_c) - s(t_{p-1} - t_c)| \tag{9}$$

Obviously, the uniform pattern could reduce the LBP dimension, thus avoid the feature vector dimension is too high, and could also reduce the negative impact of high-frequency noise. By combining these two patterns, you can get a rotation invariant uniform LBP like this:

$$LBP_{P,R}^{riu2} = \begin{cases} \sum_{i=0}^{P-1} s(t_i - t_c) & U \le 2 \\ P+1 & Otherwise \end{cases} \tag{10}$$

Table 1 shows the dimensional contrast of several patterns at different scales.

Table 1. Different dimensions in different scales

Mode	Scales (R, P)			
	(1, 8)	(2, 16)	(3, 24)	(4, 32)
$LBP_{P,R}$	256	65536	Too much	Too much
$LBP_{P,R}^{ri}$	36	4116	Too much	Too much
$LBP_{P,R}^{u2}$	59	243	555	995
$LBP_{P,R}^{riu2}$	10	18	26	34

One of the main factors that affect the extraction of texture features is the scale, so this paper considers to use the multi-scale texture to more comprehensively describe the gray scale changes in different scales in the image. We can try to change the P value and the R value, that is to calculate the LBP values of different neighborhoods of different radii. From the above description, the higher P value and R value will make the higher the computational complexity, and the experimental accuracy is not high. Therefore, in this paper, only the first three scales of the LBP operators are used to extract the multi-scale texture features, that is, scale1: R = 1, P = 8; scale2: R = 2, P = 16; scale3: R = 3, P = 24, and we will calculate the single scale LBP operator, and then calculate the two scales and three scales of integrated LBP operators.

4   Spatial-Spectral Fusion

The hyperspectral image has the characteristics of high dimension, large amount of data and high band correlation, it is more prone to huge and data redundancy. When it extract texture features with local binary pattern, huge data and computation, makes it very difficult to operate. Therefore, before extract the hyperspectral image feature, we need reduce the hyperspectral image dimension first, the principal component analysis is used in the experiment, after dimension reduction selected a bit bands to process. The following are the steps of experiment:

Step 1: Several intrinsic mode functions are obtained by two-dimensional empirical mode decomposition of the original hyperspectral image;

Step 2: Use the principal component analysis to extract the first principal component of each intrinsic mode function;

Step 3: The spatial texture features are extracted by local binary pattern for each of the first principal component;

Step 4: The spectral features of the original hyperspectral image are extracted by principal component analysis;

Step 5: Hyperspectral image spectral features and texture features need to be fused after the extraction, this paper use the feature layer fusion, and in order to facilitate the treatment, we choose a direct combination of fusion spectral features and texture features;

Step 6: Use support vector machine for classification;

Step 7: Calculate the classification accuracy.

Figure 2 is the experimental flowchart of EMD-LBP-SVM.

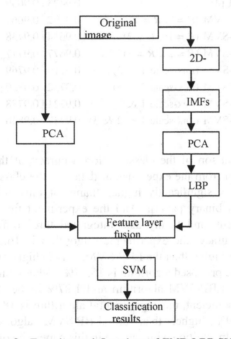

Fig. 2. Experimental flowchart of EMD-LBP-SVM.

5 Experiments

The data used in this chapter is a segment 92AV3C [11] in the hyperspectral data collected by the AVIRIS sensor in the northwestern part of Indiana state in 1992. The image size is 145 × 145 pixels and the wavelength range is 0.4–2.5 μm, contains 224 bands, the spatial resolution is 20 m, contains 16 different ground features. Twenty weather-affected bands were removed, and the remaining 200 bands of spectral images were applied to the classification.

Since the Indian Pines have a small spatial data volume, the training sample ratio is chosen to be 10%. Considering the randomness of the training sample selection, each experiment is done ten times and calculated the average of all the evaluations.

In this paper, the traditional SVM, LBP-SVM, EMD-SVM and EMD-LBP-SVM were used to classify the hyperspectral images, and we will comparing the classification results of the different methods. The experimental results are respectively expressed by the overall accuracy, average accuracy, and Kappa coefficients. The results of the classification are shown in Table 2 and Fig. 3.

Table 2. Different accuracies in different classifications

Classification methods	OA	AA	Kappa
SVM	0.8115	0.8259	0.7854
LBP-SVM [8]	0.9338	0.9672	0.9241
EMD-SVM [4]	0.9423	0.9639	0.9340
EMD-LBP-SVM ($P = 8$, $R = 1$)	0.9592	0.9669	0.9535
EMD-LBP-SVM ($P = 16$, $R = 2$)	0.9614	0.9738	0.9559
EMD-LBP-SVM ($P = 24$, $R = 3$)	0.9677	0.9777	0.9631
EMD-LBP-SVM (mix-scale 1 & 2)	0.9710	0.9769	0.9669
EMD-LBP-SVM (mix-scale 2 & 3)	0.9732	0.9820	0.9694
EMD-LBP-SVM (mix-scale 1 & 3)	0.9721	0.9788	0.9682
EMD-LBP-SVM (mix-scale 1 & 2 & 3)	0.9737	0.9826	0.9700

Table 2 is a comparison of the classification accuracy of the four classification methods. It can be seen from the experimental data that the classification accuracy of the proposed method is significantly higher than that only using empirical mode decomposition or local binary pattern. And the experiment first used three different scales of the experiment, and finally also integrated three different scales of LBP method. In overall accuracy, the experimental algorithm is 16.22% higher than the SVM algorithm, 3.99% higher than the LBP-SVM, 3.14% higher than the EMD-SVM. In average accuracy, the proposed algorithm is 15.67% higher than the SVM algorithm, 1.54% higher than the LBP-SVM algorithm and 1.87% higher than the EMD-SVM algorithm. In Kappa coefficient, the experimental algorithm is 18.46% higher than the SVM algorithm, 4.59% higher than the LBP-SVM algorithm, 3.60% higher EMD-SVM algorithm, In addition, you can see in this paper, the overall accuracy of the proposed method can basically reach more than 96%, and mixed two or three different scales of the LBP method more than 97%, average accuracy can basically reach more than 97%, mixed with two or three different scales of the LBP method even close to 98%, Kappa coefficient can basically reach more than 0.95, mixed two or three different scales of the LBP method even close to 0.97 or more, therefore, this experiment result could fully prove the effectiveness of this method.

Figure 3 is comparative simulation of the various classification methods used in the experiment. Figure 3(a) shows the ground truth, (b) is the classification simulation of the SVM method, (c) is the classification simulation of the LBP-SVM method, (d) is the experimental simulation of EMD-SVM method, (e) is the experimental simulation

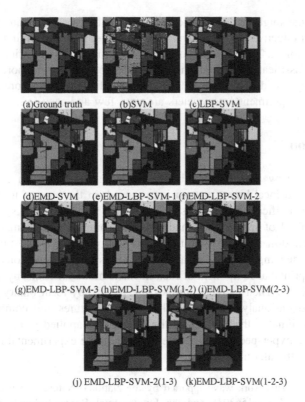

(a)Ground truth   (b)SVM   (c)LBP-SVM

(d)EMD-SVM   (e)EMD-LBP-SVM-1 (f)EMD-LBP-SVM-2

(g)EMD-LBP-SVM-3 (h)EMD-LBP-SVM(1-2) (i)EMD-LBP-SVM(2-3)

(j) EMD-LBP-SVM-2(1-3)   (k)EMD-LBP-SVM(1-2-3)

Fig. 3. Different experiment simulation graphics

of EMD-LBP-SVM when P = 8 and R = 1, (f) is the experimental simulation of EMD-LBP-SVM when P = 16 and R = 2, (g) is the experimental simulation EMD-LBP-SVM when P = 24 and R = 3, (h) is the experimental simulation of the proposed method EMD-LBP-SVM mixed scale 1 and scale 2, (i) is the experimental method EMD-LBP-SVM mixed scale 2 and scale 3, (j) is the experimental simulation of the proposed method EMD-LBP-SVM mixed scale 1 and scale 3, (k) is the experimental simulation of the proposed method EMD-LBP-SVM mixed scale 1 and scale 2 and scale 3. Some of spots in the graph are due to the misclassification of the confusion between the different objects. It can be seen from Fig. 3 that the accuracy of the method proposed in this experiment (EMD-LBP-SVM) is significantly higher than that of the traditional SVM classification method, the EMD-SVM classification method and the LBP-SVM classification method, we can see that the number of spots in this experiment is very small, especially the experimental simulation of the last four mixed different scales LBP methods, compared with the ground truth, we know the classification accuracy of the experiment is quite high.

Through the above pictures and data comparison we know the proposed method in this paper compared the SVM method, EMD-SVM method and LBP-SVM method, the classification accuracy have improved to varying degrees, and through results we found that multi-scale LBP method could improve the classification accuracy by 1–2%

compared with the single-scale LBP method. The multi-scale LBP feature could reduce the loss of local effective information at a single scale, thus confirming the prediction before the experiment. The algorithm obviously has higher classification accuracy than the traditional classification method, and the overall accuracy could more than 97%, the average accuracy could more than 98%, Kappa coefficient could more than 0.97, the key point is the experimental parameters are very few and easy set.

6 Conclusion

In this paper, a new hyperspectral image classification based on empirical mode decomposition and local binary pattern is proposed, in the case of traditional hyperspectral image classification only considering spatial information or spectral information, or the method of spatial combination is too complex and so many parameters. Firstly, we need dimensionality reduction, in this paper, we use two-dimensional empirical mode decomposition to obtain some intrinsic mode functions. Secondly we extract the first principal component of each intrinsic mode functions by using principal component analysis. Then extracted the texture feature by local binary pattern, using principal component analysis extracted the spectral features and combined with the texture features. Finally, the features of the fusion was applied to the support vector machine, and the hyperspectral images are classified, The experimental results confirm the feasibility of the algorithm.

Acknowledgment. This work was supported by the National Natural Science Foundation of China under Grant No. 61563036 and the Fundamental Research Funds for the Central Universities in China under Grant No. 2013B32514.

References

1. Tian, Y.-P., Tao, C., Zou, Z.-G.: Hyperspectral image classification based on semi-supervision of active learning and graph. Acta Geodaetica Cartogr. Sin. **44**(8), 919–926 (2015)
2. Jin, J., Zou, Z.-R., Tao, C.: High-resolution remote sensing image compression texture element classification. Acta Geodaetica Cartogr. Sin. **43**(5), 493–499 (2014)
3. Zhang, W., Du, P.-J., Zhang, H.-P.: Study on hyperspectral mixed pixel decomposition method based on neural network. Bull. Surv. Mapp. **7**, 23–26 (2007)
4. Pei-jun, D., Lin, H., Sun, D.-x.: Progress of hyperspectral remote sensing classification based on support vector machine. Bull. Surv. Mapp. **12**, 37–40 (2006)
5. Wenying, H., Jiao, Y.-m.: Remote sensing image texture information extraction method. Yunnan Geogr. Environ. Res. **3**(19), 17–20 (2007)
6. Ojala, T., Harwood, I.: A comparative study of texture measures with classification based on feature distributions. Pattern Recogn. **29**(1), 51–59 (1996)
7. Ojala, T., Pietikäinen, M., Mäenpää, T.: Multiresolution gray-scale and rotation invariant texture classification with local binary patterns. IEEE Trans. Pattern Anal. Mach. Intell. **24**(7), 971–987 (2002)

8. Masood, K., Rajpoot, N.: Texture based classification of hyperspectral colon biopsy samples using CLBP. In: IEEE International Symposium on Biomedical Imaging, Boston, MA, USA, pp. 1011–1014, 01 July 2009
9. Huang, N.E., Shen, Z., Long, S.R.: The empirical mode decomposition and the Hilbert spectrum for nonlinear and nonstationary time series analysis. In: Proceeding of Royal Society, London, vol. A454, pp. 903–995 (1998)
10. Mäenpää, T.: The Local Binary Pattern Approach to Texture Analysis-Extension and Application [EB/OL] (2006). http://herkules.oulu.fi/isbn9514270762/
11. ftp.ecn.purdue.edu/biehl/MultiSpec/92AV3C/ [OL]

Extremely Randomized Forest with Hierarchy of Multi-label Classifiers

Jinxia Li[1], Yihan Zheng[2], Chao Han[2(✉)], Qingyao Wu[2,3], and Jian Chen[2]

[1] Computer Center, Hebei University of Economics and Business,
Shijiazhuang 300000, Hebei, China
[2] School of Software Engineering, South China University of Technology,
Guangzhou 510006, China
hanchaos@163.com
[3] The State Key Laboratory of Computer Science,
Institute of Software, Chinese Academy of Sciences, Beijing, China

Abstract. Hierarchy Of multi-label classifiERs (HOMER) is one of the most popular multi-label classification approaches. However, it is limited in its applicability to large-scale problems due to the high computational complexity when building the hierarchical model. In this paper, we propose a novel approach, called Extremely Randomized Forest with Hierarchy of multi-label classifiers (ERF-H), to effectively construct an ensemble of randomized HOMER trees for multi-label classification. In ERF-H, we randomly chose data samples with replacement from the original dataset for each HOMER tree. We constructed HOMER trees by clustering labels to split each hierarchy of nodes and learns a local multi-label classifier at every node. Extensive experiments show the effectiveness and efficiency of our approach compared to the state-of-the-art multi-label classification methods.

Keywords: Multi-label classification · Random forest · Hierarchy of classifiers

1 Introduction

Compared with single-label classification, multi-label classification framework is more practical. In fact, real-world objects generally have multiple semantic meanings. For example, a patient may suffer from a few diseases; a movie can be associated with several genres; genes are associated with a number of biological functions. Therefore, multi-label classification techniques have been frequently applied in various applications, including text categorization [12], bioinformatics [22], computer vision [23], and audio emotion detection [17].

A straighforward approach for multi-label classification is to decompose the problem into a set of single-label problems, and solve the decomposed problems with existing single-label classifiers. However, this approach totally neglects the dependencies among multiple labels, which could be quite important for multi-label classification, according to [7,10]. In recent years, much effort have

© Springer International Publishing AG 2017
Y. Sun et al. (Eds.): IScIDE 2017, LNCS 10559, pp. 450–460, 2017.
DOI: 10.1007/978-3-319-67777-4_40

been devoted to exploiting the label dependency in multi-label classification algorithms. Among them, Hierarchy Of multi-label classifiERs (HOMER) [18], based on balanced clustering of labels and hierarchical multi-label classification of examples, often yields more superior experimental performance comparing to other algorithms [13].

However, HOMER is limited in its applicability to large-scale applications due to its high computational complexity. At each node, HOMER method needs to cluster the labels into disjoint subsets and train multi-label classifiers to classify the examples into children nodes. Both these two steps are computational expensive, as the parameters need to be tuning by cross-validation.

Recently, *randomization methods* has been widely used to produce an ensemble of more or less strongly diversified tree models. Many randomization methods have been proposed, such as bagging [3], random forest [4] and extremely randomized trees [8]. All these methods explicitly introduce randomization into the learning algorithm to build a different randomized version of tree model. Then the predictions of these models are aggregated for classification. It has been shown that learning performance can be significantly enhanced by using the randomization methods to combine multiple tree models.

Inspired by the recent progress in randomization methods and ensemble learning, we propose a new hierarchical tree-based ensemble algorithm, called Extremely Randomized Forest with Hierarchy of multi-label classifiers (ERF-H), which builds an ensemble of totally randomized HOMER trees for multi-label classification. Our goal is to use randomization methods to improve both accuracy and computational efficiency of HOMER on large-scale multi-label datasets. To achieve this, we explicitly introduce randomizations in the induced HOMER trees by building each tree from bootstrap examples which are selected with replacement from the original dataset. Furthermore, we randomly select the number of clusters to run the balanced clustering algorithm as well as train multilabel classifiers with randomly selected parameters. The combination with multiple HOMER trees is very attractive because of the low computational cost of growing each tree through randomization methods. This leads to a new HOMER tree-based ensemble method which eliminates the speed disadvantages of single HOMER tree.

We compare the proposed ERF-H method with 6 state-of-the-art multi-label classification algorithms on 9 benchmark multi-label datasets. The results are evaluated using the Friedman and Nemenyi tests [6]. Experimental results show that the proposed approach is superior compared to other methods.

2   Related Work

In multi-label learning problems, we have a training dataset $D = \{x_i, Y_i\}(i = 1, 2, \cdots, N)$, where N is the number of training examples. Each example consists of a feature vector x_i and a set of labels Y_i, where Y_i is a subset of a label set L that contains q possible labels. We also have a test set $U = \{x_i\}$, where the labels for each test example is unknown. Our task is to predict the class labels of $x_i \in U$.

2.1   Multi-label Classification

Zhang et al. [21] summarized the existing multi-label classification approaches into three categories (*First-order* approaches, *Second-order* approaches and *Higher-order* approaches) based on the orders of label dependencies in multi-label classification. *First-order* approaches decompose the multi-label classification task into a number of independent tasks [2,23]. The most common method is the *binary relevance* (BR) method [19], which transforms a multi-label problem into multiple separate and independent binary problems, one for each label. It is clear that the first-order methods cannot handle label dependency, which might decreases predictive performance. *Second-order* approaches consider the pairwise relations between labels, such as interaction between any pair of labels [9,15]. In general, such pairwise label dependency is estimated by the co-occurrence of labels or some other equivalent measurements. However, these approaches might over-fit the training data since these dependencies are usually inaccurate. *High-order* approaches consider even higher order of relations among labels, such as the *full-order* style of imposing all the labels [5,11]. Nevertheless, the *full-order* approaches are usually impossible when the number of labels is large and the number of possible label subset combinations can be exponentially many.

2.2   Hierarchy of Multi-label Classifier

HOMER [18] uses a top-down manner to construct a hierarchy of multi-label classifiers tree. The root node contains the label set L including all the possible labels, i.e., $L_{root} = L$, and the labels are split into k disjoint subsets by using the balanced clustering algorithm [1]. Each internal node comprises of a set of labels $L_n \subseteq L$, and L_n is a union of the label sets of its children node, i.e., $L_n = \cup L_c$, $c \in children(n)$. There are $|L|$ leaves each corresponding to a single label. Each node n contains the dataset $D_n \subseteq D$ where at leat one label of the example belongs to L_n, i.e., $D_n = \{x_i, Y_i | Y_i \cap Ln \neq \emptyset\}$.

Each internal node n also consists of a multi-label classifier h_n. Denote meta-label μ_n as the disjunction of the labels in the node n. The task of h_n is to predict one or more meta-labels of its children. The label set of h_n is $M_n = \{\mu_c | c \in children(n)\}$.

When a new example x comes, HOMER starts at h_{root}, then recursively it pass x through the multi-label classifier h_c of a child node c if μ_c is the prediction of $h_{parent(c)}$. In the end, the multi-label classifiers of the last layer will produce a prediction consists of several single-labels. The union of these single labels is considered as the predictive class labels for the unlabeled example x.

3   ERF-H Algorithm

To improve the prediction accuracy, the ERF-H algorithm employs bootstrap method to build an ensemble of randomized HOMER trees based on a classical top-down procedure. The tree constructing algorithm has two main differences with the standard HOMER method: (i) it uses bootstrap samples drawn with replacement from the original dataset to construct each tree; (ii) it splits nodes by setting parameters of the clustering algorithm and the multi-label classifier at each node fully at random. The HOMER forest constructing process is summarized as follows:

- Generate K subsets $\{D_1, D_2, ..., D_K\}$ through bagging [3]
- Grow a HOMER tree $T_i(D_i)$ for each dataset D_i.
    1. At each node, randomly select the number of clusters to run the balanced clustering algorithm.
    2. Use a multi-label classifier trained with randomly selected parameters to divide the data into children nodes.
    3. Continue this process until all data are pure or cannot be further split by the classifier.
- Combine the multi-label predictions of K unpruned HOMER trees $T_1(D_1)$, $T_2(D_2)$,..., $T_K(D_K)$ into a randomized forest to make the final multi-label prediction. The class labels with vote probabilities larger than a predefined threshold λ are considered as the predictived labels.

3.1 Complexity Analysis

We assume that the training dataset contains $|D|$ instances that are collected from the label space with $|L|$ labels. The overall training complexity of HOMER is $O(|L|(f(|L|) + |L||D| + |L|^2))$, where $f(|L|)$ is the cost of training the multi-label classifier, and $O(|L||D|+|L|^2)$ is the cost of the balanced k means clustering algorithm [18].

For the proposed ERF-H algorithm, it contains K HOMER trees. The training complexity is $O(K|L|(f(|L|) + |L||D| + |L|^2))$.

During testing, HOMER outputs a number of labels by finding a set of decision paths from the root to a set of leaves. The complexity is $O(\log |L|)$. For the ERF-H algorithm with K HOMER trees, the testing complexity is $O(K \log |L|)$.

4 Experiments

4.1 Datasets

We use 9 multi-label datasets in our experiments. The data are originally split into training and testing set. The characteristics of the datasets are summarized in Table 1. These datasets are from various domains, including biology, multimedia and text categorization. R(21) and R(90) are two subset datasets including 21 and 90 classes from the Reuters21578 collection. The multi-label datasets are available at http://mulan.sourceforge.net/datasets.html.

Table 1. Description of the multi-label datasets in terms of application domain (domain), number of training (#tr.e.) and test (#t.e.) examples, the number of features (d), the total number of labels (q) and label cardinality (l_c). The problems are ordered by their overall complexity roughly calculated as #tr.e. $\times d \times q$.

Dataset	Domain	#tr.e.	#t.e.	d	q	l_c
Emotions	Multimedia	391	202	72	6	1.87
Scene	Multimedia	1211	1196	294	6	1.07
Yeast	Biology	1500	917	103	14	4.24
Genbase	Biology	662	463	199	27	1.252
Medical	Text	333	645	1449	45	1.25
Enron	Text	1123	579	1001	53	3.38
R(21)	Text	7140	2747	500	21	1.16
R(90)	Text	7770	3019	500	90	1.24
Bibtex	Text	4880	2515	1836	159	2.4

4.2 Compared Algorithms and Parameter Instantiation

In this paper, we compare the proposed ERF-H method with 6 multi-label algorithms, namely binary relevance (BR) [19], classifier chain (CC) [16], ensemble of classifier chain(ECC) [16], RAkEL [20], calibrated label ranking(CLR) [14], and HOMER [18]. ECC, RAkEL, and CLR are ensemble of multi-label methods. Our comparative study is based on the MULAN library.

For fair comparison, we use C4.5 as base classifier for BR, and use BR + C4.5 as multi-label classifier for HOMER and ERF-H.

The parameters of the methods are instantiated with the recommendations of the literature. In particular, for the ERF-H we need three parameters: the number of clusters, number of trees K and the prediction threshold λ. We randomly select the number of clusters from 2 to 6. K is set to 30, and λ is set to 0.4 as default.

4.3 Results and Discussion

The comparative results are shown in Table 2. For clearly showing the difference among those methods, the bar figure Fig. 1. shows the accuracy of all algorithms on all datasets. These experiments reveal a number of interesting points:

- Regardless of the evaluation measures (except Macropercision), our proposed ERF-H method always results in the best performance. This result reveals the superiority of the extremely randomized forest idea in growing the HOMER trees.
- Both ERF-H and ECC are based on the ensemble techniques and achieve better performance than the other algorithms, especially the HOMER and CC methods. This suggests the effectiveness of the ensemble techniques in improving the multi-label classification performance.

Table 2. Performance of the multi-label learning approaches in terms of hamming loss, accuracy, F1 score, subset accuracy, micro F1 and macro F1.

		CC	ECC	BR	CLR	RAKELd	HOMER	ERF-H
emo.	Hamming loss	0.290(7)	0.228(2)	0.260(3)	0.263(5)	0.273(6)	0.256(4)	**0.226(1)**
	Accuracy	0.428(7)	0.469(2)	0.438(4)	0.434(6)	0.445(3)	0.438(5)	**0.531(1)**
	F1 score	0.514(7)	0.540(2)	0.540(2)	0.532(6)	0.533(5)	0.540(2)	**0.625(1)**
	Subset accuracy	0.163(4)	**0.248(1)**	0.129(5)	0.129(6)	0.168(3)	0.129(7)	0.243(2)
	Micro-F1	0.556(7)	0.617(2)	0.593(4)	0.607(3)	0.578(6)	0.593(5)	**0.665(1)**
	Macro-F1	0.551(7)	0.597(3)	0.590(4)	0.603(2)	0.581(6)	0.590(5)	**0.659(1)**
Scene	Hamming loss	0.139(4)	0.104(2)	0.140(3)	0.141(5)	0.148(7)	0.145(6)	**0.095(1)**
	Accuracy	0.595(2)	0.579(3)	0.513(5)	0.503(6)	0.517(4)	0.496(7)	**0.665(1)**
	F1 score	0.615(2)	0.591(3)	0.552(5)	0.549(6)	0.553(4)	0.528(7)	**0.692(1)**
	Subset accuracy	0.533(3)	0.543(2)	0.401(6)	0.376(7)	0.407(4)	0.403(5)	**0.588(1)**
	Micro-F1	0.615(3)	0.668(2)	0.609(5)	0.614(4)	0.590(6)	0.587(7)	**0.727(1)**
	Macro-F1	0.624(3)	0.671(2)	0.616(5)	0.623(4)	0.599(6)	0.595(7)	**0.733(1)**
Yeast	Hamming loss	0.264(5)	0.209(2)	0.259(4)	0.226(3)	0.275(7)	0.272(6)	**0.207(1)**
	Accuracy	0.432(4)	0.481(2)	0.423(5)	0.468(3)	0.410(7)	0.413(6)	**0.497(1)**
	F1 score	0.535(5)	0.589(2)	0.547(4)	0.586(3)	0.528(6)	0.520(7)	**0.603(1)**
	Subset accuracy	0.143(3)	0.155(2)	0.064(7)	0.090(5)	0.075(6)	0.101(4)	**0.165(1)**
	Micro-F1	0.558(5)	0.618(2)	0.569(4)	0.614(3)	0.547(7)	0.548(6)	**0.631(1)**
	Macro-F1	**0.398(1)**	0.369(7)	0.386(5)	0.390(4)	0.382(6)	0.392(3)	0.396(2)
med.	Hamming loss	0.013(6)	0.013(3)	0.013(4)	0.013(5)	**0.012(1)**	0.014(7)	0.012(2)
	Accuracy	0.668(4)	0.660(5)	0.657(6)	0.649(7)	0.691(2)	0.683(3)	**0.744(1)**
	F1 score	0.696(4)	0.687(6)	0.688(5)	0.679(7)	0.714(2)	0.713(3)	**0.781(1)**
	Subset accuracy	0.588(4)	0.578(5)	0.567(6)	0.558(7)	0.623(2)	0.594(3)	**0.634(1)**
	Micro-F1	0.743(7)	0.746(4)	0.750(3)	0.745(5)	0.777(2)	0.743(6)	**0.791(1)**
	Macro-F1	0.245(5)	0.218(7)	0.246(4)	0.243(6)	0.270(3)	0.271(2)	**0.323(1)**
Enron	Hamming loss	0.053(4)	**0.048(1)**	0.054(5)	0.048(2)	0.055(6)	0.064(7)	0.052(3)
	Accuracy	0.394(3)	**0.446(1)**	0.367(6)	0.388(5)	0.390(4)	0.355(7)	0.445(2)
	F1 score	0.497(3)	0.552(2)	0.473(6)	0.492(5)	0.496(4)	0.460(7)	**0.559(1)**
	Subset accuracy	0.116(3)	**0.152(1)**	0.086(7)	0.097(5)	0.112(4)	0.091(6)	0.130(2)
	Micro-F1	0.514(4)	0.560(2)	0.504(6)	0.535(3)	0.506(5)	0.469(7)	**0.561(1)**
	Macro-F1	0.157(3)	0.145(6)	0.151(5)	0.143(7)	0.155(4)	0.169(2)	**0.201(1)**
Bibtex	Hamming loss	0.015(6)	0.013(3)	0.015(5)	**0.012(1)**	0.015(7)	0.014(4)	0.013(2)
	Accuracy	0.288(5)	0.284(6)	0.299(3)	0.183(7)	0.289(4)	0.335(2)	**0.355(1)**
	F1 score	0.348(6)	0.337(7)	0.366(4)	0.417(2)	0.357(5)	0.395(3)	**0.419(1)**
	Subset accuracy	0.140(5)	0.154(4)	0.133(6)	**0.488(1)**	0.123(7)	0.183(3)	0.187(2)
	Micro-F1	0.382(6)	0.375(7)	0.397(4)	0.448(2)	0.390(5)	0.429(3)	**0.459(1)**
	Macro-F1	0.261(6)	0.205(7)	0.276(4)	0.291(3)	0.267(5)	**0.304(1)**	0.303(2)
gen.	Hamming loss	0.001(2)	0.001(2)	0.001(2)	0.001(2)	0.001(6)	0.003(7)	**0.0007(1)**
	Accuracy	0.987(2)	0.987(2)	0.987(2)	0.987(2)	0.982(2)	0.964(7)	**0.991(1)**
	F1 score	0.991(2)	0.991(2)	0.991(2)	0.991(2)	0.986(6)	0.975(7)	**0.994(1)**
	Subset accuracy	0.975(2)	0.975(2)	0.975(2)	0.975(2)	0.970(6)	0.925(7)	**0.980(1)**
	Micro-F1	0.988(2)	0.988(2)	0.988(2)	0.988(2)	0.986(6)	0.965(7)	**0.992(1)**
	Macro-F1	0.738(2)	0.738(2)	0.738(2)	0.736(6)	0.737(5)	0.595(7)	**0.770(1)**
R(21)	Hamming loss	0.016(7)	**0.012(1)**	0.015(3)	0.016(4)	0.016(5)	0.016(5)	0.014(2)
	Accuracy	0.819(3)	0.848(2)	0.813(5)	0.804(7)	0.814(4)	0.805(6)	**0.879(1)**
	F1 score	0.836(3)	0.861(2)	0.831(5)	0.834(4)	0.834(4)	0.823(7)	**0.899(1)**
	Subset accuracy	0.769(3)	0.810(2)	0.755(4)	0.744(7)	0.754(5)	0.749(6)	**0.818(1)**
	Micro-F1	0.848(7)	0.885(2)	0.856(3)	0.852(4)	0.850(5)	0.849(6)	**0.889(1)**
	Macro-F1	0.746(5)	0.798(2)	0.772(3)	0.761(4)	0.743(6)	0.740(7)	**0.816(1)**
R(90)	Hamming loss	0.006(3)	**0.005(1)**	0.006(4)	0.006(5)	0.006(6)	0.006(7)	**0.005(1)**
	Accuracy	0.777(3)	0.787(2)	0.762(5)	0.752(7)	0.762(5)	0.771(4)	**0.808(1)**
	F1 score	0.796(3)	0.804(2)	0.782(6)	0.774(7)	0.783(5)	0.791(4)	**0.829(1)**
	Subset accuracy	0.722(3)	0.739(2)	0.700(5)	0.687(7)	0.699(6)	0.714(4)	**0.743(1)**
	Micro-F1	0.794(3)	0.805(2)	0.789(4)	0.784(6)	0.787(5)	0.776(7)	**0.809(1)**
	Macro-F1	0.398(2)	0.396(3)	0.384(5)	0.370(7)	0.392(4)	0.381(6)	**0.465(1)**
Rank	Hamming loss	5	1.8	3.7	3.9	5.6	5.6	1.7
	Accuracy	3.7	2.6	4.7	5.7	4	5	1.2
	F1 score	3.9	2.9	4.5	5.1	4.6	5	1.1
	Subset accuracy	3.3	2.2	5.4	5.4	4.8	4.9	1.4
	Micro-F1	5	2.6	3.9	3.9	5.2	5.7	1.1
	Macro-F1	4.1	4	4.3	4.8	4.9	4.3	1.3

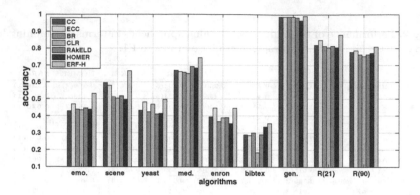

Fig. 1. Performance of the multi-label learning approaches in terms of accuracy.

- ECC performs best for the micro-precision, but ERF-H gets better result in terms of micro-recall and micro-F1. This means ERF-H is more complete but not the most accurate in compared with ECC. In general, ERF-H has a better classification performance against ECC.

4.4   Performance Measures and Statistical Evaluation

We use 4 example-based evaluation measures (hamming loss, accuracy, subset accuracy, and F1 score) and 2 label-based evaluation measures (microF1 score, and macroF1 score) [13]. The example-based measures are based on the average difference between the actual and the predicted set of labels of all examples. The label-based measures are based on the predictive performance for each label separately. If the hamming loss of an approach is smaller and the other measures are larger, the performance of the approach is better.

We use a two-step statistical test procedure to analyze the experimental results [6]. It contains a Friedman test of the null hypothesis that all learners have equal performance, and a Nemenyi test to compare all the classifiers to each other. All tests are based on the average ranks shown in the bottom row of Table 2. The results with respect to the post-hoc tests are shown in Fig. 2. A critical diagram contains an enumerated axis on which the average ranks (which are shown in the bottom row of the tables) of the algorithms are drawn. The best rank is at the right-most side of the axis. It does not differ significantly (at the significantly from each other level of $p = 0.05$) if the lines for the average ranks of the algorithms are connected.

4.5   Parameters

Our ERF-H approach has two parameters: the number of trees K and the multi-label prediction threshold λ. This experiment investigates how different values of these two parameters influence the performances of ERF-H. The performances of ERF-H with respects to different K on three datasets (emotions, scene and yeast)

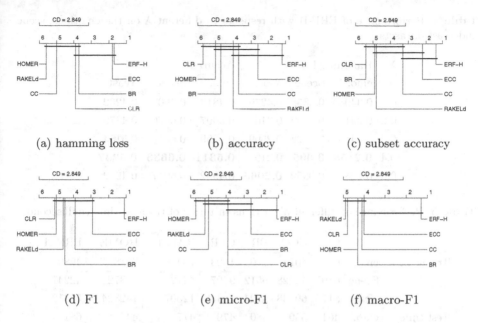

(a) hamming loss   (b) accuracy   (c) subset accuracy

(d) F1   (e) micro-F1   (f) macro-F1

Fig. 2. Graphical presentation of results from the Neminyi post hoc test at 0.05 significance level for different measures.

Table 3. Performance of ERF-H with respects to different K on the emotions, scene and yeast datasets

K	Hamming loss			Accuracy		
	emo.	Scene	Yeast	emo.	Scene	Yeast
10	0.2434	0.1105	0.2404	0.5217	0.6327	0.5018
20	0.2393	0.1034	0.2269	0.5164	0.6477	0.5184
30	0.2327	0.0967	0.2174	0.5216	0.6523	0.5189
40	**0.2261**	0.0959	0.219	**0.5311**	0.6575	0.5234
50	0.2351	**0.095**	**0.2158**	0.521	**0.6633**	**0.5237**

are given in Table 3. We can observe that the performance of ERF-H increases as the number of trees K increases. We find that ERF-H is able to achieve good performance when the number of trees K is large than 30. However, the expected training time of ERF-H will increase with the growth of K, so we use $K = 30$ as default parameter setting to provide a trade-off between accuracy and efficiency.

The performances of ERF-H with respects to different λ on the emotions, scene and yeast datasets are given in Table 4. The best performance of ERF-H is obtained when $\lambda = 0.4$ on the emotions and scene datasets, and $\lambda = 0.5$ on the yeast dataset. The optimal λ value is depended on the dataset used. We observe that $\lambda = 0.4$ is able to achieve good performances on the evaluated datasets, thus we use $\lambda = 0.4$ as default parameter setting.

Table 4. Performance of ERF-H with respects to different λ on the emotions, scene and yeast datasets

λ	Hamming loss			Accuracy		
	emo.	Scene	Yeast	emo.	Scene	Yeast
0.1	0.3498	0.2358	0.4279	0.4813	0.4909	0.4262
0.2	0.2913	0.1529	0.3109	0.5097	0.6076	0.4879
0.3	0.2483	0.1122	0.2509	0.5299	0.661	0.5154
0.4	**0.2158**	**0.095**	0.219	**0.5311**	**0.6633**	**0.5237**
0.5	0.2286	0.0959	**0.2065**	0.4992	0.6017	0.4972

Table 5. Performance of different algorithms in terms of training time (millisecond).

		CC	ECC	BR	CLR	RAkELd	HOMER	ERF-H
Traing time	emo.	629	4075	770	1323	754	9018	4089
	Scene	6197	44928	6512	9597	4412	84372	42241
	Yeast	8176	69238	9724	28432	12507	142824	58312
Test time	emo.	461	550	480	479	477	447	681
	Scene	527	964	541	798	788	513	1219
	Yeast	872	1117	914	1008	898	867	2437

4.6 Running Time

We also compare the training and testing time of different algorithms. The results are given in Tables 5. The experiments are conducted on a AMD 3.3 GHz machine with 16 GB RAM running Windows Server 2008. We can see that ERF-H requires more running time on training than CC, BR, CLR and RAKELd, but it is much faster than HOMER. Both ECC and ERF-H are ensemble methods and achieve similar running time results. Note that both ECC and ERF-H can be trained in parallel to build the ensemble, their expected running time can be reduced to almost the same as building one single singular base classifier.

5 Conclusion

In this paper, we present a new ensemble method, called Extremely Randomized Forest with Hierarchy of multi-label classifiers (ERF-H), for large scale multi-label classification. ERF-H randomly selects data examples with replacement from the original dataset to build each HOMER tree and learns the label clustering algorithm and multi-label classifier at each node with parameters chosen fully at random. We conducted extensive experiments on 10 datasets to show the effectiveness and efficiency of our approach comparing to the state-of-the-art methods.

Acknowledgments. This work was supported in part by National Natural Science Foundation of China (NSFC) under Grants 61502177, Fundamental Research Funds for the Central Universities under Grant D2172500, Special Planning Project of Guangdong Province under Grant No. 609055894069.

References

1. Banerjee, A., Ghosh, J.: Scalable clustering algorithms with balancing constraints. Data Min. Knowl. Discov. **13**(3), 365–395 (2006)
2. Boutell, M.R., Luo, J., Shen, X., Brown, C.M.: Learning multi-label scene classification. Pattern recogn. **37**(9), 1757–1771 (2004)
3. Breiman, L.: Bagging predictors. Mach. learn. **24**(2), 123–140 (1996)
4. Breiman, L.: Random forests. Mach. learn. **45**(1), 5–32 (2001)
5. Cheng, W., Hüllermeier, E.: Combining instance-based learning and logistic regression for multilabel classification. Mach. Learn. **76**(2), 211–225 (2009)
6. Demšar, J.: Statistical comparisons of classifiers over multiple data sets. J. Mach. Learn. Res. **7**, 1–30 (2006)
7. Elisseeff, A., Weston, J.: A kernel method for multi-labelled classification. In: Advances in Neural Information Processing Systems, pp. 681–687 (2001)
8. Geurts, P., Ernst, D., Wehenkel, L.: Extremely randomized trees. Mach. learn. **63**(1), 3–42 (2006)
9. Ghamrawi, N., McCallum, A.: Collective multi-label classification. In: Proceedings of the 14th ACM International Conference on Information and Knowledge Management, pp. 195–200 (2005)
10. Huang, S.J., Yu, Y., Zhou, Z.H.: Multi-label hypothesis reuse. In: Proceedings of the 18th ACM SIGKDD International Conference on Knowledge Discovery and Data Mining, pp. 525–533. ACM (2012)
11. Ji, S., Tang, L., Yu, S., Ye, J.: Extracting shared subspace for multi-label classification. In: Proceeding of the 14th ACM SIGKDD International Conference on Knowledge Discovery and Data Mining, pp. 381–389 (2008)
12. Kazawa, H., Izumitani, T., Taira, H., Maeda, E.: Maximal margin labeling for multi-topic text categorization. In: Advances in Neural Information Processing Systems (NIPS 2005), vol. 17, pp. 649–656 (2005)
13. Madjarov, G., Kocev, D., Gjorgjevikj, D., Džeroski, S.: An extensive experimental comparison of methods for multi-label learning. Pattern Recogn. **45**(9), 3084–3104 (2012)
14. Park, S.-H., Fürnkranz, J.: Efficient pairwise classification. In: Kok, J.N., Koronacki, J., de Mantaras, R.L., Matwin, S., Mladenič, D., Skowron, A. (eds.) ECML 2007. LNCS, vol. 4701, pp. 658–665. Springer, Heidelberg (2007). doi:10. 1007/978-3-540-74958-5_65
15. Qi, G., Hua, X., Rui, Y., Tang, J., Mei, T., Zhang, H.: Correlative multi-label video annotation. In: Proceedings of the 15th ACM International Conference on Multimedia, pp. 17–26. ACM (2007)
16. Read, J., Pfahringer, B., Holmes, G., Frank, E.: Classifier chains for multi-label classification. Mach. learn. **85**(3), 333–359 (2011)
17. Trohidis, K., Tsoumakas, G., Kalliris, G., Vlahavas, I.P.: Multi-label classification of music into emotions. In: ISMIR, vol. 8, pp. 325–330 (2008)
18. Tsoumakas, G., Katakis, I., Vlahavas, I.: Effective and efficient multilabel classification in domains with large number of labels. In: Proceedings of ECML/PKDD 2008 Workshop on Mining Multidimensional Data, pp. 30–44 (2008)

19. Tsoumakas, G., Katakis, I.: Multi-label classification: an overview. Int. J. Data Warehouse. Min. **3**(3), 1–13 (2007)
20. Tsoumakas, G., Katakis, I., Vlahavas, I.: Random k-labelsets for multilabel classification. IEEE Trans. Knowl. Data Eng. **23**(7), 1079–1089 (2011)
21. Zhang, M., Zhang, K.: Multi-label learning by exploiting label dependency. In: Proceedings of the 16th ACM SIGKDD International Conference on Knowledge Discovery and Data Mining, pp. 999–1008. ACM (2010)
22. Zhang, M., Zhou, Z.: Multilabel neural networks with applications to functional genomics and text categorization. IEEE Trans. Knowl. Data Eng. **18**(10), 1338–1351 (2006)
23. Zhang, M.L., Zhou, Z.H.: Ml-knn: a lazy learning approach to multi-label learning. Pattern Recogn. **40**(7), 2038–2048 (2007)

Subspace Clustering Under Multiplicative Noise Corruption

Baohua Li$^{(\boxtimes)}$ and Wei Wu

School of Mathematical Sciences, Dalian University of Technology, Dalian, China
libaoh@mail.dlut.edu.cn

Abstract. Traditional subspace clustering models generally adopt the hypothesis of additive noise, which, however, dose not always hold. When it comes to multiplicative noise corruption, these models usually have poor performance. Therefore, we propose a novel model for robust subspace clustering with multiplicative noise corruption to alleviate this problem, which is the key contribution of this work. The proposed model is evaluated on the Extend Yale B and MNIST datasets and the experimental results show that our method achieves favorable performance against the state-of-the-art methods.

Keywords: Subspace clustering · Multiplicative noise

1 Introduction

In computer vision society, some missions can be modeled as drawing high-dimensional samples from the union of multiple low-dimensional linear subspaces, that is subspace clustering. By subspace clustering, the missions such as motion trajectories segmentation [1,2], face images clustering [3], and video segmentation [4] can be reasonably interpreted and whose underlying structure of data also be well uncovered. Subspace clustering has received significant attention in recent years and many elaborate models have been developed. Thereinto, the reconstruction based clustering methods obtain good performance by employing the data *self-expressiveness* strategy which assumes that each data point in a union of subspace can be reconstructed by a linear combination of other points in the dataset. As for reconstruction based clustering, many methods have been proposed. Literature [5–8] proposed the sparse representation based algorithm to seek a sparse coefficient matrix in some sense. While, literatures [9–12] intend to find the low rank solution of the clustering models by imposing the nuclear norm constraint. Using the ridge regression technique, authors in [13] proposed the least square regression method for subspace clustering under enforced block diagonal condition and grouping effect guarantee. To balance the sparse and density, literatures [14,15] combine a quadratic data-fidelity term with trace lasso and elastic net regularization terms respectively to model the subspace clustering problem. In [16–18], the non-Gaussian noise corruption and

© Springer International Publishing AG 2017
Y. Sun et al. (Eds.): IScIDE 2017, LNCS 10559, pp. 461–470, 2017.
DOI: 10.1007/978-3-319-67777-4_41

impulsive noise corruption are considered. In addition, more general noise pollution is involved in [19] by mixture Gaussian regression. The mentioned methods perform subspace clustering issues in two stages: (1) using the estimated coefficient matrix to construct the affinity matrix which characterizes the degree of affinity among the different data points; (2) Applying the Normalized Cuts [30] to the affinity matrix to find the final clustering result. Therefore, the core of successful subspace clustering lies in how to find a suitable coefficient matrix.

Most reconstruction based subspace clustering methods are based on the assumption of additive noise, in which the captured data $A \in R^{m \times n}$ can be uniform shown as

$$A = AX + Z \tag{1}$$

where $X \in R^{n \times n}$ is the desired coefficient matrix, and $Z \in R^{m \times n}$ is the unknown noise term.

However, coherent imaging systems such as Doppler imaging, synthetic aperture radar (SAR) imaging, synthetic aperture sonar (SAS) imaging and ultrasound imaging, which play an important in medical, military, aerospace and other fields, are usually interfered by multiplicative noise. The multiplicative noise scenarios have been carefully studied by many researchers and obtained exciting performance on image restoration [20–24] etc. Usually, the degraded image F under multiplicative noise corruption is given by

$$G = (RF) \odot \Gamma \tag{2}$$

where $\odot$ denotes the Hadamard product which works in element-wise multiplication manner. R is the blur matrix, Γ denotes the noise, and G is the observed image. Here G, H, F and Γ have the agreed dimension. Especially, when R is the identity matrix, the abovementioned method (2) fails in the multiplicative noise removal problem which motivates us to consider the subspace clustering problem under multiplicative noise scenario. We contrast the Eq. (1) with (2), and proceed to provide the mathematical description of Eq. (1) with multiplicative noise flavour as

$$A = B \odot Z \tag{3}$$

where the elements of Z is not zero, A is the observed data, and B is the clean data. Note that, our ideal is to estimate a clean data matrix and then find its reconstruction coefficient. Therefore, we use B instead of AX in (3), where X is the coefficient matrix. If the elements of Z are all equal to 1, the data is clean. In this case, (3) boils down the noise free subspace clustering problem. Based on (3), a novel subspace clustering model is proposed by us.

The rest of this paper is organized as follows. Section 2 presents the necessary notations and preliminaries and based on which our proposed model is developed. Following that, we design an ADMM [26,27] based algorithm to solve the proposed model in Sect. 3. Experimental results are presented in Sect. 4. Finally, we make a conclusion in Sect. 5.

2   Notations and Preliminaries

We use capital and lowercase letters to represent matrixes and scalars respectively. Two operators will be utilized in the following pages. For a matrix C, we denote $vec\,(C)$ to return a column vector which contains all the elements of C. The operation $vec\,(C)$ is the same as the Matlab command vec. We denote $diag\,(C)$ to return a diagonal matrix, which is equal to the Matlab command $diag\,(vec\,(C))$. Let $\|\,C\,\|_F$, $\|\,C\,\|_1$, and $\|\,C\,\|_*$ denote the matrix Frobenius norm, ℓ_1 norm and nuclear norm of a matrix C respectively.

Given the corrupted data matrix $A = (A_1, \cdots, A_n)$ whose columns are data points drawn from k subspaces, different from the existing subspace clustering methods, we solve the following model

$$\min_{H,X,B} \frac{1}{2}\,\|\,H - \alpha E\,\|_F^2 + \lambda_1\,\|\,B - A \odot H\,\|_1 + \frac{1}{2}\,\|\,BX - B\,\|_F^2 + \lambda_2\,\|\,X\,\|_* \quad (4)$$

to conduct the subspace clustering problem in multiplicative noise scenario. Where H is the matrix whose entries are the reciprocal of noise, α is set to be the mean of H, and E is a matrix in which the entries are equal to 1. $\lambda_1 > 0$ and $\lambda_2 > 0$ are regularization parameters to balance the objective function. According the characteristic of subspace clustering data, we remove the total variation regularization term.

In addition, the common subspace clustering models directly employ the observed data, which are serious corrupted. It weakens the correlation of the data points within the same subspace, and leads to the poor performance. Instead of utilizing the observed data, our proposed model (4) resorts to an estimated data matrix which is obtained by preprocessing the observed data. Since the proposed model (4) involves the variance of the inverse of the noise, we need to manually input the mean of the inverse of noise before solving the object function. We list some frequently used multiplicative noises [21,24] etc. and the mean value of their inverse as bellow:

- The probability density function of Gamma noise is:

$$p\,(x) = \frac{x^{k-1}}{\Gamma\,(k)\,\theta^k}\,\exp\left(\frac{-x}{\theta}\right)$$

with $\theta > 0$ and $k > 1$. The mean value of $\frac{1}{x}$ is derived by

$$\mathbb{E}\left(\frac{1}{x}\right) = \frac{\Gamma\,(k-2)}{\theta^2\Gamma\,(k)} \int_0^{+\infty} \frac{x^{k-2}}{\Gamma\,(k-2)\,\theta^{k-2}}\,\exp\left(\frac{-x}{\theta}\right) dx$$

$$= \theta\frac{\Gamma\,(k-2)}{\theta^2\Gamma\,(k)}\,(k-2)$$

$$= \frac{1}{\theta\,(k-1)}$$

- The probability density function of Rayleigh noise is:

$$p\left(x\right) = \frac{x}{\sigma^2} \exp\left(-\frac{x^2}{2\sigma^2}\right)$$

The mean value of $\frac{1}{x}$ is derived by:

$$\mathbb{E}\left(\frac{1}{x}\right) = \int_0^{+\infty} \frac{1}{x} \frac{x}{\sigma^2} \exp\left(\frac{-x^2}{2\sigma^2}\right) dx$$

$$= \frac{\sqrt{2}}{\sigma} \int_0^{+\infty} \exp\left(-\left(\frac{x}{2\sigma}\right)^2\right) d\frac{x}{\sqrt{2}\sigma}$$

$$= \sqrt{\frac{\pi}{2\sigma^2}}$$

- Besides in additive noise situation, the Gaussian noise is also frequently used in multiplicative noise case. The probability density function of Gaussian noise is:

$$p\left(x\right) = \frac{1}{\sigma\sqrt{2\pi}} exp\left(-\frac{(x-\mu)^2}{2\sigma^2}\right)$$

Unlike the first two density functions, the mean value of $\frac{1}{x}$ in case of Gaussian

$$\mathbb{E}\left(\frac{1}{x}\right) = \frac{1}{\sigma\sqrt{2\pi}} \int_{-\infty}^{+\infty} \frac{1}{x} \exp\left(-\frac{(x-\mu)^2}{2\sigma^2}\right) dx$$

is not showed as a expression. Thanks to the numerical approximation [25] for above integral, which makes the approximate value of the inverse of Gaussian noise feasible.

3   The Numerical Algorithm

We are now in the position of numerical method section. Model (4) is reformulated as following constrained optimization problem:

$$\min_{H,X,B} \frac{1}{2} \parallel H - \alpha E \parallel_F^2 + \lambda_1 \parallel Z \parallel_1 + \frac{1}{2} \parallel BW - B \parallel_F^2 + \lambda_2 \parallel X \parallel_*$$

$$s.t. \ B - A \odot H = Z \tag{5}$$

$$W = X$$

by introducing the auxiliary variables Z and W. The augmented Lagrangian function of (5) is:

$$\mathcal{L}\left(H, X, B, W, Z, \Lambda_1, \Lambda_2\right) = \frac{1}{2} \parallel H - \alpha E \parallel_F^2 + \frac{1}{2} \parallel BW - B \parallel_F^2 + \lambda_1 \parallel Z \parallel_1$$

$$+ \lambda_2 \parallel X \parallel_* + tr\left(\Lambda_1^\top (B - A \odot H - Z)\right) + tr\left(\Lambda_2^\top (W - X)\right) \tag{6}$$

$$+ \frac{\beta}{2} \left(\parallel B - A \odot H - Z \parallel_F^2 + \parallel W - X \parallel_F^2\right)$$

Here, Λ_1, Λ_1 are multipliers, and β is the penalty parameter. Operation $tr\,(\cdot)$ denotes the trace of a square matrix. Our optimization contains two major steps: optimizing variables and optimizing multipliers. When updating a certain variable or multiplier, we always assume the rest items are fixed.

Ignoring the irrelevant terms, we update B via solving following optimization problem:

$$\min_{B} \frac{1}{2} \parallel BW - B \parallel_F^2 + \frac{\beta}{2} \parallel B - A \odot H - Z + \frac{\Lambda_1}{\beta} \parallel_F^2 \tag{7}$$

Similar to the problem (7), we find W, Z, X and H as follows: Update W

$$\min_{W} \frac{1}{2} \parallel BW - B \parallel_F^2 + \frac{\beta}{2} \parallel W - X + \frac{\Lambda_2}{\beta} \parallel_F^2 \tag{8}$$

Update Z

$$\min_{Z} \frac{1}{2} \parallel B - A \odot H - Z + \frac{\Lambda_2}{\beta} \parallel_F^2 + \parallel Z \parallel_1 \tag{9}$$

Update X

$$\min_{X} \frac{1}{2} \parallel W - X + \frac{\Lambda_2}{\beta} \parallel_F^2 + \frac{\lambda_2}{\beta} \parallel X \parallel_* \tag{10}$$

Updata H

$$\min_{H} \frac{1}{2} \parallel H - \alpha E \parallel_F^2 + \frac{\beta}{2} \parallel B - A \odot H - Z + \frac{\Lambda_2}{\beta} \parallel_F^2 \tag{11}$$

In order to decouple the $A \odot H$ and make the subproblem (11) easy to be carried out, we reformulate (11) as

$$\min_{vec(H)} \frac{1}{2} \parallel vec\,(H) - \alpha vec\,(E) \parallel_2^2$$
$$+ \frac{\beta}{2} \parallel vec\,(B) - diag\,(A)\,vec\,(H) - vec\,(Z) + \frac{vec\,(\Lambda_2)}{\beta} \parallel_F^2 \tag{12}$$

Our overall ADMM based algorithm is summarized in Algorithm 1. Since the update of the multipliers Λ_1 and Λ_2 does not need to solve optimization function, we just list their update processes in Algorithm 1. In addition, the shrinkage operator $S_\tau[x]$ will be exploited in our algorithm to simplify the expressions of optimization problems:

$$S_\tau[x] = \begin{cases} x - \tau, & if\ x > \tau \\ x + \tau, & if\ x < -\tau \\ 0, & otherwise \end{cases} \tag{13}$$

Moreover, the discussion of convergence guarantee of Algorithm 1 is similar to [28,29], we thus omit it. After calculating the reconstruction coefficient X, we construct the affinity matrix by $\frac{(|X|+|X^\top|)}{2}$. Then, the Normalized Cuts [30] is applied to perform the subspace clustering. The clustering accuracy is defined by:

$$Accuracy = 1 - \frac{missclustered\ points}{total\ data\ points} \times 100\%$$

Algorithm 1: Finding the solution by ADMM

Input: The dataset, implicative noise, regularization parameter λ_1, λ_2 and β, the mean α of inverse noise, the number of subspaces k, and tolerate error ε

Given the initial estimation of W, Z, X, H, Λ_1, Λ_2

Repeat:

1. Update B via solving (7)

$$B^{(n+1)} = \beta \left(A \odot H^{(n)} - Z^{(n)} + \frac{\Lambda_1^{(n)}}{\beta} \right) \left(\left(W^{(n)} - I \right) \left(W^{(n)} - I \right)^{\top} + \beta I \right)^{-1}$$

2. Update W via solving (8)

$$W^{(n+1)} = \left(B^{(n+1)^{\top}} B^{(n+1)} + \beta I \right)^{-1} \left(B^{(n+1)^{\top}} B^{(n+1)} + \beta X^{(n)} - \Lambda_2^{(n)} \right)$$

3. Update Z via solving (9)

$$Z^{(n+1)} = S_{\frac{\lambda_2}{\beta}} [B^{(n+1)} - A \odot H^{(n)} + \frac{\Lambda_2^{(n)}}{\beta}]$$

4. Update X via solving (10)

$$(U, \Sigma, V) = svd \left(W^{(n+1)} + \frac{\Lambda_2^{(n)}}{\beta} \right)$$

$$X^{(n+1)} = U S_{\frac{\lambda_2}{\beta}} [\Sigma] V^{\top}$$

5. Update H via solving (12)

$$vec(H)^{(n+1)} = \left(I + \beta \left(diag\left(A \right) \right)^2 \right)^{-1} \left(\alpha vec\left(E \right) + \beta vec \left(B^{(n+1)} - Z^{(n+1)} + \frac{\Lambda_2^{(n)}}{\beta} \right) \right)$$

we finish updating H by transforming $vec(H)^{(n+1)}$ in to matrix $H^{(n+1)}$

6. Update the multipliers Λ_1 and Λ_2 as follows

$$\Lambda_1^{(n+1)} = \Lambda_1^{(n)} + \beta \left(B^{(n+1)} - A \odot H^{(n+1)} - Z^{(n+1)} \right)$$

and

$$\Lambda_2^{(n+1)} = \Lambda_2^{(n)} + \beta \left(W^{(n+1)} - X^{(n+1)} \right)$$

7. Until

$\| X^{(n+1)} - X^{(n)} \|_F \leq \varepsilon$

$\| W^{(n+1)} - W^{(n)} \|_F \leq \varepsilon$

$\| Z^{(n+1)} - Z^{(n)} \|_F \leq \varepsilon$

$\| H^{(n+1)} - H^{(n)} \|_F \leq \varepsilon$

End loop. Output X

---

4   Experiments

In this section we will report the performance of our new method on Extended Yale B database and the MNIST database of hand-written digits under different commonly used multiplicative noise corruption respectively. Our method is evaluated with four state-of-the-art algorithms, that are SSC [6], LRR [11], LSR [13] and CASS [14]. The experimental results are list in tables. The parameters for each clustering algorithm are tuned to reach the best performance. The clustering accuracies of our method are shown in red color.

The extend Yale B database contains 38 objects in total and each object contains 64 images. Here, we take it's first 5, 8 and 10 classes data and corrupt them with frequently used multiplicative noises, then proceed to group these data. The data are reduced into a 30, 48, 60-dimension new data respectively before solving the object function. Figure 1 illustrates the clean data, the corrupted data and the restored data. Note that, we take its first 1 class 64 images without reducing dimensions, and show the restoring result. The restored images are closer to the clean ones, which avoids poor performance caused by serious corruption of noises. The results of clustering accuracies prove the superiority of employing restored estimated data matrix in model (4). Tables 1, 2 and 3 state the clustering result. It demonstrates that our method outperforms the SSC, LRR, LSR, and CASS. For the 5 objects face clustering issue, all the mentioned methods reach satisfied accuracies under the serous multiplicative noise corruption. As for the 8 objects and 10 objects situations, our method makes a remarkable improvement which dues to both the noise removing step and the grouping effect of nuclear norm regularization term of our model. We can see that LRR method, which adopts nuclear norm regularization term too, however, doesn't perform very well. It further proves that exploiting estimated data matrix through noise removing step is reasonable. It is also observed that the performance of each method decreases from the Tables 1, 2 and 3 as the numbers of object get larger. For the reason that the trace lasso norm is adaptive to the correlation of data, CASS performs better than SSC, LRR and LSR. The SSC enjoys the property of variable selection [31] while LSR and LRR possess the grouping effect. These elegant properties guarantee the performance of above-mentioned methods. However, the serious multiplicative noise may break the correlation of data and depress the elegant properties. Therefore, the performance of CASS, SSC, LRR, LSR are limited. Our model adopt noise removed data to estimate the coefficient matrix, which is different from the previous methods. By employing the learned data matrix which is less affected by the noise, we obtain the competitive clustering accuracies.

The MNIST database of hand-written digitals is another challenging data set in subspace clustering society. It contains 10 objects which are corresponding to 0–9 of size 28×28 pixels. We select a subset of it in which each object of 0–9 contains 50 samples. Before grouping the chosen data by all methods, we corrupt them by the three common multiplicative noises, and use PCA to project them into 12-dimensional data inspired by [32]. We evaluate the models according to different noises and report the clustering accuracies in the Table 4. As we can see, in all cases, our method performances better than the others. According to the experimental results, we know that the clustering accuracies heavily depend on the noises removal step in the multiplicative noises corruption.

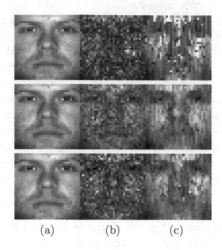

(a)   (b)   (c)

Fig. 1. (a) Clean images, (b) The noisy images corrupted by multiplicative Rayleigh, Gamma and Gaussian noise from the top down. (c) The corresponding noise removed results.

Table 1. The clustering accuracies about 5 objects

	5 objects		
	Gamma	Gaussian	Rayleigh
Our method	64.06%	61.75%	59.07%
SSC	52.64%	53.08%	50.43%
LRR	52.73%	51.77%	47.33%
LSR	50.35%	48.04%	45.17%
CASS	56.97%	55.04	52.72%

Table 2. The clustering accuracies about 8 objects

	8 objects		
	Gamma	Gaussian	Rayleigh
Our method	53.05%	51.83%	49.31%
SSC	45.69%	43.83%	40.76%
LRR	46.74%	44.01%	42.33%
LSR	41.63%	39.78%	40.02%
CASS	48.76%	46.55%	43.75%

Table 3. The clustering accuracies about 10 objects

	10 objects		
	Gamma	Gaussian	Rayleigh
Our method	41.96%	40.32%	39.31%
SSC	37.56%	34.85%	33.61%
LRR	34.07%	31.86%	32.63%
LSR	33.73%	29.06%	28.61%
CASS	38.77%	36.01%	33.88%

Table 4. The clustering accuracies of MINST data

	MINST data		
	Gamma	Gaussian	Rayleigh
Our method	39.59%	39.04%	38.88%
SSC	35.06%	33.67%	30.96%
LRR	34.66%	31.07%	28.99%
LSR	32.86%	29.43%	27.96%
CASS	36.11%	35.01%	31.96%

5   Conclusion

In this work, we propose a novel reconstruction based subspace clustering model with multiplicative noises corruption. To the best of our knowledge, it is the first work to address the subspace clustering issue under the multiplicative noises scenario. An effective numerical algorithm is designed based on ADMM. At last, the experimental results on the Extended Yale B and MNIST datasets state the effectiveness of our method.

References

1. Costeira, J.P., Kanade, T.: A multibody factorization method for independently moving objects. Int. J. Comput. Vis. **29**(3), 159–179 (1998)
2. Rao, S., Tron, R., Vidal, R., Ma, Y.: Motion segmentation in the presence of outlying, incomplete, or corrupted trajectories. IEEE Trans. Pattern Anal. Mach. Intell. **32**(10), 1832–1845 (2010)
3. Basri, R., Jacobs, D.W.: Lambertian reflectance and linear subspaces. IEEE Trans. Pattern Anal. Mach. Intell. **25**(2), 218–233 (2003)
4. Xiao, S., Tan, M., Xu, D.: Weighted block-sparse low rank representation for face clustering in videos. In: Fleet, D., Pajdla, T., Schiele, B., Tuytelaars, T. (eds.) ECCV 2014. LNCS, vol. 8694, pp. 123–138. Springer, Cham (2014). doi:10.1007/978-3-319-10599-4_9
5. Elhamifar, E., Vidal, R.: Sparse subspace clustering. In: Proceedings of the IEEE Computer Vision and Pattern Recognition, pp. 2790–2797 (2009)
6. Elhamifar, E., Vidal, R.: Sparse subspace clustering: algorithm, theory, and applications. IEEE Trans. Pattern Anal. Mach. Intell. **35**(11), 2765–2781 (2013)
7. Li, C.-G., Vidal, R.: Structured sparse subspace clustering: a unified optimization framework. In: Proceedings of the IEEE Conference on Computer Vision and Pattern Recognition, pp. 277–286 (2015)
8. Yin, M., Guo, Y., Gao, J., He, Z., Xie, S.: Kernel sparse subspace clustering on symmetric positive definite manifolds. In: Proceedings of the IEEE Conference on Computer Vision and Pattern Recognition, pp. 5157–5164 (2016)
9. Liu, G., Lin, Z., Yu, Y.: Robust subspace segmentation by low-rank representation. In: Proceedings of the 27th International Conference on Machine Learning, pp. 663–670 (2010)
10. Favaro, P., Vidal, R., Ravichandran, A.: A closed form solution to robust subspace estimation and clustering. In: Proceedings of the IEEE Computer Vision and Pattern Recognition, pp. 1801–1807 (2011)
11. Liu, G., Lin, Z., Yan, S., Sun, J., Yu, Y., Ma, Y.: Robust recovery of subspace structures by low-rank representation. IEEE Trans. Pattern Anal. Mach. Intell. **35**(1), 171–184 (2013)
12. Vidal, R., Favaro, P.: Low rank subspace clustering (LRSC). Pattern Recogn. Lett. **43**, 47–61 (2014)
13. Lu, C.-Y., Min, H., Zhao, Z.-Q., Zhu, L., Huang, D.-S., Yan, S.: Robust and efficient subspace segmentation via least squares regression. In: Fitzgibbon, A., Lazebnik, S., Perona, P., Sato, Y., Schmid, C. (eds.) ECCV 2012. LNCS, vol. 7578, pp. 347–360. Springer, Heidelberg (2012). doi:10.1007/978-3-642-33786-4_26

14. Lu, C., Feng, J., Lin, Z., Yan, S.: Correlation adaptive subspace segmentation by trace lasso. In: Proceedings of the IEEE International Conference on Computer Vision, pp. 1345–1352 (2013)
15. You, C., Li, C.-G., Robinson, D.P., Vidal, R.: Oracle based active set algorithm for scalable elastic net subspace clustering. In: Proceedings of the IEEE Conference on Computer Vision and Pattern Recognition, pp. 3928–3937 (2016)
16. Lu, C., Tang, J., Lin, M., Lin, L., Yan, S., Lin, Z.: Correntropy induced L2 graph for robust subspace clustering. In: Proceedings of the IEEE International Conference on Computer Vision, pp. 1801–1808 (2013)
17. Zhang, Y., Sun, Z., He, R., Tan, T.: Robust subspace clustering via half-quadratic minimization. In: Proceedings of the IEEE International Conference on Computer Vision, pp. 3096–3103 (2013)
18. He, R., Zhang, Y., Sun, Z., Yin, Q.: Robust subspace clustering with complex noise. IEEE Trans. Image Process. 24(11), 4001–4013 (2015)
19. Li, B., Zhang, Y., Lin, Z., Lu, H.: Subspace clustering by mixture of Gaussian regression. In: Proceedings of the IEEE Conference on Computer Vision and Pattern Recognition, pp. 2094–2102 (2015)
20. Huang, Y.-M., Lu, D.-Y., Zeng, T.: Two-step approach for the restoration of images corrupted by multiplicative noise. SIAM J. Sci. Comput. 35(6), 2856–2873 (2013)
21. Zhao, X.-L., Wang, F., Ng, M.K.: A new convex optimization model for multiplicative noise and blur removal. SIAM J. Imaging Sci. 7(1), 456–475 (2014)
22. Aubert, G., Aujol, J.-F.: A variational approach to removing multiplicative noise. SIAM J. Appl. Math. 68(4), 925–946 (2008)
23. Rudin, L., Lions, P.-L., Osher, S.: Multiplicative denoising and deblurring: theory and algorithms. In: Osher, S., Paragios, N. (eds.) Geometric Level Set Methods in Imaging, Vision, and Graphics, pp. 103–119. Springer, Heidelberg (2003). doi:10.1007/0-387-21810-6_6
24. Wang, F., Zhao, X.-L., Ng, M.K.: Multiplicative noise and blur removal by framelet decomposition and l_1-based l-curve method. IEEE Trans. Image Process. 25(9), 4222–4232 (2016)
25. Shampine, L.F.: Vectorized adaptive quadrature in matlab. J. Comput. Appl. Math. 211(2), 131–140 (2008)
26. Boyd, S., Parikh, N., Chu, E., Peleato, B., Eckstein, J.: Distributed optimization and statistical learning via the alternating direction method of multipliers. Found. Trends® Mach. Learn. 3(1), 1–122 (2011)
27. Ng, M.K., Wang, F., Yuan, X.: Inexact alternating direction methods for image recovery. SIAM J. Sci. Comput. 33(4), 1643–1668 (2011)
28. Mairal, J., Bach, F., Ponce, J., Sapiro, G.: Online dictionary learning for sparse coding. In: International Conference on Machine learning, pp. 689–696 (2009)
29. Nguyen, H., Yang, W., Sheng, B., Sun, C.: Discriminative low-rank dictionary learning for face recognition. Neurocomputing 173, 541–551 (2016)
30. Shi, J., Malik, J.: Normalized cuts and image segmentation. IEEE Trans. Pattern Anal. Mach. Intell. 22(8), 888–905 (2000)
31. Tibshirani, R.: Regression shrinkage and selection via the lasso. J. Roy. Stat. Soc. Ser. B (Methodol.) 73(3), 267–288 (1996)
32. Hastie, T., Simard, P.Y.: Metrics and models for handwritten character recognition. Stat. Sci. 13(1), 54–65 (1998)

Imaging

Geographic Atrophy Segmentation for SD-OCT Images by MFO Algorithm and Affinity Diffusion

Yubo Huang[1], Zexuan Ji[1(✉)], Qiang Chen[1], and Sijie Niu[2]

[1] School of Computer Science and Engineering,
Nanjing University of Science and Technology, Nanjing 210094, China
jizexuan@njust.edu.cn
[2] School of Information Science and Engineering, University of Jinan,
Jinan 250022, China

Abstract. Age-related macular degeneration (AMD) is a common cause of vision loss among the elderly in developed countries. Geographic atrophy (GA) appears in advanced stages of non-exudative AMD. In this paper, we present a hybrid GA segmentation model for spectral-domain optical coherence tomography (SD-OCT) images. The method first segments the layered structure of the SD-OCT scan data and produces the projection images. Then we construct the histogram of the resulting image into a probability distribution function, and use this function to fit a Gaussian mixed model (GMM) by Moth-flame optimization (MFO) algorithm. To incorporate the globe spatial information to over come the impact of noise, a robust affinity diffusion method is proposed to construct the affinity map. Finally, bias field correction process is employed to remove the intensity inhomogeneity. Two data sets, respectively consisting on 55 SD-OCT scans from twelve eyes in eight patients with GA and 56 SD-OCT scans from 56 eyes in 56 patients with GA, are utilized to quantitatively evaluate the segmentation algorithm. Experimental results demonstrate that the proposed algorithm can achieve high segmentation accuracy.

Keywords: MFO · Geographic atrophy · SD-OCT · Affinity diffusion · Bias field

1 Introduction

Age-related macular degeneration (AMD) is a disease that leads visual distortion, visual impairment and severe visual impairment [1], which is characterized by the development of two major abnormalities: Choroidal neovascularization (CNV) and Geographic atrophy(GA) [2, 3]. GA is an atrophy of the retinal pigment epithelium (RPE) with clear boundary and gradually expand [3, 4]. Accurate identification of GA regions is necessary because it is an important information for clinicians to make a treatment decision.

Most semi-automated or automated image analysis of GA are all applied to color fundus (CFP), fundus auto-fluorescence (FAF) or optical coherence tomography (OCT) modalities [5–8]. Traditional imaging techniques (CFP, FFA) are mostly based

© Springer International Publishing AG 2017
Y. Sun et al. (Eds.): IScIDE 2017, LNCS 10559, pp. 473–484, 2017.
DOI: 10.1007/978-3-319-67777-4_42

on the retinal fundus examination, in which only fundus plane structure is provided and the fundus internal fine structure cannot be observed. Therefore, it is difficult to find early pathological features [9–11]. As a high-resolution, non-invasive, noninvasive three-dimensional imaging technology, OCT can clearly show the internal fine structure and has been widely used in the eye disease examination [12]. Spectral-domain (SD-OCT), which brought the application of fourier transform theory, has produced a fundamental breakthrough in imaging speed and resolution [13]. State-of-the-art algorithms mainly segment the GA regions based on the projection image generated with the voxels between the RPE and the choroid layers. Chen et al. [14] utilized geometric active contours to produce satisfactory performances comparing with manually-defined GA regions. Niu et al. [1] proposed an automated GA segmentation method for SD-OCT images by using Chan-Vese model via local similarity factor.

Gaussian mixture model (GMM) is one of finite mixed models (FMMs) which has a wide range of applications in cluster analysis and image segmentation. The traditional mathematical method of solving GMM model is expectation-maximization (EM) algorithm. However, in the real problem, the mathematical method is often easy to fall into the local optimal solution. Thus the heuristic intelligent optimization algorithm to solve the parameters has been proposed. Tohka et al. [15] utilized real coded Genetic algorithms to solve GMM parameters to segment the MR images. Mirjalili [16] proposed a novel nature-inspired optimization paradigm called moth-flame optimization (MFO) algorithm.

In order to reduce the impact of noise pixels and the image inhomogeneity, affinity map [17, 19] and bias field [18] has been introduced in this paper. The affinity map is a matrix that can represent the similarity of any two pixels. We can consider the affinity map as global spatial information to improve noise robustness. Since intensity inhomogeneity often causes misclassification of GA regions, it is a mandatory step to remove the intensity inhomogeneity through a procedure called bias field correction. Therefore, we have proposed a hybrid model to solve the GA segmentation problem by utilizing MFO for GMM, affinity map and bias field correction.

2   Proposed Model

We have proposed a hybrid model on solving the GA segmentation problem. Firstly we segment the layered structure of the SD-OCT scan data and obtain the projection images by the method illustrated in [1, 14]. Secondly we construct the histogram of the resulting image into a probability distribution function, and use this function to fit a GMM by MFO algorithm. The original segmentation result is obtained by the GMM and it is used to correct the image bias field. Continue to iteratively segment and estimate the bias field, thus we can finally get a final segmentation result. The overall process is shown in Fig. 1.

2.1   Generation of GA Projection Image

Our method is performed on 2-D images, so we need to generate GA projection images from SD-OCT volumetric data sets. The summed-voxel projection (SVP) [20] is a

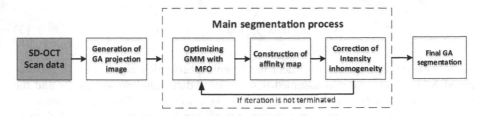

Fig. 1. The pipeline of the proposed automatic GA segmentation method

common projection method. However, it is difficult to segment the GA region in the SVP image due to the confounding influence of highly reflective retinal layers above and below GA lesions in the retina with obscure GA lesions. Therefore, the restricted summed-voxel projection (RSVP) as the projection approach is employed in [1, 14]. It is derived by restricting the sum of the voxel values to the sub-volume beneath the RPE layer to making the GA area more obvious than the SVP image. The lower boundary of the sub-volume is limited to be parallel to the top boundary (RPE layer), where the parallel distance is equal to the minimum distance between the end of the cube and the segmented RPE layer. The intensity value of the GA projection image is set as the average intensity of the sub-volume in the axial direction.

2.2 Optimizing GMM with MFO

Image intensities are denoted by $x \in \mathcal{R}^d, i = 1, \ldots, N$. All of these intensities are drawn from 1 of the K classes. We assume that K is a priori information of the image. In our task, K is set to 2 because the image has only the lesion area and the normal area. Intensities drawn from the class k follow the probability density function (pdf) $f_k(x|\theta_k)$, where $k = 1, \ldots, K$. The form of the pdf in our method is the GMM as follows:

$$f_k(x|\theta_k) = f_k(x|\mu_k, \Sigma_k) = \frac{e^{-0.5(x-\mu_k)^T \Sigma_k^{-1}(x-\mu_k)}}{\sqrt{(2\pi)^d |\Sigma_k|}} \tag{1}$$

where μ_k is the mean vector and Σ_k is the positive definite covariance matrix of the class k. Each class has a prior probability $p_k \in [0, 1]$, and satisfies $\sum_{k=1}^{K} p_k = 1$. We denote the set of all parameter values by $\Theta = \{p_k, \theta_k : k = 1, \ldots, K\}$. Combining the above models, the complete model is

$$f(x|\Theta) = \sum_{k=1}^{K} p_k f_k(x|\theta_k) \tag{2}$$

The objective is to estimate the parameters Θ given the data $\{x_i : i = 1, \ldots, N\}$. Here, the estimation is based on the maximum likelihood (ML) principle. To find the ML estimate, an optimization problem has to be solved

$$\widehat{\Theta} = \arg\max_{\Theta} \sum_{i=1}^{N} \log f(x|f_k(x_i|\Theta)) \tag{3}$$

MFO algorithm is one of efficient intelligent optimization algorithms. In the proposed MFO algorithm, it is assumed that the candidate solutions are moths and the values of the variables are the positions of moths in the space. There are n moths and n flames in one defined moth population, which can be expressed as the following matrix:

$$M = \begin{pmatrix} m_{1,1} & \cdots & m_{1,d} \\ \vdots & \ddots & \vdots \\ m_{n,1} & \cdots & m_{n,1} \end{pmatrix} F = \begin{pmatrix} F_{1,1} & \cdots & F_{1,d} \\ \vdots & \ddots & \vdots \\ F_{n,1} & \cdots & F_{n,1} \end{pmatrix} \tag{4}$$

where n is the number of moths or flames, d is the number of variables' dimensions. For all the moths and flames, a vector for storing the corresponding fitness values is introduced as follows:

$$OM = \begin{bmatrix} OM_1 \\ \vdots \\ OM_n \end{bmatrix} OF = \begin{bmatrix} OF_1 \\ \vdots \\ OF_n \end{bmatrix} \tag{5}$$

The moths play the roles of moving and searching in the solution space, while the flames store the optimal position of the moths so far. Each moth can fly around a flame and update its position matrix when a better solution is found. The MFO's specific algorithm is composed of a three-tuple as follows:

$$MFO = (\Gamma, P, T) \tag{6}$$

Γ is a function that generates a random population of moths and corresponding fitness values. The P function, which is the main function, moves the moths around the search space. This function received the matrix of M and returns its updated one eventually. The T function is used to determine whether to terminate.

Considering that Spiral's initial point should start from the moth and Spiral's final point should be the position of the flame, a logarithmic spiral is defined for the MFO algorithm as follows:

$$M_i = S(M_i, F_i) = D_i \cdot e^{bt} \cdot \cos(2\pi t) + F_j \tag{7}$$

where D_i indicates the distance of the i-th moth for the j-th flame, b is a constant for defining the shape of the logarithmic spiral, and t is a random number in $[-1, 1]$. The first moth always updates its position with respect to the best flame, whereas the last moth updates its position with respect to the worst flame in the list. The F matrix is the optimal solution so far and the moths continue to find better solutions around these flames. The complete P function is illustrated in [16].

In our model we set the classes number K to 2. Therefore the GMM parameters can be denoted as $\Theta = \{p_1, \mu_1, \Sigma_1, p_2, \mu_2, \Sigma_2\}$ according to formula (1). Θ is regarded as one row of vectors in the Moths matrix or Flames matrix in formula (4). The MFO algorithm is performed to estimate the parameters. We map the intensity histogram of the input image into a true probability density distribution through the parzen window and the estimated probability density distribution is then obtained by the estimated parameters. The Kullback–Leibler (KL) divergence, which is a measure of the non-symmetric difference between two probability distributions, is regarded as the fitness function of the MFO algorithm and a global optimal solution $\hat{\Theta}$ can be obtained. The probability that each pixel belongs to each class is easy to calculate as

$$V = \begin{bmatrix} v_{1,1} & v_{1,2} \\ \vdots & \vdots \\ v_{N,1} & v_{N,2} \end{bmatrix} \tag{8}$$

where $v_{i,j}$ is the probability that i-th pixel belongs to j-th class.

2.3 Construction of Affinity Map

In order to avoid using only intensity information, we introduce affinity maps as means of using spatial information. For an image with N pixels, we construct a weighted graph, $G = (V, E)$, by taking each pixel as a node, and connecting each pair of pixels by an edge. In other words, the graph is expressed as an $N \times N$ weight matrix W as follows:

$$W_{ij} = e^{\frac{-\left\|F_{(i)} - F_{(j)}\right\|_2^2}{\sigma_I}} \times \begin{cases} e^{\frac{-\left\|X_{(i)} - X_{(j)}\right\|_2^2}{\sigma_I}}, & \text{if } \left\|X_{(i)} - X_{(j)}\right\|_2 < r \\ 0, & \text{otherwise} \end{cases} \tag{9}$$

where W_{ij} is the similarity measure between i-th pixel and j-th pixel, $F_{(i)}$ is the brightness value of i-th pixel, $\left\|X_{(i)} - X_{(j)}\right\|_2$ is the distance of i-th pixel and j-th pixel and r is the pre-defined distance scale. We can observe that the similarity measure between a pair of pixels is determined by their spatial distance and intensity value together.

Since the local similarities radius is r, the similarity measure between pixels whose distance is greater than r are set to 0, which lost some relationship information among the pixels. A diffusion process named self-diffusion process (SD) is then introduced to improve the initial matrix to get a faithful affinity map. According to [19], the SD works directly on the input (initial) affinity and uses a regularization in each diffusion iteration. The basic algorithm is shown as follow:

1. Computing the smoothing kernel: $P = D^{-1}W$
   where D is a diagonal matrix with $D(i,i) = \sum_{k=1}^{n} W(i,k)$
2. Performing smoothing for t* steps: $W_t = W_{t-1}P + E$
3. Self-normalization: $W^* = W_t D^{-1}$

where E is the identity matrix. The convergence value is $(D - W)^{-1}$ when $t \to \infty$, which is always a degenerated (over-smoothed) result. Therefore, it also needs to conduct the iterative process to obtain a satisfactory result.

To overcome the limitations of the above method, in this paper, a robust affinity diffusion is proposed as follows:

$$W_t = \alpha W_{t-1} P + (1 - \alpha) W \text{ and } W_1 = W \tag{10}$$

where W is the neighboring similarity matrix, P is the row-normalized matrix of W, and $0 < \alpha < 1$ is the controlling parameter. The closed form of the diffusion transition matrix at step t can be written as:

$$W_t = \alpha^{t-1} W_{t-1} P^{t-1} + (1 - \alpha) \sum_{i=0}^{t-1} \alpha^i W P^i \tag{11}$$

we can derive:

$$W^* = \lim_{t \to \infty} W_t = (1 - \alpha) W (E - \alpha P)^{-1} \tag{12}$$

The probability matrix V (8) is used as the prior probability of the affinity diffusion step. We can get the posterior probability matrix by $U = W^* \times V$.

$$U = \begin{bmatrix} u_{1,1} & u_{1,2} \\ \vdots & \vdots \\ u_{N,1} & u_{N,2} \end{bmatrix} \tag{13}$$

where $u_{i,j}$ is the probability that i-th pixel belongs to j-th class after using the affinity map W. This indicates that the posterior probability is determined not only by a pixel's own intensity value, but also by all pixels information in the entire image together. It should be noted here that the W matrix represents the global spatial information, which contains the similarity measure of each pixel with all the rest of the pixels, rather than just concerned about the neighborhood pixels.

2.4   Correction of Intensity Inhomogeneity

In SD-OCT image, the intensity inhomogeneity, is an important factor in adversely affecting segmentation results. Due to the intensity inhomogeneity, there are overlaps between the ranges of the intensities of different classes, which often causes misclassification of classes. Therefore, it is often a mandatory step to remove the intensity inhomogeneity through a procedure called bias field correction.

From the formation of SD-OCT images, it has been generally accepted that an SD-OCT image I can be modeled as

$$I(x) = b(x)J(x) + n(x) \tag{14}$$

where $I(x)$ is the intensity of the observed image at voxel x, J(x) is the true image, $b(x)$ is the bias field that accounts for the intensity inhomogeneity in the observed image, and $n(x)$ is additive noise with zero-mean. As illustrated in [18], the bias field is represented by a linear combination of a given set of smooth basis functions $g_1, \cdots, g_M$, which make the bias field vary smoothly. The estimation of bias field is performed by finding the optimal coefficients $w_1, \cdots, w_M$ in the linear combination $b(x) = \sum_{k=1}^{M} w_k g_k$. We represent the coefficients $w_1, \cdots, w_M$ by a column vector $\mathbf{w} = (w_1, \cdots, w_M)^T$ and the basic functions $g_1(x), \cdots, g_M(x)$ by a column vector valued function $G(x) = (g_1(x), \cdots, g_M(x))^T$. Thus, the bias field $b(x)$ can be expressed an $b(x) = \mathbf{w}^T G(x)$

The true image $J(x)$ is approximately a constant c_i for x in the i-th class. Each class can be represented by its membership function u_i. The K classes can be represented by a binary membership functions $u_i(x)$ with $u_i(x) = 1$ for $x \in \Omega_i$ and $u_i(x) = 0$ for $x \notin \Omega_i$. Thus the true image J can be approximated by $J(x) = \sum_{i=1}^{K} c_i u_i(x)$. The energy formulation is proposed to estimate the bias field as follows:

$$F(b, J) = \int_\Omega |I(x) - b(x)J(x)|^2 dx = \int_\Omega \left| I(x) - \mathbf{w}^T G(x) \sum_{i=1}^{K} c_i u_i(x) \right|^2 dx \tag{15}$$

Here we use the above-mentioned U in (13) as $\mathbf{u}$ and use the μ in $\hat{\Theta}$ as $\mathbf{c}$, so the key question is the estimate of $\mathbf{w}$. For fixed c and u, we minimize the energy F ($\mathbf{u}, \mathbf{c}, \mathbf{w}$) with respect to the variable $\mathbf{w}$. This can be achieved by solving the equation $\frac{\partial F}{\partial \mathbf{w}} = 0$. Thus we can obtain the vector $\hat{\mathbf{w}}$ as follow:

$$\hat{\mathbf{w}} = \left(\int_\Omega G(x) G^t(x) \left(\sum_{i=1}^{K} c_i u_i(x) \right) dx \right)^{-1} \int_\Omega G(x) I(x) \left(\sum_{i=1}^{K} c_i u_i(x) \right) dx \tag{16}$$

The estimated bias field can be expressed as:

$$\hat{b}(x) = \hat{\mathbf{w}}^T G(x) \tag{17}$$

3   Experimental Results and Analysis

3.1   Experimental Data and Evaluation Studies

Two different data sets acquired with Cirrus OCT device were used to evaluate the performance of the proposed algorithm, where all the testing cases belong to the advanced stage of non-exudative AMD, which contain obvious GA. It should be noted that both data sets were described and utilized in previous work [1, 14]. The first data set consisted of 55 longtitudinal SD-OCT cube scans from twelve eyes in eight patients. The second data set consisted of 56 SD-OCT cube scans from 56 eyes in 56 patients with GA. For the first data set, manual outlines were drawn by two independent readers in the projection images in two repeated separate sessions. For the second data set, manual outlines were available as indicated in corresponding FAF images, which were considered as ground truth segmentation.

We employed four metrics to perform these quantitative comparisons: Absolute area difference (AAD), overlap ratio (OR), correlation coefficient (cc) and p-value in a U-test. Their calculation methods is illustrated in [1]. The AAD measures the absolute difference between the GA areas as segmented by two different methods. The OR is defined as the percentage of overlapped area between two segmentations.

3.2   Evaluation of GA Segmentation

The bias field estimates used in our method can be effectively segmented more robustly for SD-OCT images. Figure 2 shows the contrast of segment results. Figure 2(a) and (b) are original image and corrected image respectively. Figure 2(c) is an estimate image of the bias field for the original image (Fig. 2(a)). Figure 2(d) and (e) are the results of the segmentation of Fig. 2(a) and (b). The segmentation algorithm used here is GMM + MFO + Affinity diffusion. We can find that the segmentation result is very poor due to the intensity inhomogeneity in SD-OCT image when the bias field is not corrected. The corrected image can achieve a precise result (Fig. 2 (e)). This indicates that it is very necessary to introduce an estimate of the bias field during the segmentation process.

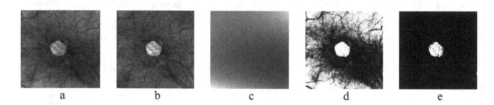

a   b   c   d   e

Fig. 2. The result of bias field correction

We evaluated the performance of the proposed method in the first data set by comparing its results to the manual segmentation gold standard and to the previously published NSJ's method [1]. Figure 3 shows a comparison of several segmentation results. The result of our method is indicated by red outlines, and NSJ's method is indicated by a green line. Blue outlines indicate the manual gold standard. We can see that both our method and NSJ's method can get quite good segmentation results, while in most cases our results are closer to the expert segmentation. It seems that our method has a higher precision.

Table 1 summarizes the results of the quantitative comparison of different method including gold standard, our method, previous QC's method [14] and previous NSJ's method. The values obtained by QC's method and NSJ's method are compared to Avg.

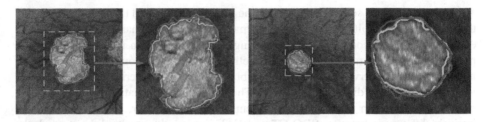

Fig. 3. Comparison of several segmentation results. The result of our method is indicated by red outlines, and NSJ's method is indicated by a green line. Blue outlines indicate the manual gold standard. (Color figure online)

Table 1. Quantitative comparison of our algorithm segmentation results and other's results

Methods compared	cc	p-value (U-test)	AAD (mm²) (mean, std)	AAD (%) (mean, std)	OR(%) (mean, std)
QC'S vs avg. expert	0.970	0.026	1.438 ± 1.26	27.17 ± 22.06	72.60 ± 15.35
Nsj's vs avg. expert	0.984	0.179	0.88 ± 0.81	12.93 ± 12.89	81.08 ± 12.80
Our seg. vs avg. expert	0.956	0.402	0.714 ± 1.24	12.27 ± 17.72	82.88 ± 13.60
Our seg. vs expert A1	0.948	0.761	0.763 ± 1.32	12.78 ± 17.82	82.18 ± 13.36
Our seg. vs expert A2	0.953	0.371	0.683 ± 1.30	11.04 ± 17.58	82.91 ± 13.30
Our seg. vs expert B1	0.959	0.286	0.782 ± 1.18	13.24 ± 17.76	81.84 ± 13.85
Our seg. vs expert B2	0.967	0.317	0.691 ± 1.04	12.24 ± 15.46	82.54 ± 12.80

Expert and displayed in the table. We also compared the differences of our method to each of the manual readers and sessions independently. Overall, our algorithm and NSJ's method performed higher similarity to the manual gold standard than QC's method substantially. The absolute area differences of our method is lower than NSJ's and QC's method(12.27% vs 12.93%, 27.17%, while the overlap ratio of our method is higher(82.88% vs 81.08%, 72.60%). Lower area differences indicate the area estimated by our method seems closer to the values measured by hand and our overlap ratio values means the precision of our segmentation is very high. We can also see that the paired U-test in measured GA area differences between our method and each of the manual segmentations was not significant (all with p-value > 0.05), while it was significant for differences between QC's method and manual segmentations (all with p-value < 0.05). All of these illustrate that the proposed hybrid model in this paper can produce a better segmentation performance than the other two methods.

Figure 4 displays several examples in the second data set and the binary images are the segmentation results of the gray-scale images. Table 2 summarizes the quantitative evaluation in this second data set, comparing each segmentation method (our method presented here, QC's method, and NSJ's method) to the manual outlines drawn in FAF images. We can observe that our method obtains the highest overlap ratio (73.02) and lowest Absolute area difference (18.32%). The quantitative evaluation indicates that the proposed method can work well in different data set. However, probably due to the intrinsic differences between SDO-CT and FAF images, the results did not reach the level in first data set.

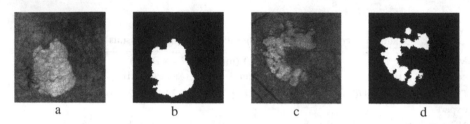

a   b   c   d

Fig. 4. Examples in the second data set

Table 2. Quantitative comparison in second data set

Methods compared	cc	p-value (U-test)	AAD (mm^2) (mean, std)	AAD(%) (mean, std)	OR(%) (mean, std)
QC'S vs FAF	0.955	0.524	0.951 ± 1.28	19.68 ± 22.75	65.88 ± 18.38
Nsj's vs FAF	0.937	0.261	1.215 ± 1.58	22.26 ± 21.74	70.00 ± 15.63
Our seg. vs FAF	0.942	0.423	1.123 ± 1.30	18.32 ± 19.78	73.02 ± 14.98

4 Conclusions

This paper proposed a hybrid model on solving the GA segmentation problem in SD-OCT images. Our approach integrates the Gaussian mixed model, the Affinity map and the bias field correction. MFO algorithm is used to estimate the parameters of the Gauss mixture model. Our method is robust to noise and is suitable for images with intensity inhomogeneity. After compared our method with previous QC's and NSJ's method, it seems the proposed model has superior to them. As summarized in Table 1, our method demonstrated very high accuracy when compared to a manual gold standard generated by two different readers and repeated at two separated sessions (mean OR = 82.88 ± 13.60; AAD = 0.714 ± 1.24 mm^2; cc = 0.956; U test p-value = 0.402). The proposed algorithm can be used in locating the GA lesions area and be helpful in the treatment of AMD.

Acknowledgments. This work was supported in part by the National Natural Science Foundation of China under Grants 61401209 & 61671242, in part by the Natural Science Foundation of Jiangsu Province, China under Grant BK20140790, in part by the Fundamental Research Funds for the Central Universities under Grant 30916011324, and in part by China Postdoctoral Science Foundation under Grants 2014T70525 & 2013M531364.

References

1. Niu, S., De, S.L., Chen, Q., et al.: Automated geographic atrophy segmentation for SD-OCT images using region-based C-V model via local similarity factor. Biomed. Opt. Express **7**(2), 581 (2016)
2. Wang, J.J., Rochtchina, E., Lee, A.J., et al.: Ten-year incidence and progression of age-related maculopathy: the Blue Mountains Eye Study. Ophthalmology **114**(1), 92–98 (2007)
3. Klein, R., Klein, B.E.K., Knudtson, M.D., et al.: Fifteen-year cumulative incidence of age-related macular degeneration: the Beaver Dam Eye Study. Ophthalmology **114**(2), 253–262 (2007)
4. Buch, H., Nielsen, N.V., Vinding, T., et al.: 14-year incidence, progression, and visual morbidity of age-related maculopathy: the Copenhagen City Eye Study. Ophthalmology **112** (5), 787–798 (2005)
5. Bindewald, A., Bird, A.C., Dandekar, S.S., et al.: Classification of fundus autofluorescence patterns in early age-related macular disease. Invest. Ophthalmol. Vis. Sci. **46**(9), 3309–3314 (2005)
6. Schmitz-Valckenberg, S., Brinkmann, C.K., Alten, F., et al.: Semiautomated image processing method for identification and quantification of geographic atrophy in age-related macular degeneration. Invest. Ophthalmol. Vis. Sci. **52**(10), 7640–7646 (2011)
7. Deckert, A., Schmitz-Valckenberg, S., Jorzik, J., et al.: Automated analysis of digital fundus autofluorescence images of geographic atrophy in advanced age-related macular degeneration using confocal scanning laser ophthalmoscopy (cSLO). BMC Ophthalmol. **5**(1), 8 (2005)
8. Lee, N., Laine, A., Barbazetto, I., et al.: Level set segmentation of geographic atrophy in macular autofluorescence images. Invest. Ophthalmol. Vis. Sci. **47**(13), 2125 (2006)
9. Allingham, M.J., Nie, Q., Lad, E.M., et al.: Semiautomatic segmentation of rim area focal hyperautofluorescence predicts progression of geographic atrophy due to dry age-related macular degeneration. Invest. Ophthalmol. Vis. Sci. **57**(4), 2283–2289 (2016)
10. Hu, Z., Medioni, G.G., Hernandez, M., et al.: Automated segmentation of geographic atrophy in fundus autofluorescence images using supervised pixel classification. J. Med. Imaging **2**(1), 014501 (2015)
11. Ramsey, D.J., Sunness, J.S., Malviya, P., et al.: Automated image alignment and segmentation to follow progression of geographic atrophy in age-related macular degeneration. Retina **34**(7), 1296–1307 (2014)
12. Jeong, Y.J., Hong, I.H., Chung, J.K., et al.: Predictors for the progression of geographic atrophy in patients with age-related macular degeneration: fundus autofluorescence study with modified fundus camera. Eye **28**(2), 209–218 (2014)
13. Yehoshua, Z., Rosenfeld, P.J., Gregori, G., et al.: Progression of geographic atrophy in age-related macular degeneration imaged with spectral domain optical coherence tomography. Ophthalmology **118**(4), 679–686 (2011)

14. Chen, Q., De, S.L., Leng, T., et al.: Semi-automatic geographic atrophy segmentation for SD-OCT images. Biomed. Opt. Express **4**(12), 2729–2750 (2013)
15. Tohka, J., Krestyannikov, E., Dinov, I., et al.: Genetic algorithms for finite mixture model based tissue classification in brain MRI. In: IFMBE Proceedings of European Medical and Biological Engineering Conference (EMBEC), pp. 4077–4082 (2005, 2013)
16. Mirjalili, S.: Moth-flame optimization algorithm: a novel nature-inspired heuristic paradigm. Knowl.-Based Syst. **89**, 228–249 (2015)
17. Shi, J., Malik, J.: Normalized cuts and image segmentation. IEEE Trans. Pattern Anal. Mach. Intell. **22**(8), 888–905 (2000)
18. Li, C., Gore, J.C., Davatzikos, C.: Multiplicative intrinsic component optimization (MICO) for MRI bias field estimation and tissue segmentation. Magn. Reson. Imaging **32**(7), 913–923 (2014)
19. Wang, B., Tu, Z.: Affinity learning via self-diffusion for image segmentation and clustering. In: 2012 IEEE Conference on Computer Vision and Pattern Recognition (CVPR), pp. 2312–2319. IEEE (2012)
20. Jiao, S., Knighton, R., Huang, X., et al.: Simultaneous acquisition of sectional and fundus ophthalmic images with spectral-domain optical coherence tomography. Opt. Express **13**(2), 444–452 (2005)

Blind Multi-frame Super Resolution with Non-identical Blur

Wei Sun[1], Jinqiu Sun[2], Xueling Chen[1], Yu Zhu[1], Haisen Li[1(✉)],
and Yanning Zhang[1]

[1] School of Computer Science, Northwestern Polytechnical University, Xi'an, China
{weisun,xuelingchen,zhuyu1986}@mail.nwpu.edu.cn, haisenli@foxmail.com,
ynzhang@nwpu.edu.cn
[2] School of Astronautics, Northwestern Polytechnical University, Xi'an, China
sunjinqiu@nwpu.edu.cn

Abstract. Real world video super resolution is an challenging problem
due to the complex motion field and unknown blur kernel. Although
multi-frame super resolution has been extensively studied in past
decades, it still remained problems and always assumed that the blur
kernels were identical in different frames. In this paper, we propose an
novel blind multi-frame super resolution method with non-identical blur.
To estimate blur kernels of different frames, we propose using salient
edges selection method for more accurate kernel estimation. The whole
process of estimation is based on Hyper-Laplacian prior, and iterative
value updating through a multi-scale process. After the kernels of different frames are estimated, the high resolution frame is reconstructed using
a cost function. The proposed method can obtain superior results, and
outperforms the state of the art in the experiments through subjective
and objective evaluation.

Keywords: Multi-frame super resolution · Non-identical kernel · Blind
estimation · Salient edges selection

1 Introduction

Multi-frame super resolution (SR) is a process of estimating a high resolution
(HR) image from a sequence of low resolution (LR) observations degraded by
various artifacts such as blurring, sampling, and noise. It is an fundamental task
in computer vision and image processing.

Although a lot of previous multi-frame SR methods have been studied in the
past 30 years [1], there were also many problems when dealt with real world SR.
Most of the previous works assumed that the point spread function (PSF) of the
camera or motion blur kernel were known, and the blur kernels of different frames
were identical [2–4]. In general, these assumptions are not realistic. Because of
the complexities of the imaging system, the blur kernel is usually arbitrary even
non-identical between different frames.

© Springer International Publishing AG 2017
Y. Sun et al. (Eds.): IScIDE 2017, LNCS 10559, pp. 485–495, 2017.
DOI: 10.1007/978-3-319-67777-4_43

Therefore, a practical SR method should be able to estimate optical flow and arbitrary even non-identical blur kernel [5], then use complementary information in adjacent frame to reconstruct the HR image. A lot of estimation methods of optical flow [6,7] and blur kernel [8–11] have been discussed in literatures, but straightforward processing of each image using existing techniques could also introduce visual artifacts.

Most works on the image and video SR are non-blind [12], only a few on blind SR. Ma *et al.* [13] proposed a multi-frame super resolution method that mainly dealt with ubiquitous motion blur. Because the blurring edges could mislead kernel estimation in multi-frame super resolution, the approach searched the sharp area in adjacent multi-frame firstly. Then an EM framework is proposed to guide residual blur kernel estimation of selected sharp areas, finally high resolution image can be reconstructed using estimated blur kernel. To suppress noise, it employed a family of sparse penalties as natural image prior. However, the approach mainly dealt with motion blurring, and it need a lot of adjacent frames to select sharp areas then obtain promising results. So it sometimes is not suit for real life high resolution images obtaining.

To solve motion field, blur kernel and noise parameter estimation simultaneously in sequence, Liu and Sun [14] proposed a Bayesian framework for adaptive video super resolution. The regularization terms for all unknowns are of type *l1* norm. An estimated noise parameter is used to update the weight of the fidelity term at each iteration of the optimization procedure. The noise level is updated at each iteration, but assumed to be identical for all pixels. Moreover, The blur kernel is assumed to be separable ($h = hx * hy$) and results are only provided with Gaussian blurs.

Inspired by a lot of successful blind deblurring methods, Faramarzi *et al.* [15] combined non-uniform interpolation SR method with blur kernel estimation, it could deal with different types of kernels and successively suppress artifacts. But the superior results are based on the assumption of identical blur kernels among different frames, it can not apply to real complex imaging system.

To deal with various types of non-identical blur kernels of different frames, in this paper, a blind SR framework dealing with non-identical blur is proposed. The method combines optical flow, blind blur kernel estimation and super resolution efficiently, and constructs the cost function including Hyper-Laplacian prior to estimate blur kernel. For robust kernel estimation, we select salient edges that suit for kernel estimation iteratively. Based on iterative algorithm, we can get non-identical blur kernels of sequence images. By applying the estimation non-identical blur kernels and motion fields of sequence to proposed blind super resolution framework finally, we obtain a practical and accurate super resolution results. The whole model of blind SR is illustrated in Fig. 1.

2   Observation Model

In the spatial domain, the linear forward observation model is a process of degeneration which illustrates how the ith LR image $l_i(x{\downarrow}, y{\downarrow}; c)$ of size $N_x^L \times N_y^L \times C$

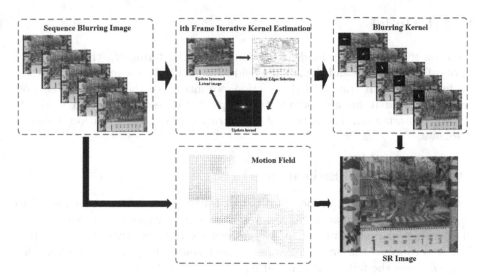

Fig. 1. Sequence frames super resolution diagram. Different blur kernels of different frame are estimated by iterative kernel updating and salient edges selecting. Finally obtaining HR image by combining motion field and blur kernel.

is generated from a HR image $h_i(x, y; c)$ of size $N_x^H \times N_y^H \times C$. Here, N_x and N_y are the number of pixel in two spatial directions and C is the color channel dimensions. The process of degeneration is defined in (1).

$$l_i(x_\downarrow, y_\downarrow; c) = [m_{i,j}(h_i(x, y; c)) * k_i(x, y)]_{\downarrow D} + n_i(x_\downarrow, y_\downarrow; c), \qquad (1)$$
$$c = 1, ..., C, i = 1, ..., P, j = i - M, ..., i + N.$$

Here, P is the total number of frames, $(x_\downarrow, y_\downarrow)$ and (x, y) are the position of LR and HR image respectively. D is the downsampling ratio, and $*$ represent two-dimensional convolution operator. The warping function between different HR frame h_i and h_j is defined with $m_{i,j}$, and the ith HR frame is blurred by the system PSF k_i. Finally, the downsampling image is corrupted by the additive noise n_i.

It is more convenient to express the whole degeneration process in matrix notion, (1) is rewritten as:

$$L_i = DK_i M_{i,j} H_i + N_i \qquad (2)$$

where H_i is the vector representation of size $N_x^H N_y^H C \times 1$ of ith HR frame, K_i and $M_{i,j}$ are the blurring and motion operators of size $N_x^H N_y^H C \times N_x^H N_y^H C$, different frames are blurred with different blur kernel K_i, D is the downsampling matrix of size $N_x^L N_y^L C \times N_x^H N_y^H C$, L_i and N_i are the vectors of the ith LR image and noise respectively, both of size $N_x^L N_y^L C \times 1$.

3   SR Model

Based on the observation model, to estimate the HR frame H_i, the blur kernel K_i of different frames and the motion field $M_{i,j}$ should be estimated firstly, then reconstruct the HR frame. In this section, we introduce the SR reconstruction framework, and it mainly include two steps: kernel estimation and final HR frame reconstruction. The details of the iterative updating algorithm is shown as follows.

3.1   Robust Kernel Estimation

Accurate blur kernel estimation of different frame is crucial for sequence SR to achieve a good performance. So we present a robust kernel estimation method based on maximum a posteriori (MAP) framework and iterative outliers selection [16]. According to the process of imaging represented in (2), the observed blur image L_i have the following formulation:

$$L_i = DK_i M_{i,j} H_i + N_i = DM_{i,j} K_i H_i + N_i = DM_{i,j} B_i + N_i \qquad (3)$$

where K_i is the kernel of ith frame and $B_i = K_i H_i$ is the ith upsampled but still blurry frame. Equation (3) suggests that we can first construct the upsampled frames B_i using an appropriate upsampling method and then apply a deblurring method to B_i to estimate K_i. The experiments in [17] have shown that using NUI [18,19] for upsampling the frames leads to better estimates of K_i compared to when K_i is estimated iteratively from the LR frames L_i using a MAP mehtod. So a blur kernel of ith frame can be estimated by solving MAP framework in (4) after upsampling the frames L_i using NUI method:

$$\min_{H_i, K_i} \|B_i - K_i H_i\|_2 + \lambda E(H_i) + \gamma E(K_i) \qquad (4)$$

where the first term imposes that the convolution output of the recovered image and the blur kernel should be similar to the observation; the second and third term $E(H_i)$, $E(K_i)$ are regularization terms on latent image and blur kernel; λ and γ are weight parameters. For getting superior kernel estimation, the update process is iterative between latent image with blur kernel estimation. Because of insignificant edges make kernel estimation vulnerable to noise, as discussed in [8,10,20,21]. For this reason,we introduce effective edge selection in the stage of iterative updating, the proposed algorithm detail is described as follow.

Intermediate Image Estimation. Based on MAP framework, the intermediate image H_i can be recovered by solving the following optimization problem when the blur kernel K_i is known:

$$H_i = arg \min_{H_i} \|K_i H_i - B_i\|_2 + \lambda E(H_i) \qquad (5)$$

For the regularization term $E(H_i)$, we use the hyper-Laplacian prior [22] to estimate the intermediate image. Thus, the intermediate latent image is estimated by:

$$H_i = arg \min_{H_i} \|K_i H_i - B_i\|_2 + \lambda \sum_{j=1}^{J} \|e_j * H_i\|^{0.8} \tag{6}$$

where e_j represent first-order derivative filter $e_1 = [1, -1]$, $e_2 = [1, -1]^T$. Because of the model (6) involve non-convex regularization that is hard to minimize directly. Thus, we adapt the variable substitution and alternating minimization scheme, which reduces the original problem into several simpler sub-problems. By introducing new auxiliary variables $u_j (j = 1, 2, ..., J)$, we can rewrite the energy function in (6) as:

$$E(H_i, u) = \|K_i H_i - B_i\|_2 + \lambda\beta \sum_{j=1}^{J} \|u_j\|^{0.8} + \beta \sum_{j=1}^{J} \|u_j - e_j * H_i\|_2 \tag{7}$$

which can be divided into two sub-problems: H-subproblem and u-subproblem. Following this scheme, the energy function will converge to a local minima by alternate updating sub-problem value in (8) and (9).

$$H - subproblem : E(H_i) = \|K_i H_i - B_i\|_2^2 + \beta \sum_{j=1}^{J} \|u_j - e_j * H_i\|_2 \tag{8}$$

$$u - subproblem : E(u) = \lambda\beta \sum_{j=1}^{J} \|u_j\|^{0.8} + \beta \sum_{j=1}^{J} \|u_j - e_j * H_i\|_2^2 \tag{9}$$

Intermediate Salient Edge Selection. Not all of Salient edges in the blurring image can improve kernel estimation, if the scale edge of an object is smaller than that of the blur kernel, the edge information can damage kernel estimation [8]. Therefore, we propose using a model to select salient informative edges from intermediate latent image H_i for kernel estimation:

$$\min_{\nabla S_i} \sum_x \frac{1}{2} \| \nabla S_i(x) - \nabla\Phi(H_i(x))\|_2 + \theta\omega(x)\| \nabla S_i(x)\|_1 \tag{10}$$

where $\Phi(\cdot)$ denotes the operation of shock filter [23]; $\nabla S_i(x) = (\partial_x S_i(x), \partial_y S_i(x))^T$ corressponds to the gradients $\nabla\Phi(H_i(x)) = (\partial_x\Phi(H_i(x)), \partial_y\Phi(H_i(x)))^T$; θ is weight parameter, and $\omega(x) = exp(-(r(x))^{0.8})$ with

$$r(x) = \frac{\|\sum_{y \in N_h(x)} \nabla B_i(y)\|_2}{\sum_{y \in N_h(x)} \| \nabla B_i(y)\|_2 + 0.5} \tag{11}$$

Here, $N_h(x)$ is a $h \times h$ neighbourhood of pixel x. A large value of $r(x)$ indicates that the area have strong image structures. Based on the shrinkage formula, (10) will convert to a closed-form solution:

$$\begin{cases} \partial_x S_i(x) = sign(\partial_x \Phi(H_i(x))) max(|\partial_x \Phi(H_i(x))| - \theta\omega(x), 0), \\ \partial_y S_i(x) = sign(\partial_y \Phi(H_i(x))) max(|\partial_y \Phi(H_i(x))| - \theta\omega(x), 0) \end{cases} \quad (12)$$

Kernel Estimation. After obtain salient edges from intermediate latent image, the blur kernel can be estimated by:

$$\min_{K_i} \|\nabla S_i(x)K_i - \nabla B_i\|_2 + \gamma\|K_i\|_2 \quad (13)$$

Similar to existing approaches [24,25], the solution of (13) can be obtained by FFT. The blur kernel of different frames can be estimated by alternative updating intermediate latent image, salient edges and kernel estimation, we apply a multi-scale pyramid [21] to the whole updating process for avoiding trapping in local minima.

3.2   Final HR Frame Reconstruction

After the non-identical blur kernel estimation of different frame is completed, the final HR frames are reconstructed through minimizing the following cost function:

$$E(H_1, ..., H_P) = \sum_{i=1}^{P}(\sum_{j=i-M}^{i+N} \|DK_iM_{i,j}H_i - L_i\|_2 + \lambda\sum_{q=1}^{2} \|e_q * H_i\|^{0.8}) \quad (14)$$

where motion field $M_{(i,j)}$ of adjacent frames are computed through optical flow method [6]. Minimizing this cost function with the variable substitution and alternating minimization based on (7), and the final HR frame estimation can be accomplished by only few updating (14).

4   Experiment Results

In this section, we conduct several experiments to evaluate the effective of the proposed method. We compare our approach with the state-of-the-art sequence image SR methods *FastUpsampler* [12] and *MFSR* [13]. Among these, the non-blind SR method *FastUpsampler* does not include a deblurring step and it can only deal with one frame, so we add the deblurring method of [9] before upsampling. The parameter λ in (5) is set to be 0.5, γ in (13) is set to be 2, θ is set to be 1, and iterative updating is used for β [22].

To measure the accuracy of our proposed blind method with non-identical kernels for different blur types, we synthetically generate LR sequence frames from selected two popular real-world video sequences: *City*, *calendar*. All these videos have 4:2:0 chroma subsampling format which means the chrominance channels have half of the horizontal and vertical resolutions of the luminance channel [26]. To test the efficient of real video which contains unknown blurring, we also do experiment on sequence that is taken by mobile camera. For the first experiment, each frame of *City* video sequence is added different scale Gaussian blurring and downsampling. Continuous four low resolution frames with estimated Gaussian PSF is shown in Fig. 2(a). The *FastUpsampel* method oversmooths the details for avoiding ringing phenomenon and produces visually less appealing results. The *MFSR* method can not recover more details in the large of edges area only using four frames. Because the proposed method first select salient edges that is benefits for kernel estimation in blind estimation stage. It also consider non-identical blurring kernel among different frames, and apply complementary information of different blurring scale frames to SR process, so it can leverage sharp structures and produce clearer results.

For the second experiment, we add different direction motion with the size of 9×9 to test the proposed method in motion blur situation (Fig. 3). The results of other methods contain visual artifacts especially in the number area, the sharp edges do not recover even bring wrong information. Our restored SR image contain more details compared to the others, and the number is more sharp. Figure 4(a) shows two frames and part detail with zoom-in of the real video sequence that is collected by handling mobile camera. Through comparison, our method can deal with real sequence that exist blurring efficiently.

For objective evaluation of the frame reconstruction performance of the proposed method, we use PSNR and SSIM scores that is shown in Table 1. Our method consistently outperforms other methods.

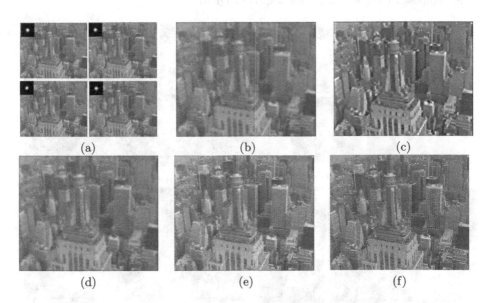

Fig. 2. Experimental results for *city* real world video sequence. (a) Continuous LR frame with different random scale Gaussian PSF: $\sigma = 1.4$, $\sigma = 1.2$, $\sigma = 1.6$, $\sigma = 1.3$ having size of 15×15, downsampling ratio of 2; (b) Upsampling to the original resolution using Bicubic; (c) Result of *FastUpsampler* [12]; (d) Result of *MFSR* [13] using 4 frames; (e) Result of our proposed method using 4 frames; (f) Original

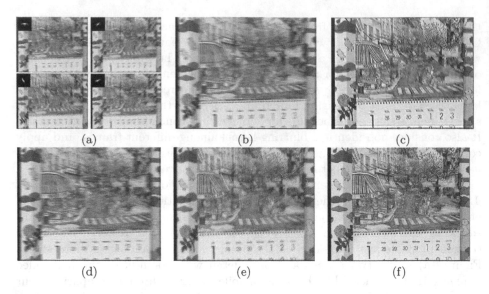

Fig. 3. Experimental results for *calendar* real world video sequence. (a) Continuous LR frame with different random direction motion blur: 0°, 45°, 120°, 210°, 9×9 motion blur, downsampling ratio of 2; (b) Upsampling to the original resolution using Bicubic; (c) Result of *FastUpsampler* [12]; (d) Result of *MFSR* [13] using 4 frames; (e) Result of our proposed method using 4 frames; (f) Original

Fig. 4. Experimental result for real video sequence. (a) Continuous LR frames selected from video sequence; (b) Result of *FastUpsampler*; (c) Result of *MFSR* using continuous 4 low resolution frames; (d) Our result using continuous 4 low resolution frames

Table 1. PSNR and SSIM scores

	City	Calendar
PSNR		
Bicubic	23.58	18.13
Fast Upsampler [12]	22.40	15.20
MFSR [13]	23.54	18.16
Proposed	25.63	19.43
SSIM		
Bicubic	0.50	0.42
Fast Upsampler [12]	0.53	0.31
MFSR [13]	0.51	0.43
Proposed	0.59	0.48

5 Conclusion

Because of real world video sequences contain complex degraded factors, it is difficult to restore high resolution frame. To solve these problems, a method of multi-frame non-identical blind super resolution is proposed in this paper. It can deal with multi-frame super resolution efficiently which degraded by non-identical blur kernels. To estimate the kernels, the method use multi-scale iterative update process. In which, the kernels is estimated by iterative selecting salient edges of intermediate latent image. After completion of kernel estimation, high resolution frames can be reconstructed through a non-blind super resolution process using the estimated non-identical kernels of continuous frames. The experiment results show that our proposed method have superior performance compared with other state of the art methods.

References

1. Wang, C.X., Hong-Qi, S.U., Fan, G.L.: Overview on super resolution image reconstruction. Comput.Technol. Dev. **54**(3), 197–212 (2011)
2. Chang, H., Yeung, D., Xiong, Y.: Super-resolution through neighbor embedding. In: IEEE Computer Society Conference on Computer Vision and Pattern Recognition, CVPR, USA, pp. 275–282 (2004)
3. Fattal, R.: Image upsampling via imposed edge statistics. ACM Trans. Graph. **26**(3), 95 (2007)
4. Zhu, Y., Zhang, Y., Yuille, A.L.: Single image super-resolution using deformable patches. In: IEEE Conference on Computer Vision and Pattern Recognition, CVPR, USA, pp. 2917–2924 (2014)
5. Riegler, G., Schulter, S., Rüther, M., Bischof, H.: Conditioned regression models for non-blind single image super-resolution. In: IEEE International Conference on Computer Vision, ICCV, Santiago, pp. 522–530 (2015)

6. Liu, C.: Beyond pixels: exploring new representations and applications for motion analysis. Massachusetts Institute of Technology (2009)
7. Sun, D., Roth, S., Black, M.J.: Secrets of optical flow estimation and their principles. In: IEEE Conference on Computer Vision and Pattern Recognition, CVPR, USA, pp. 2432–2439 (2010)
8. Xu, L., Jia, J.: Two-phase kernel estimation for robust motion deblurring. In: Daniilidis, K., Maragos, P., Paragios, N. (eds.) ECCV 2010. LNCS, vol. 6311, pp. 157–170. Springer, Heidelberg (2010). doi:10.1007/978-3-642-15549-9_12
9. Pan, J., Sun, D., Pfister, H., Yang, M.: Blind image deblurring using dark channel prior. In: IEEE Conference on Computer Vision and Pattern Recognition, CVPR, USA, pp. 1628–1636 (2016)
10. Gong, D., Tan, M., Zhang, Y., van den Hengel, A., Shi, Q.: Blind image deconvolution by automatic gradient activation. In: The IEEE Conference on Computer Vision and Pattern Recognition, CVPR, pp. 1827–1836 (2016)
11. Gong, D., Yang, J., Liu, L., Zhang, Y., Reid, I., Shen, C., Hengel, A.V.D., Shi, Q.: From motion blur to motion flow: a deep learning solution for removing heterogeneous motion blur. In: The IEEE Conference on Computer Vision and Pattern Recognition, CVPR (2017)
12. Shan, Q., Li, Z., Jia, J., Tang, C.: Fast image/video upsampling. ACM Trans. Graph. **27**(5), 153:1–153:7 (2008)
13. Ma, Z., Liao, R., Tao, X., Xu, L., Jia, J., Wu, E.: Handling motion blur in multi-frame super-resolution. In: IEEE Conference on Computer Vision and Pattern Recognition, CVPR, USA, pp. 5224–5232 (2015)
14. Liu, C., Sun, D.: A Bayesian approach to adaptive video super resolution. In: IEEE Conference on Computer Vision and Pattern Recognition, CVPR, USA, pp. 209–216 (2011)
15. Faramarzi, E., Rajan, D., Fernandes, F., Christensen, M.: Blind super-resolution of real-life video sequences. IEEE Trans. Image Process. **25**(4), 1544–1555 (2016)
16. Pan, J., Lin, Z., Su, Z., Yang, M.: Robust kernel estimation with outliers handling for image deblurring. In: IEEE Conference on Computer Vision and Pattern Recognition, CVPR, USA, pp. 2800–2808 (2016)
17. Faramarzi, E., Rajan, D., Christensen, M.P.: Unified blind method for multi-image super-resolution and single multi-image blur deconvolution. IEEE Trans. Image Process. **22**(6), 2101–2114 (2013)
18. Borman, S., Stevenson, R.: Spatial resolution enhancement of low-resolution image sequences - a comprehensive review with directions for future research (1998)
19. Park, S.C., Min, K.P., Kang, M.G.: Super-resolution image reconstruction: a technical overview. IEEE Signal Process. Mag. **20**(3), 21–36 (2003)
20. Shan, Q., Jia, J., Agarwala, A.: High-quality motion deblurring from a single image. ACM Trans. Graph. **27**(3), 73:1–73:10 (2008)
21. Cho, S., Lee, S.: Fast motion deblurring. ACM Trans. Graph. **28**(5), 145:1–145:8 (2009)
22. Krishnan, D., Fergus, R.: Fast image deconvolution using hyper-laplacian priors. In: 2009 Annual Conference on Neural Information Processing, Canada, pp. 1033–1041 (2009)
23. Osher, S., Rudin, L.I.: Feature-oriented image enhancement using shock filters. SIAM J. Numer. Anal. **27**(4), 919–940 (1990)
24. Xu, L., Zheng, S., Jia, J.: Unnatural L0 sparse representation for natural image deblurring. In: IEEE Conference on Computer Vision and Pattern Recognition, CVPR, USA, pp. 1107–1114 (2013)

25. Pan, J., Hu, Z., Su, Z., Yang, M.: Deblurring text images via L0-regularized inten-
    sity and gradient prior. In: IEEE Conference on Computer Vision and Pattern
    Recognition, CVPR, USA, pp. 2901–2908 (2014)
26. Faramarzi, E., Rajan, D., Fernandes, F.C.A., Christensen, M.P.: Blind super res-
    olution of real-life video sequences. IEEE Trans. Image Process. **25**(4), 1544–1555
    (2016)

Example-Guided Image Prior for Blind Image Deblurring

Xueling Chen, Yu Zhu$^{(\boxtimes)}$, Wei Sun, and Yanning Zhang

School of Computer Science, Northwestern Polytechnical University, Xi'an, China
zhuyu1986@mail.nwpu.edu.cn

Abstract. This paper proposes a patch-based deblurring method to leverage the unregistered sharp example image which shares global or local contents with the blurred image in a variant view. Firstly, we propose a coarse-to-fine scheme to achieve the accurate image patches matching and solve the mismatch problem caused by the blur ambiguity. Secondly, we use sharp image patches to form a patch prior which outperforms the generic prior in kernel estimation as it retain more image details which are usually filtered out using generic priors. The proposed method using unregistered sharp example image make it more practical to find examples from variant ways, *e.g.* the Internet, the succession capturing using hand-held camera, and multi-frame in videos. Experiments on real-world images show the patch prior can achieve more accurate and robust kernel estimation and outperforms state-of-the-art methods.

Keywords: Image deblurring · Unregistered image example · Patch

1 Introduction

Hand-held camera shake during the exposure commonly produces blur in photos. *Blind deblurring, i.e.* estimating both blur kernel and sharp image from a given blurred image, has been getting increasing attention [1–6]. Considering the high ill-posedness of the blind deblurring problem, the auxiliary information or priors of the desired sharp image or blur kernel is crucial for the accurate and meaningful image recovery [1,2,4,5,7,8].

Recent years witness the great progress of single-image deblurring using statistic image priors to handle the ill-posed deblurring problem, *e.g.* the heavy-tailed distribution of natural image gradients [7]. The success of this method series relies on the emphasis of the strong edges of the intermediate image during the alternative iteration process explicitly or implicitly [1]. Xu and Jia [2] pointed out that the strong edges are less ambiguous between the blurred and sharp image. This property helps to estimate the kernel more accurately. However, such image prior is generic statistical, and ignores each specific image property, usually resulting in over smooth image with less fine texture.

Some researchers try to find more specific information to alleviate the ill-posed problem. Hu *et al.* [9] utilize camera shake trajectory recorded by inertial sensors to help estimating the precise blur kernel. Others try to transfer

© Springer International Publishing AG 2017
Y. Sun et al. (Eds.): IScIDE 2017, LNCS 10559, pp. 496–507, 2017.
DOI: 10.1007/978-3-319-67777-4_44

the generic statistics to a more specific one to recover fine texture in latent images, such as learning image prior model from image local patches containing similar structure [8]. Also, using available multi-image to deblur is widely addressed in video frames [10]. As most scenes appear repeatedly in adjacent frames, which providing enough sharp image information to recover the blurred frame. Similarly, fully registered image pairs with different direction motion blur can be jointly recovered with the help of complementary information among them [11,12]. What's more, fully registered noisy/blurry image pairs can recover a sharp image from the blurred one using noisy but sharp image information [13]. However, it is hard to find a fully registered image pairs, as they are captured by special designed cameras.

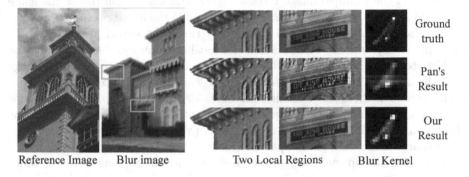

Reference Image   Blur image   Two Local Regions   Blur Kernel

Fig. 1. Our deblurring method compares with Pan's method [4]. Even given a totally different reference image with similar structures, our method can get a fairly good deblurring result.

Image deblurring using a non-elaborated registered sharp example image is more practical. But first challenge is to match the corresponding image contents between blurred sharp image. HaCohen [14] make use of non-rigid dense correspondence to register the sharp and the blurred image, and the two images are captured succession with slight view variant. Our method share the same idea to leverage the unregistered example sharp. The different is that we do not use the example image in the fidelity term to directly estimate the kernel as HaCohen [14], instead, we form a patch prior with the example image. This avoid the estimate error caused by the differences between blurred image and the sharp example and achieve more accurate kernel estimation.

The proposed method not only put no limit on the image pairs capturing, but also extend the chosen of image examples which can make use of the abundant high-quality images from the Internet. The method is especially useful for scene photo with a famous landmark, as abundant sharp example images can be found on the Internet. Figure 1 gives a comparison with the previous methods and show the effectiveness of our method.

2 Framework Overview

We firstly introduce the framework of the proposed method, and briefly explain how it works on removing image blur. The detailed implementation and definition are in Sect. 3.

The method is done by solving the following optimization problem

$$\{\bar{\mathbf{x}}, \bar{\mathbf{k}}\} = \underset{\mathbf{x}}{\operatorname{argmin}} \|\mathbf{y} - \mathbf{k} * \mathbf{x}\|_2^2 + \alpha\rho(\mathbf{R}_i\mathbf{x}, \mathbf{Q}_i\mathbf{r}) + \beta\|\mathbf{k}\|_2^2 \qquad (1)$$

where $\mathbf{x}$ is the latent sharp image, $\mathbf{y}$ is the observed blurred image and $\mathbf{k}$ is the blur kernel. The $*$ operator is the convolution which denotes the blur process of the sharp image. We refer $\mathbf{R}_i$ and $\mathbf{Q}_i$ to extract the i-th patch in the latent image $\mathbf{x}$ and the example image $\mathbf{r}$ respectively, and the reason to use the same index will be explained in Sect. 3.2. The first term is the fidelity term, the second term is the patch prior term and the third term is the kernel prior term. The patch prior plays the key role to avoid the local optimum problem, such as the 'no-blur' solution. The reason is that it provide a 'little' sharper example image than the original image in each scale which leads the optimization procedure to the approach the desired results, see Sect. 3.2. Even though the patch prior is based on the accumulation of patches, the optimization is performed on the whole image instead of patches as the patch prior term's role is to provide a whole sharp image example in optimization.

In solving the optimization problem, we use an alternating iterative minimization procedure, including three basic steps, namely, patch matching, intermediate image $\mathbf{x}$ updating and kernel $\mathbf{k}$ updating, as shown in Fig. 2. The three steps are implemented in multi-scale scheme as shown in Algorithm 1, which is fairly important for the success of the method. As the down-scale version of the image decrease the blur of the observed image [3] and it lead to the accurate patch matching and the effective patch prior. The input of the method is the

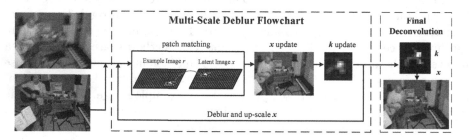

Fig. 2. The flowchart of the proposed method. The input images are a blurred image and a sharp example image. First, we do the patch matching step at certain scale, see Sect. 3.1, and second we update the latent image $\mathbf{x}$ using the patch priors that relates to the former step, see Sect. 3.2. Third, we update the kernel $\mathbf{k}$ using current latent image $\mathbf{x}$, see Sect. 3.3. And then, we up-scale the $\mathbf{x}$ and the $\mathbf{k}$ to match the next scale and do the above three steps to get the final kernel $\mathbf{k}$ estimation. Finally, we use the estimated kernel $\mathbf{k}$ to do the deconvolution to obtain the deblurred image.

observed blurred image and a sharp exmaple image. The output is the estimated kernel **k** and latent image **x**. Here, we do not use the latent image **x** as the final results, we do a final deconvolution to recover the latent image using the EPLL [15] method.

Algorithm 1. Framework of the proposed method

1: Input:blurred Image **y**, Example Image **r**
2: Initilize: Kernel **k** = *delta*, Latent Image **x** = **y**
3: Output: **k**, **x**
4: **for** scale = coarse to fine **do**
5:     Scale the **x** and **r** to the current scale
6:     $(\mathbf{R}_i, \mathbf{Q}_i)$ = patchMatching(**x**,**r**) (Sect. 3.1)
7:     **x** = Xupdate(**y**,**r**,$\mathbf{R}_i$, $\mathbf{Q}_i$,**k**) (Sect. 3.2)
8:     **k** = Kupdate(**y**,**x**) (Sect. 3.3)
9: **end for**
10: **x** = FinalDeconv(**y**,**k**)

---

3 Detailed Implementation

3.1 Accurate Blurred/Sharp Patches Matching in Multi-scale Scheme

As the result of blur ambiguity, it is difficult to find the accurate sharp image patch for the corresponding blurred image patch [3,16]. Here, we make use of the coarse-to-fine deblurring framework to iteratively solve the problem. It has been verified by Michael and Irani [3] that if blurred image **y** is down-scaled the by a factor α, the down-scaled $\mathbf{y}_\alpha$ will be α times sharper than the **y**. It motivates us to find the corresponding patch pairs from the coarsest scale to the finer scale, and reduce the ambiguity caused by blur.

The detailed implementation can be seen in the Fig. 3. First, we down-scale the example image **r** and build the multi-scale image-pyramid, as shown in Fig. 3(d)(e)(f), and then we down-scale the blurred image **y** to the coarsest scale with the same factor as (d) to obtain the (a). Second, we use subwindow of the size 9×9 to divide the (a) and (d) into overlapped patches, and for each patch in (a) we find the most similar patch in (d) to build the correspondence between the two images. Third, we deblur (a) using the intermediate estimated kernel and then up-scale it to obtain the (b). (b) is the intermediate latent image which shares the same scale factor as (e). In (b), some blur has been removed and (b) is sharper than the down-scaled blurred image which will reduce the blur ambiguity in patch matching, and this is the crucial step to achieve accurate patch matching. The above three steps are also implemented in (b), and for each patch of (b) we find the corresponding patch in (e) and then deblur (b) and up-scale it to get (c).

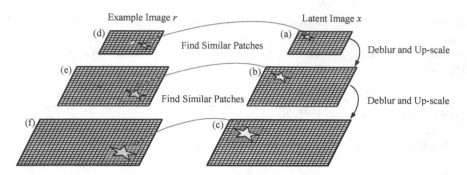

Fig. 3. Patch matching in multi-scale (a) Latent image **x** in the coarsest scale, and the coarsest latent image is initilized by the down-scaled blurred image **y**. (b) Latent image **x** in the finer scale. (c) Latent image **x** in the finest scale. (d) Example image *r* in the coarsest scale. (e) Example image *r* in the finer scale. (f) Example image *r* in the finest scale. To simplify, we only show 3 scales. The red arrows refer to find most similar example patches in for each latent image patches. The black arrows refer to deblur **x** and then up-scale it to match the next scale. (Color figure online)

One of the key to successful patch matching is to recover a sharper and ringing-free intermediate image in each scale and use it to find the corresponding sharp patches in the next finer scale. However, as the kernel **k** in the current scale is still different from the ground truth kernel, it is easily to have ringing artifact in the **x** which will lead to wrong patch matching. We solve the problem by leverage a patch prior to remove the ringing artifacts in the **x**. The detailed implementation is shown in Sect. 3.2

In order to speed up the patch matching step, we use fast approximate NN search of Olonetsky and Avidan [17] to achieve overlapped patches matching.

3.2   Latent Image Update Using Patch Prior

In this section, we elaborate our method on how to use a patch prior to assist ringing-free latent image estimation. Similar to Zoran and Weiss [15], Sun *et al.* [8] and Michael and Irani [3], we model the patch prior as the expected log likelihood (EPLL) of patches in the latent sharp image. The patch prior formular is as Eq. (2)

$$\rho(\mathbf{R}_i\mathbf{x}, \mathbf{Q}_i\mathbf{r}) = \exp(-\frac{1}{2h^2}\|\mathbf{R}_i\mathbf{x} - \mathbf{Q}_i\mathbf{r}\|_2^2) \tag{2}$$

Here, h is the bandwidth parameter. We define the $\mathbf{R}_i\mathbf{x}$ as extracting the i-th patch in latent image **x**, while the $\mathbf{Q}_i\mathbf{r}$ as extracting the most similar patch in the example image. Here, **R** and **Q** use the same index means **Q** has re-sorted the order of example patches to match the order of corresponding latent image patches.

Intensity compensate of patch pairs is also need when the example image is different from the blurred image in color and illumination. Here we use the same intensity compensate method as Zhu *et al.* [16].

So the latent image estimation using the patch prior formula is as in Eq. (3)

$$\bar{\mathbf{x}} = \underset{\mathbf{x}}{\operatorname{argmin}} \|\mathbf{y} - \mathbf{k} * \mathbf{x}\|_2^2 + \alpha \sum_i \|\mathbf{R}_i \mathbf{x} - \mathbf{Q}_i \mathbf{r}\|_2^2 \tag{3}$$

It is easily to see that, Eq. (3) is qudratic and has the closed form solution as Eq. (4). Here, we can solve the Eq. (4) using conjugate gradient (CG).

$$\mathbf{k}^T \mathbf{k} \mathbf{x} - \mathbf{k}^T \mathbf{y} + \alpha \sum_i (\mathbf{R}_i^T \mathbf{R}_i \mathbf{x} - \mathbf{Q}_i^T \mathbf{Q}_i \mathbf{x}) = 0 \tag{4}$$

Effectiveness of the Patch Prior. As ringing artifacts are undesired in the latent image esitmation, a lot of works are done to recover a ringing-free intermediate image. Xu *et al.* [18] suppress the ringing artifacts using L0 smooth method to filter out the ringing together with the image fine texture. Such method generate the over-smooth image with only strong edges. Here, the proposed patch prior has superiority to suppress the ringing artifacts without damage the fine texture.

Such ringing artifacts damage the image quality and will lead to wrong patch matching results, as shown in Fig. 4(c) and (f).

In Fig. 4, the intermediate kernels of the L0 smooth method are general less sparse than the proposed methods'. The RMSE of the proposed method is 0.0102 and 0.0107 of the L0 smooth method. The intermediate images of the L0 smooth method are unnaturally smooth, in which a lot of fine texture are lost, such as the texture on the wall, which results in less sparse kernel. In contrast, the proposed method does not generate sharp enough intermediate image, and it retain some fine texture and generate more sparse and accurate kernel. From (d) to (f) we observe the latent image changes from blur to sharp, and it is the key to avoid the trival solution (delta kernel) which guides the updated kernel follow the path from the no-blur solution to the desired solution.

3.3 Kernel Update

The sub-problem of kernel $\mathbf{k}$ update is to renew the $\mathbf{k}$ using current scale updated $\mathbf{x}$. And we use the image gradient instead of the intensity to estimate kernel which is commonly used in image deblurring works [10,14,19]. The formula is as follow

$$\bar{\mathbf{k}} = \underset{\mathbf{k}}{\operatorname{argmin}} \|\nabla \mathbf{y} - \mathbf{k} * \nabla \mathbf{x}\|_2^2 + \beta \|\mathbf{k}\|_2^2 \tag{5}$$

where $\nabla = [\nabla_r, \nabla_y]$ and ∇_r and ∇_c are the horizontal and vertical first derivatives filters. The Eq. (5) is quadratic and has closed-form solver. We use the Fourier transform as Eq. (6)

Fig. 4. Intermediate latent images and kernels of L0 smooth method and the proposed method. (a)(b)(c) are the intermediate images and kernels of L0 smooth method in the coarsest scale, middle scale and fine scale, respectively. (d)(e)(f) are the intermediate images and kernels of the proposed method in the coarsest scale, middle scale and fine scale, respectively. (g) is the ground truth image and kernel. To simplify, we only show three scales of the intermediate images and kernels and all the images and kernels are zoomed to the same size.

$$\bar{\mathbf{k}} = \mathcal{F}^{-1}\left(\frac{\bar{\mathcal{F}}(\nabla_r\mathbf{x}) \cdot \mathcal{F}(\nabla_r\mathbf{y}) + \mathcal{F}(\nabla_c\mathbf{x}) \cdot \mathcal{F}(\nabla_c\mathbf{y})}{\bar{\mathcal{F}}(\nabla_r\mathbf{x}) \cdot \mathcal{F}(\nabla_r\mathbf{x}) + \bar{\mathcal{F}}(\nabla_c\mathbf{x}) \cdot \mathcal{F}(\nabla_c\mathbf{x}) + \beta\mathcal{F}(1)} \right) \tag{6}$$

where $\mathcal{F}$ and $\mathcal{F}^{-1}$ are the Fourier transform and inverse Fourier transform, respectively.

After getting the estimated kernel $\mathbf{k}$, we use final deconvolution method to recover the latent sharp image. In order to recover a sharp image with less ringing, we employ the method of [15] to elaborately remove the ringring from the recovered image.

4   Experimantal Results

We present experimental evaluations of the proposed method using both synthetic images and real-world images. The kernel set with 12 ground truth kernels from Hu's [20,21] paper are used. As most existed blurred image sets are single blurred image without a sharp example, we construct our own image pairs dataset and use HaCohen's [14] image pairs dataset containing 13 color image pairs. Most image pairs share partial content or similar structure with different scales, intensities and views.

We compare our method with 3 state-of-art single image deblurring methods, 2 example-guided deblurring methods with full registered example [11] and non

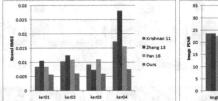 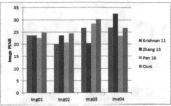

Fig. 5. Quantitative comparison on some kernels estimation (left) and deblurred images (right) of our dataset. The **x**-axis denotes the kernels and images index. The **y**-axis denotes RMSE values of estimated kernels and PSNR values of the deblurred images. We compare our results with other 3 state-of-art methods

registered example [14], respectively. For Zhang's [11] method that needs fully registered image pairs with different motion blur, we convolute the sharp image with 2 blur kernels by rotating the kernel to 90°.

The RMSE is used to quantitatively measure the accuracy of the estimated kernels. The PSNR is used to quantitatively measure the accuracy of the deblurred images. For each color image, we deblur each channel with the same estimated kernel (Fig. 5).

4.1   Compared with the Single Image Deblurring Methods

In this section, we conduct experiments on the test images downloaded from the Internet which is an easy way to acquire the corresponding feasible example images. The competitors are normalized sparse prior [19], dual blur image pairs [11] and the L_0 regularization method [4]. The results are shown in Fig. 6.

The first test image is a house image, while the corresponding example is a totally different scene. With the help of the examples, our method successfully estimate the blur kernel accurately, while the latent image is well recovered, with less artifact. Note that the text on the door is far more distinguishable than the compared methods. Pan's method [4] successfully gets the basic kernel shape but fail to recover the kernel details. Therefore the result suffers from the ringings. Meanwhile results of other methods can not make it either.

The second test image is commonly used in HDR imaging. Here the example/blurred image pairs in our method are chosen with the different exposures. Therefore illumination change challenges the methods. Also the example image is rotated with 5° to make it more practical. Results show our method outperform others with less ringing and more details. From the subregion of woman in the center, we can see our method successfully recovered her mouth region while other method can not. The recovered kernels also tell the performance.

The third test image is a low resolution and noisy image from the Internet. Pan's method [4] emphasizes the strong edges too much, leading to an oversharp image with many ringings. Zhang's method [11] fails to get the correct kernel, the final result is oversmooth. Our method and Krishnan's method [19] estimate the

Fig. 6. Visual evaluation of our dataset. From left to right: sharp example image, blurred image and ground truth kernel with 2 local regions zoomed in, Krishnan's result [19] (with estimated kernel and corresponding 2 local regions), Zhang *et al.* [11], Pan *et al.* [4], our results.

kernel well. Even though, our method outperforms this method in some details such as the people under the pavilion.

The fourth test image is a high resolution image from the Internet. With two kernels with different sizes. Only Pan *et al.* [4] and our method successfully achieve accurate kernel estimatiom. Zhang's method [11] fails to get the correct kernel in both. Krishnan's method [19] estimate the kernel only in the first image.

4.2 Compared with the Deblurring Methods Using Examples

We compare our method with two Example-Guided methods. The comparison with full registered example method [11] are shown in Fig. 6. And the comparison with non registered example method [14] are conducted on the HaCohen's [14] dataset in Fig. 7. The test three images are captured with the static background. After the blur synthesis, Gaussian noise with $\sigma = 1\%$ is added to each of the blur image. Results are shown in Fig. 7. For the first image, our method and HaCohen's method can estimate the kernel accurately. From the keyboard of the piano we can see our method can recover the distinguished keys while the other single image based method fail to make it. For the second image, all of the methods give the correct kernel structure estimation. From the figure, our method can recover the details faithfully. But some noise remains after the deblurring. For the third image which is real world, we use uniform model to deblur the image and find that ringing are less in the local region.

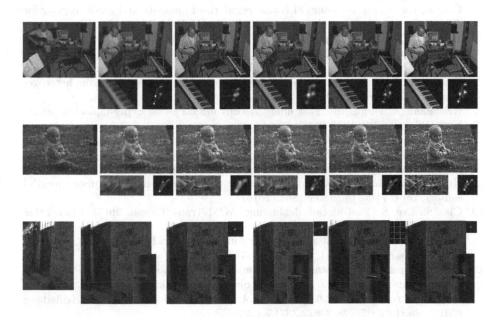

Fig. 7. Visual evaluation of HaCohen [14] dataset [14]. From left to right: sharp example image, blurred image and ground truth kernel with 2 local regions zoomed in, Xu's results [2], Krishnan's result [19] (with estimated kernel and corresponding 2 local regions), HaCohen [14], our results.

5   Conclusion

The proposed method leverage unregistered sharp example image to assist image deblurring, by using a patch-based image prior. We solve the problem of blurred/sharp patch matching iteratively by the coarse-to-fine scheme and then we propose the patch prior to make use of the example image to guide the blind image deblurring procedure to avoid the trival solution, at the same time to get a more accurate and sparse estimated kernel than the state-of-art method. The proposed method also extends the chosen of example images a lot, and make it possible to use examples with large variant.

References

1. Levin, A., Weiss, Y., Durand, F., Freeman, W.T.: Understanding and evaluating blind deconvolution algorithms. In: CVPR, pp. 1964–1971 (2009)
2. Xu, L., Jia, J.: Two-phase kernel estimation for robust motion deblurring. In: Daniilidis, K., Maragos, P., Paragios, N. (eds.) ECCV 2010. LNCS, vol. 6311, pp. 157–170. Springer, Heidelberg (2010). doi:10.1007/978-3-642-15549-9_12
3. Michaeli, T., Irani, M.: Blind deblurring using internal patch recurrence. In: Fleet, D., Pajdla, T., Schiele, B., Tuytelaars, T. (eds.) ECCV 2014. LNCS, vol. 8691, pp. 783–798. Springer, Cham (2014). doi:10.1007/978-3-319-10578-9_51
4. Pan, J., Hu, Z., Su, Z., Yang, M.H.: L0-regularized intensity and gradient prior for deblurring text images and beyond. IEEE TPAMI 99, 1–14 (2016)
5. Gong, D., Tan, M., Zhang, Y., van den Hengel, A., Shi, Q.: Blind image deconvolution by automatic gradient activation. In: CVPR, pp. 1827–1836 (2016)
6. Gong, D., Yang, J., Liu, L., Zhang, Y., Reid, I., Shen, C., Hengel, A.V.D., Shi, Q.: From motion blur to motion flow: a deep learning solution for removing heterogeneous motion blur. In: CVPR (2017)
7. Krishnan, D., Fergus, R.: Fast image deconvolution using hyper-laplacian priors. In: NIPS, pp. 1033–1041 (2009)
8. Sun, L., Cho, S., Wang, J., Hays, J.: Good image priors for non-blind deconvolution. In: Fleet, D., Pajdla, T., Schiele, B., Tuytelaars, T. (eds.) ECCV 2014. LNCS, vol. 8692, pp. 231–246. Springer, Cham (2014). doi:10.1007/978-3-319-10593-2_16
9. Hu, Z., Yuan, L., Lin, S., Yang, M.: Image deblurring using smartphone inertial sensors. In: CVPR, pp. 1855–1864 (2016)
10. Cho, S., Lee, S.: Fast motion deblurring. ACM Trans. Graph. 28(5), 145:1–145:8 (2009)
11. Zhang, H., Wipf, D.P., Zhang, Y.: Multi-image blind deblurring using a coupled adaptive sparse prior. In: CVPR, pp. 1051–1058 (2013)
12. Zhu, X., Šroubek, F., Milanfar, P.: Deconvolving PSFs for a better motion deblurring using multiple images. In: Fitzgibbon, A., Lazebnik, S., Perona, P., Sato, Y., Schmid, C. (eds.) ECCV 2012. LNCS, vol. 7576, pp. 636–647. Springer, Heidelberg (2012). doi:10.1007/978-3-642-33715-4_46
13. Yuan, L., Sun, J., Quan, L., Shum, H.: Image deblurring with blurred/noisy image pairs. ACM Trans. Graph. 26(3), 1 (2007)
14. HaCohen, Y., Shechtman, E., Lischinski, D.: Deblurring by example using dense correspondence. In: ICCV, pp. 2384–2391 (2013)
15. Zoran, D., Weiss, Y.: From learning models of natural image patches to whole image restoration (2011)
16. Zhu, Y., Zhang, Y., Yuille, A.L.: Single image super-resolution using deformable patches. In: CVPR, pp. 2917–2924 (2014)
17. Olonetsky, I., Avidan, S.: TreeCANN - k-d tree coherence approximate nearest neighbor algorithm. In: Fitzgibbon, A., Lazebnik, S., Perona, P., Sato, Y., Schmid, C. (eds.) ECCV 2012. LNCS, vol. 7575, pp. 602–615. Springer, Heidelberg (2012). doi:10.1007/978-3-642-33765-9_43

18. Xu, L., Zheng, S., Jia, J.: Unnatural L0 sparse representation for natural image deblurring. In: 2013 IEEE Conference on Computer Vision and Pattern Recognition, Portland, OR, USA, 23–28 June 2013, pp. 1107–1114 (2013)
19. Krishnan, D., Tay, T., Fergus, R.: Blind deconvolution using a normalized sparsity measure. In: CVPR, pp. 233–240 (2011)
20. Hu, Z., Yang, M.-H.: Good regions to deblur. In: Fitzgibbon, A., Lazebnik, S., Perona, P., Sato, Y., Schmid, C. (eds.) ECCV 2012. LNCS, vol. 7576, pp. 59–72. Springer, Heidelberg (2012). doi:10.1007/978-3-642-33715-4_5
21. Hu, Z., Yang, M.-H.: Learning good regions to deblur images. Int. J. Comput. Vision 115(3), 345–362 (2015)

Improved Spiking Cortical Model Based Algorithm for Multi-focus Image Fusion

Weiwei Kong[1(✉)] and Yang Lei[2]

[1] School of Computer Science and Technology,
Xi'an University of Posts and Telecommunications, Xi'an 710121, China
wwkong_xupt@163.com
[2] Department of Electronics Technology,
Engineering University of Armed Police Force, Xi'an 710086, China
surina526@163.com

Abstract. How to extract focused information as efficiently as possible is always the key issue of the fusion algorithms for multi-focus images. As an improved version of the third generation of artificial neural networks (ANN), spiking cortical model (SCM) has been proved to be an effective tool for dealing with the issues of image processing. Under the above background, a novel multi-focus image fusion algorithm based on improved SCM is proposed in this paper. Specifically, firstly, the traditional SCM is improved to be a modified version. Secondly, the concrete parameters setting is given and introduced in detail. Finally, in order to verify the effectiveness and feasibility of the proposed algorithm, several registered groups of source images are used in this paper. Experimental results demonstrate that the proposed algorithm is superior to the existing work in terms of both subjective visual presentation and objective evaluations compared with current typical ANN models.

Keywords: Image fusion · Spiking cortical model · Focused information · Artificial neural networks

1 Introduction

Since the optical system suffers from the limited focus range, thus it is often difficult and impossible to present all objects in the scene clearly. Commonly, the regions of the multi-focus source image can be categorized into two main parts including focused regions and defocused ones. Recently, the investigations on the multi-focus image fusion have become the hotspot in the area of image processing, thus more and more scholars are doing researches on the imaging mechanism of multi-focus images carefully both at home and abroad. Recently, a variety of algorithms and models have been proposed.

Generally speaking, almost all of the fusion algorithms can be classified into three categories according to the levels of the information representation, namely pixel-level, feature-level and decision-level. In pixel-level-based algorithms, each pixel is an independent unit, and its value in the final fused image comes from the pixel information of the source images. Different from pixel-level, the feature-level-based

© Springer International Publishing AG 2017
Y. Sun et al. (Eds.): IScIDE 2017, LNCS 10559, pp. 508–520, 2017.
DOI: 10.1007/978-3-319-67777-4_45

algorithms often choose the inherent features such as contours and edges rather than the pixel value to be the fusion basis. Compared with the former two kinds, the decision-level-based algorithms are considered to be the high level fusion.

From the perspective of the processing mode, the fusion algorithms can be divided into two mainstreamed ones including transform domain and spatial domain. The core idea of the former is to capture and extract the information of the edge and details of the source images as much as possible via geometric analysis. During the early stage of the image processing research, transform domain based algorithms indeed enhanced the fusion performance. However, along with the advancement of research, several of its inherent drawbacks begin to appear. For example, the problem with discrete wavelet transform (DWT) [1] is that it is merely adept at capturing point-wise singularities, and cannot be sensitive to other types of features such as lines. Therefore, DWT often causes artifacts in the final fused image. In order to overcome the drawbacks of DWT, the contourlet transform (CT) theory [2] is proposed, but CT does not have the property of shift-invariance, so the final result based on CT has the Gibbs phenomena. Compared with DWT and CT, non-subsampled contourlet transform (NSCT) [3, 4] is characterized by much better fusion performance, but the higher requirements of the computational resources prove to be their major limitations to the real-time applications. Unlike NSCT, ST [5, 6] is equipped with a rich mathematical structure similar to DWT. Its shearing filters have smaller support sizes than the directional filters used in the contourlet transform so that it can be implemented much more efficiently. Miao et al. [7] introduced ST into the field of image fusion and feature separations. In contrast to NSCT, the significant advantage of ST is the low computational complexity. However, the absence of the shift-variance property in ST which may produce some artifacts in the final fused image still can't be ignored.

Unlike the mechanism of transform domain tools, as the typical representative in the spatial domain based tools, the theory of artificial neural network (ANN) deals with the issue of image fusion from the neurons' perspective, and shows distinctive advantages compared with other categories of image fusion algorithms. In the traditional ANN model and its improved versions, pulse coupled neural network (PCNN) [8–11] has attracted sufficient attention and been widely used in the field of image fusion. PCNN is a biologically inspired ANN developed by Eckhorn [12, 13] and based on the experiment observations of synchronous pulse bursts in cat's visual cortex.

Broussard and his collaborators utilized PCNN to fuse images to improve the recognition rate, which proved that the existence of the relationship between neural firing frequency and image intensity, and confirmed that PCNN is feasible for dealing with the issue of image fusion. Even though, traditional PCNN-based fusion algorithms outperformed many current conventional ones, it also suffers from some drawbacks, such as its inherent complex structure and a great many parameters required setting. Intersecting cortical model (ICM) [14] eliminates the linking scheme of the traditional PCNN, and hence ICM has much less parameters and easier mechanism. However, the truth that absolutely ignoring the influence of the neurons in the neighborhood is adverse to the pulse synchronization of neurons. Against this background, spiking cortical model (SCM) [15, 16] developed as an improved version of PCNN owned the advantages of both PCNN and ICM. Compared with the above two ANN models, SCM has a much more reasonable structure, but only few studies address the area of image fusion based on SCM. Thus, there may be potential wide application fields for SCM.

Under the above background, it is necessary and meaningful to investigate the potential of SCM on the image fusion issue especially the fusion of multi-focus images which is of increasing importance. Meanwhile, a novel fusion algorithm for multi-focus images based on improved SCM is proposed. In this paper, the traditional SCM is modified to be ISCM with much fewer parameters and more effective function mechanism. Then, the novel model is used to deal with the issue of multi-focus image fusion. Simulation experiments are conducted to evaluate the proposed algorithm with several current popular fusion algorithms. The salient contributions of this paper can be summarized as follows.

- The classic SCM is improved to be an improved version.
- This paper presents a novel fusion algorithm for multi-focus images.
- Both qualitative and quantitative comparisons with some state-of-art and conventional ANN based algorithms have been performed on several pairs of datasets, and the superiorities of the proposed algorithm are obviously verified in this paper.
- Furthermore, the proposed algorithm has been also extended for other categories of source images.

The rest of this paper is organized as follows. In Sect. 2, a brief introduction to the traditional SCM is presented. Section 3 concretely describes the structure of ISCM and the proposed fusion algorithm for multi-focus images in detail. Experimental results and performance analysis are presented and discussed in Sect. 4. Concluding remarks and future work are given in Sect. 5.

2   Improved SCM and Its Parameters Settings

2.1   Classic SCM and Its Mathematical Model

As a novel member of ANN-based models, SCM combines the advantages of both PCNN and ICM. A SCM neuron commonly denoted by N_{ij} consists of three units: the receptive field, the modulation field, and the pulse generator. The structure of a basic SCM neuron is shown in Fig. 1.

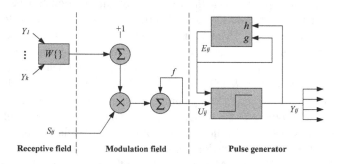

Fig. 1. The structure of the basic SCM neuron.

The discrete mathematical expressions of SCM model can be described as follows [16].

$$F_{ij}[n] = S_{ij} \tag{1}$$

$$L_{ij}[n] = V_L \sum_{kl} W_{ijkl} Y_{kl}[n-1] \tag{2}$$

$$U_{ij}[n] = f U_{ij}[n-1] + S_{ij} \sum_{kl} W_{ijkl} Y_{kl}[n-1] + S_{ij} \tag{3}$$

$$E_{ij}[n] = g E_{ij}[n-1] + h Y_{ij}[n-1] \tag{4}$$

$$Y_{ij}[n] = \begin{cases} 1, & \text{if } 1/(1 + \exp(-\gamma(U_{ij}[n] - E_{ij}[n]))) \geq 0.5 \\ 0, & \text{else} \end{cases} \tag{5}$$

As shown in Fig. 1, N_{ij} receives input signals via other neurons and external sources by two channels in the receptive field. One channel is the feeding input S_{ij} and the other one is the linking input L_{ij}, which correspond to Eqs. (1) and (2), respectively. U_{ij} is the internal activity and combines the information of the above two channels in a second order mode to form the total internal activity, as shown in Eq. (3). According to Eq. (4), E_{ij}, the dynamic threshold, will influence the firing frequency of the neurons by comparing with the output Y_{ij}. Equation (5) indicates that U_{ij} is then compared with E_{ij} to decide the value of the output Y_{ij}. If the requirement mentioned in Eq. (5) is satisfied, then the neuron N_{ij} will be activated and generate a pulse, which is characterized by $Y_{ij} = 1$, else $Y_{ij} = 0$. The result is an auto wave expanding from an active neuron to the whole region.

Apart from the parameters discussed above, there are still six other ones including V_L, W_{ijkl}, f, g, h and γ needing explaining. V_L is the magnitude scaling term; W_{ijkl} is the linking matrix; f and g denote the decay constants; h is the threshold magnitude coefficient; γ denotes a parameter of sigmoid function. Obviously, SCM has a more compact structure than tradition PCNN models in terms of the amount of parameters and operation mechanism.

2.2 Improved SCM and Its Parameters Settings

Although SCM combines the advantages of both PCNN and ICM models, its several inherent drawbacks still cannot be ignored by us, which are listed in detail as follows.

(1) The number of the parameters needing setting is large. On the one hand, by referring to the references on ANN, we can easily find that it is almost impossible to set all parameters to be the very optimal values to achieve the best fusion performance. On the other hand, in most conventional ANN based fusion algorithms, the parameters are often chosen manually on the basis of personal experience or numerous unnecessary, repeated and time consuming experiments. What is worse, the values you adjust are merely applicable in certain experiments or applications, but their performance on other occasions may be bad.

(2) The determination of the iterative number is also an onerous task for us. If n is too small, the neurons will not be activated sufficiently so that the advantage of the synchronous impulse bursting cannot be reflected enough. Of course, it is also irrational to set n as large as possible, the reason for which is that there commonly exists a comparatively desirable value or interval for the iterative number, an exceedingly large value will not only result in an adverse influence on the visual effects, but also increase the computational costs a lot. As a result, we have to develop an efficient scheme of iterative number setting.

In order to overcome the above two problems, an improved SCM (ISCM) is proposed whose concrete measures are given as follows.

Except for the iterative number n, there are six parameters in the traditional SCM altogether.

(2.1) V_L: As known to us, V_L denotes the magnitude scaling term in the linking input unit. According to Eqs. (2) and (3), if we omit the part of the former internal activity namely $fU_{ij}[n-1]$, Eq. (3) can be rewritten as follows.

$$
\begin{aligned}
U_{ij}[n] &= S_{ij} \sum_{kl} W_{ijkl} Y_{kl}[n-1] + S_{ij} \\
&= S_{ij}\Big(1 + \sum_{kl} W_{ijkl} Y_{kl}[n-1]\Big) \\
&= S_{ij}\Big(1 + \frac{1}{V_L} \cdot V_L \sum_{kl} W_{ijkl} Y_{kl}[n-1]\Big) \\
&= S_{ij}\Big(1 + \frac{1}{V_L} \cdot L_{ij}[n]\Big)
\end{aligned}
\tag{6}
$$

Comparing Eq. (6) with the internal activity in PCNN model, we can conclude that the expression $1/V_L$ corresponds to the linking strength β in PCNN. Of course, here we suppose that $V_L \neq 0$. Since the parameter V_L is counteracted in the internal activity in Eq. (3), we don't have to consider setting its value.

(2.2) W_{ijkl}: Firstly, W_{ijkl} is the linking matrix which directly decides the participation extent of the surrounding neurons in a square area centered in another certain neuron. Then, it brings the output information of the surrounding neurons into the central one to decide the value of the internal activity for further comparison between U_{ij} and the dynamic threshold E_{ij}. According to the difference of the distance between each surrounding pixel and the central pixel, suppose that the size of the square area is 3×3 and the radius is 1, the distance between the central pixel with the corner one is $2^{1/2}$, for this reason, W_{ijkl} can be set as Eq. (7) after normalization. Of course, our matrix setting does not invalidate other setting measures absolutely. But since the influence of the closer pixel is larger to some extent than the distant pixel, absorbing the distance information into the matrix setting is more direct and more convenient.

$$W_{ijkl} = \begin{bmatrix} 0.1035 & 0.1465 & 0.1035 \\ 0.1465 & 0 & 0.1465 \\ 0.1035 & 0.1465 & 0.1035 \end{bmatrix} \tag{7}$$

(2.3) f, g, h: Despite their different locations, the main functions of them are similar. f and g denote the decay constants, which decide the participation radio of the internal activity U_{ij} and the dynamic threshold E_{ij} last time, respectively. Obviously, the values of U_{ij} and E_{ij} grow with f and g increasing. Besides, h denotes the threshold magnitude coefficient to determine the transformation extent of the last output $Y_{ij}[n-1]$ into $E_{ij}[n]$. Accordingly, the value of h directly concerns E_{ij} to influence the firing threshold of each neuron. Based on the analysis and discussions above, we can easily find that the setting of these parameters is very important because of their inherent adjustment effects. On the other hand, the task of setting them is very difficult. Bad final visual performance will inevitably appear if we fail to set them to be the optimal values. In order to make the parameter setting more objective and protect the final fused result from being interfered by human factor to a great extent, meanwhile, considering the correlation between $U_{ij}[n]$ and $U_{ij}[n-1]$ is not very conspicuous, we can rewrite Eqs. (3) and (4) as follows.

$$U_{ij}[n] = S_{ij} \sum_{kl} W_{ijkl} Y_{kl}[n-1] + S_{ij} \tag{8}$$

$$E_{ij}[n] = E_{ij}[n-1] - \Delta + hY_{ij}[n-1] \tag{9}$$

In Eq. (8), the influence of the last internal activity $U_{ij}[n-1]$ is omitted. Step Δ is a positive constant. It guarantees that the dynamic threshold E_{ij} decreases linearly with the iterative number n increasing. In Eq. (9), h is usually set to be a relatively large value to ensure that the firing times of each neuron will not exceed one at most. Here we may as well set Δ and h to be 0.1 and 20, respectively.

(2.4) γ: γ denotes a parameter of sigmoid function. Let's review the requirement when $Y_{ij} = 1$.

$$\begin{aligned} & 1/(1 + \exp(-\gamma(U_{ij}[n] - E_{ij}[n]))) \geq 0.5 \\ \Leftrightarrow\ & 1 + \exp(-\gamma(U_{ij}[n] - E_{ij}[n])) \leq 2 \\ \Leftrightarrow\ & \exp(-\gamma(U_{ij}[n] - E_{ij}[n])) \leq 1 \\ \Leftrightarrow\ & -\gamma(U_{ij}[n] - E_{ij}[n]) \leq 0 \\ \Leftrightarrow\ & \gamma(U_{ij}[n] - E_{ij}[n]) \geq 0 \end{aligned} \tag{10}$$

Here we compare Eq. (10) with the pulse bursting condition in PCNN model, so we can set γ to be 1 for simplicity. Under such conditions, if $U_{ij} \geq E_{ij}$, then the neuron N_{ij} will be activated and generate a pulse, which is characterized by $Y_{ij} = 1$, whereas else $Y_{ij} = 0$.

(3) The setting of the iterative number n has great impacts on the computational costs and the final fusion effects. Inspired by reference [17], the time matrix model [17] with the same size of the source image is utilized to record the firing information of neurons. The expression of the time matrix is given as follows.

$$T_{ij}[n] = \begin{cases} n, & if \ Y_{ij} = 1 \ for \ the \ first \ time \\ T_{ij}[n-1], & else \end{cases} \tag{11}$$

where T is the time matrix, the iterative number n can be determined adaptively according to the intensity distribution of pixels in images. There are several aspects required to be noted: (i) T_{ij} will keep invariable if N_{ij} does not fire all the time; (ii) if N_{ij} fires for the first time, T_{ij} will be set to be the ordinal value of corresponding iteration; (iii) once N_{ij} has already fired, T_{ij} will not alter again. Its value relies on the ordinal number of the iteration, during which N_{ij} fired for the first time. Once all pixels have fired, the whole iteration process is over, and the value of the largest element in T indicates the total iteration times.

In conclusion, ISCM has overcome the drawbacks of the traditional SCM to a great extent, which can be listed as follows.

(1) The number of the parameters in ISCM declines a lot compared with SCM. The traditional SCM has seven variables needing setting in all, but ISCM has only four ones including W_{ijkl}, h, Δ and γ.
(2) The original function Y in ISCM is replaced by the time matrix T, which provides rich temporal and spatial information simultaneously for subsequent image processing.

3 Fusion Algorithm Based on ISCM

Here, we assume that two source images respectively denoted by A and B have been already accurately registered. The concrete steps of the fusion algorithm can be described below.

Step (a): The coefficients of A and B are both normalized between 0 and 1. Take them respectively as the feeding input to stimulate the corresponding ISCM.

Step (b): Initialize $U_{ij}[0] = T_{ij}[0] = Y_{ij}[0] = 0$, $E_{ij}[0] = 1$.

Note that all neurons don't fire at first. The reason for $E_{ij}[0]$ setting as given above is to make neurons be activated as soon as possible, which can prevent unnecessary "void" iterations and save computational costs.

Step (c): Compute $U_{ij}[n]$, $T_{ij}[n]$, $Y_{ij}[n]$ and $E_{ij}[n]$ according to Eqs. (8), (11), (5) and (9), respectively.

Step (d): Implement step (c) iteratively until all neurons are activated, namely the elements in $T_{ij}[n]$ are equal to one. Then, the coefficients of F can be chosen as follows.

$$F_{ij} = \begin{cases} A_{ij}, & if \ T_{ij,A} \leq T_{ij,B} \\ B_{ij}, & if \ T_{ij,A} > T_{ij,B} \end{cases} \qquad (12)$$

In Eq. (12), if $T_{ij,A}$ is less than $T_{ij,B}$, it means that the pixel located at (i, j) in A has much stronger feeding input than the corresponding pixel in the same place of B. Therefore, we choose A_{ij} as the corresponding pixel in F_{ij}. Conversely, B_{ij} will be chosen.

Fig. 2. Source images: (a) Leaf A (256 * 256). (b) Leaf B (256 * 256). (c) Craft A (256 * 256). (d) Craft B (256 * 256). (e) Puma A (512 * 512). (f) Puma B (512 * 512). (g) Balloon A (512 * 512). (h) Balloon B (512 * 512)

4   Experimental Results and Related Analysis

In this section, in order to test the proposed technique, extensive simulation experiments have been carried out on four groups of multi-focus images shown in Fig. 2.

Obviously, there are both in-focus and out-focus areas in each group of source images. The task confronted us is how to extract the information of the in-focus area in source images respectively, and then to fuse it into the final resultant image. In order to compare the fusion effects between our proposed technique and the current several typical ones, three techniques are chosen in this section, which are m-PCNN-based algorithm (A1) [18], SW-NSCT-PCNN-based algorithm (A2) [19], and NSST-I^{2}CM-based algorithm (A3) [20]. In order to make the final comparison result persuasive, the parameters settings in A1–A3 are still implemented as the references [18–20]. If you want to obtain more details please refer to related references mentioned above. For simplicity, we term the proposed algorithm as A4.

Subjective visual evaluation system can be utilized to provide direct comparisons. However, it is easily prone to be affected by lots of personal factors, such as eyesight level, mental state, even the mood, and so on. As a result, it is also necessary for us to evaluate the fusion effects based on the objective quality assessment. In this paper, we choose three objective quality metrics including space frequency (SF), Mutual information (MI) and Information entropy (IE) to quantitative evaluate the performance of the proposed fusion technique. The three metrics are defined as follows.

(1) Space frequency (SF)

SF reflects the whole spatial active degree of the image, which can be divided into spatial row frequency (RF) and spatial column frequency (CF), whose mathematical expressions are listed as follows.

$$RF = \sqrt{\frac{1}{M \times N} \sum_{i=1}^{M} \sum_{j=2}^{N} [P(i,j) - P(i,j-1)]^2} \tag{13}$$

$$CF = \sqrt{\frac{1}{M \times N} \sum_{i=2}^{M} \sum_{j=1}^{N} [P(i,j) - P(i-1,j)]^2} \tag{14}$$

Where P denotes the final fused image whose size is $M \times N$. The whole SF equals the root mean square between RF and CF. The larger SF is, the more the whole active degree of the final fused image is. The expression of SF can be written as follows.

$$SF = \sqrt{CF^2 + RF^2} \tag{15}$$

(2) Mutual information (MI)

MI can be used to reflect fused effects and measure the relativity between two or more images. The larger MI is, the more abundant information the fused image contains.

$$MI = \frac{\sum_{i=0}^{L-1} \sum_{k=0}^{L-1} P_{A,F}(i,k) \log \frac{P_{A,F}(i,k)}{P_A(i)P_F(k)} + \sum_{j=0}^{L-1} \sum_{k=0}^{L-1} P_{B,F}(j,k) \log \frac{P_{B,F}(j,k)}{P_B(j)P_F(k)}}{IE\_A + IE\_B} \tag{16}$$

$P_{A,F}(i,k)$ and $P_{B,F}(j,k)$ are the normalized gray histogram between source image A and the fuse image F, the normalized gray histogram between source image B and F, respectively.

(3) Information entropy (IE)

IE directly reflects the amount of average information in fused image. The larger the IE is, the more abundant the information amount of the fused image is.

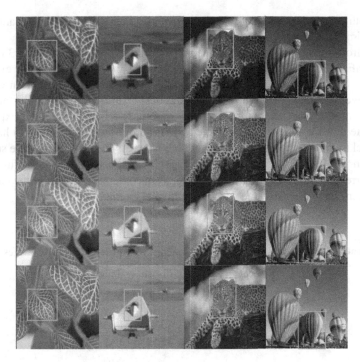

Fig. 3. Fused results of four different groups of multi-focus images based on M1–M4 (from row 1 to row 4) (Color figure online)

$$IE = - \sum_{i=0}^{L-1} P(i) \log_2 P(i) \qquad (17)$$

Where $P(i)$ indicates the probability of pixels whose gray value amount to i over the total image pixels.

The final fused results on the four groups of source images based on A1–A4 are illustrated in Fig. 3.

Observed from the fused results shown in Fig. 3, all of the four algorithms have all extracted the main features from corresponding source images and then fused them into the final fused image. However, by careful comparison, we can still find that the differences among them exist. In order to guarantee the objectivity of the conclusion, several computer experts were invited to evaluate the overall visual effects among the above four algorithms. The experts gave me the conclusion that the proposed algorithm, namely A4, is superior to other five ones. Obviously, the final fused images based on A1 don't own a reasonable contrast level. A2 and A3 suffer from the problem of blurred objects or artifacts in different degrees. Take the resultant images in the fourth column for example, compared with A1–A3, the object of the left craft based on A4 is much clearer, please see the red labels in the corresponding area. As a result, we can reach the conclusion that the proposed algorithm outperforms other current typical ones in both preserving important details information and inheriting the characteristics of source images from the subjective visual point of view.

Apart from the comparison of the visual performance, Tables 1, 2 and 3 report the values of SF, MI and IE for the four fusion algorithms in terms of objective evaluation. Please note that all the reported statistics are the average results of the four groups of source images, and the bold values indicate the best result in the same index among the mentioned algorithms. The largest values of SF for four groups of experiments demonstrate that the fuse images obtained by the proposed algorithm (A4) have more clarity level than those based on A1–A3, which is greatly consistent with the subjective comparison result. It indicates that as for the relativity between the final fused image and the corresponding source ones, A4 outperforms other ones. Only in the case of the 'Leaf' experiment, the MI value of A1 is a bit larger than that of A4.

Table 1. SF results based on A1–A4 about four groups of source images

	Leaf	Craft	Puma	Balloon
A1	16.4683	6.6817	10.1433	11.2470
A2	21.0986	5.9383	14.5112	16.6353
A3	17.7397	5.4672	13.4782	14.3517
A4	**22.1839**	**7.1731**	**15.2191**	**19.4274**

Table 2. MI results based on A1–A4 about four groups of source images

	Leaf	Craft	Puma	Balloon
A1	**5.7009**	5.7511	5.3542	5.6306
A2	5.1917	5.8573	5.9926	6.1615
A3	5.3764	5.2248	6.1425	6.2947
A4	5.5576	**5.8830**	**6.3003**	**6.4909**

Table 3. IE results based on A1–A4 about four groups of source images

	Leaf	Craft	Puma	Balloon
A1	7.3689	6.5664	6.8387	7.2099
A2	7.3572	6.4436	7.4312	7.4655
A3	7.2846	6.4692	7.3991	7.4525
A4	**7.4818**	**6.6139**	**7.5510**	**7.4821**

Based on the above analysis and comparison in terms of both subjective and objective aspects, we can find that the proposed algorithm is effective and it has obvious superiorities over other current typical algorithms on the multi-focus image fusion fields.

5 Conclusion

In this paper, a new fusion algorithm for multi-focus images based on improved SCM is proposed. Compared with the traditional models of neural networks, SCM combines the advantages of PCNN and ICM together, and much better experimental results have been obtained via several groups of multi-focus source images. The subjective visual effects and objective evaluation results both verify and prove that the proposed algorithm is superior to other typical ANN based algorithms. Next, how to further optimize and enhance the performance of the new algorithm will be the focus of our future work.

Acknowledgments. The authors would like to thank the anonymous reviewers and editors for their invaluable suggestions. The work was supported in part by the National Natural Science Foundations of China under Grant 61309008, 61309022, 61373116 and 61472302, in part by the Natural Science Foundation of Shannxi Province of China under Grant 2014JQ8349, and the Foundation of Science and Technology on Information Assurance Laboratory under Grant KJ-15-102.

References

1. Yu, B.T., Jia, B., Ding, L., et al.: Hybrid dual-tree complex wavelet transform and support vector machine for digital multi-focus image fusion. Neurocomputing **182**, 1–9 (2016)
2. Yang, L., Guo, B.L., Ni, W.: Multimodality medical image fusion based on multiscale geometric analysis of contourlet transform. Neurocomputing **72**, 203–211 (2008)
3. Yang, Y., Que, Y., Huang, S., et al.: Multimodal sensor medical image fusion based on type-2 fuzzy logic in NSCT domain. IEEE Sens. J. **16**, 3735–3745 (2016)
4. Bhatnagar, G., Wu, Q.M.J., Liu, Z.: Directive contrast based multimodal medical image fusion in NSCT domain. IEEE Trans. Multimedia **15**, 1014–1024 (2013)
5. Yang, Y., Tong, S., Huang, S.Y., et al.: Multifocus image fusion based on NSCT and focused area detection. IEEE Sens. J. **15**, 2824–2838 (2015)
6. Lim, W.Q.: Nonseparable shearlet transform. IEEE Trans. Image Process. **22**, 2056–2065 (2013)
7. Miao, Q.G., Shi, C., Xu, P.F., et al.: A novel algorithm of image fusion using shearlets. Opt. Commun. **284**, 1540–1547 (2011)
8. Kong, W.W., Zhang, L.J., Lei, Y.: Novel fusion method for visible light and infrared images based on NSST-SF-PCNN. Infrared Phys. Technol. **65**, 103–112 (2014)
9. Kong, W.W., Liu, J.P.: Technique for image fusion based on nonsubsampled shearlet transform and improved pulse-coupled neural network. Opt. Eng. **52**, 017001 (2013)
10. Shi, C., Miao, Q.G., Xu, P.F.: A novel algorithm of remote sensing image fusion based on shearlets and PCNN. Neurocomputing **117**, 47–53 (2013)
11. Yang, S.Y., Wang, M., Jiao, L.C.: Contourlet hidden Markov Tree and clarity-saliency driven PCNN based remote sensing images fusion. Appl. Soft Comput. **12**, 228–237 (2012)
12. Eckhorn, R., Reiboeck, H.J., Arndt, M., et al.: A neural networks for feature linking via synchronous activity: results from cat visual cortex and from simulations. In: Cotterill, R.M.J. (ed.) Models of Brain Function, pp. 255–272. Cambriage University Press, Cambriage (1989)
13. Eckhorn, R.: Neural mechanisms of scene segmentation: recordings from the visual cortex suggest basic circuits or linking field models. IEEE Trans. Neural Netw. **10**, 464–479 (1999)

14. Kinser, J.M.: Simplified pulse-coupled neural network. In: Proceedings of SPIE, vol. 2760, pp. 563–567 (1996)
15. Zhan, K., Zhang, H., Ma, Y.D.: New spiking cortical model for invariant texture retrieval and image processing. IEEE Trans. Neural Netw. **20**, 1980–1986 (2009)
16. Wang, N.Y., Ma, Y.D., Zhan, K.: Spiking cortical model for multifocus image fusion. Neurocomputing **130**, 44–51 (2014)
17. Ma, Y.D., Lin, D.M., Zhang, B.D.: A novel algorithm of image enhancement based on pulse coupled neural network time matrix and rough set. In: Proceedings of IEEE Conference on Fuzzy Systems and Knowledge Discovery (Institute of Electrical and Electronics Engineers, Haikou), vol. 1, pp. 86–90 (2007)
18. Wang, Z.B., Ma, Y.D.: Medical image fusion using m-PCNN. Inf. Fusion **9**, 176–185 (2008)
19. Yang, S.Y., Wang, M., Lu, Y.X., et al.: Fusion of multiparametric SAR images based on SW-nonsubsampled contourlet and PCNN. Sig. Process. **89**, 2596–2608 (2009)
20. Kong, W.W.: Multi-sensor image fusion based on NSST domain I^2CM. IET Electron. Lett. **49**, 802–803 (2013)

Robust Underwater Image Stitching Based on Graph Matching

Xu Yang[1], Zhi-Yong Liu[1,2,3](✉), Chuan Li[1,4], Jing-Jing Wang[1], and Hong Qiao[1,2,3]

[1] State Key Laboratory of Management and Control for Complex Systems, Institute of Automation, Chinese Academy of Sciences, Beijing 100190, People's Republic of China
{xu.yang,zhiyong.liu,lichuan2013,wangjingjing2014,hong.qiao}@ia.ac.cn
[2] Center for Excellence in Brain Science and Intelligence Technology, Chinese Academy of Sciences, Shanghai 200031, People's Republic of China
[3] University of Chinese Academy of Sciences, Beijing 100049, China
[4] Research Institute of Chengdu, Beijing Jingdong Century Trade Co., Ltd., Chengdu 610046, China

Abstract. Image stitching is important in intelligent perception and manipulation of underwater robots. In spite of a well developed technique, it is still challenging for underwater images because of their inevitable appearance ambiguity. For the feature based underwater image stitching, robust feature correspondence is the key because most other algorithmic parts are less directly associated with the characteristics of underwater images. Structural information between feature points may be helpful for robust feature correspondence, and based on this idea the paper proposes a robust underwater image stitching method by incorporating structural cues as additional information, whose effectiveness is validated on real underwater images. Specifically, the appearance information and structural cues are integrated by a labeled weighted graph, and the underwater image correspondence is formulated by graph matching. After geometric transformation estimation, the underwater images are finally blended into a wider viewing image.

Keywords: Underwater image · Image stitching · Feature correspondence · Graph matching · Structural information

1 Introduction

Image stitching aims at the combination of two images, or more images, with overlapped areas into a wide viewing composite, or even a panorama. It plays a key role in those robot tasks in places presenting a difficult access for human beings, such as some tasks by underwater robots, e.g. the remotely operated underwater vehicle (ROV) [1], the autonomous underwater vehicle (AUV) [2], or the autonomous remotely operated vehicle (ARV) [3]. It is because once a robot is equipped with a camera, the visual perception of its operating environment

© Springer International Publishing AG 2017
Y. Sun et al. (Eds.): IScIDE 2017, LNCS 10559, pp. 521–529, 2017.
DOI: 10.1007/978-3-319-67777-4_46

is usually of interest. Since the common camera equipped to the robots usually has a limited field of view, image stitching is thus useful to obtain a larger field of view over the operating area.

Image stitching itself has long been an important topic in image processing and computer vision. Many methods have come out, which can be roughly categorized into two types, i.e. the region based methods and the feature based methods. Early image stitching usually adopts the region based methods, which usually find a common region in two images through the region appearance information, e.g. pixel intensity. The ideas of these methods are usually straightforward and easy to implement. But at the same time these methods usually suffer from illumination changes, occlusions, geometric distortions in different images, and therefore inappropriate for real world images, especially those images obtained on a mobile robot platform. Recently most researchers in the computer vision and robot vision communities tend to use the feature based image stitching methods, because of their robustness to changing factors, such as the above mentioned changing illuminations, geometric distortions, etc. Particularly, the emergences of many excellent local features such as SIFT [4] and SURF [5] in the last fifteen years have promoted the success of the feature based image stitching, which is used in many real world camera applications, including those on intelligent mobile phones. It can be said image stitching is a solved problem for many types of images, especially the images on land.

Different from the images on land, it is still a challenging problem to stitch underwater images. The main obstacle lies in the inevitable appearance ambiguity of underwater images. It is because the limited light refracted into the water or shot from the main robot body, would further be scattered or absorbed by water molecules, plankton, or sands. Such a condition would significantly deteriorate the performance of general image stitching methods, even the feature based methods, because the ambiguous appearance of underwater images often leads to a poor discriminant ability, or effect lost, of the feature descriptor.

From the algorithmic perspective, the feature based methods mainly consist of three steps, i.e. feature correspondence, transformation estimation, image blending [6], where the combination of feature correspondence and transformation estimation is also known by the term image registration. It can be noticed that the latter two steps are less directly related with the discriminant ability of the feature descriptor, which implies that once the feature point are successfully corresponded, the stitching of the underwater images are almost the same with the stitching of the images on land. In other words, robust feature correspondence of the underwater images is the key for their stitching. Since only the appearance information is inadequate, introducing additional information or incorporating additional constraints is a intuitive way to improve the robustness of feature correspondence on underwater images. The structural information between feature points may be an effective choice [7]. Because by incorporating the structural constraints it requires that the structures extracted from the feature point sets should be consistent while maintaining the appearance similarity, which may help to avoid abnormal feature assignments.

Based on these understandings, this paper proposes a roust underwater image stitching method by introducing additional structural information. Specifically, the appearance information and structural information are integrated by a labeled weighted graph model, and the feature correspondence between two underwater images is formulated and solved by graph matching, which is followed by outlier assignment refinement. After the estimation of geometric transformation from the inlier assignments, the underwater images are finally blended into a wider viewing image.

The remaining manuscript is organized as follows: After the discussion of related works in Sect. 2, the proposed underwater image stitching method is introduced in Sect. 3, which is followed by the experimental evaluation in Sect. 4. Finally Sect. 5 concludes the paper.

2   Related Works

In this section, we first give some discussions on recent image stitching algorithms, and then introduce their applications to underwater images.

2.1   Image Stitching Algorithms

In recent years, the feature based methods have overtook the area based methods as the most common image stitching algorithms. Their success can be attributed to the robustness to changing factors, such as the geometric distortions. A benchmark algorithm is the famous scale invariant feature transform (SIFT) feature [4] based image stitching method. Then researchers have applied many types of local features to the image stitching task, of which some representative features include the speeded up robust feature (SURF) [5] feature, the binary robust independent elementary features (BRIEF) [8], the shape context feature [9], etc. Most of these algorithms are dedicated to image stitching tasks with general purposes, and get superior performance on common natural images, especially images on land. Only a few researchers generalize them to the underwater images, as introduced below.

2.2   Underwater Image Applications

In the image stitching method proposed by Leone et al. [10] for underwater environment, the Harris corner point detector with certain specific improvements is used to extract the feature points, and the texture information is used to built the feature point descriptor. Then the correspondence between two feature point sets which represents two underwater images is established by matching the feature point descriptors. The homography transformation, i.e. the translations, rotations and scaling effects, between two underwater images is estimated based on the correspondence, and then stitched image is obtained. A similar scheme is used by Elibol et al. [11] in their underwater image stitching work, or called by underwater optical mapping in their work. Differently, they adopted the SIFT feature point and descriptor extracted from the underwater images, of which the outlier assignments are refined by the famous random sample consensus (RANSAC) technique. In the real time image stitching method proposed by Ferreira et al. [12], after the binary robust independent elementary features (BRIEF) based motion estimation, the SURF is used in the feature correspondence step. Garcia-Fidalgo et al. [13] in their underwater image stitching method used a feature which is a variant of BRIEF in the framework of bags of words (BoG).

Generally, the proposed method follows a similar scheme with the above methods, but incorporates structural constraints in the feature correspondence step, which is useful against the ambiguous appearance of underwater images.

3   Underwater Image Stitching

The proposed underwater image stitching method is introduced in this section. As mentioned above, robust feature correspondence is particularly important for underwater images, which is realized by incorporating structure cues beyond the appearance information in this paper.

A feature point extracted from an underwater image is represented by a weighted labeled graph $\mathcal{G}$ to integrate the appearance information and the structural relations between feature points. Almost any of the well known local features could be adopted as the feature extractor and descriptor, e.g. Harris corner detector, SIFT extractor and descriptor, SURF extractor and descriptor, BRIEF extractor and descriptor, etc., which implies that the incorporation of structural cue lowers the demand of discriminant feature extractor and descriptor. Then it is straightforward to represent the feature set by a labeled weighted graph by representing each feature point by a graph vertex, representing the link between a pair of feature points by a graph edge, describing the vertex by a so called label using the feature descriptor, and describing the edge by a so called weight using the spatial relation measure, e.g. length and orientation of the link. Thus the feature correspondence problem can by assigning the vertices in two labeled weighted graphs, abbreviated by graph below, which problem is known as graph matching.

Mathematically, the collection of the weights in a graph $\mathcal{G}$ can be represented by weighted adjacency matrices $\mathbf{G}^i, i = 1 \cdots d$. The number of weighted adjacency matrices d depends on the weight dimension. For instance, when using the distance between feature points, i.e. the link length, as the edge weight, only one adjacency matrix $\mathbf{G}^1$ is enough for a graph, where each non-diagonal entry $\mathbf{G}^1_{ij}$ denotes the distance between the ith and j vertices in $\mathcal{G}$. The pre-calculated differences between vertex labels are stored in a label cost matrix $\mathbf{L} \in \mathbb{R}^{M \times N}$ where $\mathbf{L}_{ia}$ denotes the distance between the label of the ith vertex in $\mathcal{G}$ and that of the ath vertex in $\mathcal{H}$. Given two graphs $\mathcal{G}$ and $\mathcal{H}$ of sizes M and N respectively, their matching can be represented by an assignment matrix $\mathbf{X} \in \{0,1\}^{M \times N}$, where $\mathbf{X}_{ia} = 1$ means that the ith vertex in $\mathcal{G}$ is assigned to the ath vertex in $\mathcal{H}$. If the one-to-one matching assumption is adopted, then the assignment matrix becomes a so called partial permutation matrix, defined by

$$\mathbf{X} \in \mathcal{D} := \left\{ \mathbf{X} \Big| \sum_i \mathbf{X}_{ia} \leq 1, \sum_a \mathbf{X}_{ia} = 1, \mathbf{X}_{ia} = \{0, 1\} \right\}. \tag{1}$$

Without loss of generality, it is assumed that $M \leq N$ hereafter. Based on the above the mathematical representations, the correspondence result can be obtained by minimizing the following graph matching objective function:

$$\mathbf{X}^* = \alpha \min_{\mathbf{X}} \sum_{i=1}^{d} \|\mathbf{G}^i - \mathbf{X}\mathbf{H}^i\mathbf{X}^T\|_F + (1 - \alpha)\mathrm{tr}(\mathbf{L}^T\mathbf{X}), \tag{2}$$

$$\text{s.t } \mathbf{X} \in \mathcal{D}.$$

The optimization problem is an NP-hard high order combinatorial optimization problem with factorial computational complexity, for which the approximate method are necessary. We use the graduated nonconvexity and graduated concavity (GNCCP) [14], a continuous method based combinatorial optimization framework, to approximately solve the problem. The utilize the GNCCP, the discrete domain $\mathcal{D}$ should be relaxed to its domain $\mathcal{C}$, defined by

$$\mathcal{C} := \left\{ \mathbf{X} \Big| \sum_i \mathbf{X}_{ia} \leq 1, \sum_a \mathbf{X}_{ia} = 1, \mathbf{X}_{ia} \in [0, 1] \right\}. \tag{3}$$

And the GNCCP also makes use of the property that DD is exactly the extreme point set of $\mathcal{C}$. Specifically, the GNCCP first approximates the original optimization problem (2) by a relatively simple convex optimization problem over the continuous domain $\mathcal{C}$, and step by step implicitly transforms it to be a concave optimization problem over $\mathcal{C}$. Note by a clever design both the above convex optimization problem and concave optimization problem have exactly the same global optimum as (2) over the discrete domain $\mathcal{D}$. And the optimum point of the concave optimization problem over a convex set lies in its extreme point set, i.e. $\mathcal{D}$ by the property mentioned above. Therefore a discrete assignment matrix could be automatically obtained when the GNCCP terminates at the concave optimization problem, which usually exhibit superior performance.

In each step of the GNCCP process, the optimization problem is solved by the conditional gradient descent method [15,16], also known as the Frank-Wolfe algorithm, where the gradient of the original function (2) is needed, which is

$$\nabla = \alpha \sum_{i=1}^{d} (2\mathbf{X}(\mathbf{H}^{iT}\mathbf{X}^T\mathbf{X}\mathbf{H}^i + \mathbf{H}^i\mathbf{X}^T\mathbf{X}\mathbf{H}^{iT}) - 2(\mathbf{G}^i\mathbf{X}\mathbf{H}^{iT} + \mathbf{G}^{iT}\mathbf{X}\mathbf{H}^i)) + (1 - \alpha)\mathbf{L}. \tag{4}$$

The solution $\mathbf{X}^*$ indicate the assignments between feature points in two underwater images. As mentioned in Sects. 1 and 2, once these assignments obtained, the transformation between the images ca be estimated. Before the transformation estimation, we first employ the maximum likelyhood estimation sample consensus (MLESAC) [17] to refine the assignments, or say to remove the outlier assignments. The MLESAC is a variant of the famous RANSAC [18]. Different from RANSAC, it aims at the solution which maximizes the likelihood instead of the number of inliers, and is particularly appropriate for the estimation of complex surfaces or more general manifolds from points [17]. Then the projective matrix P between two images are estimated based on the refined inlier assignments. If a frame sequence sampled from for example a video clip are provided, the projective matrices are estimated sequentially following a similar way in [19].

Once the cascaded estimations of the projective transformation between underwater images are obtained, the final step is to warp all the images according to the transformation estimation and blending them together [6]. In order to reduce the visual influence of the seam, the stitched images are blended by rendering the overlapped area by the average intensities from both images.

4   Simulations

The proposed scheme is first evaluated on an underwater image sequence shot at the Valldemossa harbour seabed (Mallorca, Spain) [13]. This dataset contains

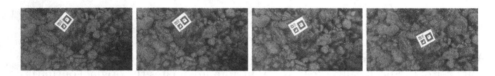

Fig. 1. Preceding 4 samples of underwater image sequence shot at Valldemossa harbour seabed.

SURF based method   The proposed method

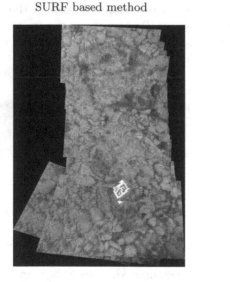
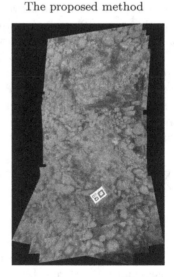

Fig. 2. Stitching result on underwater image sequence shot at Valldemossa harbour seabed.

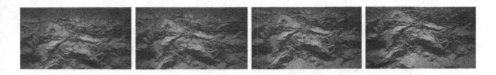

Fig. 3. Preceding 4 samples of underwater images shot by IFREMER.

SURF based method   The proposed method

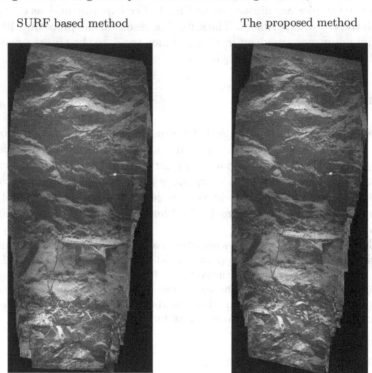

Fig. 4. Stitching result on the underwater images shot by IFREMER.

201 images with 320×180 pixels, which forms a loop around a central point. Some image samples of the dataset are shown in Fig. 1. In each image the key points are extracted by SURF [5] together with the descriptors. The graphs are constructed as described in Sect. 3. Specifically, the key points are represented by the graph vertices, with the SURF descriptor as the vertex labels. The graph structure, i.e. the set of the graph edges, are built by the Delaunay triangulation technique [20]. The length and orientation of each link in the graph structure are used as the two-dimensional edge weight. The proposed method is compared with the SURF based stitched method which does not consider the structural relation between key points. Because of the memory limit of the our computer, the preceding 40 images of the total 201 images are stitched, as illustrated in Fig. 2. It can be observed that the proposed method illustrates more smooth transition across

the images. It is directly attributed to the accurate image registration, which is essentially resulted from the robust feature correspondence by incorporating structural cues.

The proposed method is also applied to another underwater dataset[1] sampled from the video released by French research institute for exploitation of the sea (IFREMER), which is shot along the Mid-Atlantic Ridge in the North Atlantic Ocean. Some samples are illustrated in Fig. 3. The experimental setting are the same with the above experiment. The stitching result on the preceding 40 images of the total 64 images are given in Fig. 4, which validate the effectiveness of the proposed method.

5   Conclusion

This paper aims at the robust image stitching in the underwater environment, proposes to introduce the structural information to tackle the appearance ambiguity problem, a specific problem for the underwater images. Simulations witness the effectiveness of our idea. However, there may be seams or moving objects caused ghosting areas by the current version of proposed method. Therefore, we intend to make more investigations in the blending step in our future work.

Acknowledgment. This work is supported partly by the National Natural Science Foundation (NSFC) of China (grants 61503383, 61633009, U1613213, 61375005, 61210009, and 61773047), partly by the National Key Research and Development Plan of China (grant 2016YFC0300801), partly by the Beijing Municipal Science and Technology (grants D16110400140000 and D161100001416001), and Guangdong Science and Technology Department (grant 2016B090910001).

References

1. Huang, H., Tang, Q., Li, Y., Wan, L., Pang, Y.: Dynamic control and disturbance estimation of 3D path following for the observation class underwater remotely operated vehicle. Adv. Mech. Eng. **5**, 604393 (2013)
2. Huang, H., Tang, Q., Li, H., Liang, L., Li, W., Pang, Y.: Vehicle-manipulator system dynamic modeling and control for underwater autonomous manipulation. Multibody Syst. Dyn. (2017). doi:10.1007/s11044-0169538-3
3. Tang, Y., Li, S., Zhang, A.: Research on optimization design of a new type of underwater vehicle for arctic expedition. In: Proceedings of the International Symposium Underwater Technology, pp. 96–100 (2009)
4. Lowe, D.: Object recognition from local scale-invariant features. In: Proceedings of IEEE International Conference on Computer Vision, vol. 2, pp. 1150–1157 (1999)
5. Bay, H., Tuytelaars, T., Van Gool, L.: SURF: speeded up robust features. In: Proceedigs of the European Conference on Computer vision, pp. 404–417 (2006)
6. Brown, M., Lowe, D.: Automatic panoramic image stitching using invariant features. Int. J. Comput. Vis. **74**(1), 59–73 (2007)

---

[1] The dataset is named by ODEMAR, which is available at https://github.com/emiliofidalgo/bimos.

7. Yang, X., Liu, Z., Qiao, H., Song, Y., Ren, S., Zheng, S.: Underwater image matching by incorporating structural constraints. Int. J. Adv. Robot. Syst. Under Revision

8. Calonder, M., Lepetit, V., Strecha, C., Fua, P.: BRIEF: binary robust independent elementary features. In: Daniilidis, K., Maragos, P., Paragios, N. (eds.) ECCV 2010 Part IV. LNCS, vol. 6314, pp. 778–792. Springer, Heidelberg (2010). doi:10.1007/978-3-642-15561-1_56

9. Belongie, S., Malik, J.: Matching with shape contexts. In: Proceedings of the IEEE Workshop on Content-based Access of Image and Video Libraries, pp. 20–26 (2000)

10. Leone, A., Distante, C., Mastrolia, A., Indiveri, G.: A fully automated approach for underwater mosaicking. In: OCEANS, pp. 1–6 (2006)

11. Elibol, A., Garcia, R., Gracias, N.: A new global alignment approach for underwater optical mapping. Ocean Eng. **38**(10), 1207–1219 (2011)

12. Ferreira, F., Veruggio, G., Caccia, M., Zereik, E., Bruzzone, G.: A real-time mosaicking algorithm using binary features for rovs. In: Proceedings of the Mediterranean Conference on Control and Automation, pp. 1267–1273 (2013)

13. Garcia-Fidalgo, E., Ortiz, A., Bonnin-Pascual, F., Company, J.: Fast image mosaicing using incremental bags of binary words. In: Proceedings of the IEEE International Conference Robotics and Automation, pp. 1174–1180 (2016)

14. Liu, Z., Qiao, H.: GNCCP graduated nonconvexity and concavity procedure. IEEE Trans. Pattern Anal. Mach. Intell. **36**(6), 1258–1267 (2014)

15. Frank, M., Wolfe, P.: An algorithm for quadratic programming. Nav. Res. Logist. Q. **3**(1–2), 95–110 (1956)

16. Jaggi, M.: Revisiting Frank-Wolfe: projection-free sparse convex optimization. In: Proceedings of the International Conference on Machine Learning, pp. 427–435 (2013)

17. Torr, P., Zisserman, A.: MLESAC: a new robust estimator with application to estimating image geometry. Comput. Vis. Image Underst. **78**, 138–156 (2000)

18. Fischler, M., Bolles, R.: Random sample consensus: a paradigm for model fitting with applications to image analysis and automated cartography. Commun. ACM **24**(6), 381–395 (1981)

19. Li, C., Liu, Z., Yang, X., Qiao, H., Su, J.: Stitching contaminated images. Neurocomputing **214**, 829–836 (2016)

20. Loera, J., Rambau, J., Santos, F.: Triangulations: Structures for Algorithms and Applications. Springer, Heidelberg (2010)

A Fingerprint Registration Method Based on Image Field and Mean Square Error

Sheng Lan[✉] and Zhenhua Guo[✉]

Graduate School at Shenzhen, Tsinghua University, Beijing, China
q133589976@163.com, zhenhua.guo@sz.tsinghua.edu.cn

Abstract. Fingerprint registration is an important step in the technology of fingerprint identification. Image field is often used for feature extraction in fingerprint registration, however, the accuracy can not be guaranteed due to the selection of feature points. On the other hand, methods based on mutual information are widely used in the field of image registration because of the high precision, but the speed is slow owing to the complexity of calculating mutual information. By sharing the advantages and overcoming shortcomings of the image field and mutual information methods, this paper presents an improved method of fingerprint registration based on image field and mean square error (MSE). First, we calculate the image field to obtain clear ridges and valleys of the image, and then replace the mutual information with MSE, we globally search the spatial transformation parameters for image registration. The experimental results show effectiveness of the proposed algorithm.

Keywords: Fingerprint · Image registration · Image field · MSE

1 Introduction

With the requirement of identity authentication technology, automatic fingerprint identification technology has been widely used in public security, customs, banking, network security and other areas need for identification [1], as an important step in fingerprint identification technology, fingerprint image registration [2] has high theoretical and practical significance.

The image field [3] is often used for feature extraction of fingerprint [4,5]. A fingerprint image can be described by intensity field, gradient field, orientation field and frequency field. The intensity field can reflect the degree of light and dark of image, which can be used for image equalization by calculating the histogram of gray distribution and normalization. Fingerprint orientation field can reflect the direction of the trend, which can be used for image enhancement along the fingerprint ridge lines. The gradient field can reflect the degree

S. Lan—This work is partially supportted by the Natural Science Foundation of China (NSFC) (No. 61527808) and Shenzhen fundamental research fund (subject arrangement) (Grant No. JCYJ20170412170438636).

Y. Sun et al. (Eds.): IScIDE 2017, LNCS 10559, pp. 530–538, 2017.
DOI: 10.1007/978-3-319-67777-4_47

of upheaval of fingerprint, which can be used for calculating the orientation field. The frequency field can reflect the fingerprint density interval, which can be used for image enhancement along the vertical direction on the ridge lines. Methods based on image field can preferably remove noise from the original gray fingerprint, enhance the contrast of the ridge lines and get the sharp point chart for feature extraction [6,7], then feature points are used for image registration. However, due to the need of manual intervention and difficulty to select feature points, the accuracy of image registration can not be guaranteed. On the other hand, the traditional method of image registration based on mutual information [8] is a registration method based on gray level, directly process on the image gray level, which can avoid error caused by feature points selection and brings higher accuracy, however, because of the complexity of calculating mutual information, it is a rather slow algorithm.

By sharing the advantages and overcoming shortcomings of the image field and mutual information methods, this paper presents a method of fingerprint image registration based on image field and MSE. First, we calculate the image field and obtain clear ridges and valleys of the fingerprint image by preprocess of image equalization, smoothing, enhancement and image binarization. Then, instead of using feature extraction in traditional image field methods, we use mean square error (MSE [9]) instead of mutual information to globally search the spatial transformation parameters for image registration, and thus avoid errors caused by the selection of feature points. A globally search method (POWELL [10]) is used as searching strategy. On the other hand, as calculation of MSE relatively simple in comparison to mutual information, our method is able to be faster in speed. Experimental results show that our method could get more accurate registration results with less computation cost.

The rest of paper is organized as follows. Section 2 presents the proposed algorithm. Section 3 reports and analyzes the experimental results. Finally, Sect. 4 concludes our work and gives future research work.

2   Fingerprint Registration Method Based on Image Field and Mean Square Error

2.1   Pre-processing Based on Image Field

Due to the influence of various factors, the original fingerprint got by acquisition equipment contains many noise, preprocessing based on image field is to improve the quality of the input image, enhance the contrast of ridge lines and valley lines of the image for registration. As shown in Fig. 1, the image field based preprocessing includes image field calculation, image equalization, image smoothing, enhancement, and image binarization [11,12].

Based on the fingerprint image field, we can calculate the distribution histogram of gray by the intensity field, so that the image can be equalized to be evenly distributed on each intensity, image equalization can enhance the brightness contrast of the image, in order to eliminate the noise of fingerprint image

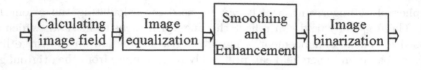

Fig. 1. Preprocessing based on image field.

and further enhance the contrast of the ridge and valley, image smoothing and enhancement are needed, here we use mean filtering [13] and image enhancement method based on orientation field [14], the Gabor function [15] is used, finally, based on the orientation field, we can calculate the value of gray by the ridge direction and the vertical direction, then carry on the image binarization.

2.2   Registration via Mean Square Error

Image registration is to find a spatial transformation T (translation, rotation and affine transformation [16]), so that the reference image R and the floating image F can reach a consensus on the space coordinate.

$$R\left(x,y\right) = T\left(F\left(x,y\right)\right). \tag{1}$$

A full image registration algorithm usually consists of four parts, namely the feature space, the search space, the search strategy and the similarity measure [17]. Feature space [18] is the part extracted from the image for registration; search space is the scope of registration and image space transformation method, this paper only consider translation, and rotation transform; search strategy is to find the optimal parameters for registration in the search space, this paper uses the POWELL method which is convenient for the global search; similarity measure is a measure of the transformation, which can provide the basis for the following search strategy. The direction acceleration method (POWELL) is a direct search method that does not need to compute the gradient of objective function, only by comparing the size of the mobile target function, the extreme point can be calculated, its basic idea is that the whole calculation process is divided into several iterations, each iteration consists of $n+1$ one-dimensional searches, during the k-th iteration, first run n times of one-dimensional searches from the initial point $x^{(k,0)}$ in n linearly independent directions along the $d^{(k,1)}, d^{(k,2)}, \ldots, d^{(k,n)}$, and get the extreme points $x^{(k,1)}, x^{(k,2)}, \ldots, x^{(k,n)}$, and then seek the best point $x^{(k,m)}$, as shown in Eq. (2).

$$f\left(x^{(k,m)} - x^{(k,m-1)}\right) = max_j\left\{f\left(x^{(k,j)} - x^{(k,j-1)}\right)\right\}. \tag{2}$$

Then run a search from the point $x^{(k,m)}$ along the direction $d^{(k,n+1)}$ of the connection of $x^{(k,0)}$ and $x^{(k,m)}$ to get the best point $x^{(k)}$ of the k-th iteration:

$$f\left(x^{(k,0)} + \mu_k d^{(k,n+1)}\right) = min_\mu f\left(x^{(k,0)} + \mu d^{(k,n+1)}\right). \tag{3}$$

$$x^{(k)} = x^{(k,0)} + \mu_k d^{(k,n+1)}. \tag{4}$$

where $d^{(k,n+1)} = x^{(k,m)} - x^{(k,0)}$, and then make the initial point $x^{(k+1,0)} = x^{(k)}$, search direction $d^{(k+1,j)} = d^{(k,j+1)}$, $j = 1,...,n$, to start a new round of iteration. In the algorithm, f acts as the similarity measure function. Mutual information [19] based registration method use mutual information as a similarity measure, which has been widely used in many fields due to its good accuracy. The concept of mutual information comes from information theory, it is used to reflect the information between two random variables correlation, the calculation can be expressed as:

$$f(R, F) = \sum_{r,f} P_{RF}(r, f) \log_2 \frac{P_{RF}(r, f)}{P_R(r) P_F(f)}. \tag{5}$$

where $P_{RF}(r, f)$ is the joint probability distribution of gray value of R and F, $P_R(r)$ and $P_F(f)$ represent the edge probability distribution:

$$P_R(r) = \sum_f P_{RF}(r, f). \tag{6}$$

$$P_F(r) = \sum_r P_{RF}(r, f). \tag{7}$$

$$P_{RF}(r, f) = hist(r, f) / \sum_{r,f} hist(r, f). \tag{8}$$

where $hist(r, f)$ is the joint histogram of gray value of R and F. We consider the image size of R and F is $M \times N$. To calculate the joint histogram, the times of the calculation is $M \times N$; to calculate the joint probability distribution by formula (8), $M \times N$ times of addition and $M \times N$ times of division are needed, which means $2 \times M \times N$ times calculation. By formulas (6) and (7) totally, the calculation of marginal probability distribution costs $M+N$ times of calculation. Finally, the calculation costs of formula (5) is $5 \times M \times N$, and the total calculation times of mutual information algorithm is $M \times N + 2 \times M \times N + M + N + 5 \times M \times N$, equal to $8M \times N + M + N$. Because of the complexity of calculating mutual information, the registration algorithm is slow. MSE (mean square error) is a more convenient method to measure the change of data, by appropriate deformation, the MSE can be calculated by:

$$f(R, F) = \log\left(255^2 / MSE\right). \tag{9}$$

$$MSE = \frac{\sum_{x=1}^{M} \sum_{y=1}^{N} (f_R(x, y) - f_F(x, y))^2}{M \times N}. \tag{10}$$

The times of calculation of Eq. (10) is $2 \times M \times N + 1$, and the times of calculation of Eq. (9) is 4, so the total times of calculations for the MSE method should be $2 \times M \times N + 5 < 8 \times M \times N + M + N$, by many experiments we find that the MSE registration algorithm can achieve similar precision to the method based on mutual information, however, because of simple calculation of MSE, speed of the algorithm can be greatly improved.

3   Experiment Result

We use the NIST Special Database 4 for image registration experiment [20], the image resolution is $512 * 512$. According to the characteristics of fingerprint, the images were classified into five groups: A = Arch, L = Left Loop, R = Right Loop, T = Tented Arch, and W = Whorl. For convenience, we reversed color of the original image. The floating image is a random translation and rotation result of the reference image, the X direction and Y direction shift range from -3 pixels to 3 pixels, the rotation angle range from $-5°$ to $5°$. Figure 2 shows an example of one reference image and one float image, Figs. 3 and 4 show an improvement in both accuracy and speed of the proposed method in comparison to image field methods and mutual information methods.

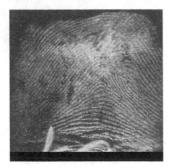   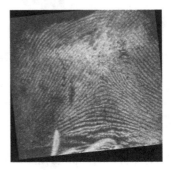

(a) A reference image   (b) A float image(H=[-3,-3],S=4)

Fig. 2. Example of one reference and one float image in NIST database for image registration experiment. The float image is a $[-3, -3]$ translation and $4°$ rotation result of the reference image, where H represents the translation and S represents the rotation.

3.1   The Experimental Results of Pre-processing

Image preprocessing based on image field is to improve the quality of the input fingerprints, enhance the contrast of ridge lines and valley lines of the image, which will affect the result of fingerprint registration. Based on the fingerprint image field, we get the results of preprocessing step by equalization, smoothing and binarization, as shown in Fig. 3. It can be seen that the fingerprint ridge lines and valley lines is more clear and contain less noise compared to the original image, which is convenient for image registration.

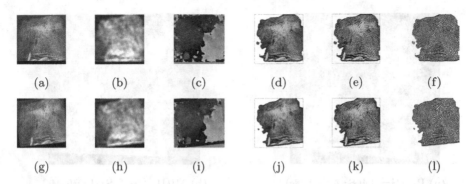

(a)   (b)   (c)   (d)   (e)   (f)

(g)   (h)   (i)   (j)   (k)   (l)

Fig. 3. Fingerprint image and result of pre-processing based on image field. a, g are the reference image and the float image, b, h are the gradient field of the image, c, i are the orientation field of the image, d, j are images after image equalization, e, k are images after image smoothing and enhancement, f, l are images after image binarization.

In Fig. 3, by comparing the original images and the preprocessed images, it's obvious that the fingerprint ridge lines and valley lines is more clear, this helps to improve the accuracy of image registration in our second step, on the other hand, due to removal of useless information, it can reduce the amount of data and improve speed of the algorithm.

3.2   The Results of Image Registration

Based on the preprocessed binary fingerprint images, we use the MSE method for image registration, and compare with the method based on mutual information and image field. As in most cases, we need to consider both the accuracy and speed of registration. Since it is hard to find a general evaluation criteria for image registration, here we use the index IRE [21] as an evaluation:

$$IRE = \frac{1}{N} \sum_{i=1}^{N} (||H_i - H_{i0}|| + ||S_i - S_{i0}||) * t_i. \tag{11}$$

where N is the number of images, H_i and S_i is the amount of translation and rotation for registration got in the experiment, while H_{i0} and S_{i0} for the actual translation and rotation, t_i for the operation time of the experiment. We can use the evaluation index to evaluate the effectiveness of registration, considering both the accuracy and speed of the algorithm, the registration results as shown in Fig. 4, in order to illustrate the effectiveness of the algorithm, we compare our method based on image field and MSE (P-MSE) with the method based on image field and mutual information (P-MI), the method based on mutual information (MI) and the method based on image field and feature points (P-F), the IRE are averaged over total runs.

We can see from Table 1, the P-MSE method has better effectiveness, it is because that P-F needs manual intervention to select proper feature points, the

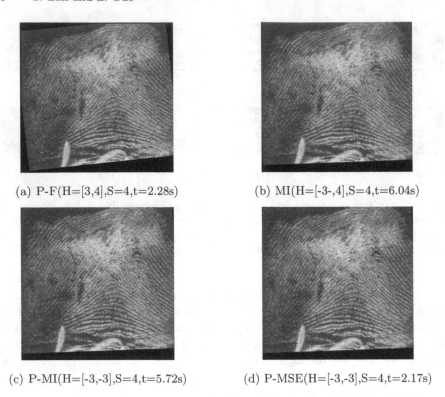

(a) P-F(H=[3,4],S=4,t=2.28s)

(b) MI(H=[-3-,4],S=4,t=6.04s)

(c) P-MI(H=[-3,-3],S=4,t=5.72s)

(d) P-MSE(H=[-3,-3],S=4,t=2.17s)

Fig. 4. Example of registered images of the float image in Fig. 2(b), where t represents the run time. (a) Result based on image field and feature points (P-F). (b) Result based on mutual information (MI). (c) Result based on image field and mutual information (P-MI). (d) Result based on image field and MSE (P-MSE).

Table 1. IRE comparison of P-MSE, P-MI, P-F and MI

Type	P-MI	P-MSE	P-F	MI
A	0.3067	**0.1659**	2.7179	0.5998
L	0.3551	**0.1243**	3.0641	0.9056
R	0.6841	**0.1532**	2.8198	0.7398
T	0.3209	**0.1177**	4.1606	0.7330
W	0.2219	**0.1112**	3.9866	0.4722
Average	0.3777	**0.1345**	3.3498	0.6901

accuracy of image registration cannot be guaranteed, although MI has a high accuracy, but the speed is too slow. As the preprocessing based on image field reduces noises and data, the P-MI method reduces the time of algorithm. By replacing MI with MSE and keeping the POWELL in the P-MI method, the P-MSE method further improves the algorithm speed and assures the precision, considering both the accuracy and speed of registration, P-MSE is more effective than the image field method and the mutual information method.

4   Conclusions

This paper presents an improved method of fingerprint image registration based on image field and MSE, which achieves good results in both speed and accuracy. The traditional method of image registration based on the image field is good in speed, but due to the difficulty of choosing feature points and other factors, the accuracy of registration results cannot be guaranteed, while the method based on mutual information is good in precision, but because of the complexity of calculating mutual information, the algorithm is slow. The proposed P-MSE method in this paper using MSE instead of mutual information based on image field for registration, reduces the amount of computation, speeds up the algorithm, and ensures the accuracy of registration. However, the proposed method is a little sensitive if the rotation and translation is large, our future work will consider how to further improve the algorithm.

References

1. Zhang, H.: Study and implementation of automatic fingerprint recognition technology. In: URKE, vol. 2, pp. 40–43 (2011)
2. Shareef, M.P., Reshma, K.V., Nair, A.T., Abraham, N.: Efficient registration via fingerprints combination. In: ICCSP, vol. 10, pp. 1540–1544 (2014)
3. Springer Science Business Media: Image Field. Springer, Heidelberg (2014)
4. Ramos, D., Krish, R.P., Fierrez, J., Garcia, J.O.: Pre-registration of latent fingerprints based on orientation field. IET Biom. 4(2), 42–52 (2015)
5. Feng, J., Zhou, J., Yang, X.: Method for registering fingerprint image (2016)
6. Srivastava, D.K., Shrivastava, A.: Fingerprint identification using feature extraction: a survey. In: IC3I, vol. 2, pp. 522–525 (2015)
7. Carneiro, F., Bessa, J.A., de Moraes, J.L., Neto, E.C., De Alexandria, A.R.: Techniques of binarization, thinning and feature extraction applied to a fingerprint system. Int. J. Comput. Appl. 103(10), 1–8 (2014)
8. Maes, F., Loeckx, D., Vandermeulen, D., Suetens, P.: Image registration using mutual information. In: Paragios, N., Duncan, J., Ayache, N. (eds.) Handbook of Biomedical Imaging, pp. 295–308. Springer, Boston, MA (2015). doi:10.1007/978-0-387-09749-7_16
9. Shamai, S., Guo, D., Verdu, S.: Mutual information and minimum mean-square error in gaussian channels. IEEE Trans. Inf. Theory 51(4), 1261–1282 (2005)
10. Powell, M.: Approximation Theory and Methods. Cambridge University Press, Cambridge (2015)
11. Singh, K., Rajiv, K.: Image enhancement using exposure based sub image histogram equalization. Pattern Recognit. Lett. 36(1), 10–14 (2014)
12. Chang, J., Chen, D.H., Guo, L., Hai, L.T.: Method for fingerprint image binarization using difference. J. Comput. Appl. 27(1), 169–171 (2007)
13. Prabhu, L., Sakthivel, N.: Meanmedian filtering for impulsive noise removal. Int. J. Basics Appl. Sci. 02(04), 47–57 (2014)

14. Zheng, H., Xu, J., Xu, L., Du, Y., Song, H.: Enhancement algorithm for structured-light stripe image based on orientation and frequency fields constraint. Comput. Sci. (2014)
15. Sastry, C.S., Ravindranath, M., Pujari, A.K., Deekshatulu, B.L.: A modified gabor function for content based image retrieval. Pattern Recognit. Lett. **28**(2), 293–300 (2007)
16. Li, S., Zeng, W., Wu, H.: Translation, rotation and scaling changes in image registration based affine transformation model. Infrared Laser Eng. **30**, 18–20 (2001)
17. Mulekar, M.S., Brown, C.S.: Distance and Similarity Measures. Springer, New York (2014)
18. Iwasaki, M.: Image classification method, image feature space displaying method, program, and recording medium (2004)
19. Atwal, G.S., Kinney, J.B.: Equitability, mutual information, and the maximal information coefficient. Proc. Natl. Acad. Sci. U.S.A. **111**(9), 3354–3359 (2014)
20. Wilson, C.L., Watson, C.I.: NIST special database 4, fingerprint database. National Institute of Standards and Technology Advanced Systems Division Image Recognition Group (1992)
21. Moghari, M.H., Ma, B., Abolmaesumi, P.: A theoretical comparison of different target registration error estimators. In: Metaxas, D., Axel, L., Fichtinger, G., Székely, G. (eds.) MICCAI 2008 Part II. LNCS, vol. 5242, pp. 1032–1040. Springer, Heidelberg (2008). doi:10.1007/978-3-540-85990-1_124

A Low Rank Regularization Method
for Motion Adaptive Video Stabilization

Huicong Wu, Hiuk Jae Shim, and Liang Xiao[⊠]

School of Computer Science and Engineering,
Nanjing University of Science and Technology, Nanjing, China
njust_wuhuicong@163.com, waitnual@gmail.com,
xiaoliang@njust.edu.cn

Abstract. Hand-held video cameras usually suffer from undesirable video jitters due to unstable camera motion. Although path optimization methods have
been successfully employed to produce stabilized videos, the methods generally
result in unintended large void areas in fast motion video sequences. To overcome this limitation, in this paper, we present a novel video stabilization
algorithm which is derived from an optimization model consisting of a motion
data fidelity term and two regularization terms: motion adaptive smoothness
term and low rank term. Particularly, we design a motion adaptive kernel to
measure neighbor motion similarity by exploiting local derivative information of
dominant motion parameter, which is incorporated into the local weighted
smoothness term to guide a motion aware regularization. Besides, the low rank
property of neighbor motions is utilized to further improve the performance of
stabilization. Experimental results show that the proposed method noticeably
stabilizes a video, and it suppresses void areas effectively in fast motion frames.

Keywords: Video stabilization · Motion smoothing · Motion adaptive kernel ·
Local weighted regularization · Low rank

1 Introduction

In recent years, with the increasing popularity of hand-held video cameras, people
prefer to take a video to record their life. Due to inevitable shaky motions, however, the
hand-held cameras are not easy to produce a satisfactory result. Therefore, video stabilization which aims to remove the unstable motions is becoming more and more
indispensable. In general, video stabilization methods consist of three steps: motion
estimation, motion smoothing and image warping [1]. The motion estimation step
estimates inter-frame motion parameters, and the motion smoothing step calculates
warping transformations using the motion parameters. Finally, the image warping step
generates stabilized frames. The first and last steps have been well studied in image
processing domain, and various smoothing methods have been proposed. In this paper,
a novel motion adaptive and low rank regularization method is proposed to estimate the
smoothed motions accurately. The main contributions of proposed method include:
(a) we model the shaky motion as a sum of smoothed motion and noisy motion, and the
smoothed motion usually has lower rank than shaky motion in a local window;

© Springer International Publishing AG 2017
Y. Sun et al. (Eds.): IScIDE 2017, LNCS 10559, pp. 539–548, 2017.
DOI: 10.1007/978-3-319-67777-4_48

(b) adjacent smoothed motions should contain similarity, thus each smoothed motion can be estimated with a local weighted optimization model; (c) in fast motion situations, calculating the neighbor weights by classical Gaussian kernel will generate large void areas (areas without pixel values) as previous methods, which causes information loss and thus affects visual experience. Therefore, we design a motion adaptive kernel to assign the weights.

The rest of this paper is organized as follows. We briefly introduce related works in Sect. 2. The proposed optimization model and motion adaptive kernel are described in Sect. 3. Experimental results are shown in Sect. 4, and finally we draw a conclusion in Sect. 5.

2 Related Work

Video stabilization methods can be roughly classified into two categories: 2-D methods and 3-D methods. 3-D methods [2–5] usually need to recover scene structures and camera motion by Structure from Motion (SFM) [6]. However, SFM is a highly non-linear optimization problem which is time-consuming, so 3-D methods have limited ability for practical application. On the contrary, 2-D methods have much less computational cost but still generate satisfactory results. Litvin et al. [7] adopted Kalman filtering to smooth the original motions. Chang et al. [8] designed a regularization method to estimate stabilized motion parameters. In [9], Gaussian smoothing method is proposed to calculate the warping transformations directly. However, it only reduces high frequency motions. Recently, Grundmann et al. [10] dramatically stabilized a video using L1 optimization method, and Zhang et al. [11] further extended the method to improve scene composition by removing distractions. Besides, feature trajectory based methods [12–14] are another part of 2-D methods, which shows powerful ability for stabilization. However, most 2-D methods leave large void areas in their resulting frames when a fast motion occurs (e.g. fast rotation motion). In [15], a bilateral filter idea is first used to control the weights of neighbor motions, however, as pointed in [16], the bilateral kernel decreases the performance of estimator. Therefore, in this paper, we estimate the smoothed motions based on the motion adaptive and low rank regularization terms, where the weights are calculated by the proposed motion adaptive kernel.

3 Optimization Model and Motion Adaptive Kernel

In the proposed method, we perform motion estimation by extracting SURF feature [17] between neighbor frames and then fit a homography model using RANSAC algorithm [18]. After obtaining warping transformations, we render the resulting frames using bicubic interpolation. Therefore, in this section, we introduce the motion smoothing step. After motion estimation step, we can obtain the original (observed shaky) motions. As Fig. 1 shows, $\mathbf{F}_i$ denotes the original motion vector obtained by stacking the columns of 3×3 homography matrix of i-th frame, and $\mathbf{H}_i$ represents the

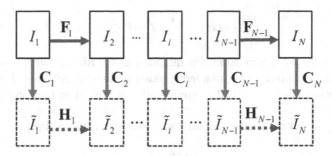

Fig. 1. The relations between the original and stabilized frames. **F**, **H**, and **C** represent original motion, smoothed motion, and warping transformation respectively.

desired smoothed motion vector. Then the shaky motion and smoothed motion are assumed to have the relationships:

$$\mathbf{F}_i = \mathbf{H}_i + \mathbf{N}_i, i = 1, 2, \dots, N - 1, \tag{1}$$

where the noisy motion $\mathbf{N}_i$ is considered as independent Gaussian random variables. The goal of video stabilization is to estimate the latent motion vectors $\{\mathbf{H}_i\}_{i=1}^{N-1}$ given observed vectors $\{\mathbf{F}_i\}_{i=1}^{N-1}$. This problem is in general ill-posed since noisy motions are random variables. Therefore, we need to make additional assumptions in order to seek a stable solution.

3.1 Optimization Model

In the proposed method, we utilize a sliding windowed manner to calculate smoothed motions. For example, in order to smooth out motion $\mathbf{F}_i$, a fixed-length window $\{\mathbf{F}_{i-r}, \dots, \mathbf{F}_i, \dots, \mathbf{F}_{i+r}\}$ is built as the input of optimization model. After obtaining the smoothed motion $\mathbf{H}_i$, the window moves one frame forward to calculate the next smoothed motion $\mathbf{H}_{i+1}$. Besides, in the algorithm, we generate the windows of first and last r frames by extending the original motions in a mirror reflection fashion.

In the following, we use the window of $\mathbf{F}_i$ as an example to show the details of proposed method. In order to stabilize a video, we need to estimate the smoothed motions. Based on the model (1), in the sliding window, we design a motion data fidelity term as

$$Q(\{\mathbf{H}_t\}) = \frac{1}{2} \sum_{t=i-r}^{i+r} \|\mathbf{H}_t - \mathbf{F}_t\|_2^2, \tag{2}$$

which means there should not be a big difference between the original and smoothed motions in order to restrict void areas. Let F and H denote the original and smoothed motion parameter matrices of the window respectively, and also each column of them is a 9×1 homography vector. Then (2) can be rewritten in a matrix form as

$$Q(\{H\}) = \frac{1}{2}\|H - F\|_F^2, \tag{3}$$

where $\|\cdot\|_F$ is the Frobenius norm. On the other hand, for the purpose of removing shakes, we present two regularization terms based on two observations. First, adjacent smoothed motions should be similar. Specifically, we calculate $\mathbf{H}_i$ using its neighbor smoothed motions in the window, and employ a classical Gaussian kernel to assign the neighbor weights. Then, a local weighted smoothness term is defined as

$$P(\{\mathbf{H}_t\}) = \frac{1}{2}\sum_{t=i-r}^{i+r} \omega_{t,i}\|\mathbf{H}_t - \mathbf{H}_i\|_2^2, \tag{4}$$

where $\omega_{t,i}$ is the weight to measure neighbor motion similarity, which is obtained by $\omega_{t,i} = G(t - i)$ and $G(\cdot)$ is the Gaussian kernel function. Similarly, in a matrix form, (4) can be rewritten as

$$P(\{H\}) = \frac{1}{2}\|H(I - B)W\|_F^2, \tag{5}$$

where I is an identity matrix; B is a square matrix with all zeros except for $(r + 1)$-th row (middle row) which contains all ones; $W = diag(\sqrt{\omega_{i-r,i}}, \ldots, \sqrt{\omega_{i,i}}, \ldots, \sqrt{\omega_{i+r,i}})$ is a diagonal matrix, and the main diagonal elements are the square root of weights. Moreover, we notice that the smoothed motion matrix H should have lower rank than original motion matrix F. Due to the existence of noisy motions, the rank of F is always 9 (full row rank). However, after reducing the noisy motions in a window, H may have lower rank especially applying aggressive smoothing. Therefore, a low rank term is expressed as

$$L(\{H\}) = \|H\|_*, \tag{6}$$

where $\|\cdot\|_*$ denotes the nuclear norm of a matrix, which is the natural convex surrogate for the rank function [19]. As a result, the final optimization model that we need to minimize is

$$\min_H Q + \lambda P + \alpha L, \tag{7}$$

where λ and α are regularization parameters to balance the three terms. In order to solve the model, we introduce a new variable A, and the model becomes a constrained optimization problem as

$$\min_{H,A} \frac{1}{2}\|H - F\|_F^2 + \frac{\lambda}{2}\|H(I - B)W\|_F^2 + \alpha\|A\|_* \text{ s.t. } A = H. \tag{8}$$

Then the augmented Lagrange function for the problem is

$$\min_{H,A,Y} \frac{1}{2}\|H - F\|_F^2 + \frac{\lambda}{2}\|H(I - B)W\|_F^2 + \alpha\|A\|_* + \langle Y, A - H\rangle + \frac{\beta}{2}\|A - H\|_F^2, \quad (9)$$

where $\langle\cdot,\cdot\rangle$ represents matrix inner product (i.e., $\langle A, B\rangle = trace(A^T B)$), Y is the Lagrange multiplier matrix, and β is a positive scalar. We utilize the alternating direction method of multipliers (ADMM) algorithm [20] to obtain the solution for each of variables. After that, we get the smoothed motion matrix H and store only $\mathbf{H}_i$ in the memory, then the window moves for the next calculation. When we obtain all $\{\mathbf{H}_i\}_{i=1}^{N-1}$, using the relation $\left(\prod_{i=1}^{n} \mathbf{F}_i\right)\mathbf{C}_{n+1} = \prod_{i=1}^{n}\mathbf{H}_i$, the warping transformation $\mathbf{C}_{n+1} = \left(\prod_{i=1}^{n}\mathbf{F}_i\right)^{-1}\left(\prod_{i=1}^{n}\mathbf{H}_i\right)$ can be obtained to generate a stabilized frame ($\mathbf{C}_1$ is set as I). However, the utilization of Gaussian (non-adaptive) kernel has a same problem as previous 2-D methods: when camera motion becomes fast, the resulting frames have large void areas, which indicates information loss. Thus, we further design a motion adaptive kernel to improve the performance of the smoothness term, which can assign the weights more reasonably.

3.2 Motion Adaptive Kernel

Inspired by steering kernel [16], which uses local gradient information to measure the similarity of two pixels. Similarly, local derivative information is exploited to determine the weights in the motion adaptive smoothness term. From the estimated original motions, we detect a dominant motion parameter to calculate the adaptive weights. Since translation can cover most of camera motions, we select horizontal, vertical translation parameter or the sum of them depending on different situations. As Fig. 2 shows, each inter-frame horizontal translation parameter f_i is the first order difference of total horizontal motion displacement at position (frame) i, which can be seen as an approximation of derivative at i. Thus, within a small window, we first calculate the local derivative at position i using Gaussian kernel in weighted average fashion:

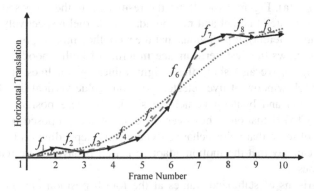

Fig. 2. Comparison of Gaussian kernel and the proposed motion adaptive kernel. The black solid arrows are the original horizontal motions, the middle frames (5 to 7) show a fast motion. The red dotted line and green dashed line are the stabilized motion using Gaussian kernel and motion adaptive kernel respectively. (Color figure online)

$\bar{f}_i = \sum_{j \in [i-c, i+c]} G(j-i) f_j$, where the sum of weights equals to 1. Then, we define a relative derivative between position i and j as $d(i,j) = \bar{f}_i - \bar{f}_j$. Finally, a relative distance can be described as $|(i-j)d(i,j)|$ to measure the motion similarity. The longer the relative distance is, the smaller the weight becomes. Mathematically, the motion adaptive weight is calculated by

$$\omega_{j,i} = \frac{\sqrt{|d(j,i)|}}{2\pi h^2} \exp\left(-\frac{((j-i)d(j,i))^2}{2h^2}\right), \tag{10}$$

where h is the bandwidth which controls the strength of weight: the weight becomes large when we choose a large h. The proposed motion adaptive kernel considers both position distance and local derivative information simultaneously, therefore, the weights are adjusted dynamically according to neighbor motions, which can generate better results than previous methods.

4   Experimental Results

We implemented and tested the proposed method in MATLAB on a desktop PC with Xeon W3520 CPU (2.66 GHz) and 4 GB memory. All test videos are from a publicly available video dataset [15]. In our algorithm, the two regularization parameters λ and α, the scalar β, the bandwidth h and two window size parameters r and c are empirically set as 1000, 1, 1, 5, 30, and 5, respectively.

4.1   Evaluation of Motion Adaptive Kernel

The first experiment demonstrates the performance of the proposed motion adaptive kernel. As Fig. 3 shows, the test video exhibits a fast horizontal motion in the middle frames. The previous method such as regularization method [8] causes excessive smoothing, thus a big difference between the original and smoothed motions is observed in Fig. 3(a). Figure 3(b)–(d) are the results using the proposed method with Gaussian kernel, bilateral kernel and motion adaptive kernel respectively (i.e. only the weights in (4) are different). It is obvious that the smoothed motion produced by motion adaptive kernel shows the best performance in terms of both smoothness and motion deviation. In Fig. 3, we also show the weight values (green lines at the bottom of graph) within each window at five selected positions (blue vertical lines). The weight values of Gaussian and bilateral kernel are similar in these positions (they behave similarly in the five situations), however, in case of the proposed motion adaptive kernel, we can observe that values change dynamically depending on given situations. Therefore, we can say that the motion adaptive kernel behaves more reasonably than the previous ones.

The comparisons of stabilized frames at the fourth position are shown in Fig. 4. Both regularization method and Gaussian kernel method display large void areas. Even though the result of bilateral kernel method shows a nearly full size picture, the scene is

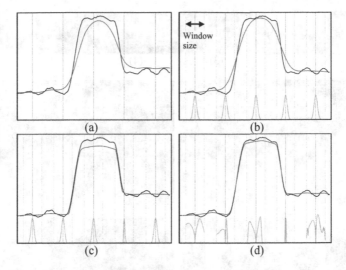

Fig. 3. Comparisons between motion adaptive kernel and other methods. The black lines and red lines are original and smoothed motions respectively, and the green lines in the bottom show the weight values within a local window at five positions. (a) Regularization method [8]. (b) Gaussian kernel. (c) Bilateral kernel. (d) Motion adaptive kernel. (Color figure online)

Fig. 4. Comparisons of stabilized frames at the fourth position. (a) Regularization method [8]. (b) Gaussian kernel. (c) Bilateral kernel. (d) Motion adaptive kernel.

tilted to some degree. However, the motion adaptive kernel method generates satisfactory frames all the time for the test video.

4.2   Comparison with Other Video Stabilization Methods

We compared the proposed method with other four popular stabilization methods: Gaussian smoothing method [9], regularization method [8], Deshaker [21] and You-Tube Stabilizer. Deshaker is a public plugin for removing motion shakes and YouTube Stabilizer is an online tool based on the state-of-the-art method in [10]. Figure 5(1)–(6) are the test videos used in this experiment. The stabilized feature trajectories by these methods are depicted in Fig. 5(a)–(f). It is obvious that the Gaussian smoothing method (b) only reduces high frequency jitters. Deshaker (c) and regularization method (d) stabilize the video one step further. Even though YouTube Stabilizer (e) shows much smoother trajectories, some low frequency motions still appear. The proposed

Fig. 5. Test videos and the smoothed trajectories using different methods. (1)–(6) are the test videos used in our experiment. (a)–(f) shows 20 feature trajectories in each stabilized videos by different methods. (a) Original. (b) Gaussian smoothing [9]. (c) Deshaker [21]. (d) Regularization method [8]. (e) YouTube Stabilizer [10]. (f) Proposed.

method (f) generates the best trajectories and is free from low frequency shakes. Besides, in order to make a quantitative evaluation, we adopted inter-frame transformation fidelity (ITF) [1] as

$$ITF = (1/N - 1) \sum_{k=1}^{N-1} PSNR(k), \tag{11}$$

where $PSNR(k)$ is the Peak Signal-to-Noise Ratio between two stabilized neighbor frames k and $k + 1$. Table 1 shows the ITF comparison results, and the proposed method can improve the original ITF about 10db, which indicates preferable performance for video stabilization. Furthermore, in order to reveal the effect of the proposed low rank regularization term, in Table 1, we also show the ITF scores of the results without low rank term. From the comparison we can find the stabilization performance is improved in most cases when low rank term is adopted.

Table 1. ITF comparison with other methods.

Video	Original	[9]	[21]	[8]	[10]	Ours (without LR)	Ours (with LR)
1	21.8	28.6	29.9	31.1	31.7	32.2	**32.5**
2	19.0	24.6	26.4	27.7	27.7	28.3	**28.3**
3	19.6	26.2	27.1	29.3	30.4	32.4	**32.8**
4	25.0	29.8	32.4	34.2	34.7	**35.5**	35.4
5	15.4	18.5	20.5	22.3	22.9	23.0	**23.1**
6	19.2	23.4	25.1	26.4	25.5	27.6	**27.9**

5 Conclusion

In this paper, we proposed a novel motion adaptive and low rank regularization method for video stabilization. The shaky motion is regarded as the sum of smoothed motion and noisy motion. And the smoothed motions should have lower rank than shaky motions, which derives the low rank term. Besides, the smoothed motions contain similarity with their neighbor motions, thus, we build a local weighted smoothness term. However, stabilized videos usually show large void areas in case of fast camera motion. Therefore, we designed a motion adaptive kernel to assign weights reasonably using the local derivative information, which largely decreases void areas in fast motion situations and provides sufficient motion stability in other ones. The proposed method has the potential to stabilize videos in real time when GPU and parallelization are adopted. Experimental results demonstrate that the proposed method brings noticeable improvement over the previous methods.

Acknowledgments. The research is supported in part by the National Key Research and Development Program of China under Grant No. 2016YFF0103604, by the National Natural Science Foundation (NSF) of China under Grant No. 61571230; and by the NSF of Jiangsu Province under Grant No. BK20161500.

References

1. Morimoto, C., Chellappa, R.: Evaluation of image stabilization algorithms. In: ICASSP (1998)
2. Buehler, C., Bosse, M., McMillan, L.: Non-metric image-based rendering for video stabilization. In: CVPR (2001)
3. Zhang, G., Hua, W., Qin, X., Shao, Y., Bao, H.: Video stabilization based on a 3D perspective camera model. Vis. Comput. **25**(11), 997–1008 (2009)

4. Liu, F., Gleicher, M., Jin, H., Agarwala, A.: Content preserving warps for 3D video stabilization. ACM Trans. Graph. **28**(3), 44:2–44:10 (2009)
5. Zhou, Z., Jin, H., Ma, Y.: Plane-based content preserving warps for video stabilization. In: CVPR (2013)
6. Hartley, R., Zisserman, A.: Multiple View Geometry in Computer Vision. Cambridge University Press, Cambridge (2003)
7. Litvin, A., Konrad, J., Karl, W.: Probabilistic video stabilization using Kalman filtering and mosaicking. In: SPIE Image and Video Communications and Processing (2003)
8. Chang, H.C., Lai, S.H., Lu, K.R.: A robust real-time video stabilization algorithm. J. Vis. Commun. Image Represent. **17**(3), 659–673 (2006)
9. Matsushita, Y., Ofek, E., Ge, W., Tang, X., Shum, H.Y.: Full-frame video stabilization with motion inpainting. IEEE Trans. PAMI **28**(7), 1150–1163 (2006)
10. Grundmann, M., Kwatra, V., Essa, I.: Auto-directed video stabilization with robust l1 optimal camera paths. In: CVPR (2011)
11. Zhang, F.L., Wang, J., Zhao, H., Martin, R.R., Hu, S.M.: Simultaneous camera path optimization and distraction removal for improving amateur video. IEEE Trans. Image Process. **24**(12), 5982–5994 (2015)
12. Liu, F., Gleicher, M., Wang, J., Jin, H., Agarwala, A.: Subspace video stabilization. ACM Trans. Graph. **30**(1), 1–10 (2011)
13. Wang, Y.S., Liu, F., Hsu, P.S., Lee, T.Y.: Spatially and temporally optimized video stabilization. IEEE Trans. Vis. Comput. Graphics **19**(8), 1354–1361 (2013)
14. Koh, Y.J., Lee, C., Kim, C.S.: Video stabilization based on feature trajectory augmentation and selection and robust mesh grid warping. IEEE Trans. Image Process. **24**(12), 5260–5273 (2015)
15. Liu, S., Yuan, L., Tan, P., Sun, J.: Bundled camera paths for video stabilization. ACM Trans. Graph. **32**(4), 78:1–78:10 (2013)
16. Takeda, H., Farsiu, S., Milanfar, P.: Kernel regression for image processing and reconstruction. IEEE Trans. Image Process. **16**(2), 349–366 (2007)
17. Bay, H., Tuytelaars, T., Van Gool, L.: SURF: speeded up robust features. In: Leonardis, A., Bischof, H., Pinz, A. (eds.) ECCV 2006 Part I. LNCS, vol. 3951, pp. 404–417. Springer, Heidelberg (2006). doi:10.1007/11744023_32
18. Fischler, M.A., Bolles, R.C.: Random sample consensus: a paradigm for model fitting with applications to image analysis and automated cartography. Commun. ACM **24**(6), 381–395 (1981)
19. Candès, E.J., Li, X., Ma, Y., Wright, J.: Robust principal component analysis? J. ACM **58**(3), 1–37 (2011)
20. Boyd, S., Parikh, N., Chu, E., Peleato, B., Eckstein, J.: Distributed optimization and statistical learning via the alternating direction method of multipliers. Found. Trends Mach. Learn. **3**(1), 1–122 (2011)
21. Deshaker. http://www.guthspot.se/video/deshaker.htm

Multi-image Deblurring Using Complement

Pei Wang[1], Jinqiu Sun[2(✉)], Haisen Li[1], Xueling Chen[1], Yu Zhu[1],
and Yanning Zhang[1]

[1] School of Computer Science, Northwestern Polytechnical University, Xi'an, China
wangpei23@mail.nwpu.edu.cn
[2] School of Astronautics, Northwestern Polytechnical University, Xi'an, China
sunjinqiu@nwpu.edu.cn

Abstract. The purpose of image restoration is to recover the latent image from the observed blurred images. In multi-image restoration, the input image's quality directly affects the final deblurring result. In this paper, a blurry images restoration method based on the complement is proposed for the selection of images in multi-image restoration. First, we give a description of complement between different images. Then, the blurred images of maximum complement are selected by image selection method which is based on the iterative maximum complement. Finally, we use an existing deblurring method to estimate the latent image. We compare the new method with traditional methods, the experimental results clearly demonstrate the efficacy of the proposed method.

Keywords: Image deblurring · Multi-image deblurring · Complement

1 Introduction

When taking photos using a hand-held camera, the camera might be moved during the exposure period, then resulting in blurred images. Although the motion blurred image restoration has been researched decades, the problem still has no better solution due to the image texture, noise, etc.

According to the number of input images, the image restoration algorithm can be divided into two types: single image and multiple images deblurring. The single type has been an active filed in recent years [4–7]. In contrast, since Rav-Acha and Peleg use two motion blurred images with different blur directions and show that restoration quality is better than when using only a single image, more and more multi-image deblurring methods have been proposed [1,2,8,10,11]. In traditional multi-image restoration method, most of them select randomly two blurry images which are used as the input of muXu2010Twolti-images deblurring method from the observation. The deblurring method we use frequently is MAP (maximum a posteriori) estimation which is based on the Bayesian inference framework. The latent image is estimated by minimizing the objective function which include "cross-blur" penalty term of MAP framework [2,3,9,10,12]. But due to the different between the images, it will be a serious problem in traditional random selection. The input image's quality directly

© Springer International Publishing AG 2017
Y. Sun et al. (Eds.): IScIDE 2017, LNCS 10559, pp. 549–558, 2017.
DOI: 10.1007/978-3-319-67777-4_49

affects the final deblurring result. In order to select the best input images, we propose a multi-image deblurring method using complement between different motion blurred images. Based on the research of image complement, we select the images with the largest complement in the sequence. The proposed complement method plays an important role in traditional multi-image restoration algorithm. The framework of the proposed method is shown in the Fig. 1:

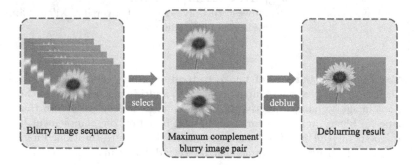

Fig. 1. The framework of the proposed method

At the first, we select the maximum complement image pair from blurry image sequence, then typical deblurring method has been used to deblur. The selection is very important in the proposed processing.

2   The Description of Complement

2.1   Complement in Blurry Images

In the process of imaging, different degraded factors resulting in varying degraded image. The motion blur exists because of the relative motion between the camera and object during the exposure. The direction and length of motion blur between the images are different in the image sequence. The difference can provide more useful information with deblurring. The different information can be named the *complement* of the blurry images. Therefore, in image processing, we can use the complement of the different sequence degraded images to improve the quality of deblurring image.

Figure 2 shows the two blurry images with different motion directions. Figure 2(a) is the horizontal blur, we can see that the information of horizontal direction has been saved while the vertical direction has been blurred. On the contrary, the vertical information has been saved more than horizontal direction in Fig. 2(b). The horizontal information in Fig. 2(a) and the vertical information in Fig. 2(b) is complementary. We can use the two images to reconstruct the latent sharp image.

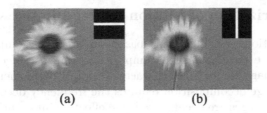

(a)   (b)

Fig. 2. Blurred images in different directions are complementary. (a) Horizontal blurred image and kernel. (b) Vertical blurred image and it's kernel.

2.2 Defining Complement Mathematically

Set is a basic concept of mathematics, it consists of one or more different data elements. Because its special property, we can use the mathematical set to describe the complement of the blurry images.

Fig. 3. Mathematical description of complementary

Figure 3 shows the relationship between two different blurry images A and B in the image sequence where A and B represent the information of the image A and B, respectively. And U represents the total information of all images in sequence. The intersection of A and B represents the correlation of the image, which means the common information in the images. The shaded area indicates that the information which is a part of the union subtracts the intersection from images. And complement between the images is described by the shaded part. The mathematical equation is defined as follows.

$$com(A, B) = \frac{\|T(A) \cup T(B) - T(A) \cap T(B)\|_1}{M \times N} \tag{1}$$

where $com(A, B)$ is the complement score of image A and B. $T(\cdot)$ represents the effective information of the image The notation '$\cup$' represents the union operation, and '$\cap$' represents the intersection operation. $\|T(A) \cup T(B)\|_1$ calculates the number of non-zero elements of the $T(A) \cup T(B)$. M and N are the length and width of the image size. Combined with the description of Sect. 2.1, the Eq. (1) can be used to compute the complement between blurry images.

3   Characterization of Motion Blurred Image Complement

The characterization of complement based on image decomposition includes the following three parts: image decomposition, image's effective information extraction and image complement characterization. Complementary representation based on image decomposition requires the frequency decomposition of the image. After the images are decomposed, the effective information of the images is extracted by the image binarization, then the complement characterization of the image can be used to describe the blurry image complement. As long as we obtain the characterization of blurry images, the image pair can be selected with the maximum complement.

3.1   Image Decomposition

The digital image signal can be decomposed into high frequency signal and low frequency signal. The edges and noises of the image correspond to the high frequency signal of the image fourier spectrum. The low frequency signal corresponds to the whole shape of the image, and does not include the image details and edges. The blur mainly affects the high frequency information of the image. The blur in different directions easily leads to the loss of different edge information of the image. Therefore, we use the high frequency component of the blurry image of different motion directions to calculate the complement score.

The blurry images B_1 and B_2 are decomposed by Gaussian filter. We use the original blurry images to minus the low frequency component, so the high component h_i of the image B_i can be extracted from the blurry image:

$$h_i = B_i - G_\sigma \otimes B_i \quad i = 1, 2 \tag{2}$$

where $\otimes$ represents the convolutional operation, and the two-dimensional Gaussian filter operator is given by:

$$G_\sigma = \frac{1}{2\pi\sigma^2} e^{-(\frac{x^2+y^2}{2\sigma^2})} \tag{3}$$

where σ represents the standard deviation of the Gaussian kernel.

3.2   Image Effective Information Extraction

Calculate Images' Gradient. Because the image's edge and texture are very sensitive to motion blur, on the basis of the decomposition of the image frequency, we need to calculate the gradient map of the high frequency component to guarantee the most effective information can be extracted.

$$\nabla h_i = \sqrt{(\nabla h_{ix})^2 + (\nabla h_{iy})^2} \quad i = 1, 2 \tag{4}$$

where ∇h_{ix} and ∇h_{iy} represent the gradient calculation along x direction and y direction, respectively. The detail calculation is shown bellow, g_x and g_y are the gradient operator generally.

$$\nabla h_{ix} = h_i \otimes g_x \,, \nabla h_{iy} = h_i \otimes g_y \quad i = 1, 2 \tag{5}$$

Image Binarization Segmentation. After calculating the gradient map of high frequency blurry images, we have the detail information of blurry image, which is insensitivity to motion blur. According to the definition of the image complement using mathematics, we need to extract the binary information of image to calculate the *union* and *intersection* between images.

$$T(B_i) - \begin{cases} 0 & \nabla h_i \leq thresh \\ 1 & \nabla h_i > thresh \end{cases} \qquad i = 1, 2 \tag{6}$$

When the binarization is carried out, the selection of the threshold *thresh* can be adjusted according to the gray value of the gradient maps.

$$thresh = \frac{1}{2}(max(\nabla h_{i_x}) - min(\nabla h_{i_x})) \qquad i = 1, 2 \tag{7}$$

where *max* and *min* represent the maximum value and the minimum value of the gradient image.

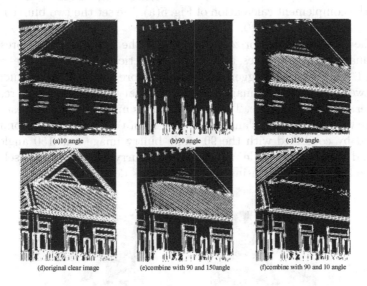

Fig. 4. Complementary information of blurry images in different directions. (a)(b)(c) are binary gradient maps of blurry images with 10, 90 and 150 angle, respectively. (e) is union image which are combined with 90 and 150 blur angle, and (f) is union of 90 and 10. (d) is the original image's gradient.

Figure 4 shows the binary high frequency gradient information which have been extracted with the different directions images and the union images which are combined by different blurry images. From the combination, compared with original clear image, it can be seen that the complementary information exists, indeed, between the degraded images at different blur directions. In the meantime, we can see that the effective information of union image are very different with different combination. Compare Fig. 4(e) with (f), it is obviously that there is more information in (e).

3.3 Image Complement Characterization

After binarization of the gradient maps, the binary image can be extracted as available information for the complementary characterization. According to the definition of Eq. (1) and the description of complement, for given blurry images pair B_1 and B_2, the equation transfer into:

$$com(B_1, B_2) = \frac{\|T(B_1) \cup T(B_2) - T(B_1) \cap T(B_2)\|_1}{M \times N} \tag{8}$$

The above equation represents the complement between the blurry image B_1 and B_2. $T(\cdot)$ represents the binary image which we mentioned above, and it's used for the image complementary characterization.

The complement between the two images with different blur angles is calculated by the Eq. (8). Figure 5 shows the complement between the two images with different blur, which Fig. 5(a) is the original image. Figure 5(b) shows the result of the complement calculation of Fig. 5(a). We set the two blurred images' motion angles are belong to [0, 180].

For the Fig. 5(b), the horizontal axis and the vertical axis represent the blurred images motion angles at different directions. For the same row, the deeper red represents the bigger complement score, and the more information exists between the two blur images. On the contrary, the deeper blue represents the smaller complement score. The green frame is marked of the largest complement score with blur images's angles. So, for Fig. 5(a), the maximum complement is achieved by combined with the 90 angle blurry image and 150 angle blurry image. And then the 90 angle and 150 angle blurry images can be selected to restore the latent image by using multiple images deblurring.

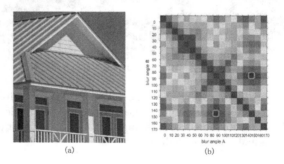

(a)   (b)

Fig. 5. The result of complement characterization. (a) Original image. (b) The result of the complement calculation of (a). (Color figure online)

3.4 Select the Maximum Complementary Images

Based on the complement definition and calculation which we mentioned above, we assume that there are s blurry images, and we need select m images from s to restore the latent image. The multi-image selection of the maximum complement information is carried out by iterative method. The specific steps of the method implementation are as follows:

Step 1: Input the s blurry images, select m images among the input, which $m < s$.

Step 2: We combine arbitrary two blurry images of s, so that there will be C_s^2 type of combination. For every combination, calculate their complement score by Eq. (8). After we obtain the C_s^2 scores which can be described as $com(B_1, B_1), \cdots, com(B_i, B_j), \cdots, com(B_s, B_s)$, we select the two images B_k and B_l which complement score is the maximum.

Step 3: Now we calculate the union set $union_{kl}$ of B_k and B_l, then according step 2, select perfect combinations which combines two images of $union_{kl}$ and one of the others $(s-2)$ blurry images, and make sure the selection's complement score is the maximum of all combination. B_t is a notation of the selected image.

Step 4: Repeat the step 3, until the selected number is m, so we have the m images $B_1, B_2, \cdots, B_i, \cdots, B_m$.

After doing all above steps, we can obtain the images which we need. Thus, we can use the selected images to restore the latent sharp image.

4   Multi-image Restoration Using Complement

The general model of image degradation is described as:

$$B_i = K_i \otimes I + N, \forall i \in \{1, \cdots, m\} \tag{9}$$

where X represents the latent image, K_i is the blurry kernel, the observed image is B_i, systematic noise is described as N.

On the basis of the complement analysis of motion blurred images, we use an existing multi-image deblurring method to reconstruct the latent image. In the paper [11], a blind image restoration algorithm based on sparse model of hidden variables is proposed. The main contribution of the method is estimation of blur kernel. The mathematical model of blind restoration is shown bellow:

$$\min_{I, \{K_l, \lambda_l\}} \sum_{l=1}^{m} \frac{1}{\lambda_l} \|B_l - K_l \otimes I\|_2^2 + g(I, \{K_l, \lambda_l\}) \tag{10}$$

where λ_l is the noise level set, m is the number of the images which is been selected. And the $g(I, \{K_l, \lambda_l\})$ is given as follow:

$$g(I, \{K_l, \lambda_l\}) \triangleq \min_{\gamma \geq 0} \sum_{l=1}^{m} \sum_{i} [\frac{I_i^2}{\gamma_i} + log(\lambda_l + \gamma_i \|\bar{K}_l\|^2)] \tag{11}$$

The above optimal model can be solved by using the coordinate descent algorithm. The specific solution can be seen in [11].

5   Experiment

The performance of the proposed method is evaluated with test images. The restoration method of motion blurry image based on the complement is used. We use image's PSNR to evaluate the performance of the deblurring algorithm.

According to the complementary characterization of the selected images, as we mentioned earlier, we already known that the largest complement between two images at the direction are 150 angle and 90 angle for image data *house*, which we used before. Therefore, the blurred image in the 150 angle direction and the blurred image in the 90 angle direction are selected as input images of deblurring algorithm. The deblurring results of motion blurred image pairs are shown in Fig. 6. The results show that the effect of multiple images restoration based on complement is better than that without using complement information's method.

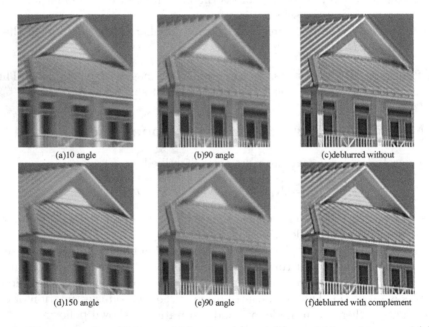

| (a)10 angle | (b)90 angle | (c)deblurred without |
| (d)150 angle | (e)90 angle | (f)deblurred with complement |

Fig. 6. The result of multi-image deblurring. (c) and (f) are deblurred image of (a)(b) and (c)(d), respectively. The PSNR value of (c) is 23.1548, and the value of (f) is 24.4553.

The more linear motion deblurring results are shown in Fig. 7. The original images are named *flower*, *academy*, *tower*, respectively. In the meantime, we calculate the PSNR and SSIM for every deblurry image, the result are shown in Fig. 8. For test images, after using complement information, the qualities of restoration are all improved obviously. The results of the experiment shown that the method using complement information approach more better performance than without complement method.

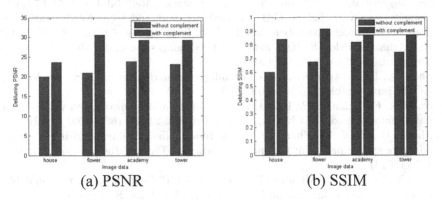

Fig. 7. The result of multi-image deblurring. From top to bottom: complement characterization, deblurring with complement, deblurring result without complement. From left to right: *flower*, *academy* and *tower*.

(a) PSNR   (b) SSIM

Fig. 8. Performance of the deblurring methods.

6   Conclusion

In order to solve the problem of the selection in image restoration, a multi-image deblurring method based on complement of motion blurred images is proposed. Based on the complement of the images, we use the iterative maximum complementary information to select the image pair with the largest complement. The blind deblurring method is used to reconstruct the latent image. Experimental results show that our method achieves better performance than others without complement deblurring methods.

Acknowledgments. This work is supported by the National Natural Science Foundation of China (No.61231016), the National 863 Program (No.2015AA016402), Seed Foundation of Innovation and Creation for Graduate Students in Northwestern Polytechnical University (Z2017184).

References

1. Cai, J.F., Ji, H., Liu, C., Shen, Z.: Blind motion deblurring using multiple images. J. Comput. Phys. **228**(14), 5057–5071 (2009)
2. Chen, J., Yuan, L., Tang, C.K., Quan, L.: Robust dual motion deblurring. In: Conference on Computer Vision and Pattern Recognition, CVPR 2008. IEEE, pp. 1–8 (2008)
3. Gong, D., Tan, M., Zhang, Y., van den Hengel, A., Shi, Q.: Blind image deconvolution by automatic gradient activation. In: Conference on Computer Vision and Pattern Recognition, pp. 1827–1836 (2016)
4. Gong, D., Tan, M., Zhang, Y., van den Hengel, A., Shi, Q.: MPGL: an efficient matching pursuit method for generalized lasso. In: AAAI Conference on Artificial Intelligence (2017)
5. Govindan, S., Saravanakumar, S.: Fast blur kernel estimation with texture preserving drd for motion deblur. Procedia Eng. **38**, 442–447 (2012)
6. Levin, A., Weiss, Y., Durand, F., Freeman, W.T.: Understanding blind deconvolution algorithms. IEEE Trans. Pattern Anal. Mach. Intell. **33**(12), 2354–67 (2011)
7. Papafitsoros, K., Schnlieb, C.B.: A combined first and second order variational approach for image reconstruction. J. Math. Imaging Vis. **48**(2), 308–338 (2014)
8. Rav-Acha, A., Peleg, S.: Two motion-blurred images are better than one. Pattern Recogn. Lett. **26**(3), 311–317 (2005)
9. Shan, Q., Jia, J., Agarwala, A.: High-quality motion deblurring from a single image. ACM Trans. Graph. **27**(3), 1–10 (2008)
10. Sroubek, F., Milanfar, P.: Robust multichannel blind deconvolution via fast alternating minimization. IEEE Trans. Image Process. **21**(4), 1687–1700 (2012)
11. Zhang, H., Wipf, D., Zhang, Y.: Multi-image blind deblurring using a coupled adaptive sparse prior. In: Computer Vision and Pattern Recognition, pp. 1051–1058 (2013)
12. Zhu, X., Šroubek, F., Milanfar, P.: Deconvolving PSFs for a better motion deblurring using multiple images. In: Fitzgibbon, A., Lazebnik, S., Perona, P., Sato, Y., Schmid, C. (eds.) ECCV 2012. LNCS, vol. 7576, pp. 636–647. Springer, Heidelberg (2012). doi:10.1007/978-3-642-33715-4_46

Biomedical Signal Processing

CAD Model Based on NN and PCA in Prostate Tumor MRI

Huiling Lu[1], Tao Zhou[1(✉)], and Hongbin Shi[2]

[1] School of Science, Ningxia Medical University,
Ningxia, Yinchuan 750004, China
lu_huiling@163.com, zhoutaonxmu@126.com
[2] Department of Urology, The Generel Hospital of Ningxia Medical University,
Ningxia, Yinchuan 750004, China
Shihongbin1124@163.com

Abstract. Aiming to feature redundancy problem in MRI Prostate Tumor ROI high dimension representation, a model, Prostate Tumor CAD Model based on NN with PCA feature-level fusion in MRI, is proposed in this paper. Firstly, geometry feature, statistical features, Hu invariant moment features, GLCM texture features, TAMURA texture features, frequency features are extracted from MRI prostate tumor ROI; Secondly PCA are used to obtain 8 dimension features in cumulative contribution rate 89.62%, and reducing the dimension of the feature vectors; Thirdly neural network is regarded as classifier to classify with BFGS, Levenberg-Marquardt, BP and GD training algorithm, Finally, MRI images of prostate patients are regarded as original data, prostate tumor CAD model based on NN with feature-level fusion are utilized to aid diagnosis. Experiment results illustrate that the ability to identify benign and malignant prostate tumor are improved at least 10% through Neural network with PCA feature-level fusion, and the strategy is effective, redundancy among features are reduces in some degree. There are positive significance for MRI prostate tumor CAD.

Keywords: Computer-aided diagnosis · Principal components analysis · Neural network prostate tumor

1 Introduction

Prostate tumors including prostate cancer and benign prostatic hyperplasia, both of which occur in the prostate, where prostate cancer is a common malignancy [1], mainly in the elderly and the recurrence rate is very high, and there is gradually dropping. Generally benign prostatic hyperplasia cannot translate into prostate cancer [2], firstly, prostatic hyperplasia occurs mainly in the central region of the prostate transitional zone, prostate cancer occurs mainly in the prostate peripheral zone (PZ), and the anatomical structure is great different between prostatic hyperplasia and prostatic cancer; secondly, there have different pathological processes, until now, only some evidence illustrate that Androgen can induce pathological prostate cancer to clinical prostate cancer, and no evidence illustrate that benign prostatic hyperplasia can be

© Springer International Publishing AG 2017
Y. Sun et al. (Eds.): IScIDE 2017, LNCS 10559, pp. 561–571, 2017.
DOI: 10.1007/978-3-319-67777-4_50

transformed into prostate cancer. However, prostatic hyperplasia and prostate cancer can exist simultaneously, clinical statistics show that a small number of prostatic hyperplasia (10%) will evolve into prostate cancer [2]. Prostate cancer is usually found by the anomaly of digital rectal examination (Digital rectal examination, DRE) and the increase of prostate specific antigen (Prostate-specific antigen, PSA) [3]. Prostate cancer, once discovered, its treatment methods and prognosis depend on the tumor histological classification and clinical staging, TNM staging and Gleason staging of prostate cancer determine the scope of the lesion and the correct staging is very important to decide whether operation, treatment and prognosis or not. MRI has many advantages such as 3D imaging, great soft tissue contrast, no biological damage, no injection of a contrast agent to show the vascular structure, and so on, which can distinguish the peripheral zone and the central gland of prostate to provide the required section in different directions, so as to facilitate understanding prostate panorama and surrounding relationship, which is conducive to determine the correct treatment guidelines by qualitative and staging, also good to the smooth implementation of surgery treatment plan and follow-up. Therefore, MRI in prostate cancer has an important role in positioning, scope of cancerous tissue, with or without penetrating the capsule, distant metastasis and other tests.

Computer aided diagnosis system (Computer-Aided, Diagnosis, CAD) is able to provide quantitative analysis to reduce the doctors' workload and provide good consistency and repeatability to doctors as diagnostic reference and suggestions to improve the diagnostic effect, diagnostic efficiency and objectivity, and to reduce the number of biopsies, at the same time lots of studies have shown that, used the CAD system could help the physician to significantly improve the quality of reading images [3]. Through the joint efforts of many researchers, computer aided diagnosis technology of prostate cancer has got significant development, the diagnostic accuracy and sensitivity have greatly improved. Doi [4] wrote a review about the history, the current status and future prospects of CAD. Niaf et al. [5] according to t-test, mutual information, minimum redundancy and maximum correlation criterion to extract statistical features Haralick features and gradient features of prostate cancer in MRI image, using four kinds of classifiers (nonlinear support vector machine, linear discriminant analysis, k-nearest neighbor and naive Bayes classifiers) to recognize them, the results shown that the computer-aided diagnosis system had the potential for the diagnosis of prostate cancer peripheral belt. Llobet et al. [6] through the analysis of 4944 cases of intestinal ultrasound images of 303 patients computer-aided diagnosis of prostate cancer, which based on k-nearest neighbor and hidden Markov model was proposed, the results shown that the urologist diagnostic capabilities with CAD improved significantly compared with no support system; Frank et al. [7] discussed the feature reduction of CAD system, analyzed three kinds of feature reduction methods which are PCC (Pearson's correlation coefficients), PCA (principal component analysis) and ICA (independent component analysis), and pointed the SVM as classifier, feature reduction of the classification sensitivity up to 85% and the classification accuracy rate up to maximum by using the PCA method. Hui [8] pointed out that the MRI in the diagnosis of liver cirrhosis was sensitive, CAD included the slice of the quantitative description and classification to provide another choice for clinicians, and presented a comprehensive description of section of CAD system, which was based on SVM method of duplicative-feature support vector machine (duplicative-feature support vector machine: DFSVM), to effectively improve the rate of correct classification.

| These literatures on the diagnosis of prostate tumor methods were discussed to some extent, but with other body organs (such as breast, brain, etc.) but compared with other organs of the body (such as breast, brain, prostate), the reports of prostate CAD were less, and these results only in a small sample, extracting features obtained in the limited conditions so false positive rate was widespread existence, and lack of more prospective study of convincing large sample. Actively seek effective feature extraction algorithm, try to improve characterization, based on the extraction of a more representative statistical characteristics, and looking good feature selection algorithm to effectively remove the redundancy between these characteristics is to improve prostate cancer Diagnostic accuracy of the CAD system is An effective channel to improve prostate cancer diagnostic accuracy of the CAD system is actively to seek effective features extraction algorithms, extract more representative statistical characteristics based on characterization, and look good feature selection algorithms to effectively remove the redundancy between these characteristics. In this paper, in order to improve the accuracy of CAD diagnosis of prostate cancer, we get 102 dimensional feature vector including 6-dimensional geometric characteristics, 6-dimensional statistical characteristics, 7-dimensional Hu moment invariant features, 56-dimensional gray co-occurrence matrix, 3-dimensional TAMURA texture features, and 24-dimensional frequency domain features of prostate ROI region in MRI image, based on feature level fusion, using classical neural networks (four kinds of training algorithm of BFGS quasi - Newton algorithm, BP algorithm, steepest descent algorithm and Levenberg-Marquardt algorithm) to verify the mode. The experimental results show that, using neural network feature level fusion to discriminate between malignant and benign lesions, the improvement is at least about 10%, this feature fusion strategy is effective, to some extent, decrease the features of irrelevant and redundant has positive significance for CAD diagnosis of prostate cancer in MRI image.

2   The Features of Prostate Tumor ROI Region in MRI Image and Its Extraction

Interesting ROI region is a sub-image of the original image which has the lesion to be detected. The lesion is split from the entire image, reducing the impact of the lesion analysis form the normal parts. In this paper, prostate tumor ROI region in MRI image is a subgraph which manual cut out from the prostate biopsy in MRI, and has stronger ability to distinguish prostate tumor (by radiologists recognized). Figure 1 shows the ROI region getting process and the extraction of characteristics.

CAD method based on medical images from a technical perspective is the target recognition technology, which has evolved from one-dimensional, two-dimensional, and three-dimensional to high-dimensional object recognition; the complexity of the target has developed from linear, nonlinear separable to complex pattern recognition; recognition theory, including classification and coverage recognition and other recognition ideas, but the existing target recognition methods are only based on a single sample of feature vectors to be divided, a large number of similar samples demonstrated

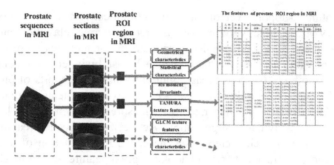

Fig. 1. Features extraction process of prostate tumor ROI region in MRI image

high-dimensional pattern characteristics to consider inadequate. Therefore, the recognition the results credibility is not high of prostate tumor ROI region in MRI image only extracting Kiwi or ten dimensional feature, it makes more meaningful only to discuss its properties in hundreds dimensional of feature description. Compared with normal tissue, prostate cancer (whether benign or malignant) physical characteristics are different, so its images in the texture, brightness, geometrical characteristics and surrounding are also different compared with normal tissue images. With this feature, in the relevant medical knowledge and the radiologist's guidance, according to the characteristics of prostate MRI image, with full consideration of features of prostate cancer in MRI, extraction of geometric features, statistical characteristics, Hu moment invariants, TAMURA texture feature, GLCM texture feature and frequency characteristics of prostate ROI region in MRI image. Because there are many literatures describes these features, just to give a brief introduction, specifically refer to the literatures.

(1) Geometrical characteristics, geometrical shape of ROI region, to some extent, reflects the pathological features of prostate. This paper selec ts six characteristics including area, perimeter, rectangularity, elongation, circularity and Euler number.

(2) Statistical characteristics, statistical methods analyze the spatial distribution of gray values, by computing local features at each point in the image, and deriving a set of statistics from the distributions of the local features. Statistical features are obtained by analysis the size and frequency of the gray value within a certain range. This paper select the following six statistical features are mean, variance skewness, kurtosis, energy and entropy.

(3) Hu moment invariants, M.K. Hu first proposed the concept of geometric moments, he used a continuous state of the second and third order central moments constructed seven invariant moments, because of good characteristics of moment invariant features with translation, rotation and scale invariance, which is widely used, usually with C1, C2, C3, C4, C5, C6, C7 to express the invariant feature of the target.

(4) TAMURA texture features, in 1978 Tamura proposed the expression of Tamura texture based on human visual perception in psychological research. Tamura texture feature six components corresponding psychology features on the 6

properties, three of them are coarseness (coarseness), contrast (contrast) and directionality (directionality), which have the good application value in the texture synthesis, image recognition and so on.

(5) GLCM texture features, the texture is composed of gray distribution in the space position appearing repeatedly, so there will exist a certain gray relationship between two pixels of a distance in the image space, called the gradation of the image gray spatial correlation properties. GLCM is a common method by studying the relevant characteristics of gray space to describe the texture feature.

(6) Frequency characteristics, based on wavelet packet texture feature extraction method is a way based on space-frequency domain characteristics of wavelet packet algorithm which translates image from the spatial domain into the frequency domain, and then extracts the image texture feature. This paper uses wavelet packet analysis of prostate MRI image for 3-order, then gets eight sub-image signal, and consists of a 24-dimensional feature vector as texture features of the image by the feature vector norm, standard deviation and energy signal of the 8 sub-image.

3 Prostate Tumor CAD Model Based on NN with PCA Feature-Level Fusion in MRI

3.1 High-Dimensional Feature Dimensionality Reduction Problem and PCA Analysis

Considering the characteristics of different categories of prostate tumor ROI region in MRI, getting the high dimensional feature can completely describe the characteristics of ROI, but also provides a lot of space for the clinical diagnosis and scientific research, at the same time it also brings two questions as following [11]:

(1) Irrelevant or redundant in the high dimensional feature. Existing research results show that irrelevant or redundant has greater impact on the accuracy of recognition algorithms, most learning algorithms required the number of training samples grow exponentially with irrelevant characteristic growth. Langley's study showed that the sample complexity of the nearest neighbor method grown exponentially with irrelevant characteristic growth [9]. From another perspective, irrelevant or redundant makes the sample points in high dimensional feature space ubiquitous sparsely distribute, and extend the distance metric function of low-dimensional space to high-dimensional space. With the increase of dimension, the distance between the data objects will disappear reducing its effectiveness, so it is particularly important to accurately find irrelevant and redundant.

(2) High-dimensional feature dimensionality reduction problem. The dimension of the feature space is too high, will produce "dimension disaster" issue, so there is an urgent need for feature selection algorithm for dimensionality reduction of high dimensional data. Two kinds of data dimension reduction methods usually are feature transformation and selection. Figure 2 shows the dimension reduction by $\varphi : R^N \to F$ transformation.

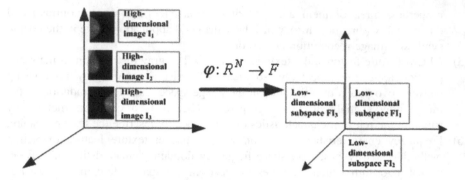

Fig. 2. High-dimensional represents the medical images dimensionality reduction

In this paper, it is difficult to give a clear conclusion to the vector space composed of the features of prostate tumor ROI region in MRI with or without the two problems and what extent, but "dimension reduction" operation is an effective means to circumvent these problems. At present, there are linear and nonlinear transformations of dimension reduction, in which linear transformation of dimension reduction method has many advantages, such as simple calculation, analytic solutions, valid of the linear structure of data, to find the best linear model under different optimization criteria, and no easy to fall into local minima. Principal component analysis PCA is one of typical nonlinear dimensionality reduction algorithm.

3.2 Prostate Tumor CAD Model Based on NN with PCA Feature-Level Fusion

MRI images of prostate tumors are regarded as the research object in this paper. Collect 13 cases MRI images of prostate tumors in clinical (5 cases of benign prostatic hyperplasia, 8 cases of prostate cancer), and extract prostate tumor ROI to pretreat, there are 102 dimension features are extracted from MRI prostate tumor ROI, including 6 dimension geometry features, 6 dimension statistical features, 7 dimension Hu invariant moment features, 56 dimension GLCM texture features, 3 dimension TAMURA texture features and 24 dimension frequency features.

8-dimension features are obtained by PCA in Cumulative contribution rate 89.62%, and reducing the dimension of the feature vectors; finally using classical neural network to test the model with BFGS, Levenberg-Marquardt, BP and GD training algorithm. Model descriptions are following:

Algorithm name: Prostate Tumor CAD Model Based on NN with PCA feature-evel fusion in MRI

Input:
 1) ROI images of prostate tumor in MRI set Xi, where i=1,2,...,180(90 images of prostate cancer , 90 images of benign prostatic hyperplasia)
 2) Sample categories n=2
Output:
 The neural network recognition accuracy of the features before and after transformation by four kinds of training algorithms.
 Steps:
Begin
 for i=1:M % M is the number of samples, extracting 102 dimension features from all samples
 T_{1i}= Geomentry(X_i); %T1 is geometric characteristics subspace formed by 6-dimensional feature vector;
 T_{2i}= Statistical(X_i); %T2 is statistical characteristic subspace formed by 6-dimensional feature vector;
 T_{3i}= Moment(X_i) ; %T3 is invariant moment features subspace formed by 7-dimensional feature vector;
 T_{4i}= GLCM (X_i) ; %T4 is GLCM texture features subspace formed by 56-dimensional feature vector;
 T_{5i}= TAMURA(X_i); %T5 is TAMURA texture features subspace formed by 3-dimensional feature vector;
 T_{6i}= Frequency(X_i) ; %T6 is frequency features subspace formed by 24-dimensional feature vector;
end
T={T_1 T_2 T_3 T_4 T_5 T_6};%T is the 102-dimensional feature space of the ROI described by combining T1 ,T2 ,T3 ,T4 ,T5 and T6
PCA_T=PCA(T); % The feature vector T is transformed by PCA, getting transform space PCA_T
 % In T and PCA_T spaces are cross-validated by neural network
for i=1:K % K-fold cross-validation
  Rec_ BFGS1(i) =NN_ BFGS (T(i));% In T (i) space using BFGS algorithm based on neural network to identify
  Rec_ BFGS2(i) =NN_ BFGS (PCA_T(i));
    % In PCA_T(i) space using BFGS algorithm based on neural network to identify
  Rec_ LM1(i) =NN_ LM (T(i));% In T (i) space using Levenberg-Marquardt training algorithm based on neural network to identify
  Rec_ LM2(i) =NN_ LM (PCA_T(i)) % In PCA_T(i) space using Levenberg-Marquardt training algorithm based on NN identify

  Rec_ BP1(i) =NN_BP(T(i)) %In T (i) space using BP algorithm based on neural network to identify
  Rec_ BP2(i) =NN_BP(PCA_T(i)) %In PCA_T(i) space using BP algorithm based on neural network to identify
  Rec_ GD1(i)=NN_ GD (T(i)) %In T (i) space using GD algorithm based on neural network to identify
  Rec_ GD2(i)=NN_ GD (PCA_T(i)) %In PCA_T(i) space using GD algorithm based on neural network to identify
End;
sum1=0; sum2=0; sum3=0; sum4=0; sum5=0; sum6=0; sum7=0; sum8=0;
for i=1:K   % Calculate the average recognition accuracy
  sum1=sum1+Rec_ BFGS1(i); sum2=sum2+Rec_ BFGS2(i)
  sum3=sum3+Rec_ LM1(i);   sum4=sum4+Rec_ LM2(i);
  sum5=sum5+Rec_ BP1(i);   sum6=sum6+Rec_ BP2(i);
  sum7=sum7+Rec_ GD1(i);   sum8=sum8+Rec_ GD1(i);
end;
  sum1=sum1/K; sum2=sum2/K; sum3=sum3/K; sum4=sum4/K;
  sum5=sum5/K; sum1=sum6/K; sum7=sum7/K; sum8=sum8/K;
end;

4 Algorithm Simulations

4.1 Experimental Environment and Data

Involved software and hardware environment in this experiment:

Software environment: windows 7, MATLAB 7.0, efilm 3.4;

Hardware environment:2G memory, 320G hard Disk, T4000 Intel Core processor;

Experiment data:Collected 13 cases MRI images of prostate tumors (5 cases of benign prostatic hyperplasia, 8 cases of prostate cancer) from department of radiology in General Hospital of Ningxia Medical University.

We obtained 180 MRI images ROI region of prostate tumors (90 MRI images of prostate cancer,90 MRI images of benign prostatic hyperplasia) and combined with the radiologist doctor's advice in experiment, taking into account the length of the article, only 5 MRI images' ROI region of prostate cancer and 5 MRI images ROI region of benign prostatic hyperplasia are given in training sample set Fig. 33-1 shows 5 MRI images ROI region of prostate cancer. Figure 33-2 shows 5 MRI images ROI region of benign prostate hyperplasia.

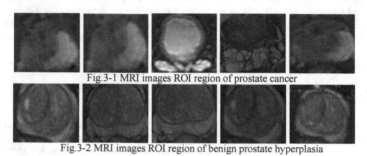

Fig.3-1 MRI images ROI region of prostate cancer

Fig.3-2 MRI images ROI region of benign prostate hyperplasia

Fig. 3. MRI images ROI region of prostate tumor

4.2   Experimental Environment and Data

(1) Feature-level fusion based on PCA

102 - dimension feature vectors are extracted from each MRI image of prostate tumor, and the feature databases of prostate cancer and hyperplasia use PCA analysis respectively, analysis each principal component contribution rate (Table 1) and the cumulative contribution rate (Fig. 4) of the feature level data fusion show that through the feature level fusion the dimension of feature vector greater than or equal 8 dimensional, and the cumulative contribution rate of the feature vector is 89.62% more than 85%, it can fully say that the 8 dimensional feature vector analysis of PCA can be expressed 102 - dimensional feature vectors before analysis, that effectively achieves the dimensionality reduction of feature vector.

(2) Prostate tumor identification based on neural network

At the last, using neural network (adopting four kinds of training algorithm of BFGS, BP,GD and Levenberg-Marquardt training algorithm) to recognize the 102 dimensional features before conversion and the 8-dimensional features after transformation, and adopting a 3-fold cross-validation method to compile statistical the results. Table 2 gives the recognition rate in different training functions with and without PCA fusion.

Table 1. Contribution rate of each principal component

Principal component	Contribution rate	Cumulative contribution rate
1	47.497	47.497
2	20.071	67.568
3	7.529	75.097
4	4.788	79.885
5	3.119	83.004
6	2.619	85.623
7	2.309	87.932
8	1.689	89.621
9	1.251	90.872
10	1.097	91.969

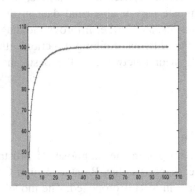

Fig. 4. The cumulative contribution rate of each principal component

Table 2. The recognition rate in different training functions with and without PCA fusion

The recognition rate		Training function			
		L-M algorithm	BFGS algorithm	BP algorithm	GD algorithm
Without PCA fusion	First time	58.33%	50%	50%	50%
	Second time	55%	60%	50%	50%
	Third time	60%	55%	50%	50%
	Average value	57.78%	55%	50%	50%
With PCA fusion	First time	60%	61.67%	66.67%	55%
	Second time	66.67%	76.67%	63.33%	60%
	Third time	68.33%	55%	53.33%	68.33%
	Average value	65%	64.45%	61.11%	61.11%
Increase rate		12.5%	17.18%	22.22%	22.22%

The experimental results show that, in the above 4 different training functions, image fusion PCA characteristics after the rate are improved obviously based on, the increase rate is respectively 12.5%, 17.18%, 22.22% and 22.22%, which shows that the method of principal component analysis in the 102 to feature in the feature space fusion is effective, not only can reduce the redundant features between, also eliminate the influence caused by the experimental results of abnormal data in feature, but also further improve the diagnosis rate of prostate MRI images, in 4 different training function, one of the best BP neural network recognition effect on prostate MRI image, and the recognition rate relatively stable.

The experimental results show that, in the above 4 different training functions, the image recognition rate of feature-level fusion based on PCA significantly improved, and its increase rate is 12.5%,17.18%, 22.22% and 22.22% respectively, indicating that the principal component analysis is an effective method to achieve the feature-level fusion of 102-dimensional feature spaces in the article, and reduce the redundancy between features, and eliminate part of abnormal data in the features decreasing the influence of experimental results, and further improve the diagnosis recognition rate of prostate MRI image. In four different training functions, the recognition effect on prostate MRI image of BP neural network is the best, and the recognition rate is relatively stable.

5   Conclusions

This article from the MRI images in the diagnosis of prostate cancer, analyzed the status quo of prostate cancer diagnosis, then Prostate Tumor CAD Model Based on NN with PCA feature-level fusion in MRI is proposed. The model is mainly description the high dimensional feature of prostate tumor ROI area in MRI, which indicates high dimensional feature can fully depict the ROI characteristics so that it can provide a lot of space to late clinical diagnosis and scientific research, but also bring some problems to the high dimensional feature such as irrelevant, redundant, and dimensionality reduction. The model uses PCA to transform in the feature level, reducing the dimension of the feature vector. Adopting classical neural network to verify the model, the experimental results show that the feature level fusion strategy is effective, and to some extent, reduces the irrelevant and redundancy between features and has a positive meaning to CAD of MRI images of prostate tumor.

Acknowledgements. This work is supported by the Natural Science Fund of China (Grant Nos. 61561040, 81160183), Scientific Research Fund of Ningxia Education Department (Grant No. NGY2016084).

References

1. Shi, F., Wei, J., Wang, Z.: High-field magnetic resonance imaging characteristics of normal and benign prostatic hyperplasia. Chin. J. Geriatr. **4**(16), 79–83 (1997)

2. Hua, L., Ju, X., Fei, W., et al.: The expression of androgen receptor in benign prostatic hyperplasia and prostate cancer. Chin. J. Geriatr. **22**(7), 405–408 (2003)
3. Yang, Z.: Computer-aided diagnosis of prostate lesions based on ultrasound images. University of Science and Technology of China Ph.D. thesis, Hefei (2009)
4. Doi, K.: Computer-aided diagnosis in medical imaging: historical review, current status and future potential. Comput. Med. Imaging Graph. **31**(4), 198–211 (2007)
5. Niaf, E., Rouvière, O., Mège-Lechevallier, F., et al.: Computer-aided diagnosis of prostate cancer in the peripheral zone using multiparametric MRI. Phys. Med. Biol. **57**(12), 3833–3851 (2012)
6. Llobet, R., Toselli, A.H., Perez-Cortes, J.C., Juan, A.: Computer-aided prostate cancer detection in ultrasonographic images. In: Perales, F.J., Campilho, A.J.C., de la Blanca, N.P., Sanfeliu, A. (eds.) IbPRIA 2003. LNCS, vol. 2652, pp. 411–419. Springer, Heidelberg (2003). doi:10.1007/978-3-540-44871-6_48
7. Zöllner, F.G., Emblem, K.E., Schad, L.R.: SVM-based glioma grading: optimization by feature reduction analysis. Z. für Med. Phys. **22**(3), 205–214 (2012)
8. Liu, H., Mei, G.D., Liu, X.: Cirrhosis classification based on MRI with duplicative-feature support vector machine (DFSVM). Biomed. Signal Process. Control **8**(4), 346–353 (2013)
9. Langley, P.: Selection of relevant features in machine learning. In: Proceedings of the AAAI Fall Symposium on Relevance, Menlo Park, CA, pp. 140–144. AAAI Press (1994)
10. Zhuo, L., Yang, M.: Using PCA algorithm arbors hyperspectral data dimensionality reduction and classification. (2013). http://www.cnki.net/kcms/detail/11.4415.P.20130603.1602.003.htm
11. Zhou, T., Lu, H.: Multi-features prostate tumor aided diagnoses based on ensemble-SVM. In: Proceedings of IEEE International Conference on Granular Computiong 2013, Beijing China, pp. 297–302 (2013)

Altered Functional Specialization in Temporal Lobe Epilepsy

Meiling Li, Qijun Zou, Jiao Li, Wei Liao, and Huafu Chen[✉]

Key laboratory for Neuroinformation of Ministry of Education,
School of Life Science and Technology, Center for Information in BioMedicine,
University of Electronic Science and Technology of China,
Chengdu 610054, China
limeilingcheng@163.com, zouqijun_ice@foxmail.com,
519985178@qq.com, weiliao.wl@gmail.com,
chenhf@uestc.edu.cn

Abstract. Temporal epileptogenic network has been extensively characterized in temporal lobe epilepsy (TLE). How epilepsy networks supporting the pathological mechanisms interact with each other via within- and across-hemisphere connections remains unclear. Here, resting state functional network specialization, depicting the degree to which brain networks preferentially connect with ipsilateral as opposed to contralateral networks, was examined in 90 TLE patients and 90 age- and education-matched healthy controls. We found decreased functional specialization in the default network and fronto-parietal network but increased functional specialization in the somatomotor network in patients with TLE. Altered functional network co-operation was further observed in the network with abnormal functional specialization. The results indicated that the altered functional specialization emphasizing the specific processing functions within hemisphere may be particularly important for seizure attack in TLE.

Keywords: Functional specialization · Temporal lobe epilepsy · Seizure · Hemisphere

1 Introduction

Temporal lobe epilepsy (TLE) is associated with the impairment of connectivity of brain regions, implicating abnormal excessive or synchronous neuronal activity in the brain [1]. Such abnormal activity typically does not occur in a single isolated neuron; rather, in large groups of neurons, which may covered several functional/structural brain networks. Current neuroimaging tools combining metabolic, electrophysiological, functional and structural profiles have been devoted to map disrupted epilepsy network [2, 3]. Among them, the default network, attention network and perceptual networks are the most important and disrupted networks in TLE [4–6]. However, such changes mainly based on the topological properties, which are subject to differences in graph construction rules, require further investigation [7].

© Springer International Publishing AG 2017
Y. Sun et al. (Eds.): IScIDE 2017, LNCS 10559, pp. 572–579, 2017.
DOI: 10.1007/978-3-319-67777-4_51

Functional specialization such as the hemispheric specialization is hypothesized to contribute to fast, efficient information processing [8, 9]. It potentially emphasizes regions crucial to specific processing functions and may provide important insights into localizing an epileptogenic zone. Investigating the functional specialization interactions among different networks may suggest how specialization arises or changed in TLE. In the current study, functional specialization in TLE was estimated using within- and cross-hemisphere functional connectivity to investigate brain vertices/networks preferentially connect with ipsilateral vertices/networks. We hypothesized that special function in distinct network will altered in patients with TLE and the network coupling with other networks may also be impaired or reorganized. The increased or decreased functional specialization and network regularity may lead to the formation of abnormal epileptogenic connections.

2 Methods

2.1 Participants

This study was approved by the Ethics Committee of Jinling Hospital, Nanjing University School of Medicine. One hundred and six TLE patients participated in this study. Diagnosis of TLE was performed according to the International League Against Epilepsy (ILAE) including: seizure history and semiology, neurologic examination, diagnostic magnetic resonance imaging, and electroencephalography records. In addition, 161 healthy controls (HC) were recruited, which were interviewed to confirm that none of them had a history of neurological or psychiatric disorder. All the participants were right-handed according to the criterion of Chinese revised-version of Edinburgh Handedness Inventory [10].

2.2 Data Acquisition

MRI data were scanned on a Siemens 3T Trio scanner (Siemens Medical Systems, Erlangen, Germany) at Jinling Hospital. During the scanning, the participants were instructed to keep still and rest with eyes closed but not to fall asleep, foam padding was used to minimize head movement. Resting-state fMRI were acquired by a single-shot, gradient-recalled echo planar imaging (EPI) sequence. The sequence parameters were as follows: repetition time = 2000 ms, echo time = 30 ms, slice thickness = 4 mm with slice gap = 0.4 mm, 30 slices, field of view = 24 cm, flip angle = $90°$, voxel size = $3.75 \times 3.75 \times 4$ mm^3. For each subject, the fMRI scanning is composed of 250 image frames.

2.3 Data Preprocessing

Functional images were preprocessed in the SPM8 (http://www.fil.ion.ucl.ac.uk/spm/) and Data Processing Assistant for Rest-State fMRI (DPARSF) [11]. To ensure magnetization equilibrium, the first 5 volumes were discarded. The remaining 245 functional scans were corrected for temporal difference among different slices and head motion. Subsequently, the functional images were normalized to the Montreal Neurological

Institute (MNI) echo-planar imaging (EPI) template in SPM8. A band filter was applied (0.01–0.08 Hz) to reduce the effect of low-frequency drifts and high-frequency noise. Next, time series were corrected for six head motion parameters, averaged signals from cerebrospinal fluid (CSF) and white matter (WM), and global brain signal [12]. The preprocessed data in MNI space were projected to the FreeSurfer template with a mesh of 40,962 vertices in each hemisphere. Then a 6-mm full-width half-maximum (FWHM) smoothing kernel was used to the preprocessed data in the surface space and warped into 4-mm mesh with 2,562 vertices in each hemisphere provided by FreeSurfer [13].

2.4   Quality Control

We exclude subjects belonging to the following cases: (1) Scanning didn't cover the whole brain; (2) The functional image registration was not successful; (3) Head motion such as the translation or rotation exceeded ±1.5 mm or ±1.5°. Finally, 40 subjects composed of 24 controls and 16 TLE subjects were excluded, leaving 137 HC and 90 TLE. We then selected 90 controls to maximum match to the 90 patients in age and gender. The Demographic information of participants used in the present study was listed in Table 1.

Table 1. Characteristics of the TLE and healthy controls

Variables (Mean + SD)	HC (N = 90)	TLE (N = 90)	p value
Age	25.97 ± 6.72	27.93 ± 9.81	0.118
Gender	M:43;F:47	M:42;F:48	0.882

The p values were obtained by a two-sample t-test for age, and a kruskal-wallis test for gender. Abbreviations: TLE: temporal lobe epilepsy; HC: healthy controls; F, female; M, male, SD, standard deviation.

2.5   Computation of Functional Specialization

Functional specialization was estimated based on the autonomy index (AI) proposed by Wang et al. [14]. Specifically, each vertex on the brain surface was taken as a seed and a seed-based functional connectivity was then estimated based on the resting state brain activity across the whole brain. For each seed, within-hemisphere degree and between-hemisphere degree was computed by counting the number of vertices that were strongly connected (e.g. $r > 0.1$) to the seed in the ipsilateral hemisphere and in the contralateral hemisphere, respectively. The degree of within- and cross-hemisphere was normalized by dividing the total number of vertices in the corresponding hemisphere. Then the AI was calculated for each vertex as the normalized degree difference between the within- and cross-hemisphere according to the following equation:

$$AI = \frac{N_i}{H_i} - \frac{N_c}{H_c} \tag{1}$$

where N_i and N_c are the numbers of vertices with the correlation values larger than the preset threshold that are connected to the seed/vertex in the ipsilateral hemisphere and contralateral hemisphere, respectively. H_i and H_c are the total number of vertices in the ipsilateral and contralateral hemisphere, respectively.

To quantify the specialization of specific functional networks, we used the functional atlas obtained by a clustering method (http://surfer.nmr.mgh.harvard.edu/fswiki/CorticalParcellation_Yeo2011) [13]. The boundaries of seven networks were projected to the surface template with 2562 vertices in each hemisphere. The seven networks include: somatomotor network (MOT), visual network (VIS), dorsal attention network (dATN), ventral attention network (vATN), Limbic network (LMB), frontoparietal network (FPN) and default network (DN). AI was then averaged within the boundary of each network in left and right hemisphere, respectively.

3 Results

Functional specialization of the human brain was quantified using the autonomy index based on the within-hemispheric and cross-hemispheric functional connectivity. The specialization distribution maps in HC and TLE both showed particularly strong difference between within-hemispheric and cross-hemispheric connectivity in some regions (Fig. 1). The strong functional specialization was mainly observed in the lateral prefrontal, inferior parietal, and temporal regions and the minimal hemispheric specialization was observed in the visual and motor cortex. Especially, in the left hemisphere, strong specialization was observed in the default network. The functional specialization distribution was similar to the results derived from a large dataset [14].

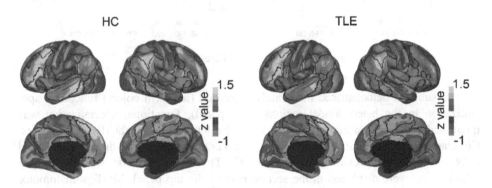

Fig. 1. Functional specialization distribution in HC and TLE.

The functional specialization maps between the two groups are visually similar. Strong specialization was observed in the association cortices including the inferior prefrontal and temporal regions (associated with language processing) in the left hemisphere and the lateral frontal and angular gyrus regions (associated with attention) in the right hemisphere. Minimal specialization was found in the sensorimotor, auditory, and visual cortices. Functional autonomy value was normalized by the total number of vertices in each hemisphere. Regions with higher within-hemisphere connectivity than cross-hemisphere connectivity are shown in warm colors. Regions with higher cross-hemispheric connectivity are shown in cold colors.

We then observed different functional specialization across seven networks between HC and TLE (Fig. 2). In the left hemisphere, decreased functional specialization was found in the LMB and DN in TLE patients. In the right hemisphere, TLE patients had decreased functional specialization in the FPN but increased functional specialization in the MOT.

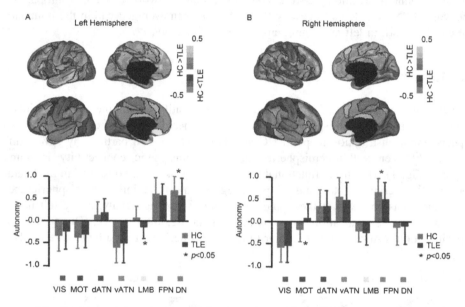

Fig. 2. Network functional specialization difference between the HC and TLE.

HC subjects demonstrated moderately increased (in warm colors) functional specialization in the hetero-modal regions (Top panel). Functional specialization was quantified across seven cerebral networks in the left (A) and right (B) hemispheres. Compared to HC, patients with TLE show weaker specialization in the DN, LMB and FPN and stronger specialization in the MOT. The seven networks with boundaries (black lines) were displayed in the second row of the top panel. MOT, somatomotor network; VIS, visual network; dATN, dorsal attention network; vATN, ventral attention network; LMB, limbic network; FPN, frontoparietal network; DN, default model network; HC: healthy controls; TLE: temporal lobe epilepsy.

To further investigate the altered functional specialization in networks, we calculated the functional specialization relation between other networks and the altered networks (Fig. 3). In the left hemisphere, patients with TLE had decreased co-operation between the MOT and LMB (HC: $p = 0.001$; TLE: $p = 0.051$), and between the VIS and DN (HC: $p = 0.004$; TLE: $p = 0.328$). In the right hemisphere, increased

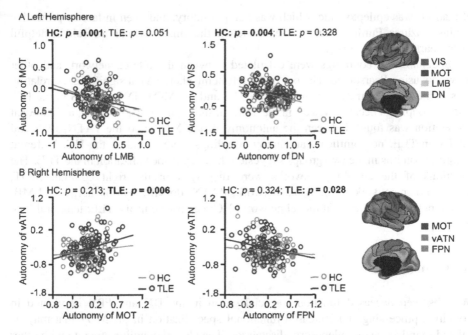

Fig. 3. Specialization coupling of the altered network with other networks.

co-operation between the vATN and MOT (HC: $p = 0.213$; TLE: $p = 0.006$), and between the vATN and FPN (HC: $p = 0.324$; TLE: $p = 0.028$) were found in patients.

4 Discussion

Decreased functional specialization was observed in the DN, FPN and LMB, whereas increased specialization was observed in the MOT in patients with TLE. DMN and FPN as two of the most important hetero-modal association regions are preferentially selected by long-range functional connectivity [15] and the rules for information processing are critical for achieving behavioral flexibility [16]. Patients with altered functional specialization probably suggested decreased in the number of long-range connections and in turn may prove abnormal development of cognitive processes in TLE. This is agree on previous studies that decreased connectivity in a large number of regions within the DMN and inferior frontal gyrus, some of which were also associated with decreased cognitive performance [17, 18]. Moreover, the increased functional specialization in the MOT may suggest a possible excessive influence of epileptiform discharges on information processing in the motor sensory system. Increased connectivity within the somato-motor network was also exhibited in children with benign epilepsy with centrotemporal spikes [19].

Specialization of the altered network also changed the coupling strength with other networks in two hemispheres. Increased or decreased interactions between networks were occurred simultaneously in the present study. It is difficult to define which alterations was epileptogenic, which was compensatory, and even in fact be protective against seizures. Further studies can evaluate whether this information would be helpful in the early epilepsy surgery.

Specialization couplings were calculated between the altered network and other networks using Pearson correlation. In the left hemisphere (A), a significant correlation was found between the autonomy in the LMB and the MOT, DN and VIS in the HC, but the couplings became weaker in the TLE. In the right hemisphere (B), a significant correlation was found between the autonomy in the MOT and the vATN, FPN and vATN in TLE, no significant autonomy couplings were found in the HC. Different coupling patterns in the two groups suggested the altered specialization in the TLE. The locations of the coupling networks were displayed in the right column. MOT, somatomotor network; VIS, visual network; dATN, dorsal attention network; LMB, limbic network; DN, default model network; HC: healthy controls; TLE: temporal lobe epilepsy.

5 Conclusions

We observed deceased functional specialization in the DN and FPN implicated in cognitive processing and increased functional specialization in the MOT that may be related to influence of epileptiform discharges. The altered coupling strength with other networks in these networks further highlights their importance in defining epileptogenic markers in TLE.

Competing Interests
The authors declare that they have no competing interests.

Acknowledgement. The work is supported by 863 project (2015AA020505) and the Natural Science Foundation of China (61533006).

Author's Contributions. Meiling Li, Wei Liao and Huafu Chen proposed and implemented the idea. Meiling Li, Qijun Zou, Jiao Li performed the data analysis. All authors drafted and approved the manuscript.

References

1. Fisher, R.S., et al.: Epileptic seizures and epilepsy: definitions proposed by the international league against epilepsy (ILAE) and the international bureau for epilepsy (IBE). Epilepsia **46** (4), 470–472 (2005)
2. Blumenfeld, H., et al.: Cortical and subcortical networks in human secondarily generalized tonic-clonic seizures. Brain **132**(Pt 4), 999–1012 (2009)
3. Laufs, H., et al.: Temporal lobe interictal epileptic discharges affect cerebral activity in "default mode" brain regions. Hum. Brain Mapp. **28**(10), 1023–1032 (2007)
4. Bettus, G., et al.: Decreased basal fMRI functional connectivity in epileptogenic networks and contralateral compensatory mechanisms. Hum. Brain Mapp. **30**(5), 1580–1591 (2009)
5. Zhang, Z., et al.: Impaired perceptual networks in temporal lobe epilepsy revealed by resting fMRI. J. Neurol. **256**(10), 1705–1713 (2009)

6. Zhang, Z., et al.: Impaired attention network in temporal lobe epilepsy: a resting FMRI study. Neurosci. Lett. **458**(3), 97–101 (2009)
7. Haneef, Z., Chiang, S.: Clinical correlates of graph theory findings in temporal lobe epilepsy. Seizure – Eur. J. Epilepsy. **23**(10), 809–818 (2014)
8. Ringo, J.L., et al.: Time is of the essence: a conjecture that hemispheric specialization arises from interhemispheric conduction delay. Cereb. Cortex **4**(4), 331–343 (1994)
9. Bullmore, E., Sporns, O.: The economy of brain network organization. Nat. Rev. Neurosci. **13**(5), 336–349 (2012)
10. Oldfield, R.C.: The assessment and analysis of handedness: the edinburgh inventory. Neuropsychologia **9**(1), 97–113 (1971)
11. Chao-Gan, Y., Yu-Feng, Z.: DPARSF: A MATLAB toolbox for "pipeline" data analysis of resting-state fMRI. Front. Syst. Neurosci. **4**, 13 (2010)
12. Fox, M.D., et al.: The global signal and observed anticorrelated resting state brain networks. J. Neurophysiol. **101**(6), 3270–3283 (2009)
13. Yeo, B.T., et al.: The organization of the human cerebral cortex estimated by intrinsic functional connectivity. J. Neurophysiol. **106**(3), 1125–1165 (2011)
14. Wang, D., Buckner, R.L., Liu, H.: Functional specialization in the human brain estimated by intrinsic hemispheric interaction. J. Neurosci. **34**(37), 12341–12352 (2014)
15. Sepulcre, J., et al.: The organization of local and distant functional connectivity in the human brain. PLoS Comput. Biol. **6**(6), e1000808 (2010)
16. Mesulam, M.M.: From sensation to cognition. Brain **121**(Pt 6), 1013–1052 (1998)
17. Liao, W., et al.: Altered functional connectivity and small-world in mesial temporal lobe epilepsy. PLoS ONE **5**(1), e8525 (2010)
18. de Campos, B.M., et al.: Large-scale brain networks are distinctly affected in right and left mesial temporal lobe epilepsy. Hum. Brain Mapp. **37**(9), 3137–3152 (2016)
19. Li, R., et al.: Epileptic discharge related functional connectivity within and between networks in benign epilepsy with centrotemporal spikes. Int. J. Neural Syst. (2017). 1750018

Resting-State Brain Activity Complexity in Early-Onset Schizophrenia Characterized by a Multi-scale Entropy Method

Xiao Wang[1], Yan Zhang[2], Shaoqiang Han[1], Jingping Zhao[3(✉)], and Huafu Chen[1(✉)]

[1] Key Laboratory for Neuroinformation of Ministry of Education, School of Life Science and Technology, Center for Information in BioMedicine, University of Electronic Science and Technology of China, Chengdu 610054, China
chenhf@uestc.edu.cn
[2] Key Laboratory for Mental Health of Hunan Province, Mental Health Institute, The Second Xiangya Hospital of Central South University, Changsha, China
[3] Mental Health Institute, The Second Xiangya Hospital of Central South University, 139, Middle Renmin Road, Changsha 410011, Hunan, China
zhaojingpinghunancsu@163.com

Abstract. Early-onset schizophrenia (EOS) is a severe mental illness associated with changes of brain's activity. However, the complexities of brain activity in EOS are still lacking. To address this issue, a multi-scale sample entropy (MSE) method was used to investigate the role of brain signal complexity in EOS. We recruited 39 patients with EOS (age from 12 to 18), 31 age- and sex-matched healthy controls. Reduced blood-oxygen-level-dependent (BOLD) complexity was observed in the superior temporal sulcus and cuneus. Increased BOLD complexity was observed in the middle frontal gyrus, superior partial lobule, precuneus and cingulate gyrus. Furthermore, we found the complexity changes in cingulate gyrus were associated with clinical symptoms of schizophrenia. These results suggested that the changes of complexity are crucial to understand the pathomechanism of schizophrenia.

Keywords: fMRI · Resting-state · Complexity · Multi-scale entropy · Early-onset schizophrenia

1 Introduction

Schizophrenia is a devastating disease characterized by abnormal activities among the brain [1–3]. There has been accumulating evidence supported that the abnormal activities of brain are associated with the change of complexity in patients with schizophrenia [4–6]. Adolescents with early-onset schizophrenia (EOS) provides a unique opportunity to explore the dynamic state alterations as they are less affected by chronic antipsychotic medication and interaction with age-related neurodegenetation [7–9]. By using multi-scales analysis, many studies have focused o'n exploring the complexity of the brain signals in schizophrenia.

© Springer International Publishing AG 2017
Y. Sun et al. (Eds.): IScIDE 2017, LNCS 10559, pp. 580–588, 2017.
DOI: 10.1007/978-3-319-67777-4_52

Some studies used electroencephalography (EEG) or magetoencephalography (MEG) technology to explore the complexity of brain activity signals in mental disorder [10, 11]. These studies suggested that the changes of complexity can be served as a valid marker for mental illness such as Alzheimer's disease, attention deficit hyperactivity disorder [12–15]. Recently, the analysis of the complexity of blood-oxygen-level-dependent (BOLD) signals obtained from resting-state functional magnetic resonance imaging (fMRI) have also been used in schizophrenia [14]. Although BOLD signals were not the direct measures of neuronal activity, they can provide time-varying effects of disease state and may provide consistent biomarker for investigating brain dynamic changes. Several studies have used entropy method to explore the complexity of brain activity in schizophrenia [16–18].

In the conventional method, the complexity was usually calculated using the single-scale and just based on the linear parameter such as the mean or variance of the time series. However, a healthy physiologic process is typically represented by a complex fluctuation with consistent entropy values over different time scales, which means the relation between the regions in the brain is not just linear. Multi-scale sample entropy (MSE) as a multi-scale method provides a profile of entropy across multiple time scales and enables extracting the complexity from physiologic signals which overcome the downside of the traditional methods as said before. Therefore, using the MSE method to explore the complexity of the EOS is very suitable.

In the current study, we aimed to assess the changes of complexity in the EOS. The MSE method was employed to investigate the complexity in the EOS patients and HCs. Two sample t test was used for statistical analysis. Furthermore, the relationship between the altered complexity at all scale and clinical symptom was investigated to test whether scale-specific complexity contributes to the clinicopathology of EOS.

2 Materials and Methods

2.1 Participants

Thirty-nine patients with EOS with mean age of 15.5 years (12–18) who met DSM-IV criteria for schizophrenia or schizophrenia form disorder were recruited from the Second Affiliated Hospital of Xinxiang Medical University. All patients were independently diagnosed by research psychiatrists and satisfied the following criteria: (1) DSM-IV-TR criteria for schizophrenia (Diagnostic and Statistical Manual of Mental Disorders, fourth edition, text revision, American Psychiatric Association, 2000); (2) no co-morbid Axis I diagnosis; (3) duration of illness is less than 2 years; (4) anti-psychotic-naive. Patients were interviewed again after six months for a final schizophrenia diagnosis. The symptoms were evaluated using the Positive and Negative Syndrome Scale (PANSS). Fifteen age-, gender-, education-, and IQ-matched healthy adolescents were included in this study. All HCs did not have any past or current neurological disorder or family history of hereditary neurological disorder. This study was approved by the Ethics Committee of the Second Affiliated Hospital of Xinxiang Medical University, and informed written consents were obtained from all subjects.

2.2 Data Acquisition

All subjects were instructed to rest with their eyes closed, not to think of anything in particular, and not to fall asleep during the resting-state fMRI (rs-fMRI) scanning [19–21]. fMRI data were collected using the 3T MRI scanner (Siemens-Trio, Erlangen, Germany) of the Second Affiliated Hospital of Xinxiang Medical University. Scanning and clinical assessments were performed in one day. Functional images were collected transversely using an echo-planar imaging (EPI) sequence with the following settings: TR/TE = 2000/30 ms, flip angle = 90°, FOV = 220 × 220 mm^2, slices = 33, matrix = 64 × 64, interslice gap = 0.6 mm, and voxel size = 3.44 × 3.44 × 4 mm^3. The scan lasted for 480 s for each subject, and 240 volumes were acquired.

2.3 Image Preprocessing

Data preprocessing was conducted using the Statistical Parametric Mapping Software (SPM8, http://www.fil.ion.ucl.ac.uk/spm8). The first 10 volumes of each participant were discarded because of the instability of the initial MRI signal and adaptation of the participants to the circumstance, leaving 230 volumes. The remaining rs-fMRI images were corrected for slice acquisition and head-motion using a least squares approach with a 6-parameter spatial transformation. Four patients and one healthy control with head motion scans exceeding 2 mm or 1° rotation were excluded. Subsequently, the corrected images were normalized according to the standard SPM8 Montreal Neurological Institute (MNI) template [22] and re-sampled to 3 × 3 × 3 mm^3 voxel size. Friston 24 motion parameters, cerebrospinal fluid, and white matter signals were included in the multiple regression model to reduce the effects of head motion and non-neuronal BOLD fluctuations [23, 24]. We also calculated the mean frame-wise displacement (FD) to further determine the comparability of head movement across groups. The largest mean FD of each subjects was less than 0.3 mm and two-sample t-test showed that there was no significant difference in the mean FD between the two groups (HC: 0.09 ± 0.05; EOS: 0.1 ± 0.03; mean ± SD, $p = 0.33$).

2.4 Analysis of the MSE of BOLD Signals

An anatomical connectivity-based template was employed to divide the brain into 246 regions of interest (ROI) [25, 26]. For each subject, the representative time series of each individual regions was obtained by averaging the fMRI signals over all voxels in this region. Finally, a 246 * 230 time series matrix was acquired.

MSE calculation can be briefly summarized in three steps: (1) constructing coarse-grained time series according to different scale; (2) calculate the sample entropy (SE) of each time series; (3) comparing the sample entropy over a range of scales [6, 14, 27].

SE (m, r, N) is precisely the negative natural logarithm of the conditional probability that a dataset of length N, having repeated itself within a tolerance of r (similarity factor) for m points (pattern length), will also repeat itself for m + 1 points, while self-matches are not included in calculating the probability [27]. Here, we used the previously parameters for MSE calculation that m = 1 and r = 0.35 and scale factors up to 5 [14].

2.5   Statistical Analysis

Statistical analysis of the MSE parametric imaging data was conducted using the MATLAB. Regional between-group differences in the whole-brain MSE mapping of each time scale were examined using the two-sample t test, and age and sex as covariates of no interest. False Discovery Rate correction theory was employed to correct multiple comparisons (q < 0.05).

Further, brain areas that showed significant effects were considered regions of interest (ROIs) for the following post-hoc analyses. Subsequently, we calculated the pearson correlation between the change of complexity in ROIs and scores of PANSS.

3   Results

3.1   Demographic Information and Clinical Symptoms

No significant differences were found in the gender, age, and years of education between the patients and HCs. For the patients group, the mean duration of illness was 16.0 months (SD = 14.4) (Table 1).

Table 1. Demographic and clinical information in the study

Demographics, Mean (SD)	EOS N = 35	Control N = 30	P value
Age (year)	15.5 (1.8)	15.3 (1.6)	0.57[a]
Gender (male/female)	20/15	13/17	0.27[b]
Education (years)	8.5 (1.48)	8.7 (1.42)	0.605[a]
Duration of psychosis (months)	16.0 (14.4)	–	–
Handedness (right/left)	35/0	30/0	–
PANSS positive symptoms	20.42 (5.72)	–	–
PANSS negative symptoms	20.91 (8.41)	–	–
PANSS general symptoms	33.28 (6.69)	–	–
PANSS total symptoms	74.62 (10.61)	–	–

Pa-value was obtained by two-sample t-test
Pb-value was obtained by two-tailed test

3.2   Between-Group Comparison of MSE Profiles in Different Time Scales

Brain regions showed significant difference in the MSE between-group in the middle frontal gyrus, superior temporal sulcus, superior parietal lobule, precuneus, cingulate Gyrus and cuneus (Fig. 1) (Table 2). Post-hoc analyses revealed that the SE in EOS

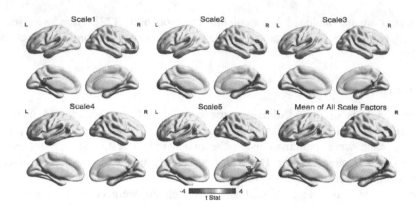

Fig. 1. Comparison of MSE difference between groups. The results were obtained by a two-sample t test. Statistical significance level is corrected for multiple comparisons using False Discovery Rate theory with (p < 0.05). MSE, multi-scale sample entropy

Table 2. Difference of MSE between groups revealed by a two-sample t test

	Brain areas	R	Peak coordinate		
			X	Y	Z
EOS > HC	Middle frontal gyrus		42	44	14
	Superior parietal lobule		19	−69	54
	Precuneus		16	−64	25
	Cingulate gyrus		9	−44	11
EOS < HC	Posterior superior temporal sulcus		−42	−24	−21
	Cuneus		−6	−94	1

was higher than that in HCs mainly in the right ventral of middle frontal gyrus, right superior partial lobule, right precuneus, and bilateral cingulate gyrus.

Moreover, a significantly lower SE was found in patients in the left superior parietal lobule and left cuneus (Fig. 2).

3.3   Demographic Information and Clinical Symptoms

Spearman correlation was calculated between the SE in the cingulate gyrus and PANSS scores after removing potential outliers. In Fig. 3, in the scale 2, the cingulate gyrus with increased SE is significantly negatively correlated with the negative and total scores of PANSS (p < 0.05). No significant correlation was found in the other scales.

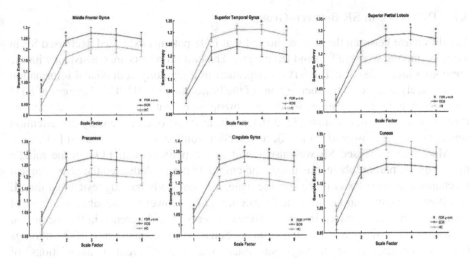

Fig. 2. Comparison of SE difference in all scales. The results were obtained by a two-sample t test. Statistical significance level is corrected for multiple comparisons using False Discovery Rate theory with (p < 0.05). SE, sample entropy

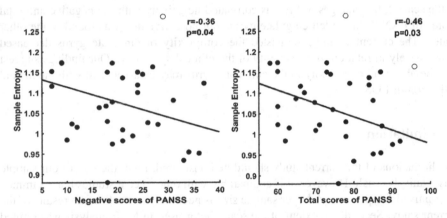

Fig. 3. Correlation between regions showing significant difference SE and clinical of symptom in patients with EOS (p < 0.05). Spearman correlations were calculated over the data after removing outliers marked by circles. SE, sample entropy

4 Discussion

In this study, we used the MSE method to investigate the changes of complexity in EOS. Several brain regions exhibited significant differences in the complexity between EOS and HC. Moreover, in the scale2, the SE in the cingulate gyrus correlated with the clinical symptoms. These results demonstrated that abnormal BOLD complexity provided new view to understand the mechanism of EOS and the changes of complexity may contribute to the syndrome domains of psychosis in EOS.

4.1   Difference in SE Between-Groups

The difference between the groups showed that EOS patients exhibited decreased SE in the left temporal gyrus (STG) and left cuneus. The current results are consistent with the previous studies [28–30]. The STG is important in integrating audiovisual information and is widely connected to other regions of the brain, the [29, 31]. The damages of STG often show abnormal hearing, which in schizophrenic patients were generally in the form of auditory hallucinations [32, 33]. Cuneus as a part of early sensory and attention neural circuit, the lower SE in it reflects the poor volitional saccade control [34].

Moreover, increased SE was also found in the patients with EOS in the middle frontal gyrus, partial lobule, precuneus and cingulate gyrus. Although the physiological mechanism of complexity in the specific scale, a recent study has suggested that the SE in different brain regions may scale-specific [14]. However, the abnormalities of complexity were still mixed in schizophrenia patients. The difference in the method of measuring the activity and the sample may contribute to inconsistency of the findings between studies. These findings suggested that the differential associations of BOLD MSE profiles may provide the clue to psychopathology of schizophrenia.

4.2   Correlation Analysis

In the scale 2, the changes of SE was correlated negatively with the negative and total scores of PANSS in the left cingulate gyrus, but no correlation was found in the other scales. The current findings suggested the complexity of cingulate gyrus decreased progressively in relation to the severity of the clinical symptoms. Our findings suggest that the abnormal complexity in the cingulate gyrus may be associated with the mental activation in EOS.

5   Limitation

The limitations of the current study should be considered. First, the size of our sample was relatively small. However, it is comparable with other recent studies that examined drug-naive EOS patients. A larger sample size is necessary to confirm the results of the current study. Second, the extent of the scale factor used in MSE analysis was limited to the length of BOLD time series (195 data points); therefore, the MSE complexity profiles based on scales of 1–5 may have captured only a portion of the underlying dynamics, unlike long resting-state fMRI signals such as those used in the Human Connectome Project.

6   Conclusion

In this study, we investigated the changes of complexity within the brains of patients with EOS and HC. Many brain areas showed significant differences in different scales. These regions mainly associated with audiovisual information processing, attention system and cognition. Moreover, the changes of SE in the cingulate gyrus were significantly associated with the clinical symptoms. Our observations may provide potential implications for exploring the Pathophysiology of schizophrenia.

Acknowledgements. The authors thank Yan Zhang, Jingping Zhao for fMRI data acquisition, as well as Shaoqiang Han and Huafu Chen for their assistance with the analyses.

Funding. The work was supported by the 863 project (2015AA020505), the Natural Science Foundation of China (61533006, 61673089) and Fundamental Research Funds for the Central Universities (ZYGX2016KYQD120, ZYGX2015J141).

References

1. Rm, H., et al.: Dynamic functional connectivity: promise, issues, and interpretations. Neuroimage **80**(1), 360 (2013)
2. Damaraju, E., et al.: Dynamic functional connectivity analysis reveals transient states of dysconnectivity in schizophrenia. Neuroimage Clin. **5**, 298–308 (2014)
3. Ma, S., et al.: Dynamic changes of spatial functional network connectivity in healthy individuals and schizophrenia patients using independent vector analysis. Neuroimage **90**, 196–206 (2014)
4. Callicott, J.H., et al.: Complexity of prefrontal cortical dysfunction in schizophrenia: more than up or down. Am. J. Psychiatry **160**(160), 2209–2215 (2003)
5. Koukkou, M., et al.: Dimensional complexity of EEG brain mechanisms in untreated schizophrenia. Biol. Psychiatry **33**(6), 397–407 (1993)
6. Takahashi, T., et al.: Antipsychotics reverse abnormal EEG complexity in drug-naive schizophrenia: a multiscale entropy analysis. Neuroimage **51**(1), 173–182 (2010)
7. Epstein, K.A., et al.: White matter abnormalities and cognitive impairment in early-onset schizophrenia-spectrum disorders. J. Am. Acad. Child Adolesc. Psychiatry **53**(3), 362–372 (2014)
8. Douaud, G., et al.: Anatomically related grey and white matter abnormalities in adolescent-onset schizophrenia. Brain **130**(9), 2375–2386 (2007)
9. Cannon, M., et al.: Evidence for early-childhood, pan-developmental impairment specific to schizophreniform disorder: results from a longitudinal birth cohort. Arch. Gen. Psychiatry **59** (5), 449–456 (2002)
10. Azami, H., et al.: Evaluation of resting-state magnetoencephalogram complexity in Alzheimer's disease with multivariate multiscale permutation and sample entropies (2015)
11. Sokunbi, M.O., et al.: Resting state fMRI entropy probes complexity of brain activity in adults with ADHD. Psychiatry Res. **214**(3), 341–348 (2013)
12. Liu, C.Y., et al.: Complexity and synchronicity of resting state blood oxygenation level-dependent (BOLD) functional MRI in normal aging and cognitive decline. J. Magn. Reson. Imaging Jmri **38**(1), 36 (2013)
13. Lai, M.C., et al.: A shift to randomness of brain oscillations in people with autism. Biol. Psychiatry **68**(68), 1092–1099 (2010)
14. Yang, A.C., et al.: Decreased resting-state brain activity complexity in schizophrenia characterized by both increased regularity and randomness. Hum. Brain Mapp. **36**(6), 2174–2186 (2015)
15. Gómez, C., et al.: Complexity analysis of resting-state MEG activity in early-stage Parkinson's disease patients. Ann. Biomed. Eng. **39**(12), 2935–2944 (2011)

16. Accardo, A., et al.: Use of the fractal dimension for the analysis of electroencephalographic time series. Biol. Cybern. **77**(5), 339–350 (1997)
17. Pincus, S.M., et al.: Hormone pulsatility discrimination via coarse and short time sampling. Am. J. Physiol. **277**(5 Pt 1), E948–E957 (1999)
18. Fernández, A., et al.: Complexity and schizophrenia. Prog. Neuro-Psychopharmacol. Biol. Psychiatry **45**, 267–276 (2012)
19. Marx, E., et al.: Eyes open and eyes closed as rest conditions: impact on brain activation patterns. Neuroimage **21**(4), 1818–1824 (2004)
20. Pang, Y., et al.: Extraversion modulates functional connectivity hubs of resting-state brain networks. J. Neuropsycho. (2015)
21. Wei, L., et al.: Specific frequency bands of amplitude low-frequency oscillation encodes personality. Hum. Brain Mapp. **35**(1), 331–339 (2014)
22. Power, J.D., et al.: Spurious but systematic correlations in functional connectivity MRI networks arise from subject motion. Neuroimage **59**(3), 2142–2154 (2012)
23. Friston, K.J., et al.: Statistical parametric maps in functional imaging: a general linear approach. Hum. Brain Mapping **2**(4), 189–210 (1994)
24. Tomasi, D., Volkow, N.D.: Aging and functional brain networks. Mol. Psychiatry **17**(5), 549–558 (2012)
25. Fan, L., et al.: The human brainnetome atlas: a new brain atlas based on connectional architecture. Cereb. Cortex **26**(8), 3508 (2016)
26. Eickhoff, S.B., Constable, R.T., Yeo, B.T.: Topographic organization of the cerebral cortex and brain cartography. Neuroimage (2017)
27. Richman, J.S., Moorman, J.R.: Physiological time-series analysis using approximate entropy and sample entropy. Am. J. Physiol.-Heart Circ. Physiol. **278**(6), H2039–H2049 (2000)
28. Bowden, N.A., Scott, R.J., Tooney, P.A.: Altered gene expression in the superior temporal gyrus in schizophrenia. BMC Genom. **9**(1), 1–12 (2008)
29. Kasai, K., et al.: Progressive decrease of left superior temporal gyrus gray matter volume in patients with first-episode schizophrenia. Am. J. Psychiatry **160**(1), 156–164 (2003)
30. Whitford, T.J., et al.: Structural abnormalities in the cuneus associated with Herpes Simplex Virus (type 1) infection in people at ultra high risk of developing psychosis. Schizophr. Res. **135**(1–3), 175 (2012)
31. Pearlson, G.D.: Superior temporal gyrus and planum temporale in schizophrenia: a selective review. Prog. Neuro-Psychopharmacol. Biol. Psychiatry **21**(8), 1203 (1997)
32. Plaze, M., et al.: Left superior temporal gyrus activation during sentence perception negatively correlates with auditory hallucination severity in schizophrenia patients. Schizophr. Res. **87**(1–3), 109–115 (2006)
33. Barta, P.E., et al.: Auditory hallucinations and smaller superior temporal gyral volume in schizophrenia. Am. J. Psychiatry **147**(11), 1457–1462 (1990)
34. Camchong, J., et al.: Common neural circuitry supporting volitional saccades and its disruption in schizophrenia patients and relatives. Biol. Psychiatry **64**(12), 1042–1050 (2008)

Component Selection in Blind Source Separation of Brain Imaging Data

Xue Wei, Ming Li, Lin Yuan, and Dewen Hu$^{(\boxtimes)}$

College of Mechatronics and Automation,
National University of Defense Technology, 109 Deya Road, Changsha 410073,
Hunan, People's Republic of China
dwhu@nudt.edu.cn

Abstract. Brain imaging technology has been wildly used in neuroscience field. Because original imaging data usually have high dimensionality and a high noise level, dimensionality reduction is generally required to remove noises and retain signals of interest. However conventional dimensionality reduction methods always carry the risk of discarding valuable signals and retaining useless noises. Here we propose a method for component identification to retain only the valuable components. This method is based on the physiological phenomenon in which the intensity of the signal of interest is changed after stimulus onset and is evaluated using stimulated data. The results indicate that the proposed method is valid to distinguish valuable components from all components, retain only the valuable components and improve the signal-to-noise (SNR) of raw imaging data.

Keywords: Blind source separation · Component identification · Dimensionality reduction

1 Introduction

Brian imaging technology, such as functional magnetic resonance imaging (fMRI) and intrinsic optical imaging (OI), has recently received a great deal of attentions because of its wide application in neuroscience [1, 2]. However, brain imaging data are generally high-dimensional and contaminated by physiological noises and random noises which are always stronger than the signals of interest [3, 4]. Thus, a valid data analysis method is required to discriminate signals of interest from noise signals, remove the noise signals and retain the signals of interest to improve the SNR of the imaging data.

Generally, signal discrimination and selection is accomplished via the dimensionality reduction of raw imaging data. In the dimensionality reduction step, raw data are decomposed into multiple ordered components, and the first several components are considered as valuable components and are retained [5, 6]. The other signals are discarded because they are considered useless. Such component selection methods based on ordering (CSO) suffer from two drawbacks: (i) the number of retained components is critical to the reduction. If the retained number is too small, some valuable components may be discarded. However the large number may result in including useless components and overfitting problems may occur [7–9]. Besides, the

© Springer International Publishing AG 2017
Y. Sun et al. (Eds.): IScIDE 2017, LNCS 10559, pp. 589–596, 2017.
DOI: 10.1007/978-3-319-67777-4_53

accurate retained number should be estimated prior to reduction, which is difficult. Researchers have proposed some criteria to estimate the retained number [10, 11]. But the hypotheses of these criteria are not always in agreement with real conditions [12–14]. Thus, accurately estimating the retained number is still a challenging problem. (ii) CSO may not be able to retain all the valuable components and remove all the useless components even with the accurate retained number, because it is difficult to distinguish valuable components from useless components based on the ordering. For example, in Principal Component Analysis (PCA), the ordering is based on the variance. however it is known that components with large variance may contain some physiological noises, and components with small variance may contain some desired signals [9, 15]. In Canonical Component Analysis (CCA), the ordering is based on the autocorrelation, however the physiological noises, such as the respiration signal and heartbeat signal, are highly autocorrelated, as well as the signals of interest [16]. Therefore, regardless of whether the number is accurate or not, there is always a possibility that the components retained in CSO do not contain all the signals of interest and contain some noises. Thus, a more sophisticated data analysis method is required that can identify whether each decomposed component is valuable, namely, stimulus-related or task-related.

In this work, a component selection method based on identification (CSI) was proposed. CSI relies on the hypothesis that the intensity of the response signal has different means before and during the stimulus. After spatially decomposing the original data, we examined the differences of the average value between the before-stimulus and during-stimulus timecourse-sections for each spatial component. If the difference was significant, the corresponding component was identified as stimulus-related and was selected. Otherwise, the component was discarded. The comparison between the proposed method (CSI) and the conventional method (CSO) showed that CSI was valid enough to discriminate and extract the desired signals and improve the SNR of the raw imaging data.

2 Methods

Recorded brain imaging data are always a sequence of images/volumes and can be described as a matrix $X_{n \times N}$, where n means the time points, and N means the number of total pixels/voxels obtained at a time point. Here, the matrix X has been centralized.

On the basis of a typical blind source separation model (BSS), recorded brain imaging data can be regarded as a linear combination of spatial source signals, and the matrix X can be described as

$$X = A \times S = a_1 s_1^T + a_2 s_2^T + \cdots + a_n s_n^T \tag{1}$$

where $X = [x_1, x_2, \cdots, x_n]^T$, $S = [s_1, s_2, \cdots, s_n]^T$, $A = [a_1, a_2, \cdots, a_n]$. Vector $x_i, i = 1, 2, \cdots n$, represents the image/volumes obtained at a time point. Vector $s_i, i = 1, 2, \cdots n$, represents the spatial component, which can be regarded as either the source image/volume, and contains response to stimulus or undesirable noises. Vector $a_i, i =$

$1, 2, \cdots, n$, contains mixing coefficients of s_i at n sample time points and can be considered as the timecourse of s_i.

Because of the changed intensity of the response signal after stimulus onset, it is inferable that for the stimulus-related component s_i, a_i is a timecourse related to the stimulus. Conversely, a_i is a timecourse independent of the stimulus for stimulus-unrelated component s_i. Thus, the identification of stimulus-related component s_i can be accomplished by testing whether a_i is a stimulus-related timecourse or not. Based on the changed intensity, we believe that the timecourse a_i of the stimulus-related component s_i is mutated at the stimulus-starting moment (data in a_i remains at almost zero before stimulus onset and changes significantly after stimulus onset), which indicates that for the stimulus-related timecourse a_i, there is a distinguished mean difference between the data in the before-stimulus section and the data in the during-stimulus section.

Based on the underlying information in the stimulus-related timecourse, two-sample t-test is applied to every vector a_i to test whether the data in the before-stimulus section and the during-stimulus section have a significant mean difference. Then, we retain the spatial components with test results showing that the corresponding time-courses have significant differences and reconstruct the image/volume sequence with the retained components. The whole method is illustrated in Fig. 1.

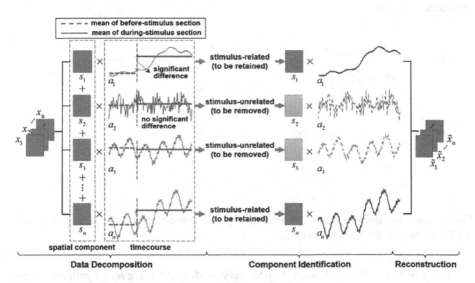

Fig. 1. The flow chart of the CSI. The CSI contains three steps, including data decomposition, component identification and reconstruction. The first step is to decompose the original imaging data into multiple spatial components by BSS methods and obtain the timecourses corresponding to each component. For stimulus-related components, the corresponding timecourses should express significant differences before and during the stimulus. And all the components are tested one by one in the second step. The spatial component with a corresponding timecourse that is significantly different is identified as a valuable component and retained. Otherwise, the component is removed. The third step includes reconstructing imaging data with the retained spatial components and corresponding timecourses.

3 Result

To demonstrate the improvement obtained from the component identification, the comparison between the conventional method (CSO) and the proposed method (CSI) was performed on stimulated data.

The stimulated data was constructed by digitally adding a known activated pattern to the image sequence of a rat somatosensory cortex with no stimulus presentation. The image sequence was acquired in an actual experiment where the rat was at rest state and severed as background noises. The details of the animal preparation and surgery were previously published [17]. The image sequence contains 300 frames and the image size is 150 × 180 pixel. The spatial distribution of the activated pattern is a rotationally symmetric Gaussian hill in a 40 × 40 pixels region (the standard deviation is 20) which is shown in Fig. 2a and its time course is shown in Fig. 2b. The activated pattern played the role of the response signal which we attempted to retain. To mimic the low SNR of real neuronal response, the activated pattern was added at an extremely low level (the SNR was set to 0.05 which was defined as the ratio between the standard deviation of the embedded response signal and the background noise.). There are 15400 different positions in the background noise $((150 - 40) * (180 - 40) = 15400)$ where the activated pattern can be embedded and we selected regularly 667 positions to embed the response signal. Therefore we obtained 667 different synthetic image sequences in total.

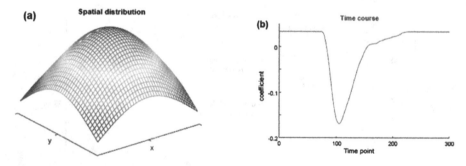

Fig. 2. The embedded activated pattern. (a) Spatial distribution of the activated pattern; (b) Timecourse of the activated pattern

The CSO and the CSI were respectively performed on every synthetic image sequence to remove the useless components and retain the stimulus-related components. Then we extracted the activated maps from the reconstructed data via first frame analysis, in which the activated pixels were greater or less than the inactivated pixels. According to the pixel value in the activated map, we could label every pixel whether it belonged to the activated region by setting a threshold value. Then we could obtain a ROC curve for every activated map due to the known actual active region of every synthetic image sequence. For the extracted map which is similar with the known activated map, the corresponding ROC curve will be close to the upper left corner of the ROC curve. Therefore the performances of the two methods can be appraised by calculating the area under the ROC curve (AUC). The larger AUC means the stimulated active map have been determined more exactly and indicated better performance of the corresponding method. In the CSO, the first 20 principal components were retained as valuable components and in first frame analysis, the first 75 time points were considered as before-stimulus section, the next 125 time points were considered as during stimulus section and the last 100 time points were considered as stimulus-interval section.

Figure 3a shows the result of the component identification in CSI. The dark diamond indicates the valuable component that should be retained and the light diamond indicates the useless component that should be removed. There are 60 rows corresponding to the 60 image sequences which were selected randomly from the 667 sequences and each row shows the first 30 principal components of an image sequence. As shown, the valuable components and useless components are crisscrossed in a row and cannot be distinguished by the PC ordering. Therefore PC ordering is not a good criterion for the selection of valuable component and regardless of the chosen retained number, the CSO method cannot guarantee that all the stimulus-related components are retained and all the noises are removed. To demonstrate the improvement brought by CSI on every single synthetic image sequence, we calculated the AUCs of the activated maps obtained from the same image sequence with different methods and then took the difference of these two AUCs. The difference greater than zero indicates that the proposed method CSI retains more valuable components and determines more accurate

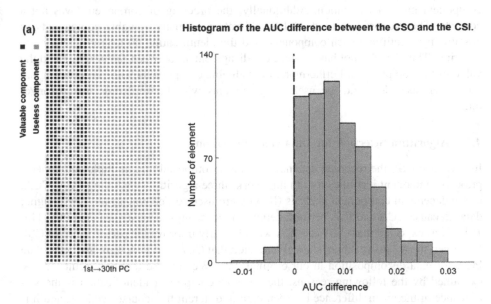

Fig. 3. Results from the CSI method. (a) Results of the component identification in CSI. Each row of diamonds shows the ordered components of an image/volume sequence (only the first 30 components are given). The mosaic of black and grey demonstrates that the stimulus-related and the stimulus-unrelated components cannot be successfully distinguished by ordering. (b) The histogram of the difference of the AUCs between the CSO and the CSI.

activated map. The histogram of the differences of the 667 synthetic image sequences was shown in Fig. 3. The mean value of the differences is 0.0084 and the number of difference greater than zero is 621.

4 Discussion

A new method is proposed to extract the signals of interest and discard noises from original brain imaging data. The greater AUC of the extracted map processed by CSI shows that the proposed method can be successfully used to improve the SNR of raw imaging data and obtain accurate brain imaging results.

4.1 The Proposed Method CSI Can Avoid Estimation of Reduced-Dimension Number

In CSO, the number of retained components is critical and there is always possibility excluding useful signals that are contained in the last-ordering components and including useless signals that are contained in the first-ordering components even with the accurate estimated number.

In our work, the inaccurate retained-component number was not a concern. We identified all the components one by one and then retained or removed components according to the results of the timecourse test. The retained-component number was not required in our method; thus, we did not have to face the problem of the retained-component number estimation. Additionally, the ordering of components was not a concern. The proposed method, CSI, could retain as many valuable components as possible by identifying each component, and the identification no longer relies on the ordering. Thus, we did not have to face the disagreement between the ordering and the value of the components. Furthermore, the valuable components with last-ordering are no longer discarded, and the useless components with first-ordering are no longer retained.

4.2 Algorithm Selection for Data Decomposition

In CSO and CSI, the recorded data needs to be decomposed into multiple components prior to component identification. In this work, three algorithms including PCA, CCA, and independent component analysis (ICA) were used to decompose the raw imaging data. Because CCA and ICA perform better at recovering source signals than PCA [17, 18], it was expected that CCA and ICA would clearly improve the performance of CSI. However, the results (data not shown) indicated that CCA and ICA were not better than PCA for data decomposition in our original data. We surmise that this result may be explained by the following reasons: the basis of component identification is the significance of the mean difference between data in different timecourse-sections, and it is effective enough to distinguish the valuable components from the useless ones. Even if PCA does not work as well as CCA and ICA do at recovering source signals, the retained components decomposed by these three methods may contain similar useful information. It also indicates that the proposed CSI method is robust to algorithm selection for data decomposition.

5 Conclusion

Valuable and useless components cannot be distinguished from each other by ordering offered by the BSS methods; thus, the CSO method carries the risk of removing the useful components and retaining noise sources. In this paper, we attempted to use the information contained in the timecourses corresponding to each spatial component to evaluate the value of the component. By testing the mean difference between the before- and during-stimulus timecourse-sections, stimulus-related components were distinguished from stimulus-unrelated components. And we retain all the response signals and remove all the physiological noises according to the test results. The comparison results showed that the CSI method was an effective tool for the extraction of the signal of interest and the imaging results processed by CSI is more accurate.

References

1. Poldrack, R.A.: The future of fMRI in cognitive neuroscience. Neuroimage **62**, 1216–1220 (2012)
2. Lin, A.J., Koike, M.A., Green, K.N., Kim, J.G., Mazhar, A., Rice, T.B., Laferla, F.M., Tromberg, B.J.: Spatial frequency domain imaging of intrinsic optical property contrast in a mouse model of Alzheimer's disease. Ann. Biomed. Eng. **39**, 1349–1357 (2011)
3. Grinvald, A., et al.: In-vivo optical imaging of cortical architecture and dynamics. In: Windhorst, U., Johansson, H. (eds.) Modern Techniques in Neuroscience Research, pp. 893–969. Springer, Heidelberg (1999). doi:10.1007/978-3-642-58552-4_34
4. Ribot, J., Tanaka, S., Tanaka, H., Ajima, A.: Online analysis method for intrinsic signal optical imaging. J. Neurosci. Methods **153**, 8–20 (2006)
5. Zheng, Y., Johnston, D., Berwick, J., Mayhew, J.: Signal source separation in the analysis of neural activity in brain. Neuroimage **13**, 447–458 (2001)
6. Kyathanahally, S.P.: Blind Source Separation Methods for Analysis and Fusion of Multimodal Brain Imaging Data. Auburn University, Auburn (2013)
7. Abou-Elseoud, A., Starck, T., Remes, J., Nikkinen, I., Tervonen, O., Kiviniemi, V.: The effect of model order selection in group PICA. Hum. Brain Mapp. **31**, 1207–1216 (2010)
8. Esposito, F., Goebel, R.: Extracting functional networks with spatial independent component analysis: the role of dimensionality, reliability and aggregation scheme. Curr. Opin. Neurol. **24**, 378–385 (2011)
9. Li, M., Liu, Y., Chen, F., Hu, D.: Including signal intensity increases the performance of blind source separation on brain imaging data. IEEE Trans. Med. Imaging **34**, 551–563 (2015)
10. Cordes, D., Nandy, R.R.: Estimation of the intrinsic dimensionality of fMRI data. Neuroimage **29**, 145–154 (2006)
11. Højen-Sørensen, P.A.D.F.R., Winther, O., Hansen, L.K.: Analysis of functional neuroimages using ICA with adaptive binary sources. Neurocomputing **49**, 213–225 (2002)
12. Li, Y., Adal, T., Calhoun, V.D.: Estimating the number of independent components for functional magnetic resonance imaging data. Hum. Brain Mapp. **28**, 1251–1266 (2007)

13. Li, X., Ma, S., Calhoun, V.D., Adali, T.: Order detection for fMRI analysis: joint estimation of downsampling depth and order by information theoretic criteria. In: IEEE International Symposium on Biomedical Imaging: From Nano to Macro, pp. 1019–1022 (2011)

14. Hui, M., Li, R., Chen, K., Jin, Z., Yao, L., Long, Z.: Improved estimation of the number of independent components for functional magnetic resonance data by a whitening filter. IEEE J. Biomed. Health Inform. **17**, 629–641 (2013)

15. Stetter, M., Schießl, I., Otto, T., Sengpiel, F., Hübener, M., Bonhoeffer, T., Obermayer, K.: Principal component analysis and blind separation of sources for optical imaging of intrinsic signals. Neuroimage **11**, 482–490 (2000)

16. Friman, O., Borga, M., Lundberg, P., Knutsson, H.: Exploratory fMRI analysis by autocorrelation maximization. Neuroimage **16**, 454–464 (2002)

17. Li, M., Liu, Y., Feng, G., Zhou, Z., De, H.: OI and fMRI signal separation using both temporal and spatial autocorrelations. IEEE Trans. Bio-med. Eng. **57**, 1917–1926 (2010)

18. De Clercq, W., Vergult, A., Vanrumste, B., Van Paesschen, W., Van Huffel, S.: Canonical correlation analysis applied to remove muscle artifacts from the electroencephalogram. IEEE Trans. Bio-Med. Eng. **53**, 2583–2587 (2006)

Gradient Vector Flow Field and Fast Marching Based Method for Centerline Computation of Coronary Arteries

Hengfei Cui[1,2] and Yong Xia[1,2(✉)]

[1] Shaanxi Key Lab of Speech and Image Information Processing (SAIIP),
School of Computer Science and Engineering, Northwestern Polytechnical University,
Xi'an 710072, PR China
[2] Centre for Multidisciplinary Convergence Computing (CMCC),
School of Computer Science and Engineering, Northwestern Polytechnical University,
Xi'an 710072, PR China
yxia@nwpu.edu.cn

Abstract. This paper develops new concept of validating centerline extraction method of coronary arteries. The approach is based on the gradient vector flow (GVF) filed of the 3D segmented coronary arteries models. It is implemented with the Gaussian based speed image. The approach was validated over 3 three-dimensional synthetic vessel models and further tested in 3 clinical coronary arteries models reconstructed from computed tomography coronary angiography (CTCA) in human patients. The results showed an excellent agreement between the proposed method and ground truth centerline in synthetic vessel models. Second, the proposed method was applicable in both left coronary arteries and right coronary arteries with average processing time of 25.7 min per case. In conclusion, the proposed gradient vector flow field and fast marching based method should have more routine clinical applicability.

Keywords: Gradient vector flow · Fast marching method · Coronary centerline

1 Introduction

Coronary centerline computation is one of the major preprocessing steps required for coronary artery stenosis quantification. In the present context, many centerline extraction approaches compute the Euclidean distance field and thus construct a medial function to extract vessel centerlines [1–3]. At each voxel location, the distance field computes the nearest distance from the vessel's boundary. However, the distance field is normally computed at voxel level, which results in that the centerline is not smooth. Post-processing steps, such as smoothing and pruning, are therefore required. In 3D, the distance field may form medial surfaces rather than medial curves for non-tubular structures.

Hassouna *et al.* [4] have proposed a new method for computing centerlines of 3D tubular structures, which combines level set method and wave propagation

© Springer International Publishing AG 2017
Y. Sun et al. (Eds.): IScIDE 2017, LNCS 10559, pp. 597–607, 2017.
DOI: 10.1007/978-3-319-67777-4_54

technique. One of the contributions of this method is the usage of the complex cost function which requires an α and β term to determine the number of branches and how the branches merge. However, these two terms are dataset dependent and must be set manually, and their interaction is nonintuitive. The speed parameters need to be selected heuristically, which prevents the algorithm from automation. When very large datasets are given as input, the processing time required for computing the distance map and performing the fast marching calculations becomes increasingly large.

This paper addresses the aforementioned disadvantages and presents a fast and accurate algorithm for computing continuous centerlines from 3D coronary artery segmentation results. The feasibility of the proposed method is demonstrated by experiments tested on both synthetic and real CTCA images. The rest of this paper is organized as follows. In Sect. 2 we describe the proposed centerline computation method. Section 3 gives the experimental results and discussions. Finally, we present our conclusions in Sect. 4.

2   Methodologies

2.1   Edge Map

The edge map [5] $f(x, y)$ of the original image $I(x, y)$ is defined as

$$f^{(1)}(x, y) = |\nabla I(x, y)| \tag{1}$$

or

$$f^{(2)}(x, y) = |\nabla[G_\sigma(x, y) * I(x, y)]| \tag{2}$$

where $G_\sigma(x, y)$ is the 2D Gaussian function. The latter one is more applicable to medical images since it is more robust in the presence of noise. There are generally some important properties of edge maps. In homogeneous regions, the gradient of an edge map ∇f is approximated to zero. At the edge locations, the ∇f vectors have large magnitudes and point toward the edges.

2.2   Gradient Vector Flow

The GVF field $\mathbf{v}(x, y) = [u(x, y), v(x, y)]$ is generally defined to minimize the following energy functional [6]

$$E_{GVF}(\mathbf{v}) = \iint \mu|\nabla\mathbf{v}|^2 + |\nabla f|^2|\mathbf{v} - \nabla f|^2 dx dy \tag{3}$$

where μ is the smoothness regularization parameter. In (3), if $|\nabla f(x)|$ is small (homogeneous region), the energy is dominated by the first term, which is used to enlarge the capture range of the force field and smooth the vector field $\mathbf{v}$. If $|\nabla f(x)|$ is large, the energy is dominated by the second term and minimized by choosing $\mathbf{v} = \nabla f(x)$.

Using the calculus of variations [7], the GVF field must satisfy

$$\mu\nabla^2\mathbf{v}(t) - (\mathbf{v} - \nabla f)|\nabla f|^2 = 0 \tag{4}$$

where ∇^2 is the Laplacian operator and t is the time variable. The intuition behind (4) is that, in a homogeneous region, $\mathbf{v}$ is determined by Laplace's equation since the second term is zero. Therefore, the resulting GVF field vectors are interpolated from the region's boundary and point into boundary concavities.

The GVF field can be found by introducing the time variable t and solving the following Euler-Lagrange equation

$$\mathbf{v}_t = \mu\nabla^2\mathbf{v}(t) - (\mathbf{v} - \nabla f)|\nabla f|^2 \tag{5}$$

where $\mathbf{v}_t$ is the partial derivative of $\mathbf{v}$ with respect to t.

It is noteworthy that the GVF magnitude $(|\mathbf{v}(x)|)$ does not form medial surfaces for non-tubular 3D objects [8]. This is because that the GVF value at each pixel is computed by diffusion process, from several local contextual boundary points. Hence, these sheets collapse to a set of medial curves. On the other hand, the distance field at a point is computed from only one boundary voxel, which may form sheets of constant distances.

2.3 Speed Image

The magnitude of the gradient of GVF field, i.e., $|\nabla\mathbf{v}(x,y)|$, can be utilized to extract centerline of an object. It is obvious that $|\nabla\mathbf{v}(x,y)|$ is nearly zero where the GVF field is smooth, and has a relatively large value in other places. Therefore, the centerlines and the edge points can be discerned from other smooth GVF field by checking the value of $|\nabla\mathbf{v}(x,y)|$. For the purpose of extracting centerlines only, the edges need to be further removed after separating the smooth area. Therefore, an edge indicator function [9] is defined

$$g(x,y) = \frac{1}{1 + f(x,y)} \tag{6}$$

which is used to multiply with the magnitude of the gradient of GVF field since the edge function is very small at the strong edge locations and equals to 1 in other places. The centerline strength function is thus defined as

$$k(x,y) = g(x,y) * |\nabla\mathbf{v}(x,y)| \tag{7}$$

then normalizing it to $[0,1]$,

$$\bar{k}(x,y) = \left(\frac{k(x,y) - \min(k)}{\max(k) - \min(k)}\right)^\gamma \tag{8}$$

where γ is the field strength in $[0,1]$.

Therefore, we propose a new speed image based on $\bar{k}(x)$ [4] for the fast marching method

$$F = e^{\beta\bar{k}} \tag{9}$$

where β is a speed parameter defined as $\beta = 1/\tau$ and τ is the parameter. It is noted that the proposed speed image F is computationally efficient since it is directly computed from the GVF and there is no extra computation required.

2.4   Branch Tracing

Based on the proposed centerline strength function $\bar{k}(x)$, the source point P_s is chosen as the point with maximal value in $\bar{k}(x)$:

$$P_s = \arg \max_x \bar{k}(x) \tag{10}$$

Then a wave front is propagated using the new speed image F from the source P_s. The resulting map T can be treated as a modified distance map from medial function $\bar{k}(x)$, using a non Euclidean metric that gives larger value in the center of the vessel.

Next, the start point m_0 is chosen as the point with maximum T value. The first branch centerline is then computed by tracking discrete points along the gradient descent in T, which is formulated as

$$\frac{dS}{dt} = -\frac{\nabla T}{|\nabla T|}, \quad S(0) = m_0 \tag{11}$$

where m_0 is the furthest geodesic point and $S(t)$ represents the discrete center-line. The classical implicit Runge-Kutta method [11] is explored to solve this stiff ordinary differential equation accurately, which is given as

$$S_{n+1} = S_n + \sum_{i=1}^{s} b_i k_i \tag{12}$$

where

$$k_i = hf(t_n + c_i h, S_n + \sum_{j=1}^{s} a_{ij} k_j) \tag{13}$$

here t_n is the time step and h is the step size.

The two stages Gauss-Legendre method (order of 4) is used to achieve high accuracy. It can be shown that [10]:

$$k_1 = hf\left(t_n + (\frac{1}{2} - \frac{1}{6}\sqrt{3})h, S_n + \frac{1}{4}k_1 + (\frac{1}{4} - \frac{1}{6}\sqrt{3})k_2\right) \tag{14}$$

$$k_2 = hf\left(t_n + (\frac{1}{2} + \frac{1}{6}\sqrt{3})h, S_n + (\frac{1}{4} + \frac{1}{6}\sqrt{3})k_1 + \frac{1}{4}k_2\right) \tag{15}$$

Given the start point m_0, initial guesses of k_1 and k_2, denoted as k_1^0 and k_2^0, were linearly interpolated from the 8 nearest neighbours of m_0 in the gradient field ∇T and then used for the first branch tracking. To improve the efficiency, an iterative method [11] was used to solve this system of nonlinear equations. All the remaining branches can be determined following the same tracking process. When the length of the branch is less than some pre-defined threshold value L, the multiple branch formation procedure terminates and this new branch is excluded from the final centerline. Algorithm 1 shows the pseudo-code of the whole procedure.

Algorithm 1. Centerline extraction algorithm

1: **Input:** Binary segmentation result V
2: **Output:** Discrete centerline S_f
3: $S_f = \emptyset, linelength = \infty$
4: $P_s = \arg \max_x \bar{k}(x)$
5: $i = 0$
6: **while** $linelength > L$ **do**
7:      $T = FastMarching(P_s, F)$
8:      $m_i = x \mid T(x) \geq T(y) \; \forall x, y \in O$
9:      $S = backtrace(m_i, P_s)$
10:     $linelength = length(S)$
11:     $P_s = P_s + S, S_f = S_f + S$
12:     $i = i + 1$
13: **end while**

3 Results and Discussion

In the experiments, the edge maps were computed by using Eq. (2). The parameter $\mu = 0.2$ and iteration $n = 10$ were chosen for GVF. The field strength parameter $\gamma = 1$ was used. Figure 1 demonstrates the computed edge maps of a 2D vessel cross section, for different values of parameter σ. It is observed that central pixel locations are discerned when σ is selected as 2. Figure 2 shows a 2D example of the edge map and the gradient vector flow field, with $\sigma = 0.5$.

Moreover, a quantitative analysis of the accuracy of the new method was performed. A total of three vessel models (Vessel 1–3) were selected, each of which was provided the ground truth centerline and the corresponding radius at each

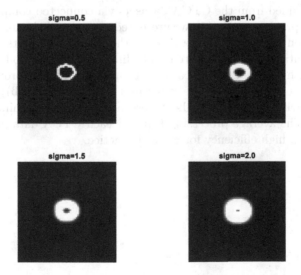

Fig. 1. Edge maps for different values of parameter σ.

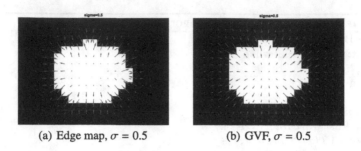

(a) Edge map, $\sigma = 0.5$ (b) GVF, $\sigma = 0.5$

Fig. 2. A 2D example of edge map and GVF field.

skeleton position [12]. The comparison of the distances and overlap scores from the ground truth centerline was also conducted for both the presented method and our previous method [1]. Figure 3 shows the obtained and the ground truth centerlines for three different synthetic vessel models for visual comparison. The final centerlines are continuous, smooth and centrally located with the vessel objects. In Table 1, we also compare the processing time, overlap measure, average error and maximum error for the computed centerlines of three different vessel models, using two different methods with step size of 0.01. The presented method is capable of reducing the processing time by 40.2%, and at the same time, achieve higher overlap measure and smaller average and maximum error.

A Hessian filter based vessel segmentation method was applied in this study into three different CTCA datasets in DICOM format (ccta 1, 14 and 22). Table 2 summarizes the dataset size, resolution and computation time of each CTCA dataset. Complete 3D segmentation results of coronary artery trees (left circumflex artery, LCX; left anterior descending artery, LAD; and right coronary artery, RCA) were obtained from the CTCA datasets via connected components based method. In our previous work [13], we presented more detailed information about the segmentation approach. Figure 4 shows the centerline results for three different CTCA datasets. The computed centerlines are observed to be continuous, naturally smooth and centrally located within all the major coronary arteries. In addition, no extra branches are generated by noise (Fig. 4(b)) because the centerline is tracked along the gradient descent in T and a termination criteria is introduced. The average processing time for each CTCA dataset is 25.7 min, which provides a high efficiency for clinical practice.

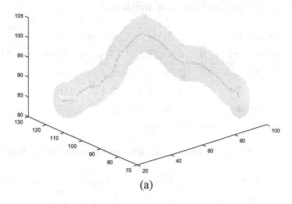

(a)

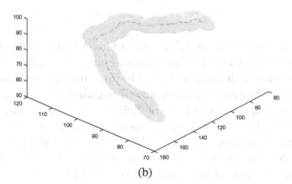

(b)

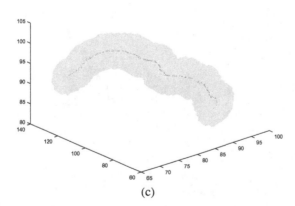

(c)

Fig. 3. Computed (*blue*) and ground truth (*red*) centerlines of 3 synthetic vessel models. (a) Vessel 1. (b) Vessel 2. (c) Vessel 3. (Color figure online)

Table 1. A comparison of processing time, overlap, average error and maximum error between the previous method and the new method.

	Previous method				New method			
	Time (s)	Overlap (%)	L_2	L_∞	Time (s)	Overlap (%)	L_2	L_∞
Vessel 1	36.2	96.1	0.58	1.33	21.4	98.6	0.47	1.19
Vessel 2	37.8	95.3	0.53	1.21	22.8	98.3	0.44	1.07
Vessel 3	36.9	94.4	0.49	1.27	22.1	97.1	0.41	1.08

Note: L_2–Average error, L_∞–Maximal error.

Table 2. Dataset size, resolution and computation time.

	Size	Resolution (mm³)	Time (min)
ccta 1	$320 \times 320 \times 460$	$0.439 \times 0.439 \times 0.25$	23
ccta 14	$400 \times 400 \times 470$	$0.31 \times 0.31 \times 0.25$	28
ccta 22	$400 \times 400 \times 426$	$0.319 \times 0.319 \times 0.3$	26

In the proposed method, the GVF based speed image is capable of driving the centerline along ridges in the time-crossing map. The algorithm is designed for tubular structures with cross sections that are roughly circular and requires no user interaction. We can vary the value of parameter γ and apply (8) on the CTCA image, to explore the effect of the field strength. It can be observed that the centerline strength will be enhanced when decrementing the field strength γ. Finally, the efficiency of this approach is higher since a lower number of computations are needed (one distance map versus two [1]), and the branch tracking procedure is more efficient by using the proposed GVF based speed image.

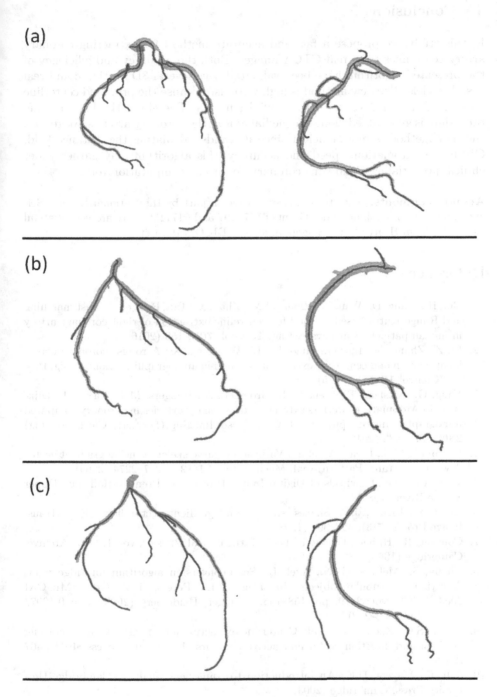

Fig. 4. Computed centerlines (*red*) for LAD and LCX (*left panel*), and RCA (*right panel*). (a) ccta 1. (b) ccta 14. (c) ccta 22. (Color figure online)

4   Conclusions

In this study, we propose a fast and accurate method for extracting coronary artery centerlines from real CTCA images. Both the accuracy and efficiency of the presented algorithm have been validated over several 3D synthetic and real vessel models. The new method is highly robust because the proposed centerline strength function does not form medial surfaces. Therefore, the back tracing procedure is guaranteed along the medial axis of the coronary arteries. Moreover, the new method is more efficient since it avoids calculating the distance field. Given the automation, speed and accuracy, this algorithm may garner wider clinical potential as a real-time coronary centerline computation tool.

Acknowledgments. The study was supported in part by the National Natural Science Foundation of China under Grants 61471297 and 61771397. We are very grateful to the National Heart Centre Singapore for the DICOM datasets.

References

1. Cui, H., Wang, D., Wan, M., Zhang, J.M., Zhao, X., Tan, R.S., et al.: Fast marching and Runge-Kutta based method for centreline extraction of right coronary artery in human patients. Cardiovasc. Eng. Technol. **7**(2), 159 (2016)
2. Li, Z., Zhang, Y., Liu, G., Shao, H., Li, W., Tang, X.: A robust coronary artery identification and centerline extraction method in angiographies. Biomed. Sig. Proces. Control **16**, 1–8 (2015)
3. Yang, G., Kitslaar, P., Frenay, M., Broersen, A., Boogers, M.J., Bax, J.J., Dijkstra, J.: Automatic centerline extraction of coronary arteries in coronary computed tomographic angiography. Int. J. Cardiovasc. Imaging (Formerly Card. Imaging) **28**(4), 921–933 (2012)
4. Hassouna, M.S., Farag, A., et al.: Variational curve skeletons using gradient vector flow. IEEE Trans. Pattern Anal. Mach. Intell. **31**(12), 2257–2274 (2009)
5. Jain, A.K.: Fundamentals of Digital Image Processing. Prentice-Hall Inc., Upper Saddle River (1989)
6. Xu, C.Y., Prince, J.L.: Snakes, shapes, and gradient vector flow. IEEE Trans. Image Proces. **7**(3), 359–369 (1998)
7. Courant, R., Hilbert, D.: Methods of Mathematical Physics, vol. 1. CUP Archive, Cambridge (1966)
8. Chang, S., Metaxas, D.N., Axel, L.: Scan-conversion algorithm for ridge point detection on tubular objects. In: Ellis, R.E., Peters, T.M. (eds.) MICCAI 2003. LNCS, vol. 2879, pp. 158–165. Springer, Heidelberg (2003). doi:10.1007/978-3-540-39903-2_20
9. Zhang, S.Q., Zhou, J.Y., et al.: Centerline extraction for image segmentation using gradient and direction vector flow active contours. J. Sig. Inf. Process. **4**(04), 407 (2013)
10. Süli, E., Mayers, D.F.: An Introduction to Numerical Analysis. Cambridge University Press, Cambridge (2003)
11. Press, W.H., Teukolsky, S.A., Vetterling, W.T., Flannery, B.P.: Numerical Recipes in C, vol. 2. Cambridge University Press, Cambridge (1996)

12. Schaap, M., Metz, C.T., van Walsum, T., van der Giessen, A.G., Weustink, A.C., Mollet, N.R., Bauer, C., Bogunović, H., Castro, C., Deng, X., et al.: Standardized evaluation methodology and reference database for evaluating coronary artery centerline extraction algorithms. Med. Image Anal. **13**, 701–714 (2009)
13. Cui, H., Wang, D., Wan, M., Zhang, J.M., Zhao, X., Tan, S.Y., Wong, A.S.L., Tan, R.S., Huang, W., Xiong, W., Duan, Y., Zhou, J., Zhong, L.: Coronary artery segmentation via hessian filter and curve-skeleton extraction. In: 2014 IEEE Conference on Biomedical Engineering and Sciences (IECBES), pp. 93–98. IEEE (2014)

EEG-Based Motor Imagery Differing in Task Complexity

Kunjia Liu$^{(\boxtimes)}$, Yang Yu, Yadong Liu, and Zongtan Zhou

College of Mechatronic Engineering and Automation,
National University of Defense Technology,
Changsha 410073, Hunan, People's Republic of China
liukunjia12@qq.com, yuyangnudt@hotmail.com,
liuyadong1977@163.com, narcz@163.com

Abstract. In this study, we explored the classification of singlehanded motor imagery (MI) EEG signals with different complexity. Eight healthy participants were asked to complete a finger-tapping task of different complexity. In signal processing, CSP features were extracted from the band-passed EEG signals. Then, these features were used to define a score using the step-wise linear discriminant analysis (SWLDA) method. The classification accuracy was evaluated by a five-fold cross-validation strategy. The experimental results showed that the average accuracy between different complexity is 79.20%, and the highest is up to 80.84%, indicating the separability of EEG-based MI tasks with different complexity. The EEG-based complexity distinction achieved in this paper would encourage further study of the realization of multiclass MI-based BCI paradigm.

Keywords: Brain-computer interface (BCI) · Electroencephalogram (EEG) · Motor imagery · Task complexity · Common spatial pattern (CSP)

1 Introduction

A BCI provides a communication and control system in which an individual sends commands to external world by generating specific patterns of brain signals instead of muscular activities. BCIs are originally developed to facilitate communication for people with limited motor function [1], but they can also be applied to improve interactions between machine and healthy individuals [2, 3]. Many researches have been done to explore various applications of BCIs in different fields, including communication, neuro-prosthetics, virtual reality, robots, robotic arms and mobility control [4–9]. These studies demonstrate the promising possibility of using BCIs to enhance the communication between human-being brains and external devices.

A prevalent type of EEG-based BCI is the motor-imagery (MI)-based BCI, which is based on the sensorimotor rhythm (SMR), where event-related desynchronization (ERD) or event-related synchronization (ERS) occurs directly after the end of a sensorimotor task or mental imagination of different types of movements. Motor imagery is usually seen as mental rehearsal of motor acts without any overt body movements,

© Springer International Publishing AG 2017
Y. Sun et al. (Eds.): IScIDE 2017, LNCS 10559, pp. 608–618, 2017.
DOI: 10.1007/978-3-319-67777-4_55

which allows self-induced brain activities to be interpreted as particular control signals that independently reflect subjective movement-related mental state of the user.

However, a normal two-class MI-based BCI has only two command options available, usually, left hand and right hand. It is still far from enough for MI-based BCI to be put into use, because massive devices such as robots and wheelchairs require more available commands to perform a series of complex actions. Therefore, extending commands is currently a much-addressed issue in the field of MI-based BCI. Many researchers have attempted to gain ground based on the somatotopic organization of the motor cortical areas. One of the typically studied variables is the effector(s) that are used in imagination, e.g., fingers, limbs, feet, elbows, shoulders, and tongue [10–13]. Another focus is mainly based on distinguishing MI tasks with different complexity, which is the main point of our work in this paper, and we believe it will facilitate potential implementations in future multiclass BCI applications.

The possibilities of discriminating MI tasks of different complexity have been well documented in previous functional Magnetic Resonance Imaging (fMRI) and functional Near-Infrared Spectroscopy (fNIRS) studies; e.g., studies by Holper and Wolf [14] showed an average classification accuracy of 81% between simple tasks and complex tasks. However, there is still a general lack of related EEG researches, because EEG data are highly non-linear and non-stationary in nature [15, 16]. Since Jasper and Penfield [17] discovered that MI could induce cortical oscillatory activity, numerous novel experimental protocols and powerful EEG signal processing techniques were presented to improve the performance of EEG-based BCIs.

In this study, we propose a novel EEG-based BCI paradigm based on two types of MI tasks associated with task complexity. It is a novel trying to use EEG signal to distinguish MI tasks with different complexity. Participants were asked to kinesthetically MI (imagine how it feels to do the movement) instead of visual MI (imagine watching the movement) because the former one resulted in detectable EEG changes, which is essential to improve MI-based BCI control [18, 19]. Simple task requires subjects to imagine a simple finger-tapping task using a single finger (index), while complex task requires subjects to perform a MI sequences with five fingers all involved.

2 Materials and Methods

2.1 Motor Imagery Tasks

The experiment consisted of two types of MI tasks, and only right hand was involved.

Simple MI: participants were asked to imagine a simple finger-tapping task by repetitively tapping their index finger with a frequency of approximately 2 Hz.

Complex MI: participants were asked to imagine a complex sequential finger-tapping task using all fingers of their right hand with a frequency of 2 Hz. The tasks are presented on the screen before the trail, including the following three pre-defined sequences: 1–3–5–2–4, 5–1–4–2–3, and 2–4–3–5–1; Numbers indicated fingers: 1 = thumb, 2 = index finger, 3 = middle finger, 4 = ring finger, and 5 = little finger. For example, the sequence 1–3–5–2–4 will be translated into thumb-middle-little-index-ring finger. The sequences were memorized by subjects at the start of each trial. The complex task is similar to that used in various fMRI and fNIRS studies [14, 20, 21].

2.2 Experiment Process

2.2.1 Experimental Setup

The subjects sat in front of a 22" LED monitor that displayed the experimental stimuli in an unshielded room at a comfortable distance of approximately 1 m from the eyes. The subject remained relaxed and avoided any physical motion during performance. The process of experiments was explained to each subject in detail beforehand, and a short familiarization session was performed before the experiments.

2.2.2 Experimental Paradigm

At the beginning of the experiment, the screen was black for 2 s. Then, an instruction cue was presented at the center of the screen to inform the specific MI type (simple or complex MI task), and the specific sequence (if it is a complex task, the sequence was randomly selected from the pre-defined sequences) for the forthcoming trial. The instruction cue remained for 2 s, followed by a 1-s blank screen (preparatory cue). When the 'Start' cue appeared at second 5, the participant did the requested motor imagine for 4 s. There was a four-second break between two sequential trials. Figure 1 illustrates the experimental paradigm.

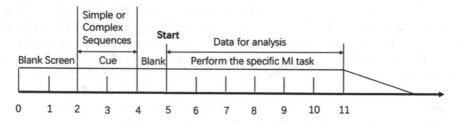

Fig. 1. The experimental procedure starts with a blank screen.

A run consisted of 5 trials of randomized MI task, followed by a 5-m rest break. Every participant performed 20 runs over two successive days. In other word, there were 100 trials of MI tasks per capita for later analysis.

2.2.3 Participants

Eight native Chinese speakers participated in this experiment (including 2 females and 6 males, aged 23–27 years, with a mean age of 25.6 years). All of the subjects were able-bodied, right handed, and did not have any known cognitive deficits. Five of the eight participants had related experience with MI BCI before the experiment. The remaining subjects were complete novices to BCI. All participants received a complete description of this study beforehand.

2.3   Data Analysis

2.3.1   EEG Data Acquisition and Preprocessing

EEG signals were recorded from the scalp by multichannel EEG amplifiers (Brain Products, GmbH, Germany) embedded with 31 electrodes around the sensorimotor cortex, based on the 64-channel modified international 10–20 system. Figure 2 depicts the electrode montage. Each channel was referenced to TP10 (the right mastoid) and grounded to AFz (the forehead). The electrode impedance was maintained below 5 kΩ. The EEG signals were amplified through a BrainAmp DC Amplifier, digitized at a rate of 200 Hz, and filtered by a 50 Hz notch filter. The stimulus presentation, data collection, offline signal processing and experimental procedures were conducted by the BCI2000 framework [22].

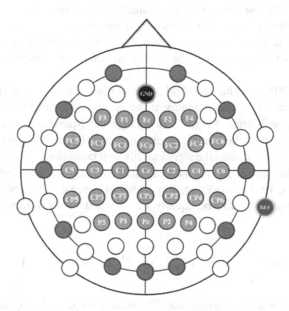

Fig. 2. EEG electrode positions based on the 64-channel modified international 10–20 system.

In this study, all of the raw EEG channels were first band-filtered between 8 and 30 Hz, a span that mainly contains alpha and beta frequency components.

2.3.2   Signal Processing

The signal processing approach is composed of CSP filtering and SWLDA classification. First the EEG signal was band-filtered to 8–30 Hz; then, CSP features are extracted from the band-passed EEG signals; subsequently, a score is defined using the SWLDA method. The block diagram illustrating the process of this experiment is shown in Fig. 3.

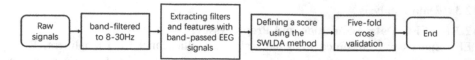

Fig. 3. Processing diagram of this experiment

2.3.3 Common Spatial Pattern (CSP) Method

Raw EEG signals usually have a poor spatial resolution because of volume conduction, but common spatial pattern (CSP) is a widely used spatial filtering technique for tackling this problem. It works well for accurate detection and recognition of brain patterns associated with MI in multichannel EEG data [23, 24]. In this study, it is applied to analyze the band-passed multichannel EEG signals.

The CSP method projects raw signals to a new time series whose variances are optimal for the classification of two classes, namely maximizing the variance of bandpass filtered EEG signals from one class while minimizing their variance from the other class [23].

The aim for CSP algorithm is to find a spatial filter matrix W, which can maximize the variance of one class while minimize the another. The method is based on the simultaneous diagonalization. Let $X^{N \times T}$ denote the band-passed EEG signal in original sensor space, where N is the number of channels and T is the number of samples per channel. $X^c (c = +, -)$ denote the signal in different conditions (simple and complex). Let Σ^+ and Σ^- be the covariance matrices of the band-passed filtered EEG signal in the two conditions. They are calculated by averaging over the trials of each group. The process to find matrix W can be equally transformed to the following optimization problem:

$$\max \sum_{i=1}^{t} W^T \Sigma^+ W, s.t. \sum_{i=1}^{t} W^T (\Sigma^+ + \Sigma^-) W = 1. \tag{1}$$

where T denotes the transpose operator. The composite spatial covariance is given as

$$\Sigma = \Sigma^+ + \Sigma^-. \tag{2}$$

Σ can also be factored as $\Sigma = U \Lambda U^T$, where U is the matrix of eigenvectors and Λ is the matrix of eigenvalues. The whitening transformation can be achieved by

$$P = \Lambda^{-\frac{1}{2}} U^T. \tag{3}$$

Define

$$S^+ = P \Sigma^+ P^T, \ S^- = P \Sigma^- P^T. \tag{4}$$

As a result of whitening transformation,

$$S^+ + S^- = I. \tag{5}$$

If (4) can be satisfied, S^+ and S^- share the same eigenvectors, i.e.

$$\text{If } S^+ = B\Lambda^+ B^T, \text{then } S^- = B\Lambda^- B^T. \tag{6}$$

where eigenvectors matrix B is orthogonal, and matrix $\Lambda^c(c = +, -)$ is diagonal. Based on Eqs. (5) and (6), it can be conducted that

$$\Lambda^+ + \Lambda^- = I. \tag{7}$$

Hence, the eigenvector with the largest eigenvalue for S^+ has the smallest eigenvalue for S^- and vice versa. This makes it convenient to classify the two conditions using eigenvectors matrix B.

Define the spatial filter matrix $W = B^T P$, the filtering of signal X is given as

$$Z = W^T X. \tag{8}$$

The Eq. (8) is used for obtaining the features for classification. W^T is spatial filter matrix, and its every column vector w_j is an independent spatial filter. While the columns of $(W^{-1})^T$ are seen as time-invariant EEG source distribution vectors, which are called the common spatial patterns. For different complexity of imagined movement, the first and last ε rows of Z are most suitable for constructing a feature vectors, because they are associated with the largest eigenvalues, in other words, variances for two conditions. Define signal $Z_p(p = 1...2\varepsilon)$ is combined by the first and last ε rows of Z. Then, the feature vector is normalized by the total variance of the projections retained and log-transformed:

$$f_p = \log(\frac{\text{var}(Z_p)}{\sum_{i=1}^{2\varepsilon} \text{var}(Z_p)}). \tag{9}$$

The feature vectors f_p of simple and complex trails are used to calculate a linear classifier [25, 26].

2.3.4 SWLDA Classification

Since step-wise linear discriminant analysis (SWLDA) is an efficient and frequently-used pattern recognition method in BCI research [27], it will be applied for signal classification. SWLDA is a technique for selecting suitable predictor variables to be included in a multiple regression model. The classification accuracy was evaluated by a five-fold cross-validation strategy.

A combination of forward and backward stepwise regression was implemented. Starting with no initial model terms, the most statistically significant predictor variable having a p-value < 0.1, was added to the model. After each new entry to the model, a backward stepwise regression was performed to remove the least significant variables, having p-values > 0.15. This process was repeated until the model includes a predetermined number of terms, or until no additional terms satisfy the entry/removal criteria [28].

3 Results

The current study presents the results of single-trial classification of MI tasks differing in task complexity using a CSP feature extraction approach. The classification results obtained along with the results of some supporting analyses are provided in this section. We briefly mathematically analyzed the classification accuracy. The simple and complex MI task is a fair extension for traditional left and right classification. We also provided CSP analysis of Subject 4, including the simple vs rest, complex vs rest, and simple vs complex filter and patterns, which is straight-forward for finding the different functional areas in the brain.

3.1 Classification Accuracy

To validate the separability of mental tasks differing in task complexity and test the feasibility of the CSP feature extraction algorithm in this study, we analyzed the classification accuracy using SWLDA. The resulting classification accuracy obtained by the five-fold cross-validation strategy for each subject is summarized in Table 1. The average accuracies distinguishing simple tasks with rest state and complex tasks with rest state are 82.16% and 82.78%, respectively, an average classification accuracy of 79.20% and highest accuracy of 80.84%.

The bar graph of cross-validation classification accuracy of each subjects is displayed as follows. It indicates that for most subjects, except subject 5, simple or complex vs rest has higher separability than simple vs complex. All subjects showed good MI control ability, as the variance of accuracy between subjects for simple vs complex is 0.397, which is mainly because most subjects (5 out of 8) have related experience with EEG MI tasks. This result shows the separability between simple, complex and rest state, which is a good supplement for traditional left and right distinction. By this means, the command options of MI are well extended (Fig. 4).

Table 1. Five-fold cross-validation classification accuracy for each subject

Subjects	Simple vs. rest	Complex vs. rest	Simple vs. complex
S1	83.61	84.87	80.72
S2	82.78	83.26	79.22
S3	81.90	80.78	77.80
S4	84.12	86.12	80.84
S5	79.44	79.03	80.01
S6	82.81	84.12	79.96
S7	80.99	82.06	76.23
S8	81.63	82.02	78.78
Mean	82.16	82.7825	79.195

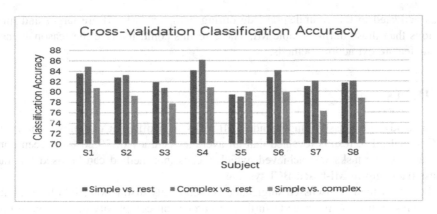

Fig. 4. Cross-validation classification accuracy

3.2   Spatial Patterns in CSP

The spatial patterns of MI tasks obtained using CSP are visualized in Fig. 5. We take Subject 4 as example, because as shown in Table 1, Subject 4 have the highest accuracy among all subjects. This figure below can be used to check the neurophysiological plausibility of different types of MI tasks [23]. The analysis shows that brain

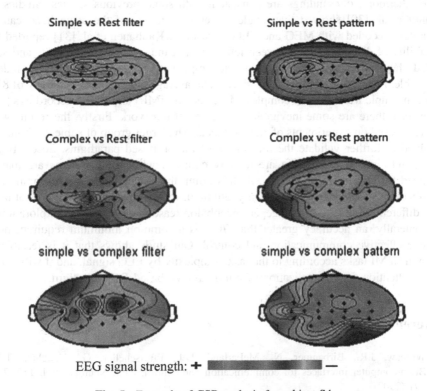

Fig. 5. Example of CSP analysis for subject S4

areas activated in the contralateral side during complex task MI are larger and more obvious than those during simple MI. These results could indicate the reasonableness of our feature extraction methods.

4 Discussion

In this study, we classify singlehanded MI tasks with different complexity based on EEG signals. The results revealed that an average accuracy of 79.20% between complex and simple tasks was achieved, which means this method can be used for multiclass paradigm in MI-based BCI systems.

Our study mainly discussed singlehanded complexity distinction (right hand), but it has the inspiration significance to further researches on complexity distinction of both hands with non-stationary EEG signals. All these efforts will eventually benefit the increase of control numbers in MI-based BCI systems. Noteworthy, there is currently no similar work to discriminate MI tasks differing in complexity by EEG signals. However, many studies have tried the various experiment paradigms, aiming at increasing the number of commands that an MI-based BCI can comprehend. For example, the study by Pfurtscheller et al. reported the distinctiveness between four different MI tasks (right hand, left hand, foot, and tongue movement) in a single trial; thus, such a paradigm may be utilized for the realization of a multiclass BCI [29].

Furthermore, our findings are consistent with some previous studies. Studies by Quandt et al. [30] showed that single finger movements of the same hand can be accurately decoded with MEG and EEG recordings. Kauhanen et al. [31] reported the possibility of discriminating between left and right index finger movements and suggested that single trial brain activity accompanying finger movements is indeed detectable. Research by Holper et al. showed an average classification accuracy of 81% between simple MI tasks and complex MI tasks with fMRI and fNIRS recordings [14].

In fact, there are some inevitable limitations of our work. Firstly, the results were reported based on experiments of eight subjects. The larger group of subjects should be involved to further validate the robustness of the proposed paradigms. Secondly, all participants in this experiment are healthy controls, while BCI systems are mainly designed for locked-in patients [1]. At this point, the effectiveness for neuromuscular disable users should be explored. Lastly, although we analyze the CSP patterns of tasks with different complexity, the deeper encephalon reasons worth further exploration.

Generally, an accuracy greater than 70% is a common minimum requirement to achieve effective communication and control. Our study shows that it is feasible to distinguish MI tasks according to the task complexity by EEG signal, and it will attract further attention for the realization of multiclass MI-based BCI paradigm.

References

1. Wolpaw, J.R., Birbaumer, N., McFarland, D.J., Pfurtscheller, G., Vaughan, T.M.: Brain-computer interfaces for communication and control. Clin. Neurophysiol. **113**, 767–791 (2002)

2. Dornhege, G., Milln, J.R., Hinterberger, T., McFarland, D.J., Muller, K.R.: Toward Brain-Computer Interfacing. The MIT Press, London (2007)
3. Wolpaw, J.R., McFarland, D.J.: Control of a two-dimensional movement signal by a noninvasive brain-computer interface in humans. Proc. Natl. Acad. Sci. 101(51), 17849–17854 (2004)
4. Pan, J., Li, Y., Gu, Z., Yu, Z.: A comparison study of two P300 speller paradigms for brain–computer interface. Cogn. Neurodyn. 7, 523–529 (2013)
5. Rebsamen, B., Guan, C., Zhang, H., Wang, C., Teo, C., Marcelo, J., Ang, H., Burdet, E.: A brain controlled wheelchair to navigate in familiar environments. IEEE Trans. Neural Syst. Rehabil. Eng. 18, 590–598 (2010)
6. Friedman, D., Leeb, R., Pfurtscheller, G., Slater, M.: Human-computer interface issues in controlling virtual reality with brain-computer interface. Hum. comput. Interact. 25(1), 67–94 (2010)
7. Millan, J., Renkens, F., Mourino, J., Gerstner, W.: Noninvasive brainactuated control of a mobile robot by human EEG. IEEE Trans. Biomed. Eng. 51, 1026–1033 (2004)
8. Chae, Y., Jeong, J., Jo, S.: Toward brain-actuated humanoid robots: asynchronous direct control using an EEG-based BCI. IEEE Trans. Robot. 25, 11131–11144 (2012)
9. Vora, B.A.J., Moore, M.: A P3 brain computer interface for robot arm control. In: Presented at the Society fo Neuroscience Abstracts. San Diego, CA, October 2004
10. Bhattacharyya, S., Konar, A., Tibarewala, D.: A differential evolution based energy trajectory planner for artificial limb control using motor imagery EEG signal. Biomed. Signal. Process. 11(1), 107–113 (2014)
11. Allison, B.Z., Pineda, J.A.: ERPs evoked by different matrix sizes: implications for a brain computer interface (BCI) system. IEEE Trans. Neural Syst. Rehabil. Eng. 11, 110–113 (2003)
12. Muller, K.R., Tangermann, M., Dornhege, G., Krauledat, M., Curio, G., Blankertz, B.: Machine learning for real-time single-trial EEG-analysis: from brain-computer interfacing to mental state monitoring. J. Neurosci. Meth. 167(1), 82–90 (2008)
13. Scherer, R., Müller, G., Neuper, C., Graimann, B., Pfurtschheller, G.: An asynchronously controlled EEG-based virtual keyboard: improvement of the spelling rate. IEEE Trans. Biomed. Eng. 51, 979–984 (2004)
14. Holper, L., Wolf, M.: Single-trial classification of motor imagery differing in task complexity: a functional near-infrared spectroscopy study. J. Neuroeng. Rehabil. 8, 34 (2011)
15. Nama, C.S., Jeon, Y., Kim, Y.-J., Lee, I., Park, K.: Movement imagery-related lateralization of event-related (de)synchronization (ERD/ERS): motor-imagery duration effects. Clin. Neurophysiol. 122, 567–577 (2011)
16. Robinson, N., Guan, C., Vinod, A.P., Ang, K.K., Tee, K.P.: Multi-class EEG classification of voluntary hand movement directions. J. Neural Eng. 10 (2013)
17. Yi, W., Qiu, S., Qi, H., Zhang, L., Wan, B., Ming, D.: EEG feature comparison and classification of simple and compound limb motor imagery. J. Neuroeng. Rehabil. 10(1), 106 (2013)
18. Allison, B., Brunner, C., Kaiser, V., Müller-Putz, G., Neuper, C., Pfurtscheller, G.: Toward a hybrid brain-computer interface based on imagined movement and visual attention. J. Neural Eng. 7(2), 26007 (2010)
19. Solis-Escalante, T., Müller-Putz, G., Brunner, C., Kaiser, V., Pfurtscheller, G.: Analysis of sensorimotor rhythms for the implementation of a brain switch for healthy subjects. Biomed. Signal Process. Control 5, 15–20 (2010)

20. Lacourse, M.G., Orr, E.L.R., Cramer, S.C., Cohen, M.J.: Brain activation during execution and motor imagery of novel and skilled sequential hand movements. NeuroImage **27**, 505–519 (2005)

21. Cao, Y., Olhaberriague, L.D., Vikingstad, E.M., Levine, S.R., Welch, K.M.A.: Pilot study of functional MRI to assess cerebral activation of motor function after poststroke hemiparesis. Stroke **29**(1), 112–122 (1998)

22. Schalk, G., McFarland, D.J., Hinterberger, T., Birbaumer, N., Wolpaw, J.R.: BCI2000: a general-purpose brain-computer interface (BCI) system. IEEE Trans. Biomed. Eng. **51**, 1034–1043 (2004)

23. Blankertz, B., Tomioka, R., Lemm, S., Kawanabe, M., Müller, K.-R.: Optimizing spatial filters for robust EEG single-trial analysis. IEEE Signal. Process. Mag. 1–12 (2008)

24. Dornhege, G., Blankertz, B., Curio, G., Müller, K.-R.: Boosting bit rates in noninvasive EEG single-trial classifications by feature combination and multi-class paradigms. IEEE Trans. Biomed. Eng. **51**(6), 993–1002 (2004)

25. Ramoser, H., Muller-Gerking, J., Pfurtscheller, G.: Optimal spatial filtering of single trial EEG during imagined hand movement. IEEE Trans. Rehabil. Eng **8**(4), 441–446 (2000)

26. Muller-Gerking, J., Pfurtscheller, G., Flyvbjerg, H.: Designing optimal spatial filters for single-trial EEG classification in a movement task. Clin. Neurophysiol. **110**, 787–798 (1999)

27. Yin, E., Zhou, Z., Jiang, J., Yu, Y., Hu, D.: A dynamically optimized SSVEP brain-computer interface (BCI) speller. IEEE Trans. Biomed. Eng. **62**(6), 1447–1456 (2015)

28. Krusienski, D.J., Sellers, E.W., McFarland, D.J., Vaughan, T.V., Wolpaw, J.R.: Toward enhanced P300 speller performance. J. Neurosci. Meth. **167**, 15–21 (2008)

29. Pfurtscheller, G., Brunner, C., Schlogl, A., Da Silva, F.H.L.: Mu rhythm (de)synchronization and EEG single-trial classification of different motor imagery tasks. NeuroImage **31**, 153–159 (2006)

30. Quandt, F., Reichert, C., Hinrichs, H., Heinze, H.J., Knight, R.T., Rieger, J.W.: Single trial discrimination of individual finger movements on one hand: a combined MEG and EEG study. NeuroImage **59**, 3316–3324 (2012)

31. Kauhanen, L., Nykopp, T., Sams, M.: Classification of single MEG trials related to left and right index finger movements. Clin. Neurophysiol. **117**, 430–439 (2006)

A Comparative Study of Joint-SNVs Analysis Methods and Detection of Susceptibility Genes for Gastric Cancer in Korean Population

Jinxiong Lv[1], Shikui Tu[1], and Lei Xu[1,2(✉)]

[1] Department of Computer Science and Engineering,
Center for Cognitive Machines and Computational Health,
Shanghai Jiao Tong University, Shanghai 200240, China
{lvjinxiong, tushikui, leixu}@sjtu.edu.cn
[2] Department of Computer Science and Engineering,
The Chinese University of Hong Kong, Hong Kong SAR, China

Abstract. Many joint-SNVs (single-nucleotide variants) analysis methods were proposed to tackle the 'missing heritability' problem, which emphasizes that the joint genetic variants can explain more heritability of traits and diseases. However, there is still lack of a systematic comparison and investigation on the relative strengths and weaknesses of these methods. In this paper, we evaluated their performance on extensive simulated data generated by varying sample size, linkage disequilibrium (LD), odds ratios (OR), and minor allele frequency (MAF), which aims to cover almost all scenarios encountered in practical applications. Results indicated that a method called Statistics-space Boundary Based Test (S-space BBT) showed stronger detection power than other methods. Results on a real dataset of gastric cancer for Korean population also validate the effectiveness of the S-space BBT method.

Keywords: GWAS · Sequence analysis · Joint-SNVs analysis test · Statistics-space Boundary based test · Gastric cancer

1 Introduction

The GWAS has made tremendous success based on the hypothesis 'Common Disease, Common Variant (CDCV)' [1], yet common variants identified via the GWAS only explained a small fraction of the heritability factors owing to two aspects. First, the traditional GWAS only focuses on the common variants to the common diseases, while the rare variants also make contributions to the common diseases in the light of 'Common Disease, Rare Variant (CDRV)' [2], and it is defined through the MAF ($1\% \leq \text{MAF} \leq 5\%$); second, it aims to detect the single genetic variants to the diseases while neglects the combined effect of SNVs [3]. The 'next generation' sequencing technologies facilitate the detection for rare variants contributing to the complex diseases. However, interesting rare variants have difficulty in being captured owing to the insufficient sample size.

In view of this, investigators have proposed many joint-SNVs analysis methods to solve them. These methods can be divided into three categories via the way to

© Springer International Publishing AG 2017
Y. Sun et al. (Eds.): IScIDE 2017, LNCS 10559, pp. 619–630, 2017.
DOI: 10.1007/978-3-319-67777-4_56

obtaining the corresponding statistics. The first road is the 'projection'. We transform the statistics vector into one statistic for simplified calculation of P-value. Thus, it is crucial to define suitable 'projection' matrix. For instance, the Hotelling's T square test transforms the difference of mean vectors for two populations into the Hotelling's T square statistic by multiplying the inverse of covariance matrix. However, accurate estimation of the covariance matrix depends on the large sample size and the low missing rate. On the basis of CDRV, some methods collapse or sum up all SNVs in a unit into a single one to discover the accumulation effect of rare variants. Here, the 'projection' matrix is diagonal. These methods can be divided into two groups according to whether the 'projection' matrix is the identity matrix or not. The two groups are named after burden test and non-burden test. The burden test assumes that SNVs contribute to the unit equally, while the non-burden test does not. The second is the 'combination' in the probability space [4]. The Fisher's method combined the P-value of each hypothesis into the Chi-square statistic in linear form, but much information is lost. And the third road is that we perform the multivariate test in the high dimensional space directly, so that it can break through the two limitations that the existing methods suffered from [5]. First, these existing methods regardless of the relationship between the dimensions; Second, the direction for each component is not taken into consideration. It is of note that S-space BBT is the representative method in the third road, and the comparative study involved it is still absent both in the large-scale simulation dataset and real-world dataset.

In this paper, we first introduced six representative methods, and then performed the comparative study in considering the varying sample size, OR, LD and MAF. The simulation results showed that (1) all the involved methods obtain stronger detection power with the sample size increasing. S-space BBT and SKATO are more sensitive than other four methods; (2) S-space BBT has stronger detection power than other methods under different OR, LD and MAF; (3) S-space BBT almost obtains smaller P-value compared with other methods in the real-world datasets. All above indicate that the S-space BBT plays an important role for joint-SNVs analysis. Thus, we applied it to a dataset of gastric cancer for Korean population and obtained a susceptibility gene list. The literature survey for selected genes was conducted to validate the effectiveness of S-space BBT. As a result, we provided the biomarker list and anticipated that it can be the reference for the gastric cancer study.

2   Representative Methods

2.1   Hotelling's T Square Test

The Hotelling's T square distribution is the generalization of the Student's t-distribution. Given two populations and they follow the independent multivariate normal distributions with same mean and covariance. The Hotelling's T square statistic is defined as:

$$t^2 = \frac{mn}{m+n}(\tilde{\mathbf{x}} - \tilde{\mathbf{y}})'\hat{\Sigma}^{-1}(\tilde{\mathbf{x}} - \tilde{\mathbf{y}}) \sim T^2(p, m+n-2)$$

$$\frac{m+n-p-1}{(m+n-2)p}t^2 \sim F(p, m+n-1-p) \tag{1}$$

where the m and n are the size of two populations, p is the number of variates, $\tilde{\mathbf{x}}$ and $\tilde{\mathbf{y}}$ are the sample means and the $\hat{\Sigma}$ indicates the covariance matrix. In order to calculate the P-value, we often transform it into F statistics.

As for the case-control study, the Hotelling's T square test obtains more accurate P-value when the sample size is large and the missing rate for genotype data is low, because both lead to the precise estimation for the covariance matrix.

2.2 Sumstat Test

The sumstat test, one kind of the burden test, treats the SNVs equally in the unit and adds all of the statistics from each SNV together to conduct the hypothesis test. It can enhance the power in considering the existence of rare variants. But it ignores the effect direction and the magnitude effect of SNVs. When the SNVs have same effect direction and the magnitude, the sumstat test obtained the better performance. The effect direction is defined via the OR, when the OR > 1, the direction is deleterious, otherwise, the direction is protective.

2.3 The Sequence Kernel Association Test (SKAT) and Its Optimal Version

For the regression model, we test whether the unit has influence on phenotype under the null hypothesis as described in the Eq. (1).

$$H_0 : \boldsymbol{\beta} = 0 \tag{2}$$

where the $\boldsymbol{\beta}$ indicates the coefficients vector and the null hypothesis means that the corresponding SNV is not associated with the phenotype. The SKAT assumes each β_j follows an arbitrary distribution with a mean of zero and a variance of $w_j\tau$ and then tests the null hypothesis H_0: $\tau = 0$ where the w_j is the prespecified weight. It obtains the variance-component score statistics which take the direction of β into consideration.

The optimal version of SKAT (SKATO) [6] combined burden statistic Q_{burden} and SKAT statistic Q_{SKAT} into the SKATO statistic Q_ρ in linear form. SKATO statistic is described as followed:

$$Q_\rho = (1 - \rho)Q_{SKAT} + Q_{burden} \tag{3}$$

where ρ indicates pair-wise correlation among β_j in Eq. (1).

Both of them belong to the non-burden test, not only taking the effect direction into account but also the magnitude of effect. So compared with the burden test, they are more robust.

2.4   Fisher's Combined Test

The Fisher's combined test combines the P-value from each test into the Chi-square statistic assuming the hypotheses are independent. The formula is defined as:

$$\chi^2_{2k} \sim -2\sum_{i=1}^{k} \ln(p_i) \qquad (4)$$

where the k is the number of the hypotheses and the p_i is the P-value obtained from the i-th hypothesis.

It suffers from the poor performance in joint-SNVs analysis owing to the information loss (e.g., LD, effect direction and so on). If there are many causal variants in the unit, the Fisher's combined method can achieve better performance.

2.5   Statistic-space Boundary Based Test

The above tests reject the H_0 as long as at least one of dimensions is rejected, and they ignore the roles of dimensions and their combination just as described in the Fig. 3(a) of [5]. The S-space BBT is one of the directional test and is described in the Fig. 6(a) of [5]. The way to achieving the combination is also given in Eq. (13)–(19) of [5].

The implementation of S-space BBT has been described in details in the Tab. 6 of [4]. Here, we give some key points of it. First, we directly use the boundary to form the rejection domain in the statistic space as followed:

$$\Gamma(\tilde{s}) = \left\{ s : (s - \tilde{s})' diag(\text{sign}(\tilde{s})) > 0 \right\} \qquad (5)$$

where $\text{sign}(s) = [sign(s_1), \cdots, sign(s_m)]'$ with $sign(v) = \frac{v}{|v|}$. Second, the P-value is calculated by the permutation test (see (65) in [4]). Third, the principle component analysis is performed to remove the second-order dependence. Forth, we adopt the posteriori version of the P-value for reduction of the background disturbance (see (93) of in [4]).

We have analyzed the application scenarios for the six methods in theory. The related computation have shown three points in the [7]. First, the six methods except for the S-space BBT swamp the significant SNVs. Second, burden test is powerful under the same effect direction. While the SKAT/SKATO is suitable for the different effect direction. Third, S-space BBT has stronger detection power in different MAF, LD and OR. In this paper, we adopted the statistic power to evaluate the six methods under the sample size, LD, OR and MAF in the simulation experiments. The detection power is defined as the proportion of true positive results. They were also evaluated on the real-world datasets.

3 Simulation Experiments

3.1 Simulation Framework

In order to compare the power of different approaches under various conditions, we use the simulation tool of PLINK software [8] to generate large simulation datasets. The number of SNVs in the joint unit is 10, which is composed of 5 causal variants and corresponding 5 observed markers. As a result, we obtained the simulation data of 10 SNVs on 100 cases vs. 100 controls, 500 cases vs. 500 controls and 1000 cases vs. 1000 controls in a stochastic way. Other parameter settings for the simulation datasets were described in the Table 1. Note that the LD in Table 1 is calculated between causal variant and corresponding observed marker, so we call it the incomplete LD. Besides, we produced 1000 replicates for each dataset for power computation and set the threshold $\alpha = 0.05$. In the [9], the detection power was estimated as the proportion of P-value $\leq \alpha$ among the 1000 replicates.

Table 1. Parameter settings for simulation datasets

Conditions	DatasetID	OR$_{het}$	OR$_{hom}$	MAF	Marker/causal variant LD
LD	Dataset1	1.2	2.4	0.05	0.4
	Dataset2	1.2	2.4	0.05	0.96
OR	Dataset3	1.1	2.2	0.05	0.8
	Dataset4	1.2	2.4	0.05	0.8
	Dataset5	1.3	2.6	0.05	0.8
MAF	Dataset6	1.2	2.4	0.01	0.8
	Dataset7	1.2	2.4	0.03	0.8

Note: OR$_{het}$ indicates the odds ratio for heterozygote causal variants.
OR$_{hom}$ indicates the odds ratio for homozygote causal variants.

Hotelling's T square test, Fisher's combined test and S-space BBT were implemented by the MATLAB. We adopted the SKATBinary function in the SKAT package of the R software to perform SKAT and SKATO. Sumstat test was performed by the PLINK/seq software with 100000 times of permutation.

3.2 Simulation Results

3.2.1 Linkage Disequilibrium

We first focus on the effect of the incomplete LD on each method. The linkage disequilibrium is the correlation between two SNVs and can be measured with the correlation coefficient [10]. The results were shown in the Fig. 1.

All of the methods achieved higher detection power with the sample size increasing. In particularly, the accurate estimation of the covariance matrix may account for the improvement for Hotelling's T square test, SKAT and SKATO. The S-space BBT obtained the best performance among the six methods. The SKATO obtained stronger detection power than sumstat test and SKAT. In conclusion, the power of the six methods is almost constant in different incomplete LD.

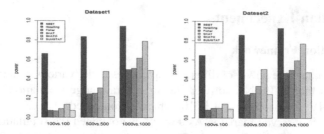

Fig. 1. Power comparison under different incomplete LD

3.2.2 Odds Ratio

The odds ratio is utilized to quantify the relationship between property A and property B in a given population. In GWAS, it quantifies the impact that one allele has on disease. When the OR > 1, the SNV is defined as deleterious one, which means that the more frequent the allele of SNV appears, the more likely to get sick. Conversely, when the OR < 1, the SNV is defined as protective one.

As shown in the Fig. 2, the power for each method enhances as the OR increasing, and SKATO is more sensitive than other methods. The sensitivity indicates the growth rate of detection power under different conditions. When the OR is large enough and the sample size is 2000, all methods achieved at least 85% power. The S-space BBT still keep the highest power in the different OR.

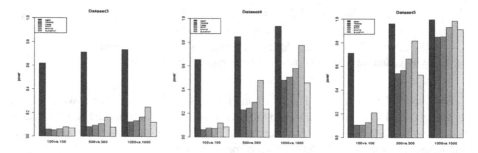

Fig. 2. Power comparison under different OR

3.2.3 Minor Allele Frequency

One site has two alleles (e.g. 'A' and 'a') in general. The frequency of second most common allele is the minor allele frequency in a given population. The rare variants are defined by the minor allele frequency. Based on the CDRV, the rare variants play a crucial role in genetic susceptibility to common diseases [2].

As described in the Fig. 3, all the six methods obtained stronger power with the MAF increasing, and each achieved greater sensitivity. It is of note that the S-space BBT kept the better performance (power$_{average}$ ≥ 60%) when the MAF = 1%, while other methods achieved less than 20% average power. The Fisher's combined method achieved 0.2% power when the sample size is 100 vs. 100 and the MAF = 1%.

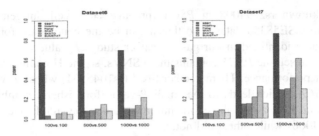

Fig. 3. Power comparison under different MAF

4 Gastric Cancer Study

4.1 Quality Control

The gastric cancer is the fifth most common malignancy in the world, especially in the Korea [11]. The selected dataset is associated with the gastric cancer for Korean population from the GEO database (Gene Expression Omnibus, ID: GSE58356). 319283 probes make up the dataset with the sample size of 683 controls and 329 cases.

As for the quality control, we took the Hardy-Weinberg's equilibrium and the missing rate into consideration. The Hardy-Weinberg law states that the allele and genotype frequency in a population will remain constant generation after generation. It is essential to regard the Hardy-Weinberg's equilibrium as one of measures in the quality control owing to the identification of questionable genotypes [12]. The threshold of the missing rate is set to 5%, and the threshold of the Hardy-Weinberg's equilibrium is set to 1.00E−04. After quality control and removing the duplicate probes, 54988 SNVs were remained. We regarded the gene as a unit and obtained 14709 units to conduct the joint-SNVs analysis via the S-space BBT.

4.2 Evaluation on the Real-World Dataset

To overcome some limitations (e.g., the LD, existence of the causal variants and so on) of the simulation experiments, we made efforts to search for the SNVs that are not only generally recognized but also can be found in the published SNVs datasets. Finally, 3 significant SNVs were found. Then all the six methods were performed for them and the results were shown in the Table 2.

Table 2. Results of three benchmarks for the six methods

Gene	SBBT	Hot	Fis	SKAT	SKATO	SUM
PSCA	8.90E−10	1.23E−06	1.65E−12	4.49E−01	4.49E−01	1.30E−03
ANK3	5.62E−08	4.83E−04	5.66E−03	7.43E−02	1.09E−01	6.85E−03
PALB2	7.96E−07	2.09E−03	1.13E−03	3.04E−01	3.28E−01	5.35E−03

Note: Hot indicates the Hotelling's T square tests; Fis means the Fisher's combined test; SUM indicates the sumstat test.

The well-known rs2294008 in *PSCA* (prostate stem cell antigen) [13, 14] is involved in the GSE58356 dataset, so that it can be the benchmark for comparative study. It was also identified in our gastric cancer study (*P*-value = 1.12E−07, OR = 1.66). In our dataset, the *PSCA* contains three SNVs, so the Hotelling's T square test obtained better performance. There is other SNV (rs1045531) whose *P*-value is 7.73E −08 in the *PSCA*, which leads to the small *P*-value for Fisher's combined method. S-space BBT and Fisher's combined method maintained the significance of the causal variant (rs2294008) while others did not.

The rs1938526 and rs420259 are found in the GSE71443 dataset consisting of 65 bipolar disorder patients and 74 controls. The dataset contains no missing value. We adopted similar quality control as the GSE58356 did.

The rs1938526 of the *ANK3* (Ankyrin 3) is the susceptibility locus for the bipolar disorder in [15]. For the *ANK3*, there are 235 SNVs involved, which may result in the inaccuracy computation of joint *P*-values owing to the small size of population. Thus, we selected SNVs located in the upstream and downstream 20 kb of rs1938526 to make up the computational unit, and 15 SNVs were remained. The smallest single locus *P*-value is 1.21E−05, while the *P*-value of rs1938526 ranks second (*P*-value = 0.06, OR = 1.58). The Hotelling's T square test obtained the smaller *P*-value compared to the other methods except for the S-space BBT owing to no missing value.

The rs420259 in *PALB2* (Partner And Localizer Of *BRCA2*) is also regarded as the meaningful SNV for the bipolar disorder [16]. For the *PALB2*, three SNVs are involved in. The rs420259 has the smallest *P*-value (*P*-value = 2.95E−03, OR = 0.42). Fisher's method, Hotelling's T square test and sumstat test achieved similar *P*-value.

In conclusion, the S-space BBT achieved smaller *P*-value compared with other methods for the *ANK3* and *PALB2*. As for the *PSCA*, the *P*-value of S-space BBT is also significant ($p_{SBBT} \leq 2.50E−06$). The SKAT and the SKATO performed worst for the three genes, which might result from the small sample size.

4.3   Literature Survey for Top 20 Associated Genes of Gastric Cancer

It is of note that we mainly focus on the combined effect of SNVs in the comparative study. Thus, some genes would be neglected owing to two points. First, smallest *P*-values of SNVs in these genes are smaller than 5.00E−08, which indicates that the traditional GWAS can detect them; second, there is only one SNV in the gene. Then, we selected top 20 genes detected via S-space BBT to conduct the literature survey. The search result was shown in the **Appendix** in detail. Further, we divided the genes into three groups. C group means those genes related to gastric cancer, B group indicates those related to other kinds of cancers and A group is other cases. In summary, there are 65% genes associated with the gastric cancer and other kinds of cancers. It indicates that the S-space BBT is reliable in the joint-SNVs analysis.

5 Conclusion

We conducted a comparative study on the main threads of joint-SNVs analysis methods in considering the sample size, LD, OR and MAF. The simulation experiments were designed to show that the S-space BBT has stronger detection power compared with other involved methods in different conditions. The simulation results showed that the S-space BBT plays an crucial role in detection of the susceptibility genes. More generally, we evaluated them on the real-world dataset and reached the same conclusion. Thus, we applied the S-space BBT to the dataset of gastric cancer for Korean population and obtained 20 significant genes. In order to validate the efficiency of the S-space BBT, we conducted literature survey for the top 20 genes, of which 65% are associated with the gastric cancer and other kinds of cancers. The prevalence of many diseases is low, which leads to the sample disequilibrium problem in statistical tests. The reactions of different joint-SNVs analysis methods to the problem might be an interesting issue, further, it is essential for investigators to propose novel methods to solve it.

Acknowledgements. This work was supported by the Zhi-Yuan chair professorship start-up grant (WF220103010) from Shanghai Jiao Tong University.

Appendix

As the Table 3 showed, 6 genes belong to the group C. The somatic mutation of *DCTN1* was discovered both in the primary cancer and the metastatic cancer [17]. The *PSCA* had been detected via the GWAS for Japanese and Korean populations [18] and conferred susceptibility to urinary bladder cancer in US and European populations [31]. While the *GAL3ST1* was identified via the whole exome sequencing for people from same family [20]. The *FRS2/FRS3* is related to the autophosphorylated FGFRs in the FGF signaling pathway, which is associated with the later stage for gastric cancer [21]. Hasegawa et al. found the altered expression of *PPP2R1B* in the lymph node metastasis for intestinal-type gastric cancer [22]. The expression of *B4GALNT1* plays an crucial role in the molecular mechanisms underlying the regulation of cancer-associated GM2 expression in stomach and colon [23]. The group B consists of 7 genes. The *FBXO11* induces the *BCL6* degradation to suppress the tumorigenicity for the diffuse large B-cell lymphomas [28]. The *MYLK2* is associated with the colorectal cancer [30]. These genes in group B might be common pathogenic genes for both gastric cancer and other kinds of cancers. As for the group A, to our limited knowledge, no literature clarified whether they are associated with cancers or not, which might play a potential role in the occurrence of gastric cancer on the genetic level.

In summary, there are 65% genes associated with the gastric cancer and other kinds of cancers. It indicates that the S-space BBT is reliable in the joint-SNVs analysis.

Table 3. Literature survey for the top 20 genes

Categories	Gene_Symbol	Annotation	Description	Reference
C	DCTN1	Dynactin Subunit 1	Somatic mutation	[17]
	PSCA	Prostate Stem Cell Antigen	Relate to the gastric cancer	[18, 19]
	GAL3ST1	Galactose-3-O Sulfotransferase 1	One of the 12 novel non-synonymous single nucleotide variants for the gastric- and rectal cancer in a family	[20]
	FRS3	Fibroblast Growth Factor Receptor Substrate 3	Relate to the FGF signaling pathway	[21]
	PPP2R1B	Protein Phosphatase 2 Scaffold Subunit Abeta	Relate to the lymph-node metastasis in gastric cancer	[22]
	B4GALNT1	eta-1,4-N-Acetyl Galactosaminyltransferase 1	Relate to gastric cancer	[23]
B	ACTR3C	ARP3 Actin-Related Protein 3 Homolog C	Downregulation under the cadmium treatments in the HepG2 cells	[24]
	ABCA12	ATP Binding Cassette Subfamily A Member 12	Upregulation in colorectal cancer	[25]
	C9orf152	Chromosome 9 Open Reading Frame 152	Biomarker for classification of endometrial carcinoma	[26]
	ZNF574	Zinc Finger Protein 574	Related to the colorectal cancer	[27]
	FBXO11	F-Box Protein 11	Tumor suppressor genes for the diffuse large B-cell lymphomas	[28]
	AIFM3	Apoptosis Inducing Factor, Mitochondria Associated 3	Related to the human cholangiocarcinoma	[29]
	MYLK2	Myosin Light Chain Kinase 2	Related to the colorectal cancer	[30]
A	CCT8L2	Chaperonin Containing TCP1 Subunit 8 Like 2		
	JRK	Jrk Helix-Turn-Helix Protein		
	PSTPIP2	Proline-Serine-Threonine Phosphatase Interacting Protein 2		
	CNBD2	Cyclic Nucleotide Binding Domain Containing 2		
	OTUD3	OTU Deubiquitinase 3		
	BTN2A1	Butyrophilin Subfamily 2 Member A1		
	MAN1B1	Mannosidase Alpha Class 1B Member 1		

References

1. Hindorff, L.A., Sethupathy, P., Junkins, H.A., Ramos, E.M., Mehta, J.P., Collins, F.S., Manolio, T.A.: Potential etiologic and functional implications of genome-wide association loci for human diseases and traits. Proc. Natl. Acad. Sci. U.S.A. **106**(23), 9362–9367 (2009)
2. Schork, N.J., Murray, S.S., Frazer, K.A., Topol, E.J.: Common vs. Rare allele hypotheses for complex diseases. Curr. Opin. Genet. Dev. **19**(3), 212–219 (2009)
3. Manolio, T.A., Collins, F.S., Cox, N.J., Goldstein, D.B., Hindorff, L.A., Hunter, D.J., McCarthy, M.I., Ramos, E.M., Cardon, L.R., Chakravarti, A., Cho, J.H., et al.: Finding the missing heritability of complex diseases. Nature **461**(7265), 747–753 (2009)
4. Xu, L.: Bi-linear matrix-variate analyses, integrative hypothesis tests, and case–control studies. Appl. Inform. **2**(1), 1–39 (2015)
5. Xu, L.: A new multivariate test formulation: theory, implementation, and applications to genome-scale sequencing and expression. Appl. Inform. **3**(1), 1–23 (2016)
6. Lee, S., Wu, M.C., Lin, X.: Optimal tests for rare variant effects in sequencing association studies. Biostatistics **13**(4), 762–775 (2012)
7. Lv, J.X., Huang, H.C., Chen, R.S., Xu, L.: A comparison study on multivariate methods for joint-SNVs association analysis, pp. 1771–1776
8. Purcell, S., Neale, B., Todd-Brown, K., Thomas, L., Ferreira, M.A., Bender, D., Maller, J., Sklar, P., De Bakker, P.I., Daly, M.J.: Plink: a tool set for whole-genome association and population-based linkage analyses. Am. J. Hum. Genet. **81**(3), 559–575 (2007)
9. Wu, M.C., Lee, S., Cai, T., Li, Y., Boehnke, M., Lin, X.: Rare-variant association testing for sequencing data with the sequence kernel association test. Am. J. Hum. Genet. **89**(1), 82–93 (2011)
10. Slatkin, M.: Linkage disequilibrium–understanding the evolutionary past and mapping the medical future. Nat. Rev. Genet. **9**(6), 477–485 (2008)
11. Ferlay, J., Soerjomataram, I., Dikshit, R., Eser, S., Mathers, C., Rebelo, M., Parkin, D.M., Forman, D., Bray, F.: Cancer incidence and mortality worldwide: sources, methods and major patterns in GLOBOCAN 2012. Int. J. Cancer **136**(5), E359–E386 (2015)
12. Gomes, I., Collins, A., Lonjou, C., Thomas, N.S., Wilkinson, J., Watson, M., Morton, N.: Hardy-weinberg quality control. Ann. Hum. Genet. **63**(Pt 6), 535–538 (1999)
13. Sakamoto, H., Yoshimura, K., Saeki, N., Katai, H., Shimoda, T., Matsuno, Y., Saito, D., Sugimura, H., Tanioka, F., Kato, S.: Genetic variation in psca is associated with susceptibility to diffuse-type gastric cancer. Nat. Genet. **40**(6), 730–740 (2008)
14. Song, H.R., Kim, H.N., Piao, J.M., Kweon, S.S., Choi, J.S., Bae, W.K., Chung, I.J., Park, Y. K., Kim, S.H., Choi, Y.D., Shin, M.H.: Association of a common genetic variant in prostate stem-cell antigen with gastric cancer susceptibility in a korean population. Mol. Carcinog. **50** (11), 871–875 (2011)
15. Ferreira, M.A., O'Donovan, M.C., Meng, Y.A., Jones, I.R., Ruderfer, D.M., Jones, L., Fan, J., Kirov, G., Perlis, R.H., Green, E.K., Smoller, J.W., et al.: Collaborative genome-wide association analysis supports a role for ANK3 and CACNA1C in bipolar disorder. Nat. Genet. **40**(9), 1056–1058 (2008)
16. Wellcome Trust Case Control C: Genome-wide association study of 14,000 cases of seven common diseases and 3,000 shared controls. Nature **447**(7145), 661–678 (2007)
17. Zhang J, Huang JY, Chen YN, Yuan F, Zhang H, Yan FH, Wang MJ, Wang G, Su M, Lu G, Huang Y *et al*: Whole genome and transcriptome sequencing of matched primary and peritoneal metastatic gastric carcinoma. *Scientific reports* (2015) 5(13750

18. Sakamoto, H., Yoshimura, K., Saeki, N., Katai, H., Shimoda, T., Matsuno, Y., Saito, D., Sugimura, H., Tanioka, F., Kato, S.: Genetic variation in psca is associated with susceptibility to diffuse-type gastric cancer. Nature genetics **40**(6), 730–740 (2008)

19. Song, H.R., Kim, H.N., Piao, J.M., Kweon, S.S., Choi, J.S., Bae, W.K., Chung, I.J., Park, Y. K., Kim, S.H., Choi, Y.D., Shin, M.H.: Association of a common genetic variant in prostate stem-cell antigen with gastric cancer susceptibility in a korean population. Molecular carcinogenesis **50**(11), 871–875 (2011)

20. Thutkawkorapin, J., Picelli, S., Kontham, V., Liu, T., Nilsson, D., Lindblom, A.: Exome sequencing in one family with gastric-and rectal cancer. BMC genetics **17**(1), 1 (2016)

21. Katoh, M.: Dysregulation of stem cell signaling network due to germline mutation, snp, helicobacter pylori infection, epigenetic change, and genetic alteration in gastric cancer. Cancer biology & therapy **6**(6), 832–839 (2007)

22. Hasegawa, S., Furukawa, Y., Li, M., Satoh, S., Kato, T., Watanabe, T., Katagiri, T., Tsunoda, T., Yamaoka, Y., Nakamura, Y.: Genome-wide analysis of gene expression in intestinal-type gastric cancers using a complementary DNA microarray representing 23,040 genes. Cancer research **62**(23), 7012–7017 (2002)

23. Yuyama, Y., Dohi, T., Morita, H., Furukawa, K., Oshima, M.: Enhanced expression of gm2/gd2 synthase mrna in human gastrointestinal cancer. Cancer **75**(6), 1273–1280 (1995)

24. Cartularo, L., Laulicht, F., Sun, H., Kluz, T., Freedman, J.H., Costa, M.: Gene expression and pathway analysis of human hepatocellular carcinoma cells treated with cadmium. Toxicology and applied pharmacology **288**(3), 399–408 (2015)

25. Hlavata, I., Mohelnikova-Duchonova, B., Vaclavikova, R., Liska, V., Pitule, P., Novak, P., Bruha, J., Vycital, O., Holubec, L., Treska, V., Vodicka, P., et al.: The role of abc transporters in progression and clinical outcome of colorectal cancer. Mutagenesis **27**(2), 187–196 (2012)

26. Sung CO, Sohn I: The expression pattern of 19 genes predicts the histology of endometrial carcinoma. *Scientific reports* (2014) 4(5174

27. Wang, H., Liang, L., Fang, J.Y., Xu, J.: Somatic gene copy number alterations in colorectal cancer: New quest for cancer drivers and biomarkers. Oncogene **35**(16), 2011–2019 (2016)

28. Duan, S., Cermak, L., Pagan, J.K., Rossi, M., Martinengo, C., di Celle, P.F., Chapuy, B., Shipp, M., Chiarle, R., Pagano, M.: Fbxo11 targets bcl6 for degradation and is inactivated in diffuse large b-cell lymphomas. Nature **481**(7379), 90–93 (2012)

29. Chua-On, D., Proungvitaya, T., Techasen, A., Limpaiboon, T., Roytrakul, S., Wongkham, S., Wongkham, C., Somintara, O., Sungkhamanon, S., Proungvitaya, S.: High expression of apoptosis-inducing factor, mitochondrion-associated 3 (aifm3) in human cholangiocarcinoma. Tumour biology : the journal of the International Society for Oncodevelopmental Biology and Medicine **37**(10), 13659–13667 (2016)

30. Parsons, D.W., Wang, T.L., Samuels, Y., Bardelli, A., Cummins, J.M., DeLong, L., Silliman, N., Ptak, J., Szabo, S., Willson, J.K., Markowitz, S., et al.: Colorectal cancer: Mutations in a signalling pathway. Nature **436**(7052), 792 (2005)

31. Wu, X., Ye, Y., Kiemeney, L.A., Sulem, P., Rafnar, T., Matullo, G., Seminara, D., Yoshida, T., Saeki, N., Andrew, A.S., Dinney, C.P., et al.: Genetic variation in the prostate stem cell antigen gene psca confers susceptibility to urinary bladder cancer. Nature genetics **41**(9), 991–995 (2009)

The Prognostic Role of Genes with Skewed Expression Distribution in Lung Adenocarcinoma

Yajing Chen[1], Shikui Tu[1], and Lei Xu[1,2(✉)]

[1] Center for Cognitive Machines and Computational Health, and Department of Computer Science and Engineering, Shanghai Jiao Tong University, Shanghai, China
{cyj907,tushikui,leixu}@sjtu.edu.cn
[2] Department of Computer Science and Engineering, The Chinese University of Hong Kong, Hong Kong SAR, China

Abstract. Many studies assumed gene expression to be normally distributed. However, some were found to have left-skewed distribution, while others have right-skewed distribution. Here, we investigated the gene expression distribution of five lung adenocarcinoma data sets. We assumed that samples in the tail and non-tail of a skewed distribution were drawn from different populations with different survival outcomes. To investigate this hypothesis, skewed genes were detected to build a tail indicator matrix comprising of binary values. Survival analysis revealed that patients with more skewed genes in their tails had worse survival. Hierarchical clustering of the tail indicator matrices discovered a gene set with similar tail configurations for either left or right skewed genes. The two gene sets divided patients into three groups with different survivals. In conclusion, there is a direct association between genes with skewed distribution and the prognosis of lung adenocarcinoma patients.

Keywords: Skewed distribution · Gene expression · RNA-sequencing · Microarray · Survival · Lung adenocarcinoma

1 Introduction

Gene expression profiling measures the activity of a large number of genes at a time. DNA microarray technology and RNA-sequencing are two main approaches to obtain gene expression data, though the latter is taking place of the former nowadays. RNA-sequencing uses next-generation sequencing to measure the RNA quantity in a sample, while microarray is based on hybridization of the predesigned probes and RNAs. Despite the difference in the techniques, the obtained expression values are highly correlated, which implies that the analysis approaches and the results from one might be applicable to the other [4,16,20]. Many gene expression profiling studies assume the distribution of gene expression to be Gaussian. In this case, statistical methods like t-test can be applied to detect genes with the differential expression between patient groups. However,

© Springer International Publishing AG 2017
Y. Sun et al. (Eds.): IScIDE 2017, LNCS 10559, pp. 631–640, 2017.
DOI: 10.1007/978-3-319-67777-4_57

Thomas et al. showed that gene expression is not always normally distributed [15]. Some genes display heavy-tailed distributions [6]. The reason why these genes have non-Gaussian distribution remains unknown.

Lung cancer is the most frequent cancer in the world, covering 13% of the total cancer incidence. It can be further divided into small cell lung carcinoma (SC) and non-small cell lung carcinoma (NSCLC). Lung adenocarcinoma is a histological type of NSCLC. 40% of the lung cancers in the US are adenocarcinomas. Patients with lung cancer have a 5-year survival of 10–15% [13]. The poor prognosis of this disease urges the discovery of new reliable and effective therapeutic approaches.

Here, we focused on the genes with skewed distribution in lung adenocarcinoma, and hypothesized that the tail and non-tail of a skewed distribution indicated distinct populations that might form subtypes of the disease. We investigated the survival of patients in tail and non-tail of skewed distribution, and see if different prognostic groups were formed. The data sets used in this study included a RNA-sequencing and four microarray data sets. We computed skewness to detect genes with heavy-tailed distributions, and obtained tail indicator matrices by labelling the tails. Survival analysis showed that patients with more tails in the skewed genes had poorer overall survival, regardless of the tail direction. Hierarchical clustering of the tail indicator matrices helped discover and select genes with similar tail configurations for either left or right skewed genes. They classified patients into three groups, i.e., one with both left and right tails, one with either left or right tail, and the other with no tail. Kaplan-Meier plots showed that patients with both left and right tails had worst survival, while those with no tail had the best prognosis. Literature review on the genes in *L-list* and *R-list* demonstrated their potential roles as therapeutic targets.

In conclusion, genes of skewed distribution are correlated to the prognosis of lung adenocarcinoma patients. A high number of tails in the skewed genes predicts a lower survival rate. Patients can be divided into three prognostic groups according to the tail configuration in *L-list* and *R-list*. The genes in the *L-list* and *R-list* can provide reliable therapeutic targets for the disease.

2   Methods

2.1   Analysis Procedures

Shown in Fig. 1 are the two main analysis procedures applied in this study. Details of these procedures are explained in the following sections.

2.2   Tail Detector

After obtaining a list of left-tail and right-tail genes, their tails were labelled to compute tail indicator matrices (see Fig. 3(b) as an example). The tail indicator matrix comprises of binary values, with 0 as non-tail and 1 as tail. To label the tails in a right-tail gene, samples were removed one by one according to the

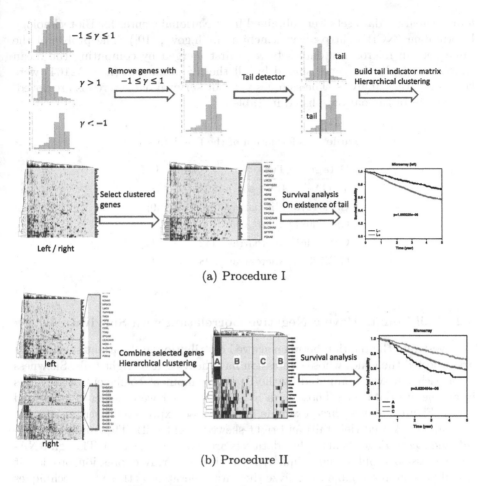

(a) Procedure I

(b) Procedure II

Fig. 1. Illustration of analysis procedures.

decreasing order of expression levels until the skewness of the remaining samples was smaller than or equal to zero. The samples removed were labelled as tail, while the remaining were labelled as non-tail. A similar procedure was applied to determine the tails for left-tail genes, though the samples were removed according to an increasing order of expression instead.

3 Experiments and Results

3.1 Data Preprocessing

The data sets we used in this study included a RNA-sequencing and four microarray data sets. The RNA-sequencing data set was generated by the Cancer Genome Atlas (TCGA) (https://cancergenome.nih.gov/) and downloaded from the UCSC Xena browser (https://xenabrowser.net/datapages/, [3]). The four microarray data sets were obtained from National Center for Biotechnology Information (NCBI) (https://www.ncbi.nlm.nih.gov/, [10]). The probes of the same gene in microarray data sets were first merged by computing the mean expression. Only the genes shared by all the five data sets ($n = 12164$) were kept for further analysis. The samples without survival data were also removed. The details of the data are listed in Table 1.

Table 1. Information of the five data sets.

Data set	Type	#Sample	Ref
TCGA	RNA-seq	503	[8]
GSE31210	Microarray	226	[9]
GSE50081	Microarray	130	[1]
GSE68465	Microarray	442	[12]
GSE72094	Microarray	398	[11]

3.2 Tail Counts Have a Negative Correlation with Survival

To investigate the relation between skewed distribution and patient survival, we first selected the genes whose expression have heavy left or right tails. Skewness γ for each gene was computed by the Pearson's moment coefficient of skewness in all the five data sets. Those with $\gamma > 1$ or $\gamma < -1$ were detected as skewed genes. Figure 2 shows three example genes whose expression distributions are normal, left-skewed (left-tail) and right-skewed (right-tail). The number of *left-tail* and *right-tail* genes in the five data sets are shown in Fig. 3(a). Though RNA-seq expression profiles share similarities with microarray expression profiles, it would be more reasonable to analyze the data generated by these two techniques separately. First, we integrated the resulting tail genes of the four microarray data sets. Figure 3(a) shows that the number of right-tail genes is much larger than the left-tail genes in the microarray data sets in general. Therefore, to

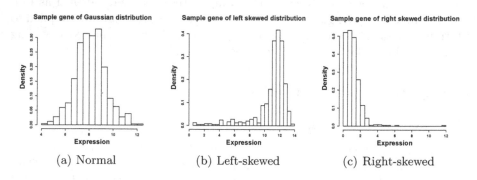

(a) Normal   (b) Left-skewed   (c) Right-skewed

Fig. 2. Histogram of three example gene distributions.

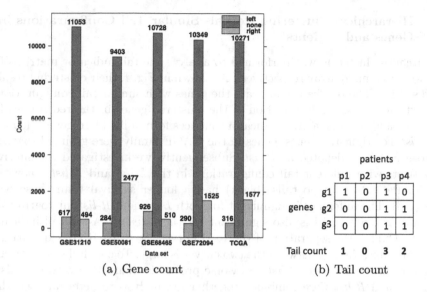

(a) Gene count   (b) Tail count

Fig. 3. Barplots for skewed genes in the five data sets and the illustration for counting tails.

obtain two balanced sets of reliable left- and right-tail genes, we selected the right-tail genes ($n = 129$) that were detected in all four microarray data sets, and the left-tail genes ($n = 209$) that were detected in more than one data set.

The intersection of the genes generated by the RNA-seq data and microarray data were computed, resulting in 50 left-tail and 100 right-tail genes. Subsequently, the tail labelling approach was used to compute tail indicator matrices for the shared left-tail ($n = 50$) and right-tail ($n = 100$) genes in the five data sets. The four matrices computed from the microarray data were merged. The total number of tails in the selected left or right tail genes was counted for each patient (see Fig. 3(b) for details). The results of Cox regression showed that a higher number of tail counts leads to a lower survival rate in both RNA-seq and microarray data (shown in Table 2).

Table 2. Results from Cox regression for survival versus tail counts. **Left** means left-tailed; **Right** means right-tailed. **p** means p-value obtained from likelihood-ratio test. **HR** means hazard ratio.

Data set	Left (p)	Left (HR)	Right (p)	Right (HR)
RNA-seq	0.000178	1.0595	0.347	1.00645
Microarray	1.23e-05	1.04205	0.000107	1.01822

3.3 Hierarchical Clustering Reveals Similar Tail Configurations in Genes and Patients

Hierarchical clustering was performed to analyze the tail indicator matrices for RNA-seq and microarray respectively. As shown in Fig. 4, their clustering results were similar. The red frames indicate the genes with similar tail configurations. For left-tail genes, the intersection of the clustered genes in the red frames for RNA-seq and microarray tail indicator matrices form a list of 16 genes, denoted as *L-list*. For right-tail genes, genes in the GAGE family were grouped together in both matrices, denoted as *R-list*. Subsequently, we investigated the survival of patients with different tail configurations in the *L-list* and *R-list* (shown in Fig. 5). Patients with no tails (black) had a longer survival than those with tails (red). The difference is significant for both *L-list* and *R-list* microarray tail indicator matrices, and is also significant for the *L-list* RNA-seq tail indicator matrix. Though the separation of the K-M curves for *R-list* RNA-seq matrix is not significant, the gap between the two curves is large. To conclude, the patients with tails in the *L-list* or *R-list* had worse prognosis than those with no tails.

L-list and *R-list* were combined together for analysis to determine whether similarities exist between both sets of data. Figure 6(a), (b) display the hierarchical clustering results for their RNA-seq and microarray tail indicator matrices. Patients were clustered into three distinct groups, where one showed a large number of tails in both *L-list* and *R-list* (A), one showed a high number of tails

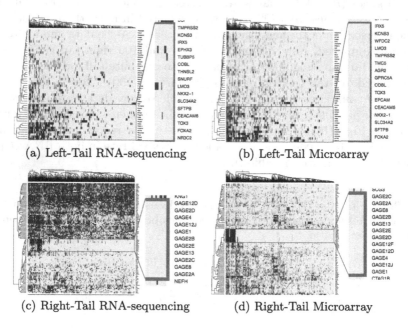

   (a) Left-Tail RNA-sequencing   (b) Left-Tail Microarray

   (c) Right-Tail RNA-sequencing   (d) Right-Tail Microarray

Fig. 4. Hierarchical clustering of left-tail and right-tail indicator matrices, where yellow represents non-tail and blue represents left-tail or right-tail. The red frames were added manually to indicate the clustered genes. (Color figure online)

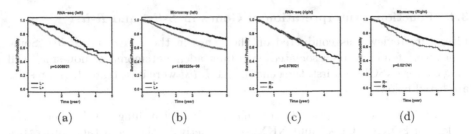

(a)   (b)   (c)   (d)

Fig. 5. Kaplan-Meier plots for different tail configurations. *L-list* in RNA-seq, $p = 0.0089$ (a). *L-list* in microarray, $p = 1.89e - 06$ (b). *R-list* in RNA-seq $p = 0,57$ (c). *R-list* in microarray, $p = 0.022$ (d). $L+$ represents patients with at least one tail in *L-list*; $L-$ represents patients with no tail in *L-list*. $R+$ represents patients with at least one tail in *R-list*; $R-$ represents patients with no tail in *R-list*. The p-values in the plots were computed by log-rank test. (Color figure online)

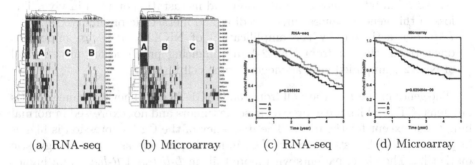

(a) RNA-seq   (b) Microarray   (c) RNA-seq   (d) Microarray

Fig. 6. Hierarchical clustering for *L-list* and *R-list* (a,b). Kaplan Meier plots for different tail configuration of genes in *L-list* and *R-list* (c,d). **A** represents patient group with tails in *L-list* and *R-list*. **B** represents patient groups with tails in either *L-list* or *R-list*. **C** represents patient groups with no tails in the two list. The p-values in the plots ($p = 0.066$ for RNA-seq, $p = 3.62e - 06$ for microarray) were computed by log-rank test. (Color figure online)

in merely *L-list* (B), and the other showed almost no tails (C). Based on the clustering results, we divided the patients into three groups by their tail configurations, i.e., one with tails in *L-list* and *R-list* (A), one with tails in either list (B), and the other with no tails (C). Figure 6(c), (d) show the survival curves for the three groups. Group A (black) has the worst survival, group B (red) has intermediate survival, and group C (green) has the best prognosis. Though only the microarray data show significant difference with $p < 0.05$, there is clear separation of the three survival curves for both RNA-seq and microarray data. These results indicated that *L-list* and *R-list* can serve as prognostic markers for lung adenocarcinoma.

3.4   Functional Interpretation of Genes in *L-list* and *R-list*

Literature review was conducted on the genes in the *L-list* and *R-list*. Some genes were found to be associated with lung adenocarcinoma or non-small cell lung cancer (NSCLC). First, the genes in the *L-list* were investigated, and results are listed below.

- *FOXA2* is a transcription factor that is involved in lung development. The loss of FOXA2, CDX2 and NKX2-1 can activate the metastatic process of lung adenocarcinoma [5].
- The non-detectable status of *SFTPB* is related to high risk of lung cancer [14].
- *SLC34A2* is an important gene during the fetal lung development, which was suppressed in the lung adenocarcinoma cell line A549. Its up-regulation inhibits cell invasion, tumor growth and metastasis ability [17].
- *NKX2-1* inhibits tumor differentiation and metastatic potential in vivo. The loss of this gene enhances tumor seeding and metastatic proclivity [19].
- *LMO3* is activated by the amplification of NKX2-1. It is an important downstream effector from NKX2-1 in enhancing proliferation and survival of NKX2-1-amplified lung adenocarcinoma cell lines [18].

The genes in the *R-list* are all from the GAGE family. They are cancer/testis antigens (CTA) which are expressed in some tumors and not expressed in normal lung tissue except for the testis. The frequency of the CTA expression is higher in patients of higher stages in NSCLC, indicating its role as a poor prognostic marker [2]. Therefore, patients with more tails in *L-list* and *R-list* had a higher probability of tumor metastasis and invasion of lung carcinoma, leading to poor prognosis.

4   Conclusion

Genes with skewed distribution in expression data were investigated and a correlation with survival was discovered. Patients with more tails in the left or right-tail genes had worse survival. Furthermore, RNA-seq and microarray data shared many siimlar tail configurations in some left and right tail genes, which were denoted as *L-list* and *R-list*. These genes helped to classify the patients into three prognostic groups, indicating three possible subtypes of the disease. Kaplan-Meier plots displayed that the patient group with tails in both *L-list* and *R-list* suffer from poorer prognosis than those with tail in either list. Those with no tail had the best survival among the three groups. Literature survey revealed that some genes in the *L-list* and *R-list* were reported to be related to lung adenocarcinoma. Patients with more tails in the *L-list* and *R-list* were more likely to have tumor metastasis, which explained why more tail counts predicted worse survival. The genes in *L-list* and *R-list* might provide potential therapeutic targets for lung adenocarcinoma. We think that the analysis procedures illustrated here can also serve as a biomarker detector and survival predictor for other diseases.

Acknowledgments. This work was supported by the Zhi-Yuan chair professorship start-up grant (WF220103010) from Shanghai Jiao Tong University.

References

1. Der, S.D., Sykes, J., Pintilie, M., Zhu, C.Q., Strumpf, D., Liu, N., Jurisica, I., Shepherd, F.A., Tsao, M.S.: Validation of a histology-independent prognostic gene signature for early-stage, non-small-cell lung cancer including stage IA patients. J. Thorac. Oncol. **9**(1), 59–64 (2014)
2. Gjerstorff, M.F., Pøhl, M., Olsen, K.E., Ditzel, H.J.: Analysis of GAGE, NY-ESO-1 and SP17 cancer/testis antigen expression in early stage non-small cell lung carcinoma. BMC Cancer **13**(1), 466 (2013)
3. Goldman, M., Craft, B., Swatloski, T., Cline, M., Morozova, O., Diekhans, M., Haussler, D., Zhu, J.: The UCSC cancer genomics browser: update 2015. Nucleic Acids Res. **43**, D812–D817 (2014)
4. Guo, Y., Sheng, Q., Li, J., Ye, F., Samuels, D.C., Shyr, Y.: Large scale comparison of gene expression levels by microarrays and RNAseq using TCGA data. PLoS one **8**(8), e71462 (2013)
5. Li, C.M.C., Gocheva, V., Oudin, M.J., Bhutkar, A., Wang, S.Y., Date, S.R., Ng, S.R., Whittaker, C.A., Bronson, R.T., Snyder, E.L., et al.: Foxa2 and Cdx2 cooperate with NKX2-1 to inhibit lung adenocarcinoma metastasis. Genes devel. **29**(17), 1850–1862 (2015)
6. Marko, N.F., Weil, R.J.: Non-gaussian distributions affect identification of expression patterns, functional annotation, and prospective classification in human cancer genomes. PLoS one **7**(10), e46935 (2012)
7. Meyer, D., Dimitriadou, E., Hornik, K., Weingessel, A., Leisch, F., Chang, C.C., Lin, C.C., Meyer, M.D.: Package e1071 (2017)
8. Network, C.G.A.R., et al.: Comprehensive molecular profiling of lung adenocarcinoma. Nature **511**(7511), 543–550 (2014)
9. Okayama, H., Kohno, T., Ishii, Y., Shimada, Y., Shiraishi, K., Iwakawa, R., Furuta, K., Tsuta, K., Shibata, T., Yamamoto, S., et al.: Identification of genes upregulated in ALK-positive and EGFR/KRAS/ALK-negative lung adenocarcinomas. Cancer Res. **72**(1), 100–111 (2012)
10. Sayers, E.W., Barrett, T., Benson, D.A., Bolton, E., Bryant, S.H., Canese, K., Chetvernin, V., Church, D.M., DiCuccio, M., Federhen, S., et al.: Database resources of the national center for biotechnology information. Nucleic Acids Res. **39**(suppl 1), D38–D51 (2011)
11. Schabath, M.B., Welsh, E.A., Fulp, W.J., Chen, L., Teer, J.K., Thompson, Z.J., Engel, B.E., Xie, M., Berglund, A.E., Creelan, B.C., et al.: Differential association of STK11 and TP53 with KRAS mutation-associated gene expression, proliferation and immune surveillance in lung adenocarcinoma. Oncogene **35**, 3209 (2015)
12. Shedden, K., Taylor, J.M., Enkemann, S.A., Tsao, M.S., Yeatman, T.J., Gerald, W.L., Eschrich, S., Jurisica, I., Giordano, T.J., Misek, D.E., et al.: Gene expression-based survival prediction in lung adenocarcinoma: a multi-site, blinded validation study. Nat. Med. **14**(8), 822–827 (2008)
13. Stewart, B., Wild, C.P., et al.: World cancer report 2014 (2014)
14. Taguchi, A., Hanash, S., Rundle, A., McKeague, I.W., Tang, D., Darakjy, S., Gaziano, J.M., Sesso, H.D., Perera, F.: Circulating pro-surfactant protein B as a risk biomarker for lung cancer. Cancer Epidemiol. Prev. Biomark. **22**(10), 1756–1761 (2013)

15. Thomas, R., de la Torre, L., Chang, X., Mehrotra, S.: Validation and characterization of DNA microarray gene expression data distribution and associated moments. BMC Bioinform. **11**(1), 576 (2010)
16. Trost, B., Moir, C.A., Gillespie, Z.E., Kusalik, A., Mitchell, J.A., Eskiw, C.H.: Concordance between RNA-sequencing data and DNA microarray data in transcriptome analysis of proliferative and quiescent fibroblasts. Roy. Soc. Open Sci. **2**(9), 150402 (2015)
17. Wang, Y., Yang, W., Pu, Q., Yang, Y., Ye, S., Ma, Q., Ren, J., Cao, Z., Zhong, G., Zhang, X., et al.: The effects and mechanisms of SLC34A2 in tumorigenesis and progression of human non-small cell lung cancer. J. Biomed. Sci. **22**(1), 52 (2015)
18. Watanabe, H., Francis, J.M., Woo, M.S., Etemad, B., Lin, W., Fries, D.F., Peng, S., Snyder, E.L., Tata, P.R., Izzo, F., et al.: Integrated cistromic and expression analysis of amplified NKX2-1 in lung adenocarcinoma identifies LMO3 as a functional transcriptional target. Genes Dev. **27**(2), 197–210 (2013)
19. Winslow, M.M., Dayton, T.L., Verhaak, R.G., Kim-Kiselak, C., Snyder, E.L., Feldser, D.M., Hubbard, D.D., DuPage, M.J., Whittaker, C.A., Hoersch, S., et al.: Suppression of lung adenocarcinoma progression by NKX2-1. Nature **473**(7345), 101–104 (2011)
20. Zhao, S., Fung-Leung, W.P., Bittner, A., Ngo, K., Liu, X.: Comparison of RNA-Seq and microarray in transcriptome profiling of activated T cells. PloS one **9**(1), e78644 (2014)

Survival-Expression Map and Essential Forms of Survival-Expression Relations for Genes

Yajing Chen[1], Shikui Tu[1], and Lei Xu[1,2(✉)]

[1] Center for Cognitive Machines and Computational Health, and Department of Computer Science and Engineering, Shanghai Jiao Tong University, Shanghai, China
{cyj907,tushikui,leixu}@sjtu.edu.cn
[2] Department of Computer Science and Engineering, The Chinese University of Hong Kong, Hong Kong SAR, China

Abstract. The relation between survival and gene expression has been investigated in many studies. Some used a univariate Cox model to detect genes with expression significantly related to survival. Some built a multivariate Cox model to analyze the influence of multiple genes on death risk. The original Cox model assumes a linear relation between survival and expression. But some evidence implied the existence of non-linear relation. Whether the survival-expression relations for different genes share some particular forms remain unknown. Here, we clustered the survival-expression (S-E) relations by k-means. We also developed a survival-expression (S-E) map to display the S-E relations for each cluster and summarized four essential forms of relations. We believe that the four essential S-E forms might assist the discovery of therapeutic targets and enhance the understanding of mechanisms in cancers.

Keywords: Cox regression · Spline · Survival-expression map · Essential survival-expression relation

1 Introduction

Many studies performed survival analysis to investigate the relation between survival and gene expression. Some built a multivariate Cox model to predict survival from the expression of several genes [2,9–11]. Some aimed to find prognostic biomarkers and used a univariate Cox model to discover the significant survival-related genes [2,3,6]. As the Cox model is linear, these studies implicitly assumed a linear survival-expression (S-E) relation. Others applied non-linear models and found evidence of non-linear relations [5,7,8]. However, whether the S-E relations for different genes share some particular forms and what these forms look like remain unknown.

Here, we aimed to find out the essential forms of S-E relations and clarify their patterns. After computing the survival rates for each gene using Cox regression with natural splines, we applied K-means on the resulting S-E relations, namely, the survival rates arranged by increasing expression. We proposed

© Springer International Publishing AG 2017
Y. Sun et al. (Eds.): IScIDE 2017, LNCS 10559, pp. 641–649, 2017.
DOI: 10.1007/978-3-319-67777-4_58

a survival-expression (S-E) map to display S-E relations for multiple genes simultaneously. It is a heat map whose rows are genes and columns are samples. Each row display the S-E relation for a gene. We analyzed the gene expression data for breast cancer, lung adenocarcinoma, lung squamous cell carcinoma. For all of them, the S-E relations can be clustered into four groups. The S-E maps for the four clusters revealed the differences in changing rates and tendencies of S-E relations. We claimed that each cluster represents an essential form of S-E relations, which might play a distinct role in biological systems. Careful investigation in the S-E maps uncovered the variation in the proportion and detailed configurations of the essential S-E forms for different cancers, indicating the heterogeneity of cancers. Surprisingly, even though lung adenocarcinoma and lung squamous cell carcinoma are both non-small-cell lung carcinoma, their S-E maps display significant difference.

In conclusion, we discovered four essential S-E forms by S-E maps for different cancers. We believe that the discovery might assist the finding of therapeutic targets to improve prognosis.

2   Methods

2.1   Cox Regression with Natural Splines

To analyze the relation between survival and gene expression, we applied Cox regression with natural splines (degree of freedom is 2). Cox regression model is a proportional hazard model which defines hazard rate λ as:

$$\lambda(t|x) = \lambda_0(t) exp(\beta^T x) \tag{1}$$

where $\lambda(t|x)$ represents the hazard rate at time t with covariate x, $\lambda_0(t)$ is the baseline hazard function which is irrelevant to x. Cox regression does not have to explicitly specify the form of baseline hazard function $\lambda_0(t)$, as it is cancelled out during the computation of β. The survival rate $S(t|x)$ is defined as:

$$S(t|x) = exp(-\Lambda(t|x)) \tag{2}$$

where $\Lambda(t|x)$ is the cumulative hazard function:

$$\Lambda(t|x) = \int_0^t \lambda(T|x)\, dT \tag{3}$$

However, the computation of survival rate requires us to specify the baseline hazard $\lambda_0(t)$. Thus, we define cumulative baseline hazard $\Lambda_0(t)$ as the Nelson-Aalen estimator [1]:

$$\Lambda_0(t) = \sum_{t_i \le t} \frac{d_i}{n_i} \tag{4}$$

The $\beta^T x$ in the hazard function is replaced by the natural splines $s(x)$. The hazard function becomes:

$$\Lambda(t|x) = exp(s(x))\Lambda_0(t) \tag{5}$$

The algorithm is implemented in the R language. The Cox model is built by the *coxph* function in the *survival* package, and natural splines is given in the *ns* function in the *splines* package. Finally, we obtained the survival rates for each gene and patient.

2.2 Survival-Expression (S-E) Map

The S-E map is a heat map that combines multiple S-E relations as its rows. Each S-E relation is the survivals sorted by increasing expression. Thus, the survivals in the same column but different rows of the S-E map might be computed for different patients. The colors indicate the magnitude of survival rates, where green represents low survival and red represents high survival. Rows of the S-E map are rearranged according to the results of hierarchical clustering, so as to display similar patterns in the S-E relations.

3 Experiments and Results

3.1 Data Preprocessing

We analyzed three cancers in this study, i.e., breast cancer (BRCA), lung adenocarcinoma (LUAD) and lung squamous cell cancer (LUSC). They were all RNA-sequencing expression profiles downloaded from the UCSC Xena browser (https://xenabrowser.net/datapages/, [4]). The RNA-sequencing expression data revealed the gene activity inside specific tissues. They were generated by the Cancer Genome Atlas (TCGA) research network (http://cancergenome.nih.gov/). The survival data include information about the overall and disease-free survival for each patient, which can also be obtained from TCGA. We selected breast cancer (BRCA) because it has the largest sample size in the TCGA hub. We selected lung adenocarcinoma (LUAD) and lung squamous cell carcinoma (LUSC) because they have large sample size and both of them are subtypes of non-small-cell lung cancer. The statistics of these data sets are listed in Table 1. The number of genes in each data set is 20530.

Table 1. Statistics of the three data sets.

Cancer	Abbreviation	Sample size
Lung squamous cell carcinoma	LUSC	494
Lung adenocarcinoma	LUAD	503
Breast cancer	BRCA	1080

3.2 Computation of the Overall Survival

We applied Cox regression with natural splines ($df = 2$) to analyze the influence of gene expression on survival. As the prognosis of BRCA is better than LUSC and LUAD, we computed the 5-year overall survival for LUAD and LUSC, and the 10-year overall survival for BRCA. Then, the survival rates were sorted according to the increasing expression for each gene, resulting in a survival-expression (S-E) relation. We performed likelihood ratio tests to evaluate the reliability of the fitting Cox models. Genes with small p-values were thought to display reliable S-E relations, which were selected for further investigation. The resulting p-value distributions for the three cancers are diffcrent. We chose the genes with $p < 0.001$ for LUAD, $p < 0.02$ for LUSC and $p < 0.01$ for BRCA, so that the number of selected genes ranged between 500 and 1000, i.e., 651, 445 and 793 respectively for LUAD, LUSC and BRCA.

3.3   Discovery of Four Essential Forms of Survival-Expression Relations

After applying k-means with $k = 4$ on the S-E relations, we made the S-E maps for the four clusters. Figs. 1, 2 and 3 display the results for BRCA, LUAD and LUSC respectively. The four clusters for each cancer mainly show two patterns. One is increasing survival with increasing expression (e.g. Fig. 1(b), (d)), while the other is decreasing survival with increasing expression (e.g. Fig. 1(a), (c)), each of which covers two clusters. The two clusters with the same changing tendency display different changing rates at different expression. One has a fast changing rate at low expression (e.g. Fig. 1(a), (b)), while the other has a fast changing rate at high expression (e.g. Fig. 1(c), (d)).

We claim that these four clusters represent four essential forms of S-E relations in biological systems. We named them as $I+$, $I-$, $D+$ and $D-$ according to the changing rates and tendencies. Details are shown in Table 2.

Table 2. Characteristics of four essential S-E forms.

S-E form	Changing tendency	Changing rate
$I+$	Increasing	Slow at low expression, fast at high expression
$I-$	Increasing	Fast at low expression, low at high expression
$D+$	Decreasing	Slow at low expression, fast at high expression
$D-$	Decreasing	Fast at low expression, low at high expression

After careful investigation in the four S-E forms, we found some bell-shape S-E relations. For example, at the bottom of Fig. 1(a), the colors of some S-E relations display patterns as orange-yellow-green-yellow, indicating the changing tendencies as high-low-high. Similar patterns can also be found in Fig. 1(b), (c) and (d). The opposite tendency in the tail of a bell-shape S-E relation seems to locate in the region where survivals change slowly. For example, at the bottom of Fig. 1(a) for $D-$, the bell-shape S-E relations show a slight increase at the high expression region, where survivals change slowly. At the bottom of Fig. 1(d), though the major tendency is increasing, we can see a slight decreasing at the low expression region, where survivals also change slowly. Similar results are shown in Figs. 2 and 3.

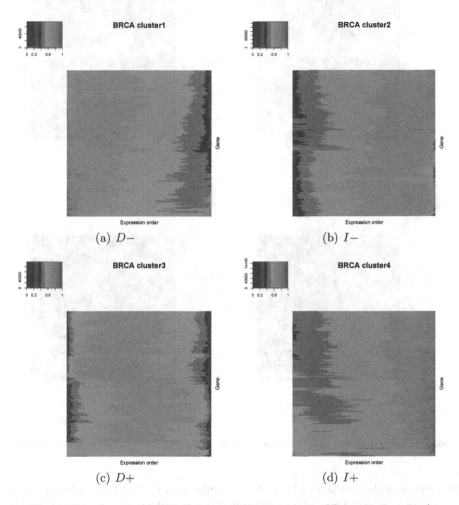

(a) $D-$

(b) $I-$

(c) $D+$

(d) $I+$

Fig. 1. The four essential S-E forms for breast cancer (Color figure online)

3.4   Analysis of the S-E Maps Between Cancers

The above section showed the similarities between the same S-E forms for three cancers. In this section, we focused on difference in the S-E forms between different diseases. We mainly compared the S-E maps for LUAD and LUSC, which are both non-small-cell lung cancer. First of all, the color configurations of the S-E maps for LUAD and LUSC are different. For LUAD, the major colors are green and blue, while for LUSC, the major colors become yellow and green. The

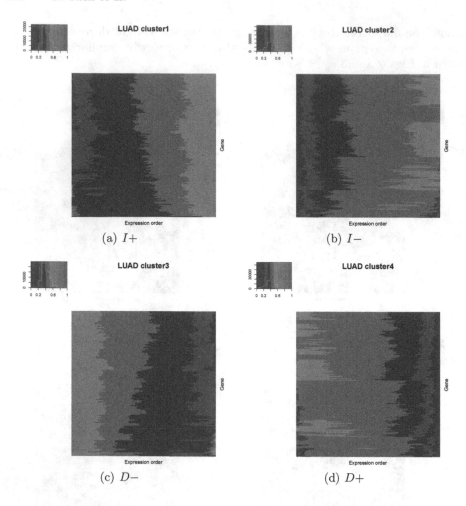

Fig. 2. The four essential S-E forms for lung adenocarcinoma (Color figure online)

difference in colors indicates that the patients with LUAD have a lower 5-year overall survival rate than LUSC. Second, the proportion of bell-shape S-E relations in LUSC are larger than LUAD. In LUSC, they cover almost half of the $D-$, $I-$ and $I+$ forms (Fig. 3(a), (b) and (d)). But we cannot see such a large proportion in LUAD. Third, the ratios of each S-E form between LUAD and LUSC are different, as shown in Table 3. However, the difference is not significant by chi-square test ($p = 0.2133$). These results indicate that the mechanisms and characteristics of LUSC and LUAD might be disease-specific, though they both belong to non-small-cell lung cancer.

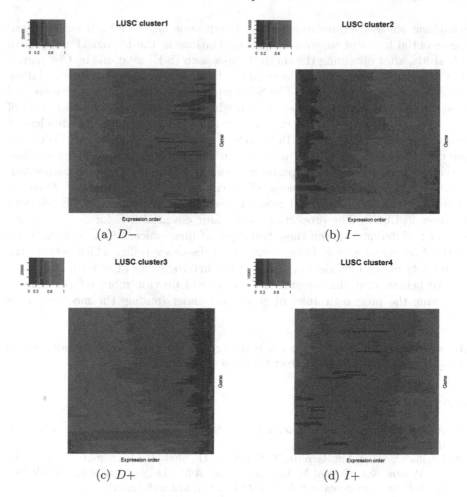

Fig. 3. The four essential S-E forms for lung squamous cell cancer (Color figure online)

Table 3. The number of genes in each S-E form for LUSC and LUAD

Cancer	I+	I−	D+	D−
LUSC	94 (21%)	51 (11%)	167 (38%)	133 (30%)
LUAD	131 (20%)	221 (34%)	180 (28%)	119 (18%)

4 Conclusion

The relation between survival and gene expression is implicitly assumed to be linear in many studies. Univariate Cox regression is often applied to evaluate the significance of the association between survival and gene expression. Some researchers performed Cox regression with splines to analyze the non-linear survival-expression relation, but no one has ever bothered to reveal the essential forms of survival-expression relations in the biological systems. In this study, after obtaining the survival-expression (S-E) relations by Cox regression with natural splines, we clustered the S-E relations into 4 groups and drew the S-E maps for each group. The S-E maps showed that the four clusters had their special S-E configurations, which might demonstrate the essential forms of S-E relations. Different S-E forms have various changing rates and tendencies in survival versus expression. Bell-shape S-E relations exist in each form and the opposite tendencies tend to appear in tail of the expression region where survivals change slowly. Comparisons between S-E maps for lung adenocarcinoma (LUAD) and lung squamous cell carcinoma (LUSC) showed the difference in their 5-year overall survival rates. The coverage of bell-shape S-E relations is larger in LUSC. The proportion of the four essential S-E forms are not significantly different between these two types of lung cancers. The configurations of the four essential S-E forms seem to be disease-specific, which reflects the complexity of the biological system and the heterogeneity of cancers.

We believe that the essential S-E forms will provide more information for analyzing the prognostic roles of genes and understanding the mechanisms of cancers.

Acknowledgments. This work was supported by the Zhi-Yuan chair professorship start-up grant (WF220103010) from Shanghai Jiao Tong University.

References

1. Aalen, O.: Nonparametric inference for a family of counting processes. Ann. Stat. **6**, 701–726 (1978)
2. Dhanasekaran, S.M., Barrette, T.R., Ghosh, D., Shah, R., Varambally, S., Kurachi, K., Pienta, K.J., Rubin, M.A., Chinnaiyan, A.M.: Delineation of prognostic biomarkers in prostate cancer. Nature **412**(6849), 822–826 (2001)
3. Diamandis, E.P., Scorilas, A., Fracchioli, S., Van Gramberen, M., De Bruijn, H., Henrik, A., Soosaipillai, A., Grass, L., Yousef, G.M., Stenman, U.H., et al.: Human kallikrein 6 (hK6): a new potential serum biomarker for diagnosis and prognosis of ovarian carcinoma. J. Clin. Oncol. **21**(6), 1035–1043 (2003)
4. Goldman, M., Craft, B., Swatloski, T., Cline, M., Morozova, O., Diekhans, M., Haussler, D., Zhu, J.: The UCSC cancer genomics browser: update 2015. Nucleic Acids Res. **43**, D812–D817 (2014)
5. Li, H., Luan, Y.: Boosting proportional hazards models using smoothing splines, with applications to high-dimensional microarray data. Bioinformatics **21**(10), 2403–2409 (2005)
6. Luo, L.Y., Katsaros, D., Scorilas, A., Fracchioli, S., Bellino, R., van Gramberen, M., de Bruijn, H., Henrik, A., Stenman, U.H., Massobrio, M., et al.: The serum concentration of human kallikrein 10 represents a novel biomarker for ovarian cancer diagnosis and prognosis. Cancer Res. **63**(4), 807–811 (2003)
7. Rini, B., Goddard, A., Knezevic, D., Maddala, T., Zhou, M., Aydin, H., Campbell, S., Elson, P., Koscielny, S., Lopatin, M., et al.: A 16-gene assay to predict recurrence after surgery in localised renal cell carcinoma: development and validation studies. Lancet Oncol. **16**(6), 676–685 (2015)

8. Rockova, V., Abbas, S., Wouters, B.J., Erpelinck, C.A., Beverloo, H.B., Delwel, R., van Putten, W.L., Löwenberg, B., Valk, P.J.: Risk stratification of intermediate-risk acute myeloid leukemia: integrative analysis of a multitude of gene mutation and gene expression markers. Blood **118**(4), 1069–1076 (2011)
9. Sotiriou, C., Neo, S.Y., McShane, L.M., Korn, E.L., Long, P.M., Jazaeri, A., Martiat, P., Fox, S.B., Harris, A.L., Liu, E.T.: Breast cancer classification and prognosis based on gene expression profiles from a population-based study. Proc. Nat. Acad. Sci. **100**(18), 10393–10398 (2003)
10. Van't Veer, L.J., Dai, H., Van De Vijver, M.J., He, Y.D., Hart, A.A., Mao, M., Peterse, H.L., van der Kooy, K., Marton, M.J., Witteveen, A.T., et al.: Gene expression profiling predicts clinical outcome of breast cancer. Nature **415**(6871), 530–536 (2002)
11. Wang, Y., Klijn, J.G., Zhang, Y., Sieuwerts, A.M., Look, M.P., Yang, F., Talantov, D., Timmermans, M., Meijer-van Gelder, M.E., Yu, J., et al.: Gene-expression profiles to predict distant metastasis of lymph-node-negative primary breast cancer. Lancet **365**(9460), 671–679 (2005)

Fast Vein Pattern Extraction Based on a Binary Filter

Shidong Li, Shuang Sun, and Zhenhua Guo[(⊠)]

Graduate School at Shenzhen, Tsinghua University, Shenzhen, China
Shidong_li@163.com, sun-sl6@mails.tsinghua.edu.cn,
Zhenhua.guo@sz.tsinghua.edu.cn

Abstract. Vein pattern extraction from Near-infrared (NIR) images is essential in most vein recognition algorithms. Among many vein recognition systems, the matched filter has been widely applied because it works well in enhancing vein images. However, the matched filter is time-consuming as it uses multi-scale and multi-orientation Gauss or Gabor filters to generate the Matched Filter Response (MFR) images. In this paper, we propose a binary filter for vein pattern extraction which can achieve similar results as the Gauss or Gabor filter but with fewer parameters and faster processing speed. The proposed method could process 27 images with image resolution of 320 * 240 per second, which is about three times faster than Gauss or Gabor filter.

Keywords: Vein extraction · Binary filter · Matched filter · Fast convolution

1 Introduction

Vein recognition, because of its advantages compared to other biometric recognition technologies (such as fingerprint, iris, face, etc.), has received more attentions for the past decade [1]. The relevant advantages can be summarized as follows:

(1) "Liveness" detection: NIR vein imaging is based on the NIR light absorption difference between the blood in vein vessel and the surrounding skin. So, NIR vein images could be acquired from the live body.
(2) Difficulty to forgery: Veins are underneath the skin rather than the outside bio-metrics such as fingerprint, which imposes difficulties to forge.
(3) Non-contactness: NIR vein images can be acquired from a hand without touching.
(4) High security: Vein pattern of each people is unique and stable.

Hand vein exploited for recognition can be divided into three categories: hand-dorsal vein [2, 3], palm vein [4] and finger vein [5], while this paper mainly focuses on the hand-dorsal vein.

Many methods have been investigated to obtain a binary image for vein pattern. For instance, Miura et al. [6] extracted finger vein pattern based on repeated line tracking. Wang et al. [7] exploited a local dynamic threshold method to segment the hand-dorsal vein area from the background. Gupta and Gupta [8], Huang et al. [9], and Yang and Shi [10] adopted the matched filter to enhance NIR vein images at first and then

© Springer International Publishing AG 2017
Y. Sun et al. (Eds.): IScIDE 2017, LNCS 10559, pp. 650–661, 2017.
DOI: 10.1007/978-3-319-67777-4_59

segmented the vein area. Song et al. [11] proposed a local mean curvature algorithm to extract centerlines of vein-networks.

However, the existing algorithms do not have satisfied results. Figure 1 shows pattern extraction results of the above mentioned algorithms (Fig. 1(a)–(h) are from the literature [11]). Figure 1(a) is a vein-like image and Fig. 1(b) is a noisy image by adding noise to Fig. 1(a). Figure 1(c) and (d) show results of binary vein images obtained using the repeated line tracking which consist of many thin lines leading to an unsatisfied visual effect. The local threshold based techniques prove to be effective in simple cases, but cannot work well in NIR vein images acquired from low-cost imaging device with high-level noise, low contrast and non-uniform illumination conditions, as shown in Fig. 1(e) and (f). The local mean curvature algorithm has a good performance on vein pattern extraction but this algorithm aims at obtaining centerlines of vein-networks, as shown in Fig. 1(g) and (h), so it may lose some discriminant information.

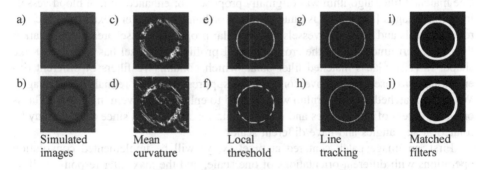

Simulated images	Mean curvature	Local threshold	Line tracking	Matched filters

Fig. 1. Comparisons of different vein pattern extraction algorithms. (a) and (b) are test patterns; (c)–(j) are patterns extracted by different algorithms.

In this paper, we aim at obtaining the binary vein image rather than the skeleton image due to the following reasons:

(1) A binary vein image presents more characteristics of vein-networks including the orientation and width of the vessels. However, the skeleton does not contain the width information.
(2) A skeleton of vein-networks can be obtained from a binary image by a thinning algorithm, while a binary image cannot be acquired from a skeleton one because width information is missing.

Figure 1(i) and (j) show the binary images obtained by the matched filter algorithm. Compared to previous mentioned vein extraction algorithms, the matched filter algorithm has a more satisfied result. In addition, it can also, to a certain extent, remain width information of vein-networks. However, the multi-scale and multi-orientation Gauss or Gabor filter bank [8–10], which are widely chosen as the matched filter in previous works, suffer the problems of being time-consuming.

In this paper, we propose a binary filter for NIR vein images enhancement, which is very simple and has fewer parameters. The experiments show that the binary filter can achieve similar vein images enhancement results as the Gauss or the Gabor filter. Meanwhile, the binary filter can perform rapid convolution based on the integral images. The proposed algorithm extracts the vein patterns in a speed of 27 frames per second for a 320 * 240 image resolution, which is about three times faster than the Gauss or Gabor filter.

Remaining of this paper is organized as follows. Section 2 gives a simple intro-duction of the matched filter algorithm. Section 3 presents the proposed algorithm. Section 4 shows experimental results and discussions. Finally, Sect. 5 concludes this paper.

2 Related Work

The matched filter algorithm was originally proposed for enhancement of blood vessels in retinal images [12], and it is quite suitable for hand vein pattern extraction because retinal vessels and hand vein vessels share similar properties: vessel areas appear darker than their surroundings and the cross-sectional profile of a vessel has a Gaussian-like shape. In [12], Gauss matched filter bank, which contains 12 filters of different ori-entations, was used to improve contrast and suppress noise in retinal vessel images. When the matched filter algorithm was adopted to enhance NIR vein images, the Gauss or Gabor filters of multi-scales and multi-orientations were used since the vein may be rotated at any angles and have different scales.

First, an image patch centered at pixel (x, y) will be implemented convolution operations with different orientations of one scale, and the max filter response will be retained to the pixel (x, y). The single-scale MFR image is generated when all the pixels finish up the previous procedure. If the filter bank includes N orientations and M scales, $K_{n,m}$ denotes the filter of n-th orientation and m-th scale and the single-scale MFR image at m-th scale is obtained by

$$S_m(x, y) = \max_{1 \leq n \leq N} (P(x, y) \otimes K_{n,m})$$

(1)

where $P(x, y)$ presents the image patch centered at pixel (x, y) with the same size of the filter, and the symbol "$\otimes$" denotes convolution. Then single-scale MFR images of different scales will perform pixel-wise product operations to get the final enhanced vein image:

$$E(x, y) = \prod_{m=1}^{M} S_m(x, y)$$

(2)

where $E(x, y)$ denotes the enhanced vein image.

The enhanced NIR vein image is a grayscale image, so we need to segment the vein area from the background. Usually, the binary vein image will be generated by just adopting a threshold algorithm on the multi-scale and multi-orientation MFR image. In this paper, we chose the OTSU adaptive threshold algorithm [13] since it is very convenient and can generate a satisfied binary image.

3   Vein Pattern Extraction

3.1   The Matched Filter Algorithm

Chaudhuri et al. [12] firstly adopted the matched filter to extract retinal vessels and they thought that an optimal filter must have a same shape as the cross-sectional profile of a vein. So they chose the Gauss filter as the optimal filter. Under this circumstance, when this method was applied in vein image enhancement, the Gauss and Gabor filter are considered as the optimal filters. The ideal goal of a vein enhancement algorithm is to maximize filter output at vein pixels and minimize filter output at non-vein pixels. However, the optimal filter defined by Chaudhuri et al. [12] maximizes the output signal-to-noise ratio at a given vein pixel location, but may get higher response for non-pixels due to noise or strong edge.

Subtracting mean value of the filter is necessary to the matched filter algorithm and it was highlighted in many literatures [8, 10, 12], which is important to achieve satisfied performance.

The filter with zero mean value will eliminate the direct current (DC) component of the input signal, which could be proved as follow:

$$
\begin{aligned}
R &= \sum_{x=1}^{k}\sum_{y=1}^{k} K(x,y) \times I(x,y) \\
&= \sum_{x=1}^{k}\sum_{y=1}^{k} K(x,y) \times (I_0(x,y) + m) \\
&= \sum_{x=1}^{k}\sum_{y=1}^{k} K(x,y) \times I_0(x,y) + m \times \sum_{x=1}^{k}\sum_{y=1}^{k} K(x,y) \\
&= \sum_{x=1}^{k}\sum_{y=1}^{k} K(x,y) \times I_0(x,y)
\end{aligned}
\tag{3}
$$

where R is the filter output, $K(x,y)$ is the filter with size of $k * k$, $I(x,y)$ is the input image patch with mean value of m and $I_0(x,y)$ is obtained from $I(x,y)$ by subtracting m. A cross-sectional profile of vein with zero mean value and a Gauss filter are presented in Fig. 2(a).

From Fig. 2(a), we can see that a cross-sectional profile of a vein with zero mean value and a Gauss filter have similar polarity. In this case, all pixel-wise products between a vein cross-sectional profile and a Gauss filter are positive and the sum of them will be a lager number. Therefore, the filter output of the other signals, which do not have similar polarity as the Gauss filter, will be small. Even though Gauss or Gabor matched filters have a satisfied result in most cases. However, such filters are complicated with too many parameters which are difficulties to optimize and are

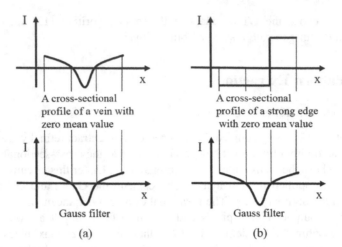

Fig. 2. The relationship between a cross-sectional profile of a vein or a strong edge with zero mean value and a Gauss filter.

time-consuming. Assuming the background of a NIR vein image have constant intensity with white noise, the filter output will ideally be zero. So, the filter used to enhance the NIR vein images does not need to share the same shape as the local vein patch shapes but share the same polarity. It motives us to use a binary filter rather than a real-value filter for vein extraction.

Meanwhile, the magnitude of the matched filter output is related to input signal's energy, which is decided by the amplitude of the signal rather than the filter. For example, a local patch with a strong edge may have larger energy than a vein area patch (see Fig. 2(b)). So, local normalization is necessary. If we conduct a normalization operation to transform the pixel intensity into a uniform scale, then the filter response will be limited into a certain range. In this paper, contrast limited adaptive histogram equalization (CLAHE) [14] method was used to achieve this effect. Comparisons of the proposed algorithm with and without the CLAHE method will be presented in Sect. 4.

3.2   A Binary Filter

Based on the discussion in Sect. 3.1, we propose a novel binary filter to replace Gauss or Gabor filter, which is presented in Fig. 3. Value of the white area of the filter is one and the dark area is a negative constant, which keeps zero mean for the filter. The filter has only three parameters: w and h denote the width and height of the dark region inside the filter; l denotes the size of the filter (In our work, the filter is fixed as a square.). Therefore, it is easier to tune the parameters than Gauss and Gabor filters as the latter have four parameters.

The binary filter is simple, but its NIR vein image enhancement effect is nearly same as the Gauss or Gabor filter. More experimental details are presented in the Sect. 4.

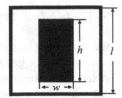

Fig. 3. The proposed binary filter.

3.3 Fast Convolution Based on the Integral Image

Besides easy to tune the parameters, the binary filter can implement fast convolution operation based on the integral image [15, 16]. There are two types of integral image, upright integral image and rotated integral image (see Fig. 4), which were originally proposed for rapidly extracting the Haar features and Haar-like features. Upright integral image and rotated integral image can be obtained by:

$$I_U(x,y) = \sum_{x_0 \le x, y_0 \le y} I(x_0, y_0) \tag{4}$$

$$I_R(x,y) = \sum_{y_0 \le y, |x-x_0| \le (y-y_0)} I(x_0, y_0) \tag{5}$$

where $I_U(x,y)$ and $I_R(x,y)$ denote upright integral image and rotated integral image separately, and $I(x_0, y_0)$ is the original grayscale image.

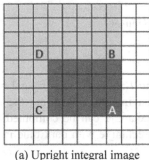

(a) Upright integral image

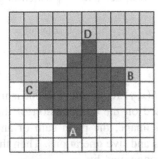

(b) Rotated integral image

Fig. 4. Two types of integral image.

Using integral image, the pixel sum of any horizontal rectangles or any 45° rotated rectangles can be computed in four lookups from the upright integral image or rotated integral image. The pixel sum of rectangle *ABDC* in Fig. 4 can be computed by:

$$I_{sum}(ABDC) = I_A + I_D - I_B - I_C \tag{6}$$

where I means upright integral image or rotated integral image.

However, by previous formulation, we cannot get accurate pixel sum of some 45° rotated rectangles with a center pixel from the rotated integral image, which is demonstrated in the Fig. 5. The pixel intensities in two lines (the green one and the yellow one) will be erroneously added into the pixel sum of the rectangle in dark blue by using Eq. (6).

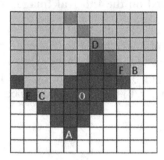

Fig. 5. The pixel sum of rectangle in dark blue cannot be computed using the rotated integral image. (Color figure online)

In this paper, we propose a novel linear integral image to solve the above mentioned problem. Each value of the linear integral image is the sum of pixels located on the line of upper left 45°, which is given by:

$$L(x, y) = \sum_{\substack{x - x_0 = y - y_0 \\ x, y \le x_0, y_0}} I(x_0, y_0) \tag{7}$$

where $L(x, y)$ represents the linear integral image and $I(x_0, y_0)$ is the original image.

Based on the combination of the proposed linear integral image and the rotated integral image, the pixel sum of any 45° rotated rectangle can be computed by six lookups and five operations. Then the pixel sum of rectangle with dark blue in Fig. 5 can be computed by:

$$I_{sum} = I_A + I_D - I_B - I_C - L_E - L_F \tag{8}$$

where I means the rotated integral image and L means the linear integral image.

As the proposed binary filter is consist of two rectangles and each rectangle shares the same value, we can implement fast convolution to speed up feature extraction.

4 Experimental Results

This section has two parts. The first part evaluates the results of NIR vein image enhancement based on the binary filter, while the second part evaluates the speed of our approach.

4.1 NIR Vein Images Enhancement Using the Binary Filter

The experiment are conducted on a database consisting of NIR hand-dorsal vein images acquired under 850 nm NIR light, which is created in our lab. It includes 2000 images acquired from 200 users (five images per hand), each of size 224 * 224 pixels. All the images have been manually labeled.

We use precision, recall, F1 score and intersection over union (IOU) as the metrics to evaluate the results. n_T and n_F represent the number of true and false predicted vein pixels respectively, and n_L is the number of labeled vein pixels. These four metrics are computed by:

$$\text{Precision} = n_T/(n_T + n_F) \tag{9}$$

$$\text{Recall} = n_T/n_L \tag{10}$$

$$\text{F1} = 2 \times \text{Precision} \times \text{Recall}/(\text{Precision} + \text{Recall}) \tag{11}$$

$$\text{IOU} = n_T/(n_L + n_F) \tag{12}$$

The first experiment presents the results of multiple groups of different numbers of scales and orientations, which is demonstrated in Table 1. The "Proposed_S1_O8" means the filter bank including one scale and eight orientations.

Table 1. Comparison of the results of different filter banks. "Proposed_S1_O8" means the proposed method with one scale and eight orientations.

Methods	Precision (%)	Recall (%)	F1 score (%)	IOU (%)
Proposed_S1_O8	68.7316	74.2442	71.3817	54.3955
Proposed_S2_O4	66.2933	82.4923	73.5109	57.4192
Proposed_S2_O6	65.9083	85.4749	74.4271	58.586
Proposed_S2_O8	65.5259	86.8182	74.6841	58.9427
Proposed_S2_O10	65.1772	87.616	74.749	59.0396
Proposed_S2_O12	65.1134	87.7334	74.7496	59.0492
Proposed_S3_O8	68.0713	75.5068	71.5965	55.8136

From Table 1, we can see that results of the binary filters of two scales are better than that of one scale or three scales. The product operation between two different single-scale MFR images can improve contrast and suppress noise, while the product operation of three scales will suppress weak filter response of slight vessels.

The binary filters of multi-orientation are used because a vein vessel may be rotated at any angle. Therefore, a higher precision can be achieved by adopting more filters in different orientations, which can be validated by the results in Table 1. The result of the binary filters of eight different orientations outperforms that of six or four by a large margin. However, improvement by using ten or twelve filters are not worthy as they will lead to higher computation complexities. Based on this finding, we chose the binary filter bank with two scales and cight orientations, and Gauss or Gabor filter bank were set the same number of scales and orientations in the following experiments.

Table 2 presents metrics of the results of three filters. Some vein patterns extracted based on the three filters are shown in Fig. 6.

Table 2. Performance of the three filters.

Methods	Precision (%)	Recall (%)	F1 score (%)	IOU (%)
Gabor filter	67.9886	83.0029	74.7493	59.1872
Gauss filter	65.5497	86.1951	74.4679	58.7135
Binary filter	65.5259	86.8182	74.6841	58.9427
Binary filter (no CLAHE)	64.9027	79.1036	71.3030	55.0642

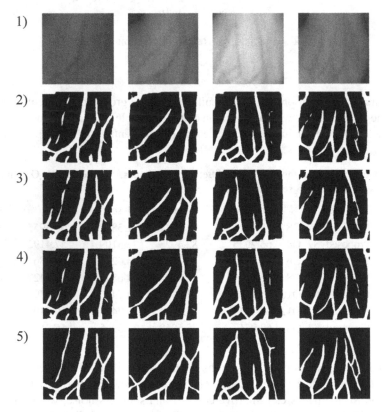

Fig. 6. Comparison of the vein segmentation results of three filters: (1) original images, (2) the binary filter, (3) Gauss filter, (4) Gabor filter, and (5) the ground truth of the original images.

Figure 6 shows that the vein enhancement and segmentation results of three filters are similar with each other. The F1 score and IOU of three filters are almost identical. The experiments show that the proposed binary filter could achieve similar performance of vein pattern extraction as the Gauss or Gabor filter.

The results of the proposed algorithm without the CLAHE method are presented in the last row in Table 2, which demonstrates that matched filters with CLAHE will improve results.

4.2 Speed Experimental Results

The binary filter bank has two scales and eight orientations. Among the eight binary filters with different orientations, only four binary filters of orientation of $0°$, $45°$, $90°$, and $135°$ can implement fast convolution operation based on the integral image directly. And the other four filters with $22.5°$, $67.5°$, $112.5°$, and $157.5°$ can also exploit the integral image to speed up convolution when the original image is rotated by $22.5°$. All the experiments are implemented using VS2013 on a desktop PC which is equipped with a Core i5-4570 CPU 3.2 GHz and 12 GB memory. The size of the three filters is 57 * 57. Table 3 shows the time cost of four convolutions with $0°$, $45°$, $90°$ and $135°$.

From Table 3, it is clear that the running speed of convolution with the binary filter has improved obviously.

Table 3. The time cost of four convolutions with $0°$, $45°$, $90°$ and $135°$.

Methods	Resolution: 224 * 224	Resolution: 320 * 240
Binary filter	3 ms	4 ms
Gabor filter	17 ms	34 ms
Gauss filter	16 ms	33 ms

The last experiments is conducted to compare the time cost of the whole vein extraction algorithm, including preprocessing, vein enhancement and segmentation, and the results are given in Table 4.

Table 4. Comparison of the time cost of the whole algorithm based on three filters (The number in the parentheses is the number of frames per second).

Methods	Resolution: 224 * 224	Resolution: 320 * 240
Binary filter	27 ms (37)	37 ms (27)
Gabor filter	72 ms (14)	138 ms (7)
Gauss filter	69 ms (14)	132 ms (7)

From Table 4, it is clear that the processed time drops by a large margin. The proposed algorithm can process 37 images per second when the size of image is 224 * 224 pixels. Meanwhile, the proposed algorithm can also extract vein pattern in real time even when the image size increases to 320 * 240.

5 Conclusion

In this paper, we proposed a binary filter for vein pattern extraction which can achieve similar results as the Gauss and Gabor filter but with fewer parameters and faster processing speed. Our approach could achieve a real time effect when the resolution of the vein image is smaller than 320 * 240, which is about three times faster than Gauss or Gabor filter. However, the proposed method may not be fast enough for a large image, for example, 640 * 480. How to get real time performance for larger resolution will be studied in our future work.

Acknowledgments. This work is partially supported by Shenzhen fundamental research fund (subject arrangement) (Grant No. JCYJ20170412170438636).

References

1. Huang, D., Zhang, R., Yin, Y., Wang, Y., Wang, Y.: Local feature approach to dorsal hand vein recognition by centroid-based circular key-point grid and fine-grained matching. Image Vis. Comput. **58**, 266–277 (2017)
2. Lee, J.C., Lo, T.M., Chang, C.P.: Dorsal hand vein recognition based on directional filter bank. Sig. Image Video Process **10**, 145–152 (2016)
3. Li, X., Huang, D., Zhang, R., Wang, Y., Xie, X.: Hand dorsal vein recognition by matching width skeleton models. In: 2016 IEEE International Conference on Image Processing (ICIP), pp. 3146–3150. IEEE (2016)
4. Lee, J.C.: A novel biometric system based on palm vein image. Pattern Recognit. Lett. **33**, 1520–1528 (2012)
5. Liu, F., Yang, G., Yin, Y., Wang, S.: Singular value decomposition based minutiae matching method for finger vein recognition. Neurocomputing **145**, 75–89 (2014)
6. Miura, N., Nagasaka, A., Miyatake, T.: Feature extraction of finger-vein patterns based on repeated line tracking and its application to personal identification. Mach. Vis. Appl. **15**, 194–203 (2004)
7. Wang, K., Zhang, Y., Yuan, Z., Zhuang, D.: Hand vein recognition based on multi supplemental features of multi-classifier fusion decision. In: Proceedings of 2006 IEEE International Conference on Mechatronics and Automation, pp. 1790–1795. IEEE (2006)
8. Gupta, P., Gupta, P.: A vein biometric based authentication system. In: Prakash, A., Shyamasundar, R. (eds.) ICISS 2014. LNCS, vol. 8880, pp. 425–436. Springer, Cham (2014). doi:10.1007/978-3-319-13841-1_24
9. Huang, D., Zhu, X., Wang, Y., Zhang, D.: Dorsal hand vein recognition via hierarchical combination of texture and shape clues. Neurocomputing **214**, 815–828 (2016)
10. Yang, J., Shi, Y.: Finger-vein network enhancement and segmentation. Pattern Anal. Appl. **17**, 783–797 (2014)
11. Song, W., Kim, T., Kim, H.C., Choi, J.H., Kong, H.J., Lee, S.R.: A finger-vein verification system using mean curvature. Pattern Recognit. Lett. **32**, 1541–1547 (2011)
12. Chaudhuri, S., Chatterjee, S., Katz, N., Nelson, M., Goldbaum, M.: Detection of blood vessels in retinal images using two-dimensional matched filters. IEEE Trans. Med. Imaging **8**, 263–269 (1989)
13. Otsu, N.: A threshold selection method from gray-level histograms. Automatica **11**, 23–27 (1975)

14. Zuiderveld, K.: Contrast limited adaptive histogram equalization. In: Graph Gems, pp. 474–485 (1994)
15. Viola, P., Jones, M.: Rapid object detection using a boosted cascade of simple features. In: Proceedings of 2001 IEEE Computer Society Conference on Computer Vision and Pattern Recognition, CVPR 2001, p. I. IEEE (2001)
16. Lienhart, R., Maydt, J.: An extended set of haar-like features for rapid object detection. In: Proceedings of 2002 International Conference on Image Processing, p. I. IEEE (2002)

Recommendation

Geographical and Overlapping Community Modeling Based on Business Circles for POI Recommendation

Man-Rui Li, Ling Huang, and Chang-Dong Wang[✉]

School of Data and Computer Science, Sun Yat-sen University, Guangzhou, China
limr3@mail2.sysu.edu.cn, huanglinghl@hotmail.com,
changdongwang@hotmail.com

Abstract. Point-of-interest (POI) recommendation is a challenging task since check-in data is extremely sparse and the social relationships in traditional recommendation have a limited effect. To solve this challenge, we propose a new geographical model with social influence and user preference. More specifically, we firstly propose a business circle conception which is more suitable for the modern consumption pattern in an urban city in POI recommendation. Then we decompose the user-location matrix into two geographical latent factors and integrate them into our business circle framework. Besides, we incorporate the user preference as a regularization of matrix factorization framework into our model by means of aggregating overlapping interest communities of users via their check-ins categories. Extensive experiments are conducted on two real-world datasets and the experimental results demonstrate that our model outperforms other existing algorithms.

Keywords: POI · LBSN · Overlapping community · Business circles

1 Introduction

Early recommender systems have been widely used in online website to promote the sales volume of the online consumption [1,2], while the recent recommender systems began to focus on recommending the offline consumption by using the information provided through integrated devices. Using the information of Internet of things is a growing trend [3] and online location-based social networks (LBSNs) is one of the typical representatives [4]. Because the information of real-time locations becomes easier to be acquired by GPS, LBSNs have undergone a rapid development, such as Foursquare, Gowalla and Facebook Places, in which users can share the experience in the physical world by checking in a POI. LBSNs can help people discovering the interesting places, especially in an urban city, where the POIs are in enormous quantity and hard to choose for users.

LBSNs can benefit users outdoor activities and bridge the gap between the physical world and online social network, but POI recommendation in LBSNs encounters more challenging issues than recommendation in traditional online

© Springer International Publishing AG 2017
Y. Sun et al. (Eds.): IScIDE 2017, LNCS 10559, pp. 665–675, 2017.
DOI: 10.1007/978-3-319-67777-4_60

websites for the following reasons. **(a) Extremely sparse check-in data.** POI recommendation is facing a huge challenge caused by extremely sparse check-in data. An individual user usually checks in a limited number of locations for two factors. On one hand, though there are numerous locations in a city, users only check in a small section for the restriction of distance. On the other hand, users prefer to check in their favorite locations repetitively, leading to the number of check-ins in different locations is limited. **(b) The limited effect of social influence.** The traditional social relationship has limited effect in LBSNs for the reason that there is only a very small portion of overlapping check-ins among social friends [5]. In other words, we should explore more appropriate implicit social relationships in LBSNs to improve the accuracy of recommendation. **(c) Complex and diverse factors.** In POI recommender systems, relatively speaking, there are more factors to be considered, such as geographical factor, temporary factor, category attributes and so on. Among these factors, geographical and temporary factors are intrinsic properties in POI recommendation for the check-ins in specific time and location. In other words, POI recommendation is an online-to-offline recommendation.

In this paper, we explore the geographical limits and overlapping interest community information to improve the recommendation accuracy. Firstly, we propose a business circle conception which is more suitable for the real modern consumption pattern in an urban city in POI recommendation. Besides, we propose two geographical latent factors inspired by these previous work [6–8] and integrate them into our business circle framework. Moreover, we incorporate the user preference into our model by means of aggregating overlapping interest communities of users via their check-ins categories. In experiments, we analyze the effect of parameters in our algorithm and then compare our CBGeoMFC model with three baseline algorithms to evaluate the performance our method.

2   The Proposed Model

2.1   Business Circle Conception

In urban cities, instead of a specific shop, people would often choose an area (such as a business circle) when they go outside. For example, when you have a date with your friends, you may expect a series of dating activities and all that dating locations are geographically close. So you will consider a prosperous region containing all that required venues, and we call that region as a business circle. In addition, in order to integrate the situation that people check in near the home or working place, we can also regard the surrounding area of the home and working place as a business circle.

Business circle is highly correlated to the geographical features, and we use the Haversine Formula [9] to get the great circle distance between two locations using their geo-co-ordinates as

$$d = 2\mathbf{R}\arcsin\left(\left(\sin^2(\frac{\varphi_2 - \varphi_1}{2}) + \cos(\varphi_1)\cos(\varphi_2)\sin^2(\frac{\lambda_2 - \lambda_1}{2})\right)^{\frac{1}{2}}\right) \quad (1)$$

where φ is latitude, λ is longitude, and $\mathbf{R}$ is the earth's radius.

And we define the negative of great-circle distance is the geographical similarity between two locations $geoSim(l_j, l_{j'}) = -d(l_j, l_{j'})$, where l_j ad $l_{j'}$ are two locations in the location set $L = \{l_1, ... l_n\}$. By applying $f(x) = \frac{x - \min(x)}{\max(x) - \min(x)}$ to map the similarity to the range of $[0, 1]$, which is taken as the input of the Affinity Propagation (AP) clustering algorithm [10] to cluster all the check-in locations into groups, the business circles can be obtained with each group representing one business circle and the representative location being the centroid of the business circle.

2.2   Geographical Modeling

Matrix Factorization. Low rank matrix factorization is widely used in POI recommendation, which approximates a user-location frequency matrix R by the multiplication of two k-dimensional latent matrices, namely user check-in preference latent matrix $U \in \mathbb{R}^{m \times k}$ and location characteristic latent matrix $L \in \mathbb{R}^{n \times k}$, where m and n are the numbers of users and locations respectively.

$$\min_{U,L} ||I \odot (R - UL^T)||_F^2 + \lambda_u ||U||_F^2 + \lambda_l ||L||_F^2 \tag{2}$$

where $I \in \mathbb{R}^{m \times n}$ is a $[0, 1]$ matrix with $I_{i,j} = 1$ indicating that user i has visited location j and $I_{i,j} = 0$ otherwise. The tuning parameters λ_u, λ_l is non-negative to avoid overfitting and $|| \cdot ||_F^2$ is the Frobenius norm.

CBGeoMF. In this section, we redefine the users' activity and POI influence and incorporate them into our business circle conception, and hence propose a new geographical matrix factorization model based on business circles (CBGeoMF).

The business circle is a geographical priority concept. In other words, users always concern more about which consumption centers they want to go to instead of a specific POI. To abstract this conception, first of all, we give the definition of user's activity on business circles and attraction of POI in the business circle.

User's Activity on Business Circles

1. User is more active on the business circles where he/she has checked in more times.
2. The business circles which are geographically near the activity business circles have more potential to be active.

For the foregoing considerations, the problem is to evaluate the density distribution on each business circle, which can be modeled by using the two-dimensional kernel density estimation approach. Assume that we have obtained a business circle set $C = \{c_1, ... c_w\}$, the estimated density of user u_i in a business circle c_q can be defined as

$$x_{i,c_q} = \frac{1}{P_i} \sum_{c_{q'} \in P_i} \frac{n_i^{c_{q'}}}{\delta} K\left(\frac{d(c_q, c_{q'})}{\delta}\right) \tag{3}$$

where $K(\cdot)$ is standard normal distribution with δ being the standard deviation, P_i is the set of business circles that user u_i has visited and $n_i^{c_{q'}}$ is the check-in number of u_i on business circle c'_q. And $d(c_q, c_{q'})$ denotes the distance of business circles c_q and $c_{q'}$, i.e. the distance between the corresponding centroids returned by AP.

In this way, the activity of user on circle is represented by a non-negative matrix X, which is estimated by both the check-in frequency counts and geographical information.

Attraction of POI in the Business Circle. Considering that users are more likely attracted by the popular locations, the locations in a dense region are more attractive. Similar to user's activity, the attraction of location l_j in a business circle c_q, denoted as $y_{l_j \in c_q}$, can be computed in the same way using kernel density estimation approach, $y_{l_j \in c_q} = \frac{1}{\delta n_{c_q}} \sum_{l_{j'} \in c_q} K\left(\frac{d(l_j, l_{j'})}{\delta}\right)$, where n_{c_q} is the number of locations in c_q, and $l_{j'}$ is the other locations in the same business circle c_q as l_j. In this way, the attraction of POIs in their corresponding circles can be represented by the non-negative matrix Y.

The preference of users on locations can be decomposed into two low rank w-dimensional latent factors, where w is the number of business circles. One is user's activity on business circles $X \in \mathbb{R}^{m \times w}$, and the other is the attraction of POI in the corresponding circle $Y \in \mathbb{R}^{n \times w}$. By augmenting the matrix of the two latent geographical matrices X and Y with the original latent matrices U and L, the objective function can be defined as follows

$$\min_{U,L} ||I^R \odot (R - UL^T - XY^T)||_F^2 + \lambda_u ||U||_F^2 + \lambda_l ||L||_F^2 + \lambda_x ||X||_F^2 \tag{4}$$

In the above objective function, the user's activities on circles X are initially initialized by using Eq. (3), and then they are iteratively updated to meet the fact that users activity behavior would change to reflect their real check-in behavior.

2.3 Overlapping Community Modeling

Social network is often taken as a good side information in recommender systems to improve accuracy. However, it works relatively less effective in a location-based recommendation for two reasons: **(a) Geographical Limitation.** Unlike the traditional recommendation, LBSN has an intimate relations with both online and offline, that is to say, we have to consider the geographical limitation in real life. For example, though two users have similar following relationship or preference, it is still not suitable to recommend one's check-in locations to another if their frequently visited places are far away. **(b) Sparse Social Relationship.**

Another reason is that compared to a traditional online social relationship, users in LBSN have fewer direct social connections and 90% user's overlap check-ins to his/her friends' check-ins is less than 20%, which is mentioned in the previous studies [5,11]. We incorporate the category influence and geographical social network information as a regularization term of matrix factorization based model to consider both the user's preference and geographical limitation, which is inspired from the relevant studies [12–14].

As aforementioned, the AP algorithm is used to cluster all the check-ins into a set of clusters, and each cluster represents a business circle. Considering the geographical limitation, the users who have similar frequently visited business circles are defined to be geographical social friends. We can effectively settle the defects that there are few overlap check-ins of a user to his/her friends' check-ins by aggregating the check-ins on locations into the check-ins on business circles. The common check-ins on locations are infrequent, even though the users have the similarity routines, however, their frequently visited business circles can be the same. In this way, the geographical feature can be enhanced. Furthermore, compared with [11,15,16] which believed users are more geographical similar when their 'homes' are near, our approach don't need to build a multi-center gaussian model for each user to estimate the rough home location.

In the first place, the Pearson Correlation Coefficient (PCC) is used to calculate the similarity matrix S_{if} of users. Considering the number of check-ins of each user differs greatly, we first normalize the visit frequency of each user from 0 to 1 by $r_{i,c_q}^c = \frac{fre_{i,c_q}}{\max\{fre_i\}}$. Then the similarity S_{if} is defined as $S_{if} = \frac{\sum_{c \in C_{if}} (r_{i,c_q}^c - \bar{r_i^c})(r_{f,c_q}^c - \bar{r_f^c})}{\sqrt{\sum_{c \in C_{if}} (r_{i,c_q}^c - \bar{r_i^c})^2} \sqrt{\sum_{c \in C} (r_{f,c_q}^c - \bar{r_f^c})^2}}$, where fre_{i,c_q} is the check-in frequency of user i on business circle c_q, r_{i,c_q}^c is the rating (normalized visited frequency) of user i on business circle c_q, $\bar{r_i^c}$ is the average rating of user i on all the circles, and C_{if} is the business circles set that both user u_i and user u_f have visited.

Table 1. The statistics of data sets.

Statistics	User	Venue	Check-ins	Category	Sparsity
Foursquare (NYC)	1083	38330	226970	251	0.22%
Jiepang (HongKong)	130	1153	5124	130	1.17%

Then we cluster the users into overlapping communities $M = \{m_1, ... m_h\}$ according to corresponding category information. And the regularization term is $\frac{\lambda_h}{2} \sum_{i=1}^m \sum_{p=1}^h I_{ip}^M Z_{ip} \sum_{u_f \in m_{ip}} S_{if} \|U_i - U_f\|_F^2$, where m_{ip} denotes the community m_p containing the users in the same community as u_i, I_{ip}^M equals 1 if u_i belongs to m_p or equals 0 otherwise, Z_{ip} is the preference of u_i on community m_p, which is defined as $Z_{ip} = \frac{fre_{i,m_p}}{\max_{\forall m}\{fre_i^m\}}$, where fre_{i,m_p} is the check-in frequency of user i on community m_p.

In this way, we solve the problem that common check-ins of users on locations are insufficient by incorporating both the user preference (category information) and geographical social network information.

2.4   The Overall Model and Optimization

Accordingly, the objective function of the overall model is as follows

$$E = \frac{1}{2} \sum_{i=1}^{m} \sum_{j=1}^{n} I_{ij}^{R}(R_{ij} - U_i L_j^T - X_i Y_j^T)^2 + \frac{\lambda_u}{2}||U||_F^2 + \frac{\lambda_l}{2}||L||_F^2 + \frac{\lambda_x}{2}||X||_F^2$$

$$+ \frac{\lambda_h}{2} \sum_{i=1}^{m} \sum_{p=1}^{h} I_{ip}^{M} Z_{ip} \sum_{u_f \in m_{ip}} S_{if}||U_i - U_f||_F^2 \qquad (5)$$

Stochastic Gradient Descent (SGD) is used to find a local minimum of our objective function Eq. (5), and the gradient descent w.r.t. U_i, L_j and X_i are

$$\frac{\partial E}{\partial U_i} = -\sum_{j=1}^{n} I_{ij}^{R}(R_{ij} - U_i L_j^T - X_i Y_j^T)L_j + \lambda_u U_i + \lambda_h \sum_{p=1}^{h} I_{ip}^{M} Z_{ip} \sum_{u_f \in m_{ip}} S_{if}(U_i - U_f)$$

$$\frac{\partial E}{\partial L_j} = -\sum_{i=1}^{m} I_{ij}^{R}(R_{ij} - U_i L_j^T - X_i Y_j^T)U_i + \lambda_l L_j \qquad (6)$$

$$\frac{\partial E}{\partial X_i} = -\sum_{j=1}^{n} I_{ij}^{R}(R_{ij} - U_i L_j^T - X_i Y_j^T)Y_j + \lambda_x X_i$$

3   Experiments

3.1   Dataset Description

We use the Foursquare and Jiepang datasets to evaluate the performance of our model which are widely used in previous studies in the location-based recommender system [6,17]. The Foursquare dataset contains a long-term (about 10 months) check-in data in New York City (NYC) collected from Foursquare, ranging from April 2012 to February 2013, and Jiepang dataset contains check-ins in HongKong from December 2011 to September 2012. Each check-in contains a user ID, a location ID, a category ID and a category name, latitude and longitude coordinates. Table 1 lists the general statistics of our two datasets. To split the dataset into training set and testing set, we firstly aggregate the check-ins for each location of each user. After that, we select 80% data randomly as the training set to train the model and the remaining 20% are held out for testing.

3.2   Evaluation Metrics

Considering only limited locations will be recommended to users in location-based recommendation, we use the Top-N recommendation approach to evaluate our model, including Precision@N, Recall@N, and F1@N metrics. F1@N is the harmonic mean of precision and recall.

3.3 Parameters Analysis

There are six parameters in our algorithm, namely λ_h, λ_x, λ_u, λ_l, N, and k. We set the dimension of latent matrices k as 2 and the regularization coefficients λ_u, λ_l equal to 0.01 in all experiments. Then we analyze the effect of social regularization coefficient λ_h, user activity regularization coefficient λ_x, and the length of recommend list N respectively by fixing other parameters in this section.

Parameter Analysis on λ_h. Parameter λ_h represents the impact on community social relationship. On the Foursquare dataset, we fix the parameters $N = 5, \lambda_x = 0.01$ and increase the λ_h from 0 to 0.07 unevenly. It's obvious that the accuracy is sensitive to the community social regularization. Compared with the situation that we don't consider the impact of community influence (i.e. $\lambda_h = 0$), the accuracy can be promoted 44.8% when $\lambda_h = 0.005$ on Foursquare. Besides, on the Jiepang dataset, we fix the parameters $N = 5, \lambda_x = 40$, and the accuracy can be promoted by 6.3% at the best performance when setting $\lambda_h = 0.07$ as shown in Table 2.

Table 2. Parameter analysis: the influence of λ_h on the proposed CBGeoMFC model.

Foursquare	λ_h	0	0.0001	0.0005	0.001	0.005	0.01	0.05	0.07
	Precision@5	0.0404	0.0404	0.0408	0.0421	0.0585	0.0550	0.0430	0.0440
	Recall@5	0.0120	0.0121	0.0121	0.0125	0.0174	0.0164	0.0128	0.0131
	F1@5	0.0185	0.0185	0.0187	0.0193	0.0268	0.0252	0.0197	0.0202
Jiepang	λ_h	0	0.01	0.05	0.07	0.1	0.3	0.5	0.7
	Precision@5	0.0492	0.0492	0.0508	0.0523	0.0523	0.0492	0.0446	0.0431
	Recall@5	0.0732	0.0732	0.0755	0.0778	0.0778	0.0732	0.0664	0.0641
	F1@5	0.0589	0.0589	0.0607	0.0626	0.0626	0.0589	0.0534	0.0515

Parameter Analysis on λ_x. Parameter λ_x is the Frobenius norm coefficient of user activity. On Foursquare, we fix the parameters $N = 5, \lambda_h = 0.005$ and then test λ_x from 0.0001 to 150 as shown in Table 3. The accuracy has almost no improvement when $\lambda_x < 1$ while it increases rapidly when $\lambda_x > 1$ and achieves the best performance as $\lambda_x = 100$. As for the Jiepang dataset, the best result is achieved as $\lambda_x = 40$. The experiment result shows that user activity on most circles is unimportant, for the large value of λ_x generates a rather strong constraints. When the parameter λ_x continues to increase, the constraints become stronger so that it removes some obvious features and the value begins to decrease. This is consistent with the real world situation that users always check-in in a limited number of circles. The accuracy is promoted 9.9% and 19.2% at the best performance on the Foursquare and Jiepang datasets respectively.

Table 3. Parameter analysis: the influence of λ_x on the proposed CBGeoMFC model.

Foursquare	λ_x	0	0.0001	0.001	0.01	0.1	1	10	50	100	150
	Precision@5	0.0504	0.0504	0.0504	0.0504	0.0504	0.0508	0.0517	0.0547	0.0554	0.0547
	Recall@5	0.0150	0.0150	0.0150	0.0150	0.0151	0.0152	0.0154	0.0163	0.0165	0.0163
	F1@5	0.0231	0.0231	0.0231	0.0231	0.0231	0.0233	0.0237	0.0251	0.0254	0.0251
Jiepang	λ_x	0	10	20	30	40	50	60	7	80	90
	Precision@5	0.0400	0.0400	0.0431	0.0462	0.0477	0.0462	0.0462	0.0462	0.0431	0.0415
	Recall@5	0.0595	0.0595	0.0641	0.0686	0.0709	0.0686	0.0686	0.0686	0.0641	0.0618
	F1@5	0.0478	0.0478	0.0515	0.0552	0.0570	0.0552	0.0552	0.0552	0.0515	0.0497

Parameter Analysis on N. Parameter N is the length of recommend list in the top-N recommendation. Setting $\lambda_x = 0.01, \lambda_h = 0.01$ and $\lambda_x = 40, \lambda_h = 0.7$ on the Foursquare and Jiepang datasets respectively, we change the parameter N from 1 to 20 and find that precision has a decline trend while recall continues growing when parameter N increases. F_1 measuring result considers both precision and recall and arrives the best performance as $N = 20$ on Foursquare and $N = 7$ on Jiepang as shown in Fig. 1.

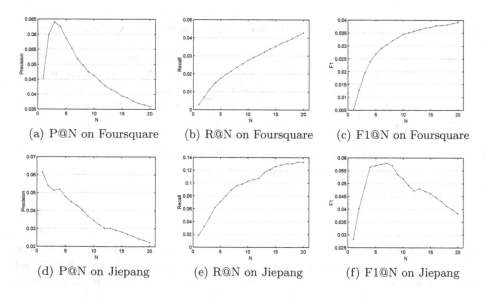

(a) P@N on Foursquare   (b) R@N on Foursquare   (c) F1@N on Foursquare

(d) P@N on Jiepang   (e) R@N on Jiepang   (f) F1@N on Jiepang

Fig. 1. The influence of N on the CBGeoMFC model.

3.4   Comparison Experiments

In this section, comparison results will be reported to compare the proposed CBGeoMFC method with some existing methods, namely UBCF, PMF [18] and MFC [12]. For all comparison methods, the parameters are set as suggested in the corresponding papers. In particular, for the sake of fairness, we set the same random initial value for all the methods, and set the learning rate as 0.0001 and regularization coefficients as 0.01 for every algorithms. We use parameter $\lambda_x = 40, \lambda_h = 0.1$ and $\lambda_x = 100, \lambda_h = 0.005$ on the Jiepang and Foursquare datasets in our CBGeoMFC model.

The comparison results are shown in Fig. 2. Our algorithm achieve the best performance among all the three comparison algorithms in terms of all evaluation metrics for considering both social influence and geographical limits. From the results, we can find that UBCF is not appropriate in POI recommendation for the reason that the check-in data is too sparse and users have few overlapping check-ins. Compared with PMF, our model improves the recommendation precision, recall and F1 by 93.9%, 63.7%, 80.5% and 64.1%, 24.9%, 34.8% respectively on Jiepang and Foursquare. Compared with MFC, our model improves the recommendation precision, recall and F1 by 39.1%, 46.2%, 40.5% and 34.9%, 15.2%, 20.7% respectively on Jiepang and Foursquare. Overall, our CBGeoMFC model outperforms the compared algorithms.

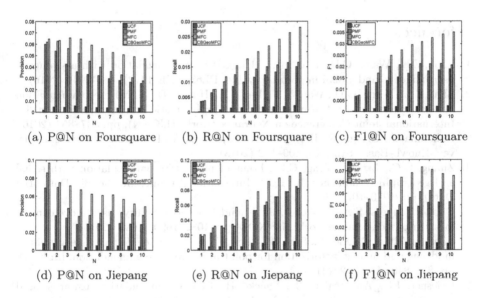

(a) P@N on Foursquare   (b) R@N on Foursquare   (c) F1@N on Foursquare

(d) P@N on Jiepang   (e) R@N on Jiepang   (f) F1@N on Jiepang

Fig. 2. Performance comparison.

4   Conclusions

POI recommendation can benefit users outdoor activities in an urban city and bridge the gap between the physical world and online social network, but it encounters more challenges for the reasons that the check-in data is too sparse, various factors have to be considered and the effect of social influence is limited. To solve these challenges, we propose a business circles conception and use AP algorithm to cluster all the check-ins into business circles. By aggregating the check-ins on locations into the check-ins on business circles, we can effectively settle the defects that there are few overlap check-ins of a user to his/her friends' check-ins. Then we compute the user's activity and attraction of the POI on each circle to evaluate the user preference on location, which is more consistent with the modern consumption pattern in an urban city. In addition, we cluster the users into overlapping communities according to the corresponding category information and social relationships. Finally, we incorporate all above factors into a matrix factorization framework. The experimental results have confirmed that our algorithm remarkably outperforms other existing algorithms.

Acknowledgments. This work was supported by NSFC (61502543), Guangdong Natural Science Funds for Distinguished Young Scholar (2016A030306014), Tip-top Scientific and Technical Innovative Youth Talents of Guangdong special support program (No. 2016TQ03X542), and the Fundamental Research Funds for the Central Universities (16lgzd15).

References

1. Zhao, Z.L., Wang, C.D., Lai, J.H.: AUI&GIV: recommendation with asymmetric user influence and global importance value. PLoS ONE **11**(2), e0147944 (2016)
2. Zhao, Z.L., Wang, C.D., Wan, Y.Y., Lai, J.H., Huang, D.: FTMF: recommendation in social network with feature transfer and probabilistic matrix factorization. In: International Joint Conference on Neural Networks (IJCNN), pp. 847–854 (2016)
3. Bobadilla, J., Ortega, F., Hernando, A., Gutiérrez, A.: Recommender systems survey. Knowl.-Based Syst. **46**, 109–132 (2013)
4. Zhang, D.C., Li, M., Wang, C.D.: Point of interest recommendation with social and geographical influence. In: IEEE International Conference on Big Data (Big Data), pp. 1070–1075 (2016)
5. Cho, E., Myers, S.A., Leskovec, J.: Friendship and mobility: user movement in location-based social networks. In: KDD, pp. 1082–1090. ACM (2011)
6. Lian, D., Zhao, C., Xie, X., Sun, G., Chen, E., Rui, Y.: GeoMF: joint geographical modeling and matrix factorization for point-of-interest recommendation. In: KDD, pp. 831–840. ACM (2014)
7. Griesner, J.B., Abdessalem, T., Naacke, H.: POI recommendation: towards fused matrix factorization with geographical and temporal influences. In: Proceedings of the 9th ACM Conference on Recommender Systems, pp. 301–304. ACM (2015)
8. Baral, R., Wang, D., Li, T., Chen, S.C.: GeoTeCS: exploiting geographical, temporal, categorical and social aspects for personalized POI recommendation. In: Proceedings of the 17th International Conference on Information Reuse and Integration (IRI), pp. 94–101. IEEE(2016)
9. Inman, J.: Navigation and Nautical Astronomy, for the Use of British Seamen. F. & J. Rivington, London (1849)
10. Frey, B.J., Dueck, D.: Clustering by passing messages between data points. Science **315**(5814), 972–976 (2007)
11. Cheng, C., Yang, H., King, I., Lyu, M.R.: Fused matrix factorization with geographical and social influence in location-based social networks. In: AAAI, vol. 12, p. 1 (2012)

12. Li, H., Wu, D., Tang, W., Mamoulis, N.: Overlapping community regularization for rating prediction in social recommender systems. In: Proceedings of the 9th ACM Conference on Recommender Systems, pp. 27–34. ACM (2015)

13. Ma, H., Zhou, D., Liu, C., Lyu, M.R., King, I.: Recommender systems with social regularization. In: WSDM, pp. 287–296. ACM (2011)

14. Jamali, M., Ester, M.: A matrix factorization technique with trust propagation for recommendation in social networks. In: Proceedings of the Fourth ACM Conference on Recommender Systems, pp. 135–142. ACM (2010)

15. Li, H., Hong, R., Zhu, S., Ge, Y.: Point-of-interest recommender systems: a separate-space perspective. In: ICDM, pp. 231–240. IEEE (2015)

16. Li, H., Ge, Y., Zhu, H.: Point-of-interest recommendations: learning potential check-ins from friends. In: KDD (2016)

17. Yang, D., Zhang, D., Zheng, V.W., Yu, Z.: Modeling user activity preference by leveraging user spatial temporal characteristics in LBSNs. IEEE Trans. Syst. Man Cybern. Syst. **45**(1), 129–142 (2015)

18. Salakhutdinov, R., Mnih, A.: Probabilistic matrix factorization. In: NIPS, vol. 1, pp. 1–2 (2007)

Event Recommendation via Collective Matrix Factorization with Event-User Neighborhood

Mei Li[1], Dong Huang[2], Bin Wei[3], and Chang-Dong Wang[1(✉)]

[1] School of Data and Computer Science, Sun Yat-sen University, Guangzhou, China
limei931122@hotmail.com, changdongwang@hotmail.com
[2] College of Mathematics and Informatics, South China Agricultural University,
Guangzhou, China
huangdonghere@gmail.com
[3] School of English and Education, Guangdong University of Foreign Studies,
Guangzhou, China
pypzengqs@126.com

Abstract. Event-based social networks (EBSNs) recently emerge as a new type of social network and have been growing rapidly. Because of the very large volume of various events, the demand of event recommendation becomes increasingly important. In this paper, we propose a novel approach called Collective Matrix Factorization with Event-User Neighborhood (CMF-EUN) model to handle this problem. CMF-EUN combines the strengths of matrix factorization and neighborhood based methods. Due to the fact that RSVP matrix is generally extremely sparse, it is difficult to find similar neighborhoods using the widely adopted similarity measures. To address this, we calculate the similarities based on some specific features of events and users in EBSNs. The heterogeneous social relationships are also taken into consideration. Experimental results conducted on real datasets collected from DoubanEvent show that the proposed model provides superior performance and outperforms several baseline methods.

Keywords: Event recommendation · Event-based social networks · Context · Neighborhood method

1 Introduction

With the rapid development of event-based applications such as Meetup and DoubanEvent, a new type of social network called event-based social networks (EBSNs) have recently appeared. In such platforms, users can attend, organize and participate social events. The goal of EBSNs is to use events as the medium to extend the social relationships in virtual network to real life. A large number of new events are created every day. For an upcoming event, the event organizer hopes to maximize the attendees. For users, they are eager to enhance user experience. Therefore, event recommendation is essential.

© Springer International Publishing AG 2017
Y. Sun et al. (Eds.): IScIDE 2017, LNCS 10559, pp. 676–686, 2017.
DOI: 10.1007/978-3-319-67777-4_61

Different from traditional recommendation problems [1–5], event recommendation encounters heterogeneous online + offline social relationships [6,7] and implicit feedback data. Besides, in EBSNs, event content (which provides introduction of the event theme), event location (where the event will be held), event time (when the event will start) and event organizer (who launches and organizes the event) are the major components of an event which affect user decisions in attending the event [8]. Furthermore, the events in EBSNs are short-lived, and always have little or no feedback which may cause a bad cold start problem [9].

Aiming to address the above issues, in this work, a novel event recommendation approach called Collective Matrix Factorization with Event-User Neighborhood model is proposed. To incorporate online social relationships into the model, collective matrix factorization [10] is used. The user latent factor and item latent factor are learned by employing an online social relationship matrix and a user-item RSVP[1] matrix. To find similar neighborhoods [11,12], we calculate the similarities based on the specific features of events and users in EBSNs. If two events are similar on event content, event location, event time and event organizers, we can consider these two events as similar events. Similarly, we consider two users are similar when they have the similar tastes about the four aspects. For finding the similar users, both the offline social relationships and the users' tastes about the event features are taken into consideration. We have collected real datasets from DoubanEvent and conducted comprehensive experiments. The results show that our proposed model provides excellent performance and outperforms several alternatives.

2 The Proposed CMF-EUN Model

2.1 Model Overview

In this work, a novel approach called Collective Matrix Factorization with Event-User Neighborhood (CMF-EUN) model is proposed. CMF-EUN consists of three parts: collective matrix factorization, event neighborhood discovery and user neighborhood discovery. The overall model is shown as follows:

$$\widehat{R_{ue}} = p_u^T q_e + |N^{k1}(u,e)|^{-\frac{1}{2}} \sum_{e' \in N^{k1}(u,e)} w_{ee'}(R_{ue'} - \overline{R_u}) + |N^{k2}(e,u)|^{-\frac{1}{2}} \sum_{u' \in N^{k2}(e,u)} w_{uu'}(R_{u'e} - \overline{R_e})$$

(1)

where p_u and q_e represent the feature vector of user u and event e obtained by collective factorization matrix respectively, $\overline{R_u}$ denotes the average score for user u, $\overline{R_e}$ denotes the average score for event e, $N^{k1}(u,e)$ denotes the set of the events that are similar with event e, $N^{k2}(e,u)$ denotes the set of users that are similar with user u, and $k1$ and $k2$ represent the number of event neighborhoods and user neighborhoods respectively. The neighborhood set are determined by the similarity measure $Sim(e,e_j)$ and $Sim(u,u_j)$, which will be introduced later. $w_{ee'}$ and $w_{uu'}$ are parameters which control the influence weights of event e' to e and user u' to u respectively.

---

[1] RSVP stands for the French expression "répondez s'il vous plaît", meaning "please respond".

2.2   Collective Matrix Factorization

Given an online social network A^{on}, $U = \{u_1, u_2, ...u_m\}$ represents the users in EBSNs. Let $S = \{w_{ij}^{on}\}$ denote the $m \times m$ matrix of A^{on}, which is also called the online social network matrix in this paper. For each user pair u_i and u_j, let $w_{ij}^{on} \in (0, 1]$ denote the weight associated with an edge from u_i to u_j. The online social relationship between u_i and u_j can be defined as $w_{ij}^{on} = \frac{|G_{u_i} \cap G_{u_j}|}{|G_{u_i}|}$, where $G(u_i)$ represents all social groups that u_i attend, $|G_{u_i} \cap G_{u_j}|$ denotes the number of nodes in the set $G_{u_i} \cap G_{u_j}$. Obviously, S is an asymmetric matrix, since in the real scene, u_i trusting u_j does not necessary indicate u_j trusts u_i.

The idea of collective matrix factorization is to derive a high-quality d-dimensional feature representation P of users by reconstructing online social relation matrix S and user-event RSVP matrix R, where users are the shared elements. The parameters are determined by:

$$\min_{p_*, q_*} \sum_{(u,e) \in \mathbb{K}} (R_{ue} - p_u^T q_e)^2 + \sum_{(u,f) \in \mathbb{S}} (S_{uf} - p_u^T o_f)^2 + \lambda(||p_u||^2 + ||q_e||^2 + ||o_f||^2) \quad (2)$$

where p_u represents the feature vector of user u, q_e represents the feature vector of event e and o_f represents the feature vector of user f respectively. The value of $\widehat{R_{ue}}$ can be predicted as $\widehat{R_{ue}} = p_u^T q_e$.

2.3   Event Neighborhood Discovery

In traditional recommendation problems, when calculating the similarities between two users or two events, the Pearson correlation Coefficient (PCC) [13] and the cosine similarity are often used. However, in EBSNs the RSVP matrix is extremely sparse, so we calculate the similarities based on the specific features of events and users in EBSNs. If two events are similar on event content, event location, event time and event organizers, we can consider these two events as similar events.

Event Content. The content of an event plays a major role in determining users' decisions in attending an event. Usually an event is characterized by three parts: title, category and description. We put these three parts together as a whole text, and remove stop words and short words. Latent Dirichlet Allocation (LDA) [14] model is used to obtain the event distribution on latent topics. And then to compute the content similarity the Jensen-Shannon (JS) distance is used.

According to LDA, the document d_e of event e is represented as a probability topic distribution, denoted as θ_{d_e}, and each topic $z_{d_e} \in \{1, 2, ...T\}$ is represented as a probability distribution over words, denoted as $\phi_{z_{d_e}}$. For each word $w_{d_e,n}$ in document d_e, draw a topic $z_{d_e,n} \sim Mult(\theta_{d_e})$, draw a word $w_{d_e,n} \sim Mult(\phi_{z_{d_e},n})$.

The joint distribution of an event document is defined as:

$$p(w_{d_e}, z_{d_e}, \theta_{d_e}, \phi | \alpha, \beta) = \prod_{n=1}^{N_e} p(w_{d_e,n} | \phi_{z_{d_e,n}}) p(z_{d_e,n} | \theta_{d_e}) p(\theta_{d_e} | \alpha) p(\phi | \beta) \quad (3)$$

where $p(\theta_{d_e} | \alpha)$ and $p(\phi | \beta)$ are derived from Dirichlet distribution parameterized by α and β respecitvely. N_e represents the number of words of event e.

Finally, all events' document-topic distribution Θ and topic-word distribution Φ can be obtained by utilizing Gibbs sampling [15]. Given the topic distributions $\theta_{d_{e_i}}$ and $\theta_{d_{e_j}}$ of event e_i and e_j, the Jensen-Shannon divergence is defined as follows:

$$JS(\theta_{d_{e_i}}, \theta_{d_{e_j}}) = \frac{1}{2}[D_{KL}(\theta_{d_{e_i}}, \frac{\theta_{d_{e_i}} + \theta_{d_{e_j}}}{2}) + D_{KL}(\theta_{d_{e_j}}, \frac{\theta_{d_{e_i}} + \theta_{d_{e_j}}}{2})] \quad (4)$$

where $D_{KL}(.)$ is the Kulllback Libeibler (KL) divergence and can be calculated as $D_{KL}(\theta_1, \theta_2) = \sum_{j=1}^{T} \theta_{1j} \log_2 \frac{\theta_{1j}}{\theta_{2j}}$.

Finally, the content similarity between two events u_i and u_j can be defined as $Sim_{content}(e_i, e_j) = 1 - JS(\theta_{d_{e_i}}, \theta_{d_{e_j}})$.

Event Location. Event location denotes the physical place where the event is held. We calculate the distance between all pairs of events with different locations. We assume that two events with closer locations are more similar to each other. The location similarity between e_i and e_j is defined as $Sim_{distance}(e_i, e_j) = e^{\left\{\frac{Distance(e_i, e_j)}{2}\right\}}$, where $Distance(e_i, e_j)$ is the distance between e_i and e_j calculated by latitude and longitude.

Event Time. Another important factor that might affect users' decision on attending an event is when the event occurs. The event time is often given by the form of year/month/date/hour: minute: seconds.

Firstly, we calculate the event similarity based on the day of the week factor $Sim_{week}(e_i, e_j) = \begin{cases} 0 & W(e_i) \neq W(e_j) \\ 1 & W(e_i) = W(e_j) \end{cases}$, where $W(e_i)$ denotes which day of the week the event occurs, and $W(e_i) \in \{1, 2, 3, 4, 5, 6, 7\}$ corresponds to the day of the week.

Secondly, we calculate the event similarity based on the hour of the day factor $Sim_{hour}(e_i, e_j) = e^{\left\{-\frac{(t_{e_i} - t_{e_j})}{2}\right\}}$, where t_{e_i} denotes which hour of the day the event occurs, and $t_{e_i} \in \{1, 2, 3...24\}$.

Event Organizer. In EBSNs, event organizers are the person who create the events. Therefore, we assume that if two events are held by the same organizer, the two events are similar. The organizer similarity between e_i and e_j is defined as a binary value function. $Sim_{host}(e_i, e_j) = \begin{cases} 0 & H(e_i) \neq H(e_j) \\ 1 & H(e_i) = H(e_j) \end{cases}$, where $H(e_i)$ represents the organizer of event e_i.

Similarity Between Events. The similarity between events can be calculated as follows:

$$Sim(e_i, e_j) = \gamma_1 * Sim_{content}(e_i, e_j) + \gamma_2 * Sim_{week}(e_i, e_j) + \gamma_3 * Sim_{hour}(e_i, e_j)$$
$$+ \gamma_4 * Sim_{distance}(e_i, e_j) + \gamma_5 * Sim_{host}(e_i, e_j)$$

(5)

where γ_1, γ_2, γ_3, γ_4, γ_5 represents the weights of different features, respectively. To evaluate the contributions of these features, we adopt each feature separately to make recommendation. The weight of each feature can be set based on recommendation performance on test datasets. The detailed settings of these relative weights are listed in experiments.

2.4   User Neighborhood Discovery

Similarly, we consider two users to be similar when they have the similar tastes about the four aspects.

User's Preference on Event Content. In this section, each user is represented as a vector of words extracted from the past events the user attended. Each event is characterized by event title, event description and event category. Then the next steps are similar to the steps for calculating the content similarity between events. LDA is used to obtain the user distribution on latent topics. At last, to compute the similarity between users on user's preference on event content, the Jensen-Shannon (JS) distance is used $Sim_{content}(u_i, u_j) = 1 - JS(\theta_{d_{u_i}}, \theta_{d_{u_j}})$, where $\theta_{d_{u_i}}$ and $\theta_{d_{u_j}}$ represent the distribution of user u_i and u_j obtained by LDA.

User's Preference on Event Location. While most users tend to go to events located close to their home, fewer users are open to participate in events located farther away. Moreover, users tend to go to events located within a limited number of regions. Here we use K-means clustering to find the center of past events the user attended. The similarity between users on user's preference on event location is defined as $Sim_{distance}(u_i, u_j) = e^{\left\{\frac{Distance(u_i(center), u_j(center))}{2}\right\}}$, where $Distance(u_i(center), u_j(center))$ is the distance calculated by the center of past events attended by user u_i and user u_j.

User's Preference on Event Time. When an event is created, the event time is an important factor determining who can participate in this event. When a user makes decision, he/she will consider whether the event time complies with his personal schedule.

To calculate the similarity between users on user's preference on the day of the week factor, each user u is represented as a 1-dimensional vector, with a vector component set to 1 whenever the event happened at that particular day.

$$Sim_{week}(u_i, u_j) = \frac{\sum_{w=1}^{7}(W_{u_i,w} - \overline{W_{u_i}})(W_{u_j,w} - \overline{W_{u_j}})}{\sqrt{\sum_{w=1}^{7}(W_{u_i,w} - \overline{W_{u_i}})^2}\sqrt{\sum_{w=1}^{7}(W_{u_i,w} - \overline{W_{u_i}})^2}} \tag{6}$$

where $\overline{W_{u_i}}$ means the number of events on average user u_i attends in a week.

To calculate the similarity between users on user's preference on the hour of the day factor, each user u is represented as a 1-dimensional vector, with a vector component set to 1 whenever the event happened at that particular hour.

$$Sim_{hour}(u_i, u_j) = \frac{\sum_{t=1}^{24}(T_{u_i,t} - \overline{T_{u_i}})(T_{u_j,t} - \overline{T_{u_j}})}{\sqrt{\sum_{t=1}^{24}(T_{u_i,t} - \overline{T_{u_i}})^2}\sqrt{\sum_{t=1}^{24}(T_{u_i,t} - \overline{T_{u_i}})^2}} \tag{7}$$

where $\overline{T_{u_i}}$ means the number of events on average user u_i attends in a day.

User's Preference on Event Organizer. Whether a user attends an event is also affected by the event organizer [16]. Assume there are t organizers in total. Each user u is represented as a $1 \times t$-dimensional vector $H_{u,o}$. The value of $H_{u,i}$ is the number of events user u attends which are hosted by organizer i. The similarity between users on user's preference on the event organizer is defined as follows:

$$Sim_{host}(u_i, u_j) = \frac{\sum_{o=1}^{t}(H_{u_i,o} - \overline{H_{u_i}})(H_{u_j,o} - \overline{H_{u_j}})}{\sqrt{\sum_{o=1}^{t}(H_{u_i,o} - \overline{H_{u_i}})^2}\sqrt{\sum_{o=1}^{t}(H_{u_i,o} - \overline{H_{u_i}})^2}} \tag{8}$$

where $\overline{H_{u_i}}$ means the number of events on average user u_i attends hosted by each organizer.

Similarity Between Users. The similarity between users can be calculated as $Sim(u_i, u_j) = \epsilon_1 * Sim_{content}(u_i, u_j) + \epsilon_2 * Sim_{week}(u_i, u_j) + \epsilon_3 * Sim_{hour}(u_i, u_j) + \epsilon_4 * Sim_{distance}(u_i, u_j) + \epsilon_5 * Sim_{host}(u_i, u_j)$, where ϵ_1, ϵ_2, ϵ_3, ϵ_4, ϵ_5 represents the weights of different features, respectively. To evaluate the contributions of these features, we adopt each feature separately to make recommendation. The weight of each feature can be set based on recommendation performance on test datasets. The detailed settings of these relative weights are shown in experiments.

We further update the similarity between users by taking offline social relationships into consideration.

Assuming the offline social network is A^{off}, $U = \{u_1, u_2, ...u_m\}$ represents the users in EBSNs. Let $S' = \left\{w_{ij}^{off}\right\}$ denote the $m \times m$ matrix, which is called the offline social network matrix in this paper. For each user pair, u_i and u_j, let $w_{ij}^{off} \in (0,1]$ denote the weight associated with an edge from u_i to u_j. The offline social relationship between u_i and u_j can be defined as $w_{ij}^{off} = \frac{|E_{u_i} \cap |E_{u_j}|}{|E_{u_i}|}$, where $E(u_i)$ represents all events that u_i attends, $|E_{u_i}| \cap |E_{u_j}|$ denotes the number of nodes in the set $E_{u_i} \cap E_{u_j}$.

Then the similarity between users can be updated as $Sim(u_i, u_j)' = \vartheta * Sim(u_i, u_j) + (1 - \vartheta) * S'_{i,j}$, where $\vartheta \in (0,1)$ is the parameter for balancing the two parts and can be tuned based on recommendation performance on test datasets.

2.5   Optimization Approach

The overall objective function is as follows:

$$
\min \sum_{(u,e) \in \mathbb{K}} \left(R_{ue} - p_u^T q_e - |N^{k1}(u,e)|^{-\frac{1}{2}} \sum_{e' \in N^{k1}(u,e)} w_{ee'}(R_{ue'} - \overline{r_u}) \right.
$$

$$
\left. - |N^{k2}(e,u)|^{-\frac{1}{2}} \sum_{u' \in N^{k2}(e,u)} w_{uu'}(R_{u'e} - \overline{r_e}) \right)^2 \qquad (9)
$$

$$
+ \sum_{(u,f) \in \mathbb{S}} (S_{uf} - p_u^T o_f)^2 + \lambda(||p_u||^2 + ||q_e||^2 + ||o_f||^2 + \sum_{e' \in N^{k1}(u,e)} w_{ee'}^2 + \sum_{u' \in N^{k2}(e,u)} w_{uu'}^2)
$$

where λ determines the extent of regularization.

The stochastic gradient descent algorithm is used to optimize the objective function given by Eq. (9). The gradient for all parameters can be computed as follows, where L denotes the loss function:

$$
\frac{\partial L}{\partial p_u} = -2 \sum_{e=1}^{n} (R_{ue} - \widehat{R_{ue}}) \cdot q_e + 2 \sum_{f=1}^{m} (S_{uf} - \widehat{S_{uf}}) \cdot o_f + 2\lambda p_u
$$

$$
\frac{\partial L}{\partial q_e} = -2 \sum_{u=1}^{m} (R_{ue} - \widehat{R_{ue}}) \cdot p_u + 2\lambda q_e \qquad \frac{\partial L}{\partial o_f} = -2 \sum_{f=1}^{m} (S_{uf} - \widehat{S_{uf}}) \cdot p_u + 2\lambda o_f
$$

$$
\frac{\partial L}{\partial w_{ee'}} = -2 \sum_{u=1}^{m} (R_{ue} - \widehat{R_{ue}}) \cdot |N^{k1}(u,e)|^{-\frac{1}{2}} \cdot (R_{ue'} - \widehat{R_u}) + 2\lambda w_{ee'}
$$

$$
\frac{\partial L}{\partial w_{uu'}} = -2 \sum_{e=1}^{n} (R_{ue} - \widehat{R_{ue}}) \cdot |N^{k2}(e,u)|^{-\frac{1}{2}} \cdot (R_{u'e} - \widehat{R_e}) + 2\lambda w_{uu'}
$$

$$
(10)
$$

The parameters are updated as follows:

$$
p_u \leftarrow p_u - \eta_1 \cdot \frac{\partial L}{\partial p_u}, \ q_e \leftarrow q_e - \eta_1 \cdot \frac{\partial L}{\partial q_e}, \ w_{ee'} \leftarrow w_{ee'} - \eta_2 \cdot \frac{\partial L}{\partial w_{ee'}}, \ w_{uu'} \leftarrow w_{uu'} - \eta_2 \cdot \frac{\partial L}{\partial w_{uu'}}
$$

3   Experiments

3.1   Datasets

In our experiments, we collected real datasets from Douban Event. To simulate real scenarios, we select users from Beijing and Shanghai, the two largest cities in China. The event attendance histories of these users are from 2016/01/01 to 2017/01/01. For each event, we get event content (category, title and description), event time, event location (latitude and longitude), event organizer and a list of attending users. For each user, we get user id, the social groups he/she joins. To filter noisy data, we remove users who attended less than five events. Then, to evaluate the effectiveness of the recommendations, we divide the dataset into a training set and a testing set. Approximately 70% of the attendance histories are assigned to the training set and the rest 30% to the testing set. The statistic of the dataset is shown in Table 1.

Table 1. Statistics of experimental datasets.

Datasets	User	Event	Organizer	Group	User-event	User-group
Beijing	2255	8652	1854	197	24183	6653
Shanghai	1837	7390	1565	204	21491	5335

3.2   Evaluation Settings

We adopt Precision@N and Recall@N to evaluate the effectiveness of the recommendation lists produced by different methods, which are two widely used evaluation metrics [17]. We compare the effectiveness of the proposed method with the following event recommendation methods: Singular Value Decomposition (SVD) [17], Collective Matrix Factorization (CMF) [18], Singular Value Decomposition with Multi-Factor Neighborhood (SVD-MFN) [16], Singular Value Decomposition with Event-User Neighborhood (SVD-EUN) (a method that similar to our method but replace CMF by SVD). For the compared methods, the best parameters are tuned as suggested in the corresponding papers. In our method, the weight of each feature can be set based on recommendation performance on test datasets. The detailed settings are shown in Table 2.

Table 2. The weight of different features.

Datasets	γ_1	γ_2	γ_3	γ_4	γ_5	ϵ_1	ϵ_2	ϵ_3	ϵ_4	ϵ_5
Beijing	0.365	0.010	0.010	0.255	0.360	0.510	0.055	0.110	0.080	0.245
Shanghai	0.440	0.010	0.020	0.220	0.310	0.480	0.060	0.065	0.130	0.265

3.3   Comparison Results

According to the results shown in Table 3, on both Shanghai and Beijing datasets, CMF outperforms SVD, this is because CMF integrates the online social network into recommendation. The comparison between SVD and SVD-MFN shows that the performance of SVD-MFN is better than SVD. The reason may be that, to obtain event neighborhood, SVD-MFN takes the features of EBSNs into consideration. This reveals that the memory-based and model-based collaborative filtering approaches can enrich each other. In most of experiments, SVD-EUN performs better than SVD-MFN. This is because SVD-EUN incorporates both the event and user neighborhood methods into matrix factorization. Besides, SVD-EUN integrates offline social network to find similar users. It shows that the performance of event recommendation can be improved by integrating social network information into the model.

Table 3. Comparison results on the Shanghai and Beijing datasets.

Methods	Beijing				Shanghai			
	P@3	*P@5*	*R@3*	*R@5*	*P@3*	*P@5*	*R@3*	*R@5*
SVD	0.0006	0.0005	0.0006	0.0006	0.0007	0.0008	0.0006	0.0009
CMF	0.0034	0.0031	0.0032	0.0048	0.0054	0.0050	0.0047	0.0071
SVD-MFN	0.0782	0.0631	0.0729	0.0981	0.0412	0.0351	0.0352	0.0499
SVD-EUN	0.0844	0.0667	0.0787	0.1037	0.0445	0.0364	0.0385	0.0518
CMF-EUN	**0.0885**	**0.0719**	**0.0820**	**0.1118**	**0.0797**	**0.0682**	**0.0681**	**0.0971**

The proposed CMF-EUN model achieves the best performance in terms of both *Precision@N* and *Recall@N*. This owes to that CMF-EUN combines the strengths of matrix factorization and neighborhood based methods. Besides RSVPs, CMF-EUN fully considers the factors in EBSNs which may affect users' decisions in attending the event.

4   Conclusions

With the popularity of event-based social networks, there is an increasing demand for event recommendation. In this paper, we propose a novel approach incorporating both event-based and user-based neighborhood methods into a matrix factorization model. To incorporate online social relationships into the model, collective matrix factorization is used. Due to the fact that RSVP matrix is extremely sparse, we calculate the similarities based on the specific features of events and users in EBSNs. Experimental results have shown the effectiveness of our method.

Acknowledgments. This work was supported NSFC (61502543 & 61602189), Guangdong Natural Science Funds for Distinguished Young Scholar (2016A030306014), the PhD Start-up Fund of Natural Science Foundation of Guangdong Province, China (2016A030310457), and Tip-top Scientific and Technical Innovative Youth Talents of Guangdong special support program (2016TQ03X542).

References

1. Breese, J.S., Heckerman, D., Kadie, C.: Empirical analysis of predictive algorithms for collaborative filtering. In: Proceedings of the Fourteenth Conference on Uncertainty in Artificial Intelligence, pp. 43–52. Morgan Kaufmann Publishers Inc. (1998)
2. Balabanović, M., Shoham, Y.: Fab: content-based, collaborative recommendation. Commun. ACM **40**(3), 66–72 (1997)
3. Pennock, D.M., Horvitz, E., Lawrence, S., Giles, C.L.: Collaborative filtering by personality diagnosis: a hybrid memory-and model-based approach. In: Proceedings of the Sixteenth Conference on Uncertainty in Artificial Intelligence, pp. 473–480. Morgan Kaufmann Publishers Inc. (2000)
4. Zhao, Z.L., Wang, C.D., Lai, J.H.: AUI&GIV: recommendation with asymmetric user influence and global importance value. PLoS One **11**(2), e0147944 (2016)
5. Zhao, Z.L., Wang, C.D., Wan, Y.Y., Lai, J.H., Huang, D.: FTMF: recommendation in social network with feature transfer and probabilistic matrix factorization. In: International Joint Conference on Neural Networks (IJCNN), pp. 847–854 (2016)
6. Chin, A., Tian, J., Han, J., Niu, J.: A study of offline events and its influence on online social connections in douban. In: Green Computing and Communications (GreenCom), 2013 IEEE and Internet of Things (iThings/CPSCom), IEEE International Conference on and IEEE Cyber, Physical and Social Computing, pp. 1021–1028. IEEE (2013)
7. Zhang, D.C., Li, M., Wang, C.D.: Point of interest recommendation with social and geographical influence. In: IEEE International Conference on Big Data (Big Data), pp. 1070–1075 (2016)
8. Wang, Z., He, P., Shou, L., Chen, K., Wu, S., Chen, G.: Toward the New Item Problem: Context-Enhanced Event Recommendation in Event-Based Social Networks. In: Hanbury, A., Kazai, G., Rauber, A., Fuhr, N. (eds.) ECIR 2015. LNCS, vol. 9022, pp. 333–338. Springer, Cham (2015). doi:10.1007/978-3-319-16354-3_36
9. Yuan, Q., Cong, G., Ma, Z., Sun, A., Thalmann, N.M.: Time-aware point-of-interest recommendation. In: Proceedings of the 36th International ACM SIGIR Conference on Research and Development in Information Retrieval, pp. 363–372. ACM(2013)
10. Ma, H., Yang, H., Lyu, M.R., King, I.: SoRec: social recommendation using probabilistic matrix factorization. In: Proceedings of the 17th ACM Conference on Information and Knowledge Management, pp. 931–940. ACM (2008)
11. Sarwar, B., Karypis, G., Konstan, J., Riedl, J.: Item-based collaborative filtering recommendation algorithms. In: Proceedings of the 10th International Conference on World Wide Web, pp. 285–295. ACM (2001)
12. Song, S., Wu, K.: A creative personalized recommendation algorithm—user-based slope one algorithm. In: 2012 International Conference on Systems and Informatics (ICSAI), pp. 2203–2207. IEEE (2012)

13. Shi, Y., Larson, M., Hanjalic, A.: Exploiting user similarity based on rated-item pools for improved user-based collaborative filtering. In: Proceedings of the Third ACM Conference on Recommender Systems, pp. 125–132. ACM (2009)
14. Blei, D.M., Ng, A.Y., Jordan, M.I.: Latent dirichlet allocation. J. Mach. Learn. Res. 3(Jan), 993–1022 (2003)
15. Heinrich, G.: Parameter estimation for text analysis. University of Leipzig, Technical report (2008)
16. Du, R., Yu, Z., Mei, T., Wang, Z., Wang, Z., Guo, B.: Predicting activity attendance in event-based social networks: content, context and social influence. In: Proceedings of the 2014 ACM International Joint Conference on Pervasive and Ubiquitous Computing, pp. 425–434. ACM (2014)
17. Zhang, S., Wang, W., Ford, J., Makedon, F., Pearlman, J.: Using singular value decomposition approximation for collaborative filtering. In: Seventh IEEE International Conference on E-Commerce Technology, CEC 2005, pp. 257–264. IEEE (2005)
18. Singh, A.P., Gordon, G.J.: Relational learning via collective matrix factorization. In: Proceedings of the 14th ACM SIGKDD International Conference on Knowledge Discovery and Data Mining, pp. 650–658. ACM (2008)

Author Index

Printed in the United States
By Bookmasters